VOLUME

II

Biology

The Web of Life

Daniel D. Chiras

University of Denver

West Publishing Company

Minneapolis/St. Paul

New York

Los Angeles

San Francisco

To my wife Kathleen,
and my delightful sons, Skyler and Forrest,
with love and affection

Interior and cover design Diane Beasley
Copyediting Bill Waller
Artwork Carlyn Iverson, Elizabeth Morales-Denney, Publications Services, Pat Rossi, Cyndie Wooley, J/B Woolsey & Associates (Individual art credits follow Index)
Photo research John Cunningham/Visuals Unlimited (Photo credits follow Index)
Composition Graphic World
Page layout Diane Beasley
Indexing Sandi Schroeder
Cover image White-tailed deer mother and 5-minute old young. © William J. Weber/Visuals Unlimited.
Production, prepress, printing and binding by West Publishing Company

610 Opperman Drive
P.O. Box 64526
St. Paul, MN 55164-0526

Printed in the United States of America

00 99 98 97 96 95 94 93 8 7 6 5 4 3 2 1 0

Library of Congress Cataloging-in-Publication Data

Chiras, Daniel D.
Biology, the web of life / Daniel Chiras.
p. cm.
Includes bibliographical references (p.) and index.
ISBN 0-314-01251-6 (hardcover). —ISBN 0-314-01343-1 (paperback). —ISBN 0-314-01344-X (Volume I). —ISBN 0-314-01345-8 (Volume II). —ISBN 0-314-01346-6 (Volume III).
1. Biology. I. Title.
QH308.2.C45 1993
574—dc20

92-34101
CIP

Brief Contents

PART IV

Homeostasis, Integration, and Control: Organs and Organ Systems of Animals 225

PART V

The Continuation of Life: Reproduction and Development 537

About the Author

Daniel D. Chiras received his Ph.D. in reproductive biology in 1976 from the University of Kansas Medical School, where his research on ovarian physiology earned him the Latimer Award. In September 1976, Dr. Chiras joined the Biology Department at the University of Colorado in Denver in a teaching and research position. Since 1976, he has taught numerous undergraduate and graduate courses, including general biology, cell biology, histology, endocrinology, and reproductive biology. Dr. Chiras also has a strong interest in environmental issues and has taught a variety of courses on the environment.

Currently an adjunct professor at the University of Denver and at the University of Colorado in Denver, he has also been a visiting professor at the University of Washington.

Dr. Chiras is the author of numerous technical publications on ovarian physiology, critical thinking, sustainability, and models for environmental education, which have appeared in the *American Biology Teacher, Biology of Reproduction,* the *American Journal of Anatomy,* and other journals. He has also written numerous articles for newspapers and magazines on environmental issues, and is the author of the environment section for World Book Encyclopedia's *Science Year 1993*. Dr. Chiras has also published five college and high school textbooks, including *Human Biology: Health, Homeostasis, and the Environment* and *Environmental Science: Action for a Sustainable Future*. He is the coauthor of *Natural Resource Conservation: An Ecological Approach* (with Oliver S. Owen). He currently serves as an editor of *Environmental Carcinogenesis and Ecotoxicology Reviews*.

Dr. Chiras has also written books for a general audience, including *Beyond the Fray: Reshaping America's Environmental Response*, and, most recently, *Lessons from Nature: Learning to Live Sustainably on the Earth*. Besides his active scientific pursuits, he is an avid kayaker, skier, bicyclist, and organic gardener.

Contents

PART IV

Homeostasis, Integration, and Control: Organs and Organ Systems of Animals 225

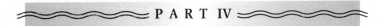

genetic engineering. This book presents a number of modern-day controversies in Point/Counterpoint sections. Some address social and political issues that require a good biological background, and others focus on scientific debates.

Each Point/Counterpoint consists of two brief essays written by distinguished writers and thinkers. These lively essays present opposing views on such important issues as genetic engineering, fetal cell transplantation, cancer, and global warming. Point/Counterpoints also offer students a chance to practice critical thinking skills.

Critical Thinking Skills

As noted earlier, Chapter 1 presents students with a number of guidelines for improving their critical thinking skills. These will help students become more discerning thinkers, an ability that will prove useful in this and many other college courses—not to mention later in life.

Additional emphasis is placed on critical thinking throughout the text. Each chapter, for example, contains a section on Exercising Your Critical Thinking Skills, which calls on students to use their critical thinking skills. These exercises include case studies, hypothetical scenarios, or summaries of news or scientific reports. They ask students to analyze situations or reports and give alternative interpretations or find flaws in reasoning. Each exercise emphasizes one or two critical thinking rules presented in the first chapter. Critical thinking questions are also included after each Point/Counterpoint.

Environment and Health

The health of the Earth's organisms and the health of the environment in which they live are closely connected. To help illustrate these connections, each chapter ends with an Environment and Health section. These brief sections illustrate some of the ways in which the physical and chemical environments affect our health and the health of other species.

In-Text Summaries

Chapter section heads are written as summary statements that capture key concepts or facts presented in the material that follows. These in-text summaries provide students a way to review major concepts in preparation for exams.

End-of-Chapter Summaries

Unlike most other biology textbooks, this one contains a fairly detailed summary at the end of each chapter. These summaries cover key concepts and ideas. Students can use them for a quick review after reading the chapter and as they prepare for exams. The in-text summaries, detailed end-of-chapter summaries, and extensive questions (discussed below) provide an excellent study guide.

Summary Tables

To help students summarize key concepts, processes, and systems, I have included summary tables in many chapters. Students can use these tables to prepare for exams or to review material after reading the chapter.

Test of Terms

To help students review the key terminology in each chapter, a Test of Terms has been included at the end of each chapter. These tests contain fill-in-the-blank questions and can be used by students to assess their grasp of the main terms and concepts presented in the chapter. They may also prove useful when preparing for exams. Students can fill in the blanks immediately after reading the chapter or after they have spent some time studying the material.

Test of Concepts

Each chapter also contains a number of brief essay questions that enable students to assess their understanding of the material.

Art Program

This book contains a remarkable collection of drawings and photographs. These colorful illustrations supplement the text, helping make the more complex concepts and processes understandable.

SUPPLEMENTARY ITEMS FOR INSTRUCTORS

To help you teach this course, an extensive ancillary package has been developed. It includes an instructor's guide, test bank (also available in a computerized test-generation program), study guide, transparency acetates, slides, laboratory manual, videotapes, and videodisk, among other items.

Instructor's Guide

An instructor's guide with a 2000-question test bank is available from West Publishing Company. For each chapter, the instructor's guide contains lecture outlines with tips and suggested enrichment topics, page-referenced key terms found in the chapter, food-for-thought questions that can be used to provide students with something to think about as they read the next assignment from the book, and a list of film and video sources that might prove valuable in lectures.

Computerized Testing

The test bank is also available in a form that allows tests to be computer generated. Contact your West sales representative for details.

Study Guide

A study guide with helpful review items is available to supplement the text's built-in study guide material. The study guide is also available in electronic form on disk for learning laboratory usage.

Transparency Acetates and Slides

A set of transparency masters of important diagrams and acetates of key full-color art pieces will be available from the publisher for adopters to use in classes. Many combine photos and art. A slide set with other important pieces of art and photographs from the text will also be available.

Laboratory Manual

A laboratory manual with 33 class-tested lab exercises for introductory biology classes is available.

Videotapes

West offers a video library of films from which adopters of the text may choose. Contact your local West sales representative for further information.

Videodisk

A videodisk is available from West with animations, some of which are art from the book, still images (photos and art), and film clips on relevant topics.

ACKNOWLEDGEMENTS

A project of this magnitude is the fruit of a great many people. I wish to thank the thousands of scientists and teachers who have contributed to our understanding of the web of life. A special thanks to the extraordinary teachers who have made tremendous contributions to my education, especially the late Dr. Weldon Spross, the late Dr. H. T. Gier, Dr. Gilbert Greenwald, Dr. Floyd Foltz, and Dr. Douglas Poorman.

I am also deeply indebted to many people for their assistance during the writing of this book. A special thanks to my editors, Jerry Westby and Theresa O'Dell, for their exceptional patience, guidance, perserverance, and inspiration. Thanks to my outstanding production editor, Tom Hilt, for his calmness in the midst of turmoil, cordiality, attention to detail, and diligence. My appreciation goes to our copyeditor, Bill Waller, for a skillful job of editing. A word of thanks also to my colleague Dr. John Cunningham of Visuals Unlimited who supplied the excellent photographs. Thanks also go to my team of artists: Carlyn Iverson, Elizabeth Morales-Denney, Publications Services, Pat Rossi, Cyndie Wooley, and J/B Woolsey & Associates. Thanks also to Richard Shippee for his excellent review of the artwork. I greatly appreciate West's marketing team, Ann Hillstrom and Amelia Jacobson, and the many sales representatives who have helped make this book a success. It has been a pleasure and an honor to have worked with such a fine and talented group of people.

Thanks also to the many authors who contributed the Point/Counterpoints in this book. Your work will make this a more exciting journey, and will help students see the practical applications of the study of biology.

Throughout this extraordinarily difficult time, my wife, Kathleen, and our two sons, Skyler and Forrest, have offered considerable support and a much-needed counterbalance to the stresses and strains of a project of this scope. A very special thanks to Kathleen for acquiring the Point/Counterpoints in this volume and making all of the copyediting changes on disk. Thanks to all of you for seeing me through this project and helping me along.

Finally, a special thanks to all the reviewers who offered many useful comments throughout this project. Their insights and attention to detail have been greatly appreciated. A list of those who have reviewed the manuscript is found on the next page.

Reviewers

D. DARYL ADAMS
Mankato State University

JAMES S. BACKER
St. Vincent College

WILLIAM E. BARSTOW
University of Georgia

LAWRENCE J. BELLIPANNI
University of Southern Mississippi

CLYDE E. BOTTRELL
Tarrant County Junior College

KATHLEEN BURT-UTLEY
University of New Orleans

ROY B. CLARKSON
West Virginia University

BARBARA CRANDALL-STOTLER
Southern Illinois University

KENNETH J. CURRY
University of Southern Mississippi

PETER DALBY
Clarion University

GARY E. DOLPH
Indiana University at Kokomo

JAMIN EISENBACH
Eastern Michigan University

SALLY K. FROST-MASON
University of Kansas

BERNARD L. FRYE
University of Texas-Arlington

ROBERTA GIBSON
Glendale Community College

MADELINE M. HALL
Cleveland State University

LASZLO HANZELY
Northern Illinois University

JOHN P. HARLEY
Eastern Kentucky University

ANN T. HARMER
Orange Coast College

JAMES G. HARRIS
Utah Valley Community College

WILLIAM R. HAWKINS
Mt. San Antonio College

RONALD K. HODGSON
Central Michigan University

ROBERT R. HOLLENBECK
Metropolitan State University

HARRY R. HOLLOWAY
University of North Dakota-Main Campus

GEORGE A. HUDOCK
Indiana University, Bloomington

ARTHUR B. JANTZ
Western Oklahoma State University

TOM KANTZ
California State-Sacramento

ARNOLD J. KARPOFF
University of Louisville

GEORGE KLEE
Kent State University

MARK LEVINTHAL
Purdue University

JAMES LUKEN
Northern Kentucky University

ANN S. LUMSDEN
Florida State University

KATHY MARTIN
Central Connecticut State University

PATRICIA MATTHEWS
Grand Valley State University

PRISCILLA MATTSON
University of Lowell

HEATHER McKEAN
Eastern Washington University

MICHAEL P. McKINLEY
Glendale Community College

JAN MERCER
Tarrant County Junior College

JAMES E. MICKLE
North Carolina State University

GLENDON R. MILLER
Wichita State University

STEVEN N. MURRAY
California State University—Fullerton

DEBRA K. PEARCE
Northern Kentucky University

CHRIS E. PETERSEN
College of Dupage

BARBARA Y. PLEASANTS
Iowa State University

DAVID J. PRIOR
Northern Arizona University

RODNEY A. ROGERS
Drake University

EDWIN DANIEL SCHREIBER
Mesa Community College

JAY TEMPLIN
Widener University

ROBERTA WILLIAMS
University of Nevada

STEPHEN WILLIAMS
Glendale Community College

LARRY D. WILSON
Miami Dade Community College

DANIEL D. CHIRAS
Evergreen, Colorado

Study Skills

College is a demanding time in the lives of many students. Term papers, tests, reading assignments, and classes require a new level of commitment and intellectual activity. The work load can become overwhelming and frustrating. Fortunately, there are ways to lighten the load, to manage your time efficiently, to increase your chances of getting good grades, and to improve your knowledge and understanding.

This section offers some suggestions to help you manage your studies and improve your mastery of the subjects you study. If you already are adept and efficient at studying and taking tests, you may still benefit from reading this section. Every suggestion you can use to your advantage will help.

To begin, read over the suggestions listed below. Pick a few that seem right for you under each category, then put them into action. Applying these ideas could pay huge dividends—not just in college, but throughout your entire life. Learning is a lifetime endeavor, and those that learn fastest seem to get the most out of life.

Mastering basic study skills will require some work at first and may require that you break some bad habits. In the long run, the additional time investment could save you lots of time, help you get better grades, and become a more efficient learner. Most important, it could help you improve your knowledge and understanding.

GENERAL STUDY SKILLS

- Study in a quiet, well-lighted space.
- Work at a desk or table. Don't lie on a couch or bed.
- Establish a specific time each day to study and stick to your schedule.
- Study when you are most alert. Many people find they retain more if they study in the evening a few hours before bedtime.
- Let your friends and family know when you study and ask them to respect that time.
- Take frequent breaks—one every hour or so. Exercise or move around during your study breaks to help you stay alert.
- Reward yourself after a study session with an ice cream cone or a mental pat on the back.
- Study each subject every day to avoid cramming for tests. Some courses may require more hours than others, so adjust your schedule accordingly.
- Look up new terms or words whose meaning is unclear to you in the glossaries in your textbooks or in a dictionary.

IMPROVING YOUR MEMORY

You can improve your memory by following the PMC method. The PMC method involves three simple learning steps: (1) paying attention, (2) making information memorable, and (3) correlating new information with facts you already know.

Step 1. Paying attention means taking an active role in your education—taking your mind out of neutral. Eliminate distractions when you study. Review what you already know and formulate questions about what you are going to learn *before* a lecture or *before* you read a chapter in the text. Reviewing and questioning help prime the mind.

Step 2. Making information memorable means finding ways to help you retain information in your memory. Repetition, mnemonics, and rhymes are three examples.

- Repetition can help you remember things. The more you hear or read something, the more likely you are to remember it. Scribble notes while you read or study.
- You can also use learning tools, mnemonics, to help remember lists. For example, *k*eep *p*iling *c*hocolate *o*n *f*or *g*oodness *s*akes helps you remember the taxonomic classification scheme: kingdom, phylum, class, order, family, genus, and species.
- Rhymes and sayings are also helpful. If you are having trouble remembering a list of facts, try making up a rhyme.
- If you're having trouble remembering key terms, look up their roots in the dictionary. This helps them stick in your memory. Use the list of prefixes, suffixes, and roots on the back endsheets of this book.
- Draw pictures and diagrams of processes.

Step 3. Correlating with things you know means tying facts together or making sense of the bits and pieces of information you are learning and have learned previously.

- Instead of filling your mind with disjointed facts and figures, try to see how they relate to previous information you have learned. Stop and scan your memory for similar facts. Correlating facts with previous knowledge enables you to comprehend the big picture. The end-of-chapter questions will assist you in this function.
- After studying your notes or reading a chapter in your textbook, determine the main points. How does the new information you have learned fit into your view of life or the general subject under discussion? How can you use the information?

BECOMING A BETTER NOTE TAKER

- Spend 5–10 minutes before each lecture reviewing the material you learned in the previous lecture. This is extremely important for learning.
- Know the topic of each lecture *before* you enter the class. Spend a few minutes *before* each class reflecting on facts you already know about the subject that is to be discussed.
- If possible, read the text *before* each lecture. If not, at least look over the main headings in the chapter, read the topic sentences, and look over the figures.
- Develop a shorthand system of your own. Symbols such as = (equals), w/o (without), w (with), > (greater than), < (less than), ↑ (increase), and ↓ (decrease) can save you time.
- Develop special abbreviations. For example, if you find yourself writing the word human over and over again in your notes, abbreviate it to H. Muscle could be abbreviated as m or mm. Species is sometimes abbreviated sp.
- Omit vowels and abbreviate words to decrease writing time (for example: omt vwls shrten wrtng tme). This takes some practice.
- Don't take down every word your professor says, but be sure your notes contain the main points, supporting information, and important terms.
- Watch for signals from your professor indicating important material ("This is an extremely important point. . . ").
- If possible, sit near the front of the class to avoid distractions.
- Review your notes soon after lecture while they're still fresh in your mind. Be sure to leave room in your notes during class to add material you missed. Recopy your notes if you have the time.
- Compare your notes with those of your classmates to be sure you understood everything and did not miss anything important.
- Attend lecture regularly.
- Use a tape recorder, if necessary and if it's acceptable to your professor, if you have trouble catching all the points.
- If your professor talks too quickly, politely ask him or her to slow down.
- If you are unclear about a point, ask during class. Chances are other students are confused as well. If you are too shy, go up after lecture and ask, or visit your professor during his or her office hours.

HOW TO GET THE MOST OUT OF WHAT YOU READ

- Before you read a chapter or other assigned readings, preview the material by reading the main headings or chapter outline to see how the material is organized.
- Pause over each heading and ask a question or two about each main heading.
- Next, read the first sentence of each paragraph. When you have finished, turn back to the beginning of the chapter and read it thoroughly.
- Take notes in the margin or on a separate sheet of paper. Underline or highlight key points.
- Don't skip terms that are confusing to you. Look them up in the glossary in the back of your textbook or in a dictionary. Make sure you understand each term before you move on.
- Use the study aids in your textbook, including end-of-chapter questions and summaries. Don't just look over the questions and say, "Yeah, I know that." Write out the answer to each question as if you were turning it in for a grade and save your answers for later study. Look up answers to questions that confuse you. This book has questions that test your understanding of the terms, concepts, and processes. Critical thinking questions are also included to help you sharpen your critical thinking skills.

PREPARING FOR TESTS

- Don't fall behind on your reading assignments and review lecture notes as frequently as possible.
- If you have the time, you may want to outline your notes and your assigned readings. Try to prepare the outline with your book and notes closed. Determine weak areas, then go back to your text or class notes to study those areas.
- Space your study to avoid cramming. One week before your exam, go over all of your notes. Study for two nights, then take a day off. Study again for a couple of days. Take another day off, then make one final push before the exam, being sure to study not only the facts and concepts, but also how the facts are related. Unlike cramming, which puts a lot of information into your brain for a one-time event, spacing will help you retain information for the test and for the rest of your life.

- Be certain you can define all terms and give examples of how they are used.
- Draw key structures over and over until they stick in your memory.
- You may find it useful to write flash cards to review terms and concepts.
- After you have studied your notes and learned the material, look at the big picture—the importance of the knowledge and how the various parts fit together.
- You may want to form a study group to discuss what you are learning and to test one another.
- Attend review sessions offered by your instructor or by your teaching assistant. Study before the review session and go to the session with questions.
- See your professor or class teaching assistant with questions as they arise.
- Take advantage of free or low-cost tutoring offered by your school or, if necessary, hire a private tutor to help you through difficult material. Get help quickly, though. Don't wait until you are way behind. Remember that learning is a two-way street. A tutor won't help unless you are putting in the time.
- If you are stuck on a concept, it may be that you may have missed some important previous material. Look back over your notes or ask your tutor or professor what facts might be missing and causing you to be confused.
- If you have time, write and take your own tests, including all types of questions.
- Study tests from previous years, if they are available legally.
- Determine how much of a test will come from notes and how much will come from the textbook.

TAKING TESTS

- Eat well and get plenty of exercise and sleep before tests.
- Remain calm during the test by deep breathing.
- Arrive at the exam early or on time.
- If you have questions about the wording of a test question, ask your professor. Don't be shy.
- Look over the entire test first so you can budget your time.
- Skip questions you can't answer right away and come back to them at the end of the session if you have time.
- Read each question carefully and be sure you understand its full meaning before you answer it.
- For essay questions and definitions, organize your ideas on a piece of scrap paper or the back of the test *before* you start writing.

Now take a few moments to go back over the list. Check off those things you already do. Then, mark the new ideas you want to incorporate into your study habits. Make a separate list, if necessary, and post it by your desk or on the wall and keep track of your progress. Good luck!

West's Commitment to the Environment

In 1906, West Publishing Company began to recycle gold shavings, sheepskin scraps, and paper trimmings left over from the production of books. This began a tradition of efficient and responsible use of resources that has continued longer than most printers and publishers have been in existence.

Today, West Publishing's operation is environmentally conscious in almost every aspect of its operation. West was one of the first publishers to use recycled paper in books. Today, up to 95% of our legal books and 65 to 70% of our college texts are printed on recycled, acid-free stock. In addition, West recycles virtually all scrap paper left over from production. In the production plant, over 22 million pounds of scrap paper is recycled annually—the equivalent of 182,000 trees.

Being a responsible publisher and printer goes beyond paper. Since the 1960s, West has also recycled steel, sheet metal, and other metals. We have devised ways to capture and recycle waste inks, solvents, oils, and vapors created in the printing process and have eliminated the use of styrofoam book packaging. Today, we also recycle plastics of all kinds, wood, glass, corrugated cardboard and batteries.

Employees at West's home office and production plant are provided with containers for recycling aluminum, glass, and paper. Employees are encouraged not only to recycle, but also to suggest additional ways to use our resources more efficiently. It is employee initiative and involvement that has contributed to West's long-standing tradition of environmental responsibility.

When you order a West book, you get more than a token commitment to a clean environment. We thought you would appreciate knowing this.

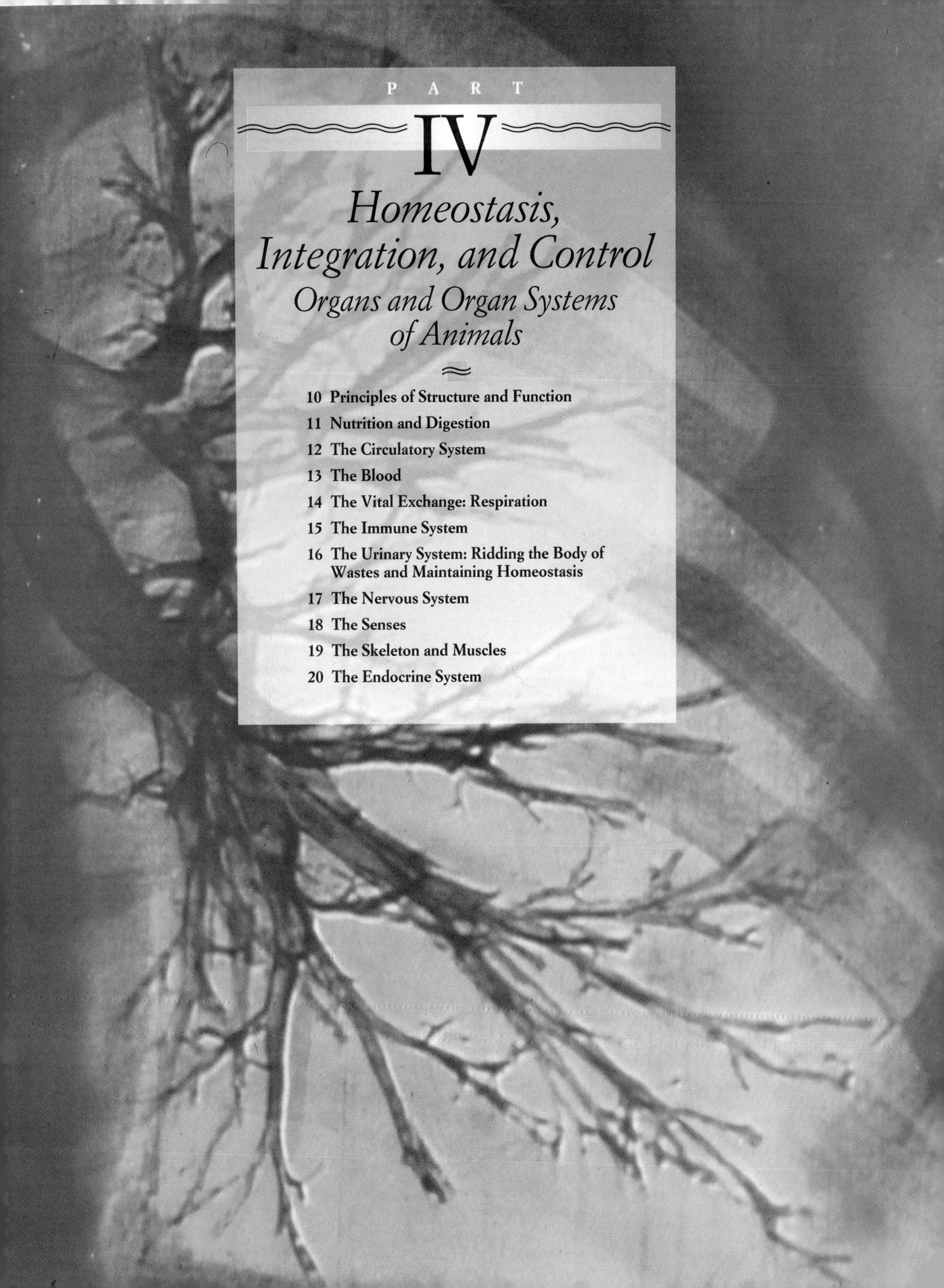

PART

IV

Homeostasis, Integration, and Control

Organs and Organ Systems of Animals

CHAPTER

10

Principles of Structure and Function

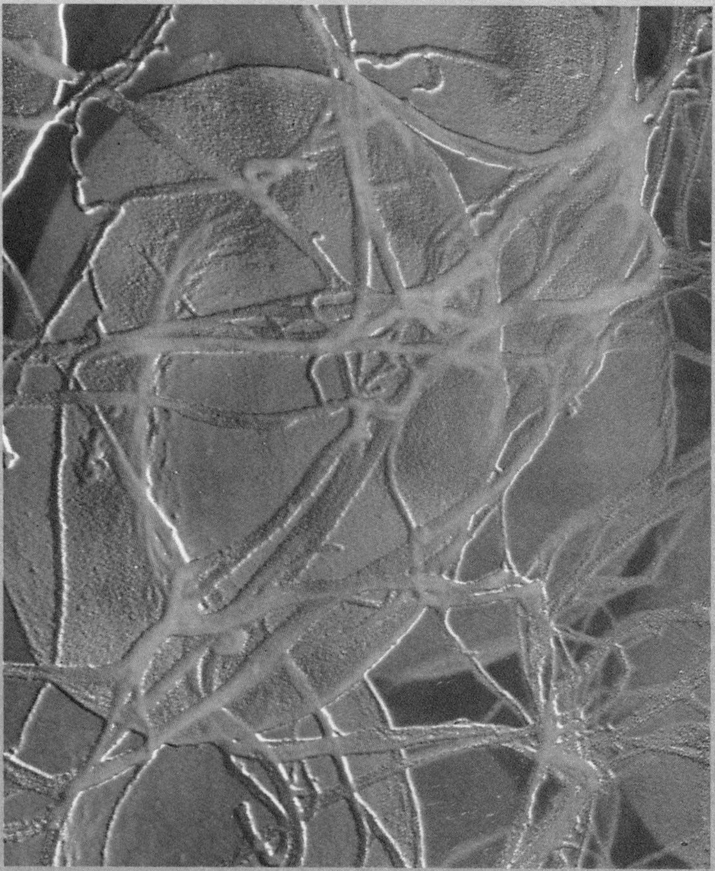

Colorized electron micrograph of red blood cells trapped in fibrin network of a blood clot.

A human being, a horse, and a bald eagle are strikingly different organisms. One walks on two feet, the second on four hooves, and the third depends on powerful wings to soar above the landscape. Despite their marked differences, these three organisms share several features. First, they are all members of the animal kingdom, a large and diverse group of multicellular organisms. Second, the cells of each of these uniquely adapted organisms are arranged in tissues, which, in turn, combine to form various organs, such as the heart, stomach, and brain. Third, each organism contains a number of homeostatic mechanisms that maintain internal constancy, permitting survival in an ever-changing world.

This chapter focuses on the second feature, the organizational pattern common to many multicellular animals, and the third feature, homeostasis. Throughout this chapter, and in the remaining chapters of Part IV, we will focus on organ systems and their role in homeostasis, studying representative members of each major group of animals.

FROM CELLS TO ORGAN SYSTEMS

To begin our study of the structure and function of animals, we look at the beginning of life in sexually reproducing vertebrate (backboned) animals to put things in context. Life begins at fertilization. The fertilized ovum contains all of the information needed to develop into a fully functional adult and all of the information needed to control complex life functions, such as growth, reproduction, and homeostasis.

Cells Unite to Form Tissues, and Tissues Combine to Form Organs

During embryonic development, the fertilized ovum divides many times, first producing a ball of cells called the **morula** (▻ Figure 10–1). Soon thereafter, the cells undergo a process called **differentiation,** during which they become structurally and functionally specialized. The first sign of differentiation is the emergence of three distinct cell types: ectoderm, mesoderm, and endoderm. **Ectoderm** lies on the outside of the embryo and gives rise to the skin, the eyes, and the nervous system. **Mesoderm** lies in the middle and forms muscle, bone, and cartilage. **Endoderm** is the innermost layer and forms the lining of the intestinal tract and several digestive glands.

During embryonic development in most animals, ectodermal, mesodermal, and endodermal cells give rise to a variety of highly differentiated cell types, each of which carries out very specific functions. These specialized cells are often bound together by extracellular fibers and other extracellular materials, forming **tissues** (from the Latin "to weave"). Extracellular materials may be liquid (as in blood), semisolid (as in cartilage), or solid (as in bone). Tissues, in turn, combine to form **organs,** discrete structures in the body that carry out specific functions that benefit the entire organism. Organs, like their counterparts in cells, the organelles, provide for a division of labor.

Plants undergo a similar process during development, forming specialized cells and tissues. In addition, the cells and tissues of a plant give rise to highly specialized structures, such as leaves and roots, that carry out specific functions much as the organs of animals do. (The structure and function of plants are described in Chapter 26.)

Cells Combine to Form Four Primary Tissues

Four major tissue types are found in vertebrate animals: epithelial, connective, muscle, and nervous. These are called **primary tissues.** Table 10–1 lists the primary tissues and shows that each has two or three subtypes. Muscle tissue, for instance, comes in three varieties: cardiac, skeletal, and smooth. Cardiac muscle cells are found exclu-

▻ **FIGURE 10–1 Early Embryonic Development in Humans** The fertilized ovum (*a*) divides mitotically, eventually forming a solid ball of cells, the morula (*b*), which later becomes a hollow blastocyst (*c*). The inner cell mass shown here becomes the embryo, differentiating into ectoderm, endoderm, and mesoderm in later stages of development (*d*).

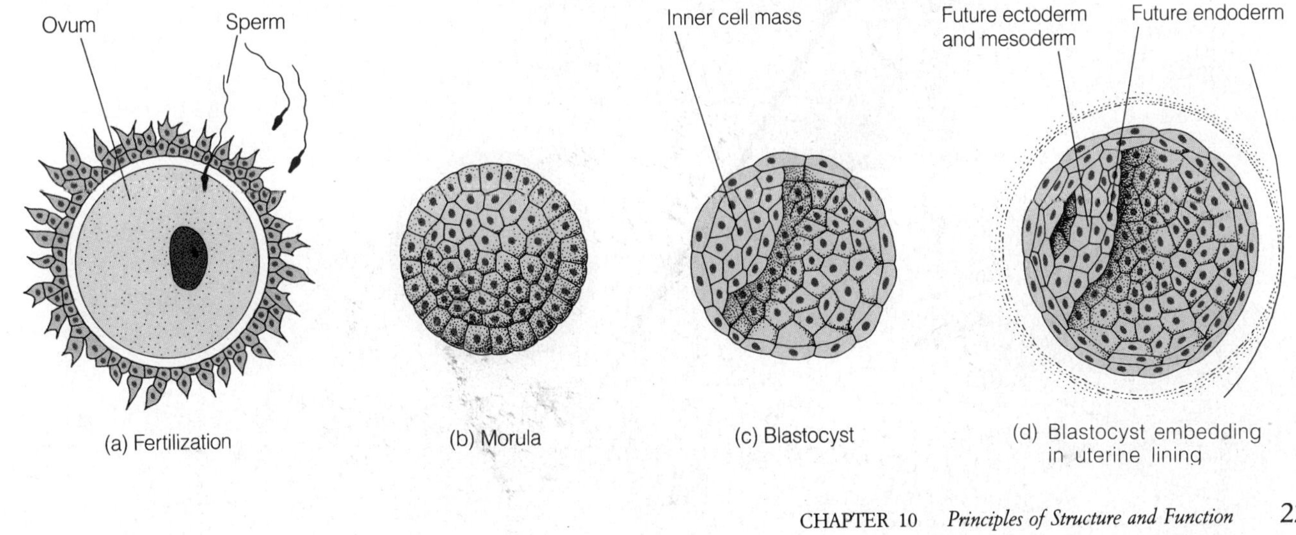

sively in the heart. Skeletal muscle cells are located chiefly in body muscles. Smooth muscle cells are found in the walls of the stomach, intestinal tract, and blood vessels.

The primary tissues exist in all organs in varying amounts. The lining of the stomach, for example, consists of a single layer of epithelial cells called the surface epithelium (▻ Figure 10–2). Just beneath the lining is a layer of connective tissue. A thick sheet of smooth muscle cells forms the bulk of the stomach wall. Smooth muscle cells are also found in blood vessels supplying the tissues of the stomach. Nerves enter with the blood vessels and control the flow of blood.

TABLE 10–1 The Primary Tissues and Their Subtypes

Epithelial tissue	Connective tissue
Membranous	Connective tissue proper
Glandular	Specialized connective tissue
Muscle tissue	Blood
Cardiac	Bone
Skeletal	Cartilage
Smooth	
Nervous tissue	
Conductive	
Supportive	

Epithelium Forms the Lining or External Covering of Organs and Also Forms Glands

Epithelial tissue exists in two basic forms: glandular and membranous. The **membranous epithelia** consist of sheets of cells tightly packed together, forming the external coverings or linings of organs. ▻ Figure 10–3 shows some of the remarkable variety in epithelial membranes and illustrates the presence of two basic types of membranous epithelium: **simple epithelia,** consisting of a single layer of cells, and **stratified epithelia,** consisting of many layers of cells.

Glandular epithelia are clumps of cells that form many of the glands of the body. Epithelial glands arise during embryonic development from tiny invaginations of surface epithelia, as illustrated in ▻ Figure 10–4. Some glands remain connected to the epithelium by hollow ducts and are called **exocrine glands** (glands of external secretion); products of the exocrine glands flow through ducts into some other body part. In humans, sweat glands in the skin are exocrine glands that produce a clear, watery fluid that is released onto the surface of the skin by small ducts. This fluid evaporates from the skin, helping cool the body. Salivary glands are another type of exocrine gland. Located around the oral cavity, the salivary glands produce saliva, a fluid that is released into the mouth via small ducts.

Some glandular epithelial cells break off completely from their embryonic source, as shown in Figure 10–4, to form **endocrine glands** (glands of internal secretion). The endocrine glands produce hormones that are released from

▻ **FIGURE 10–2 Human Stomach** (*a*) The stomach, like all organs, contains all four primary tissues. (*b*) These are shown here in cross section.

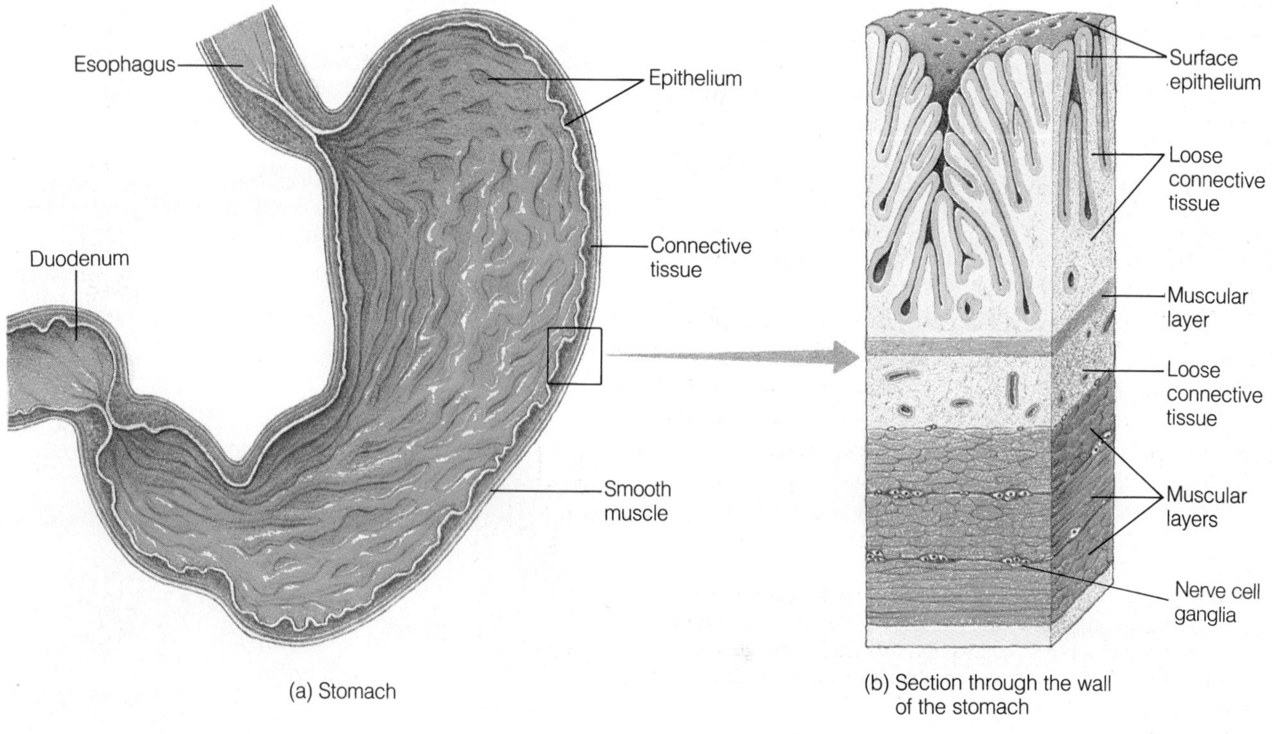

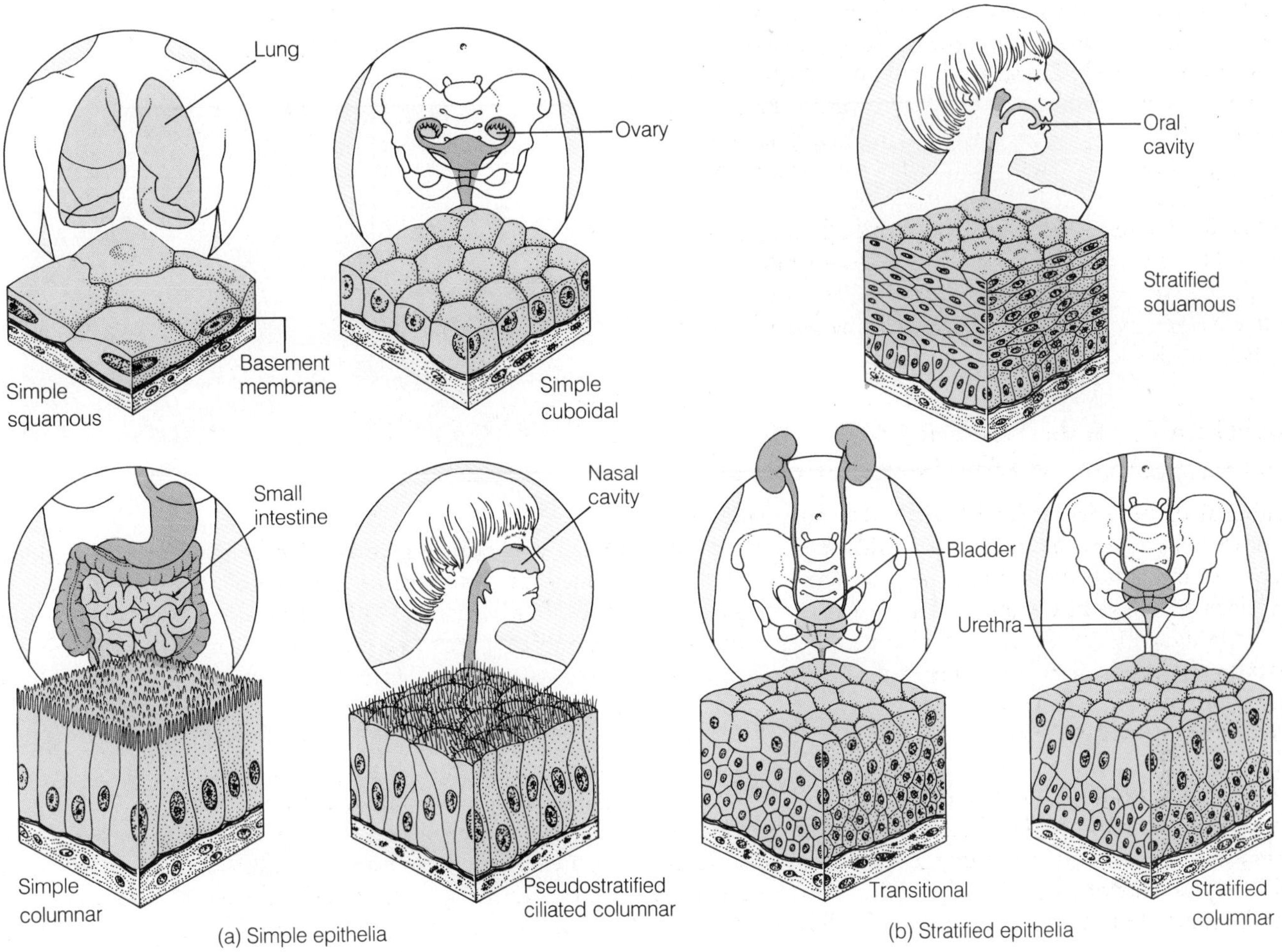

▻ **FIGURE 10–3 Membranous Epithelia** Single-celled (simple) epithelia (*a*) and stratified epithelia (*b*) exist in different parts of the body.

the cell and diffuse into the bloodstream, where they travel to other parts of the body (Chapter 20).

Epithelial Tissues Illustrate the Basic Biological Principle That Structure Correlates with Function. One of the basic rules of architecture is that form (the structure of a building) often follows function—in other words, architectural design reflects underlying function. Animals exhibit a similar relationship achieved through evolution.

The membranous epithelia provide many examples of the correlation between structure and function. A good example is the outer layer of the human skin, the **epidermis,** which, among other things, protects us humans from water loss. Shown in ▻ Figure 10–5, the epidermis con-

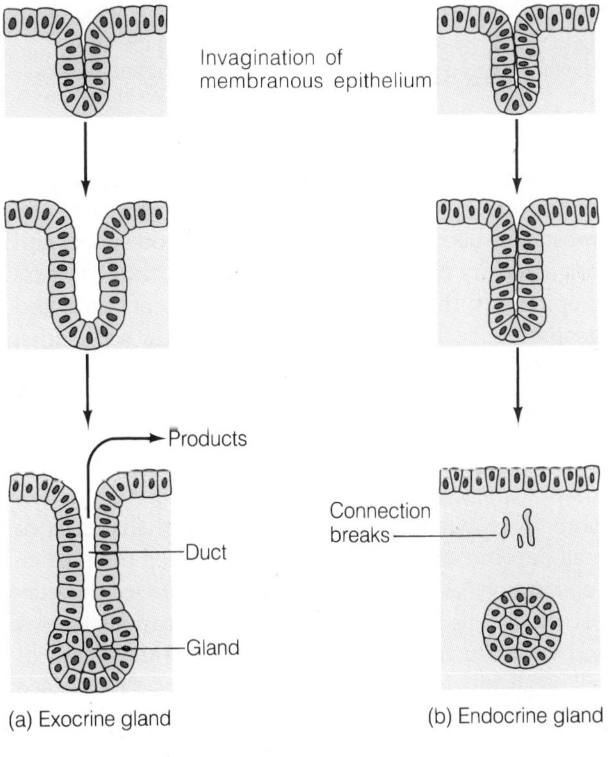

▻ **FIGURE 10–4 Formation of Endocrine and Exocrine Glands** (*a*) Exocrine glands arise from invaginations of membranous epithelia that retain their connection. (*b*) Endocrine glands lose this connection and must secrete their products into the bloodstream.

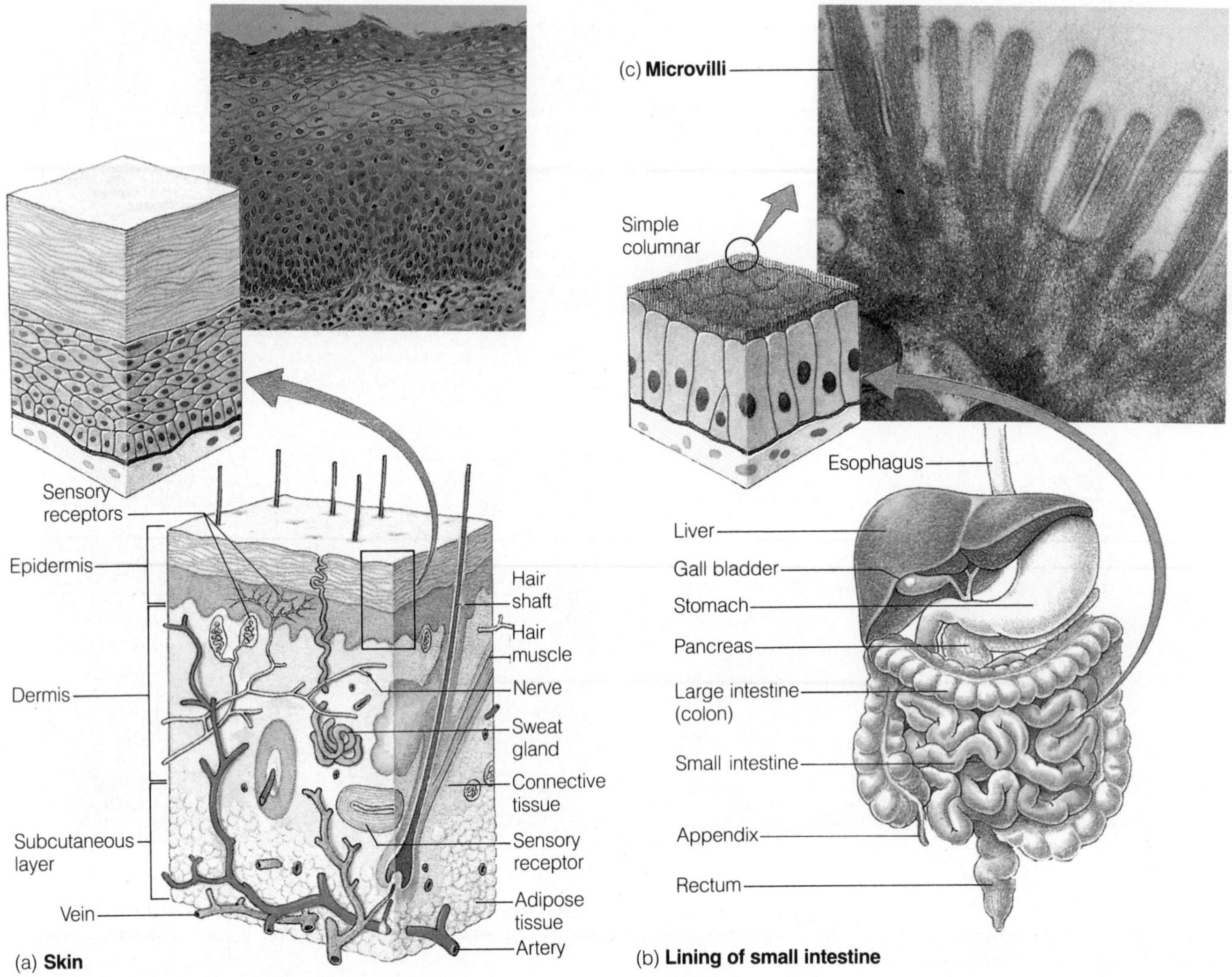

▻ **FIGURE 10–5 Comparison of Two Epithelia with Markedly Different Functions** (*a*) A cross section of the skin showing the stratified squamous epithelium of the epidermis (above), which protects underlying skin from sunlight and dessication, and (*b*) the simple columnar epithelium of the lining of the small intestine, which is specialized for absorption. (*c*) The plasma membranes of the cells lining the intestine are thrown into folds (microvilli) that greatly increase the surface area for absorption.

sists of numerous cell layers. The cells flatten toward the surface and are tightly joined by special connections. The outermost cells become isolated from the blood supply and die, forming a dry protective layer. Together, the thickness of the epidermis, the adhesion of one cell to another, and the dry protective layer of dead cells impede water loss. They also present a formidable barrier to microorganisms.

Another example of the relationship between structure and function is the epithelium of the small intestine. As shown in Figure 10–5b, the lining of the small intestine consists of a single layer of columnar cells, uniquely suited to absorb food materials. The columnar epithelial cells of the small intestine are also structurally modified to enhance food absorption. As illustrated in Figure 10–5c, the surfaces of these cells have numerous tiny protrusions known as **microvilli,** which markedly increase the surface area of the cell available for absorption. The larger the surface area, the more efficient is food absorption.

Connective Tissue Binds the Cells and Organs of the Body Together

As the name implies, **connective tissue** is the body's glue. Connective tissues bind cells and other tissues together and are present in all organs in varying amounts.

The body contains several types of connective tissue, each with specific functions (▻ Figure 10–6). Despite the differences, all connective tissues consist of two basic components: cells and varying amounts of extracellular material. Two types of connective tissue will be discussed here: connective tissue proper and the specialized connective tissues—bone, cartilage, and blood.

Connective Tissue Proper Consists of Two Subtypes, Determined by the Relative Proportion of Fibers and Cells. **Connective tissue proper** is an important structural component of the vertebrate body and is

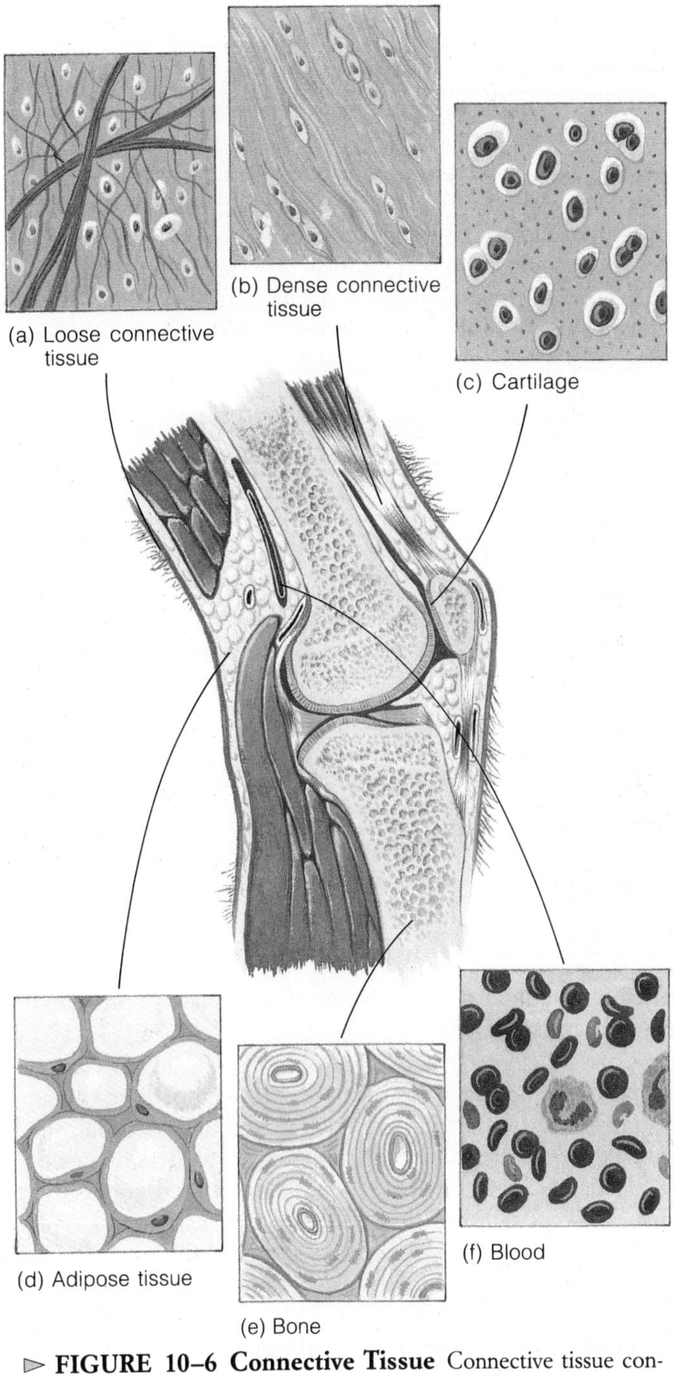

▻ **FIGURE 10–6 Connective Tissue** Connective tissue consists of many diverse subtypes.

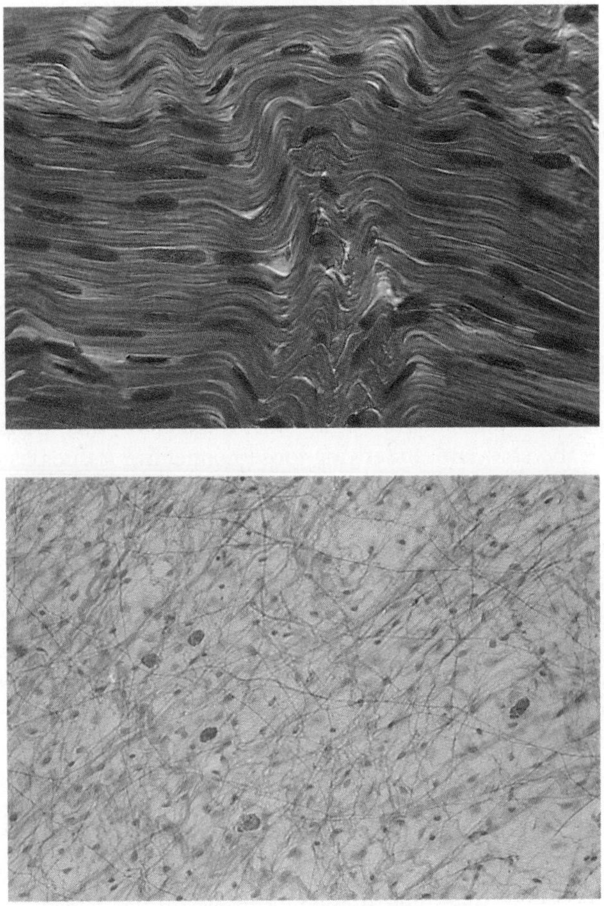

▻ **FIGURE 10–7 Light Micrographs of Connective Tissue** (*a*) Dense connective tissue. (*b*) Loose connective tissue.

composed of dense connective tissue and loose connective tissue. **Dense connective tissue (DCT)** consists primarily of densely packed fibers produced by cells interspersed between them. DCT is found in ligaments and tendons (▻ Figure 10–7a). Ligaments join bones to bones at joints and provide support for joints; tendons join muscle to bone and aid in body movement.[1] The layer of the skin underlying the epidermis, the **dermis,** is also dense connective tissue, although the fibers are less regularly arranged than those in ligaments and tendons (see Figure 10–5a). The dermis binds the epidermis to underlying muscle and bone.

Loose connective tissue (LCT) is the body's packing material. As shown in Figure 10–7b, LCT contains cells in a loose network of collagen and elastic fibers. Both of these fibers are made of protein. Loose connective tissue forms around blood vessels in the body and in skeletal muscles, where it binds the muscle cells together. Loose connective tissue also lies beneath epithelial linings of the intestines and trachea, anchoring them to underlying structures.

The chief difference between dense and loose connective tissues lies in the ratio of cells to extracellular fibers. As you can see in Figure 10–7, dense connective tissue has far more fibers than loose connective tissue.

The extracellular fibers found in dense and loose connective tissue are produced by a connective tissue cell known as the **fibroblast.** In addition to producing the extracellular fibers that hold many tissues together, fibroblasts repair damage created by cuts or tears in body tissues. When the skin is cut, for example, fibroblasts in the dermis migrate into the injured area, where they begin producing large quantities of collagen. The collagen fibers

[1] I use the mnemonic *LBJ* to remember that ligaments connect bones at joints.

fill the wound, closing it off. The epidermis soon begins to grow over the damaged area, helping repair the damage and restoring the integrity of the skin. When the cut is small, the epidermis covers the damaged area completely, leaving no scar, but in larger wounds, the epidermal cells are often unable to grow over the entire wound, leaving some of the underlying collagen exposed and producing a scar.

Besides helping to hold the body together, loose connective tissue gives refuge to several cell types that protect us against bacterial and viral infections. One of the most important of the protective cells is the **macrophage** ("big eater"). Containing numerous lysosomes to digest foreign material, macrophages engulf microorganisms that penetrate the skin and underlying loose connective tissues as a result of injury. This helps prevent bacteria from spreading to other parts of the body and producing a systemic infection—that is, one that affects the entire organism. Macrophages also play a role in immune protection, which will be discussed in Chapter 15.

Loose connective tissues also contain lymphocytes and neutrophils, two types of blood cells that play an important role in protecting the body from invaders described in later chapters.

Some loose connective tissues contain large, conspicuous fat cells. The **fat cell** is one of the most distinctive of all body cells. When fully formed, it contains huge fat globules that occupy virtually the entire cell, pressing the cytoplasm and the nucleus to the periphery. Fat cells occur singly or in groups of varying size. Large numbers of fat cells in a given region form a modified type of loose connective tissue known as **adipose tissue,** or, less glamorously, fat. Adipose tissue is an important storage depot for lipids, particularly triglycerides, which are used as an energy source under certain conditions. Fatty deposits also provide a degree of heat-conserving insulation for humans and many other mammals and some birds, especially those that live in aquatic environments where heat loss can be substantial.

For humans, fatty deposits often become unsightly. To rid the body of these deposits, individuals can begin exercise programs and reduce their intake of food. The combination of the two can prove quite effective. Others opt for a surgical measure called **liposuction** (▻ Figure 10–8). In this procedure, a small incision is made in the skin through which surgeons insert a suction device to aspirate fat deposits under the skin in various locations, such as the thighs, buttocks, and abdomen. The fat cells extracted from one region can even be transferred to other regions, such as the breast, to resculpt the human body. Liposuction is a relatively safe technique, but it is not free from risk.

▻ **FIGURE 10–8 Liposuction** During liposuction surgery, the physician aspirates fat from deposits lying beneath the skin, helping to reduce unsightly accumulations.

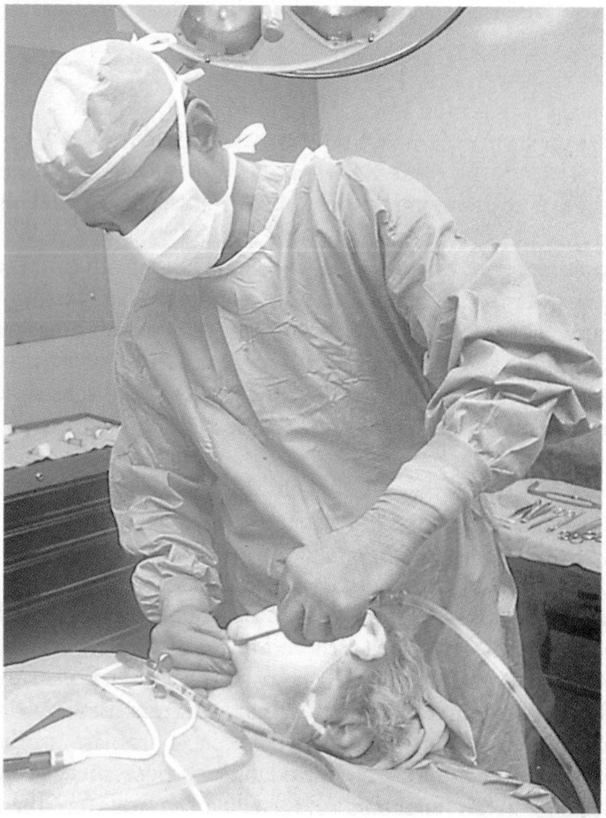

The Specialized Connective Tissues Are Structurally and Functionally Modified to Perform Specific Functions Essential to Homeostasis

The body contains three types of specialized connective tissue: cartilage, bone, and blood.

Cartilage. Cartilage consists of cells embedded in an abundant and rather impervious extracellular material, the **matrix** (▻ Figure 10–9). Surrounding virtually all types of cartilage is a layer of dense, irregularly packed connective tissue, the **perichondrium** ("around the cartilage"). This layer contains the blood vessels that supply nutrients to cartilage cells through diffusion. No blood vessels penetrate the cartilage itself. Because cartilage cells are nourished by diffusion from perichondral capillaries, damaged cartilage heals very slowly. For this reason, joint injuries that involve the cartilage often take years to repair or may not heal at all.

Three types of cartilage are found in vertebrates: hyaline, elastic, and fibrous. Each serves a special purpose, which is reflected in its underlying structure. The most prevalent type of cartilage is **hyaline cartilage** (Figure 10–9a). Hyaline cartilage contains numerous collagen fibers, which appear white to the naked eye. Found on the ends of many bones, hyaline cartilage greatly reduces friction, so bones move over one another with ease. It also makes up the bulk of the nose and is found in the larynx (voice box) and the rings of the trachea, which you can feel below the larynx. The ends of the ribs that join to the

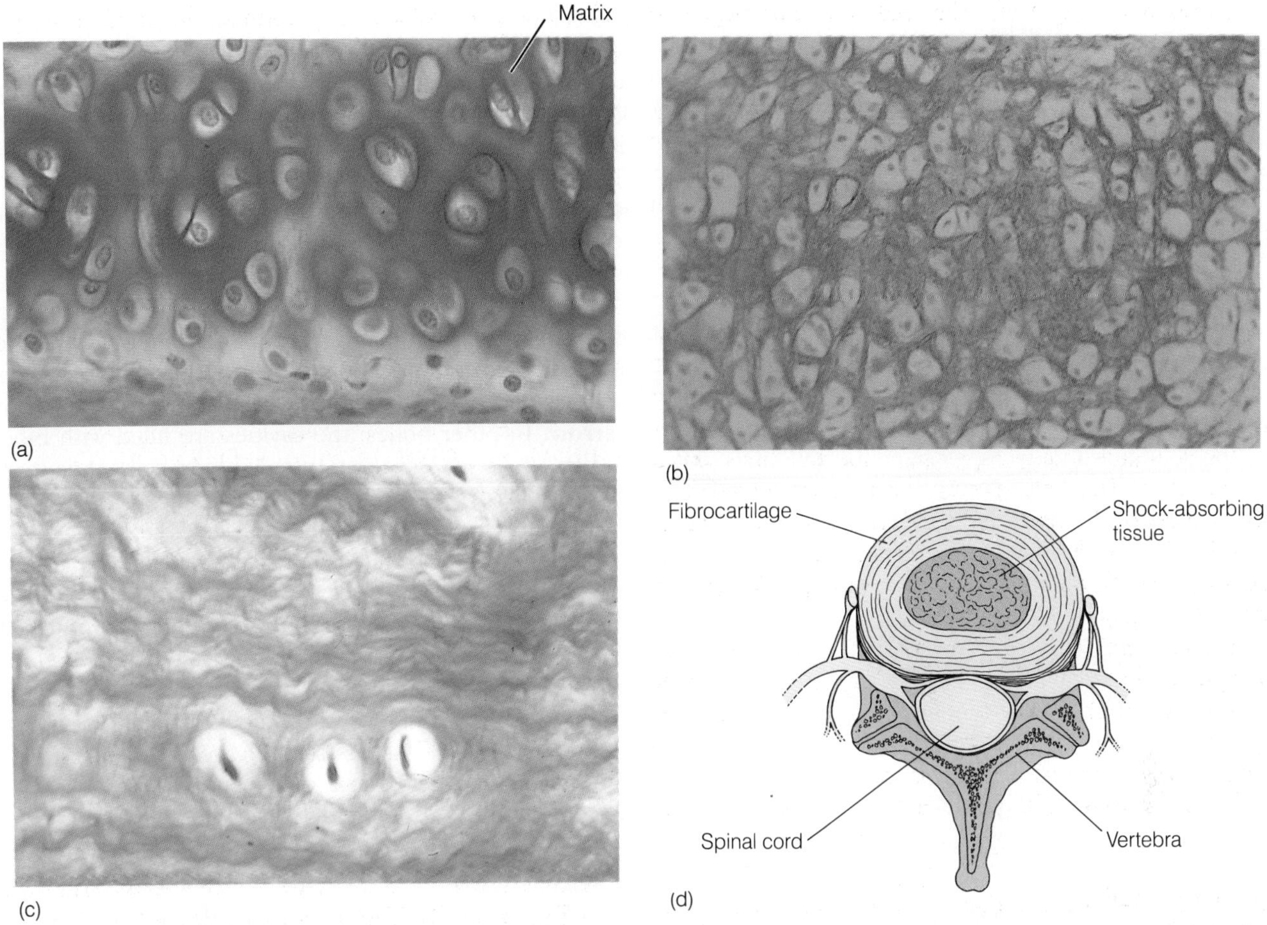

▻ **FIGURE 10–9 Light Micrographs of Cartilage** (*a*) Hyaline. (*b*) Elastic. (*c*) Fibrous. (*d*) Intervertebral disk showing arrangement of fibrocartilage in a protective ring around the soft, spongy part of the disk that absorbs shock.

sternum (breast bone) are composed of hyaline cartilage. In embryonic development, the first skeleton is hyaline cartilage, much of which is later converted to bone.

Elastic cartilage is similar to hyaline cartilage but contains many wavy elastic fibers, which give it much greater flexibility (Figure 10–9b). Elastic cartilage is found in regions where support and flexibility are required—for example, the external ears and eustachian tubes, cartilaginous ducts that help equalize pressure in the inner ear (Chapter 18).

Fibrocartilage is the rarest of all cartilage. Like hyaline cartilage, it consists of an extracellular matrix containing numerous bundles of collagen fibers. As shown in Figure 10–9c, however, fibrocartilage contains far fewer cells than either hyaline or elastic cartilage.

Fibrocartilage is found in the outer layer of **intervertebral disks,** the shock-absorbing tissue between the vertebrae of the spine (Figure 10–9d). An intervertebral disk consists of a soft, cushiony central region that absorbs shock. Fibrocartilage forms a ring around the central portion of the disk, holding it in place. Over time, the fibrocartilage ring may weaken or tear, so that the central part of the disk bulges outward (herniates). Referred to as a slipped or herniated disk, this condition may result in a significant amount of pain in the neck, back, or one or both legs, depending on the location of the damaged disk. Pain is generated when the disk presses against nearby nerves. Surgeons can correct the problem by removing the herniated portion of the disk. You can reduce your chances of "slipping" a disk in the first place by watching your weight, sitting upright (not slouching) in a chair, and lifting heavy objects carefully (▻ Figure 10–10).

Bone. Bone is another form of specialized connective tissue. Contrary to what many might think, bone is a dynamic living tissue. Besides providing internal support and protecting internal organs such as the brain, heart, and lungs, bone plays an important role in maintaining optimal blood calcium levels and is therefore a homeostatic organ. Calcium is required for many body functions: muscle contraction, normal nerve functioning, and even blood clotting.

Like all connective tissues, bone consists of cells embedded in an abundant extracellular matrix (▻ Figure

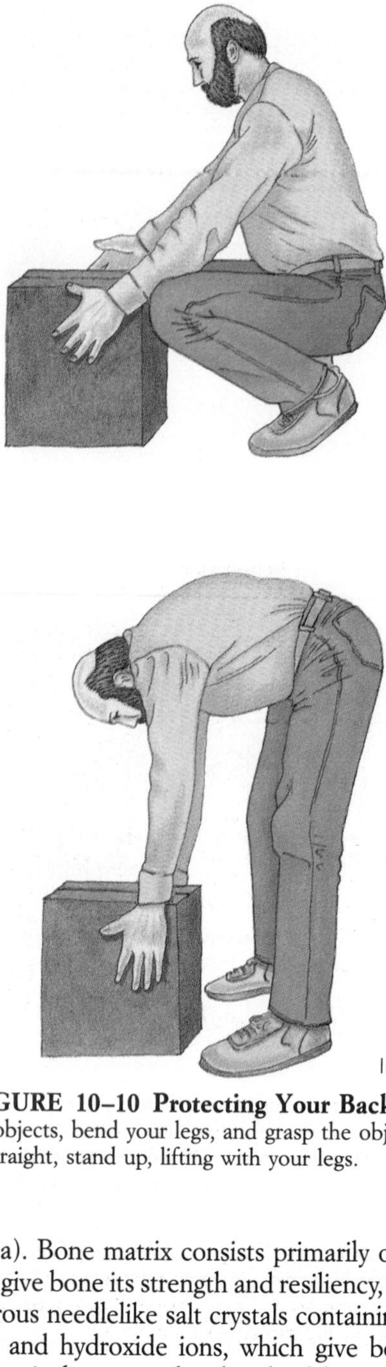

▻ **FIGURE 10–10 Protecting Your Back** When lifting heavy objects, bend your legs, and grasp the object. With your back straight, stand up, lifting with your legs.

10–11a). Bone matrix consists primarily of collagen fibers, which give bone its strength and resiliency, interspersed with numerous needlelike salt crystals containing calcium, phosphate, and hydroxide ions, which give bone its hardness. Calcium in bone can be dissolved by weak acids, leaving behind a collagen replica that can be turned into a thick paste, called **demineralized bone matter (DBM).** DBM is rather remarkable stuff, and it is being used to repair severe bone damage, as described in Health Note 10–1.

Two types of bone tissue are found in the body: compact bone and spongy bone (Figure 10–11a). Compact bone is dense and hard. As illustrated in the photomicrograph in Figure 10–11b, the cells in compact bone **(osteocytes)** are located in concentric rings of calcified matrix. These surround a **central canal** through which the blood vessels and nerves pass. Each osteocyte is endowed with numerous processes that course through tiny canals in the bony matrix, known as **canaliculi** (literally, "little canals"). The canaliculi provide a route for nutrients and wastes to flow to and from the osteocytes and the central canal.

Inside most bones of the body is a tissue known as spongy bone. **Spongy bone** consists of an irregular network of calcified collagen spicules (Figure 10–11b). As illustrated in Figure 10–11c, on the surface of the spicules are numerous **osteoblasts,** bone cells that produce collagen, which later becomes calcified. Once these cells are surrounded by calcified matrix, they are referred to as osteocytes.

Between the spicules are numerous cavities, which often communicate with much larger cavities in the center of bones. In most of the bones of an adult, the large and small cavities are filled with fat cells and form the **yellow marrow.** In other bones, the cavities are filled with blood cells and cells that give rise to new blood cells, thus forming **red marrow,** described in later chapters.

On the surfaces of many bony spicules of spongy bone are large, multinucleated cells called **osteoclasts** ("bone breakers") (Figure 10–11d). These cells are part of a homeostatic system that ensures proper blood calcium levels. When calcium levels in the blood fall, osteoclasts are activated by the parathyroid hormone produced by the thyroid gland. So activated, the osteoclasts digest small portions of the spongy bone. This in turn releases calcium into the bloodstream and helps restore proper levels.

Spongy bone is also remodeled when bones are subjected to new stresses. For example, the leg bones of a desk-bound executive from Atlanta who goes on a skiing vacation in Jackson Hole, Wyoming, are remodeled as his legs are subjected to the rigors of skiing. This adjustment accommodates for the new stresses and strains. During this process, osteoclasts tear down some of the spongy bone, while osteoblasts rebuild new bone in other areas to meet the new stresses. By the end of the two-week ski trip, the executive's bones have been considerably refashioned and are much stronger than when he left home. When he is back at his desk, his bones revert to their previous, weaker state.

Blood. Blood is also a specialized form of connective tissue and consists of two components: (1) the formed elements (cells and platelets) and (2) a large amount of extracellular material, a fluid called **plasma** (▻ Figure 10–12).

The formed elements of blood consist of the red and white blood cells and the platelets. **Red blood cells,** or **erythrocytes,** transport oxygen and small amounts of carbon dioxide to and from the lungs and body tissues. **White blood cells,** or **leukocytes,** are involved in fighting infections. **Platelets** are fragments of large cells (megakaryocytes) that are located in the red bone marrow, the principal site of blood cell formation. Platelets play a key role in blood clotting.

Muscle Tissue Consists of Specialized Cells That Contract When Stimulated

Muscle, the third in our series of primary tissues, is found in virtually every organ in the body. Muscle gets its name

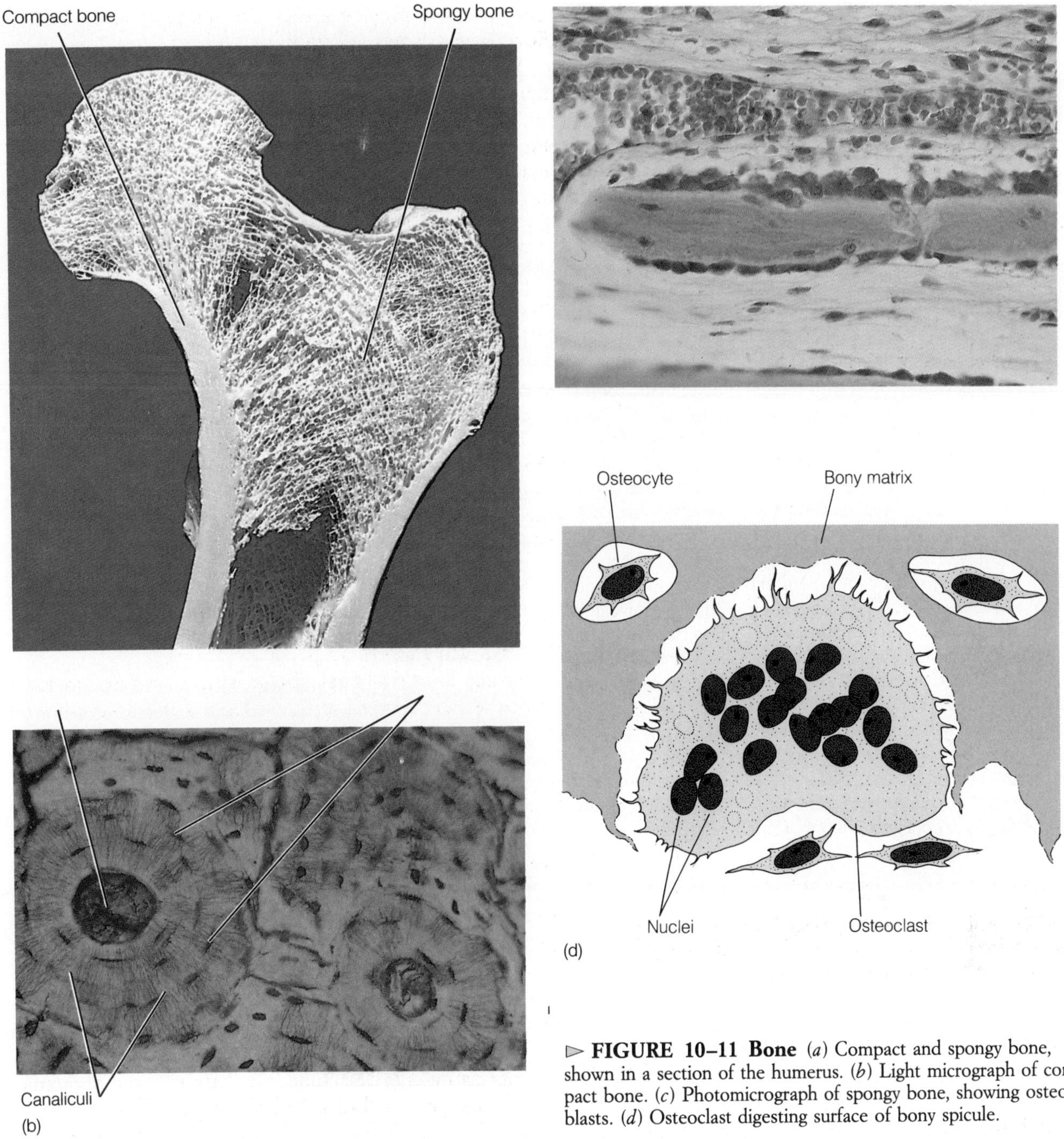

▻ **FIGURE 10–11 Bone** (*a*) Compact and spongy bone, shown in a section of the humerus. (*b*) Light micrograph of compact bone. (*c*) Photomicrograph of spongy bone, showing osteoblasts. (*d*) Osteoclast digesting surface of bony spicule.

from the Latin word for "mouse" *(mus).* Early observers likened the contracting muscle of the biceps to a mouse moving under a carpet.

Like nervous tissue, muscle demonstrates a characteristic of life described in the first chapter, irritability—that is, the ability to perceive and respond to stimuli. Muscle is therefore an excitable tissue that, when stimulated, is capable of contracting, producing mechanical force. Muscle cells working in large numbers can create enormous forces. Muscles of the jaw, for instance, create a pressure of 200 pounds per square inch, forceful enough to snap off a finger. (Don't try this at home.) Muscle also moves body parts, propels food along the digestive tract, and expels the fetus from the uterus during birth. Muscle of the heart contracts and pumps blood through the 50,000 miles of blood vessels in the human body. Acting in smaller numbers, muscle cells may be responsible for more intricate movements, such as those required to play the piano or move the eyes.

As mentioned earlier, three types of muscle are found in humans: skeletal, cardiac, and smooth. The cells in each type of muscle contain two types of contractile protein filaments, **actin** and **myosin.** These very same fibers were first encountered in your study of biology in the microfila-

HEALTH NOTE 10–1

REMAKING THE HUMAN BODY

David Eastland was driving home from work late one evening when he fell asleep at the wheel. His car swerved off the road, striking a tree. The impact of the accident crushed Eastland's right leg, destroying a large segment of his femur (thigh bone).

In earlier times, this accident would have cost Eastland his leg, because surgeons would have had no way of repairing such extensive bone damage. Even bone grafts were insufficient in such instances. Today, however, thanks to advances in medical science, surgeons can literally remake the human skeleton using a material called demineralized bone matter (DBM). Demineralized bone matter consists primarily of a protein known as collagen. DBM is produced by immersing the bones from human cadavers and other animals in a weak acid solution. As noted in the text, this treatment dissolves the mineral matter of the bone, leaving behind a thick, rubbery paste, which can be used to repair severe bone damage.

After replacing the missing bone with DBM, surgeons actually inject bone cells into it. These cells, usually derived from small fragments taken from a patient's hip bone, grow and divide in the DBM, eventually converting the implant into a functional bone, practically indistinguishable from normal bone.

Demineralized bone matter can also be used to repair birth defects and to replace segments of bone that have been surgically removed because of bone cysts, tumors, or severe infections. One of the most remarkable success stories is that of a boy born with a disease known as cloverleaf syndrome. In this disease, growth in the bones of the skull fails to keep pace with the brain's growth, eventually killing its victims. Dr. John Mulliken, a surgeon at Boston's Children's Hospital, removed the skullof a 6-year-old boy suffering from the disease. He then fashioned a new and larger skull from DBM. Five years later, the boy was alive and well.

Artificial Skin

Another promising development is artificial skin. Researchers at the Massachusetts Institute of Technology have developed an artificial skin that can be applied to burned areas, helping to reduce fluid loss, prevent infection, and promote faster recovery.

Artificial skin consists of two layers that resemble the body's natural skin. The lower layer, the artificial dermis, is made of collagen fibers extracted from cowhide and a substance extracted from the cartilaginous skeleton of sharks. The upper layer, the artificial epidermis, is made of plastic.

In severely burned patients, physicians clean away the burned flesh, then sew the artificial skin in place. Within a short time, fibroblasts and blood vessels from the surrounding region invade the artificial dermis. Over a period of several months, these cells produce new collagen fibers, replacing the fibers of the implant.

The plastic epidermis is peeled off within a few weeks after the dermis has revascularized, and surgeons reconstruct a new cellular epidermis using some of the patient's own epidermal cells, which are grafted in thin sheets onto the underlying dermis.

Like DBM, artificial skin is relatively

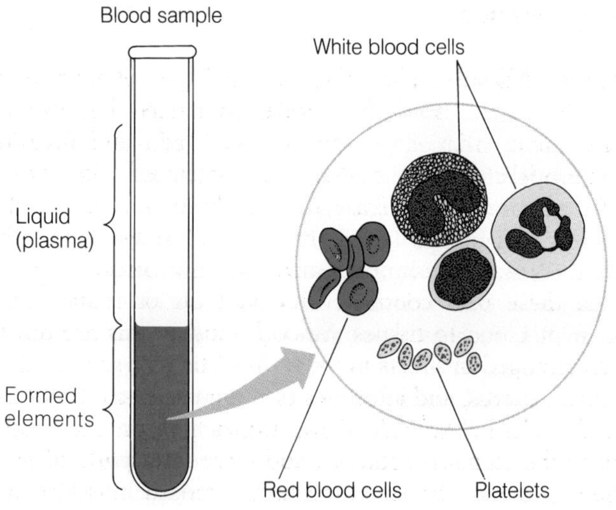

▻ **FIGURE 10–12 Blood** Blood is about 55% liquid (plasma) and 45% formed elements: red blood cells, white blood cells, and platelets.

mentous network lying beneath the plasma membrane of animal cells that is responsible for cytokinesis, the division of the cytoplasm. When stimulated, actin and myosin filaments of muscle cells slide over one another, shortening them and causing contraction.

Skeletal Muscle. The majority of the body's muscle is called **skeletal muscle**—so named because it is frequently attached to the skeleton. When skeletal muscle contracts, it causes body parts (arms and legs, for instance) to move. Most skeletal muscle in the body is under voluntary, or conscious, control. Signals from the brain brought about by thoughts cause the muscle to contract. A notable exception is the skeletal muscle of the upper esophagus, which contracts automatically during swallowing.

Skeletal muscle cells are long cylinders formed during embryonic development by the fusion of many embryonic muscle cells. Because of this, skeletal muscle cells are usually referred to as **muscle fibers.** Each muscle fiber contains many nuclei (▻ Figure 10–13a). Because of the easy to make and transplant. New skin grows in without the severe scarring that accompanies more conventional burn treatment.

Rebuilding Ailing Hearts

Each year, 400,000 people in the United States develop end-stage heart failure, aserious weakening of the heart. This disease results from the death of cardiac muscle cells caused by excess stress on a heart whose arteries are clogged with cholesterol. Nearly all of these people die within a year.

Many patients with end-stage heart failure could benefit from heart transplants, but because of a lack of donors, only 1400 heart transplants were performed in 1987. Artificial hearts may provide some hope, but they are costly and plagued with problems, not the least of which is that they strap patients to a machine for the rest of their lives.

One promising technique that could benefit thousands of patients each year is known as the skeletal muscle wrap. As ▻ Figure 1 illustrates, in this procedure a segment of the latissimus dorsi, a large muscle of the back, is cut loose, then wrapped around the diseased heart. The muscle is stapled in place, forming a contractile basket. Then it is attached to a pacemaker, which senses the heart's natural electrical pulses and delivers bursts of electricity to the skeletal muscle, causing it to contract in step with the heart. This technique leaves the blood supply and nerve supply of the muscle intact, so there is little worry over graft survival. Skeletal muscle wraps improve cardiac function in patients by about 20%, a change that means the difference between living as an invalid and leading a fairly normal life.

Using a patient's own skeletal muscle has many benefits. First, it is readily available, and in most cases, patients are more than willing to "donate" tissue for their own survival. Second, because the graft is taken from a person's own body, surgeons do not have to worry about the immune system rejecting the graft. Promising as this technique is, it is no substitute for prevention of heart disease.

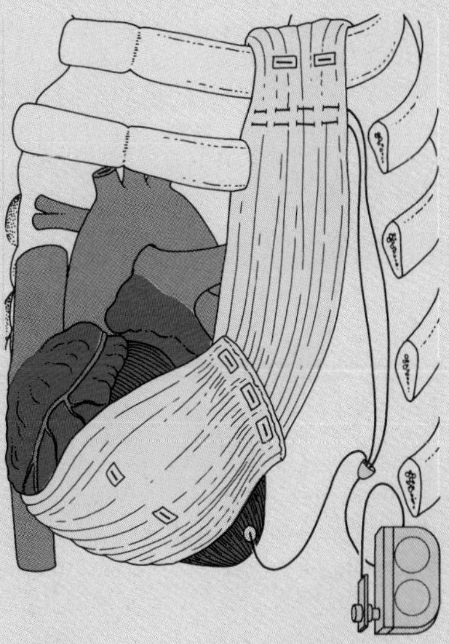

▻ **FIGURE 1 Skeletal Muscle Wrap** Patients with ailing hearts may be aided by skeletal muscle wrapped around the failing organ. Pacemakers can stimulate the muscle to beat in step with the heart muscle, thus increasing cardiac output.

dense array of contractile fibers in the cytoplasm of the muscle fiber, the nuclei are generally pressed against the plasma membrane. As you might suspect, this highly specialized cell cannot divide, and damaged muscle cells cannot be replaced.

Skeletal muscle fibers appear banded, or **striated,** when viewed under the light microscope, as illustrated in Figure 10–13a. The striations result from the unique arrangement of actin and myosin filaments inside muscle cells, a topic discussed in Chapter 19.

Cardiac Muscle. Like skeletal muscle, **cardiac muscle** is striated (Figure 10–13b). Unlike skeletal muscle, cardiac muscle is involuntary; that is, it contracts without conscious control. Found only in the walls of the heart, cardiac muscle cells contain a single nucleus. They also branch and interconnect freely, but individual cells are tightly connected to one another, an adaptation that helps maintain the structural integrity of the heart, an organ subject to incredible strain as it pumps blood through the body day and night for years on end. The points of connection also provide pathways for electrical impulses to travel from cell to cell, allowing the heart muscle to contract uniformly when stimulated.

Smooth Muscle. The third and final type of muscle is smooth muscle. **Smooth muscle,** so named because it lacks visible striations, is involuntary. Actin and myosin filaments are present but are not organized like those found in other types (Figure 10–13c). Smooth muscle cells may occur singly or in small groups. Small rings of smooth muscle cells, for example, surround tiny blood vessels. When these cells contract, they shut off or reduce the supply of blood to tissues. Smooth muscle cells are most often arranged in sheets in the walls of organs, such as the stomach, uterus, and intestines (see Figure 10–2). Smooth muscle cells in the wall of the stomach churn the food, mixing the stomach contents, and force tiny spurts of liquified food into the small intestine. Smooth muscle contractions also propel the food along the intestinal tract.

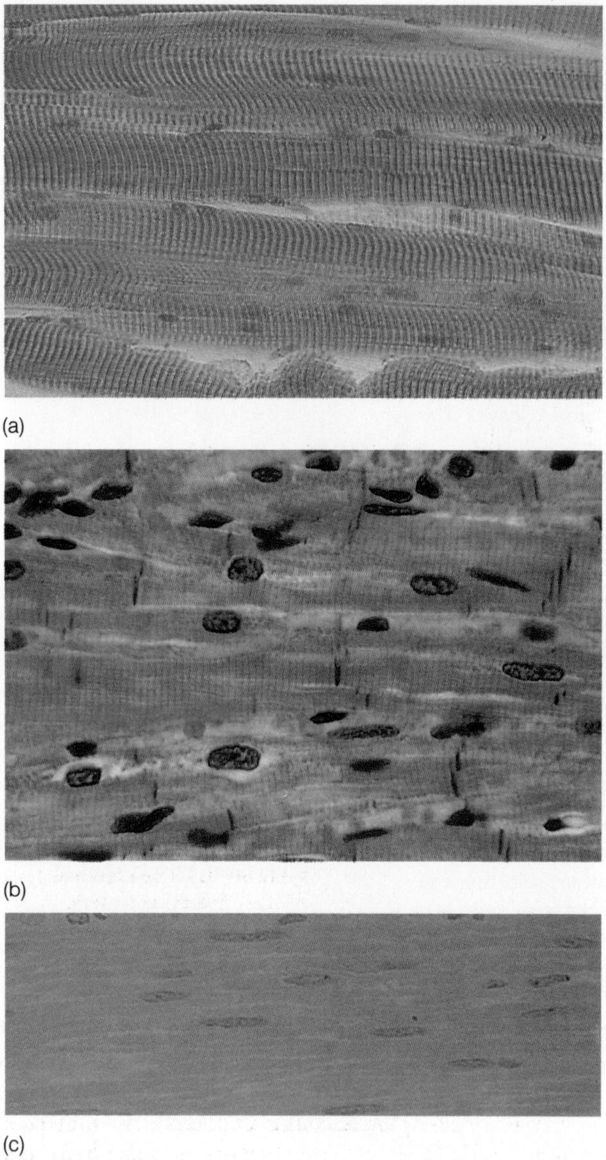

▻ **FIGURE 10–13 Light Micrograph of the Three Types of Muscle** (*a*) Skeletal. (*b*) Cardiac. (*c*) Smooth.

Nervous Tissue Contains Specialized Cells Characterized by Irritability and Conductivity

Last but not least of the primary tissues is nervous tissue. **Nervous tissue** consists of two types of cells: conducting cells and supportive cells. Many of the conducting cells, known as **neurons,** are modified to respond to specific stimuli, like pain or temperature. Stimulation results in bioelectric impulses, which the neuron transmits from one region of the body to another. As noted above, the ability to respond to stimuli is a characteristic of all living things and is called irritability. The ability to transmit an impulse is called **conductivity.** The properties of irritability and conductivity in neurons allow animals to be aware of their environment and to respond to a variety of internal and external stimuli. The evolution and refinement of the nervous system has been vitally important to the emergence of the more complex organisms in the animal kingdom.

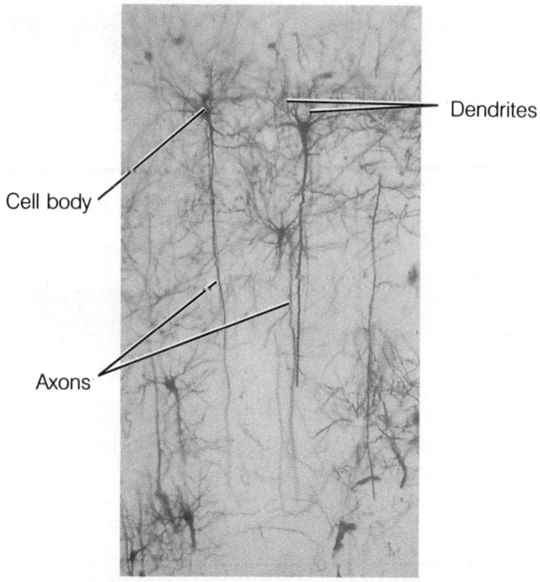

▻ **FIGURE 10–14 Multipolar Neuron** Attached to the cell body of the multipolar neuron are many highly branched dendrites, which deliver impulses to the cell body. Multipolar neurons have one long, unbranching fiber called the axon, which transmits impulses away from the cell body.

The supportive cells of the nervous system are a kind of nervous system connective tissue. These cells are incapable of conducting impulses, but they help transport nutrients from blood vessels to neurons and help guard against toxins by creating a barrier to many potentially harmful substances. As you will see in later chapters, the supportive cells also help increase the rate of conduction in neurons with which they are associated. Together, the neurons and their supporting cells combine to form the brain, spinal cord, and nerves of the nervous system, which are described in more detail in Chapters 17 and 18.

Although Three Types of Neurons Can Be Found in the Body, They Share Many Characteristics. At least three distinct types of neurons are found in the body. Despite their differences, they have several common features. We will study these similarities by looking at one of the most common nerve cells, the **multipolar neuron,** shown in ▻ Figure 10–14. The multipolar neuron contains a prominent, multiangular **cell body** to which are attached several short, highly branched processes, known as **dendrites.** (It is the presence of these processes that give this neuron its name, multipolar.) The dendrites receive impulses from receptors or other neurons and transmit them to the cell body. Also attached to the cell body of the multipolar neuron is a large, fairly thick process, the **axon,** which transports bioelectric impulses away from the cell body.

Like muscle cells, nerve cells are highly differentiated and cannot divide. When a cell is destroyed, it degenerates and cannot be replaced by cell division. A cut nerve axon, however, may partially regenerate. A new axon may grow

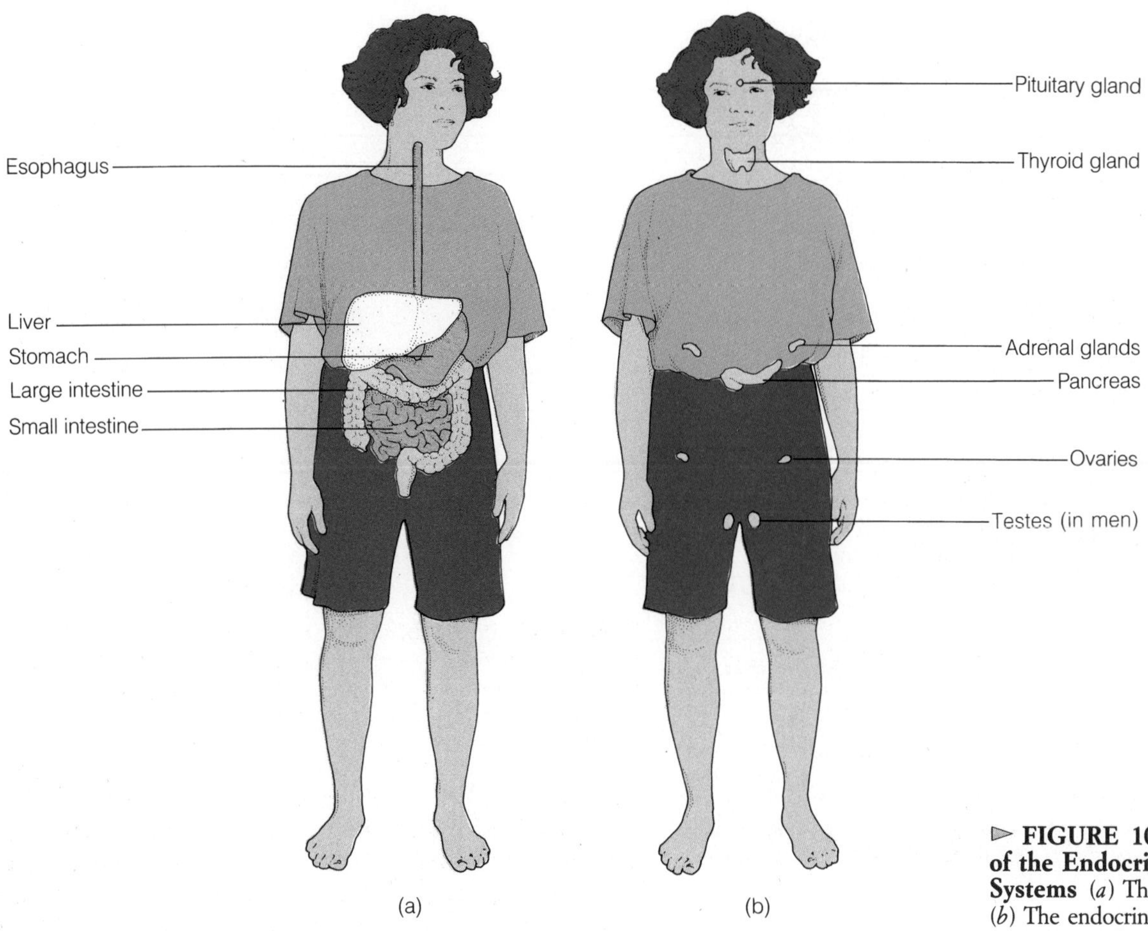

▻ **FIGURE 10–15 Comparison of the Endocrine and Digestive Systems** (*a*) The digestive system. (*b*) The endocrine system.

from the damaged end, reestablishing previous connections and restoring some sensation or control over muscles. New research may someday provide ways to stimulate regeneration of nerve cells artificially, thus helping physicians more fully restore nerve function to victims of accidents.

Organs Often Function in Groups Called Organ Systems

The cell contains organelles ("little organs") that carry out many of its functions in isolation from the biochemically active cytoplasm. Compartmentalization is an evolutionary adaptation that also occurs at the organismic level in organs. Discrete structures called **organs** evolved to perform specific functions, such as digestion, enzyme production, and hormone synthesis. Most organs, however, do not function alone. Instead, they are part of a group of cooperative organs, called an **organ system.** The brain, spinal cord, and nerves are part of the organ system known as the nervous system.

As you will see in the upcoming chapters, components of an organ system are sometimes connected—as in the digestive system—and are sometimes dispersed throughout the body—as in the endocrine system (▻ Figure 10–15). Some organs belong to more than one system. For example, the pancreas produces digestive enzymes that are secreted into the small intestine, where they break down food materials. The pancreas is therefore part of the digestive system. The pancreas also contains cells that produce insulin and glucagon, two hormones that help control blood glucose levels. The pancreas therefore also belongs to the endocrine system.

The following 10 chapters describe the major organ systems and the functions they perform, paying special attention to their role in homeostasis and highlighting important aspects of their evolution. ▻ Figure 10–16 summarizes the functions of each organ system and lists the role each plays in the overall economy of the human organism. You may want to take a moment to read the descriptions in the light blue boxes before proceeding.

As Figure 10–16 shows, organ systems are the functional machinery of the body. In sickness and in health, the 10 major organ systems remain interconnected and mutually dependent.[2] This interdependence is often apparent when one system "breaks down." Such an event can have catastrophic effects on other organ systems and the organism

[2]You will note that this figure includes the integumentary system as an organ system. I prefer to discuss the skin in Chapter 9 and others, rather than devote a full chapter to this important system.

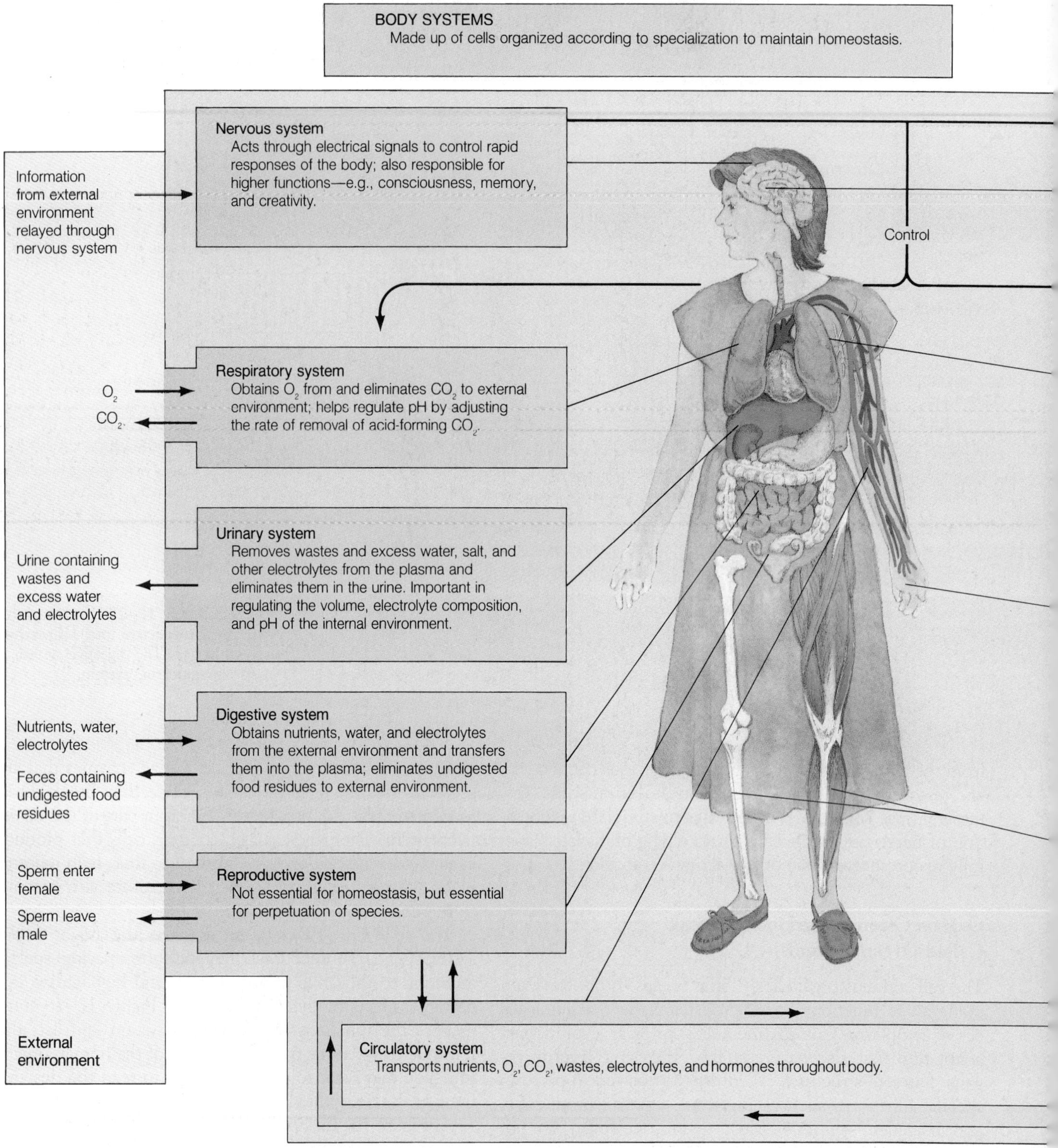

BODY SYSTEMS
Made up of cells organized according to specialization to maintain homeostasis.
Information from external environment relayed through nervous system
Nervous system
Acts through electrical signals to control rapid responses of the body; also responsible for higher functions—e.g., consciousness, memory, and creativity.
Control
O_2
CO_2
Respiratory system
Obtains O_2 from and eliminates CO_2 to external environment; helps regulate pH by adjusting the rate of removal of acid-forming CO_2.
Urine containing wastes and excess water and electrolytes
Urinary system
Removes wastes and excess water, salt, and other electrolytes from the plasma and eliminates them in the urine. Important in regulating the volume, electrolyte composition, and pH of the internal environment.
Nutrients, water, electrolytes
Feces containing undigested food residues
Digestive system
Obtains nutrients, water, and electrolytes from the external environment and transfers them into the plasma; eliminates undigested food residues to external environment.
Sperm enter female
Sperm leave male
Reproductive system
Not essential for homeostasis, but essential for perpetuation of species.
External environment
Circulatory system
Transports nutrients, O_2, CO_2, wastes, electrolytes, and hormones throughout body.

▻ **FIGURE 10–16 Role of the Body Systems in Maintaining Homeostasis**

as a whole. When defective kidneys fail to remove toxins from the blood, for instance, other cells and other organ systems become poisoned. If the disease is untreated, death may occur in a matter of days.

PRINCIPLES OF HOMEOSTASIS

Homeostasis is one of the primary themes of this book. Defined in Chapter 1 as the internal constancy of the body, homeostasis occurs on a variety of levels—in cells, tissues,

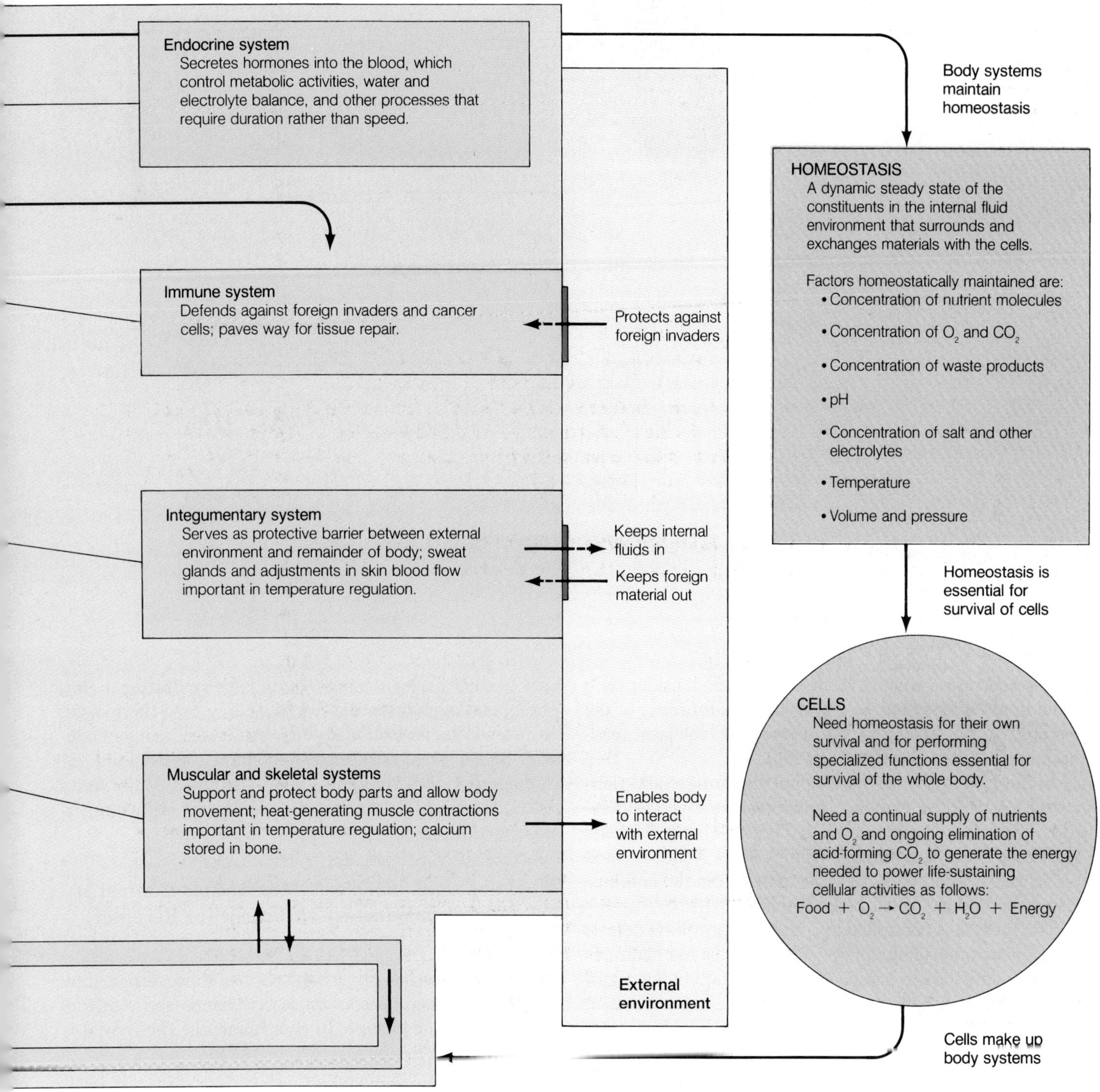

organs, organ systems, organisms, and even the environment. Homeostatic systems at all levels of biological organization have several common features worth noting before we begin our study of organ systems.

Endocrine system
Secretes hormones into the blood, which control metabolic activities, water and electrolyte balance, and other processes that require duration rather than speed.
Body systems maintain homeostasis
HOMEOSTASIS
A dynamic steady state of the constituents in the internal fluid environment that surrounds and exchanges materials with the cells.
Factors homeostatically maintained are:
• Concentration of nutrient molecules
• Concentration of O_2 and CO_2
• Concentration of waste products
• pH
• Concentration of salt and other electrolytes
• Temperature
• Volume and pressure
Immune system
Defends against foreign invaders and cancer cells; paves way for tissue repair.
Protects against foreign invaders
Integumentary system
Serves as protective barrier between external environment and remainder of body; sweat glands and adjustments in skin blood flow important in temperature regulation.
Keeps internal fluids in
Keeps foreign material out
Homeostasis is essential for survival of cells
CELLS
Need homeostasis for their own survival and for performing specialized functions essential for survival of the whole body.
Need a continual supply of nutrients and O_2 and ongoing elimination of acid-forming CO_2 to generate the energy needed to power life-sustaining cellular activities as follows:
Food + $O_2 \rightarrow CO_2$ + H_2O + Energy
Muscular and skeletal systems
Support and protect body parts and allow body movement; heat-generating muscle contractions important in temperature regulation; calcium stored in bone.
Enables body to interact with external environment
External environment
Cells make up body systems

Homeostatic Systems Maintain Constancy Chiefly Through Negative Feedback Mechanisms

Feedback mechanisms were briefly described in Chapter 4. To expand your understanding of the most common feed-

▷ **FIGURE 10–17 Homeostasis and the House** (*a*) Heat is maintained in a house by a furnace, which produces heat to balance heat loss. The thermostat monitors the internal temperature and switches the furnace on and off in response to temperature changes. (*b*) A hypothetical temperature graph showing temperature fluctuation around the operating point.

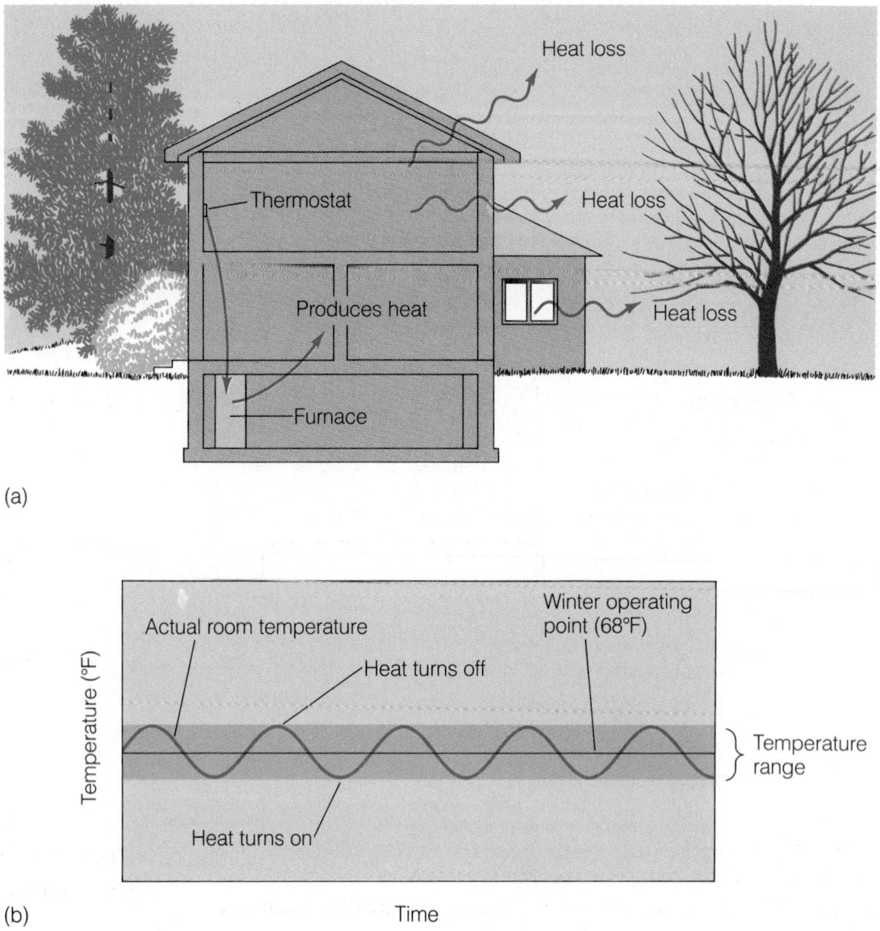

back mechanism, negative feedback, we will consider once again the heating system in a typical house, for many of the features of this system are also present in biological feedback mechanisms (▷ Figure 10-17a).

In the winter, the heating system (the furnace and thermostat) of a house maintains a constant internal temperature, even though the outside temperature may fluctuate markedly. Heat lost through ceilings, walls, windows, and tiny cracks is replaced by heat generated from the combustion of natural gas or oil in the furnace.

The furnace is controlled by a thermostat, which detects changes in room temperature. When indoor temperatures fall below the desired setting, the thermostat sends a signal to the furnace, turning it on. Heat from the furnace is then distributed through the house, raising the room temperature. When the room temperature reaches the desired setting, the thermostat shuts the furnace off. Like all negative feedback mechanisms, the product of the system (heat) "feeds back" on the process, shutting it down.

A graph of a hypothetical house temperature is shown in Figure 10–17b and illustrates another important principle of homeostasis: *homeostatic systems do not maintain absolute constancy.* Rather, they maintain conditions (such as body temperature) within a given range.

All homeostatic mechanisms in vertebrates operate in a similar fashion, maintaining conditions by negative feedback within a narrow range around the **operating point.** The operating point is akin to the setting on a thermostat. As you will see in later chapters, vertebrates maintain constant levels of a great many chemical components, including hormones, nutrients, wastes, and ions. Many vertebrates also maintain physical parameters like body temperature, blood pressure, blood flow, and others.

All Homeostatic Feedback Mechanisms Contain at Least Two Components—A Sensor (or Receptor) and an Effector

Biological homeostatic mechanisms, like those designed by engineers, contain sensors to detect change and effectors to correct for the change. In your home, the thermostat is the sensor, and the furnace is the effector.

In your body, temperature receptors, special modified nerve cell endings in the skin, are one type of sensor. They detect changes in the ambient (outside) temperature and send signals to the brain, alerting it when the temperature rises or falls. When the ambient temperature falls, the brain sends signals to the body to reduce heat loss and increase heat output. The options for generating more heat are many. Put another way, the body contains several different types of effectors.

One of the main sources of heat is the catabolism of

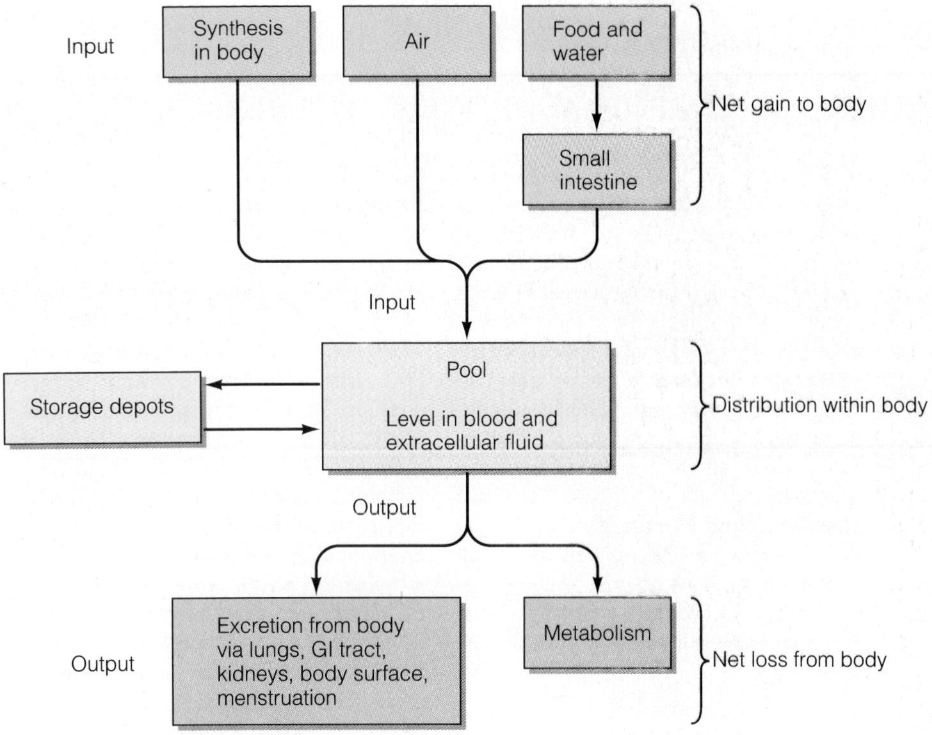

▷ **FIGURE 10–18 Generalized View of a Homeostatic System** Inputs and outputs are balanced to maintain more or less constant levels of chemical and physical parameters.

glucose and other molecules. In earlier chapters, you learned that the cells of the body break down glucose to make ATP. During this process, heat is given off. Each cell, then, is a tiny furnace whose heat radiates outward and is distributed throughout the body by the blood. Unlike the furnace in your home, the cellular "furnaces" cannot be turned up very quickly. They respond much more slowly than a furnace and are part of a delayed response to low temperature. As winter progresses, energy catabolism increases, and the body produces more heat.

In order to respond to sudden changes in outside temperature, the body must rely on other, more rapid mechanisms. If you walked outdoors on a cold winter night dressed only in a light sweater and blue jeans, for example, receptors in your skin would sense the cold and send signals to the brain. The brain in turn would send signals to blood vessels in the skin, causing them to constrict, reducing the flow of blood in the skin. The restriction of blood flow through the skin reduces heat loss. If it is cold enough outside, the brain may also send signals to the muscles, causing them to undergo rhythmic contractions, commonly known as shivering. Shivering burns additional glucose, releasing extra heat. (It's not unlike adding additional logs to a fire. The more fuel that's burned, the more heat you get.) For a discussion on how some animals respond to seasonal changes, see Spotlight on Evolution 10–1.

Many voluntary actions may also be "ordered" by the brain to help reduce heat loss or generate more heat. For example, you might put on a hat or turn around and go back inside. In humans, conscious acts are often crucial components of homeostasis.

Homeostasis is Maintained by Balancing Inputs and Outputs

The generalized diagram in ▷ Figure 10–18 illustrates the many ways the body regulates the level of various chemicals. Let's consider the input side first. As illustrated, ions (for example, calcium) and other chemical substances (for example, glucose) are ingested in food or water. Other essential chemicals like oxygen enter our bodies in the air we breathe. Lastly, some substances like certain amino acids are produced in the body. All three routes tend to raise internal concentrations.

Excretion and metabolism tend to lower concentrations. **Excretion** is the loss of materials through a number of specialized body systems. Carbon dioxide, for example, is excreted by the lungs, as is water. Water is also excreted by the kidneys. Metabolism, chemical reactions occurring in the body cells, also removes certain substances from the body. Blood glucose levels, for instance, are decreased by cellular catabolism.

Internal concentrations are kept in balance, in large part by input and output. However, internal storage depots (regions where chemicals are stored) also participate in homeostatic balance. For example, as you learned in previous chapters, glucose is stored in the liver, and blood glucose levels are therefore also determined by the input and output occurring in this important organ.

The relative importance of the various homeostatic pathways depends on the substance in question. Water, for example, enters the body primarily through ingestion (the food and liquids we consume) and is removed by the

SPOTLIGHT ON EVOLUTION 10–1

HIBERNATION AND TORPOR: ADAPTIVE HYPOTHERMIA

Stress is a daily occurrence in the lives of animals. Predators, cold spells, hot spells, and shortages of food, among other factors, all contribute to stress. In many regions, stressful conditions may be seasonal; for example, periods of plenty may alternate with periods of great scarcity. Such cycles are conspicuous in the Arctic, with its long summer days. During the Arctic summer, plants grow rapidly, and many species make their home there, feeding off the abundance of life. But Arctic summers are followed by prolonged winters in which bitter cold brings production to a halt and sends many species to warmer grounds. Surprisingly to many, some tropical rain forests also exhibit annual cycles with distinct rainy (monsoon) seasons punctuated by hot, dry seasons.

Animals respond to seasonal stress in several ways, including hibernation (adaptive hypothermia), aestivation, torpor, and migration.

Some mammals, like chipmunks, ground squirrels, woodchucks (groundhogs), skunks, and bears, for instance, enter a period of greatly reduced activity called winter hibernation. Thus, during the coldest months of the year when food is scarce, these animals hole up in their dens or burrows and enter into a state of suspended animation. Metabolism, breathing, and heart rate all decrease. During hibernation, for instance, a black bear's heart and breathing rates may slow to as little as 10% of the active rates. That would be like your pulse dropping to 7 beats per minute and your respiration rate to one or two breaths per minute. The bear's body temperature drops by about 5°C, and metabolism drops by about 50%—an adaptation that results in a dramatic energy savings.

Mammals prepare for hibernation in at least two ways—by storing large amounts of body fat, which they live off during the long months of hibernation, and by growing thick winter pelts, which reduce heat loss. As days become shorter, internal biological clocks trigger changes in feeding behavior, which in turn result in increases in food consumption and body fat content. The internal clock also triggers growth of the animal's fur coat. These adaptive changes, found in animals that remain active in the winter as well, serve all species during lean days of winter.

Another interesting adaptation that helps conserve energy is known as torpor. Torpor is a temporary reduction of the metabolic set point and is found in small animals, such as bats and hummingbirds (▷Figure 1). These species enter a state of decreased metabolism every day during periods of inactivity. The reason is that animals with a small body size in relation to body surface area lose heat faster than it can be generated through metabolism. Thus, at night or during periods of rest, a hummingbird's body temperature drops relative to that of its environment, saving enormous amounts of energy. Resting on a branch on a cool spring day, the hummingbird's temperature plummets, but when it is time to move again, the bird flexes its wings, shivers a little, and increases its metabolic heat production, raising its body temperature to about 40°C, a temperature at which it can perform.

▷ **FIGURE 1 The Hummingbird** This lovely bird saves energy in temperate climates by reducing metabolic activity during periods of inactivity.

Bats exhibit both hibernation and torpor. During the daylight hours, bats repair to caves or dark places, where they remain inactive, falling into a state of torpor. During the winter months, they enter hibernation for many weeks.

Some animals undergo a summer dormancy, known as aestivation, which permits them to cope with temperature extremes or seasonal shortages of food and water. Like hibernation, aestivation is characterized by slow metabolism and inactivity, which permits animals to survive long periods of elevated temperatures and water shortages. Still others migrate long distances, moving from warm spring and summer habitat to warm wintering grounds, finding in each location ample food and suitable temperature for survival.

All of these adaptations save energy and promote survival and reproduction. They are but a handful of the evolutionary strategies for survival.

kidneys, lungs, and skin. For iron, the rate of absorption by the intestinal tract is a key determinant of its levels in the blood. When iron levels decrease—say, because of a decrease in iron intake—the body reestablishes the balance by increasing the rate of absorption in the small intestine.

Homeostasis Can Be Upset by Changes in the Input, Output, or Storage Occurring in the Body's Complex Homeostatic Network

Consider output first. On a hot day, water escapes our

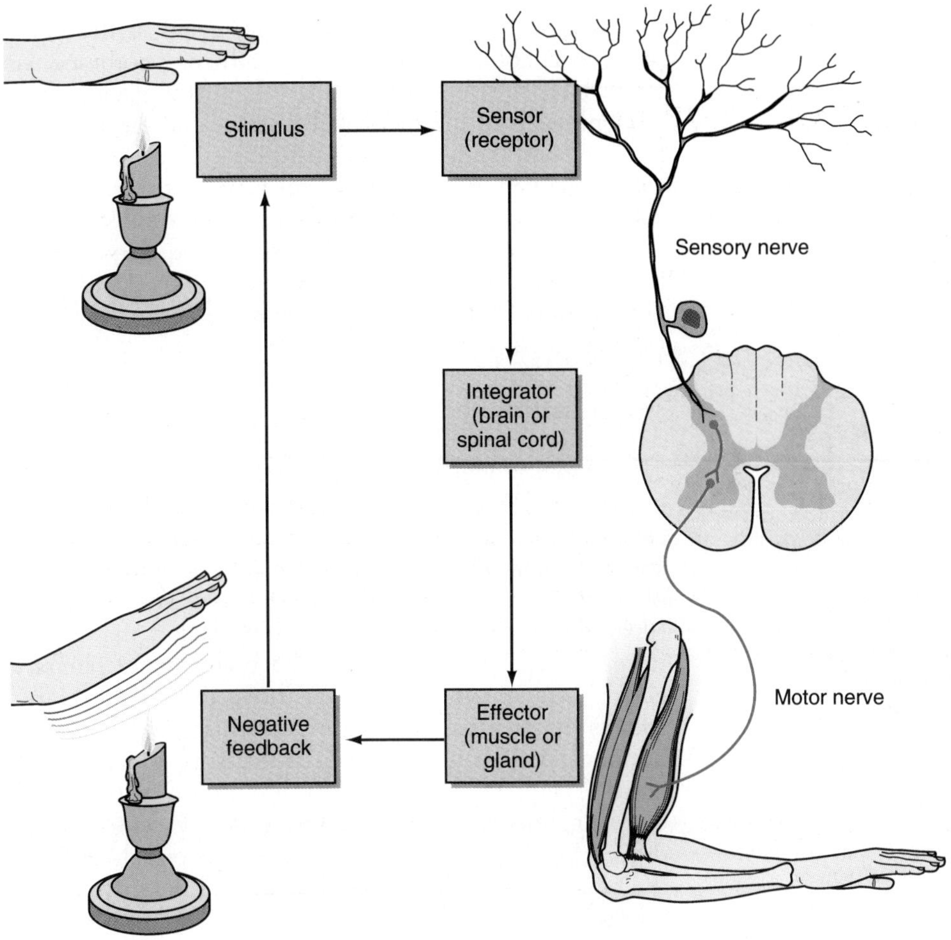

▷ **FIGURE 10–19 Nervous System Reflex** Reflexes involve some kind of stimulus, a sensor, an integrator, and an effector. The sensor detects a change and sends a signal to the integrator, the brain or spinal cord, which elicits a response in the effector organs. Negative feedback from the effector eliminates the stimulus.

bodies very rapidly as a result of perspiration, an adaptation that helps cool us down. Heavy perspiration can result in a severe decrease in water volume (dehydration) that may upset homeostasis—so much so, in fact, that death may occur. Severe diarrhea—a drastic increase in water output—also results in a dangerous depletion of body fluids that can be so severe that it results in death. In fact, dehydration resulting from severe diarrhea kills millions of Third World children each year.

Just as these changes in the output of a substance drastically alter homeostasis, so do changes in input. For example, going without water for a prolonged period (approximately three days) can kill a human being. Excess input can also prove dangerous. Excess salt intake, for example, can result in hypertension (high blood pressure) in some people.

Whatever the cause, imbalances in homeostasis can have dramatic impacts on human health.

Homeostatic Control Requires the Action of Nerves, Hormones, and Various Chemicals that Operate Over Short Distances

Homeostatic mechanisms are **reflexes**—that is, automatic physiological responses triggered by certain stimuli. Reflexes occur without conscious control. Homeostatic reflexes in the body involve two main mechanisms: nervous and chemical. Chapter 31 describes homeostatic mechanisms that operate in the environment.

The Nervous System Reflex. The line between cause and effect in a homeostatic system that involves the nervous system is fairly easy to trace. As shown in ▷ Figure 10–19, changes are detected by a sensor, or **receptor,** a nerve cell ending or a special structure that responds to various stimuli in the internal or external environment. In vertebrates, the sensor sends nerve impulses to the brain or spinal cord by a nerve. The brain and spinal cord, in turn, direct an appropriate response to counterbalance the change, maintaining or restoring balance.

At any one time, the brain and spinal cord receive thousands of signals from receptors in the body. These signals alert the nervous system to a variety of internal and external conditions. The brain and spinal cord are centers of integration able to make sense of information and generate a meaningful response. They control the activity of many **effectors,** generally muscles and glands. On a cold winter night, for example, nerve impulses travel from temperature sensors in your skin to your brain. The brain, in turn, sends signals to the blood vessels in the skin, causing them to constrict and reduce blood flow through the skin, thus conserving heat. As noted earlier, if it is cold enough, the

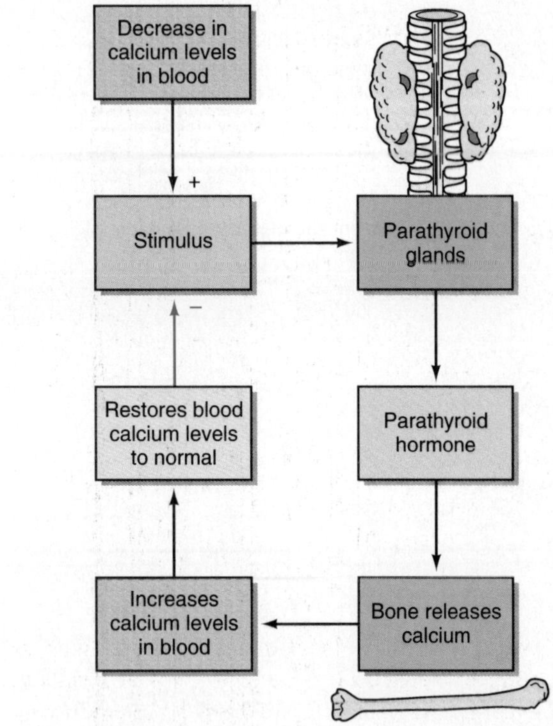

▻ **FIGURE 10–20 Endocrine Reflex Not Involving the Nervous System** This reflex operates through the bloodstream. A decrease in calcium in the blood stimulates a series of reactions that helps restore normal blood calcium levels. A plus sign indicates that the stimulus increases parathyroid gland activity; a minus sign indicates the opposite effect.

brain may also send impulses to the muscles, causing shivering. The muscles surrounding the blood vessels of the skin and those involved in shivering are effectors in this reflex.

Chemical Control. Hormones also participate in reflexes. Consider an example. The parathyroid glands, four tiny glands embedded in the thyroid gland in the neck, produce a hormone known as parathyroid hormone, or PTH. PTH is released from the cells of the glands when calcium levels in the blood fall. It travels in the blood to the bone and there stimulates osteoclasts, bone-destroying cells described earlier. These cells cause the release of calcium and help raise the calcium level of the blood.

In this reflex, the cells of the parathyroid gland are the receptors, which detect the drop in blood calcium. The bone is the effector, which produces the desired response (▻ Figure 10–20).

Some endocrine reflexes operate through the nervous system, making the lines of cause and effect a bit more difficult to follow. A good example is the thyroid gland. The thyroid produces a hormone called thyroxine (▻ Figure 10–21). This hormone increases the metabolic rate of cells and thus increases heat production. Thyroxine levels are monitored by cells in the brain (the receptors). When blood levels fall, these cells release a chemical substance (thyroid-stimulating hormone releasing factor) that travels from its site of production in the brain to the pituitary gland, located just beneath the brain. The pituitary responds by releasing another hormone, thyroid-stimulating hormone, or TSH, which travels in the blood to the thyroid gland. There, TSH steps up the production of thyroxine, correcting the deficiency. This reflex has one sensor, which detects levels of thyroxine, but three effectors: the cells of the pituitary that release TSH, the cells of the thyroid that release thyroxine, and body cells that respond to thyroxine by increasing their rate of metabolism.

Local Chemical Control. Nervous and endocrine mechanisms generally occur over considerable distances. In the nervous system, for example, nerves carry impulses from the brain and spinal cord to distant parts of the body, often several feet away. In the endocrine system, the bloodstream carries the messages from endocrine glands to

▻ **FIGURE 10–21 A Neuroendocrine Reflex** This reflex involves both the endocrine and nervous systems. Sensors are located in the brain.

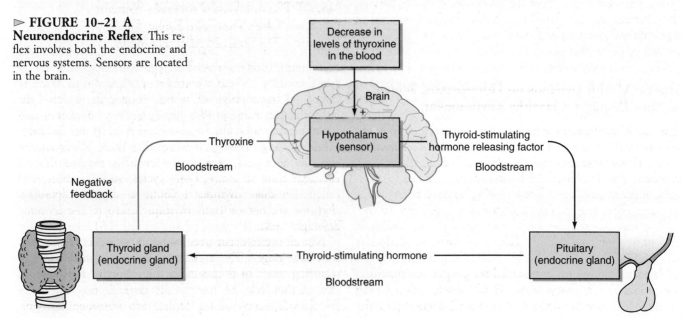

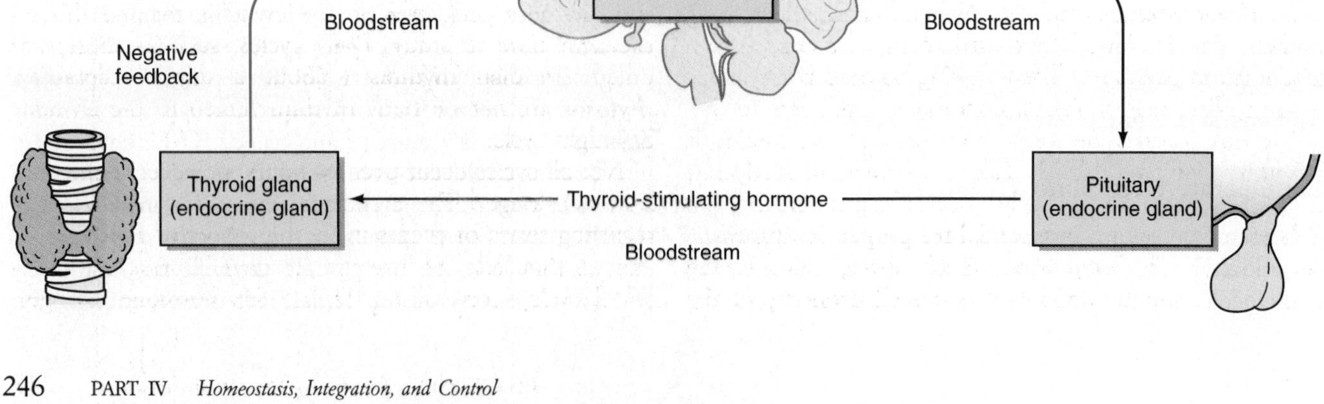

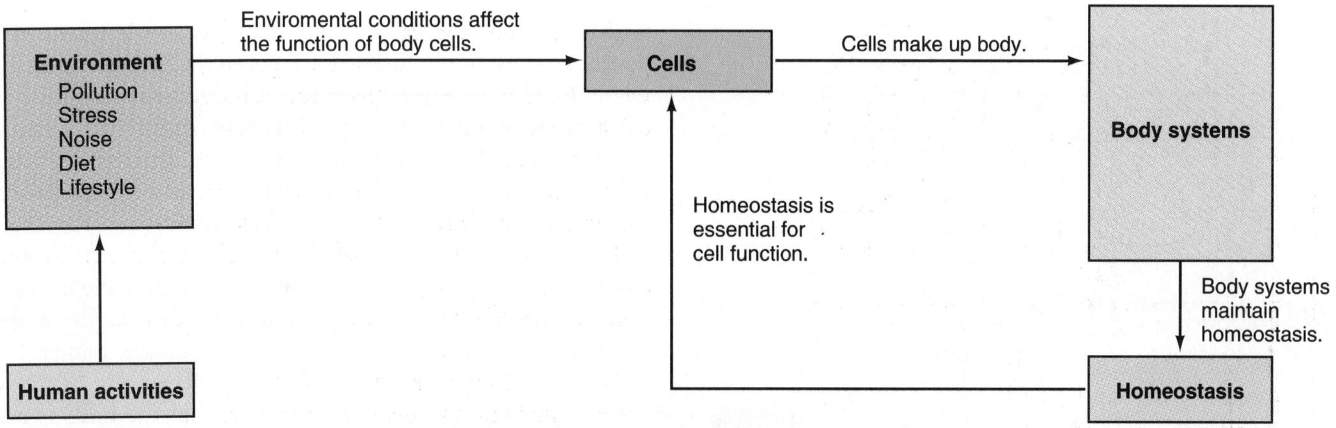

▻ FIGURE 10–22 Health, Homeostasis, and the Environment Human health is dependent on maintaining homeostasis. Homeostasis, however, is affected by the condition of our environment. Stress, pollution, noise, and other environmental factors upset the function of cells and body systems, thus upsetting homeostasis and human health. As a result, human health is dependent on a healthy environment.

distant effectors. But not all systems require messages that are transported over long distances. In some cases, chemi cal control is exerted through the extracellular fluid only a cell or two away. Chemicals that elicit local effects are called **paracrines.** Produced by individual cells, paracrines diffuse to neighboring cells via the extracellular fluid. Epidermal growth factor produced by skin cells is one example. These molecules stimulate cell division when skin cells are damaged or lost, providing a degree of local control over cell growth.

One of the best-known paracrines is a group of chemical substances known as **prostaglandins.** Prostaglandins comprise a rather large group of molecules with diverse functions. Some help stimulate blood clotting; others stimulate smooth muscle contraction.

An interesting evolutionary adaptation akin to paracrines are the **autocrines,** molecules produced by a cell that affect the function of that cell. Autocrines are part of the simplest chemical reflexes in the body. Some prostaglandins, for example, function as autocrines as well as paracrines.

Human Health Depends on Homeostasis, Which in Turn Requires a Healthy Environment

This book emphasizes a theme that is important to all of us notably, that human health is dependent on homeostasis. Homeostasis, in turn, requires a healthy, clean environment. The health of the environment, of course, is also dependent in part on well-functioning homeostatic systems in nature that help maintain conditions conducive to life.

The relationship between homeostasis and health is shown in ▻ Figure 10–22. Take a moment to study this diagram. Notice that the body systems help maintain overall homeostasis, which is essential for proper cell function. Cells also contain homeostatic mechanisms, that are vital to their own function and to the overall economy of the organism. Notice too that environmental factors, such as pollution, affect the function of body systems and cells in ways that can upset the internal balance, thus altering human mental and physical health.

Just as human health is affected by the condition of the environment, so too is the health of many other species. Numerous examples of this connection will be pointed out in this book.

BIOLOGICAL RHYTHMS

The previous discussion may have given the impression that homeostasis establishes an unwavering condition of stability that remains more or less the same, day after day, year after year. In truth, many physiological processes undergo rhythmic change.

Not All Physiological Processes Remain Constant Over Time but Those That Fluctuate Do So in Predictable Ways

Body temperature varies during a 24-hour period by as much as 0.5° C. Blood pressure may change by as much as 20%, and the number of white blood cells, which fight infection, can vary by 50% during the day. Alertness also varies considerably. About 1:00 P.M. each day, for instance, most people go through a slump. For most of us, activity and alertness peak early in the evening, making this an excellent time to study. Daily cycles, such as these, are called circadian rhythms ("about a day"). **Circadian rhythms** are natural body rhythms linked to the 24-hour day-night cycle.

Not all cycles occur over 24 hours, however. Some can be much longer. The **menstrual cycle,** for instance, is a recurring series of events in the reproductive functions of women that lasts, on average, 28 days. During the menstrual cycle, levels of the female sex hormone estrogen

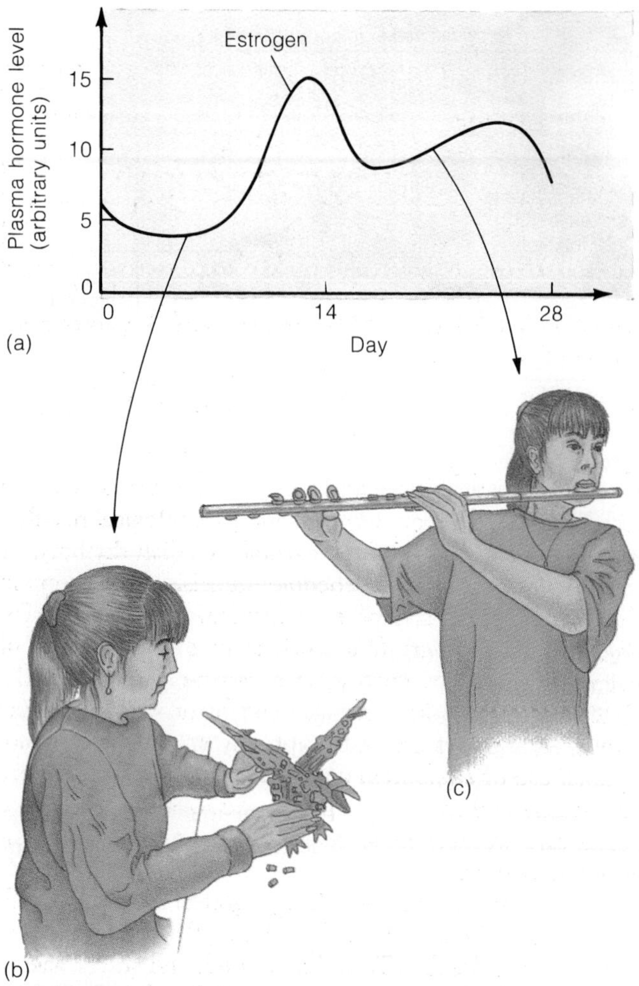

▻ **FIGURE 10–23 Estrogen Levels during the Menstrual Cycle** (*a*) Note that over a 28-day period, estrogen levels vary considerably. The variation is part of a finely tuned reproductive cycle that prepares the woman for pregnancy. (*b*) During the early part of the cycle when estrogen levels are low, studies show that women excel at tasks involving spatial relationships, but they do less well at tasks involving fine motor skills. (*c*) During the later half of the cycle when estrogen levels are high, women excel at tasks involving motor skills, like playing a musical instrument, but do less well at tasks involving spatial relationships.

undergo dramatic shifts. As illustrated in ▻ Figure 10–23a, estrogen concentrations in the blood are low at the beginning of each cycle and peak on day 14, when ovulation normally occurs. Throughout the remaining 14 days, estrogen levels are rather high. Then they drop off again when a new cycle begins.

Estrogen follows this cycle month after month in women of reproductive age. Interestingly, one recent study suggests that estrogen levels may exert profound influences on neural functions in women. Early in the cycle when estrogen levels are low, the study showed, women excel at tasks involving spatial relations—for example, solving three-dimensional puzzles. When estrogen levels are elevated in the second half of the cycle, many women may find such tasks more difficult (Figure 10–23b).

Motor skills follow a different pattern. When estrogen levels are low, some women are less able to perform complex motor skills, such as the finger movements required to play a musical instrument, than they are during the remainder of the cycle (Figure 10–23c). When estrogen concentrations in the blood are high, motor skills become easier.

Many hormones follow daily rhythmic cycles. The male sex hormone testosterone, for example, follows a 24-hour cycle. The highest levels occur in the night, particularly during dream sleep. Dream sleep occurs primarily in the early morning hours—the later the hour, the longer the periods of dream sleep (▻ Figure 10–24).

The important point of all of this is that the body is not static. *Although many chemical substances are held within a fairly narrow range by homeostatic mechanisms, others fluctuate widely in normal and quite essential cycles.* Just as the seasons change, altering the face of the landscape, so do the internal body seasons. But over the long run, these changes are predictable. They also occur within prescribed physiological limits; they do not run out of control. They are part of the body's dynamic balance, just as the weather changes throughout the year are part of the dynamic balance of the planet's climate.

In Humans, Internal Biological Rhythms Are Apparently Controlled by the Brain

Just how the body controls its many internal rhythms remains a question. Research suggests that the brain controls many biological cycles. One region in particular, the **suprachiasmatic nucleus (SCN),** is thought to play a major role in coordinating several key rhythms. The SCN, a clump of nerve cells in the base of the brain in a region called the hypothalamus, may also regulate other control centers. As a result, the suprachiasmatic nucleus is often referred to as the "master clock."

Like a clock, the SCN ticks off the minutes, faithfully imposing its control on the body, turning body functions on and off like the master control in an automated factory. Even in complete darkness, the master clock ticks on, directing many circadian rhythms.

In humans, research suggests, the master clock operates on a 25-hour cycle. That is, if isolated in a dark room, many of us would fall into a 25-hour sleep-wake cycle. But as with most other biological phenomena, there is considerable variability. Some individuals, for example, operate on 28-hour sleep-wake cycles. Some are shorter than 24 hours.

Sleep researchers believe that in the real world the natural 25-hour clock is modulated by the 24-hour day-night cycle. In other words, the environment alters the clock and maintains many biological rhythms on a 24-hour cycle. Control of the clock is thought to reside in a gland in the brain known as the **pineal gland.** It secretes a hormone thought to keep the suprachiasmatic nucleus in sync with the day-night cycle.

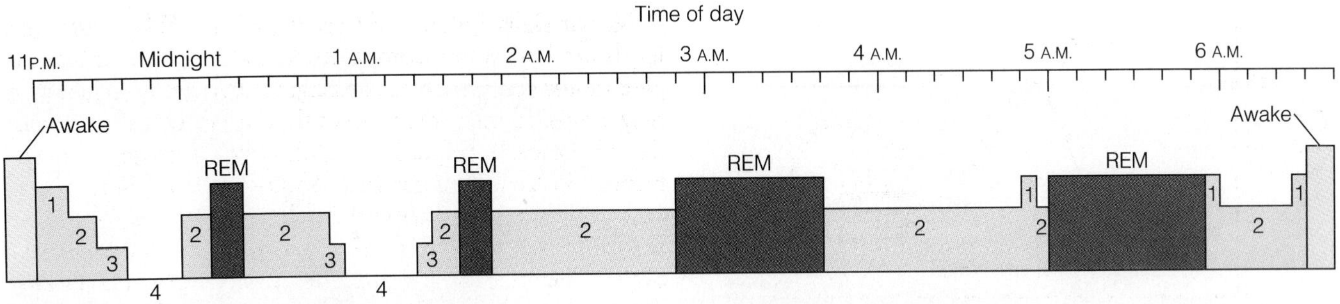

▻ **FIGURE 10–24 Stages of Sleep** Numbers indicate the stages of sleep: the higher the number, the deeper the sleep. Note that around midnight sleep is deepest. As morning approaches, a person's sleep is lighter and REM sleep, or dream sleep, occurs in longer increments.

The study of biological rhythms is a fascinating field that has yielded some important information and insights. One practical application is a better understanding of jet lag, that drowsy, uncomfortable feeling people get from the disruption of sleeping patterns caused by long-distance airplane travel. Studies suggest that jet lag occurs when the body's biological clock ticks out of synchrony with the day-night cycle of a traveler's new surroundings. A businesswoman who travels from Los Angeles to New York, for instance, may be weary and irritable the first few days in the new time zone. At 10:00 P.M. New York time, when New Yorkers are heading to bed, she is wide awake because her body is still on Los Angeles time—three hours earlier. When the alarm goes off at 6:00 A.M. New York time, our weary traveler crawls out of bed exhausted, because as far as her body is concerned, it is 3:00 A.M. Los Angeles time. To avoid the ill feelings, some sleep researchers suggest that you abide by your internal clock, following "home time" when in a new time zone, or, barring that, try to reset your biological clock before you get there. You can do this by going to bed an hour or two earlier when you are traveling from west to east. Do just the opposite for east-to-west travel.

Research on body rhythms has also shown that people respond to drugs differently at different times of the day and night. By administering drugs at the body's most receptive times, physicians may be able to reduce the doses, lowering toxic side effects and fighting diseases more effectively.

ENVIRONMENT AND HEALTH: TINKERING WITH OUR BIOLOGICAL CLOCKS

Modern life with its stress, noise, pollution, hectic pace, and weird work schedules can deal a blow to our internal biological rhythms, with serious consequences. Dr. Richard Restak, a neurologist and author, notes that the "usual rhythms of wakefulness and sleep . . . seem to exert a stabilizing effect on our physical and psychological health." The greatest disrupter of our natural circadian rhythms, he says, is the variable work schedule, surprisingly common in the United States and other industrialized countries. Today, one out of every four working men and one out of every six working women is on a variable work schedule, shifting frequently between day and night work. In many industries, workers are at the job day and night to make optimal use of equipment and buildings. As a result, more restaurants and stores are open 24 hours a day, and more health-care workers must be on duty at night to care for accident victims.

What American business has forgotten is that for millions of years, humans have slept during the night and been awake during the day. Turn that around, and you're asking for trouble. Make a person work at night when he or she normally sleeps, say sleep experts, and you can expect more accidents and lower productivity. Consider an example.

At 4:00 o'clock in the morning in the control room of the Three Mile Island nuclear reactor in Pennsylvania, three operators failed to notice warning lights and to note that a valve in the system had remained open. When the morning-shift operators entered the control room, they quickly discovered the problem, but it was too late. Pipes in the system had burst, sending radioactive steam and water into the air and into two buildings. John Gofman and Arthur Tamplin, two radiation health experts, estimate that the radiation released from the accident, the worst nuclear accident in U.S. history, will result in at least 300 and possibly as many as 900 additional fatal cases of cancer in the residents living near the troubled reactor, although other experts (especially in the nuclear industry) contest these projections, saying that the accident will not have a noticeable effect. Whatever the outcome of this debate, one thing is certain: the 1979 accident at Three Mile Island will cost several billion dollars to clean up.

Late in April 1986, another nuclear power plant ran amok. This crisis, in the former Soviet Union, was far more severe (▻ Figure 10–25). In the wee hours of the morning, two engineers were testing the reactor. They deactivated key safety systems in violation of standard operational pro-

▶ **FIGURE 10–25 Disaster at Chernobyl Nuclear Power Plant** This costly accident was caused primarily by operator error. What price do we pay to have industries running 24 hours per day to keep production levels high?

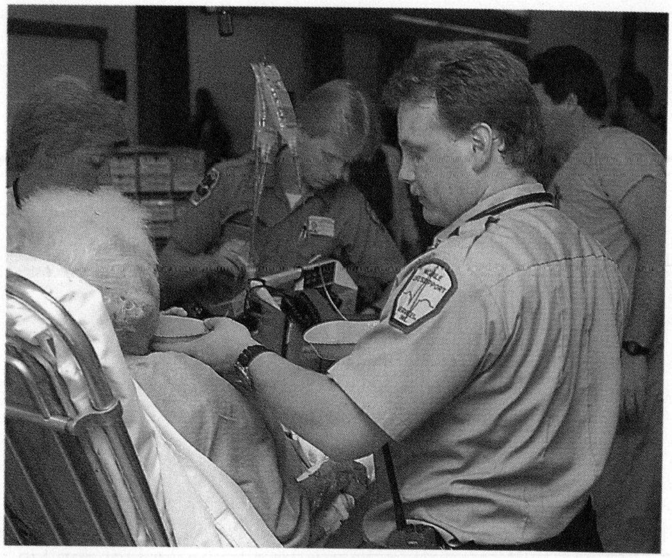

▶ **FIGURE 10–26 Late Night in the Emergency room.** Increasing numbers of men and women working the night shift may increase the number of judgment error accidents on the job. This causes an increase in the need for medical care. However, late-night medical care workers may also suffer from errors in judgment.

tocol. This single error in judgment (possibly due to fatigue) led to the largest and most costly nuclear accident in history. Steam built up inside the reactor and blew the roof off the containment building. A thick cloud of radiation rose skyward and then dissipated throughout Europe and the world. Workers battled for days to cover the molten radioactive core, which spewed radiation into the sky while the world watched in horror.

Although no one will ever know for sure, the Chernobyl disaster, like the accident at Three Mile Island, may have been the result of workers operating at a time unsuitable for clear thinking. One has to wonder how many plane crashes, auto accidents, and acts of medical malpractice can be traced to judgment errors resulting from our insistence on working against the natural body rhythms.

Making matters worse, many companies that maintain shifts round the clock spread the burden evenly among employees. One week, workers are on the day shift; the next week, they are switched to the "graveyard shift" (midnight to 8:00 A.M.). The next week, they are put on the night shift (4:00 P.M. to midnight). Many workers subject to such disruptive changes report that they often feel run down and have trouble staying awake at the job. Their work performance suffers. When employees who have been on the graveyard shift arrive home, they are physically exhausted but cannot sleep, because they are trying to sleep at a time when the body is trying to wake them up. What is more, weekly changes in schedule never permit workers' internal clocks to adjust. Studies show that most people require 4 to 14 days to adjust to a new schedule.

Not surprisingly, workers on alternating shifts suffer more ulcers, insomnia, irritability, depression, and tension than workers on unchanging shifts. Many suffer from impaired judgment on the job, and in some circumstances they may pose a threat to society (Figure 10–26).

Thanks to studies of biological rhythms, researchers are finding ways to reset the body's clock. These efforts could help lessen the misery and suffering of shift workers and could improve the performance of those on the graveyard shift. For instance, one simple measure is to place all shift workers on three-week cycles instead of weekly cycles. This gives their biological clocks time to adjust. And instead of shifting workers from daytime to a graveyard shift, transfer them forward, rather than backward (for example, from a daytime to an evening shift and from an evening shift to the graveyard shift). Studies suggest that moving forward is far easier for workers than shifting backward.

For reasons not well understood, special treatment with bright lights helps reset the biological clock. Patients who are suffering from insomnia, because their biological clocks are out of phase, for example, receive daily doses of bright light, which somehow reset their internal clocks. Similar treatments are being tested for shift workers. They are a small price to pay for a healthy work force and a safer society.

SUMMARY

FROM CELLS TO ORGAN SYSTEMS

1. The basic structural unit of animals (and all other organisms) is the cell. Cells and extracellular material form tissues, and tissues, in turn, combine to form organs.
2. Four primary tissues are found in the bodies of most multicellular animals such as humans: epithelial tissue, connective tissue, muscle tissue, and nervous tissue. Most of these tissues have two or more subtypes.
3. Organs contain all four primary tissues in varying proportions.
4. Epithelial tissues consist of two types: membranous epithelia, which form coverings or linings of organs, and glandular epithelia, which form exocrine and endocrine glands.
5. Connective tissues bind other tissues together, provide protection, and support body structures. All connective tissues consist of two basic components: cells and extracellular fibers.
6. Two types of connective tissue are found in the body: connective tissue proper (tendons, ligaments, and loose connective tissue) and specialized connective tissue (bone, cartilage, and blood).
7. Cartilage consists of specialized cells embedded in a matrix of extracellular fibers and other extracellular material. The cells are supplied with nutrients from blood vessels in the periphery of the cartilage, which is why damaged cartilage is repaired slowly.
8. Bone is a dynamic tissue that provides internal support, protects organs such as the brain, and helps regulate blood calcium levels.
9. Bone consists of bone cells (osteocytes) and a calcified cartilage matrix. Two types of bone tissue exist: spongy and compact.
10. The osteoclast, one type of bone cell, plays a major role in reshaping bone to meet the changing demands of the body and in releasing calcium to help maintain blood calcium levels.
11. Blood is another form of specialized connective tissue. It consists of numerous blood cells and platelets and an extracellular fluid, called plasma.
12. Muscle is an excitable tissue that contracts when stimulated. Three types of muscle tissue are found in the human body: cardiac, skeletal, and smooth muscle.
13. Cardiac muscle is found in the heart and is involuntary. Skeletal muscle is under voluntary control, for the most part, and forms the muscles that attach to bones. Smooth muscle is involuntary and forms sheets of varying thickness in the walls of organs and blood vessels.
14. Nervous tissue is the fourth primary tissue. It consists of two types of cells: conducting and nonconducting (supportive). The conducting cells, called neurons, transmit impulses from one region of the body to another. The nonconducting cells are a type of nervous system connective tissue.
15. Tissues combine to form organs in which specialized functions are carried out. Most organs are part of an organ system, a group of organs that cooperate to carry out some complex function.

PRINCIPLES OF HOMEOSTASIS

16. The nervous and endocrine systems coordinate the functions of organ systems, helping the body achieve homeostasis, or internal constancy. Homeostasis also occurs at the level of the cell.
17. All homeostatic systems maintain constancy by balancing inputs and outputs as well as movement to and from storage depots. Homeostatic systems do not maintain absolute constancy and can be upset by alterations in input or output or by changes in movements in or out of storage depots. Imbalances can have serious effects on an individual's health.
18. Homeostatic mechanisms are reflexes, occurring without conscious control.
19. Homeostatic reflexes occur at all levels of biological organization and involve two main mechanisms: nervous and hormonal.
20. Nervous and endocrine mechanisms generally occur over considerable distances. In some cases, control is exerted locally through the extracellular fluid.

BIOLOGICAL RHYTHMS

21. Many physiological processes undergo definite rhythmic changes. These natural rhythms may take place over a 24-hour period or over much longer or shorter periods.
22. The brain controls many biological cycles. A clump of nerve cells (the suprachiasmatic nucleus) located in the hypothalamus is thought to play a major role in coordinating several key functions and several other control centers. It is therefore often referred to as the "master clock."

ENVIRONMENT AND HEALTH: TINKERING WITH OUR BIOLOGICAL CLOCKS

23. The hectic pace of modern life, the shifting work schedules that many people follow, and the stressful environments in which we live can upset internal rhythms, with disastrous consequences.
24. The greatest disrupter of our natural circadian rhythms is the variable work schedule, which is surprisingly common in industrialized nations.
25. Workers on alternating shifts suffer from a higher incidence of ulcers, insomnia, irritability, depression, and tension than workers on regular shifts. Making matters worse, tired, irritable workers whose judgment is impaired by fatigue pose a threat not only to themselves but also to society.
26. Researchers are finding ways to reset the biological clock, which could help lessen the misery and suffering of shift workers and could increase the performance of those on the graveyard shift.

EXERCISING YOUR CRITICAL THINKING SKILLS

1. In the Environment and Health section, I noted that workers on night shifts are more accident-prone and suffer from lower productivity. I then cited two possible examples: the Three Mile Island and Chernobyl nuclear reactor accidents, both of which occurred at night. One of the academic reviewers of this text wrote, "I am certain I can cite terrible worker error in daytime circumstances." The examples, she contended, are "biased and used for effect."

   Looking at this issue more carefully, and using your critical thinking skills, how can you determine whether my statement that night shift workers are more accident-prone is valid? Is there a way to determine whether the accidents at the nuclear power plants were the result of the late hour? If so, how? If not, why not? Is there a way to determine if accidents at nuclear plants are more frequent at night or during the day? Is the reviewer's statement "I am certain I can cite terrible worker error in daytime circumstances" necessarily a valid refutation of the points I made? What is wrong with this line of reasoning?
2. Health care costs are skyrocketing, and many critics are arguing that our health care dollars are being misused. In particular, they argue that expensive procedures, such as organ transplants for indigent people costing taxpayers $100,000 or more each, are draining dollars that could be invested in preventive medicine, such as prenatal care. Prenatal care is medical care and advice on nutrition and other matters that are given to pregnant women and are crucial to the development of a healthy fetus. By various estimates, the cost of one organ transplant could supply medical care for 1000 pregnant women, assistance that could prevent many of the problems that result in ailments in newborns that necessitate costly transplant procedures.

   What is your opinion on this matter? Should Americans spend more on prevention and less on dramatic life-saving measures, such as organ transplants? Why? Can you think of ways to provide both types of health care without raising taxes?

   After you have given your opinion on this matter, put yourself in the position of a parent whose child needs a kidney transplant to survive. How does this perspective affect your position?

TEST OF TERMS

1. The first three cell types to emerge during embryonic development are ____________, ____________, and ____________.
2. All tissues are made of cells and a variable amount of ____________ ____________.
3. The four primary tissues are ____________, ____________, ____________, and ____________.
4. Glands that secrete their products into ducts that, in turn, carry the products to some other part of the body are called ____________ glands.
5. The outer layer of the skin is called the ____________; it is a type of ____________ tissue.
6. Tendons are a form of ____________ tissue.
7. The cell that produces collagen in tendons and other similar primary tissues is the ____________.
8. The soft tissue forming most of the nose is made of ____________ ____________.
9. The intervertebral disk contains a ring of ____________ that holds the soft, cushiony center of the disk in place.
10. The blood vessels of bone are found in tunnels called ____________ canals. Nutrients flow out of these canals through even smaller tunnels in the bony matrix called ____________, providing the bone cells, or ____________, with nourishment.
11. The dense outer layer of most bones is called ____________ bone.
12. The fluid component of blood is known as ____________.
13. The blood contains two components that do not have nuclei or organelles; they are the ____________ ____________ ____________ and the ____________.
14. The involuntary muscle found in the lining of the small intestine and stomach is called ____________ muscle. It contains two contractile proteins, ____________ and ____________, which are also present in other muscle types.
15. A conducting nerve cell, or ____________, is a specialized cell that generates and transmits bioelectric impulses from one part of the body to another.
16. A group of organs that cooperate to perform some key function is called a(n) ____________ ____________.
17. All homeostatic systems maintain chemical and physical parameters within a narrow range, called the ____________ ____________.
18. Endocrine and neural ____________ are a series of events that begins with a stimulus and ends with a response and is crucial for maintaining homeostasis.
19. The process of making sense of various chemical and nervous inputs is called ____________.
20. Organs and glands that carry out the instructions of the brain and spinal cord are called ____________.
21. A chemical released into the bloodstream that travels to distant sites where it elicits some response is called a(n) ____________; a chemical that acts on cells close to its point of origin is called a(n) ____________.
22. Body temperature and some hormone levels vary over a 24-hour period; these and other similar cycles are called ____________ rhythms.
23. A clump of nerve cells called the ____________ nucleus in the hypothalamus of the brain is thought to be the master clock, controlling several key internal body rhythms and several other clocks.
24. Travelers suffer from ____________ ____________ because their biological clocks are out of synchrony with the 24-hour day-night cycle.

Answers to the Test of Terms are found in Appendix B.

TEST OF CONCEPTS

1. Define the following terms: tissues, extracellular material, organs, and organ systems.
2. List the four primary tissues and their subtypes.
3. Describe the similarities and differences in the embryonic origin of endocrine and exocrine glands. Explain why the two secrete differently.
4. From your study of biology, discuss some examples in which structure reflects function.
5. Describe the two types of connective tissue and how they function.
6. Why do cartilage injuries repair so slowly? Bone is repaired much more easily than cartilage. Why do you think this is true?
7. In what way is bone part of a homeostatic system?
8. Describe the chief differences among cardiac, skeletal, and smooth muscle.
9. What is an organ system? List some examples.
10. Define homeostasis, and describe the major principles of homeostasis presented in this chapter. Use an example to illustrate your points.
11. How do storage depots enter into the input-output concept of homeostasis?
12. Homeostatic mechanisms are largely reflexes, involving chemical or nervous impulses. Describe each type of reflex, giving an example. You may find it helpful to include drawings of the systems to help explain these reflexes.
13. Are biological rhythms an exception to the principle of homeostasis?
14. Describe the biological clock, and explain how it is synchronized with the 24-hour day-night cycle.
15. How does shift work upset the biological clock? How can these problems be mitigated?

CHAPTER

11

Nutrition and Digestion

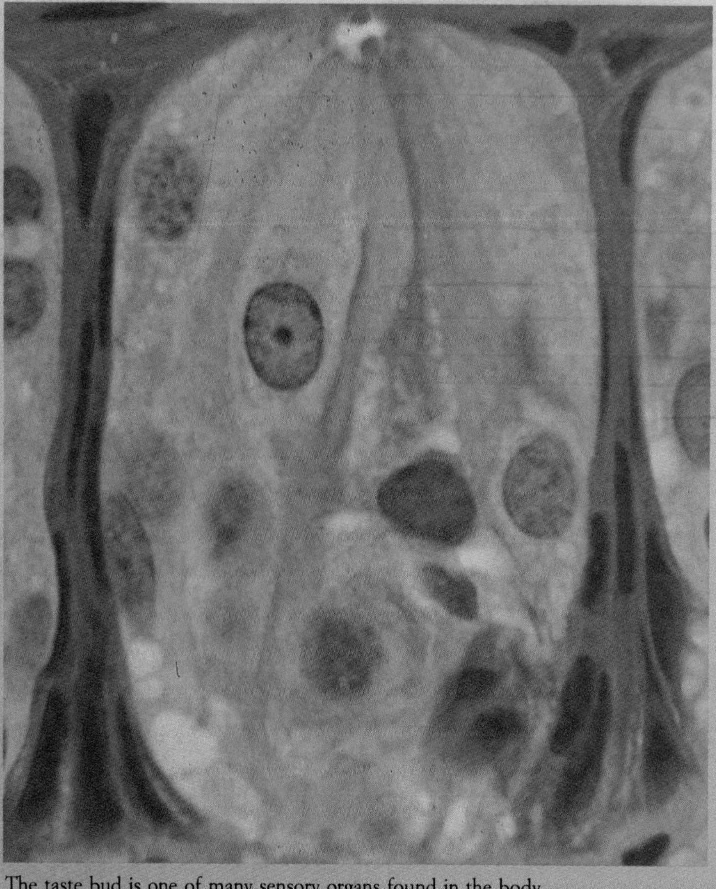

The taste bud is one of many sensory organs found in the body.

It must be a law of human nature. Ask almost any couple, and they will tell you: she lies shivering under the covers on a cold winter night while he bakes. Out on a hike in winter, he stays warm in a light jacket while she bundles up in stocking cap, down coat, and gloves. What causes this difference between many men and women?

Part of the answer may lie in iron—not pumping iron, but dietary iron. Quite simply, many American women do not consume enough iron to offset losses that occur during **menstruation,** the monthly discharge of blood and tissue from the lining of the uterus. Iron deficiencies in women may reduce internal heat production.[1]

John Beard, a researcher at Pennsylvania State University, recently published a study that supports this conclusion. In his research, Beard compared two groups of women, one with low levels of iron in the blood and another with normal levels. Beard found that body temperature dropped more quickly in iron-deficient women exposed to cold than in those with normal iron levels. He also found that iron-deficient women generated 13% less body heat.

Adding credence to the hypothesis that iron deficiency reduces body heat and makes women colder, Beard gave the iron-deficient women iron supplements for 12 weeks, after which they responded normally to cold.

At least two hypotheses can explain these results. The first is that iron deficiencies reduce the amount of oxygen carried in the blood. Iron is a vital component of the hemoglobin molecule, a protein in the red blood cells, which binds to oxygen. Because red blood cells transport oxygen from the lungs to body tissues, a decrease in iron in the hemoglobin molecule may reduce the amount of oxygen available for cellular respiration. Because cellular respiration produces ATP and heat, a shortage of oxygen may lower body heat.

The second hypothesis is based on the role of iron in energy production. Iron is a vital component of the electron transport proteins found in the mitochondria. As noted in Chapter 6, the electron transport system produces most of the ATP in animal cells. If iron levels are low, energy production could be impaired. Because ATP production also releases waste heat, iron deficiencies could reduce internal heat production.

The research on iron may help explain why many women are colder than their male partners and also why many women who take iron supplements during pregnancy report a marked improvement in their response to cold. These findings, and others presented in this chapter, illustrate how important proper nutrition is to normal physiological function.[2]

To further your understanding of nutrition, this chapter begins with a discussion of the major dietary requirements of humans, then describes the process of digestion.

A PRIMER ON NUTRITION

Some people eat to live, but many others seem to live to eat. No matter what your orientation, food probably occupies a central part of your life. You plan your daytime activities around meals. You spend a good part of your life shopping for, preparing, and eating meals. And, depending on your income, 10% to 20% of the money you earn goes to buy food. Thus, one and a half to three hours of each workday goes to providing money for food. (The rest goes to income tax!)

If you live to be 65, you will consume over 70,000 meals. Because foods affect your body in many ways, what you eat will determine how you feel in your later years. That's how important nutrition is to your health. Despite the increased emphasis on nutrition today, studies suggest, most Americans still pay little attention to their diet. To perform our very best, though, we must eat well, acquiring energy and nutrients required by our cells, tissues, and organs.

A **balanced diet** provides proper levels of both nutrients and energy and contains a variety of foods. In 1992, the U.S. Department of Agriculture released a food pyramid to supplant the long-standing four food groups. The food pyramid places foods in six major groups and prescribes allotments from each group necessary for proper nutrition (▶ Figure 11–1).

Table 11-1 lists the basic nutrients required by humans and includes some of the foods and beverages that provide them. As indicated in the table, nutrients are divided into two groups: **macronutrients** and **micronutrients.** Take a

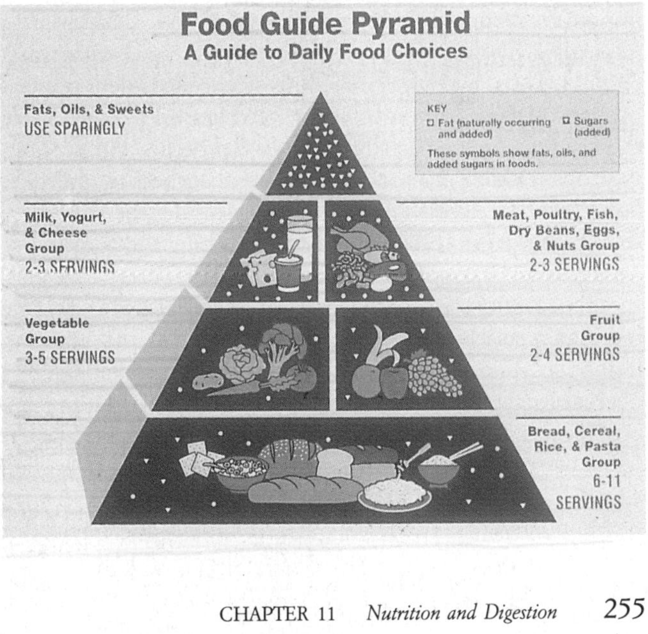

▶ **FIGURE 11–1 Food Guide Pyramid** Replacing the four food groups, the food pyramid divides food into six groups. Recommended daily intake is shown for each group.

[1] Differences in body temperature between men and women may also result from other factors, such as body mass and surface area. Low blood pressure or poor circulation would also make the extremities feel cold.

[2] Because iron can be toxic if ingested in excess, people considering supplementing their diets with iron tablets should consult a physician.

TABLE 11–1 Macronutrients and Micronutrients

NUTRIENTS	FOODS CONTAINING THEM
Macronutrients	
Water	All drinks and many foods
Amino acids and proteins	Milk and milk products, meats, eggs
Lipids	Milk and milk products, meats, eggs, nuts, oils
Carbohydrates	Breads, pastas, cereals, sweets
Micronutrients	
Vitamins	Many vegetables, meats, and fruits
Minerals	Many vegetables, meats, fruits, nuts, and seeds

few minutes to study the table. If you would like to assess your eating habits, take the test on page 266.

The following section describes the macronutrients and micronutrients more fully and explains why they are important to the proper functioning of the human body and the maintenance of health.

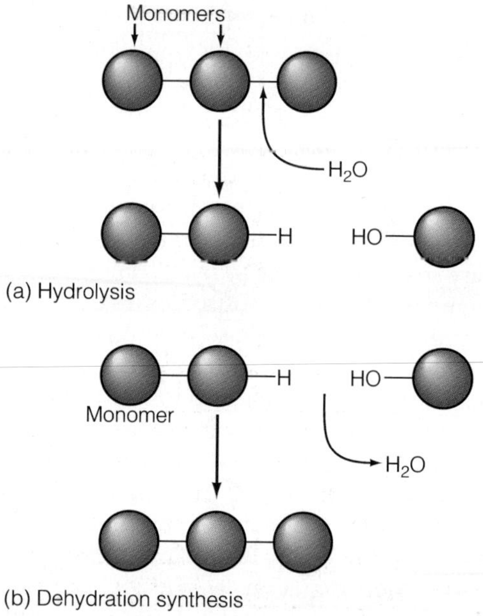

▻ **FIGURE 11–2 Hydrolysis and Dehydration Synthesis** (*a*) Hydrolysis is a reaction in which water is added across a covalent bond, causing the bond to split. (*b*) Dehydration synthesis is the opposite reaction, in which water is lost from two reacting molecules.

Macronutrients Are Required in Relatively Large Quantities and Include Four Substances: Water, Carbohydrates, Lipids, and Proteins

Water. Water is one of the most important of all the substances we ingest. Without it, a person can survive only about three days. Despite its importance, water rarely shows up on nutrition charts. In part, that is because it is supplied in so many different ways. For example, water constitutes the bulk of the liquids we drink and is present in virtually all of the solid foods we eat; water is even produced internally during cellular metabolism (Chapter 6).

Maintaining the proper level of water in the body is important for several reasons. First, water participates in many chemical reactions in the body. One of the most important reactions is known as **hydrolysis.** Discussed in Chapter 2, hydrolysis means to break up *(lyse)* with water *(hydro).* Thus, hydrolytic reactions occur anytime a bond is broken by the addition of water (▻ Figure 11–2). Most of the food we eat consists of polymers (proteins and polysaccharides), which are too large to be absorbed by the cells lining the small intestine. Enzymes inside the intestinal tract break down the polymers by catalyzing the addition of water across the bonds that hold the monomers together. In this reaction, a hydrogen from water attaches to one monomer, and a hydroxyl group attaches to the adjacent monomer. Because of water's importance in metabolism, a decrease in its level can impair metabolism, including energy production. According to some studies, athletic performance may drop significantly when water level falls even slightly.

Maintaining an adequate water volume is also important because it helps stabilize body temperature. A decline in the amount of water in your body, for example, decreases the volume of your blood and your extracellular fluid. These declines cause your body temperature to rise, because the heat normally produced by body cells is being absorbed by a smaller volume of water. A rise in body temperature can impair cellular function and can, if high enough, lead to death. One reason physicians tell you to drink plenty of water when you are suffering from a fever is to help control the increase in your body temperature.

Maintaining proper water levels also helps individuals maintain normal concentrations of nutrients and toxic waste products in the blood and extracellular fluid. If you don't drink enough liquid, your urine will become more concentrated; the rise in the concentration of chemicals in the blood and urine increases your chances of developing a kidney stone, a deposit of calcium and other materials that can block the flow of urine, causing extensive damage to this organ.

Maintaining proper water levels is so important that animals have evolved a variety of homeostatic mechanisms to help carry out this function, described in Chapter 16. As the previous discussion suggests, however, water homeostasis also plays a role in maintaining homeostasis in other areas—for example, energy production, body temperature, and blood levels of various ions and molecules.

Carbohydrates. Carbohydrates serve many purposes in the biological world. One of the most important is providing a source of energy.

All active organisms, humans included, need a continuous

TABLE 11–2 Energy Sources in the Average American Diet

SOURCE	PERCENTAGE
Fat	42
Sucrose	24
Starch	22
Protein	12

supply of energy. Cells, for instance, require energy to carry out thousands of functions needed to maintain homeostasis, to grow, to divide, and to transport molecules across their membranes. Surprisingly, 70% to 80% of the total energy required by fairly sedentary humans (individuals who get little exercise) goes to perform basic functions—metabolism, food digestion, absorption, and so on. The remaining energy is used to power body movements, such as walking, talking, and turning on the television. All waste energy (heat), of course, helps maintain body temperature. For more active people, these percentages shift considerably. A basketball player, for example, uses proportionately more energy for vigorous muscular activity than a sedentary office worker.

At rest, the body relies on nearly equal amounts of carbohydrate (mostly glucose) and fat (triglycerides) to supply basic body needs. As you may recall from Chapter 6, glucose is catabolized during cellular respiration.

Glucose comes from a variety of sources. The most common source is the molecule known as starch. Produced by plants, starch is present in grains and their by-products (pasta and bread). It is also provided by many vegetables. Starch contributes on average about 22% of the total energy requirements of most Americans (Table 11–2). Glucose also comes from glycogen, a polysaccharide found in small amounts in meat.

In the digestive system of humans and other vertebrates, glycogen and starch are broken down into glucose molecules, which are absorbed into the bloodstream and distributed to body cells. The cells break down some glucose immediately to produce energy. Most of the rest is converted into glycogen in muscles and the liver and stored for later use (▻ Figure 11–3). Because most of us eat only a few times a day, the liver plays a crucial role in maintaining homeostasis. Between meals and during exercise, the liver breaks down glycogen to form glucose, which is released into the bloodstream. Glycogen in muscle is catabolized to provide energy for muscular contraction but is not released into the bloodstream.

Another important carbohydrate, not involved in energy production, is fiber. Dietary **fiber** is nondigestible polysaccharide (such as cellulose) found in fruits, vegetables, and grains. As noted in Chapter 2, humans cannot digest cellulose, because the body lacks the enzymes needed to break the covalent bonds joining the monosaccharide units (glu-

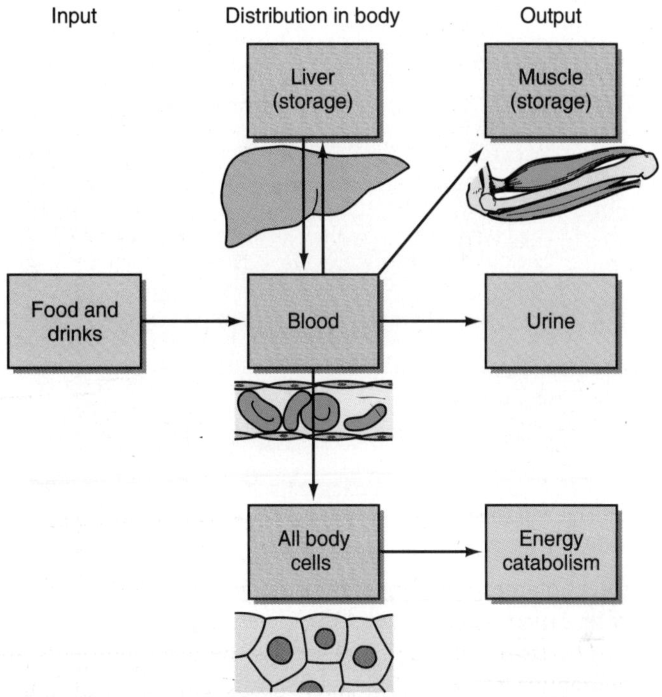

▻ **FIGURE 11–3 Glucose Balance** Glucose levels in the blood result from a balance of input and output.

cose) in the molecule. Consequently, fiber passes through the intestine largely unaffected by stomach acidity or digestive enzymes.

Fiber exists in two basic forms: water-soluble and water-insoluble. Water-soluble fibers are gummy polysaccharides (for example, pectins) in fruits, vegetables, and some grains—including apples, bananas, carrots, barley, and oats. Several studies suggest that water-soluble fiber helps lower blood cholesterol by acting as a sponge that absorbs dietary cholesterol inside the digestive tract, thus preventing it from being absorbed into the bloodstream. Some water-soluble fiber may change the pH of the intestine, making cholesterol insoluble and more difficult to absorb. (The chemical action of water-soluble fibers is discussed in more detail in the section of this chapter on the liver.)

In contrast, water-insoluble fibers are rigid cellulose molecules in celery, cereals, wheat products, and brown rice. Some foods, such as green beans and green peas, contain a mixture of both types. Water-insoluble fiber increases the water content of the **feces,** the semisolid waste produced by the large intestine. This makes the feces softer and facilitates their transport through the large intestine. Increasing the water content also reduces constipation and helps prevent pressure from building up in the large intestine. In some cases, pressure from constipation causes small pouches, or diverticulae, to form in the wall of the large intestine, resulting in a condition known as **diverticulosis** (▻ Figure 11–4). The pouches in the intestinal wall can become inflamed and infected with bacteria, producing a condition known as **diverticulitis.** Bacteria from the local

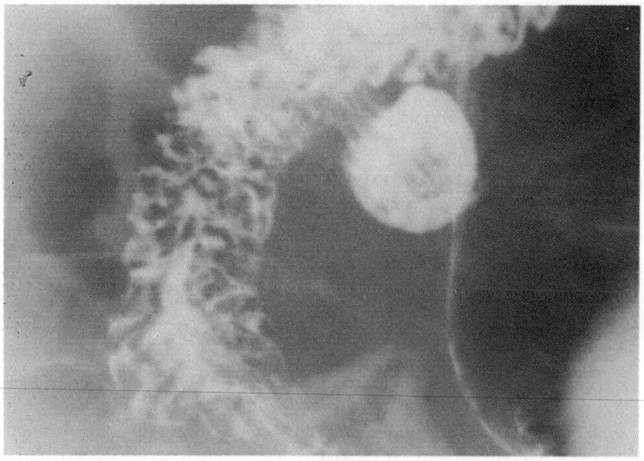

▻ **FIGURE 11–4 Diverticulosis** X-ray showing outpocketing of large intestine.

infection may enter the bloodstream, causing a dangerously high fever. Occasionally, the diverticulae burst, releasing feces into the abdominal cavity. Because feces contain billions of bacteria, their release into the abdominal cavity results in a massive infection that is difficult to treat and is sometimes fatal.

The incidences of diverticulosis and diverticulitis, although not great, have increased substantially in the United States since the introduction of white flour made from wheat. Why? The wheat grain consists of three parts: (1) the germ, the vitamin- and mineral-rich portion that becomes a new plant; (2) the endosperm, the starchy portion that provides nutrients for the growing plant; and (3) the bran, or shell, which contains most of the fiber (▻ Figure 11–5a). When whole-wheat grain is ground, it produces whole-wheat flour, a brown powder containing the entire wheat grain (Figure 11–5b). In contrast, white flour is produced from grain whose bran and germ have been removed (Figure 11–5b). Removing the bran eliminates most, if not all, of the water-insoluble fiber (cellulose). Removing the germ eliminates most of the vitamins and minerals. White flour, therefore, consists mostly of starch.

The widespread use of white flour in the United States and elsewhere has increased the incidence of diverticulosis and diverticulitis because of the elimination of water-insoluble fiber from our diets. Not surprisingly, these conditions are rare in areas, such as Africa, where fiber intake is higher.

Research suggests that water-insoluble fiber, such as that present in the bran, also reduces the incidence of colon cancer, which afflicts about 3% of the U.S. population. How does fiber function? A recent study showed that some bacteria in the large intestine produce a potent chemical mutagen that may cause cancer. Researchers believe that by accelerating the transport of wastes through the intestine, water-insoluble fiber either reduces the formation of the mutagen or reduces the time in which the intestinal cells are exposed to it. Interestingly, in rural Africa, colon

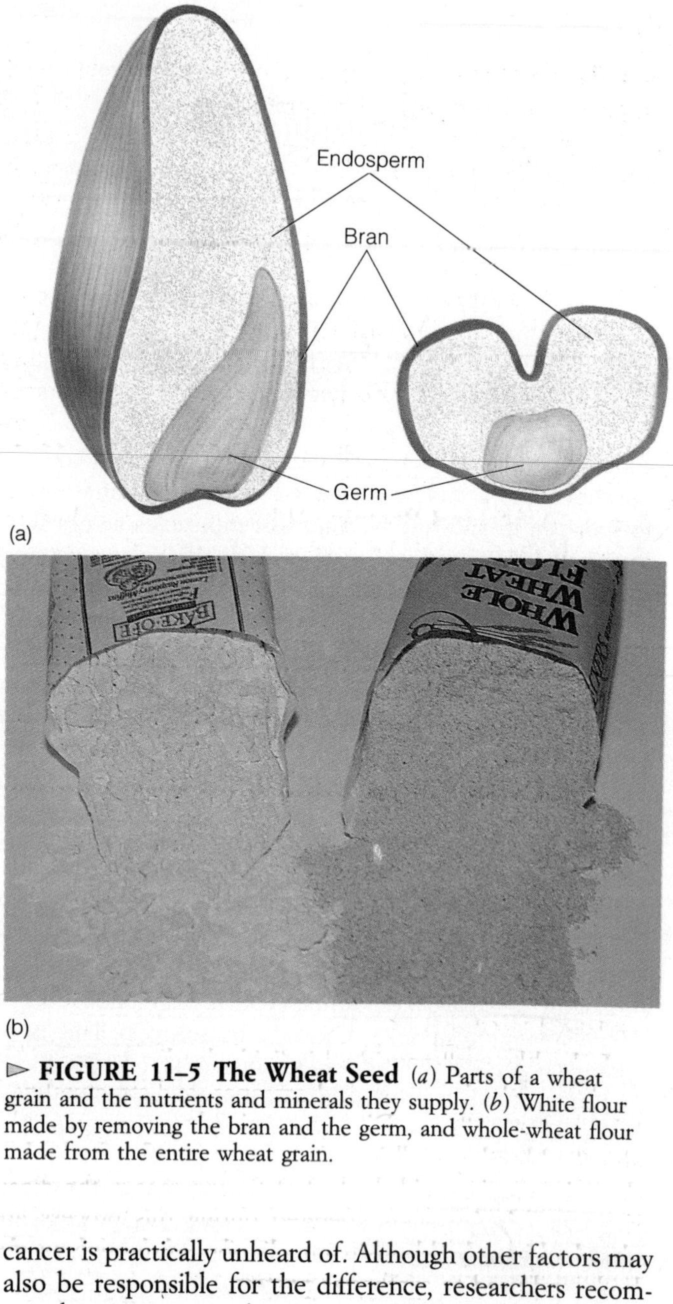

▻ **FIGURE 11–5 The Wheat Seed** (*a*) Parts of a wheat grain and the nutrients and minerals they supply. (*b*) White flour made by removing the bran and the germ, and whole-wheat flour made from the entire wheat grain.

cancer is practically unheard of. Although other factors may also be responsible for the difference, researchers recommend an increase in the amount of fiber we eat.

Lipids. When most of us think about fats, or lipids, in our diets, we think about their harmful effects. In truth, lipids are absolutely essential to proper nutrition and, as noted earlier, provide about half of the body energy during rest. Fatty acids are broken down in a series of cellular chemical reactions whose products enter the biochemical pathways of cellular respiration, as noted in Chapter 6. The complete breakdown of fat yields slightly more than twice as much ATP as the complete breakdown of carbohydrates (about 9 calories per gram).

Although lipids provide cellular energy, they serve other functions as well. The layer of fat beneath the skin in humans, for instance, helps insulate the body from heat loss and helps maintain homeostasis. Fatty deposits around certain organs also cushion them from damage. Horseback riders, motorcyclists, and runners, for instance, can engage in their sports for hours without damage to internal organs in part because of nature's natural cushions.

Certain lipids are also needed by body cells to synthesize steroid hormones, and lipids are a principal component of the plasma membrane of all body cells. In the intestine, lipids increase the uptake of the fat-soluble vitamins (A, D, E, and K).

Most Americans consume lipids in excess. On average, fats provide about 42% of the dietary caloric intake (see Table 11-2), but to lower the risk of heart attack, fat intake should only be about 30%, perhaps even lower, and animal fat should be eaten in small quantities. (For more on lipids and their effects on heart disease, see Health Note 11–1.)

Amino Acids and Protein. Although many people commonly think of proteins as a source of energy, they are only used for this purpose under two conditions: either when dietary intake of carbohydrates and fats is severely restricted or when protein intake far exceeds demand. Dietary deficiencies occur in millions of children throughout the world. The arms and legs of children deprived of protein are often emaciated because their muscle cells break down protein in an effort to provide energy (▻ Figure 11–6).

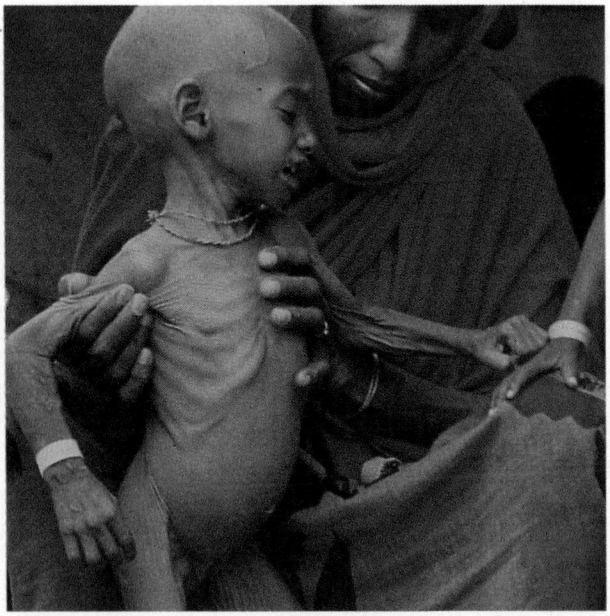

▻ **FIGURE 11–6 A Protein-Deficient Child** This child suffers from severe protein deficiency (kwashiorkor). His arms and legs are emaciated because muscle protein has been broken down to supply energy. His belly is swollen because of a buildup of fluid in the abdomen.

On the other end of the spectrum are the overfed populations of the world. Many Americans, for example, eat far more food than they need. In fact, the average American consumes twice the daily requirement for protein. Amino acids derived from the surplus dietary protein are broken down in the body to produce energy and fat, as explained in Chapter 6.

In healthy, well-nourished individuals, dietary protein is used to synthesize enzymes, hormones, and structural proteins such as collagen. Dietary proteins, however, cannot be absorbed by the small intestine; they must first be broken down into amino acids by hydrolytic enzymes in the digestive tract. Amino acids liberated during this process are absorbed into the bloodstream, distributed throughout the body, and used to synthesize protein.

Proteins in the human body contain 20 different amino acids—all of which can be provided from the diet. The body, however, is capable of synthesizing 11 of the amino acids it needs from nitrogen and smaller molecules derived from carbohydrates and fats. Thus, if they are not present in the diet, they can be made. The remaining 9 amino acids cannot be synthesized by body cells; that is, they must be provided by the diet. These amino acids are appropriately called **essential amino acids.** A deficiency of even one of the essential amino acids can cause severe physiological problems. As a result, nutritionists recommend a diet containing many different protein sources. In this way, individuals can receive all of the amino acids they need.

Proteins are divided into two major groups by nutritionists: complete and incomplete. **Complete proteins** contain ample amounts of all of the essential amino acids. Complete proteins are found in milk, eggs, meat, fish, poultry, and cheese. **Incomplete proteins** lack one or more essential amino acids and include those found in many plant products: nuts, seeds, grains, most legumes (peas and beans), and vegetables. Vegetarians who avoid all animal products, including milk and eggs, must acquire the essential amino acids they need by combining two incomplete protein sources, as shown in ▻ Figure 11–7 (page 262).

Micronutrients are Substances Needed in Small Quantities and Include Two Broad Groups: Vitamins and Minerals

Vitamins. Vitamins are a diverse group of organic compounds present in very small amounts in many of the foods we eat. These molecules are absorbed by the lining of the digestive tract without being broken down. Some vitamins occur in foods in a form known as precursors, or **provitamins.** Inside the body, these precursor molecules are chemically altered to one or more active forms.

The 13 known vitamins play an important role in many metabolic reactions. Because vitamins are recycled many times during metabolic reactions, they are needed only in very small amounts. In fact, 1 gram of vitamin B-12, about 1 teaspoon, would supply over 300,000 people for a day.

Most vitamins are not synthesized in the cells of the body or, if they can be made, are not produced in sufficient amounts to satisfy cellular demands, making dietary intake essential. Vitamin D, for instance, is manufactured by the skin when exposed to sunlight. However, most Americans spend so much time indoors that dietary input is absolutely essential to good health. Table 11–3 (page 263) lists the vitamins and their functions.

HEALTH NOTE 11–1

LOWERING YOUR CHOLESTEROL

Let there be no doubt about it: diseases of the heart and arteries are leading causes of death in the United States.

Atherosclerosis, the accumulation of plaque on artery walls, and the problems it creates are responsible for nearly two of every five deaths in the United States each year. Thanks to improvements in medical care and diet, however, the death rate from atherosclerosis has been falling steadily in recent years, but it is still a major concern. New research, in fact, shows that atherosclerotic plaque is present even in children.

Researchers believe that atherosclerotic plaques begin to form after minor injuries to the lining of blood vessels. High blood pressure, they think, may damage the lining, causing platelets and cholesterol in the blood to adhere to the injured site. The blood vessel responds by producing cells that grow over the fatty deposit. This thickens the wall of the artery, reducing blood flow. Additional cholesterol is then deposited in the thickened wall, forming a larger and larger obstruction.

Cholesterol deposits impair the flow of blood in the heart and other organs, cutting off oxygen to vital tissues. Blood clots may form in the restricted sections of arteries, further reducing blood flow. When the oxygen supply to the heart is disrupted, cardiac muscle cells can die, resulting in heart attacks and death. Blood clots originating in other parts of the body may also lodge in diseased vessels, obstructing blood flow. Oxygen deprivation can weaken the heart, impairing its ability to pump blood. When the oxygen supply to the heart is restricted, the result is a type of heart attack known as a **myocardial infarction**. Victims of a myocardial infarction feel pain in the center of the chest and down the left arm. If the oxygen-deprived area is extensive, the heart may cease functioning altogether.

Atherosclerotic plaque also impairs the flow of blood to the brain. Blood clots catch in the restricted areas and block the flow of blood to vital regions of the brain. Victims may lose the ability to speak or to move limbs. If the damage is severe enough, they may die. Thanks to advances in medical treatment, however, many victims can be saved. And over time, they can recover lost functions as other parts of the brain take over for the damaged regions. Atherosclerosis also affects other organs, such as the kidneys.

Atherosclerosis and cardiovascular disease result from nearly 40 factors, some more important than others. Several of these risk factors, such as old age and sex (being male), cannot be changed. Other factors are controllable.These include high blood pressure, high blood cholesterol, stress, smoking, inactivity, and excessive food intake. Of all the risk factors, three emerge as the primary contributors to cardiovascular disease: elevated blood cholesterol, smoking, and high blood pressure.

Consider cholesterol. Cholesterol is essential to normal body function. It is part of the plasma membrane and is needed to synthesize certain hormones. Interestingly, in most people, the majority of the cholesterol in your blood is produced by the liver. The liver synthesizes and releases about 700 milligrams of cholesterol per day. Only about 225 milligrams of cholesterol are derived from the food we eat each day. Normally, the concentration of cholesterol in the blood is constant. If dietary input falls, the liver increases its output. If the amount of cholesterol in the diet rises, the liver reduces its production. So what's all the fuss about cholesterol in a person's diet?

Even though the liver regulates cholesterol levels, it cannot work fast enough. That is, it may simply be unable to absorb, use, and dispose of cholesterol quickly enough. Consequently, excess cholesterol circulates in the blood after a meal and is deposited in arteries.

Cholesterol is carried in the bloodstream bound to protein. These complexes of protein and lipid fall into two groups: **high-density lipoproteins (HDLs)** and **low-density lipoproteins (LDLs).** HDLs and LDLs function very differently. LDLs, for example, transport cholesterol from the liver to body tissues. In contrast, HDLs are scavengers, picking up excess cholesterol and transporting it to the liver, where it is removed from the blood and excreted in the bile. Research shows that the ratio of HDL to LDL is an accurate predictor of cardiovascular disease. The higher the ratio, the lower the risk of cardiovascular disease.

A high cholesterol level (or hypercholesterolemia) tends to run in families. Thus, if a parent has died of a heart attack or suffers from this genetic disease, his or her offspring are more likely to have high cholesterol levels.

For many years, physicians have advised patients to cut back on high-cholesterol foods, especially eggs, to reduce blood levels of cholesterol. Interestingly, though, a reduction in dietary cholesterol in one individual may result in very little decline in total blood cholesterol, but in another a reduction may result in a much larger drop. The difference in response can be attributed to

Because vitamins are needed in almost all cells of the body, a dietary deficiency in just one vitamin can cause wide-ranging effects. Vitamins also interact with other nutrients. Vitamin C, for example, increases the absorption of iron in the small intestine. Large doses of vitamin C, however, decrease copper utilization by the cells. Consequently, maintaining good health requires the proper balance of vitamins and other nutrients.

Vitamins fall into two broad categories: water-soluble and fat-soluble. **Water-soluble vitamins** include vitamin C and eight different forms of vitamin B. Water-soluble vitamins are transported in the blood plasma unassisted and

▷ **FIGURE 1** (*a*) Fatty foods like these increase cholesterol levels. (*b*) Foods like these will help you reduce cholesterol levels.

exercise, genetics, initial cholesterol levels, and age.

To reduce your chances of atherosclerosis, the American Heart Association recommends (1) limiting dietary fat to less than 30% of the total caloric intake, (2) limiting dietary cholesterol to 300 milligrams per day, and (3) acquiring 50% or more of one's calories from carbohydrates, especially polysaccharides (notably, starches found in potatoes, rice, and other foods). Reductions in saturated fats (animal fats) can also help lower cholesterol levels, for reasons explained in Chapter 2. You can cut back on saturated fat by reducing your consumption of red meat and trimming the fat off all meats before cooking. You can also increase your consumption of fruits, vegetables, and grains, letting these low-fat foods displace some of the fatty foods you might otherwise have eaten (▷ Figure 1).

New and still controversial research also indicates that a diet rich in fish oils can help reduce blood cholesterol. Fish oils contain polyunsaturated fatty acids called omega-3 fatty acids. These fatty acids stimulate the release of prostaglandins, which increase the flexibility of the red blood cells and reduce their stickiness, which is essential for blood clotting. Research on mice shows that a diet rich in omega-3 fatty acids doubles their life span. One of the conclusions of the study, though, is that omega-3 fatty acids, which are extremely susceptible to oxidation, are effective only if oxidation can be prevented. Mincing fish prior to cooking increases oxidation and lowers the level of omega-3 fatty acids.

One's cholesterol level can be lowered with drugs, diet, and exercise. Research spanning several decades shows that a lower cholesterol level translates into a decline in cardiovascular disease. Unfortunately, experts disagree on several key issues. One is exactly who will benefit from a reduction in cholesterol. Some researchers say that only high-risk people with a cholesterol level over 250 milligrams per 100 milliliters of blood should take steps to cut back. Because two-thirds of the American adult population has a blood cholesterol level over 200 milligrams, some experts believe that the entire adult population should take steps to reduce cholesterol.

High cholesterol is also surprisingly common in children, leading many health experts to believe that steps should be taken to prevent problems later in life. In children under the age of 2, however, diets should not be restricted. A diet that is too restrictive may actually impair physical growth and development. What children need is a well-balanced diet, low in fats, especially animal fat, with sufficient calories from other sources. If nothing else, it could help create the eating habits necessary for good health throughout adult life.

are eliminated by the kidneys. Because they are water-soluble, most are not stored in the body in any appreciable amount.

Water-soluble vitamins generally work in conjunction with enzymes, promoting the cellular reactions that supply energy or synthesize cellular materials. Contrary to common myth, the vitamins themselves do not provide energy.

Many proponents of vitamin use believe that megadoses of water-soluble vitamins are harmless because these vitamins are excreted in the urine and do not accumulate in the body. That is, many people think that we can ingest unlimited amounts of water-soluble vitamins with impu-

▻ **FIGURE 11–7**
Complementary Protein Sources
By combining protein sources, a vegetarian who consumes no animal byproducts can be assured of getting all of the amino acids needed. Legumes can be combined with foods made from grains or nuts and other seeds.

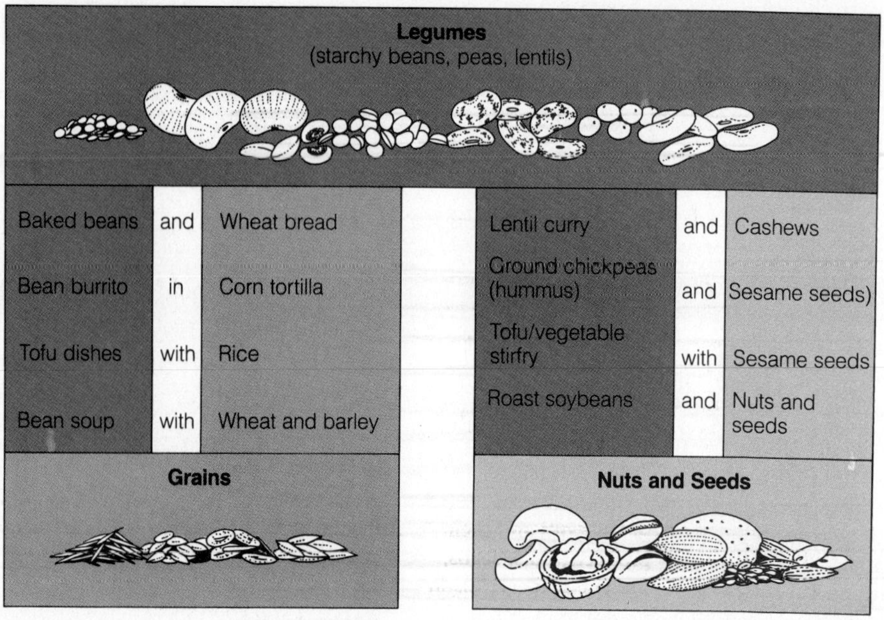

nity. New research, however, shows that this is not entirely true. Some water-soluble vitamins, such as vitamin C, when ingested in excess, can be toxic. (See Table 11–3 for some examples.)

The **fat-soluble vitamins** are vitamins A, D, E, and K. They perform many different functions. Vitamin A, for example, is converted to light-sensitive pigments in receptor cells of the retina, the light-sensitive layer of the eye. These pigments play an important role in vision. Another member of the vitamin A group removes harmful chemicals (oxidants) from the body. Unlike water-soluble vitamins, the fat-soluble vitamins are stored in body fat and accumulate in the fat reserves. The accumulation of fat-soluble vitamins can have many adverse effects (Table 11–3). An excess of vitamin D, for example, can cause weight loss, nausea, irritability, kidney stones, weakness, and other symptoms. Large doses of vitamin D taken during pregnancy can cause birth defects.

Vitamin excess is encountered largely in the developed countries and usually occurs only in people taking vitamin supplements. Each year, in fact, approximately 4000 Americans are treated for vitamin supplement poisoning. To avoid problems from excess vitamins, nutritionists recommend eating a balanced diet that provides all of the vitamins the body needs, rather than taking vitamin pills. Megadoses, they say, should be avoided.

Vitamin deficiencies, like dietary excesses, can lead to serious problems. A deficiency of vitamin D, for example, can produce rickets, a disease that results in bone deformities. Vitamin K deficiencies can result in severe bleeding on injury and internal hemorrhaging. Vitamin C deficiency can result in delayed wound healing and reduced immunity, making people more susceptible to infectious disease.

Most people afflicted by vitamin deficiencies are those who fail to get enough to eat, although even apparently well-fed individuals may suffer from a vitamin deficiency if they are not eating a well-rounded diet. Sadly, about one of every six people living in the nonindustrialized nations of the world, or about 700 million to 800 million people, go to bed hungry; most of these people suffer from multiple vitamin deficiencies and exhibit many symptoms. People suffering from vitamin deficiency typically complain of weakness and fatigue. Children with insufficient vitamin intake fail to grow. All told, about 10 million children under the age of 5 suffer from extreme malnutrition in the less-developed nations of the world. Another 90 million under the age of 5 are moderately malnourished.

Deficiencies of vitamin A afflict over 100,000 children worldwide each year. If not corrected, vitamin A deficiency causes the eyes to dry. Ulcers may form on the eyeball and can rupture, causing blindness (▻ Figure 11–8).

Vitamin deficiencies are also surprisingly common in the industrialized world, where poverty is a way of life for many millions of people. In the United States, one of every five children is born into poverty and is a candidate for vitamin deficiency.

Minerals. On average, an adult contains about 5 pounds of **minerals,** naturally occurring inorganic substances vital to many life processes. Humans require about two dozen minerals, such as calcium, sodium, iron, and potassium, to carry out normal body functions.

Minerals, like vitamins, are micronutrients and are derived from the food we eat and the beverages we drink. Minerals are divided into two groups: the **major minerals** and the **trace minerals** (Table 11–4). The major minerals are present in the body in amounts larger than 5 grams; the trace minerals are found in lesser quantities.

Calcium and phosphorus, for example, are major minerals, and make up three-fourths of the total mineral content

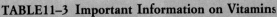

TABLE 11–3 Important Information on Vitamins

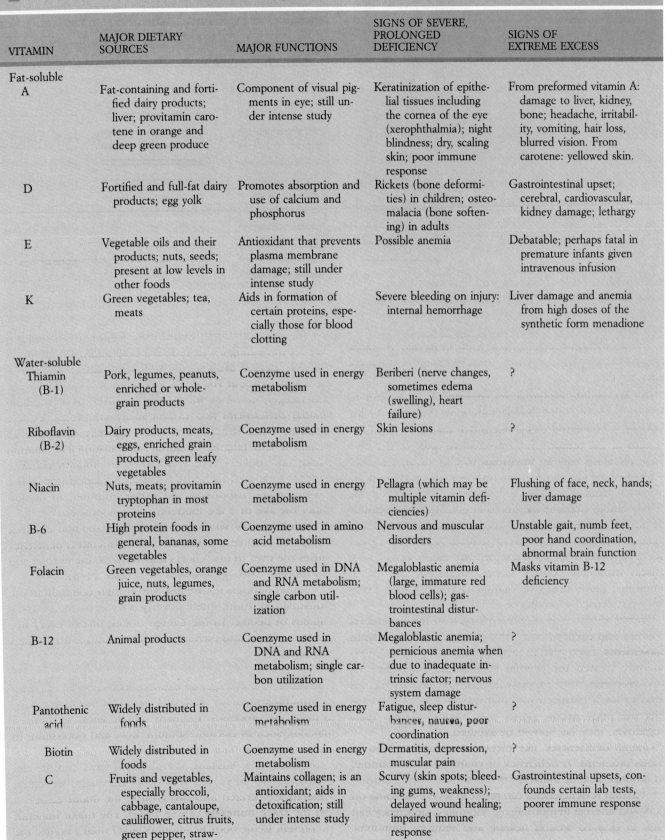

VITAMIN	MAJOR DIETARY SOURCES	MAJOR FUNCTIONS	SIGNS OF SEVERE, PROLONGED DEFICIENCY	SIGNS OF EXTREME EXCESS
Fat-soluble				
A	Fat-containing and fortified dairy products; liver; provitamin carotene in orange and deep green produce	Component of visual pigments in eye; still under intense study	Keratinization of epithelial tissues including the cornea of the eye (xerophthalmia); night blindness; dry, scaling skin; poor immune response	From preformed vitamin A: damage to liver, kidney, bone; headache, irritability, vomiting, hair loss, blurred vision. From carotene: yellowed skin.
D	Fortified and full-fat dairy products; egg yolk	Promotes absorption and use of calcium and phosphorus	Rickets (bone deformities) in children; osteomalacia (bone softening) in adults	Gastrointestinal upset; cerebral, cardiovascular, kidney damage; lethargy
E	Vegetable oils and their products; nuts, seeds; present at low levels in other foods	Antioxidant that prevents plasma membrane damage; still under intense study	Possible anemia	Debatable; perhaps fatal in premature infants given intravenous infusion
K	Green vegetables; tea, meats	Aids in formation of certain proteins, especially those for blood clotting	Severe bleeding on injury: internal hemorrhage	Liver damage and anemia from high doses of the synthetic form menadione
Water-soluble				
Thiamin (B-1)	Pork, legumes, peanuts, enriched or whole-grain products	Coenzyme used in energy metabolism	Beriberi (nerve changes, sometimes edema (swelling), heart failure)	?
Riboflavin (B-2)	Dairy products, meats, eggs, enriched grain products, green leafy vegetables	Coenzyme used in energy metabolism	Skin lesions	?
Niacin	Nuts, meats; provitamin tryptophan in most proteins	Coenzyme used in energy metabolism	Pellagra (which may be multiple vitamin deficiencies)	Flushing of face, neck, hands; liver damage
B-6	High protein foods in general, bananas, some vegetables	Coenzyme used in amino acid metabolism	Nervous and muscular disorders	Unstable gait, numb feet, poor hand coordination, abnormal brain function
Folacin	Green vegetables, orange juice, nuts, legumes, grain products	Coenzyme used in DNA and RNA metabolism; single carbon utilization	Megaloblastic anemia (large, immature red blood cells); gastrointestinal disturbances	Masks vitamin B-12 deficiency
B-12	Animal products	Coenzyme used in DNA and RNA metabolism; single carbon utilization	Megaloblastic anemia; pernicious anemia when due to inadequate intrinsic factor; nervous system damage	?
Pantothenic acid	Widely distributed in foods	Coenzyme used in energy metabolism	Fatigue, sleep disturbances, nausea, poor coordination	?
Biotin	Widely distributed in foods	Coenzyme used in energy metabolism	Dermatitis, depression, muscular pain	?
C	Fruits and vegetables, especially broccoli, cabbage, cantaloupe, cauliflower, citrus fruits, green pepper, strawberries	Maintains collagen; is an antioxidant; aids in detoxification; still under intense study	Scurvy (skin spots; bleeding gums, weakness); delayed wound healing; impaired immune response	Gastrointestinal upsets, confounds certain lab tests, poorer immune response

SOURCE: Adapted from J. L. Christian and L. L. Greger, *Nutrition for Living*, 2d ed., copyright © 1988 by the Benjamin/Cummings Publishing Company. Used with permission.

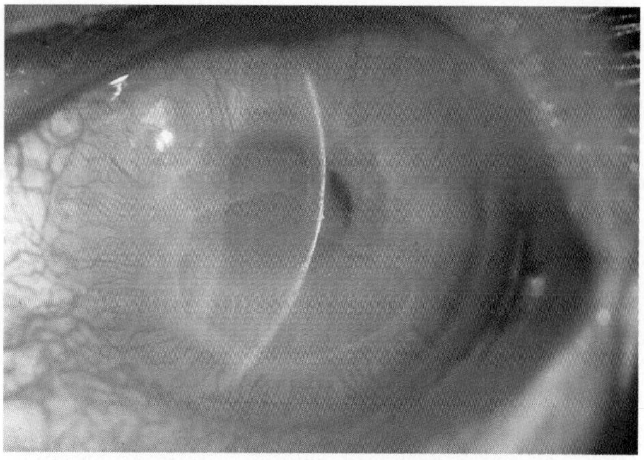

▻ **FIGURE 11–8 Vitamin A Deficiency** Vitamin A deficiency causes the cornea of the eye to dry and become irritated. If the deficiency is not corrected, ulcers may form and break, causing blindness.

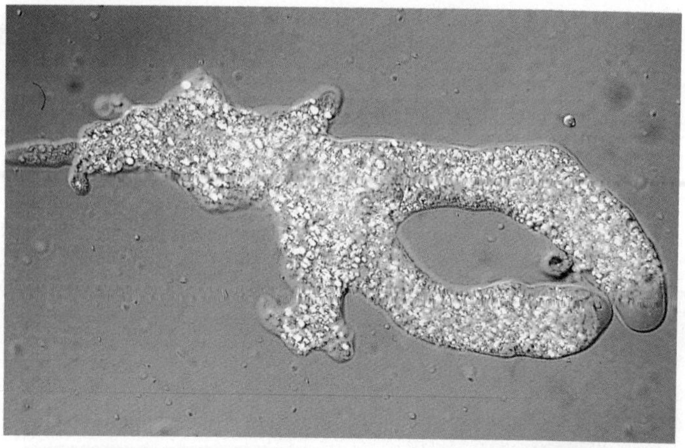

▻ **FIGURE 11–9 The Simplest Form of Digestion** Amoebae and other protists phagocytize materials, then digest them intracellulary within food vacuoles.

of the human body. These two minerals form part of the dense extracellular matrix of bone and are required in a much greater quantity than zinc or copper, two trace minerals that are components of some enzymes.

The distinction between major and trace minerals is not meant to imply that one group is more important than the other. In fact, a daily deficiency of a few micrograms (a microgram is 1/1000 of a gram, and a gram is 1/454 of a pound) of iodine is just as serious as a deficiency of several hundred milligrams of calcium.

Table 11–4 lists some of the minerals, their function, and problems that arise when they are ingested in excess or deficient amounts.

If you'd like to assess your eating habits and learn how to eat a more healthy diet, complete the survey in Table 11–5.

THE DIGESTIVE SYSTEM

The human digestive system is a product of millions of years of evolution. During the course of evolution, strategies for digesting food have generally increased in complexity. In single-celled protists, such as *amoebae,* food is phagocytized, then digested intracellularly in food vacuoles (▻ Figure 11–9). This is the simplest form of digestion and, no doubt, the earliest to evolve. Its inefficiency probably accounts for its relatively rare occurrence within the animal world.

Much more efficient modes of digestion are provided by various types of digestive tracts, tubular canals that run through organisms. Digestive tracts provide a site for extracellular digestion in a wide number of vertebrates and invertebrates alike.

The simplest digestive tract is found in flatworms and a group of animals known as the **cnidarians** (ny-dar-ee-uns), which includes jellyfish, hydras, sea anemones, and coral (▻ Figure 11–10). In these creatures, the gut consists of a saclike or branched tube with only a single opening. Food enters this opening and is digested by enzymes secreted into the gut. Waste is ejected through the same opening.

Most other animals have a complete gut, consisting of a tube opened at both ends. Food enters the mouth and is digested by enzymes secreted into the gut. Waste is excreted at the opposite end, the **anus** (▻ Figure 11–11,

▻ **FIGURE 11–10 The Incomplete Gut** Food enters the gut, or coelenteron, of the jellyfish and other cnidarians through an opening that also serves to rid these organisms of undigested waste. Enzymes released by some of the cells lining the sac digest the food inside the gut. (*a*) Schematic and (*b, next page*) photograph.

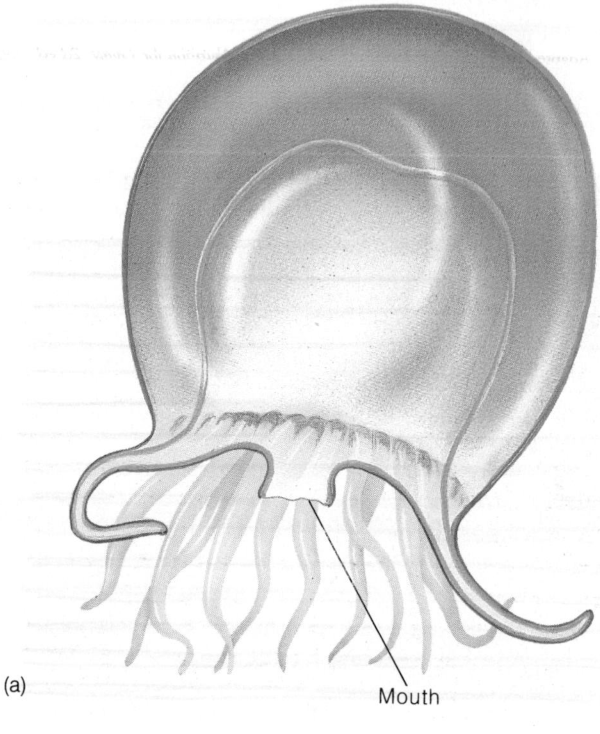

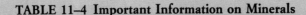

TABLE 11–4 Important Information on Minerals

MINERAL	MAJOR DIETARY SOURCES	MAJOR FUNCTIONS	SIGNS OF SEVERE, PROLONGED DEFICIENCY	SIGNS OF EXTREME EXCESS
Major minerals				
Calcium	Milk, cheese, dark green vegetables, legumes	Bone and tooth formation; blood clotting; nerve transmission	Stunted growth; maybe bone loss	Depressed absorption of some other minerals
Phosphorous	Milk, cheese, meat, poultry, whole grains	Bone and tooth formation; acid-base balance; component of coenzymes	Weakness; demineralization of bone	Depressed absorption of some minerals
Magnesium	Whole grains, green leafy vegetables	Component of enzymes	Neurological disturbances	Neurological disturbances
Sodium	Salt, soy sauce, cured meats, pickles, canned soups, processed cheese	Body water balance; nerve function	Muscle cramps; reduced appetite	High blood pressure in genetically predisposed individuals
Potassium	Meats, milk, many fruits and vegetables, whole grains	Body water balance; nerve function	Muscular weakness; paralysis	Muscular weakness; cardiac arrest
Chloride	Salt, many processed foods (as for sodium)	Plays a role in acid-base balance; formation of gastric juice	Muscle cramps; reduced appetite; poor growth	Vomiting
Trace minerals				
Iron	Meats, eggs, legumes, whole grains, green leafy vegetables	Component of hemoglobin and enzymes	Iron-deficiency anemia, weakness, impaired immune function	Acute: shock, death; chronic: liver damage, cardiac failure
Iodine	Marine fish and shellfish; dairy products; iodized salt; some breads	Component of thyroid hormones	Goiter (enlarged thyroid)	Iodide goiter
Fluoride	Drinking water, tea, seafood	Maintenance of tooth (and maybe bone) structure	Higher frequency of tooth decay	Mottling of teeth; skeletal deformation

SOURCE: Adapted from J. L. Christian and L. L. Greger, *Nutrition for Living,* 2d ed., copyright © 1988 by the Benjamin/Cumming Publishing Company. Used with permission.

(b)

page 268). In the course of evolution, the various parts of the digestive tract have evolved to perform specific functions, as shown in the figure. Some regions may store food; others grind and mix the food. Still others play a role in digestion and absorption. The earthworm digestive tract is shown in ▻ Figure 11–12 (page 268), along with the functions of each region.

In humans and most animals, food digestion requires physical and chemical processes that take place in the digestive tract (▻ Figure 11–13, page 269). In the mouth, for example, food is sliced, crushed, and torn by the teeth into smaller particles, greatly increasing the surface area presented to the digestive enzymes in the stomach and small intestine and therefore increasing the efficiency of digestion. Enzymes participate in a chemical breakdown of food. In the small intestine, amino acids, monosaccharides, and other small molecules produced by enzymatic digestion are absorbed into the bloodstream for distribution to body cells.

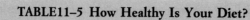

TABLE 11–5 How Healthy Is Your Diet?

This quiz enables you to assess your eating habits. The more points you get, the better your nutritional health is likely to be.

PART I

1. I usually limit my meat, fish, poultry, or egg servings to once or twice a day. Yes/No
2. I eat red meats (beef, ham, lamb, or pork) not more than about three times a week. Yes/No
3. I remove fat or ask that fat be trimmed from meat before cooking. Yes/No
4. I eat about three or four eggs per week, including those cooked with other foods. Yes/No
5. I sometimes have meatless days and eat such protein-rich foods as legumes and nuts. Yes/No
6. I usually broil, boil, bake, or roast meat, fish, or poultry; I usually don't fry it. Yes/No

Total "YES" answers: _____

PART II

7. I have two or more cups of milk or the equivalent in milk products every day. Yes/No
8. I drink low-fat or nonfat milk (2% or less butterfat) rather than whole milk. Yes/No
9. I eat ice cream or ice milk only twice a week or less. Yes/No
10. I seldom have more than about 3 tsp of margarine or butter per day. Yes/No

Total "YES" answers: _____

PART III

11. I usually have one serving (½ c) of citrus fruit or juice (oranges, grapefruit, etc.) each day. Yes/No
12. I have at least one serving of dark green or deep orange vegetables each day. Yes/No
13. I eat fresh fruits and vegetables when I can get them. Yes/No
14. I cook vegetables without fat (if I use margarine, it's measured and added after cooking). Yes/No
15. I eat fresh fruit for dessert more often than pastries. Yes/No

Total "YES" answers: _____

PART IV

16. I generally eat whole-grain breads. Yes/No
17. Most of the cereals I use are whole-grain and good sources of fiber. Yes/No
18. The cereals I use have little or no sugar added. Yes/No
19. I use brown rice in preference to white rice. Yes/No
20. I generally have at least four servings of bread or cereal grain products each day. Yes/No

Total "YES" answers: _____

PART V

21. I am usually within 5 to 10 lb of the weight considered appropriate for my height. Yes/No
22. I drink no more than 1½ oz of alcohol (one to two drinks) per day. Yes/No
23. I do not add salt to food after preparation and prefer foods salted lightly or not salted at all. Yes/No
24. I try to avoid foods high in refined sugar and use sugar sparingly. Yes/No
25. I always eat a breakfast of at least cereal and milk, egg and toast, or other protein-carbohydrate combination with fruit or fruit juice. Yes/No

Total "YES" answers: _____

For each "yes" answer, give yourself 1 point. How are your points distributed among the various areas of nutrition?

		Excellent	Good	Fair	Poor	Your score
Part I	Meat and meat alternate choices	5–6	4	3	0–2	_____
Part II	Dairy choices	4	3	2	0–1	_____
Part III	Fruit and vegetable choices	5	4	3	0–2	_____
Part IV	Grain choices	5	4	3	0–2	_____
Part V	Weight control, other choices	5	4	3	0–2	_____

How is your over-all nutrition score? **Total points earned:** _____

(24–25 is excellent; 19–23 is good; 14–18 is fair; 13 or lower is poor.)

SOURCE: Adapted with permission from Roger Sargent, *Have a Good Life Series*, Greenville, S.C.: Liberty Life.

SPOTLIGHT ON EVOLUTION 11–1

INSECT ADAPTATIONS FOR FEEDING

Insects are one of the most successful groups of animals that inhabit the Earth. In fact, there are more insect species than all other forms of life combined! What is more, some scientists believe that hundreds of thousands—perhaps millions—of additional species are yet to be discovered. Most of these will probably be discovered in tropical rain forests, where one recent study showed that as many as 20,000 species may be found on a single tree!

Adapted to a wide range of habitats, insects have been on the planet for about 400 million years. They owe their great success in large part to the evolution of flying. This ability helps them escape predators, find food, locate mates, and move to new habitats more quickly than most other animals.

Insects occupy nearly every terrestrial and aquatic habitat known to humankind, and they are rare only in the seas. With so many species occupying so many habitats, this remarkable class of animals displays a wide range of adaptations. Here we focus our attention on a few adaptations related to the ingestion and digestion of food.

As anyone who has ever walked through the woods or tended a garden knows, many insects feed on plant or animal juices. These species come equipped with a tiny, needlelike proboscis (pro-bos-is). These hollow structures can penetrate flowers, stems, leaves, or even skin (▻ Figure 1) and can be used to suck nutrient-rich fluids from plants and animals.

Perhaps the most familiar example is the mosquito. The female mosquito requires blood for successful egg production. Besides having a sharp proboscis, which she inserts into the skin, the female mosquito has tiny glands that release a chemical anticoagulant. This substance prevents blood from clotting, thus ensuring a steady flow while she dines on an animal's blood.

Some insects have evolved strong mandibles composed of chitin (a type of polysaccharide). The mandibles are used for biting into wood, leaves, or prey. Leaf-cutting ants, for instance, have extremely strong chitinous mandibles, which aid in the chewing of leaves, which they deposit in their underground tunnels. Here the leaves support a healthy colony of fungi, which are eaten by the ants.

A number of insects have auxiliary food storage capacity—that is, sites where they can store ingested foodstuffs for later use. This permits many insects to feed intermittently. The mosquito comes to mind as a good example. It contains a small, thin-walled sac, that can store blood for up to a week before it passes into the digestive tract. Honeybees also have a storage site for nectar (sugar secreted by the base of a flower's petals). Nectar is stored in a special segment of the gut where it is converted to honey. When the bee returns to its hive, it regurgitates the honey into a waxy honeycomb.

The grasshopper's digestive system—like that of so many other insects—contains three segments, the foregut, the midgut, and the hindgut. The foregut contains a region known as the crop, which functions primarily in food storage. In some insects, the crop has a roughened surface that helps grind food, which then passes on to the midgut, where digestion and absorption occur.

Most insects digest their food within the digestive tract, but the common housefly has a unique mode of digestion. As a fly walks across the icing of cake, it tastes the sweet substance with its feet. (In fact, many insects have their taste receptors in their feet.) As it wanders around, the fly disgorges enzymes, which it then stirs into the food with its mouth. These enzymes digest food molecules. But don't worry, says the biologist Robert Wallace. When it's done, the housefly sucks any mess it has made with its tubelike mouthparts, leaving the rest for you.

▻ FIGURE 1 Proboscis of the mosquito.

The Mouth Is the Site of the Physical Breakdown of Food

The mouth is a complex structure in which food is broken down mechanically and, to a much lesser degree, chemically. Food taken into the mouth is sliced into smaller pieces by the sharp teeth in front; it is then ground into a pulpy mass by the flatter teeth toward the back of the mouth. As the food is pulverized in the mouth, it is liquefied by **saliva,** a watery secretion released by the **salivary glands,** three sets of exocrine glands located around the oral cavity (▻ Figure 11–14). The release of saliva is triggered by the smell, feel, taste, and sometimes even the thought of food.

Saliva (1) liquifies the food, making it easier to swallow, (2) kills or neutralizes some bacteria via the enzymes and antibodies it contains, (3) dissolves substances so they can

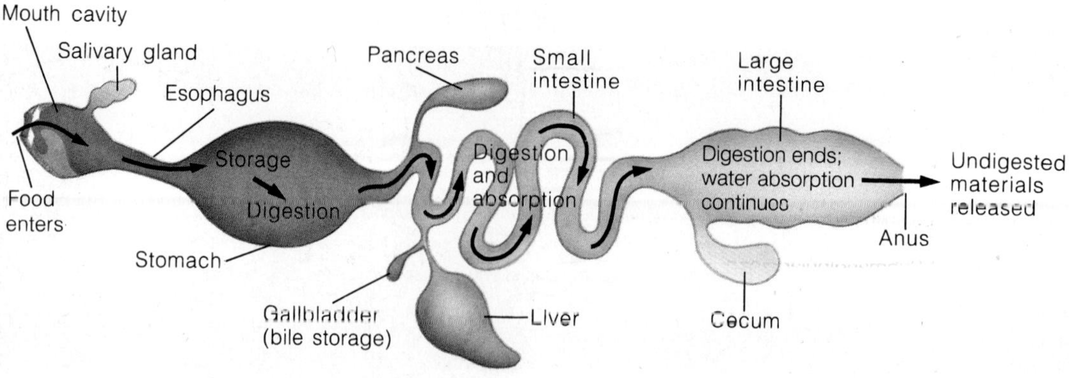

▻ **FIGURE 11–11 The Complete Gut** This diagram illustrates the general anatomy of the tubular digestive tract present in many animals. Food enters the mouth and is excreted at the opposite end through the anus. Along the way, it is chemically and physically broken down, then transported into the bloodstream.

be tasted, and (4) begins to break down starch molecules with the aid of the enzyme **amylase.** Saliva also cleanses the teeth, washing away bacteria and food particles. Because the release of saliva is greatly reduced during sleep, bacteria tend to accumulate on the surface of the teeth, where they break down microscopic food particles, producing some foul-smelling chemicals that give us "dragon breath," or "morning mouth."

Controlling the bacteria that live on the teeth by regular brushing is important, because these organisms secrete a rather sticky material called **plaque.** Plaque adheres to the surface of the teeth, trapping bacteria. Entombed in their own secretions, these bacteria release small amounts of a weak acid that dissolves the hard outer coating of our teeth, the **enamel.** This acid forms small pits in the enamel, commonly referred to as **cavities.** Continued acid secretion may cause the cavities to deepen into the softer layer beneath the enamel. If the cavity is left untreated, an entire tooth can be lost to decay.

Brushing helps remove plaque and helps reduce the incidence of cavities. Most toothpastes also contain small amounts of fluoride, which hardens the enamel and helps reduce cavities. In most cities and towns, small amounts of fluoride are added to drinking water as a preventive measure. However, one recent study suggests that excess fluoride may be carcinogenic, and health officials are reexamining this practice.

A recent study showed that chewing sugarless gum increases the flow of saliva and cleanses the teeth. By chewing sugarless gum within 5 minutes after a meal, and for at least 15 minutes, you can reduce the incidence of cavities.

After food is chewed, it must be swallowed. The tongue plays a key role in swallowing by pushing food to the back of the oral cavity into the **pharynx,** a funnel-shaped chamber that connects the oral cavity with the **esophagus,** a long muscular tube that leads to the stomach (see Figure 11–13).

The tongue, which also aids in speech, contains taste receptors, the **taste buds,** on its upper surface. Taste buds are stimulated by four basic flavors: sweet, sour, salty, and bitter (▻Figure 11–15). Various combinations of these flavors (combined with odors we smell) give us a rich assortment of tastes.

Food propelled from the pharynx into the esophagus is prevented from entering the trachea, or windpipe, which carries air to the lung and lies in front of the esophagus, by the epiglottis (▻ Figure 11–16). The **epiglottis** is a flap of tissue that acts like a trap door, closing off the trachea during swallowing. (Details of epiglottis function are found in Chapter 14.)

Swallowing begins with a voluntary act—the tongue pushing food into the back of the oral cavity. Once food enters the pharynx, however, the process becomes automatic. Food in the pharynx stimulates stretch receptors in the wall of this organ, which, in turn, trigger the **swallowing reflex,** an involuntary contraction of the muscles in the wall of the pharynx. This forces the food into the esophagus.

▻ **FIGURE 11–12 The Digestive System of the Earthworm** This valuable organism has a complete gut, which consists of a single tubular structure with a mouth and anus.

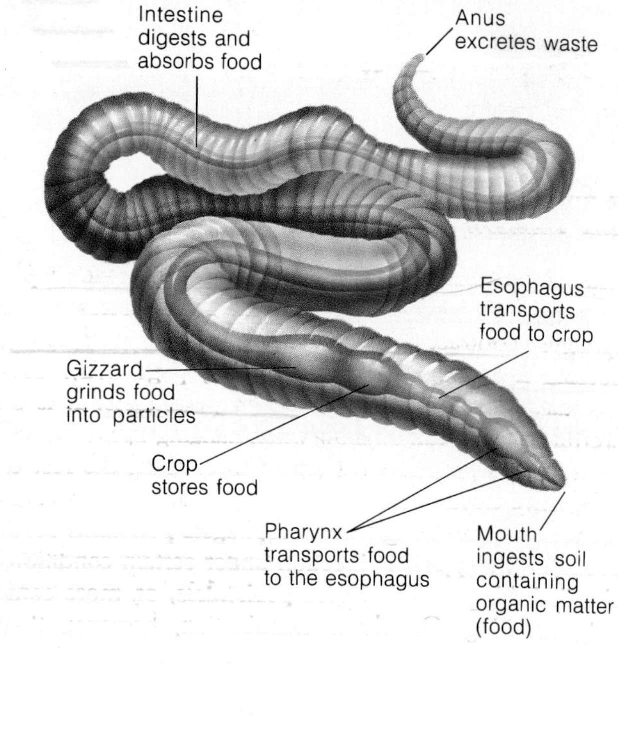

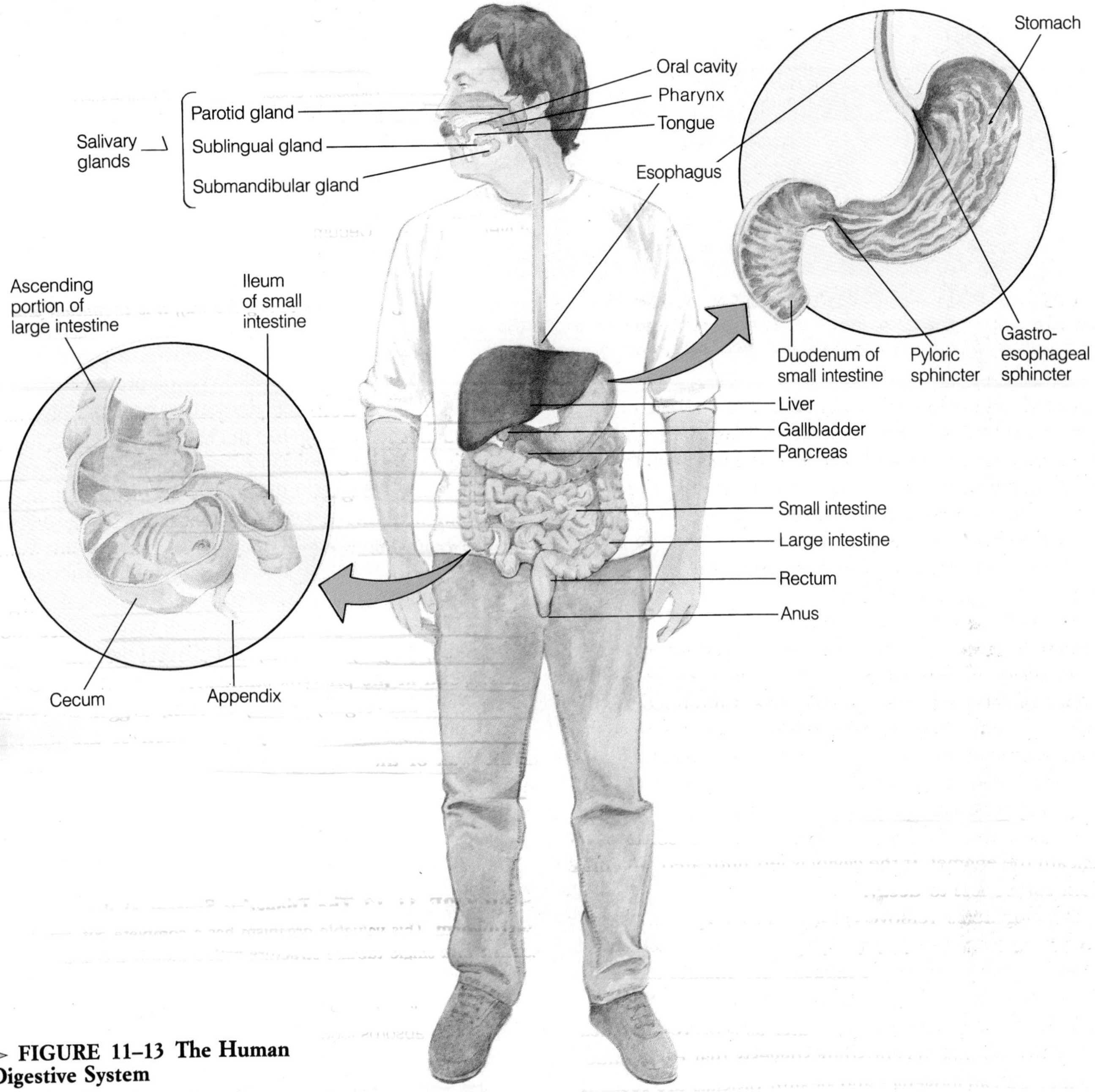

▻ **FIGURE 11–13 The Human Digestive System**

The Esophagus Transports Food to the Stomach via Peristalsis

Involuntary contractions of the muscular wall of the esophagus propel the food to the stomach. The muscles of the esophagus contract above the swallowed food mass, squeezing it along (▻ Figure 11–17a, page 272). This involuntary muscular action is called **peristalsis.** It is so powerful that you can swallow when hanging upside down. Peristalsis also propels food (and waste) along the rest of the digestive tract.

Scientists once thought that esophageal peristalsis could proceed in the opposite direction under certain conditions and called this process **reverse peristalsis,** or, more commonly, vomiting. On closer examination, however, they found that the stomach and esophagus are both relaxed during vomiting. Food is expelled from the stomach as a result of contractions in the muscles of the abdomen and the diaphragm, which separates the thoracic and abdominal cavities and plays an important role in breathing.

Vomiting is a reflex action that occurs when irritants are present in the stomach. It is, therefore, a protective adaptation that allows the body to rid the stomach of bad food and harmful viruses and bacteria, so important to vertebrates. Vomiting may also be caused by (1) emotional factors, such as stress, (2) rotation or acceleration of the head, as in motion sickness, (3) stimulation of the back of the throat, and (4) elevated pressure inside the brain caused by injury.

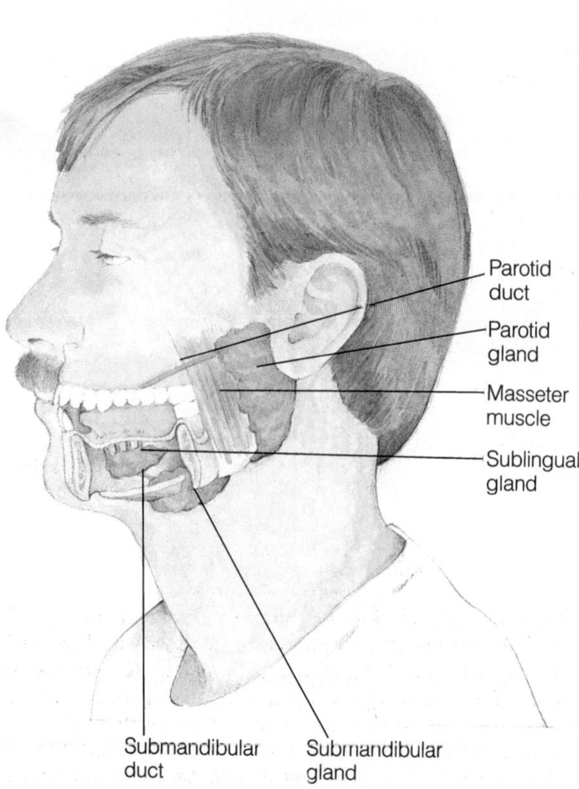

▻ **FIGURE 11–14 Salivary Glands** Three salivary glands (parotid, submandibular, and sublingual) are located in and around the oral cavity and empty into the mouth via small ducts.

The Stomach Stores Food, Releasing It into the Small Intestine in Spurts

In humans, the stomach, as shown in Figure 11–13, lies on the left side of the abdominal cavity, partly under the protection of the rib cage. Food enters the stomach via the esophagus. The opening to the stomach, however, is closed off by the **gastroesophageal sphincter,** a thickened layer of smooth muscle at the juncture of the esophagus and stomach (▻ Figure 11–18). As food enters the lower esophagus, the gastroesophageal sphincter opens, allowing the food to enter the stomach. The sphincter then promptly closes, preventing food and acid from percolating upward. If the sphincter fails to close, acid rising in the esophagus causes irritation, a condition commonly known as "heartburn" (see Figure 11–17b).

Inside the stomach, food is liquified by acidic secretions of tiny glands in the wall of the stomach, the **gastric glands.** These glands produce a watery secretion called **gastric juice.** It contains hydrochloric acid (HCl), pepsin, and a proteolytic (protein-digesting) enzyme precursor known as **pepsinogen** (discussed in more detail below). These substances come from two distinctly different cell types in the gastric gland.

Inside the stomach, food is churned by peristalsis and mixed with gastric juice. The churning action of the stomach's muscular walls may help break down large pieces of food. Combined with the liquid from salivary glands and the gastric glands, the food becomes a rather thin, watery paste known as **chyme.** The stomach can hold 2 to 4 liters (2 quarts to 1 gallon) of chyme, which is gradually released into the small intestine at a rate suitable for proper digestion and absorption. The periodic release maximizes the efficiency of digestion.

Contrary to what many think, very little enzymatic digestion occurs in the stomach. The stomach's role is largely to prepare most food for enzymatic digestion that will occur in the small intestine. There are some exceptions to this rule, however. Protein is one of them.

Proteins are denatured by hydrochloric acid (HCl) in the stomach. Hydrochloric acid also acts on pepsinogen, converting it to the active form **pepsin,** a proteolytic enzyme that catalyzes the breakdown of proteins into large fragments. These fragments are further broken down in the small intestine, the next stop in the digestive system. Interestingly, pepsinogen molecules are also activated by pepsin. Thus, once a few pepsin molecules are formed, they assist HCl in activating the remaining pepsinogen molecules.

Hydrochloric acid creates an acidic environment (pH 1.5–3.5) in the stomach. Besides activating pepsinogen and denaturing protein (rendering it digestible), HCl also breaks down connective tissue fibers in meat and kills most bacteria, helping protect the body from infection.

Also, contrary to popular belief, the stomach does not absorb foodstuffs. Only a few substances, such as alcohol, can actually penetrate the lining of the stomach to enter the bloodstream. Alcohol consumed on an empty stomach passes quickly into the bloodstream, often producing a rather immediate dizzying effect. The presence of food in the stomach, however, retards alcohol absorption.

The stomach lining is protected from destruction by an alkaline secretion known as **mucus,** produced by certain cells in the lining. Mucus coats the epithelium, protecting it from acid and pepsin. The tissues beneath the epithelium are protected from acid leakage by the cells of the epithelium, which are tightly joined to one another, forming a leak-proof barrier.

Unfortunately, the stomach's protective mechanisms can break down. A number of factors—stress, coffee (caffeine), excess aspirin, nicotine, and alcohol or combinations of them—can increase acid levels in the stomach, overwhelming the mucous layer. When this happens, hydrochloric acid and pepsin come in contact with the epithelial cells and may digest parts of the wall of the stomach, forming painful **ulcers** (▻ Figure 11–19).[3] When detected early, most ulcers can be treated by reducing stress and changing one's diet—for example, eating smaller amounts of food and reducing coffee, aspirin, and alcohol. Stress reduction and dietary changes, such as these, reduce acid

[3] Ulcers also occur in the esophagus and, more commonly, in the small intestine. Excessive acid percolating into the esophagus and the release of excess acid into the small intestine can overwhelm the protective mucous layer found in both organs.

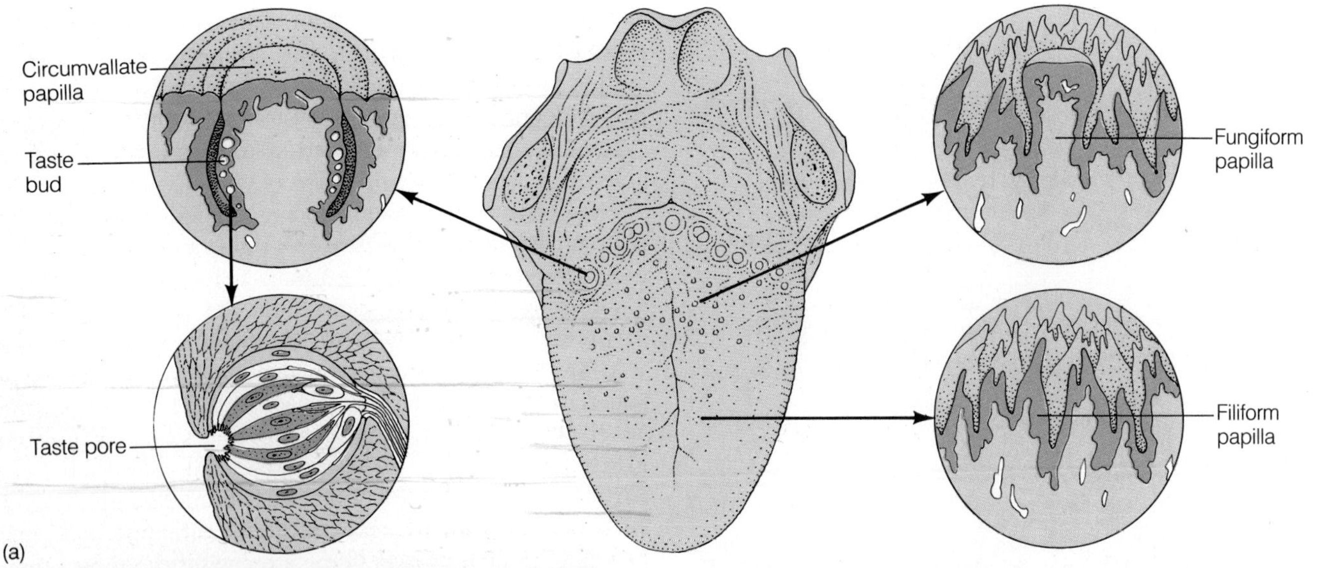

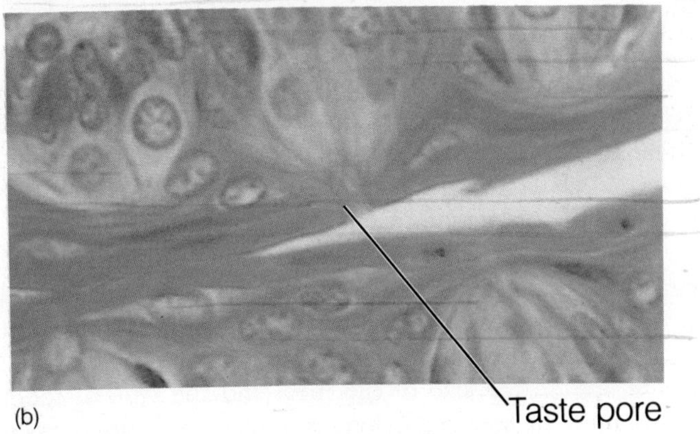

▻ **FIGURE 11–15 The Tongue and Taste Buds** (*a*) The tongue is a muscular organ that aids in swallowing and phonation (producing sounds). Its upper surface is dotted with protrusions called papillae. Three types are present: the fungiform, filiform, and circumvallate. Taste buds are located on the fungiform and circumvallate papillae. Four types of taste buds are found: those that detect salt, bitter, sweet, and sour flavors. Each type is found on a specific region on the tongue. Food molecules dissolved in saliva enter the taste pore and stimulate these cells, which in turn trigger nerve impulses to the brain. (*b*) A photomicrograph of taste buds; the arrow indicates a taste pore.

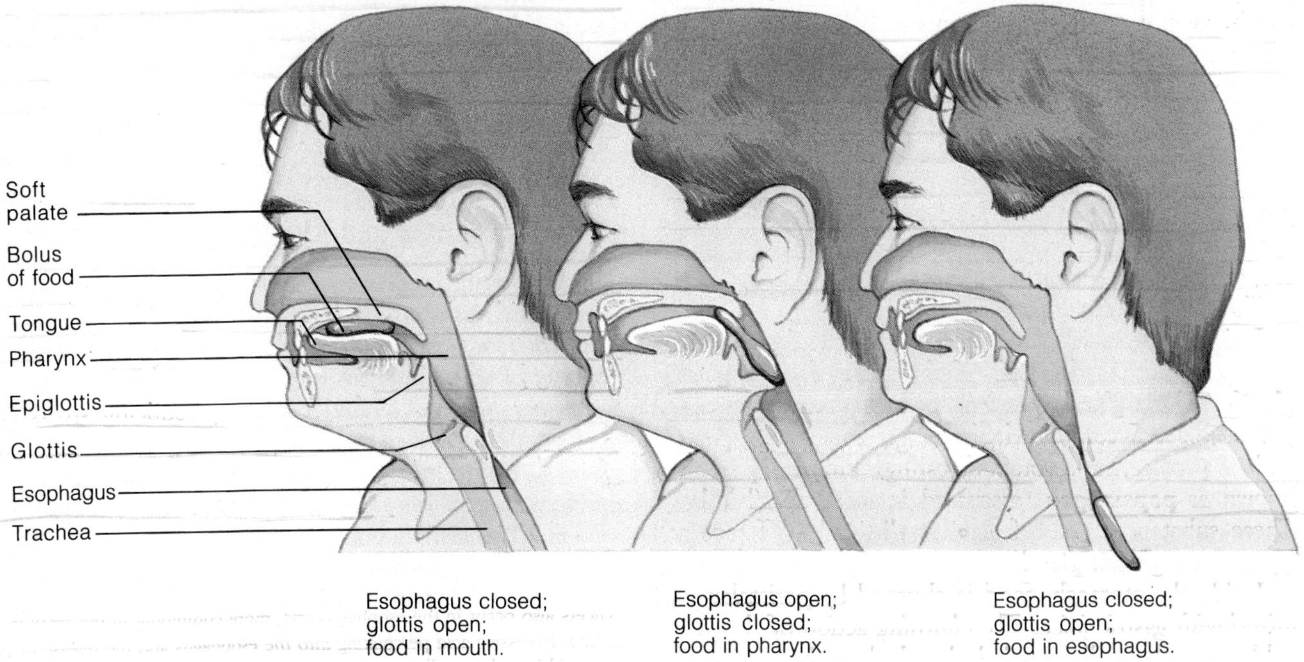

▻ **FIGURE 11–16 The Epiglottis** This trapdoor prevents food from entering the trachea during swallowing. As illustrated, the trachea is lifted during swallowing, pushing against the epiglottis, which bends downward.

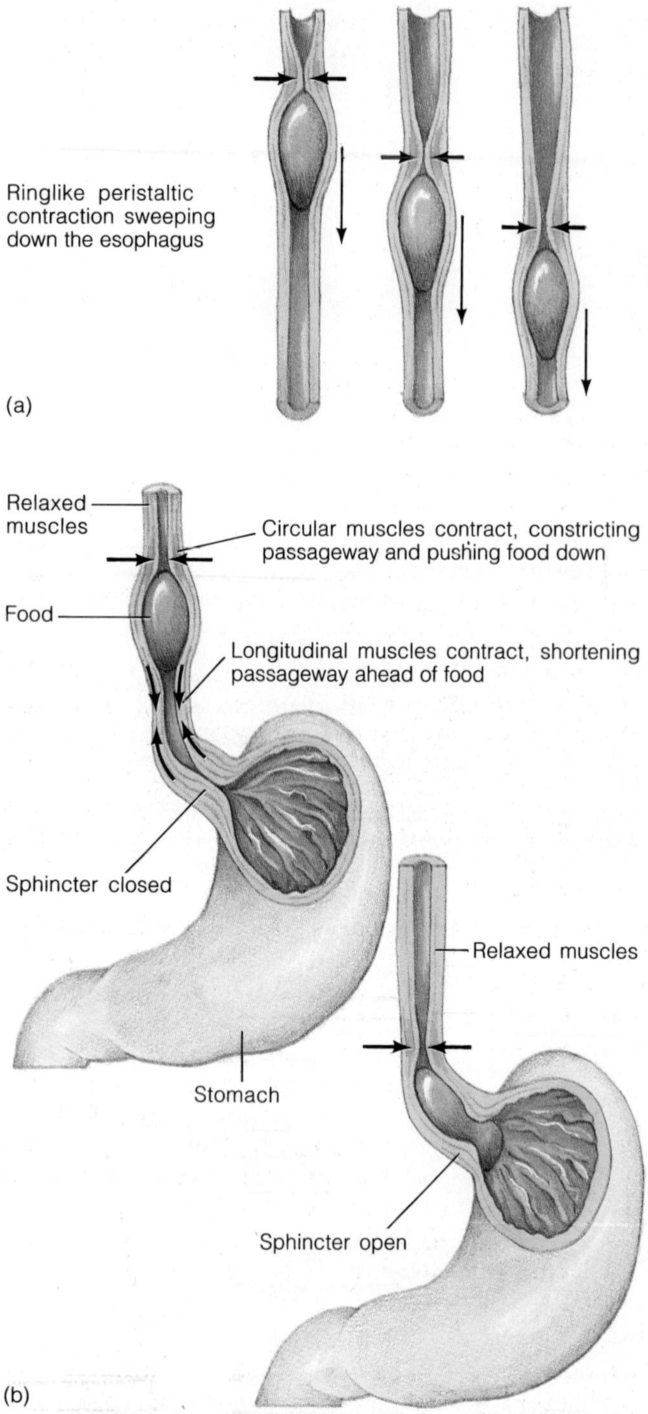

▻ **FIGURE 11–17 Peristalsis**
(*a*) Peristaltic contractions in the esophagus propel food into the stomach. (*b*) When food reaches the stomach, the gastroesophageal sphincter opens, allowing food to enter.

secretion in the stomach (see Health Note 1-1). When ulcers are not detected early and when damage is severe, parts of the stomach may have to be removed surgically.

The Stomach's Function Is Regulated by Nerves and Hormones. The stomach does not produce hydrochloric acid continuously. Were that to occur, continuous production might endanger the stomach lining. Instead, HCl is produced on demand. Its secretion is controlled by the endocrine and nervous systems, sometimes working together. These highly evolved systems provide a remarkable degree of control over a very complex process.

Activation begins when food is smelled, tasted, or simply seen (▻ Figure 11–20). The sight, smell, and taste of food activate centers of the brain, which, in turn, transmit nerve impulses to the stomach via the **vagus nerve.** The vagus nerve terminates in the stomach wall and activates HCl production by the gastric glands. Nerve impulses also stimulate the synthesis and release of the hormone gastrin, which is produced by the lining of the stomach. **Gastrin** increases the output of gastric juice, containing HCl and pepsinogen. Acid production is also stimulated by the presence of proteins and peptides in the stomach and small intestine. In fact, protein in the stomach is the most potent stimulus of gastric gland secretion.

Chyme Leaves the Stomach and Enters the Small Intestine. Chyme in the stomach is ejected into the small intestine by peristaltic muscle contractions. A peristaltic wave travels across the stomach every 20 seconds. When the wave of contraction reaches the far end of the stomach, the **pyloric sphincter** (a ring of smooth muscle at the juncture of the small intestine and stomach) opens, and chyme squirts into the small intestine.

The stomach contents are emptied in two to six hours, depending on the size of the meal and the type of food. The larger the meal, the longer it takes to empty. Solid foods (meat) empty faster than liquid foods (a milkshake). Peristaltic contractions continue after the stomach is empty and are felt as hunger pangs.

As chyme leaves the stomach, the major stimulus for enhanced gastric-gland secretion, the presence of protein in the stomach, is removed. For this and other reasons, acid and pepsinogen release decline, shutting down the stomach's HCl production until the next meal.

The Small Intestine Serves as a Site of Food Digestion and Absorption

The small intestine is a coiled tube in the abdominal cavity about 6 meters (20 feet) long in adults (see Figure 11–13). So named because of its small diameter, the small intestine consists of three parts in the following order: the duodenum, jejunum, and ileum. Inside the small intestine, macromolecules are digested with the aid of numerous enzymes—that is, broken down into smaller molecules—that are transported into the bloodstream and the lymphatic system. The lymphatic system, discussed more fully in Chapter 12, is a network of vessels that carries extracellular fluid from the tissues of the body to the circulatory system. In addition, tiny lymphatic vessels in the wall of the small intestine, known as **lacteals,** absorb fats and transport them to the bloodstream.

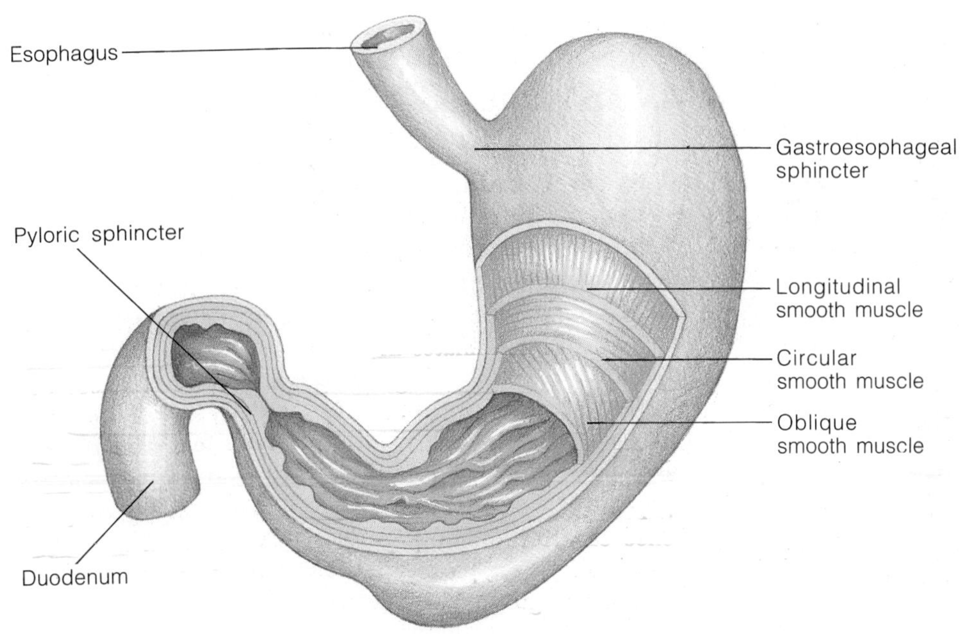

▻ **FIGURE 11–18 The Stomach** The stomach lies in the abdominal cavity. In its wall are three layers of smooth muscle that help mix the food and force it into the small intestine, where most digestion occurs. The gastroesophageal and pyloric sphincters control the inflow and outflow of food, respectively.

Digestion in the Small Intestine Requires Enzymes from Two Major Sources. The digestion of food molecules inside the small intestine requires enzymes produced from two distinctly different sources: the pancreas, an organ that lies beneath the stomach, and the lining of the small intestine itself.

The **pancreas** is nestled in a loop formed by the first portion of the small intestine, the duodenum (▻ Figure 11–21). The pancreas is a dual-purpose organ; it has endocrine and exocrine functions. As an exocrine gland, it produces enzymes and sodium bicarbonate essential for the digestion of foodstuffs in the small intestine. Its endocrine function is fulfilled by special cells that produce hormones that help regulate blood glucose levels and thus help maintain homeostasis.

▻ **FIGURE 11–19 Ulcer**

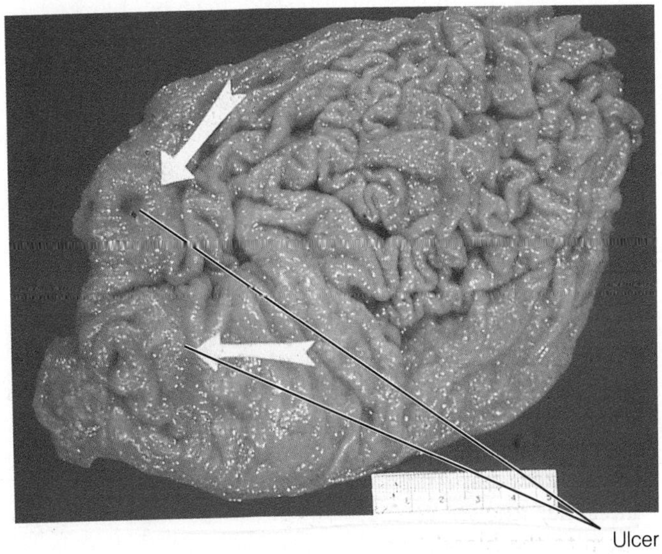

The digestive enzymes of the pancreas are produced in small glandular units. These glands empty into many small ducts that converge to form the large pancreatic duct. The pancreatic duct joins with a duct draining the gallbladder and liver and then empties into the duodenum, the first segment of the small intestine. Each day approximately

▻ **FIGURE 11–20 Pathway Leading to the Release of HCl by the Stomach**

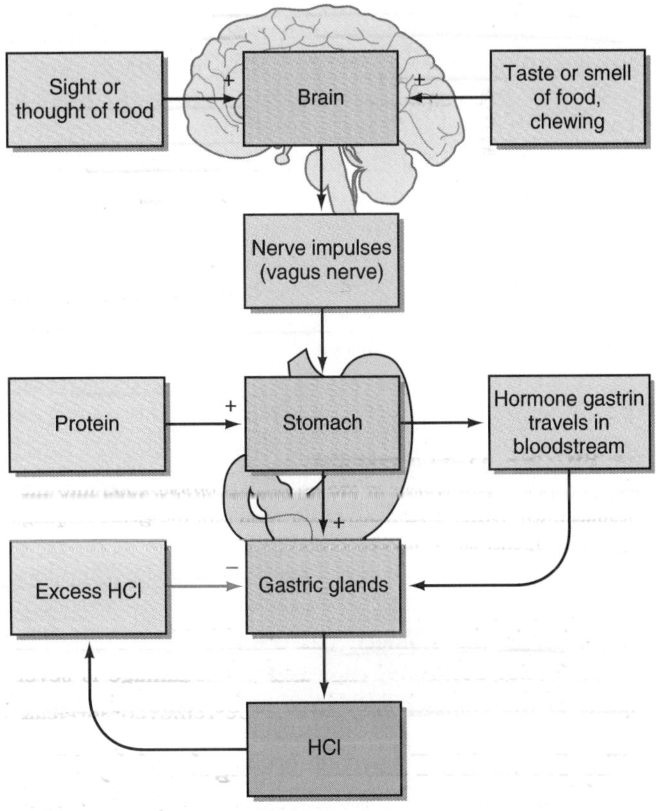

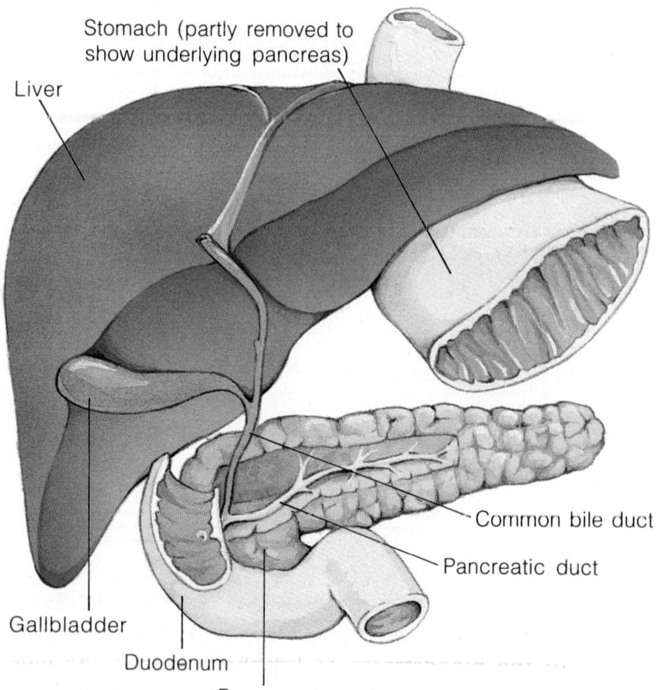

▻ **FIGURE 11–21 Organs of Digestion** The liver, gallbladder, and pancreas all play key roles in digestion and empty by the common bile duct into the small intestine, in which digestion takes place.

1200 to 1500 milliliters (1.0 to 1.5 quarts) of pancreatic juice is produced and released into the small intestine. This liquid is composed of water, sodium bicarbonate, and several important digestive enzymes (Table 11–6).

Sodium bicarbonate neutralizes the acidic chyme released by the stomach and thus helps protect the small intestine from the harmful effects of stomach acid. It also gives the pancreatic juice a pH of about 8, creating an environment optimal for the function of the pancreatic digestive enzymes.

Pancreatic enzymes released into the lumen of the small intestine act on the large molecules in food (proteins, starches, and so on), as shown in Table 11–6. As a result of pancreatic enzymatic activity, fats are completely reduced to monoglycerides (glycerol attached to one fatty acid) and fatty acids, which can be absorbed without further action. Proteins are broken into small peptide fragments and some amino acids. Carbohydrates are broken down into disaccharides and some monosaccharides.

Like the stomach, the pancreas secretes its enzymes in an inactive form. This protects the gland from self-destruction. **Trypsinogen,** for example, is the inactive form of the protein-digesting enzyme trypsin. Trypsinogen is activated by a substance on the epithelial lining of the small intestine. Trypsin, in turn, activates other digestive enzymes.

The final stage of digestion occurs with the aid of enzymes produced by the epithelial cells of the small intestine. These enzymes are embedded in the membranes of the epithelial cells (Table 11–6). As a result, the final phase of digestion occurs just before the nutrient is absorbed into the cell.

The Liver Produces an Emulsifying Agent, Bile, Which Plays a Key Role in the Digestion of Fat.

The **liver** is one of the largest and most versatile organs in the body, performing many functions essential to homeostasis. By various estimates, the liver performs as many as 500 different functions. Situated in the upper-right quadrant of the abdominal cavity, under the protection of the rib cage, the liver is one of the body's storage depots for glucose (Figure 11–13). It also stores fats, iron, copper, and many vitamins. By storing glucose and lipids and releasing them as they are needed, the liver helps ensure a constant supply of energy-rich molecules to body cells. The liver also synthesizes some key blood proteins involved in clotting and is an efficient detoxifier of potentially harmful chemicals, such as nicotine, barbiturates, and alcohol. These functions contribute significantly to homeostasis.

The liver also plays a key role in the digestion of fats through the production of a fluid called **bile,** which contains water, ions, and molecules such as cholesterol, fatty acids, and bile salts. **Bile salts** are steroids that emulsify fats—that is, break fat globules into smaller ones. This process is essential for lipid digestion, because lipid-digesting enzymes in the small intestine do not work well on large fat globules. Unless fat globules are broken into smaller ones, fat digestion will be impaired.

Produced by the cells of the liver, bile is first transported to the **gallbladder,** a sac attached to the underside of the liver (see Figure 11–13). The gallbladder concentrates the bile by removing water from it. (This action also helps conserve water.) Bile is stored in the gallbladder until needed. When chyme is present in the small intestine, the gallbladder contracts, causing bile to flow out through the duct system and into the small intestine (Figure 11–21).

Bile flow to the small intestine may be blocked by **gallstones,** deposits of cholesterol and other materials that form in the gallbladder of some individuals. Gallstones may lodge in the ducts draining the organ, thus reducing—even completely blocking—the flow of bile to the small intestine. The lack of bile salts greatly reduces lipid digestion. Because lipid digestion is reduced, fat is not absorbed. Passed on with the undigested food mass into the large intestine of a person with gallstones, these fats are then decomposed by bacteria in the large intestine, but they are not absorbed. The decomposition of these fats often gives the feces a foul odor. The higher percentage of fat also makes the feces quite buoyant and difficult to flush.

Approximately 1 of every 10 American adults has gallstones, although many (30% to 50%) of these people exhibit no symptoms whatsoever. Gallstones occur more frequently in older, overweight individuals, and the incidence

TABLE 11–6 Digestive Enzymes

SITE OF PRODUCTION	ENZYME	ACTION
Salivary glands	Amylase	Digests polysaccharides in oral cavity
Stomach	Pepsin	Breaks proteins into peptides
Pancreas	Trypsin	Cleaves peptide bonds of polypeptides and proteins
	Chymotrypsin	Same as trypsin
	Carboxypeptidase	Cleaves peptide bonds on carboxy end of polypeptides
	Amylase	Breaks starches into smaller units (maltose)
	Phospholipase	Cleaves fatty acids from phosphoglycerides to form monoglycerides
	Lipase	Cleaves two fatty acids from triglycerides
	Ribonuclease	Breaks RNA into smaller nucleotide chains
	Deoxyribonuclease	Breaks DNA into smaller nucleotide chains
Epithelium of small intestine	Maltase	Breaks maltose into glucose subunits
	Sucrase	Breaks sucrose into glucose and fructose subunits
	Lactase	Breaks lactose into glucose and galactose subunits
	Aminopeptidase	Breaks down peptides into amino acids

in the elderly is about 1 in every 5. When they cause problems, gallstones are usually removed surgically. This procedure requires that the entire gallbladder be removed. Interestingly, bile continues to be produced in these patients but is stored in the common bile duct, which becomes distended to accommodate the liquid bile. Scientists are now testing a new drug that dissolves gallstones in many patients, hoping that it may someday eliminate the need for surgery in many instances.

The Intestinal Epithelium Is Specially Modified for Absorption. Virtually all food digestion occurs in the duodenum and jejunum. Once food molecules are digested, they must be absorbed—that is, transported across the epithelial lining of the first two portions of small intestine into the bloodstream or lymphatic system. In these regions, absorption is facilitated by three structural modifications of the small intestine. ▻ Figure 11–22a shows, for instance, that the lining of the small intestine is thrown into circular folds, which increase the overall surface area. As shown in Figure 11–22b, on the surfaces of the circular folds are many fingerlike projections known as **villi,** which also increase the surface area available for the absorption of food molecules. The intestinal surface area is further increased by **microvilli,** tiny protrusions of the plasma membranes of the epithelial cells lining the villi, which are located on the surface facing the lumen, or cavity (Figure 11–22d). Each epithelial cell that functions in absorption contains an estimated 3000 microvilli. Two hundred million microvilli occupy a single square millimeter of intestinal lining. Together, the circular folds, villi, and microvilli result in a dramatic increase in the surface area of the lining of the small intestine, making it 600 times greater than if it were merely a flat layer, lined by epithelial cells.

Numerous mechanisms are involved in absorption—too numerous to be discussed in this book. Three of the most common are diffusion, osmosis, and active transport, described in Chapter 3. Many nutrients enter the epithelial cells via active transport, then diffuse out of the intestinal cells into the bloodstream or lymphatic vessels. As illustrated in ▻ Figure 11–23, each villus of the small intestine is endowed with a rich supply of blood and lymph capillaries. These tiny vessels absorb nutrients that have passed through the lining of the small intestine, then transport these nutrients elsewhere.

Although most nutrients diffuse from the epithelial cells into the blood capillaries, fatty acids and monoglycerides produced by the enzymatic digestion of triglycerides follow a different route. As shown in ▻ Figure 11–24 (page 278), these molecules diffuse into the cells lining the villi and inside the cells they combine to reform triglycerides. The triglycerides, in turn, combine to form fat droplets within the cytoplasm of the epithelial cells of the small intestine. These tiny fat droplets are then coated with a layer of lipoprotein (lipid bound to protein), which is produced by the endoplasmic reticulum. This treatment renders the fat droplets, or **chylomicrons,** water-soluble. The coated fat droplets are then released by the epithelial cells by exocytosis into the interstitial fluid (Figure 11–24). Because blood capillaries are relatively impermeable to chylomicrons, most of the lipid globules enter the more porous lymph capillaries in the villi. You will note in Figure 11–24 that small or medium chain fatty acids not incorporated in triglycerides pass directly into blood capillaries in the villi.

The Large Intestine Is the Site of Water Resorption

The **large intestine** is about 1.5 meters (5 feet) long and consists of four basic regions, the cecum, appendix, colon, and rectum (▻ Figure 11–25, page 279). The **cecum** is simply apouch that forms below the juncture of the large and small intestine. A small, wormlike structure, the **appendix,** attaches to the bottom of the cecum. Most of the large

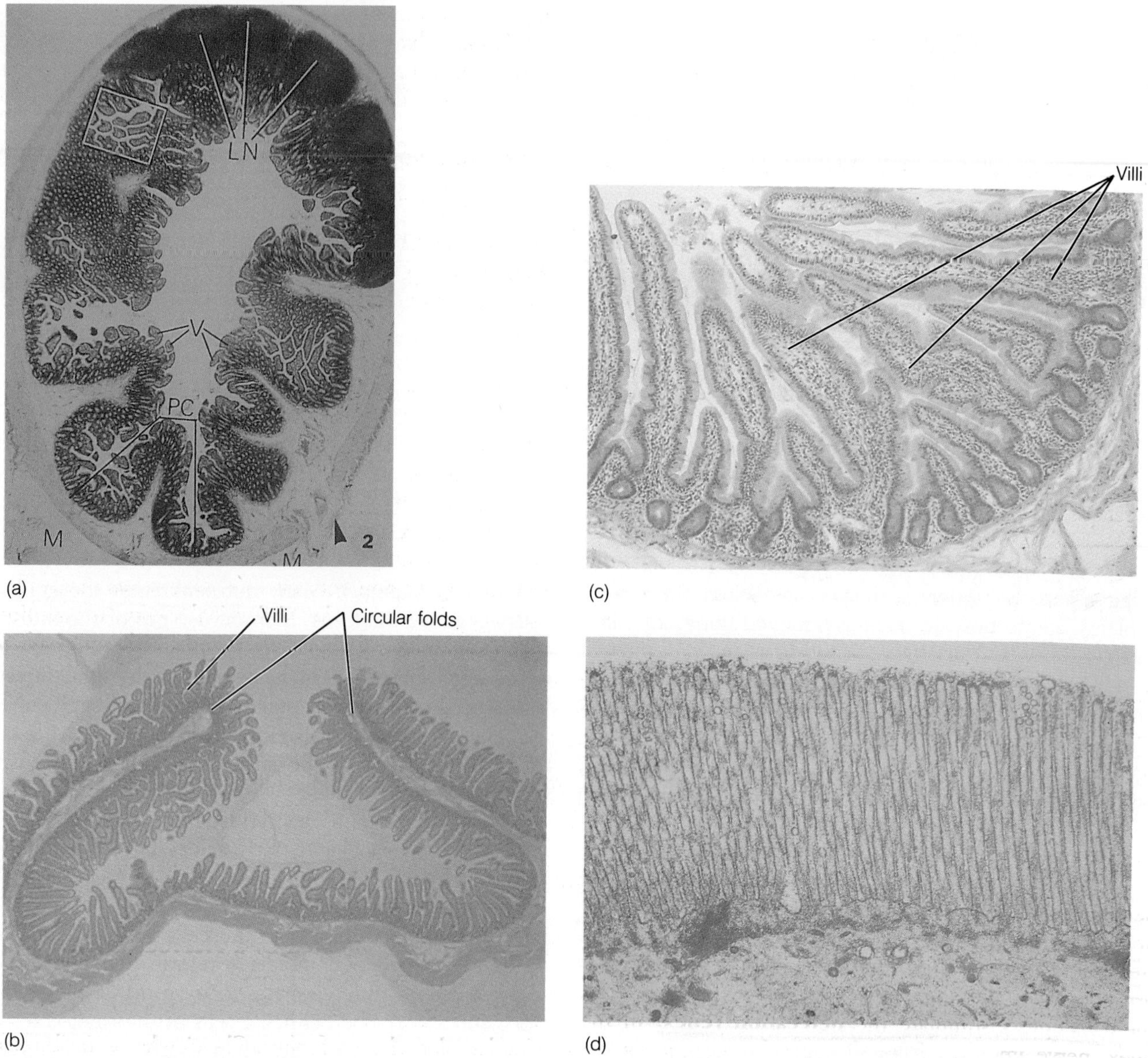

▻ **FIGURE 11–22 The Small Intestine** The small intestine is uniquely "designed" to increase absorption. (*a*) A cross section showing the folds. LN means lymph nodules (aggregation of lymphocytes); V means villi; PC means plica circulares (arculan fold). (*b*) A light micrograph of folds and villi. (*c*) Higher magnification of villi. (*d*) An electron micrograph of the apical surface of the absorptive cells showing the microvilli.

intestine consists of the **colon.** Unlike the small intestine, which is coiled and packed in a small volume, the colon consists of three relatively straight portions, the **ascending colon,** the **transverse colon,** and the **descending colon.** The colon empties into the **rectum.**

So named because of its large diameter, the large intestine receives materials from the small intestine. The material entering the large intestine consists of a mixture of water, undigested or unabsorbed food molecules, and undigestible food residues, such as cellulose. It also contains sodium and potassium ions.

The colon absorbs approximately 90% of the water and sodium and potassium ions that enter it. The undigested or unabsorbed nutrients feed a rather large population of bacteria in the large intestine. These bacteria synthesize several key vitamins: B-12, thiamine, riboflavin, and, most importantly, vitamin K, which is often deficient in the human diet. These vitamins are absorbed by the large intestine.

The contents of the large intestine (after water and salt have been removed) are known as the **feces.** The feces consist primarily of undigested food, indigestible materials, and bacteria. Bacteria, in fact, account for about one-third of the dry weight of the feces.[4]

[4]The feces are about 30% dead bacteria, 10% to 20% fat, 10% to 20% inorganic matter, 2% to 3% protein, and about 30% undigested roughage (cellulose).

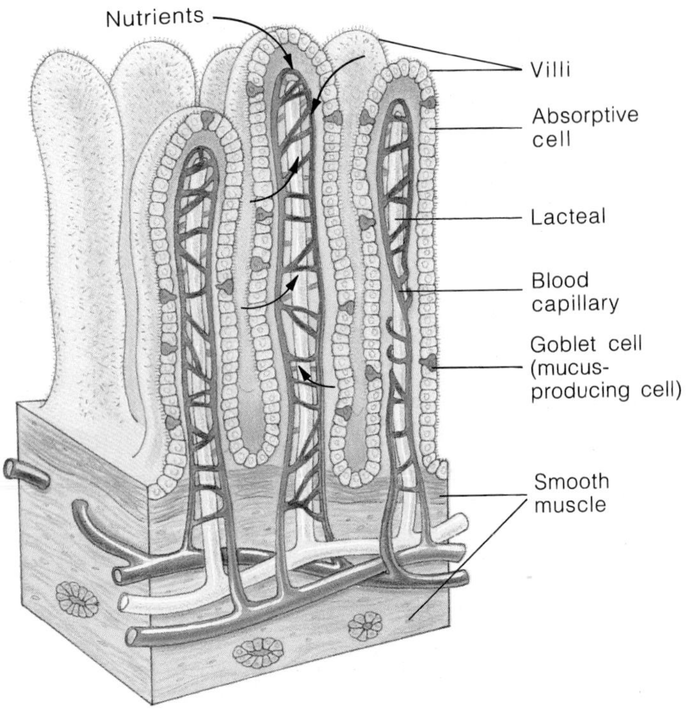

▷ **FIGURE 11–23 The Villus** Each villus contains a loose core of connective tissue, a lacteal, or lymph capillary, and a network of blood capillaries. Nutrients pass from the lumen through the epithelium and into the interior of the villi, where they are picked up by the lymph and blood capillaries.

The feces are propelled by peristaltic contractions along the colon until they reach the rectum, the last segment of the large intestine (Figure 11–25). As fecal matter accumulates in the rectum, it distends the organ. This action, in turn, stimulates stretch receptors in the wall of the rectum. These receptors stimulate the **defecation reflex.** In this reflex, nerve impulses from the stretch receptors in the rectum travel via sensory nerves to the spinal cord. In the spinal cord, the nerve impulses stimulate nerve cells that supply the smooth muscle in the wall of the rectum, causing them to contract and expel the feces. Other nerve impulses from the spinal cord travel to the **internal anal sphincter,** a ring of smooth muscle that holds the feces back. The internal anal sphincter normally keeps fecal matter from entering the anal canal. In the defecation reflex, however, the sphincter relaxes and the fecal matter can escape. Defecation does not occur until the **external anal sphincter** relaxes. This sphincter is composed of skeletal muscle and is under conscious control. If the time and place are appropriate, the external anal sphincter is relaxed, and defecation can occur. If circumstances are inappropriate, voluntary tightening of the external anal sphincter prevents defecation, despite prior activation of the reflex. When defecation is delayed, the muscle in the wall of the rectum eventually relaxes, and the urge to defecate subsides—at least until the next movement of feces into the rectum occurs. Defecation is usually assisted by voluntary contractions of the abdominal muscles and a forcible exhalation, which increase intra-abdominal pressure.

If defecation is delayed too long, a person may become constipated. Constipation results from excess water absorption, which makes the feces hard and dry. Besides being uncomfortable, constipation may result in a dull headache, loss of appetite, and depression.

Constipation is generally caused by (1) ignoring the urge to defecate; (2) decreases in colonic contractions, which occur with age, emotional stress, or low-bulk diets; (3) tumors or colonic spasms that obstruct the movement of feces; and (4) nerve injury, which impairs the defecation reflex.

Constipation is not only uncomfortable, it can create serious medical problems. Hardened fecal material, for instance, may become lodged in the appendix, mentioned above. This, in turn, may lead to inflammation of the organ, a condition called **appendicitis.** When this occurs, the appendix becomes swollen and filled with pus and must be removed surgically to prevent the organ from bursting and spilling its contents into the abdominal cavity. Fecal matter leaking into the abdominal cavity as mentioned earlier, introduces billions of bacteria and can result in a deadly infection.

≈ CONTROLLING DIGESTION

Digestion is a complex and varied process that is controlled largely by nerves and hormones. This section discusses some of the key events involved in the control of digestion, building on material you have learned earlier.

Salivation Is Stimulated by a Nervous Reflex

Digestion begins in the oral cavity. As noted earlier, the sight, smell, taste, and sometimes even the thought of food stimulate the release of saliva. Chewing has a similar effect. The secretion of saliva is therefore controlled by the nervous system and is largely a reflex (▷ Figure 11–26). For a discussion of the discovery of this phenomenon see the Scientific Discoveries 11–1.

Gastric-Gland Secretion Is Controlled by a Variety of Different Mechanisms

Besides activating salivary production, these stimuli also cause the brain to send nerve impulses along the vagus nerve to the stomach (Figure 11–20). As described earlier, these nerve impulses initiate the secretion of hydrochloric acid and pepsinogen from cells of the gastric glands. Nerve impulses also stimulate the secretion of the hormone gastrin. Gastrin is released into the blood and acts on the gastric glands, stimulating additional HCl and pepsinogen secretion.

The most potent stimulus for gastric secretion is the presence of protein in the stomach. Protein stimulates chemical receptors inside the stomach that activate net-

▻ **FIGURE 11–24 Fat Digestion and Absorption**

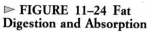

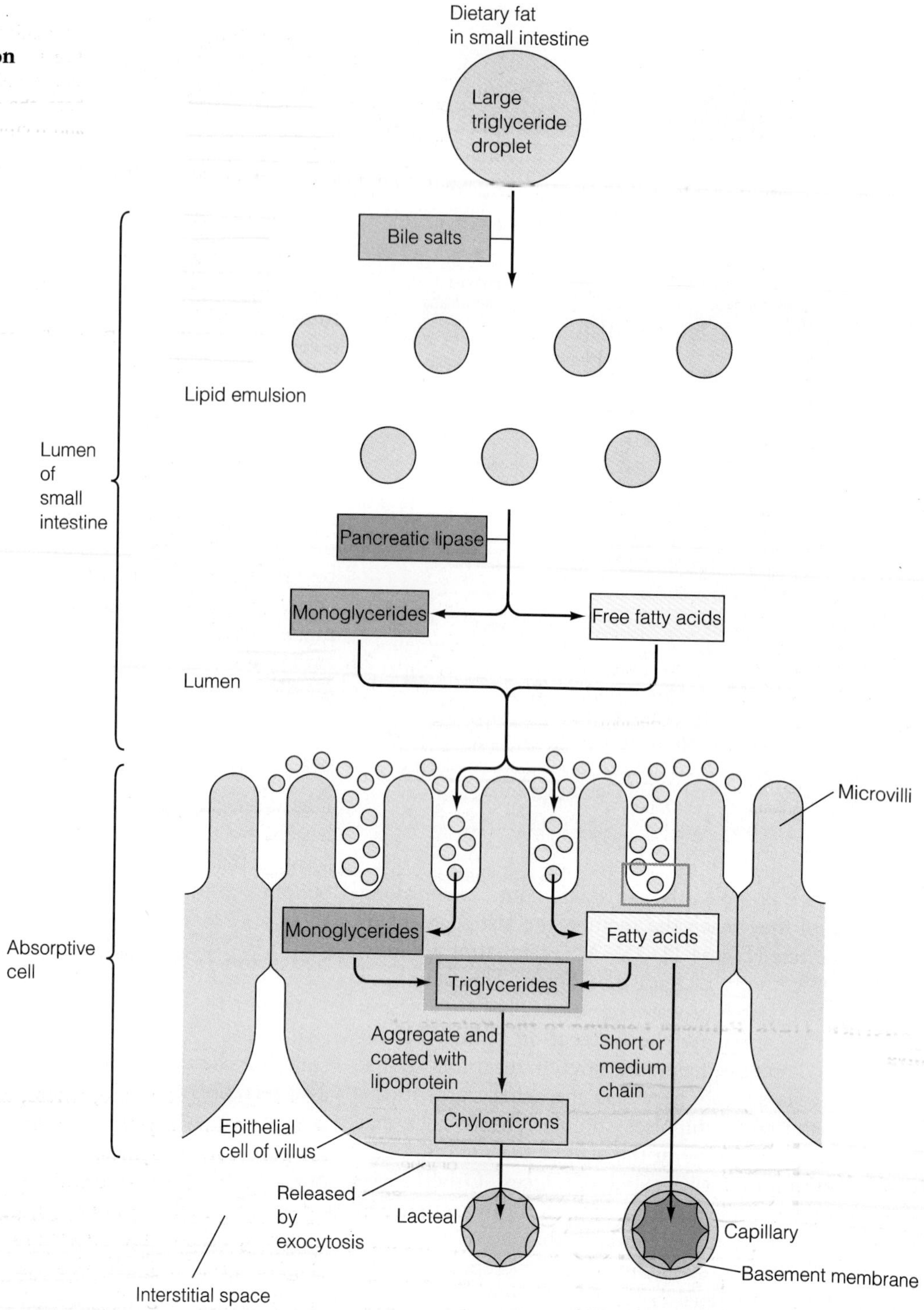

works of nerves in the stomach wall. These nerves, in turn, stimulate the gastric glands to secrete HCl and pepsinogen. Protein also stimulates gastrin secretion directly, in ways beyond the scope of this book.

The concentration of HCl in the stomach is also regulated by a negative feedback mechanism. When the acid content rises too high, it inhibits gastrin secretion, thus shutting off HCl production (see Figure 11–20).

Pancreatic Secretions Released into the Small Intestine Are Stimulated by Two Intestinal Hormones

Acidic chyme leaves the stomach and enters the small intestine, where it stimulates the release of a hormone known as secretin. **Secretin** is produced by the cells of the duodenum and travels in the bloodstream to the pancreas. Here it stimulates the release of sodium bicarbonate ▻ Figure 11–27). Sodium bicarbonate, in turn, is excreted

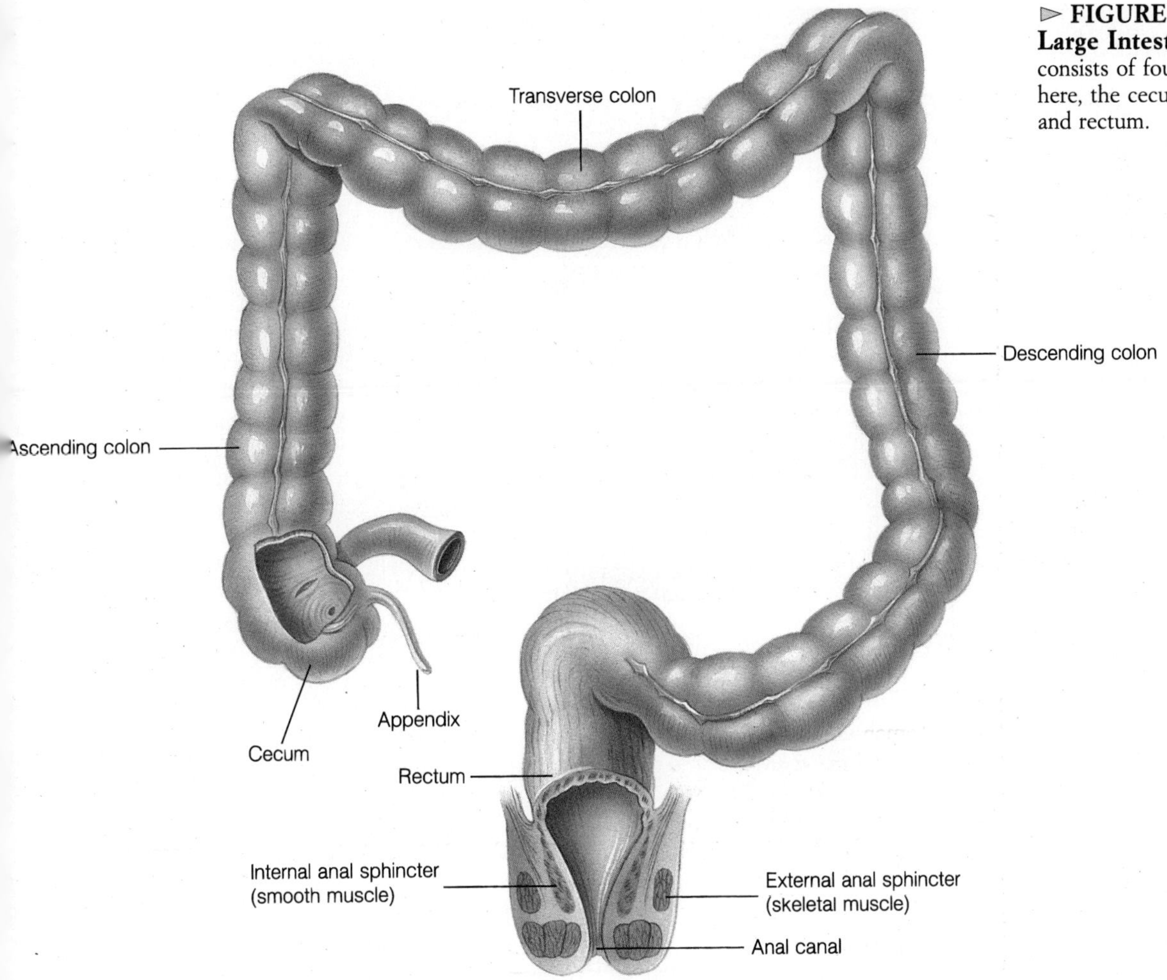

▻ **FIGURE 11–25 The Large Intestine** This organ consists of four basic parts, shown here, the cecum, appendix, colon, and rectum.

▻ **FIGURE 11–26 Pathway Leading to the Release of Saliva**

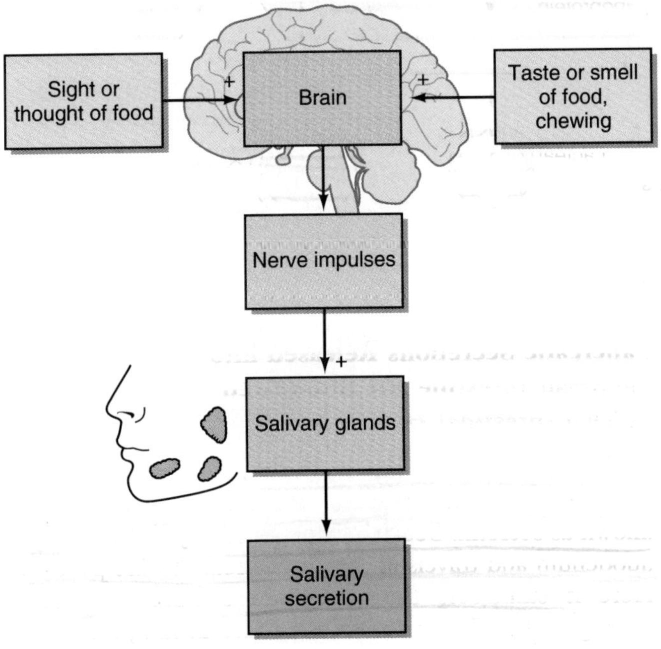

into the small intestine, where, as noted earlier, it neutralizes the acidic chyme and creates an environment optimal for pancreatic enzymes.

The release of pancreatic enzymes is triggered by another intestinal hormone, **cholecystokinin (CCK),** produced by cells of the duodenum in the presence of chyme (Figure 11–27). CCK also stimulates the gallbladder to contract, releasing bile into the small intestine.

Interestingly, recent evidence links abnormally low secretion of CCK to **bulimia,** an eating disorder characterized by recurrent binge eating followed by vomiting. CCK has been found in the brain's hypothalamus with other hormones and may be involved in a range of behaviors, including bulimia. Approximately 4% of America's young women, and a far smaller fraction of men, suffer from this disorder. Bulimia is thought to have both biological and psychological roots, but researchers had failed to identify a biochemical cause until recently. Although no single chemical is likely to control a complex behavior like appetite, it appears that CCK plays an important role.

SCIENTIFIC DISCOVERIES THAT CHANGED THE WORLD 11–1

DISCOVERING THE NATURE OF DIGESTION

Featuring the work of van Helmont and Beaumont

The foundation for the modern understanding of digestion was laid by a number of pioneering scientists. Two noteworthy contributors were a 17th-century Flemish physician, Jan Baptista van Helmont, and a 19th-century American, William Beaumont.

Van Helmont was a man of great originality and ingenuity. In his time, most people thought that the digestion of food in the human was akin to cooking—that is, that food inside the stomach was broken down by body heat. Van Helmont, however, dismissed this theory with the simple observation that digestion occurs in cold-blooded animals such as fish. The lack of heat in such animals suggested to him that an alternative process was at work. But what was it?

The road to discovery began in a rather bizarre way. One day, a tame sparrow that used to visit the scientist attempted to bite van Helmont's tongue. The physician detected a slight acidity in the bird's throat and hypothesized that acid must digest food.

To test this hypothesis, van Helmont tried to digest meat with a vinegar solution, but he could not. Consequently, he hypothesized that the body must contain "ferments," chemical substances in the stomach and small intestine that are specific for different types of food. Today, we call these "ferments" the digestive enzymes produced by the stomach lining, pancreas, and small intestine.

Little insight into digestion was forthcoming in the period between van Helmont's work and the studies of the American scientist William Beaumont. Beaumont was the adventurous son of a Connecticut farmer who left his home in 1806. Soon thereafter, he became a schoolmaster in New York, where he studied medicine and associated sciences in his spare time. Several years later, he became an apprentice to a physician in Vermont and two years later he received a license to practice medicine.

Soon after becoming a licensed physician, Beaumont joined the army, where he would make his important discoveries about digestion. He owes his great work to an accident. In 1822, a French-Canadian porter and servant in the army was wounded in the abdomen by a musket that had accidentally discharged at close range. The servant, Alexis St. Martin, who was only 18, was brought to Beaumont for treatment. The doctor found that the bullet had ripped a hole in St. Martin's stomach, out of which poured food the man had eaten.

As was the custom of the time, Beaumont bled the victim, removing a substantial amount of his blood. Despite this treatment, the patient survived, but healing became a protracted process. For over eight months, Beaumont tried in vain to close the hole in the stomach. During this time, it dawned on the young physician that he had stumbled across a golden opportunity to study digestion.

Beaumont's experiments consisted of feeding the patient various types of food, then studying what took place inside the poor man's stomach. In the next nine years, Beaumont studied St. Martin's stomach contents with a wide assortment of foods, looking at the rate and temperature of digestion and the chemical conditions that favored different stages of the process. During his work, he noticed that the stomach lining could be injured by drinking alcohol.

Beaumont's studies helped settle a controversy over the nature of digestion. Some of his contemporaries contended that gastric juice was a kind of chemical solvent. Others believed that it merely liquified food and that digestion occurred as a result of a vital force present in living organisms. By showing that some digestion could take place outside the stomach in the presence of gastric juice, Beaumont demonstrated that gastric secretions did more than moisten food and that no "vital force" was at work. Digestion, Beaumont asserted, was a purely chemical phenomenon.

Beaumont did not address one question that could have been answered with existing techniques—that is, what caused gastric juice to be secreted? Was it the presence of food in the stomach?

The answer to that question came in 1889 from the work of Ivan Pavlov, another notable scientist, whose experiments on behavior are discussed in Scientific Discoveries 29–1. In studies on dogs, Pavlov showed that the secretion of gastric juice was mediated by the nervous system. How did he make this determination?

In one of many experiments, the Russian scientist surgically connected a fold of a dog's stomach to an opening in the animal's side, which permitted him to examine the production of gastric secretions. He next cut and tied off the esophagus of the dog, so that food the animal swallowed could not enter the stomach. Following the surgery, Pavlov gave the dog food and found that as soon as food entered the dog's mouth, gastric juice began to be secreted. These secretions continued as long as food was present, thus suggesting nervous system involvement. Since that time, a great deal has been learned about digestion and its control. For example, as you learned in this chapter, hormones are also involved in the control of many digestive processes.

The small intestine also produces the hormone, **gastric inhibitory peptide (GIP).** This hormone is released in response to fatty acids and sugars in chyme. GIP inhibits HCl production and peristalsis in the stomach, slowing down the rate at which chyme is discharged from the stomach and therefore enhancing digestion and absorption of food already in the small intestine.

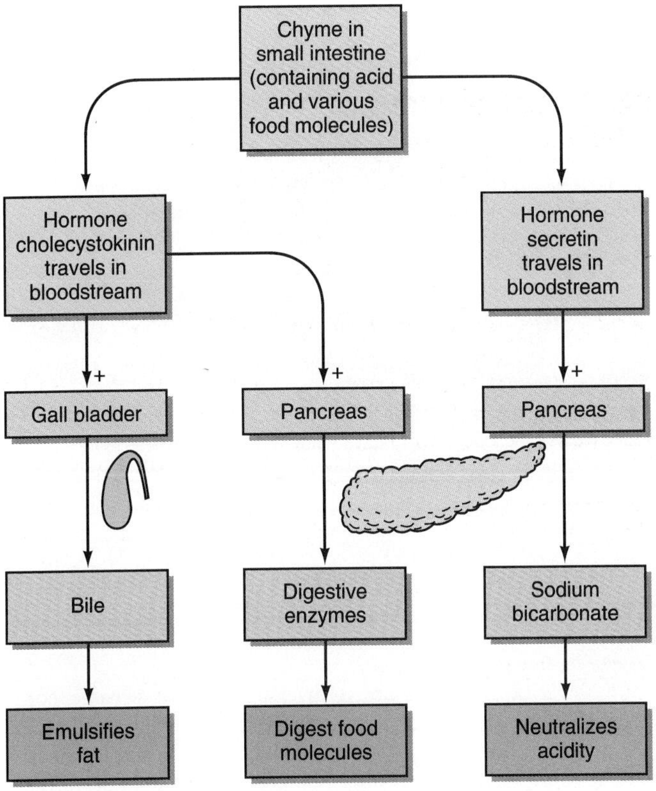

▻ **FIGURE 11–27 Control of the Pancreas and Gallbladder**

ENVIRONMENT AND HEALTH: EATING RIGHT/LIVING RIGHT

In few places is the relationship between homeostasis and human health as evident as in human nutrition. Because homeostasis requires an adequate supply of nutrients, nutritional deficiencies can have noticeable impacts on body functions and human health. Supporting this idea, many studies show that an unhealthy diet can increase the risk of contracting certain diseases, including cancer, heart disease, hypertension, and others.

Consider a few examples. Magnesium, one of the major minerals, is routinely ingested in insufficient amounts. New research suggests that such deficiencies may underlie a number of medical conditions, including diabetes, high blood pressure, pregnancy problems, and cardiovascular disease.

Research shows that adding magnesium to the drinking water of rats with hypertension can eliminate high blood pressure. Studies in rabbits show that magnesium reduces lipid levels in the blood and also reduces plaque formation in blood vessels. Rabbits on a high-cholesterol, low-magnesium diet, for example, have 80% to 90% more atherosclerotic plaque than rabbits on a high-cholesterol, high-magnesium diet.

Researchers have found that in humans, magnesium deficiencies during pregnancy result in migraine headaches, high blood pressure, miscarriages, stillbirths, and babies with low birth weight. Research suggests that magnesium deficiency causes spasms in the blood vessels of the placenta, which reduce blood flow to the fetus. This can retard fetal growth and may even kill a fetus in utero. Magnesium supplements greatly reduce the incidence of these problems.

Researchers believe that 80% to 90% of the American public may be magnesium deficient. One reason the American diet may be deficient in magnesium is that phosphates in many carbonated soft drinks bind magnesium in the intestine, preventing it from being absorbed into the blood. Magnesium deficiencies can be reversed by eating more green leafy vegetables, seafood, and whole grain cereals. Mineral supplements could help as well, but they should be used with extreme caution, because excess magnesium can cause neurological disorders.

Zinc is a trace mineral that has also been implicated in a wide range of health problems. Rats fed diets severely deficient in zinc, for example, have more birth defects, are often stunted, and reach sexual maturation later than their counterparts fed normal diets. Concerned that less severe zinc deficiencies may cause problems in humans, researchers followed 10 monkeys from birth through adolescence. One group was fed a diet low in zinc. The other received far more than it required. Monkeys fed the zinc-deficient diet showed several curious symptoms. Their immune function was suppressed 20% to 30%, making them more susceptible to disease. Significant learning impairments were also observed. The monkey studies substantiate studies in rodents and suggest concern for people in less developed countries, who often subsist on low-zinc diets consisting primarily of cereals.[5]

Over the years, numerous dietary recommendations have been issued to help Americans live healthier lives and reduce the risk of cancer and heart attack. Nutritionists recommend that we daily consume (1) fruits and vegetables, especially cabbage and greens, (2) high-fiber foods, such as whole-wheat bread and celery, and (3) foods high in vitamins A and C. A healthy diet also minimizes the consumption of animal fat; red meat; and salt-cured, nitrate-cured, smoked, or pickled foods, including bacon and lunch meat.

As I noted earlier in the chapter, the American public has not taken these recommendations to heart. A survey of the eating habits of nearly 12,000 subjects conducted from 1976 to 1980, in fact, showed that the diets of both black and white Americans were typically deficient in the very foods that nutritionists recommend, such as fruits and vegetables.[6] When asked to recall everything they had

[5]To determine if you are receiving an adequate supply of micronutrients, you can undergo a blood test or an analysis of your diet at a nutritional clinic. If there are problems, a trained nutritionist will be able to make recommendations to correct the problem.

[6]Critical thinking suggests caution in interpreting these results. Studies of the dietary habits of people in 1990 would be more informative.

eaten in the previous 24-hour period, fewer than one in five people in the study reported having eaten any of these foods. In sharp contrast, many of the people surveyed had eaten red meat, bacon, and lunch meat.

A healthy diet results largely from habit and circumstance—that is, our social environment. How does our environment affect our nutrition and health? In the hustle and bustle of modern society, many of us ignore proper nutrition, grabbing fat-rich foods when we are hungry because we haven't the time to sit down to a nutritionally balanced meal (▻ Figure 11–28). Our fast-paced world places a high premium on saving time, often ignoring the importance of eating right. Fast food may allow us to hurry on to our next appointment, but unless it is nutritionally sound, it probably decreases the quality of your lives in the long run.

▻ **FIGURE 11–28 Fast Food Restaurants** The hectic pace of modern life leads many of us to eat at places whose food is rich in fat and rarely provides a balanced meal.

SUMMARY

1. Studies suggest that iron levels in the body may affect heat production in women, a relationship that, if substantiated by additional research, helps illustrate the importance of diet for physiological processes.

A PRIMER ON NUTRITION

2. Humans acquire energy and nutrients from the food they eat. There are two types of nutrients: macronutrients, substances needed in large quantity, and micronutrients, substances required in much lower quantities.
3. The four major macronutrients are water, carbohydrates, lipids, and proteins.
4. Water is in the liquids we drink and the foods we eat. Maintaining adequate water intake is important, because water is involved in many chemical reactions in the body. It also helps maintain body temperature and a constant level of nutrients and wastes in body fluids, so vital to homeostasis.
5. Carbohydrates and lipids are major sources of cellular energy; 70% to 80% of all energy required by the body goes for basic functions.
6. Contrary to popular myth, protein does not supply much energy, except when lipid and carbohydrate intake is low or when protein intake exceeds daily requirements. Dietary protein is chiefly a source of amino acids for building proteins.
7. Some amino acids can be synthesized in the body. These are known as nonessential amino acids. Others cannot be synthesized in the body and must be supplied in the food we eat. These are known as essential amino acids.
8. To ensure an adequate supply of all amino acids, individuals should eat complete proteins, such as those found in milk or eggs, or combine lower quality protein sources.
9. Lipids provide energy during rest and aerobic activity. They serve many other functions in the body, such as insulation.
10. Besides providing energy, carbohydrates serve other important functions. Dietary fiber, for example, increases the liquid content of the feces, reducing constipation, the incidence of diverticulosis, and the risk of colon cancer.
11. Micronutrients are needed in much smaller quantities and include two groups: vitamins and minerals.
12. Vitamins are a diverse group of organic compounds that are required in relatively small quantities for normal metabolism. A deficiency or surplus of one or more vitamins may alter homeostasis, with serious effects on human health.
13. Human vitamins fit into two categories: water-soluble and fat-soluble. The water-soluble vitamins include vitamin C and the B-complex vitamins. The fat-soluble vitamins include vitamins A, D, E, and K.
14. Minerals fit into one of two groups: trace minerals, those required in very small quantity, and major minerals, those required in greater quantity. Deficiencies and excesses of both types of minerals can lead to serious health problems.

THE DIGESTIVE SYSTEM

15. Food is physically and chemically broken down in the digestive system. Small molecules produced during digestion are absorbed by the intestinal tract into the bloodstream or lymphatic system and circulated throughout the body for use by the cells.
16. Food digestion begins in the mouth. The teeth mechanically break down the food. Saliva liquifies it, making it easier to swallow. Salivary amylase begins to digest starch molecules.
17. Food is pushed by the tongue to the pharynx, where it triggers the swallowing reflex. Peristaltic contractions propel the food down the esophagus to the stomach.
18. The stomach is an expandable organ that stores and further liquifies the food. The churning action of the stomach, brought about by peristaltic contractions, mixes the food, turning it into a paste referred to as chyme. The stomach releases food into the small intestine in timed pulses, ensuring efficient digestion and absorption. Very limited chemical digestion and absorption occur in the stomach.
19. The stomach produces hydrochloric acid, which denatures protein, allowing it to be acted on by enzymes. The stomach also produces a proteolytic enzyme called pepsin, which breaks proteins into peptides. The lining of the stomach is protected from acid by mucus. When the mucous protection fails, however, the lining may be eroded by acids, creating an ulcer.

CHAPTER

12

The Circulatory System

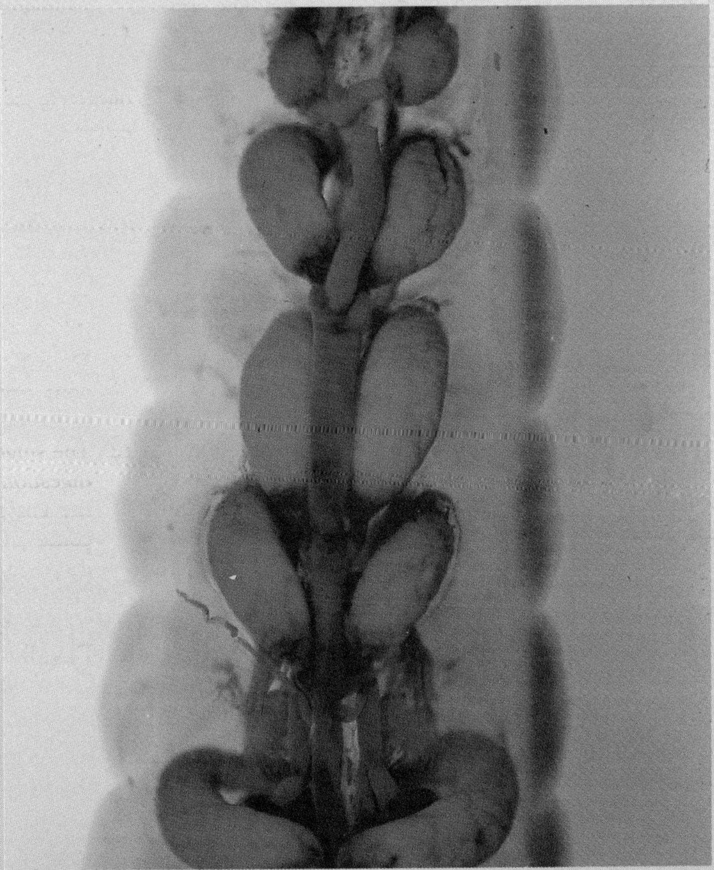

Earthworm hearts shown here help propel the blood through a closed system of blood vessels.

David McMahon, a 48-year-old New York attorney, collapsed in his office one morning. Dazed and in great pain, he was rushed to a nearby hospital. There a team of cardiologists discovered a blood clot lodged in a narrowed section of his right coronary artery, which supplies the heart. The physicians began immediate action to prevent further damage to the oxygen-starved heart muscle. Through a small incision in McMahon's groin, they inserted a tiny plastic catheter into the large artery that carries blood to the leg. They then threaded the catheter up through the arterial system to his heart (▻ Figure 12–1). Once they reached the heart, physicians guided the catheter into the clogged coronary artery, where they injected an enzyme, called streptokinase, through the catheter. Streptokinase dissolved the blood clot, restoring blood flow to McMahon's heart muscle.

David McMahon survived. Unfortunately, many others who suffer similar attacks are not so lucky. They either arrive at the hospital too late to be treated or, if they do make it to the hospital, do not receive blood-clot-dissolving agents or other treatments in time to avoid extensive damage to their heart muscle.

Heart attacks strike thousands of Americans each year. They are one of a handful of diseases of the circulatory system caused by the stressful conditions of modern life, poor eating habits, and a host of other factors. This chapter examines the circulatory system of animals. This chapter also takes a look at the lymphatic system, which is involved in both circulation and immune protection.

▻ **FIGURE 12–1 The Circulatory System** (*a*) The circulatory system consists of a series of vessels that transport blood to and from the heart, the pump. (*b*) The circulatory system has two major circuits, the pulmonary circuit, which delivers blood to the lungs, and the systemic circuit, which delivers blood to the rest of the body.

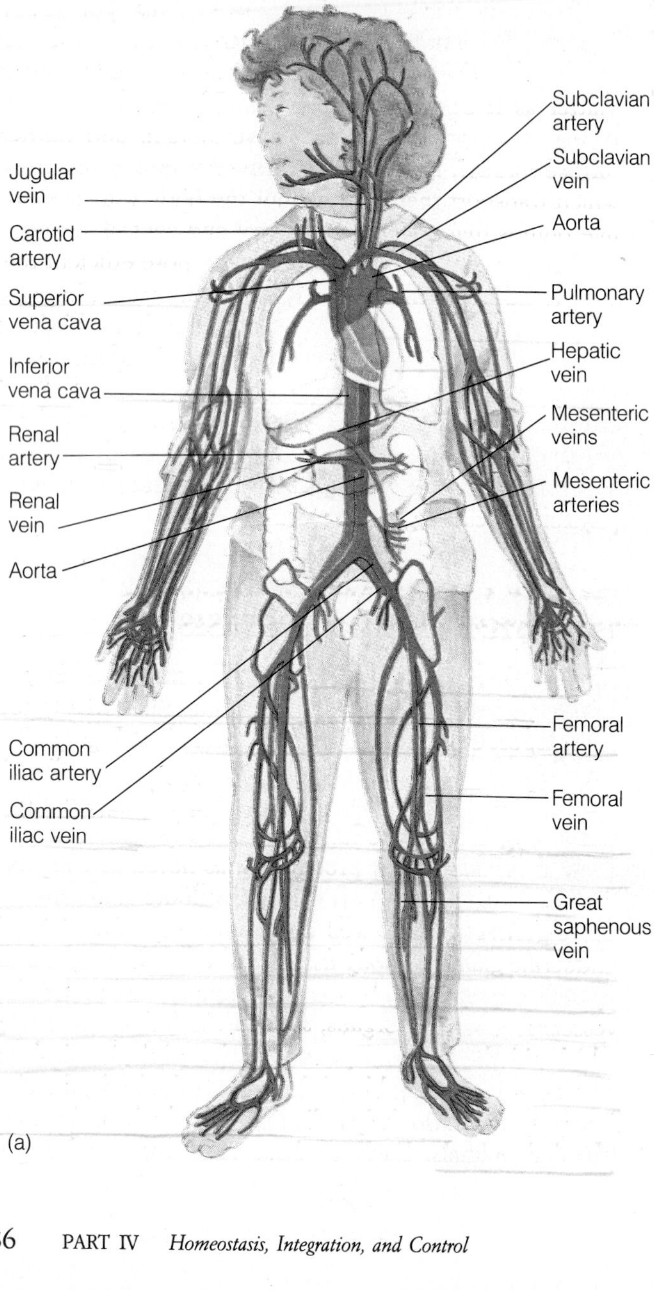

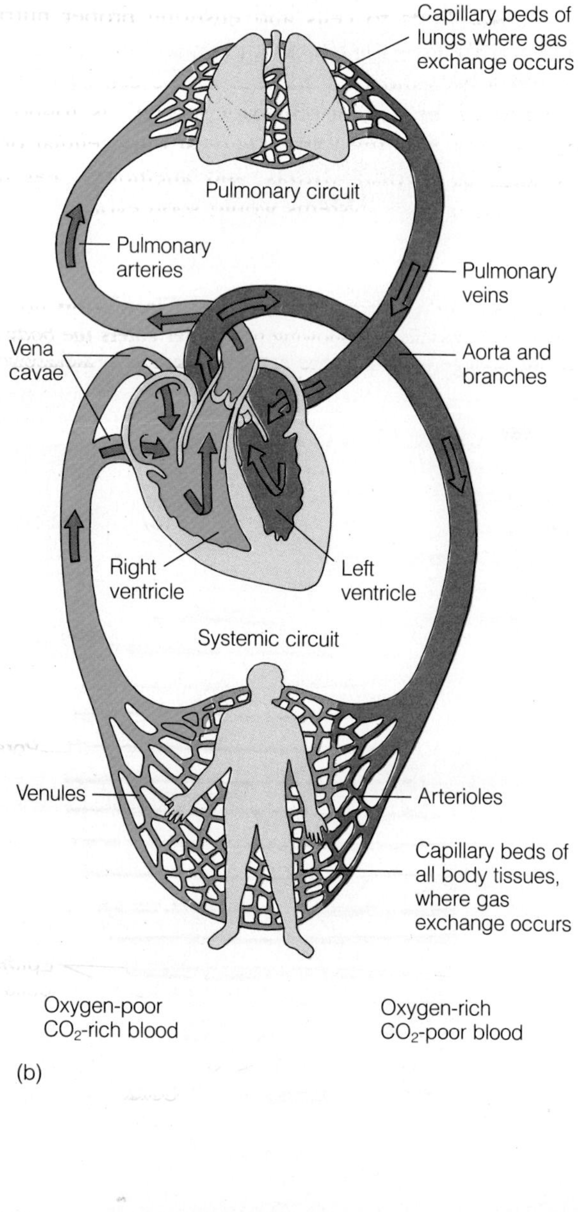

≈ CIRCULATORY SYSTEMS: AN EVOLUTIONARY IMPERATIVE

In the early days of evolution, life was a rather simple affair. For the first 2.5 billion years, in fact, all that existed were single-celled monerans (bacteria) living in the shallow coastal waters where life presumably arose. The monerans were later joined by protists, single-celled eukaryotes, and together these two life forms dominated the scene for many more millions of years.

Monerans and protists acquired nutrients from the seawater. Seawater also carried wastes away. In a sense, then, the watery environment bathing the cells served as a huge external circulatory system. But with the advent of multicellularity, organisms began to evolve elaborate ways to circulate nutrients to and remove wastes from cells that had become increasingly distant from the external environment.

The earliest form of internal circulation is seen in the sponge, one of the earliest animals (▻ Figure 12–2). Seawater enters through tiny pores in the sponge's body and then flows through a network of internal channels. This series of channels forms a primitive, but effective, means of transporting fluids to cells and ensuring proper nutrition and waste disposal. As in the protists and monerans, nutrients and wastes move in and out of the cell by diffusion.

This basic type of "circulatory system" is found in a wide assortment of the Earth's earliest multicellular organisms, such as jellyfish, hydras, and anemones. Far more elaborate circulatory systems would soon evolve.

▻ **FIGURE 12–2 Primitive Internal Circulation** (*a*) Sponge. (*b*) Cross section showing how water enters the body cavity of the sponge to nourish the cells and wash away metabolic wastes.

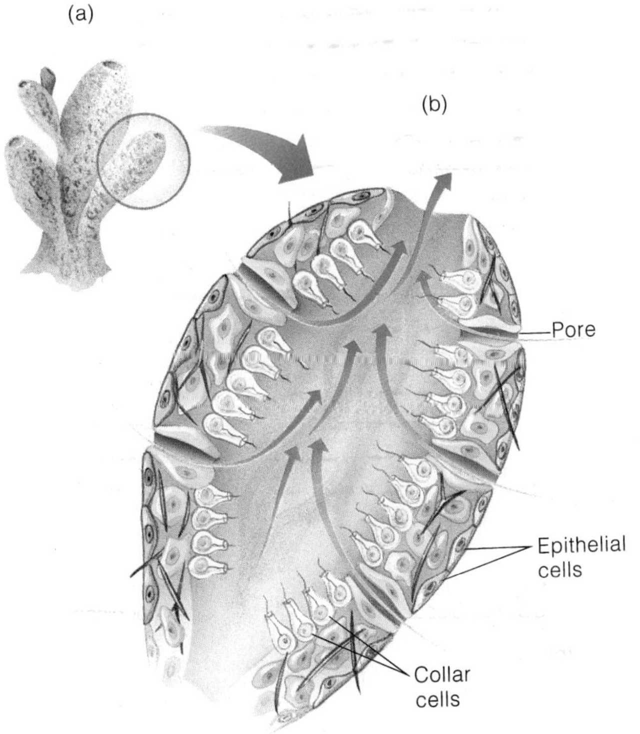

These more complex systems fall into two major categories: open and closed. An **open circulatory system** is one in which the blood vessels empty into the body cavity, where the organs are located. A good example is the grasshopper (▻ Figure 12–3). The grasshopper's organs receive ample nutrition and are adequately rid of wastes by this crude but effective system. Blood is removed from the body cavity through tiny openings (ostia) in the dorsal blood vessel. The walls of the dorsal vessel are thickened and serve as pumps. When the muscles are relaxed, the holes in the wall open, permitting blood to enter. When the muscles contract, the holes close, and blood is propelled forward in the dorsal vessel.

Open circulatory systems are rather inefficient and are generally found in organisms that do not exhibit high rates of metabolism: arthropods (insects, spiders, and crustaceans) and molluscs (clams and snails). **Closed circulatory systems** transport blood through a closed system of vessels, delivering it more directly to cells through thin-walled capillaries.

One of the first closed circulatory systems appears in the segmented worms (annelids) such as the earthworm, shown in ▻ Figure 12–4. The earthworm feeds on organic matter as it burrows through soil. The food it ingests is digested in the primitive digestive system, and nutrients diffuse through the wall of the intestine into blood vessels, which transport them throughout the body with the aid of five pumps that connect the dorsal and ventral vessels.

Closed circulatory systems permit a more efficient transport of nutrients than open systems and are an important adaptation in warm-blooded species, such as birds and mammals, which require a sufficient supply of nutrients to maintain body temperature and the high metabolic demand of cells. The human circulatory system, described in this chapter, is an excellent example of a closed system and a testament to the remarkable improvements in animal physiology resulting from evolution.

≈ THE CIRCULATORY SYSTEM'S FUNCTION IN HOMEOSTASIS

The circulatory system of humans, like that of other vertebrates, consists of a muscular pump and a network of vessels to transport blood to and from the heart (see Figure 12–1). The functions of the human circulatory system are listed in Table 12–1. The blood coursing through the system transports oxygen from the lungs to the body cells, where it is used in cellular energy production, as noted in Chapter 6. The circulatory system also helps distribute nutrients from the digestive tract as well as hormones produced by the endocrine glands to body tissues and cells. Waste produced by cells is swept away by the blood and carried in the blood vessels to excretory organs, such as the kidneys and lungs, which rid the body of potentially harmful chemical substances. By pumping blood throughout the body, the circulatory system also helps distribute body heat in warm-blooded animals. Each of these functions—and a great

▻ **FIGURE 12–3 Open Circulatory System**
In circulatory systems such as those in insects, the heart (actually a series of hearts) pumps blood into a series of vessels that empty into a body cavity, or hemocoel. Each organ of the body is directly bathed with blood in the hemocoel. When the hearts relax, blood is sucked back into them through openings guarded by one-way valves. When the hearts contract, the valves are pressed shut, forcing the blood to travel through the vessels.

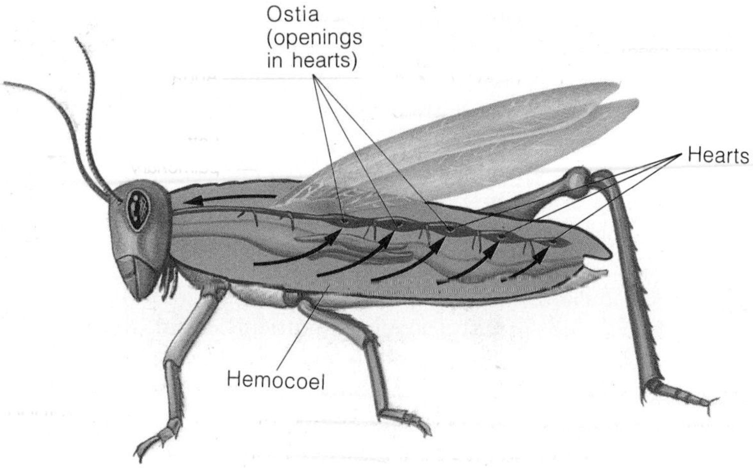

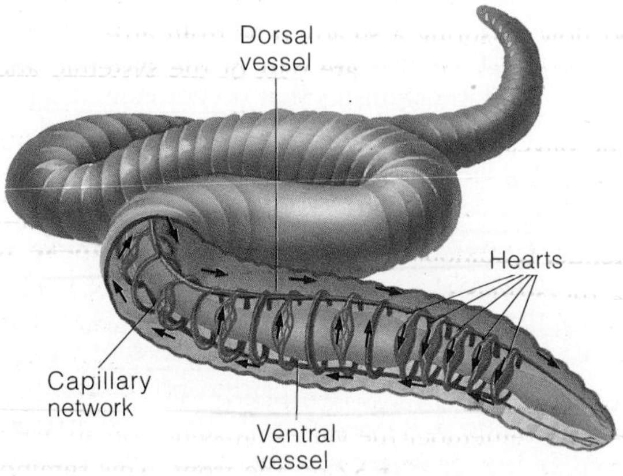

▻ **FIGURE 12–4 Circulatory System of the Earthworm**
In this closed circulatory system, blood remains within the vessels and hearts. Nutrients must be transported out of the vessels to reach body cells.

TABLE 12-1 Functions of the Circulatory System

Transports oxygen to body cells
Transports nutrients from the digestive system to body cells
Transports hormones to body cells
Transports wastes from body cells to excretory organs
Distributes body heat

many more—serves a higher purpose, the maintenance of homeostasis; that is, it helps maintain the relatively constant internal conditions necessary for cellular function.

THE HEART

The heart is a muscular pump in the thoracic (chest) cavity. Sometimes referred to as the workhorse of the cardiovascular system, the heart propels blood through the 50,000 miles of blood vessels in the human body. Each day, the heart beats approximately 100,000 times, adjusting its rate to meet the changing needs of the body. If you had a dollar for every heartbeat, you would be a millionaire in 10 days. Over a 70-year lifetime, you would collect $2.5 billion for your heart's work.

The heart, shown in ▻ Figure 12–5, is a fist-sized organ whose walls are composed of three layers, the pericardium, the myocardium, and the endocardium. The **pericardium** forms a thin, closed sac surrounding the heart and the bases of large vessels that enter and leave the heart. The pericardial sac is filled with a clear, slippery fluid that reduces friction produced by the heart's repeated contraction. The middle layer, the **myocardium,** is the thickest part of the wall and is composed chiefly of cardiac muscle cells, briefly described in Chapter 10. The inner layer, the **endocardium,** is the endothelial layer, which forms the lining of the heart chambers.

The Circulatory System Has Two Distinct Circuits through Which Blood Flows

The circulatory system, shown in simplified form in Figure 12–1b, consists of two distinct circuits, the **pulmonary circuit,** which carries blood to and from the lungs, and the **systemic circuit,** which transports blood to and from the rest of the body. The central driving force in both of these circuits is the heart. As shown in Figure 12–1b, the heart consists of four hollow chambers—two on the right side of the heart and two on the left.

Blood is pumped through the pulmonary circuit by the right side of the heart—that is, the right atrium and right ventricle. As noted above, the pulmonary circuit delivers blood to the lungs, where it loses most of its carbon dioxide and replenishes its supply of oxygen. The oxygenated blood is then returned to the heart and distributed to body tissues via the systemic circulation, which is served by the left atrium and left ventricle. In the tissues of the body, the blood releases much of the oxygen it gained in the lungs and takes up carbon dioxide released by cells during cellular respiration. The blood is then returned to the heart, where it enters the pulmonary circuit once again.

Figure 12–5 illustrates the course that blood takes through the heart. Drawn in blue, blood low in oxygen (and rich in carbon dioxide) enters the right side of the heart from the systemic circuit through the **superior** and

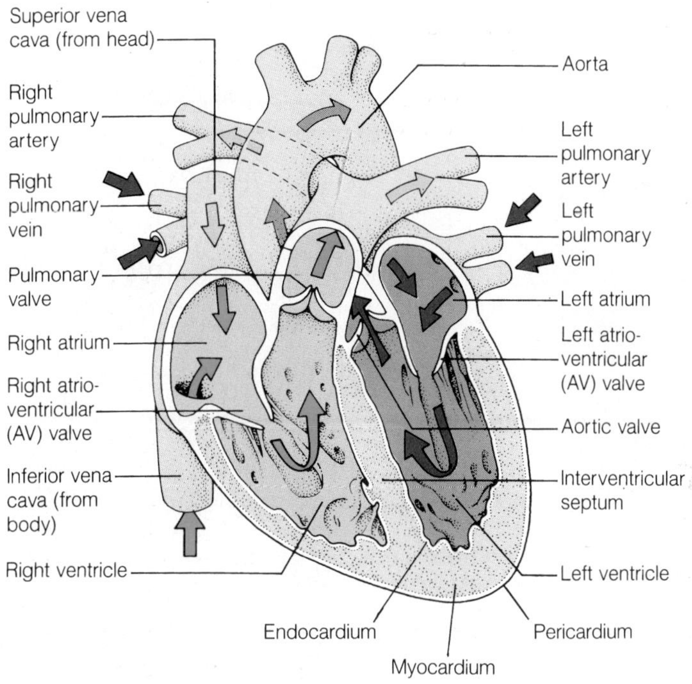

▻ **FIGURE 12–5 Blood Flow through the Heart** Deoxygenated (carbon-dioxide-enriched) blood (blue) flows into the right atrium from the systemic circulation, then is pumped into the right ventricle. The right ventricle, in turn, pumps the blood into the pulmonary artery, which delivers it to the lungs, where carbon dioxide is released and oxygen is picked up. Reoxygenated blood (red) is returned to the left atrium, then flows into the left ventricle, which pumps it to the rest of the body through the systemic circuit.

inferior vena cavae. These large veins empty directly into the **right atrium,** the uppermost chamber on the right side of the heart. The blood is pumped from here into the **right ventricle,** the lower chamber on the right side. When the right ventricle is full, the muscles in its wall contract, forcing blood into the pulmonary arteries, which lead to the lungs.

Blood whose oxygen supply has been restored, drawn in red in Figure 12–5, flows back to the heart from the lungs in the **pulmonary veins.** The pulmonary veins, in turn, empty directly into the **left atrium,** the upper chamber on the left side of the heart. From here, the oxygen-rich blood is pumped to the **left ventricle.** When the left ventricle is full, its thick, muscular walls contract and propel the blood into the aorta. The **aorta** is the largest artery in the body. It carries the oxygenated blood away from the heart, delivering it to the cells and tissues of the body via many smaller branches (discussed below).

The flow of blood I have just described presents a slightly misleading view of the way the heart really works. As shown in ▻ Figure 12–6, both atria actually fill and contract simultaneously, delivering blood to their respective ventricles. The right and left ventricles also fill simultaneously, and when both ventricles are full, they too contract in unison, pumping the blood into the systemic and pulmonary circulations. The coordinated contraction of heart muscle is brought about by an internal timing device, or pacemaker (described later).

Heart Valves are Located Between the Atria and Ventricles and Between the Ventricles and the Large Vessels into Which They Empty

In the evolution of circulatory systems, one of the most important features was the emergence of valves to help regulate the flow of blood through the heart. The human heart contains four valves that control the direction of blood flow, ensuring a steady flow from atria to ventricles to the large vessels that are part of the systemic and pulmonary circuits (▻ Figure 12–7a).

The valves between the atria and ventricles are known as **atrioventricular valves.** Each valve consists of two or three flaps of tissue anchored to the inner walls of the ventricles by slender tendinous cords, the **chordae tendineae,** resembling the strings of a parachute (Figure 12–7a). The right atrioventricular valve, between the right atrium and right ventricle, is called the **tricuspid valve,** because it contains three flaps. The left atrioventricular valve is the **bicuspid valve.**[1] To remember the valves, imagine you are wearing a jersey with the number 32 on the front. This reminds you that the tricuspid valve is on the right side and the bicuspid valve is on the left.

Between the right and left ventricles and the arteries into which they pump blood (pulmonary artery and aorta, respectively) are the **semilunar valves** (Figures 12–7a and 12–7b). The semilunar valves (literally, "half moon") consist of three semicircular flaps of tissue (Figures 12–7b and 12–7c).

The atrioventricular and semilunar valves are one-way valves, opening when blood pressure builds on one side and closing when it increases on the other, much like the purge valves in scuba diving masks, which allow divers to force water out of their masks, or the ball valves in snorkels, which operate similarly. When the ventricles contract, blood forces the semilunar valves open. Blood flows out of the ventricles into the large arteries. The backflow of blood causes the valve to close, preventing blood from draining back into the ventricles. The atrioventricular valves function in like fashion.

Heart Sounds Result from the Closing of Various Heart Valves

When physicians listen to your heart, they are actually listening to sounds of the heart valves closing. The noises they hear are called the **heart sounds** and are often described as "LUB-dupp." The first heart sound (LUB) results from the closure of the atrioventricular valves. It is longer and louder than the second heart sound (dupp),

[1] The bicuspid valve is also called the mitral valve, because it resembles a miter, a hat worn by the Pope and Catholic bishops.

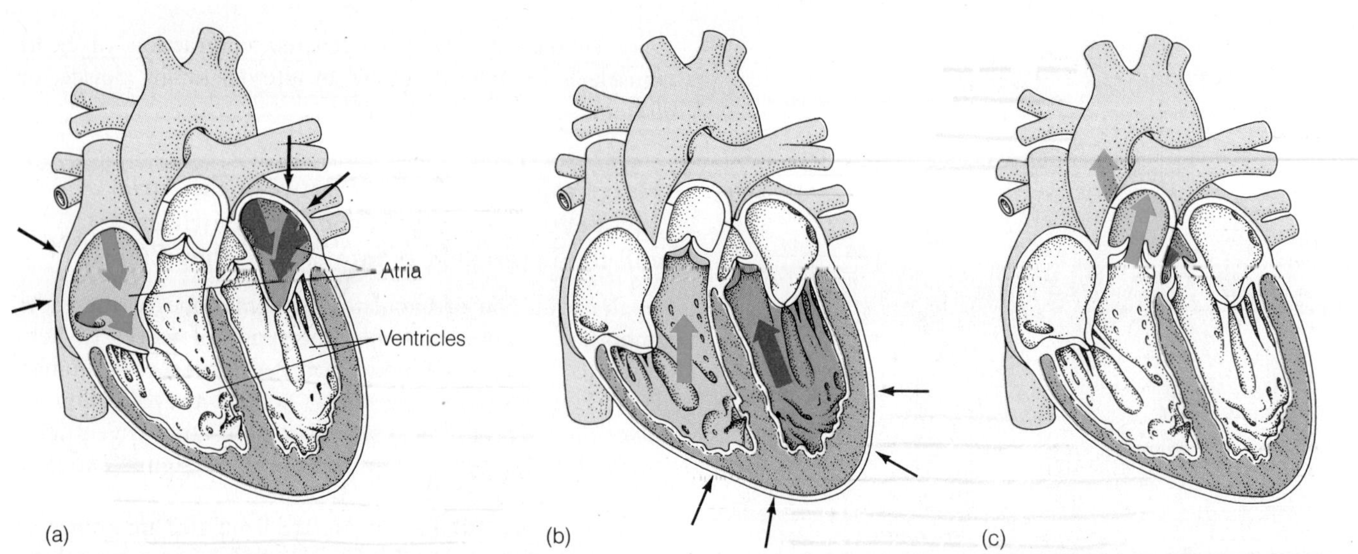

▻ **FIGURE 12–6 Blood Flow through the Heart**
(*a*) Blood enters both atria simultaneously from the systemic and pulmonary circuits. When full, the atria pump their blood into the ventricles. (*b*) When the ventricles are full, they contract simultaneously, (*c*) delivering the blood to the pulmonary and systemic circuits.

produced when the semilunar valves snap shut. Interestingly, the right and left atrioventricular valves do not close at precisely the same time. Nor do the semilunar valves. Thus, by careful placement of the stethoscope, a physician can listen to each valve individually to determine if it is functioning well.

For most of us, our heart valves function flawlessly throughout life. However, some diseases can alter the function of the valves, greatly affecting the efficiency of the heart and the circulation of blood. Rheumatic fever, for example, is caused by a bacterial infection and affects many parts of the body, including the heart. Although relatively rare in developed countries, rheumatic fever is still a significant problem in the Third World. Rheumatic fever begins as a sore throat caused by certain types of **streptococcus** bacteria. The sore throat—known as strep throat—is usually followed by general illness. During this infection, antibodies (proteins made by cells of the immune system) to the streptococcus bacteria, which circulate in the blood, damage the heart valves, preventing them from closing completely. This damage causes blood to leak back into the atria and ventricles after contraction and results in a distinct "sloshing" sound, commonly called a **heart murmur.** This condition, generally referred to as **valvular incompetence,** reduces the efficiency of the heart and causes the organ to work harder than usual to make up for the inefficient pumping. Increased activity, in turn, causes the walls of the heart to enlarge. In severe cases, valvular incompetence can result in heart failure. To prevent heart failure, damaged valves can be replaced by artificial implants.

Tumors (benign and malignant) and scar tissue have an opposite effect; that is, they can obstruct the valve, reducing blood flow through the heart. This condition is known as **valvular stenosis** (from the Greek word *steno,* meaning "narrow"). Valvular stenosis prevents the ventricles from filling completely. As in valvular incompetence, the heart must beat faster to ensure an adequate supply of blood to the body's tissues. This acceleration also puts additional stress on the organ.

Heart Rate Is Controlled by an Internal Pacemaker

The activity level of multicellular animals is highly variable. For this reason, they need a way to control the rate at which the heart beats and, consequently, the rate at which blood flows to body tissues. That is to say, they need a way to accelerate heart rate and blood flow when tissues demand more oxygen and decelerate it when demand falls.

During the evolution of multicellular animals, several mechanisms evolved to control heart rate. One of them is an internal pacemaker, known as the **sinoatrial (SA) node** (▻ Figure 12–8). Located in the wall of the right atrium, the SA node is composed of a clump of specialized cardiac muscle cells. The cells of the SA node contract spontaneously and rhythmically. Each contraction produces a bioelectric impulse, akin to those produced by nerve cells. The impulse given off by the SA node spreads rapidly from muscle cell to muscle cell in both atria. Because cardiac muscle cells are tightly joined and because the impulse travels quickly, the two atria contract simultaneously and uniformly.

Left to their own devices, cardiac muscle cells would contract independently and in a disorderly way, creating an ineffective pumping action. The SA node, therefore imposes a single rhythm on all of the atrial heart muscle cells.

The electrical impulse next passes from the atria to the ventricles; however, its passage is briefly slowed by a barrier of unexcitable tissue that separates the atria from the ventricles. The impulse is delayed approximately 1/10 of a second. After this brief holdover, the impulse is channeled

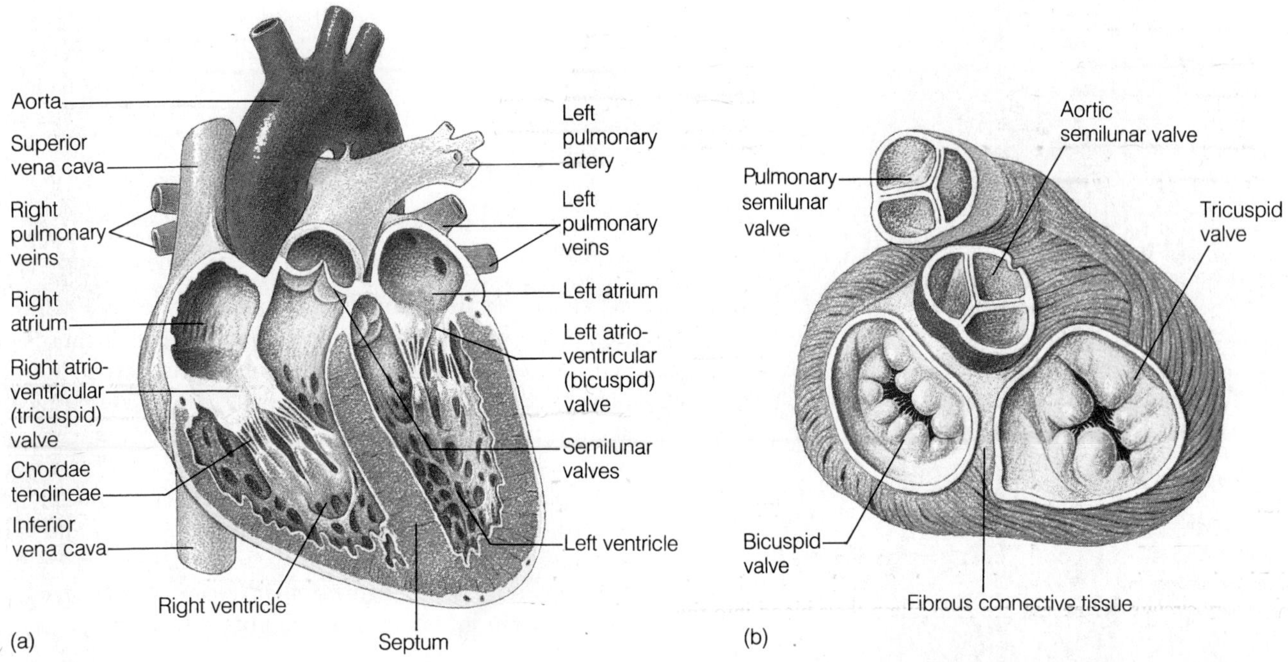

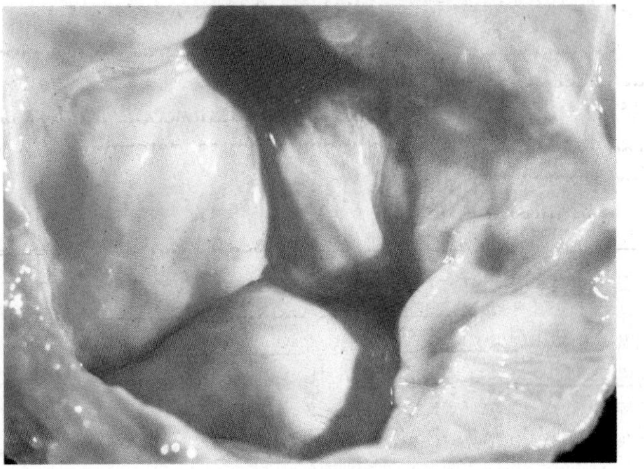

▻ **FIGURE 12–7 Heart Valves** (*a*) A cross section of the heart showing the four chambers and the location of the major vessels and valves. (*b*) A view of the heart from above, with the major vessels removed to show the valves. (*c*) Pulmonary semilunar valve.

through a second mass of specialized muscle cells, the **atrioventricular (AV) node,** shown in Figure 12–8. From the AV node, the impulse travels along a tract of specialized cardiac muscle cells, known as the **atrioventricular bundle.** As shown in Figure 12–8, the atrioventricular bundle divides into two branches (called the bundle branches) that travel on either side of the wall separating the two ventricles. The bundle branches give off smaller branches that terminate on **Purkinje fibers,** specialized muscle cells that come in contact with cardiac muscle cells of the ventricles.

The slight delay created when the impulse travels from the atria to the ventricles gives the blood-filled atria ample time to contract and empty into the ventricles. It also provides the ventricles plenty of time to fill before they are stimulated to contract. Unlike the muscle cells of the atria, the cardiac muscle cells of the two ventricles do not contract in unison, in large part, because the impulse is not transmitted as quickly and as uniformly through the ventricles as it is through the atria. Contraction begins at the bottom of the heart and proceeds upward, squeezing the blood out of the ventricles into the aorta and pulmonary arteries.

The SA node of the human heart produces a steady rhythm of about 100 beats per minute when isolated from outside influences, but this is much too fast for most human activities. To bring the heart rate in line with body demand, the SA node must be dampened. This dampening is brought about by nerves that supply the heart with impulses from a control center in the brain. At rest or during nonstrenuous activity, these impulses slow the heart to about 70 beats per minute, thus aligning heart rate with body demands. (They dampen much as the brakes reduce a car's speed.) During exercise or stress, when the heart rate must increase to meet body demands, the decelerating impulses from the brain are reduced. (In other words, the body lets up on the brakes, permitting the heart rate to increase.) Other nerves also influence heart rate. These nerves carry impulses that accelerate the heart rate even further, allowing the heart to attain rates of 180 beats or more when body demand for oxygen is great. (Continuing with the car analogy, these nerves are akin to the accelerator pedal, which increases the RPMs of the engine.)

Several hormones also play a role in controlling heart rate. One of these hormones is **epinephrine,** commonly known as **adrenalin.** During stress or exercise, the adrenal glands, located on top of the kidneys, secrete this hor-

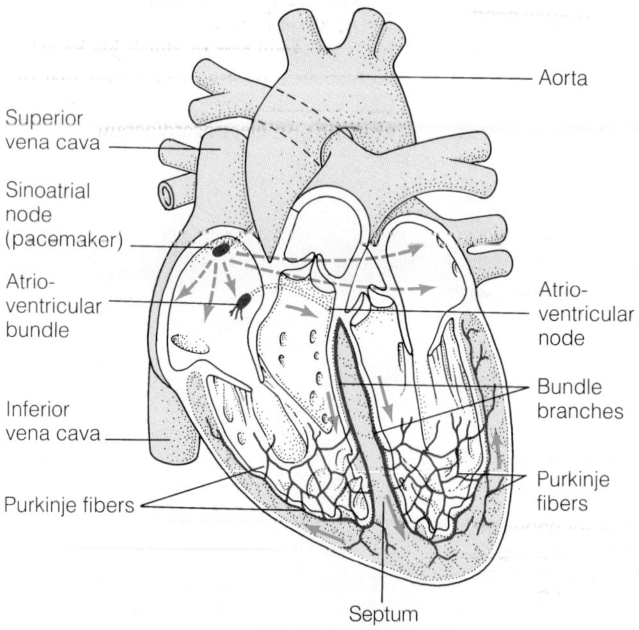

▻ **FIGURE 12–8 Conduction of Impulses in the Heart** The sinoatrial node is the heart's pacemaker. Located in the right atrium, it sends timed impulses into the atrial heart muscles, coordinating muscle contraction. The impulse travels from cell to cell in the atria, then passes to the atrioventricular node and into the ventricles via the atrioventricular bundle and its two branches, which terminate on the Purkinje fibers.

mone. Epinephrine stimulates the heart rate, increasing the flow of blood through the heart.

The nervous and endocrine system mechanisms described above are important evolutionary adaptations that help vertebrates cope with the changing demands of their lives.

Heart Muscle Cells Unleashed from Their Control Beat Independently, Greatly Reducing the Heart's Effectiveness

A patient is wheeled into the emergency room of a major hospital, where the doctors find that his heart is in a kind of cardiac anarchy known as **fibrillation.** Individual cells are beating on their own accord, and little blood is actually moving in and out of the heart chambers.

This patient, like thousands of others each year, is suffering one kind of heart attack. Having lost control from the SA node, the patient's heart has been converted into an ineffective, quivering mass of muscle tissue with individual cardiac muscle cells contracting at their own rate. During fibrillation, the heart pumps very ineffectively and may cease pumping altogether, a potentially fatal condition known as **cardiac arrest.**

Physicians treat fibrillation by applying a strong electrical current to the chest, a procedure appropriately known as **defibrillation.** The electrical current passes through the wall of the chest and is often sufficient to restore normal electrical activity and heartbeat. A normal heartbeat can also be restored by **cardiopulmonary resuscitation (CPR),** a procedure in which the heart is "massaged" externally by applying pressure to the sternum (breastbone).

Electrical Activity in the Heart Can Be Measured on the Surface of the Chest

When the electrical impulse that stimulates muscle contraction in the heart reaches a cardiac muscle cell, it causes the cell to contract. Normally, the outside surface of the cardiac muscle cell is slightly positive. The inside surface is slightly negative. When the impulse arrives, it causes a rapid change in the permeability of the cardiac muscle cell's plasma membrane to sodium ions. Sodium ions flow inward, changing the polarity of the membrane, and temporarily making the inside of the cell more positive than the outside. This change in polarity causes the release of calcium from internal storage depots, which in turn causes the cell to contract.

The shift in cardiac muscle cell polarity, or **depolarization,** can be detected by surface electrodes, small metal plates connected to wires and a voltage meter (▻ Figure 12–9a). The electrodes are placed on a person's chest. The resulting reading on a voltage meter is called an **electrocardiogram (ECG),** or sometimes also **EKG** (Figure 12–9b).

For a normal person, the tracing produced on the voltage meter has three distinct waves (Figure 12–9b). The first wave, the **P wave,** represents the electrical changes occurring in the atria of the heart. The second wave, the **QRS wave,** is a record of the electrical activity taking place during ventricular contraction, and the third wave, the **T wave,** is a recording of electrical activity occurring as the ventricles relax.

Diseases of the heart may disrupt one or more waves of the ECG. As a result, an ECG is often a valuable diagnostic tool for cardiologists. Bear in mind, though, that the ECG detects only those diseases that alter the heart's electrical activity.

Cardiac Output Varies from One Person to the Next, Depending on Activity and Conditioning

The total amount of blood pumped by the ventricles each minute is called the **cardiac output.** Cardiac output is a function of two variables: **heart rate,** or the number of contractions the heart undergoes per minute, and **stroke volume,** or the amount of blood pumped by each ventricle during each contraction. At rest, the heart beats approximately 70 times per minute, and the stroke volume is about 70 milliliters. This produces a cardiac output of 5000 milliliters, or 5 liters per minute.

Cardiac output varies among individuals, depending on their physical condition and their level of activity. The heart of a trained athlete, for example, can pump 35 liters of blood per minute (seven times the cardiac output at rest).

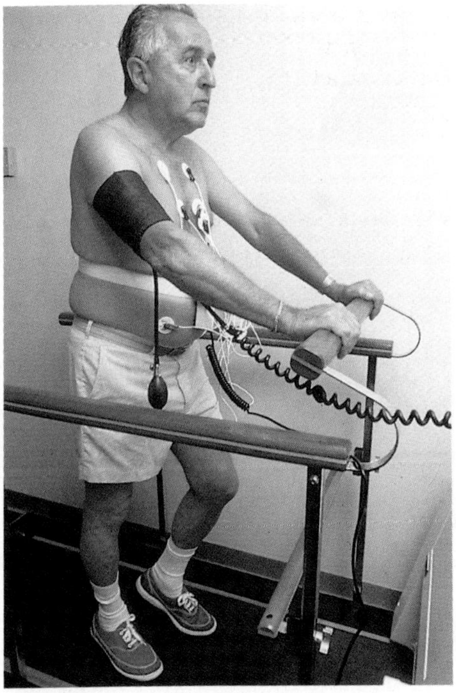

(a)

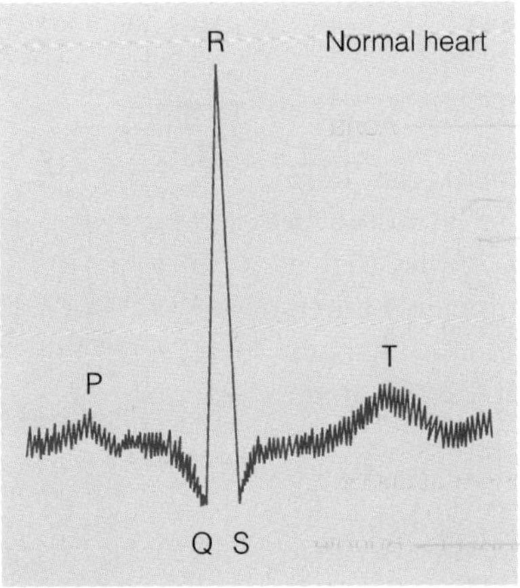

P = atrial depolarization, which triggers atrial contraction.

QRS = depolarization of AV node and conduction of electrical impulse through ventricles. Ventricular contraction begins at R.

T = repolarization of ventricles.

P to R interval = time required for impulses to travel from SA node to ventricles.

(b)

▻ **FIGURE 12–9 The Electrocardiogram** (*a*) This patient taking a treadmill test to check his heart's performance is wired to a meter that detects electrical activity produced by the heart. (*b*) An electrocardiogram.

Most nonathletes, however, can increase the cardiac output to only about 20 liters per minute.

HEART ATTACKS: CAUSES, CURES, AND TREATMENT

Heart attacks come in several varieties. The most common heart attack is called a **myocardial infarction** (mentioned in Chapter 11 and described below) and is caused by a **thrombosis,** a clot blockage of one or more of the **coronary arteries,** small vessels that supply blood to the heart. A blood clot lodged in a coronary artery restricts the flow of blood to the heart muscle, cutting off the supply of oxygen and nutrients. Depriving the heart muscle of oxygen can damage and even kill the cells. The region damaged by a thrombosis is called an **infarct,** hence the name myocardial infarction.

Myocardial Infarctions Usually Occur When Blood Clots Lodge in Arteries Narrowed by Atherosclerosis

As noted in Health Note 11–1, the formation of atherosclerotic plaque results from a combination of stress, poor diet, lack of exercise, smoking, heredity, and several other factors. Narrowing of a coronary artery by plaque does not necessarily result in a heart attack, however, unless the narrowing is quite severe. Nonetheless, less severe narrowing does makes the vessel more susceptible to blood clots. When a clot forms in the vessel at the site of narrowing or when a clot originating elsewhere in the body lodges in the narrowed vessel, trouble begins.

The outcome of heart attacks varies. If the size of the infarct (damaged area) is small and if the change in electrical activity of the heart is minor and transient, a heart attack is usually not fatal. If the damage is great or electrical activity is severely disrupted, however, myocardial infarctions can prove fatal.

Heart attacks can occur quite suddenly, without warning, or may be proceeded by several weeks of **angina,** pain that is felt when the supply of oxygen to the myocardium is reduced. Anginal pain appears in the center of the chest and can spread to a person's throat, upper jaw, back, and arms (usually just the left one). Angina is a dull, heavy, constricting pain, that appears when an individual is active, then disappears when he or she ceases the activity.

Angina attacks may also be caused by stress and exposure to carbon monoxide, a pollutant that reduces the oxygen-carrying capacity of the blood. Angina begins to show up in men at age 30 and is nearly always caused by coronary artery disease. In women, angina tends to occur at a much later age. Interestingly, about 90% of all "chest pain" turns out to be unrelated to the heart. Patients go to their doctors thinking they are suffering from angina, when, in fact, what they are feeling is tension and pain in the wall of the chest, usually in the muscles between the ribs. Deep breathing, relaxation, and stress reduction are effective in reducing, even eliminating, this pain.

Prevention is the Best Cure, but in Cases Where Damage has Already Occurred, Medical Science has a Great Deal to Offer

Proper diet, exercise, and stress management can help reduce the risk of heart problems later in life, as noted in Health Notes 1–1 and 11–1). Research also shows that a daily dose of aspirin (half a tablet a day is sufficient) taken over long periods can help reduce an individual's chances of having a heart attack. A recent study of 22,000 male physicians in the United States, in fact, showed that aspirin in small doses reduced the risk of first heart attacks by nearly half.[2] Studies suggest that aspirin reduces heart attacks by impairing blood clotting.

Prevention should be the first line of attack against heart disease. It could save Americans hundreds of millions of dollars each year in medical bills, lost work time, and decreased productivity. But given human nature, the fast pace of modern life, and our inattentiveness to exercise and proper diet, heart disease will probably be around for a long time. To reduce the death rate, physicians therefore also look for ways to treat patients after they have had a heart attack.

One promising development is the use of blood-clot-dissolving agents like streptokinase, mentioned at the beginning of the chapter. When administered within a few hours of the onset of a heart attack, streptokinase can greatly reduce the damage to heart muscle and accelerate a patient's recovery.

Ironically, streptokinase is an enzyme derived from the bacterium that causes rheumatic fever. Because streptokinase is a foreign substance, it evokes an immune reaction. In some people, the reaction is quite severe and may even cause death. The immune reaction to this drug has inspired a search for similar chemicals without the dangerous side effects. One promising chemical is urokinase, a clot-busting enzyme produced by human cells that does not trigger an immune reaction.

Scientists are also testing another naturally occurring clot dissolver, called TPA (tissue plasminogen activator). Tests in humans suggest that TPA may also be free of the dangerous side effects of streptokinase. Consequently, TPA has been approved for use in humans since 1987. Nevertheless, its use is not without its problems. Two of the most significant are (1) the recurrence of blood clots in many patients and (2) the high cost of the drug.

In cases where the coronary arteries are completely blocked by atherosclerotic plaque, it is necessary to reestablish full blood flow to the heart muscle. To restore blood flow, physicians often perform **coronary bypass surgery,** in which they transplant segments of intact vein from the leg into the heart (▻ Figure 12–10). These venous bypasses transport blood around the clogged coronary arteries, restoring blood flow. Once hoped to be a long-term solution to a widespread problem, coronary bypass surgery may provide only temporary relief with little, if any, long-term benefits. Research studies show that bypass patients have a significantly higher rate of survival in the five years following surgery than patients who just receive drugs. In the next seven years, however, these studies show, long-term survival from coronary bypass surgery is about the same as that of patients treated with diet and medications. In the long run, say researchers, bypass surgery is only slightly more effective than nonsurgical medical treatments.

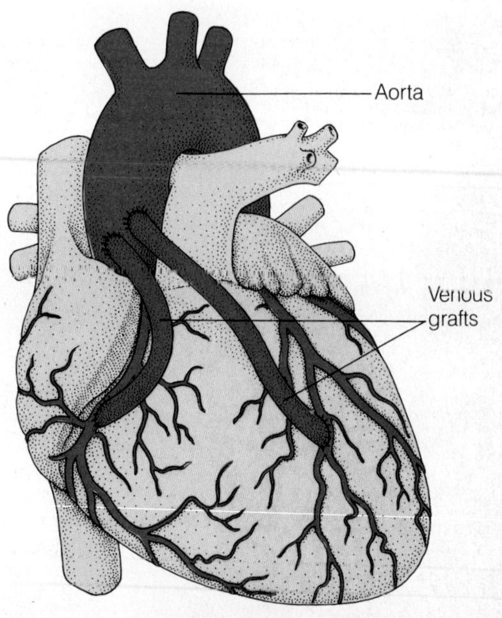

▻ **FIGURE 12–10 Coronary Bypass Surgery** Venous grafts bypass coronary arteries occluded by atherosclerotic plaque.

The problem with coronary bypasses is that the venous grafts often fill fairly quickly with cholesterol plaque, and the heart becomes starved for oxygen once again, especially when forced to work harder (for example, during exercise). The recurrence of plaque in venous grafts has led researchers to turn to marginally important arteries, such as the internal mammary artery, for coronary bypass surgery. Researchers believe that these arteries will prove more resistant to plaque buildup than veins. Physicians can also clean clogged blood vessels by using a small catheter with a tiny balloon attached to its tip in conjunction with the clot-dissolving agents. After the chemical clot dissolvers are administered to a patient, the balloon is inflated. This procedure, called **balloon angioplasty,** forces the artery open and apparently loosens the plaque from the wall, allowing it to be washed away by the blood.

Scientists are also experimenting with lasers to burn away plaque in artery walls. Catheters containing fine glass fibers can be inserted into the blocked arteries during open-heart surgery. They can also be inserted through the artery in the thigh and snaked through the arterial system until they reach the clot. Laser beams transmitted through the glass fibers burn away the plaque. Unfortunately, as in other techniques, cholesterol builds up again in the walls of arteries within a few months.

[2] You should consult your physician if you are thinking about taking aspirin as a preventive measure.

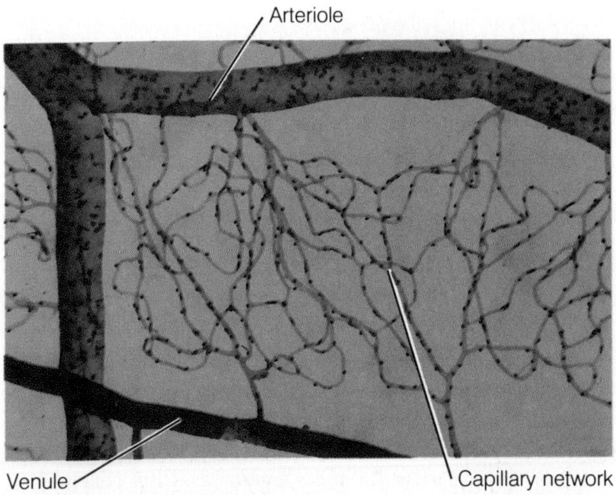

▻ **FIGURE 12–11 Capillary Network** A network of capillaries between the arteriole and the venule delivers blood to the cells of body tissues, not shown.

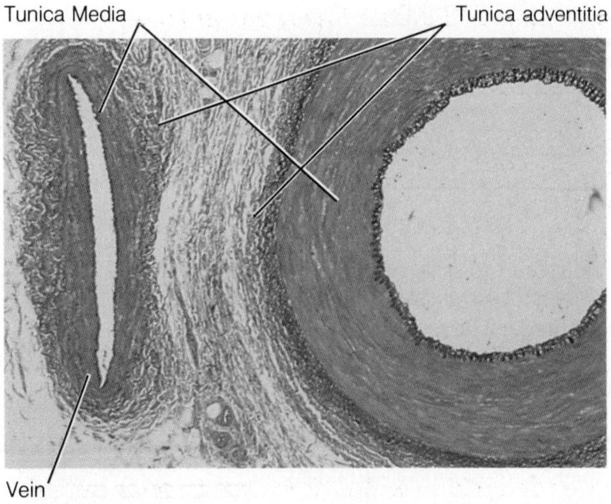

▻ **FIGURE 12–12 Artery and Vein** A cross section through a vein shows that the muscular layer, the tunica media, is much thinner than it is in an artery. Veins typically lie alongside arteries and in histological sections such as these usually have irregular lumens.

≈ THE BLOOD VESSELS

The circulatory system can be divided into four functional parts. The first is the heart, which pumps blood throughout the body. The second is the arteries, which form a delivery system that carries blood from the heart to the body tissues. The third is an exchange system, consisting of networks of tiny vessels known as capillaries, found in body tissues. The fourth is the return system, consisting of veins that carry oxygen-depleted and waste-enriched blood back to the heart from the body tissues. (For a discussion of the discoveries that led to our understanding of circulation, see the Scientific Discoveries 12–1.)

Arteries, which transport blood away from the heart, branch many times, forming smaller and smaller vessels. The smallest of all arteries is the **arteriole.** As illustrated in ▻ Figure 12–11 arterioles empty into **capillaries,** tiny, thin-walled vessels that permit wastes and nutrients to pass through with relative ease. Capillaries form extensive, branching networks in body tissues, referred to as **capillary beds**, which provide an avenue for exchange between the blood and the tissue fluid surrounding the cells of the body.

Blood flows out of the capillaries into the smallest of all veins, the **venules**. Venules, in turn, converge to form small veins, which unite with other small veins, in much the same way that small streams unite to form a river.

▻ Figure 12–12 shows a cross section of an artery and a vein. As illustrated, these two vessels are structurally different. Veins, for example, tend to be smaller and to have thinner walls. Despite their obvious differences, arteries and veins have a common architecture. For example, both consist of three layers: (1) an external layer of connective tissue, the **tunica adventitia,** which binds the vessel to surrounding tissues; (2) a middle layer, the **tunica media,** which is primarily made of smooth muscle; and (3) an internal layer, the **tunica intima,** which is composed of a layer of flattened cells, the **endothelium,** and a thin, nearly indiscernible layer of connective tissue, which binds the endothelium to the tunica media (▻ Figure 12–13).

▻ **FIGURE 12–13 General Structure of the Blood Vessel** The artery shown here consists of three major layers, the tunica intima, tunica media, and tunica adventitia.

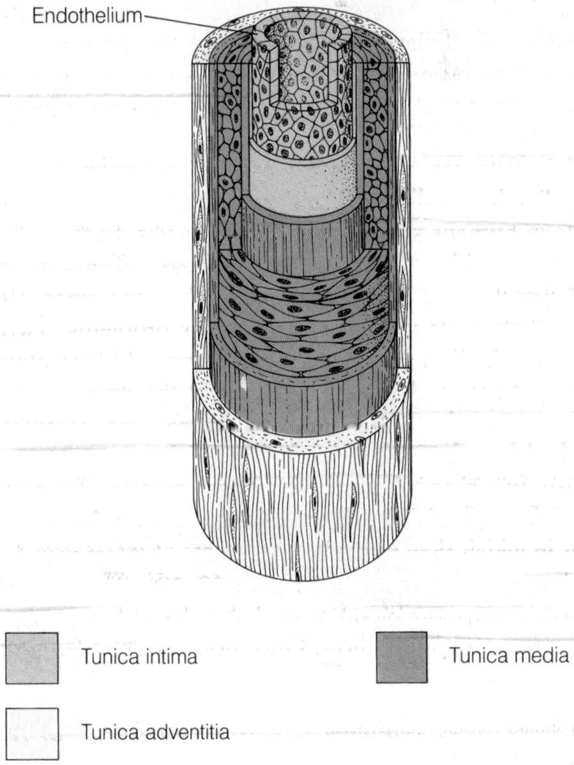

SCIENTIFIC DISCOVERIES THAT CHANGED THE WORLD 12–1

THE CIRCULATION OF BLOOD IN ANIMALS

Featuring the Work of William Harvey and Stephan Hales

The 17th-century British physician William Harvey is generally credited with the discovery of the circulation of blood in animals. Harvey is known as a scientist with a short temper who wore a dagger in the fashion of the day, which he reportedly brandished at the slightest provocation (▻ Figure 1). He was probably not the kind of professor you might "argue" with over grades.

Temperament aside, Harvey is generally honored for his work on the role of the heart and the flow of blood in animals and is often praised as a pioneer of scientific methodology. His application of quantitative procedures to biology, some say, ushered in the modern age of this science.

In Harvey's medical school days, anatomists thought the intestines produced a substance called chyle, a kind of lymph that passed from the intestines to the liver. Chyle was, they thought, derived from the food people ate. The liver converted the chyle to venous blood and then distributed the blood through arteries and veins. As a medical student, Harvey learned that blood oozing through arteries and veins supplied organs and tissues with nourishment. He also learned that there was no real circulation, that blood merely ebbed back once in a while to the heart and lungs, where impurities were removed.

▻ **FIGURE 1 William Harvey**
This important scientific figure greatly advanced our knowledge on circulation in animals.

These ideas were proposed by the Greek physician Galen 14 centuries earlier and had persisted nearly without challenge until Harvey's time.

As a teacher in the Royal College of Physicians in 1616, Harvey began to describe the circulation of blood, based on the results of experiments and observations on animals. Apparently rebuked for his ideas by some of his colleagues, Harvey engaged in many years of research to provide supporting evidence. In these studies, he described the muscular character of the heart and the origin of the heartbeat in the right atrium. He also demonstrated that the pulse felt in arteries resulted from the impact of blood pumped by the heart. Furthermore, he described the pulmonary and systemic circuits and proposed that blood flowed to the tissues and organs of the body via the arteries and returned via the veins.

A brief examination of just one of his experiments illustrates that even though Harvey is a key figure in the history of biological science and played a key role in promoting quantitative study, some of

Arteries and Arterioles Deliver Oxygen-Rich Blood to Body Tissues and Organs

The largest of all arteries is the aorta. As noted earlier, this massive vessel carries oxygenated blood from the left ventricle of the heart to the rest of the body. The aorta loops over the back of the heart, then descends through the chest and abdomen, giving off large branches along its way. These branches carry blood to the head, the extremities (arms and legs), and major organs, such as the stomach, the intestines, and the kidneys (see Figure 12–1). The very first branches of the aorta are the coronary arteries, which supply the heart muscle.

The aorta and many of its chief branches are **elastic arteries,** so named because they contain numerous wavy elastic fibers interspersed among the smooth muscle cells of the tunica media (▻ Figure 12–14a). As blood pulses out of the heart, the elastic arteries expand to accommodate the blood. Like a stretched rubber band, though, the elastic fibers in the tunica media cause the arterial walls to recoil. This provides an ancillary pump that helps push the blood along the arterial tree and also helps maintain an even flow of blood through the capillaries.

The elastic arteries branch to form smaller vessels, the **muscular arteries** (Figure 12–14b). Muscular arteries contain fewer elastic fibers but can still expand and contract with the flow of blood. You can feel this expansion and contraction in the arteries lying near the skin's surface in your wrist and neck. It's the pulse that health-care workers use to measure your heart rate.

The smooth muscle of the tunica media of muscular arteries responds to a number of stimuli, including nerve impulses, hormones, carbon dioxide, and lactic acid. These stimuli cause the blood vessels to open or close to varying degrees. This, in turn, allows the body to adjust blood flow through its tissues to meet increased demands for nutrients and oxygen. Arterioles in muscles, for instance, dilate when an animal is threatened by danger. This opening increases blood flow to the muscle, allowing the animal to flee or to meet the danger head on. At the same time, vessels in the digestive system constrict, temporarily reducing the digestive process and increasing the amount of blood available his work was less than exceptional, based as it was on some poor assumptions and inaccurate observations. As an example, consider the work he used to rebut Galen's hypothesis that the blood was produced by the food people ate.

Harvey first approximated the amount of blood the heart ejected with each heartbeat (stroke volume), then determined the pulse rate. He called on earlier observations of a heart from a human cadaver to determine stroke volume. At that time, he had noted that the left ventricle contained more than two ounces of blood and then, for reasons not entirely clear to historians of science, had hypothesized that the ventricle ejected "a fourth, a fifth, a sixth or only an eighth" of its contents. (Today, studies indicate that the heart ejects nearly all of its contents.) Harvey estimated that the stroke volume was about 3.9 grams of blood per beat. Modern estimates put it at 89 grams per beat. Harvey also made a grave error in determining pulse. His value of 33 beats per minute is about half of the actual rate in humans. No one knows how he could have been so wrong. Armed with two erroneous measurements, Harvey derived a figure for the amount of blood that circulated through the body that was 1/36 of the lowest value accepted today.

Regardless, Harvey "proved" his point that each half-hour the blood pumped by the heart far exceeds the total weight of blood in the body. From this he concluded that blood must be circulated. It is not, as Galen proposed, produced by the food we eat. The amount of food one eats could not produce blood in such volume.

Harvey debunked another falsehood perpetrated through the centuries—the Galenic myth that blood flowed into the extremities in both arteries and veins. Harvey first wrapped a bandage around an extremity. This obstructed the flow of blood through the veins but not the arteries. He noted that the veins swelled because, as he conjectured, blood was being pumped into them via underlying arteries and there was nowhere for the blood to go. Tightening the bandage further cut the blood flow in the arteries as well and thus prevented the veins from swelling. From these observations, Harvey correctly surmised that the arteries deliver blood to the extremities, and the veins return it to the heart.

Harvey's work laid the foundation of modern cardiovascular physiology, but it left many questions unanswered. Many of these questions were addressed by the highly industrious English biologist Stephan Hales, who was born a century after Harvey. In a long series of rather gruesome but scientifically rigorous experiments on horses, dogs, and frogs, Hales explored many aspects of the cardiovascular system. Benefiting from more modern methods of study, he charted blood pathways and examined blood flow and blood pressure in different parts of the circulatory system. After settling many of the unanswered questions left by Harvey, Hales went on to study plant physiology and is perhaps best known for his work on the circulation of sap in plants, discussed in Scientific Discoveries 26–1.

to the muscles. Regulating the flow of blood to body tissues is required to control body temperature as well, as noted in earlier chapters.

Blood Pressure. The force that blood applies to the walls of a blood vessel is known as the **blood pressure.** Like many other physical conditions in the human body, blood pressure varies from time to time. For example, it changes in relation to one's activity and stress levels. In a given artery, it rises and falls with each heart beat. Blood pressure also varies throughout the cardiovascular system, being the highest in the aorta and dropping considerably as the arteries branch. When blood reaches the capillaries, the flow of blood and blood pressure are greatly reduced, enhancing the rate of exchange between the blood and the tissues.

Blood pressure is measured by using an inflatable device with the tongue-twisting name of **sphygmomanometer** (sfig-mo-ma-nom-a-ter), or, more commonly, blood pressure cuff (▻ Figure 12–15a). The blood pressure cuff is first wrapped around the upper arm. A stethoscope is positioned over the artery just below the cuff. Air is pumped into the cuff until the pressure stops the flow of blood through the artery (Figure 12–15b). The pressure in the cuff is then gradually reduced as air is released. When the blood pressure in the artery exceeds the external pressure of the cuff, the blood starts flowing through the vessel once again. This point represents the **systolic pressure,** the peak pressure at the moment the ventricles contract. Systolic pressure is detected by a thump heard through the stethoscope and is the higher of the two numbers in a blood pressure reading (120/70).[3] The pressure at the moment the heart relaxes to let the ventricles fill again is the **diastolic pressure** and is the lower of the two readings. It is determined by continuing to release air from the cuff until no arterial pulsation is audible. At this point, blood is flowing continuously through the artery.

A typical reading for a young, healthy adult is about 120/70, although readings vary considerably from one individual to the next. Thus, what is normal for one person

[3]Blood pressure is measured in millimeters of mercury (mm Hg).

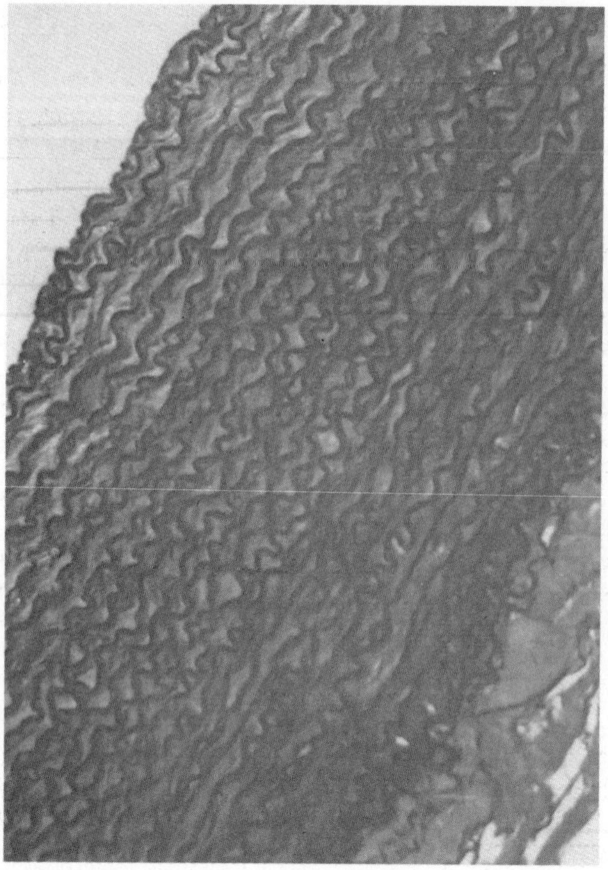

(a)

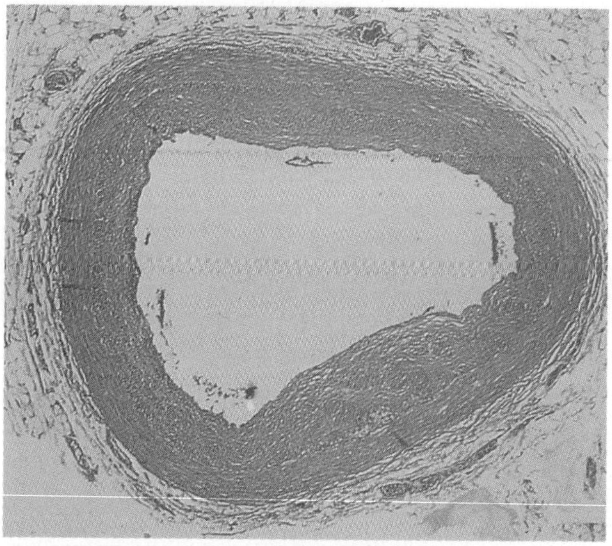

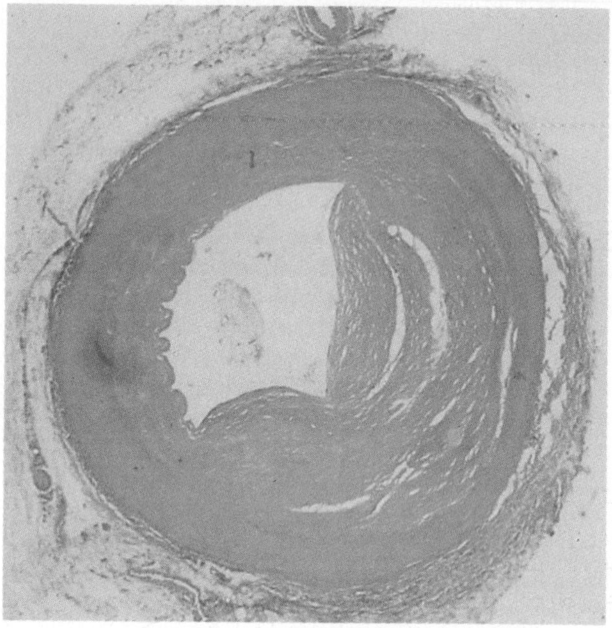

(c)

▷ **FIGURE 12–14 Arteries** (*a*) A cross section of the wall of the aorta showing numerous wavy elastic fibers common in all elastic arteries. (*b*) The muscular artery has a thick tunica media as well but fewer elastic fibers. (*c*) Muscular artery partly occluded by atherosclerotic plaque.

may be abnormal for another. As a person ages, blood pressure tends to rise. Therefore, a healthy 65-year-old might have a blood pressure reading of 140/90. In a 30-year-old, this reading might be cause for alarm.

Hypertension is a prolonged elevation in blood pressure. It has many causes, including kidney disease, high salt intake, and obesity. Nearly always a symptomless disease, hypertension is often characterized by a gradual increase in blood pressure over time. A person may feel fine and display no physical problems whatsoever for years. Symptoms, such as headaches, palpitations (rapid, forceful beating of the heart), and a general feeling of ill health, usually occur only when blood pressure is dangerously high. Consequently, early detection and treatment are essential to prevent serious problems, including heart attacks. (For more on hypertension, see Health Note 12–1.)

Capillaries Permit the Exchange of Nutrients and Wastes Between Blood and Body Cells

As described above, the heart, arteries, and veins form an elaborate system that propels and transports blood to and from the capillaries. As also noted earlier, capillaries form branching networks, known as capillary beds, among the cells of body tissues (▷ Figure 12–16a). It is in these extensive networks of vessels that wastes and nutrients are exchanged between the cells of the body and the blood.

As illustrated in Figure 12–16b, the walls of the capillaries consist of flattened endothelial cells; they permit dissolved substances to pass through them with great ease—and provide another illustration of the correlation between structure and function in the body. In some places, such as the kidneys and small intestine, small windows, or **fenestrae,** are present in the cells of capillary walls (Figure 12–16c). Fenestrae permit even greater movement of molecules to and from the capillary.

If you could remove all of the capillaries from the body and line them up end to end, they would extend over

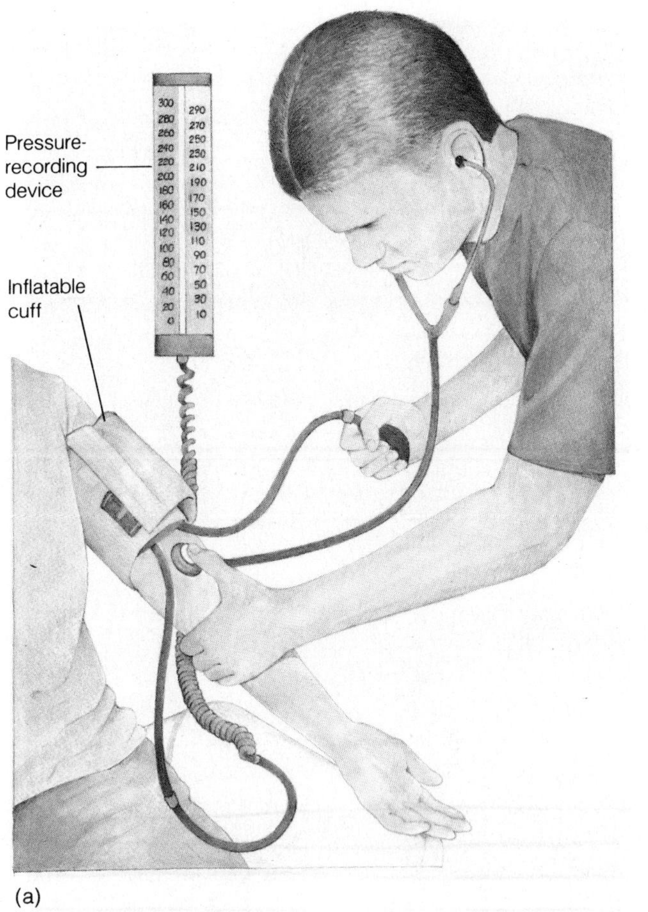

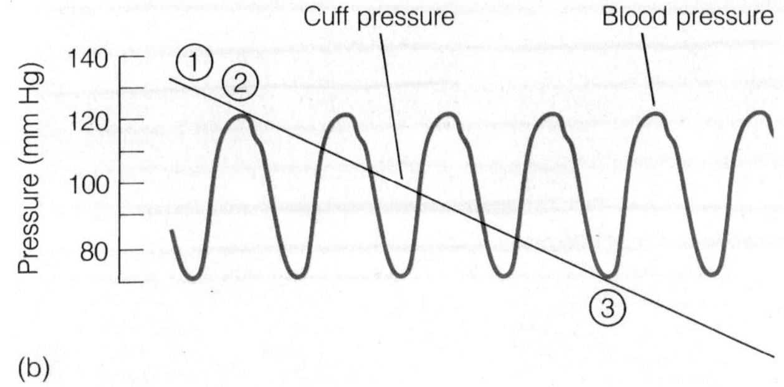

▻ **FIGURE 12–15 Blood Pressure Reading** (*a*) A sphygmomanometer (blood pressure cuff) is used to determine blood pressure, indicated by the red line. As shown, the blood pressure in a given region rises and falls with each contraction of the heart. (*b*) When the pressure in the cuff exceeds the arterial peak pressure, blood flow stops ①. No sound is heard. Cuff pressure is gradually released. When pressure in the cuff falls below the arterial pressure, blood starts flowing through the artery once again. This is the systolic pressure ②. The first sound will be heard. Cuff pressure continues to drop. When cuff pressure is equal to the lowest pressure in the artery, the artery is fully open and no sound is heard ③. This is the diastolic pressure.

80,000 kilometers (50,000 miles)—enough to circle the globe at the equator two times. The extensive branching of capillaries brings them in close proximity to body cells but also slows the rate of blood flow through capillary networks and decreases pressure, both of which increase the efficiency of capillary exchange.

A "typical" capillary network is shown in ▻ Figure 12–17a. As illustrated, most capillary beds contain **thoroughfare channels,** or **metarterioles,** circulatory "shortcuts" that connect the arterioles with the venules. The metarterioles give off smaller branches, the **true capillaries,** the site of the exchange of nutrients and wastes.

The flow of blood through the capillary bed, and therefore the supply of nutrients to tissues, is largely controlled by the constriction and dilation of the metarteriole. When the metarteriole is open, blood flows into the capillary network, providing ample nutrients. When it constricts, blood passes by, traveling on to other tissues in need of nutrients.

The constriction and dilation of the metarteriole also help regulate body temperature. On a cold winter day, for example, the metarterioles of the capillary networks in the skin close down, restricting blood flow and conserving body heat. Just the reverse happens on a warm day: The flow of blood through the skin increases, releasing body heat and often creating a pink flush.

Blood flow through capillary networks is also controlled at another level. As Figure 12–17a shows, tiny rings of muscle surround the capillaries as they branch from the metarterioles. These muscle rings are called **precapillary sphincters.** They open and close like floodgates in response to local chemical signals (such as carbon dioxide levels) from nearby tissues. The relaxation of the precapillary sphincters causes blood to rush into the capillary bed. The precapillary sphincters therefore provide a means of delivering blood on demand to cells that need it or reducing the flow to cells that do not.

As blood flows *into* a capillary bed, nutrients, gases, water, and hormones immediately begin to diffuse out of the tiny vessels. As the blood flows *through* the capillary bed, water-dissolved wastes, such as carbon dioxide, begin to flow into the capillaries by diffusion. Wastes and nutrients can travel (1) between the endothelial cells of the capillary, (2) through the endothelial cells by diffusion, (3) through fenestrae, and (4) through the endothelial cells in minute pinocytotic vesicles (Chapter 3). The forces that control movement across the capillary wall are explained in ▻ Figure 12–18 (page 302).

Veins and Venules Transport the Oxygen-Poor and Waste-Laden Blood Back to the Heart

Blood leaves the capillary beds stripped of its nutrients and loaded with cellular wastes. Blood draining from the capillaries first enters the smallest of all veins, the venules. The venules converge with others, forming small veins. **Veins** carry blood toward the heart. Unlike the arteries, the veins start off small and converge with other veins, forming larger and larger vessels. Eventually, all blood returning to the heart from the systemic circulation enters either the superior or inferior vena cavae, the two main veins that

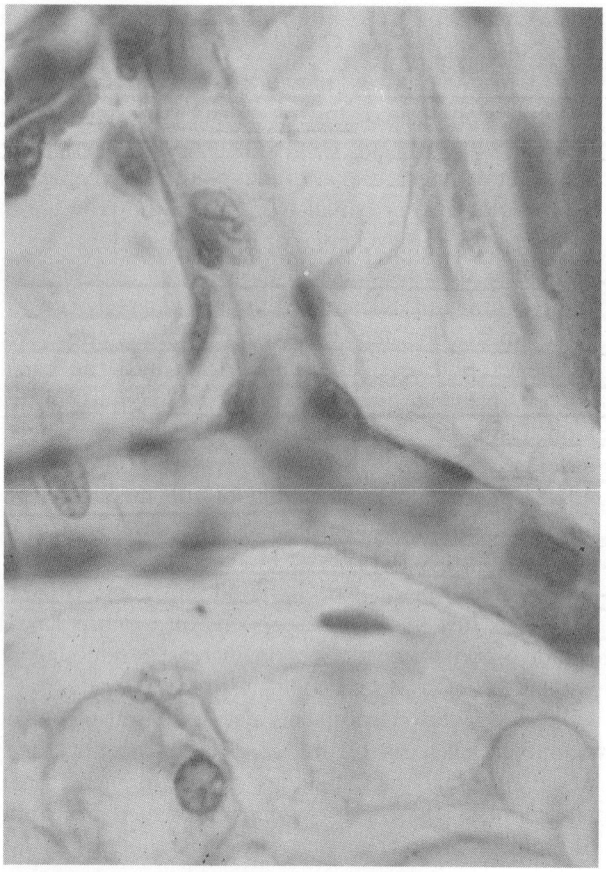

(a)

(b)

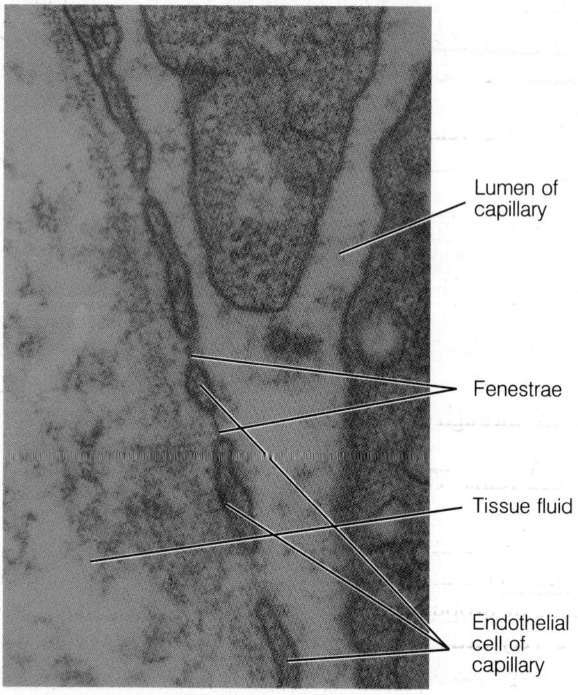

(c)

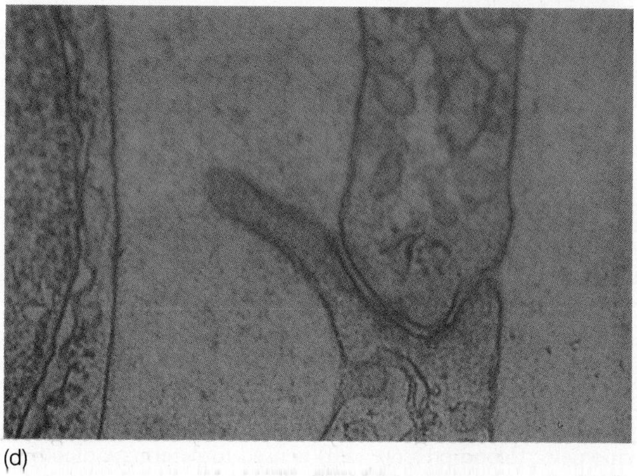

(d)

▻ **FIGURE 12–16 The Capillary** (*a*) A light micrograph of a capillary showing the endothelial cells that make up the wall of this vessel. (*b*) A cross section of a capillary showing the nucleus of an endothelial cell and capillary lumen. (*c*) A section through the wall of a highly porous capillary showing the fenestrae. Note that each window is spanned by a thin membrane that permits rapid movement of molecules into and out of the capillary. (*d*) A cross section through the wall of a capillary showing how materials flow through via endocytosis.

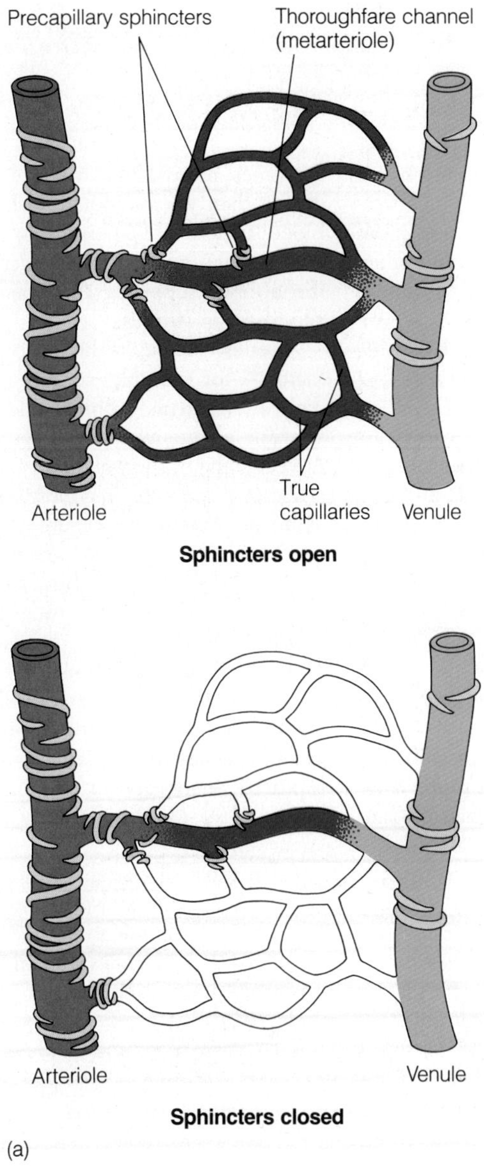

▻ **FIGURE 12–17 Control of Blood Flow through Capillaries** (*a*) The flow of blood into a capillary bed is controlled by the metarterioles and the precapillary sphincters. When the metarterioles are open, they allow blood to flow into the capillary network. Precapillary sphincters must open to permit further influx. When precapillary sphincters close, they reduce blood flow through the capillary bed. (*b*) A scanning electron micrograph of an arteriole showing the smooth muscle cells that contract and relax, controlling flow through these vessels.

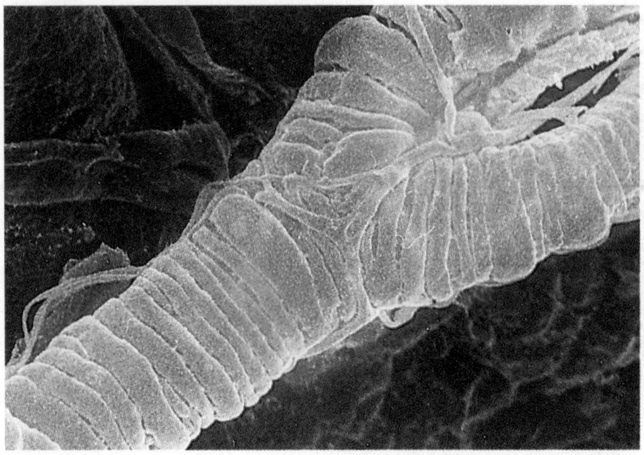

drain into the right atrium of the heart (see Figure 12–5). These vessels drain the upper and lower parts of the body, respectively.

Veins and arteries generally run side by side throughout the body much like opposing lanes on a freeway. The arteries take blood away from the heart and toward body tissues, and the veins return blood to the heart.

Blood pressure in the veins is low, and veins have relatively thin walls with fewer smooth muscle cells (see Figure 12–12). Because the veins' walls are so thin, obstructions can cause them to balloon, in much the same way that a tree down across a small stream can cause water to pool upstream. Blood pools in the obstructed veins, forming rather unsightly bluish bulges called **varicose veins** (▻ Figure 12–19, page 304).

Some people inherit a tendency to develop varicose veins, but most cases can be attributed to factors that reduce the flow of blood back to the heart. Abdominal tumors, pregnancy, obesity, and even sedentary lifestyles often result in varicose veins in the lower extremities. The restriction of blood flow may result in muscle cramps and the buildup of fluid, **edema** (swelling), in the ankles and legs.

Varicose veins may also form in the wall of the anal canal. The veins in this region are known as the internal hemorrhoidal veins. A swelling of the internal hemorrhoidal veins results in a condition known as **hemorrhoids.** Because the internal hemorrhoidal veins are supplied by numerous pain fibers, this condition can be quite painful.

How Do the Veins Work? ▻ Figure 12–20 (page 304) shows that blood pressure declines fairly rapidly as blood travels through the arteries and that it continues to decrease, but at a slower rate, as it flows through the capillaries and veins. The lowest blood pressure readings occur in the superior and inferior vena cavae. With such low pressure, how do the veins return blood to the heart?

For blood in veins above the heart, gravity is the chief means of propulsion. But for veins below the heart, which have very little pressure to force the blood along, return flow depends on the movement of body parts, which "squeezes" the blood upward. As you walk to class, for example, the contraction of muscles in your legs pumps the blood in the veins, like the hands of a person milking a cow. This forces the blood upward, slowly and surely causing it to move against the force of gravity. Even the nervous muscle contractions that occur during study help move the blood back to the heart.

HEALTH NOTE 12–1

HYPERTENSION AND ANEURYSMS: CAUSES AND CURES

The stresses of modern life and the unhealthy diets of many people take their toll on the heart and blood vessels. One of the most common problems of modern times is atherosclerosis, the buildup of cholesterol plaque in the walls of arteries, discussed in previous chapters. Arteries clogged with cholesterol force the heart to work harder, putting strain on this organ. Perhaps the most significant problem arises from blood clots that lodge in narrowed coronary arteries, reducing the flow of blood to the heart muscle, often with devastating consequences.

Hypertension, or high blood pressure, is another altogether-too-common problem of our times. In some individuals, hypertension is hereditary, passed from parent to offspring. Hypertension is also thought to result from stress and diet, especially high salt intake in some individuals. In other instances, it is caused by disorders such as kidney failure or hormonal imbalances. Pregnancy can lead to hypertension, and so can use of oral contraceptives. Low levels of cadmium in food, air, and water may also contribute to this disease, as noted in this chapter.

A few facts about hypertension are clear, though. People who are overweight when they are young are more likely to suffer from hypertension when they reach adulthood. An adult who is hypertensive and overweight, however, can often control the disease by losing weight and by reducing salt intake.

If this disease is untreated, blood pressure rises steadily over the years. As noted in the chapter, hypertension is a nearly symptomless disease for many people. Individuals feel fine for many years, despite their gradually rising blood pressure, and may not exhibit any signs of the disease until it has progressed to the dangerous stage. Thus, it is important for people over the age of 40 to have their blood pressure checked each year.

Hypertension is more common in men than in women and is more common in African Americans than in Caucasians. The disease is dangerous because the increased pressure in the circulatory system forces the heart to work harder. Elevated blood pressure may also damage the lining of arteries, creating a site at which atherosclerotic plaque forms. Atherosclerosis increases the risk of heart attack: a hypertensive person is six times more likely to have a heart attack than an individual with normal blood pressure. Hypertension also increases the chances for an occlusion in the arteries supplying the brain, which can result in strokes.

Arteries can be weakened by various factors—for example, certain infectious diseases (such as syphilis) and atherosclerotic plaque as well as hypertension. A weakening of the wall of an artery may cause it to balloon, a condition known as an **aneurysm** (▷ Figure 1). Like a worn spot on a tire, an aneurysm can give way when pressure builds inside or when the wall becomes too thin.

When an aneurysm breaks, blood pours out of the circulatory system. Because they happen so quickly, most aneurysms lead to death. An estimated 30,000 Americans die each year from ruptured aneurysms in the brain, and nearly 3000 die from ruptured aortic aneurysms.

As in most diseases, the first line of defense against aneurysms is prevention. By reducing or eliminating the two main causes—atherosclerosis and high blood pressure—individuals can greatly lower their risk.

Physicians recommend a number of

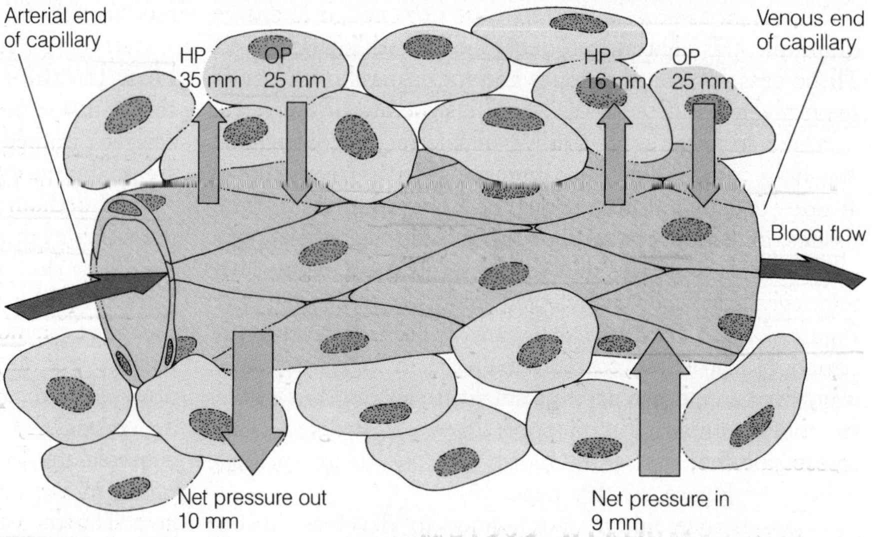

▷ **FIGURE 12–18 Capillary Exchange** Fluid begins to flow out of the capillary as soon as it enters the vessel from the arteriole. A slight hydrostatic pressure (HP), resulting from blood pressure, creates a slight outward force of 35 mm Hg. Osmotic pressure (OP), the tendency for water to flow inward because of a higher internal concentration of dissolved solute, is only 25 mm Hg. As a result, there is a net outward force on the arterial side of the capillary network of 10 mm Hg. The hydrostatic pressure declines as the blood flows through the capillary bed, and therefore the osmotic pressure forcing water inward exceeds the outward pressure. This drives water back into the capillary. Notice that the net movement inward does not equal the net movement out. As a result, excess fluid accumulates in the tissue and must be drained by the lymphatic system.

steps to reduce your chances of developing atherosclerosis and hypertension. If you smoke, stop. If you are overweight, exercise and lose weight. If you ingest excess salt, gradually cut back on your intake. If you are stressed at work and at home, find ways to reduce stress levels. If you drink alcohol, drink in moderation, or quit. If you eat foods rich in fats and cholesterol, cut down on them and consume more water-soluble fiber like that found in apples, bananas, citrus fruits, carrots, barley, and oats, which reduces cholesterol uptake by the intestine, as explained in Chapter 11.

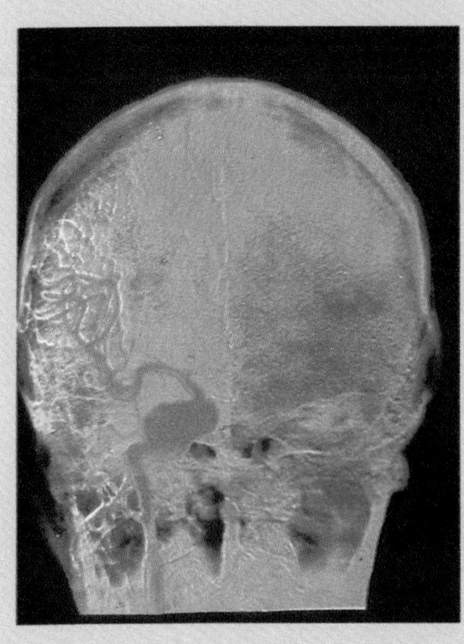

▻ **FIGURE 1 Aneurysm** This X-ray shows a ballooning of one of the arteries in the brain. If untreated, an aneurysm can break, causing a stroke.

The second line of defense against cardiovascular disease is early detection and treatment. Hypertension can be detected by regular blood pressure readings. Atherosclerosis can be discovered by blood tests to measure cholesterol. Aneurysms can be detected by X-ray. Pain also alerts patients and physicians that something is wrong. Once any of these diseases is detected, physicians have many options.

In the event of an aneurysm, surgeons can remove the weakened section of the artery and replace it with a section of a vein. In other instances, where venous grafts are more difficult (for example, in the brain), surgeons can clamp or tie off the artery just before the bulge, preventing blood flow through the damaged section. This works only when other arteries provide adequate blood flow to the area served by the damaged artery. In larger arteries, pieces of Dacron or other synthetic materials can be sewn into the wall of the artery, protecting it from breaking. Researchers are also experimenting with an alloy of nickel and titanium called nitinol. Nitinol is a "metal with a memory." When a fine nitinol wire is heated and wrapped around a cylinder, it forms a tightly coiled spring. When the spring is cooled, it reverts to a straight wire. When reheated, the metal returns to a coil.

In experiments with dogs, scientists create a wire coil that corresponds to the internal diameter of an artery. Next they cool the wire, causing it to revert to the straight form, and then push it through a catheter inserted in the artery. As the wire emerges from the cooled catheter inside the artery, body heat causes it to coil again. In place, the coil adds strength to the wall of the artery, preventing rupture. Experiments with dogs show that the endothelial cells of the tunica intima soon grow over the implant, making it a permanent part of the artery's wall. If successful in humans, this procedure could help save hundreds, perhaps thousands, of lives each year. It is, however, no substitute for a healthy diet and a healthy environment.

Important as muscles are, they are not enough to ensure return flow. Evolution has "provided" an additional factor: valves. **Valves** are flaps of tissue that span the veins and prevent the backflow of blood. The structure of the valves is shown in Figure 12–19a. As illustrated, the semilunar flaps of the veins resemble those found in the heart. Just as in the valves of the heart, blood pressure, however slight, pushes the flaps open (Figure 12–19a). This allows the blood to move forward. As the blood fills the segment of the vein in front of the valve, it pushes back on the valve flaps and forces them shut. You can locate the valves in the superficial veins on your forearm by pressing gently on a vein, then running your finger toward your wrist. You will see that the vein will collapse behind your finger until it crosses a valve.

≈ THE LYMPHATIC SYSTEM

The lymphatic system is functionally related to two systems: the circulatory system and the immune system. This section examines the role of the **lymphatic system,** an extensive network of vessels and glands, in circulation (▻ Figure 12–21).

You may recall from Chapter 10 that the cells of the body are bathed in a liquid called interstitial fluid. Interstitial fluid is in equilibrium with the blood plasma and provides a medium through which nutrients, gases, and wastes can diffuse between the capillaries and the cells.

Tissue fluid is replenished by water that diffuses out of the capillaries. The flow of water out of the capillaries, however, normally exceeds the return flow by about 3 liters per day. The "excess" water is picked up by small **lymph capillaries** in tissues. Like the capillaries of the circulatory system, these vessels have thin, highly permeable walls through which water and other substances pass with ease. As illustrated in ▻ Figure 12–22, the cells in the walls of the lymph capillaries overlap, creating a number of one-way valves, which resemble swinging doors. The accumulation of interstitial fluid in the tissues forces the "doors" to open, causing the fluid to flow into

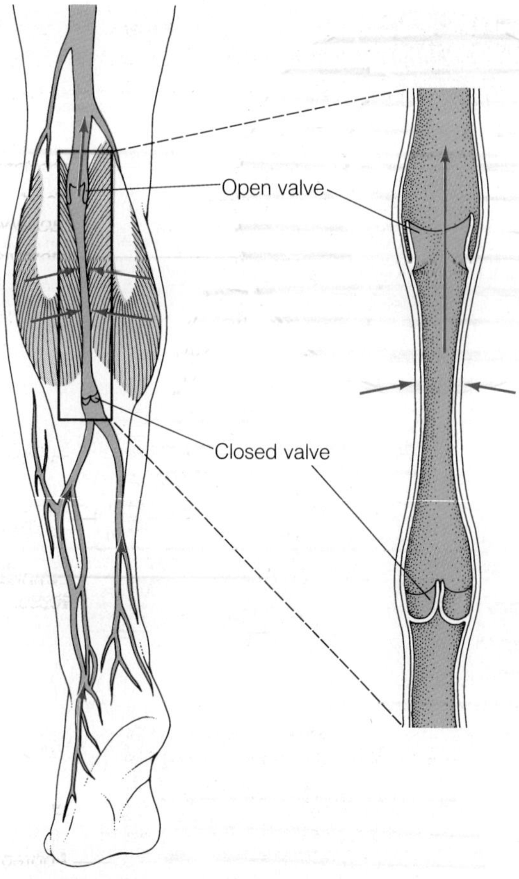

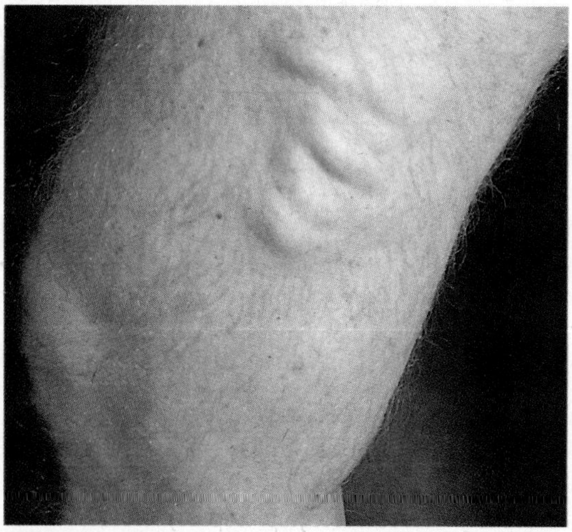

▻ **FIGURE 12–19 Valves in Veins** (*a*) The slight hydrostatic pressure in the veins and the contraction of skeletal muscles propel the blood along the veins back toward the heart. The one-way valves stop the blood from flowing backward. (*b*) Any restriction of venous blood flow to the heart causes veins to balloon out, creating bulges commonly known as varicose veins.

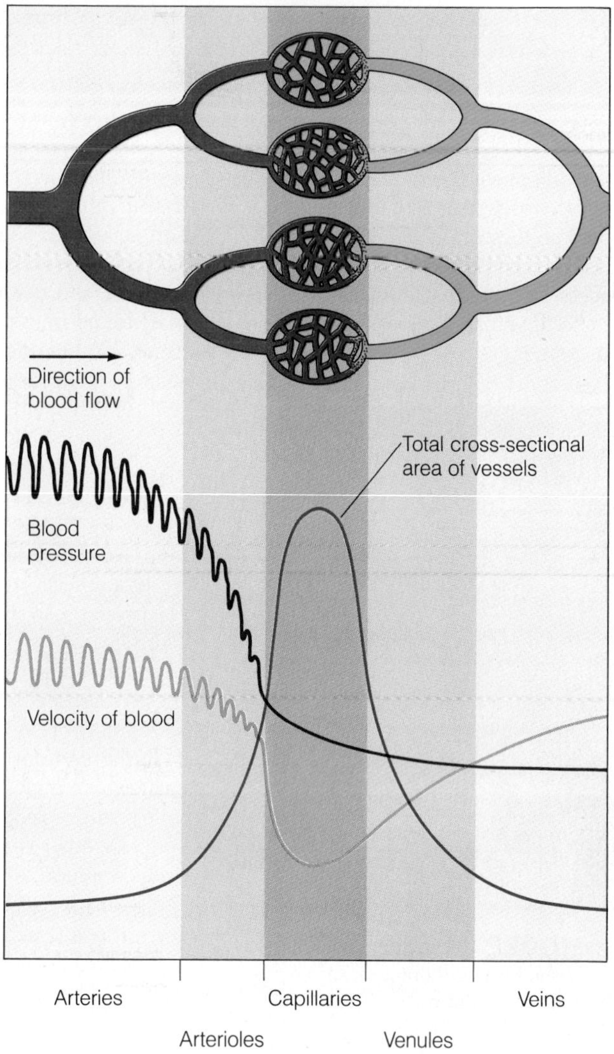

▻ **FIGURE 12–20 Blood Pressure in the Circulatory System** Blood pressure declines in the circulatory system as the vessels branch. Arterial pressure pulses because of the heartbeat, but pulsation is lost by the time the blood reaches the capillary networks, creating an even flow through body tissues. Blood pressure continues to decline in the venous side of the circulatory system.

the lymphatic capillaries. Once inside the fluid is called **lymph.**

Lymph drains from the capillaries into larger ducts, the collecting vessels. These vessels, in turn, merge with others, creating larger and larger ducts, eventually forming the **thoracic duct** and the **right lymphatic duct,** which empty into the large veins at the base of the neck (Figure 12–22).

Lymph moves through the vessels of the lymphatic system in much the same way that blood is transported in veins. In the upper parts of the body, lymph flows by gravity. In regions below the heart, lymph is propelled largely by muscle contraction. Breathing and walking, for example, pump the lymph out of the extremities. Lymphatic flow is also assisted by valves akin to those found in the veins.

The lymphatic system also consists of several lymphatic organs: the lymph nodes, the spleen, the thymus, and the tonsils. The lymphatic organs function primarily in immune protection and are discussed in more detail in Chapter 15.

In 1966, Dr. Leland C. Clark of the University of Cincinnati's College of Medicine immersed a live laboratory mouse in a clear fluorocarbon solution saturated with oxygen. To the amazement of the audience, the mouse did just fine, "breathing in" the oxygen-rich liquid as if it were air. Some time later, Clark extracted the mouse from the solution, and after a moment, the rodent began to move about, apparently unharmed by the ordeal.

The point of this demonstration was not to show that an animal could "breathe" this fluid into its lungs and survive but, rather, to show that the solution held enormous amounts of oxygen, so much so that it could possibly be used as a substitute for blood. Clark and his colleagues hoped that their "artificial blood" would someday be a boon to medical science, helping paramedics keep accident victims who have lost substantial amounts of blood alive while they were being transported to the hospital.[1] In rural America, where a trip to the hospital takes considerable time, artificial blood could save thousands of lives a year.

BLOOD: ITS COMPOSITION AND FUNCTIONS

Animal blood is a far cry from Clark's artificial substitute. The **blood** in the circulatory system of vertebrates, including humans, for example, is a water-based fluid, not a fluorocarbon, and consists of two basic components: plasma and formed elements. **Plasma** is the liquid portion of the blood and is about 90% water. It contains many dissolved substances. Three types of formed elements are suspended in the plasma: (1) white blood cells (or leukocytes), (2) red blood cells (or erythrocytes), and (3) platelets (or thrombocytes).

In a man weighing 70 kilograms (150 pounds), the cardiovascular system contains 5 to 6 liters of blood, or 1.3 to 1.5 gallons. On average, women have about a liter less. Blood accounts for about 8% of our total body weight.

▻ Figure 13–1 shows that the blood plasma makes up about 55% of the blood volume and formed elements about 45%. Technically speaking, the volume of the blood occupied by blood cells is referred to as the **hematocrit.** To determine the hematocrit, blood is withdrawn from a patient and placed in a test tube, then inserted into a centrifuge, a device that spins the samples at high speeds (Figure 13–1). Centrifugation causes the formed elements to separate from the plasma and "settle" to the bottom of the test tube. The white blood cells and platelets settle out on top of the RBCs. They make up slightly less than 1% of the total blood volume. Therefore, the hematocrit is largely determined by the RBCs.

The hematocrit varies in individuals inhabiting different altitudes. In people living in the thin air of the Mile High City of Denver, for example, hematocrits are typically about 5% higher than in people living at sea level. The slight increase in the hematocrit (RBC concentration) in Denver residents compensates for the slightly lower level of

[1] "Artificial blood," of course, is a misnomer. The fluorocarbon solution takes over the role of the plasma and the red blood cells, which transport oxygen and carbon dioxide, but not the role of the white blood cells, which play a role in immune protection.

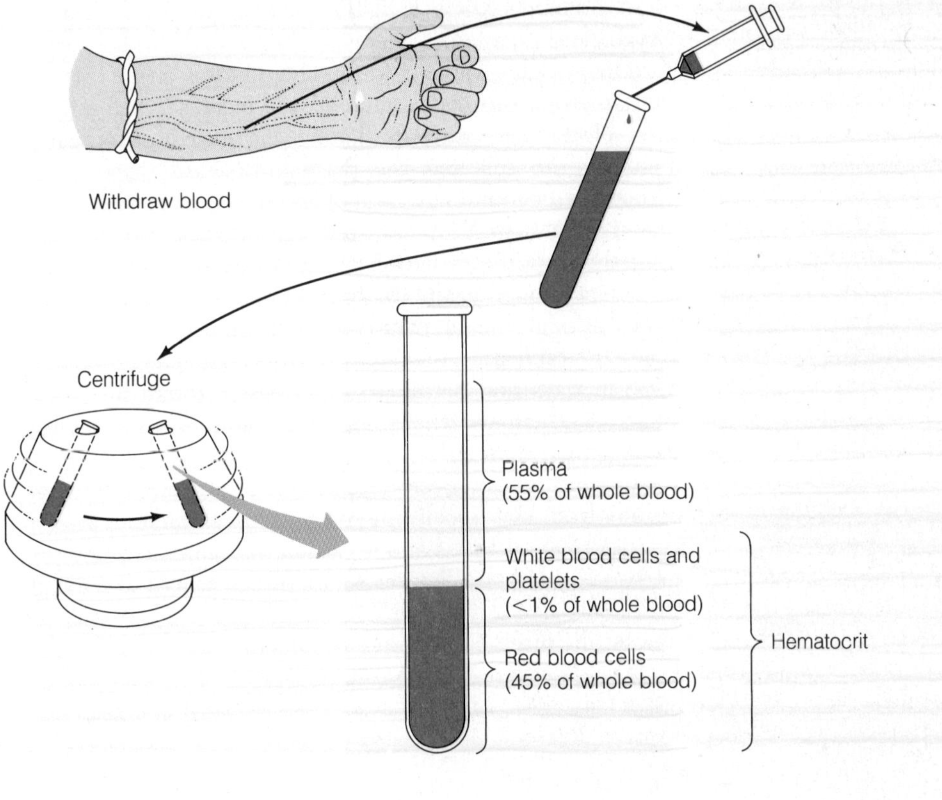

▻ **FIGURE 13–1 Blood Composition** Blood removed from a person can be centrifuged to separate plasma from the cellular component. Red blood cells constitute about 45% of the blood volume, except at higher altitudes where they make up about 50% of the volume to compensate for the lower oxygen levels.

oxygen in the atmosphere at the higher altitude and is the result of a homeostatic mechanism that helps maintain normal oxygen delivery to body cells. The effect of elevation on hematocrit is one reason why the U.S. summer Olympic team has its headquarters in Colorado Springs. Officials think that the higher hematocrit may help many athletes outperform their lowland competitors.

The Blood Plasma Is a Watery Transport Medium

The plasma is a light yellow (straw-colored) fluid. Dissolved in the plasma are: (1) gases, such as nitrogen, carbon dioxide, and oxygen; (2) ions, such as sodium, chloride, and calcium; (3) nutrients, such as glucose and amino acids; (4) hormones; (5) proteins; and (6) various wastes. Lipid molecules are also found in the plasma, but they are either suspended in tiny globules or bound to certain plasma proteins, which transport them through the bloodstream.

Plasma proteins are the most abundant of all dissolved substances in the blood plasma. Three major types of protein are found in the blood plasma: (1) albumins, (2) globulins, and (3) fibrinogen (Table 13–1).

All of these proteins contribute to osmotic pressure (described briefly in Chapter 3 in the section on membrane transport), which is essential to capillary exchange and, therefore, the proper distribution of wastes and nutrients, a function vital to homeostasis.[2] Blood proteins also augment the effects of bicarbonate, a buffer described in Chapter 2, and therefore help regulate the pH of the blood. They do it by binding to hydrogen ions, preventing the hydrogen ion concentration in the blood from rising.

In addition to these common functions, many plasma proteins serve very specific functions. Albumins and two types of globulins, for example, bind to hormones, ions, and fatty acids, helping transport these molecules through the bloodstream. These carrier proteins are large, water-soluble molecules. Their binding to much smaller lipid molecules renders the latter water-soluble, and facilitates their transport through the largely aqueous bloodstream. Carrier proteins also protect smaller molecules from destruction by the liver.

Another group of globulins, known as the gamma globulins, are **antibodies,** proteins that "neutralize" viruses and bacteria or target them for destruction by phagocytic cells in body tissues known as macrophages, a topic discussed in more detail in Chapter 15.

Still another important blood protein is **fibrinogen.** This unique protein is converted into **fibrin,** which forms blood clots when blood vessels are injured. Blood clots prevent blood loss and thus also help maintain homeostasis. (More on blood clotting later in the chapter.)

[2] Osmotic pressure is the force responsible for the movement of water across a selectively permeable membrane and is created by the difference in solute concentrations on either side of the membrane. The greater the concentration difference, the greater the osmotic pressure.

TABLE 13–1 Summary of Plasma Proteins

PROTEIN	FUNCTION
Albumins	Maintain osmotic pressure and transport smaller molecules, such as hormones and ions
Globulins	Alpha and beta globulins transport hormones and fat-soluble vitamins; gamma globulins (antibodies) bind to foreign substances.
Fibrinogen	Converted into fibrin network that help form blood clots

Red Blood Cells Are Flexible, Highly Specialized Cells that Transport Oxygen and Small Amounts of Carbon Dioxide in the Blood

The **red blood cell (RBC),** or **erythrocyte,** is the most abundant cell in the blood of vertebrates. In fact, in humans, 20 drops of blood equal 1 milliliter and contain approximately 5 billion RBCs.[3] If RBCs were people, a single milliliter of blood would contain the entire world population.

RBCs are highly specialized "cells." In humans and other mammals, in fact, RBCs lose their nuclei and organelles during cellular differentiation (▷ Figure 13–2a). RBCs found in all other vertebrates, however, retain their nuclei (Figure 13–2b).

In terrestrial (land-dwelling) vertebrates, the bone marrow continuously produces new RBCs to replace the billions that die each day. In fish, RBCs are made in the kidney.

In humans, RBCs are biconcave disks that transport oxygen and, to a lesser degree, carbon dioxide in the blood (Figure 13–2c; Table 13–2). The unique shape of the human RBC, shown in Figure 13–2, increases the surface area of the cell, thus facilitating the exchange of gases between the cell and the plasma. Like so many other structures encountered in biology, the human RBC is a remarkable example of the marriage of form and function forged during the evolution of life on Earth.

Swept along in the bloodstream, RBCs travel many times through the circulatory system each day. When they reach the capillaries, the highly flexible cells often bend and twist in ways that permit them to pass through the many miles of channels whose internal diameters are slightly smaller than theirs.

Roughly one of every 500 to 1000 African Americans suffers from **sickle-cell anemia.** Sickle-cell anemia affects the flexibility of RBCs and is caused by a genetic mutation, as you learned in Chapter 8. In places where malaria is prevalent, like Africa, this mutation provides protection against the disease. However, protection exacts a high cost.

[3] Many texts note the concentration per cubic millimeter, which is about 5 million RBCs. There are, of course, 1000 cubic millimeters in 1 cubic centimeter, or 1 milliliter.

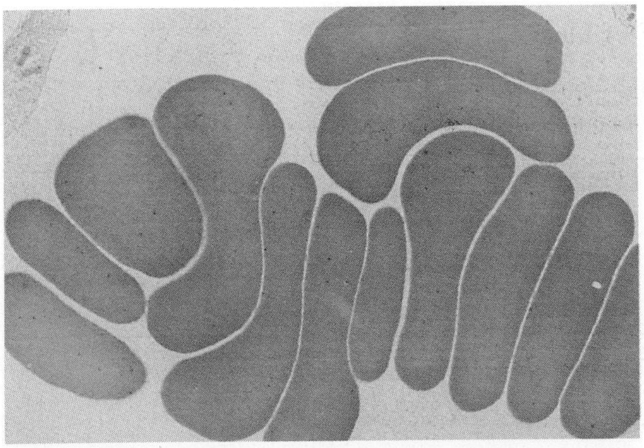

(a) Transmission electron micrograph of human RBC

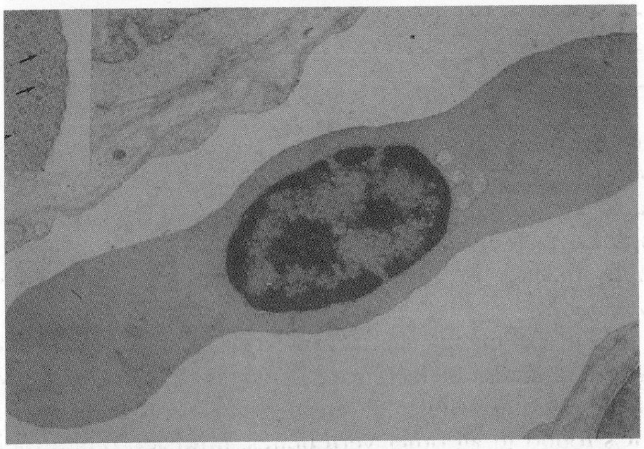

(b) Nucleated RBC from a bird

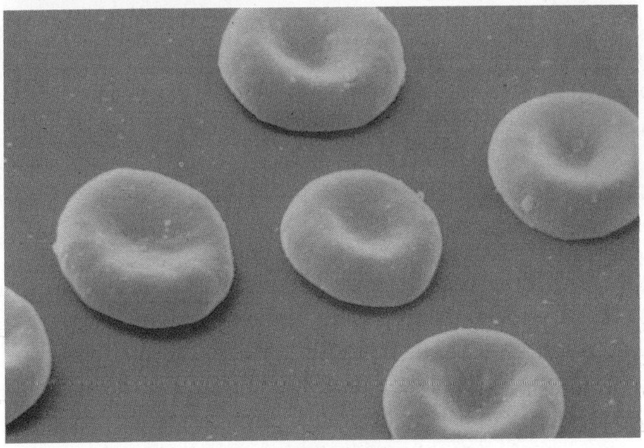

(c) Scanning electron micrograph of human RBC

▻ **FIGURE 13–2 Red Blood Cells**

Why? The RBC hemoglobin of those who carry the gene for sickle-cell anemia contains one incorrect amino acid. This defect alters the tertiary structure of the hemoglobin molecule, causing RBCs to transform from biconcave disks into sickle-shaped cells when they encounter low levels of oxygen in capillaries (▻ Figure 13–3). The sickle-shaped cells are considerably less flexible and unable to bend and twist. As a result, the RBCs collect at branching points in capillary beds much like logs in a logjam. Here, they block capillary blood flow, disrupting the influx of nutrients and oxygen to tissues. This reduces oxygen levels in body tissues and results in a condition known as **anoxia.** Anoxia causes considerable pain and can kill body cells. Blockages in the lungs, heart, and brain can even be life-threatening and often lead to heart attacks and brain damage. Many people who have sickle-cell anemia die in their late 20s and 30s; some die even earlier.

On average, RBCs live about 120 days before the liver and spleen remove them from circulation. The iron contained in the hemoglobin, however, is recycled by these organs—as many nutrients are—and is used to produce new RBCs in the red bone marrow. The recycling of iron is not 100% efficient, however, so small amounts of iron must be ingested each day in the diet. Loss of blood from an injury or, in women, during menstruation increases the body's demand for dietary iron. Without adequate intake, oxygen transport may become impaired.

In the red bone marrow, RBCs are produced by **stem cells,** undifferentiated cells that trace back to embryonic development. These cells give rise to 2 million RBCs per second! (For a discussion of some exciting efforts to manipulate stem cells to fight cancer and other diseases, see Health Note 13–1.)

In infants and children, almost all of the bone marrow is dedicated to the production of RBCs. As growth slows, though, the red marrow of many bones becomes inactive and gradually fills with fat cells, becoming **yellow marrow,** a fat storage depot. By the time an individual reaches adulthood, only a few bones, such as the hip bones, sternum (breastbone), ribs, and vertebrae are engaged in RBC production. In severe, prolonged anemia, however, yellow marrow can be converted back into red marrow to produce RBCs.

The number of RBCs in the blood remains more or less constant over long periods. Maintaining a constant concentration of RBCs is essential to homeostasis and is controlled by the hormone erythropoietin. The kidney secretes **erythropoietin** when blood oxygen levels decline—for example, when a person moves to high altitudes or loses a significant amount of blood. Erythropoietin is transported throughout the bloodstream. In the red bone marrow, it stimulates the stem cells to multiply, increasing RBC production. As the RBC concentration increases, oxygen supplies increase. When oxygen levels return to normal, erythropoietin levels fall, reducing the rate of RBC formation in a classical negative feedback mechanism so common among homeostatic mechanisms.

Hemoglobin Is an Oxygen-Transporting Protein Found in RBCs. **Hemoglobin** is a large protein molecule composed of four protein subunits. Found exclusively in the RBCs, it accounts for about a third of the RBC's weight. As shown in ▻ Figure 13–4, each hemoglobin subunit contains a **heme group,** which consists of a large, organic ring structure, called a **porphyrin ring.** In the center of the ring is an iron ion.[4]

[4] The iron is Fe^{++}, the ferrous ion. To be effective, iron supplements should contain ferrous iron.

TABLE 13–2 Summary of Blood Cells

NAME	LIGHT MICROGRAPH	DESCRIPTION	CONCENTRATION (NUMBER CELLS/mm^3)	LIFE SPAN	FUNCTION
Red blood cells (RBCs)		Biconcave disk; no nucleus	4 to 6 million	120 days	Transports oxygen and carbon dioxide
White blood cells					
Neutrophil	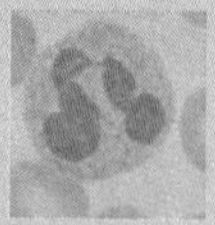	Approximately twice the size of RBCs; multilobed nucleus; clear-staining cytoplasm	3000 to 7000	6 hours to a few days	Phagocytizes bacteria
Eosinophil		Approximately same size as neutrophil; large pink-staining granules; bilobed nucleus	100 to 400	8 to 12 days	Phagocytizes antigen-antibody complex; attacks parasites
Basophil		Slightly smaller than neutrophil; contains large, purple cytoplasmic granules; bilobed nucleus	20 to 50	Few hours to a few days	Releases histamine during inflammation
Monocyte	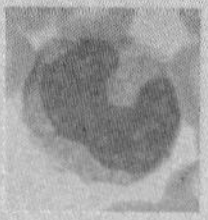	Larger than neutrophil; cytoplasm grayish-blue; no cytoplasmic granules; U- or kidney-shaped nucleus	100 to 700	Lasts many months	Phagocytizes bacteria, dead cells, and cellular debris
Lymphocyte	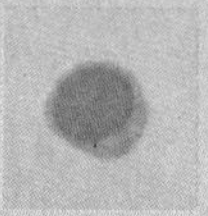	Slightly smaller than neutrophil; large, relatively round nucleus that fills the cell	1500 to 3000	Can persist many years	Involved in immune protection, either attacking cells directly or producing antibodies
Platelets		Fragments of megakaryocytes; appear as small darkstaining granules	250,000	5 to 10 days	Play several key roles in blood clotting

When blood from the pulmonary arteries flows through the capillary beds of the lungs, oxygen diffuses into the blood, then into the RBCs. Inside the RBC, oxygen binds to the iron in hemoglobin molecules for transport through the circulatory system. The importance of this process is underscored by the fact that 98% of the oxygen in the blood is actually transported bound to iron in hemoglobin. The remaining 2% is dissolved in the blood plasma.

Carbon dioxide also binds to hemoglobin, but to a much lesser degree. Most carbon dioxide molecules react with water in the RBC to form bicarbonate ions (HCO_3^-). This reaction is catalyzed by carbonic anhydrase inside the RBCs. Most of the bicarbonate ions then diffuse out of the RBCs and are transported in the plasma.

Disease of RBCs Can Alter Homeostasis. Homeostasis requires the normal operation of the heart and blood vessels. It also requires that the blood absorb a sufficient amount of oxygen as it passes through the lungs. Unfortunately, the oxygen-carrying capacity of the blood may decrease for one reason or another. A reduction in the oxygen-carrying capacity of the blood is generally known as **anemia** and may result from (1) a decrease in the number of circulating RBCs, (2) a reduction in the hemoglobin content of the RBCs, or (3) the presence of abnormal hemoglobin in RBCs. The causes of these conditions are many. Consider just a few. First, the number of RBCs in the blood may decline because of excessive bleeding or because of the presence of a tumor in the red bone marrow that reduces RBC production. Several infectious diseases (such as malaria) also decrease the RBC concentration in the blood. A reduction in the amount of hemoglobin in RBCs may be caused by iron deficiency or a deficiency in vitamin B-12, protein, and copper in the diet. Abnormal hemoglobin is produced in sickle-cell anemia and other genetic disorders.

Anemia generally results in weakness and fatigue. Individuals are often pale and tend to faint or become short of breath easily. People suffering from anemia often have an increased heart rate, because the heart beats faster to com-

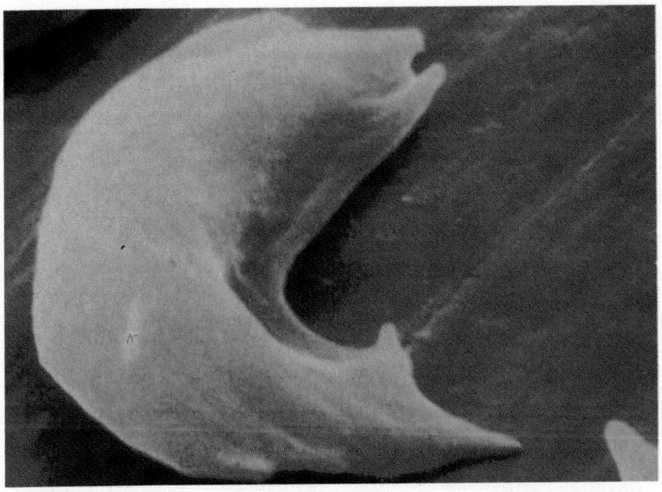

▻ **FIGURE 13–3 Sickle-Cell Anemia** Scanning electron micrograph of a sickle cell.

pensate for a reduction in the oxygen-carrying capacity of the blood.

As a rule, anemia is not a life-threatening condition. However, it does weaken one's resistance to other diseases or injuries and also limits a person's productivity and energy level. Therefore, no matter what the cause, anemia should be treated quickly.

White Blood Cells Are a Diverse Group That Helps Protect the Body from Infections

White blood cells (WBCs), or **leukocytes,** are nucleated cells that are part of the body's protective mechanism, which combats harmful microorganisms, such as bacteria and viruses. White blood cells are produced in the bone marrow and circulate in the bloodstream, but as noted earlier, they constitute less than 1% of the blood volume. White cells do most of their work outside of the bloodstream, in the tissues. The bloodstream, therefore, serves as a transport vehicle, delivering WBCs to sites of infection. When WBCs arrive at the scene, they escape through the walls of the capillaries by squeezing between the endothelial cells. This process is known as **diapedesis** (from the Greek *dia,* meaning "through" and *pedesis,* meaning "leaping") (▻ Figure 13–5).

Table 13–2 lists and describes the five types of WBCs found in the blood. The three most numerous, which are discussed here, are neutrophils, monocytes, and lymphocytes.

Neutrophils are the most abundant WBC. Approximately twice the size of the RBC, these cells are distinguished by their multilobed nuclei. So named because their cytoplasm has a low affinity for stains, neutrophils circulate in the blood like a cellular police force awaiting microbial invasion. Attracted by chemicals released from infected tissue, neutrophils escape from the bloodstream, then migrate to the site of infection by amoeboid movement.

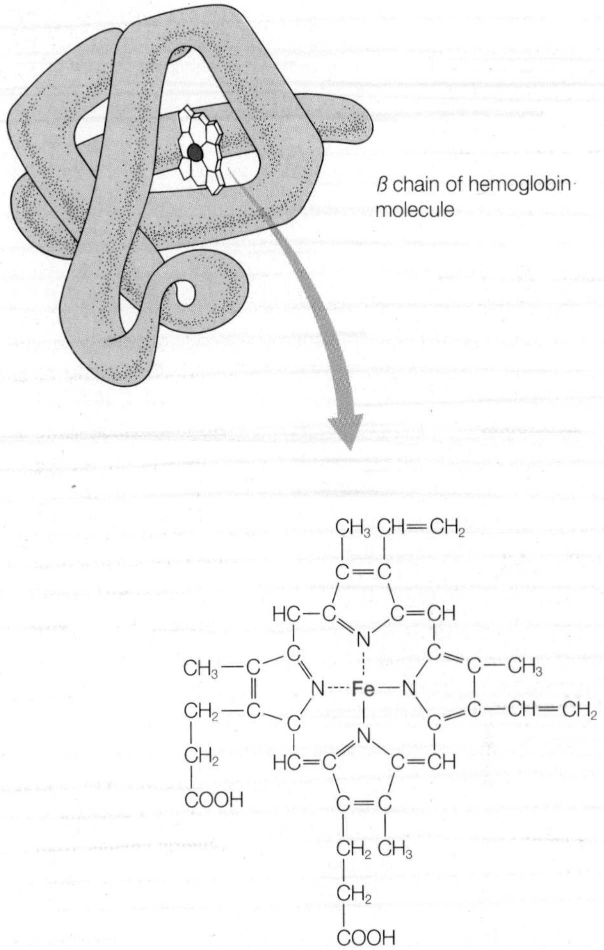

▻ **FIGURE 13–4 Porphyrin Ring** The porphyrin ring of the hemoglobin molecule contains an iron ion that binds to oxygen and carbon monoxide.

Neutrophils are usually the first WBCs to arrive on the scene. When they arrive, they immediately begin to phagocytize (engulf) microorganisms, helping prevent the spread of bacteria and other organisms from the site of invasion.

▻ **FIGURE 13–5 Diapedesis** White blood cells (leukocytes) escape from capillaries by squeezing between endothelial cells.

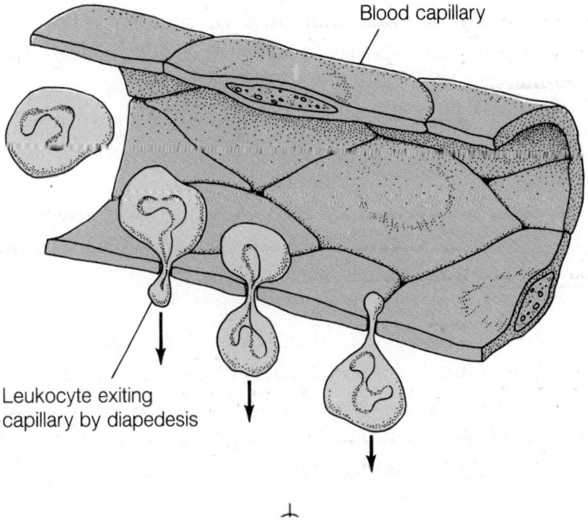

HEALTH NOTE 13-1

MANIPULATING THE STEM CELL

A baby is born into the world. After the nurse-midwife cuts the umbilical cord, which supplies oxygen and nutrient-rich blood to the fetus, technicians whisk the cord off to the lab, where they carefully extract the remaining blood. From the blood sample they extract stem cells, bone marrow cells that routinely escape into the blood (▻ Figure 1). These cells are frozen and placed in a storage bank for future use. If in the ensuing years the donor develops cancer or any of a handful of diseases affecting the blood cells or immune system, medical workers will thaw the stem cells they have put into cold storage and place them in a tissue culture, where they will be treated in one of a handful of ways to produce normal, healthy body cells that can be transfused into the patient.

Although this scenario is science fiction, if researchers at a number of universities throughout the world succeed, it may become commonplace in the coming years.

Stem cells in bone marrow transplants are currently used to restore the blood-producing cells and immune systems of people exposed to high levels of radiation (like those exposed at the Chernobyl nuclear reactor in the former Soviet Union). They are also used to restore these systems in patients treated with chemotherapy, a systemic drug treatment for cancer that kills cancer cells but also wipes out all actively dividing-cells, among them the cells of the bone marrow. Bone marrow transplants are also performed in patients subjected to radiation treatment for cancer and individuals with leukemia, a type of cancer that results from the proliferation of lymphocytes in the bone marrow, which, in turn, crowds out other cells, greatly reducing the production of red blood cells, platelets, and other cells vital to homeostasis.

The stem cell's desirability for medical treatment results from its ability to form a whole host of useful cells. As noted in this chapter, the bone marrow stem cell gives rise to five different white blood cells, red blood cells, and platelets. If medical researchers could find a way to manipulate the cell to replace diseased cells, they could cure an array of diseases.

One example is sickle-cell anemia, the genetic disorder described in the chapter. If immunologically compatible stem cells (that is, cells that would not be rejected by the recipient's immune system) could be transplanted into a person with sickle-cell anemia, researchers say, they could repopulate the bone marrow, replacing the person's own stem cells, which give rise to sickle cells, with ones that give rise to normal RBCs.

Chronic enzyme defects in white blood cells, which are genetic in origin, could also be treated, as could osteopetrosis, a bone disorder resulting in overly dense bone. Osteopetrosis is caused by a metabolic abnormality in macrophages. (You will recall that macrophages are derived from monocytes, which are derived from bone marrow stem cells.)

Some researchers believe that stem cells could be used in the treatment of cancers such as leukemia. First, bone marrow would be removed from a patient about to undergo radiation or chemotherapy. Normal stem cells would be isolated from the malignant ones. (This procedure has not been perfected.) After treatment, presumably when all malignant cells in the body had been destroyed, normal stem cells would be reinjected into the patient. Setting up residence in the marrow, the cells would begin to proliferate and would eventually restore blood formation and immune functions. This procedure would use the person's own cells and, if successful, bypass immune reactions that result when bone marrow is transplanted from another person.

Researchers see all kinds of intriguing possibilities for stem cell manipulation. With further research, they hope that stem cells could someday be intentionally altered in the laboratory in ways that cause them to produce lymphocytes specifically targeted to fight cancer or viral infections, such as AIDS. Although a great deal of basic research remains to be done, many scientists believe that they are on the road to a radically new form of treatment for many heretofore untreatable diseases.

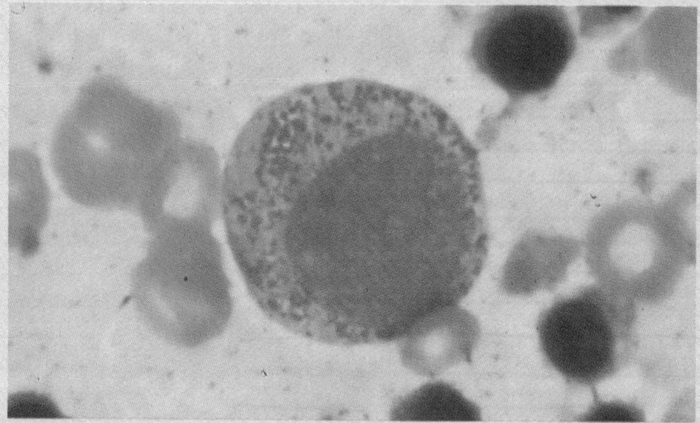

▻ **FIGURE 1 Stem Cell from Human Bone Marrow**

When a neutrophil's lysosomes are used up, the cell dies and becomes part of the yellowish liquid, or **pus,** that exudes from wounds. Pus is a mixture of dead neutrophils, cellular debris, and bacteria, both living and dead.

Monocytes are also phagocytic cells. Slightly larger than neutrophils, monocytes contain distinctive U-shaped or kidney-shaped nuclei. Like neutrophils, monocytes leave the bloodstream to do their "work." They migrate through body tissues via amoeboid motion. Once on the scene, they begin phagocytizing microorganisms, dead cells, and dead neutrophils. Thus, neutrophils are the "first-line" troops, and monocytes are something of a mop-up crew.

Monocytes also take up residence in connective tissues of the body, where they are referred to as **macrophages.** These cells remain more or less stationary, like watchful soldiers ever ready to attack and phagocytize invaders.

The second most abundant WBC is the **lymphocyte.** Most lymphocytes exist outside the circulatory system in **lymphoid organs,** such as the spleen, thymus, and lymph nodes, and **lymphoid tissue,** aggregations of lymphocytes beneath the lining of the intestinal and respiratory tracts. It is in these locations that they attack microbial intruders (▻ Figure 13–6).

Two types of lymphocytes are found in the body, both of which play a vital role in immune protection (Chapter 15). The first type, the **T lymphocyte,** or **T cell,** attacks foreign cells, such as fungi, parasites, and tumor cells.[5] T lymphocytes are thus said to provide "cellular immunity." The second type is called the **B lymphocyte,** or **B cell.** When activated, it actually transforms into another kind of cell, known as the **plasma cell.** Plasma cells, in turn, synthesize and release antibodies. **Antibodies** are proteins that circulate in the blood and bind to foreign substances, neutralizing them or just "marking" microorganisms and tumor cells for destruction by macrophages. (More on this in Chapter 15.)

The WBCs, like the other formed elements of blood, are involved in homeostasis. Their numbers can increase or decrease greatly during a microbial infection and other diseases. In fighting off a disease, they help return the body to normal function once again.

An increase in the number of WBCs, called **leukocytosis,** is a normal homeostatic response to intruders. It ends when the microbial invaders have been destroyed. Increases and decreases in various types of blood cells can be used to diagnose many medical disorders. For example, an increase in eosinophils, not discussed here, is sometimes an indication of an allergy. A dramatic increase in lymphocytes and lower abdominal pain are usually signs of appendicitis (discussed in Chapter 11). Because variations in the WBC count accompany many diseases, a blood test is a standard procedure for patients undergoing diagnostic testing.

[5]T lymphocytes usually attack large eukaryotic cells, such as fungi and parasites. Most bacteria are controlled by antibodies. Only the few bacteria that are intracellular parasites, such as *M. tuberculosis,* are attacked by T cells, but even then, the lymphocyte attacks the host cell, not the bacterium directly.

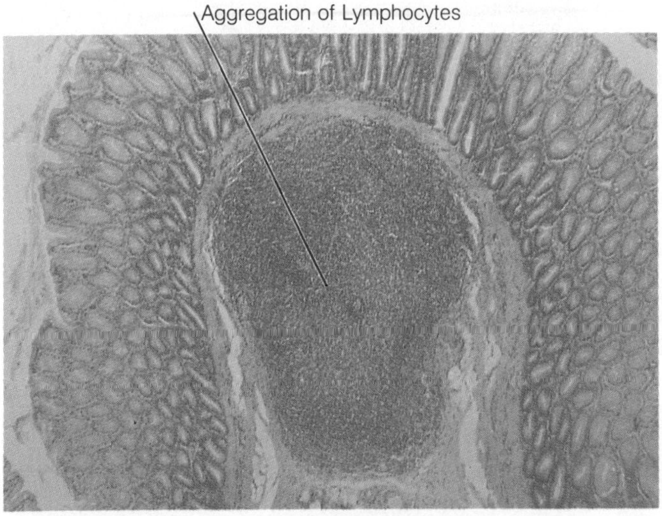

▻ **FIGURE 13–6 Lymphoid Tissue** The loose connective tissue beneath the lining of the large intestine and other sites is often packed with lymphocytes that have proliferated in response to invading bacteria.

Diseases Involving White Blood Cells Also Affect the Body's Internal Balance. By protecting the body, WBCs are part of one of the most important homeostatic mechanisms. WBCs, however, can run amok. For example, some white blood cells can become cancerous, dividing uncontrollably in the bone marrow, then entering the bloodstream. A cancer of WBCs is called **leukemia** (literally, "white blood"). The most serious type of leukemia is **acute leukemia,** so named because it kills victims quickly. Children are the primary victims of this disease.

In acute leukemia, WBCs fill the bone marrow and crowd out the cells that produce RBCs and platelets. A decline in the production of RBCs leads to anemia. A reduction in platelet production reduces clotting, increasing internal bleeding. Making matters worse, the cancerous WBCs produced in leukemia are often incapable of fighting infection. Because of this defect and because the production of platelets is reduced, victims of acute leukemia typically succumb to infections and internal bleeding.

Leukemia can be treated by irradiating the bone marrow and by administering a drug (vincristine) that stops mitosis. Twenty years ago, only one of every four children with leukemia survived. Today, thanks to vincristine, the odds have changed dramatically: three of every four children

▻ **FIGURE 13–7 Rosy Periwinkle** Many tropical plants contain chemicals that provide extraordinary medical benefits. One substance from the rosy periwinkle has helped physicians treat leukemia. Unfortunately, the tropical rain forests are being cut at an alarming rate, reducing our chances of finding other cures.

with the disease survive! Interestingly, vincristine was derived from a plant known as the rosy periwinkle, which is found in tropical rain forests (▻ Figure 13–7). Thousands of other drugs have a similar origin, underscoring the importance of protecting the rain forests from decimation (▻ Figure 13–8 a and b).

Another common disorder of the WBCs is **infectious mononucleosis,** commonly called "mono" or "kissing disease." Mono is caused by a virus transmitted through saliva and may be spread by kissing; by sharing silverware, plates, and drinking glasses; and possibly even through drinking fountains. The virus spreads through the body and affects many organs.

Although the virus infects only lymphocytes in the bloodstream, the number of monocytes and lymphocytes in the blood increases rapidly during an infection. Individuals suffering from mono complain of fatigue, aches, sore throats, and low-grade fever. Physicians recommend that victims get plenty of rest and drink lots of liquids while the immune system eliminates the virus. Within a few weeks, symptoms generally disappear, although weakness may persist for two more weeks.

Platelets Are Fragments of Cells Found in the Bone Marrow and Are a Vital Component of the Blood-Clotting Mechanism

The circulatory system is a delicate structure. Even minor bumps and scrapes can cause it to leak. Leakage is normally prevented by **blood clotting,** surely one of the most intricate homeostatic systems to have evolved in the living world. A simplified version is discussed here.

Among the most important agents of blood clotting are the platelets, tiny formed elements produced in the bone marrow by fragmentation a huge cell known as the **megakaryocyte** (▻ Figure 13–9). Platelets lack nuclei and organelles and therefore are not true cells. Like RBCs, they cannot divide. Platelets are carried passively in the bloodstream. Coated by a layer of a sticky material, platelets tend to adhere to irregular surfaces, such as tears in blood vessels or cholesterol-containing plaque.

The process of blood clotting and the role of platelets are summarized in ▻ Figure 13–10. As illustrated, the process begins when cells in the damaged tissue release a substance into the bloodstream called **thromboplastin.** Thromboplastin is a lipoprotein that converts an inactive plasma enzyme, **prothrombin** (produced by the liver) into its active form, **thrombin.** Thrombin, in turn, acts on another blood protein, **fibrinogen,** also produced by the liver. When activated, fibrinogen is converted into **fibrin,** creating long, branching fibers that produce a weblike network in the wall of the damaged blood vessel (◂ Figure 13–11). The fibrin web traps RBCs and platelets, forming a plug that cuts off the flow of blood to the tissue. Platelets captured by the fibrin web release additional thromboplastin, known as platelet thromboplastin, which causes more

▻ **FIGURE 13–8 Forest Destruction** (*a*) Rampant deforestation leaves the tropics denuded but also robs humankind of many potential cures for disease. (*b*) The loss of forests and other habitats in the industrial countries is also of grave importance. The bark of this tree, the Pacific yew, contains a chemical that may prove effective in fighting breast cancer. However, each year thousands of acres of old-growth forest, where the tree lives, are being cut down.

(a)

(b)

▻ **FIGURE 13–9 Megakaryocyte** A light micrograph of a megakaryocyte, a large, multinucleated cell found in bone marrow, which fragments, giving rise to platelets.

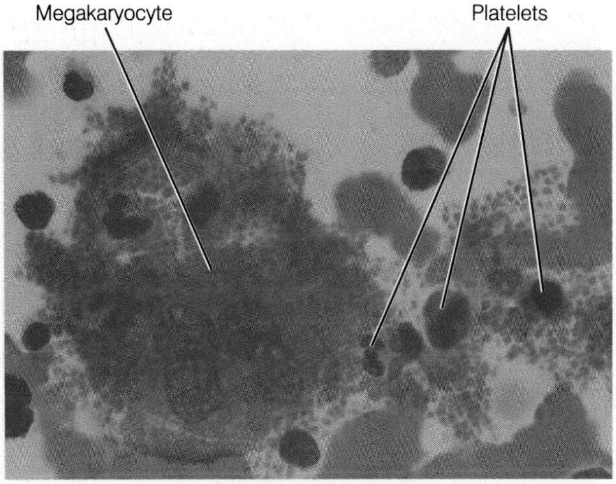

fibrin to be laid down and therefore serves to reinforce the fibrin network.

Blood clotting occurs fairly quickly. In most cases, a damaged blood vessel is sealed by a clot within 3 to 6 minutes of an injury; 30 to 60 minutes later, platelets in the clot begin to draw the clot inward, stitching the wound together. How? Platelets contain contractile proteins like those in muscle cells. Contraction of the protein fibers draws the fibrin network inward, pulling the edges of the

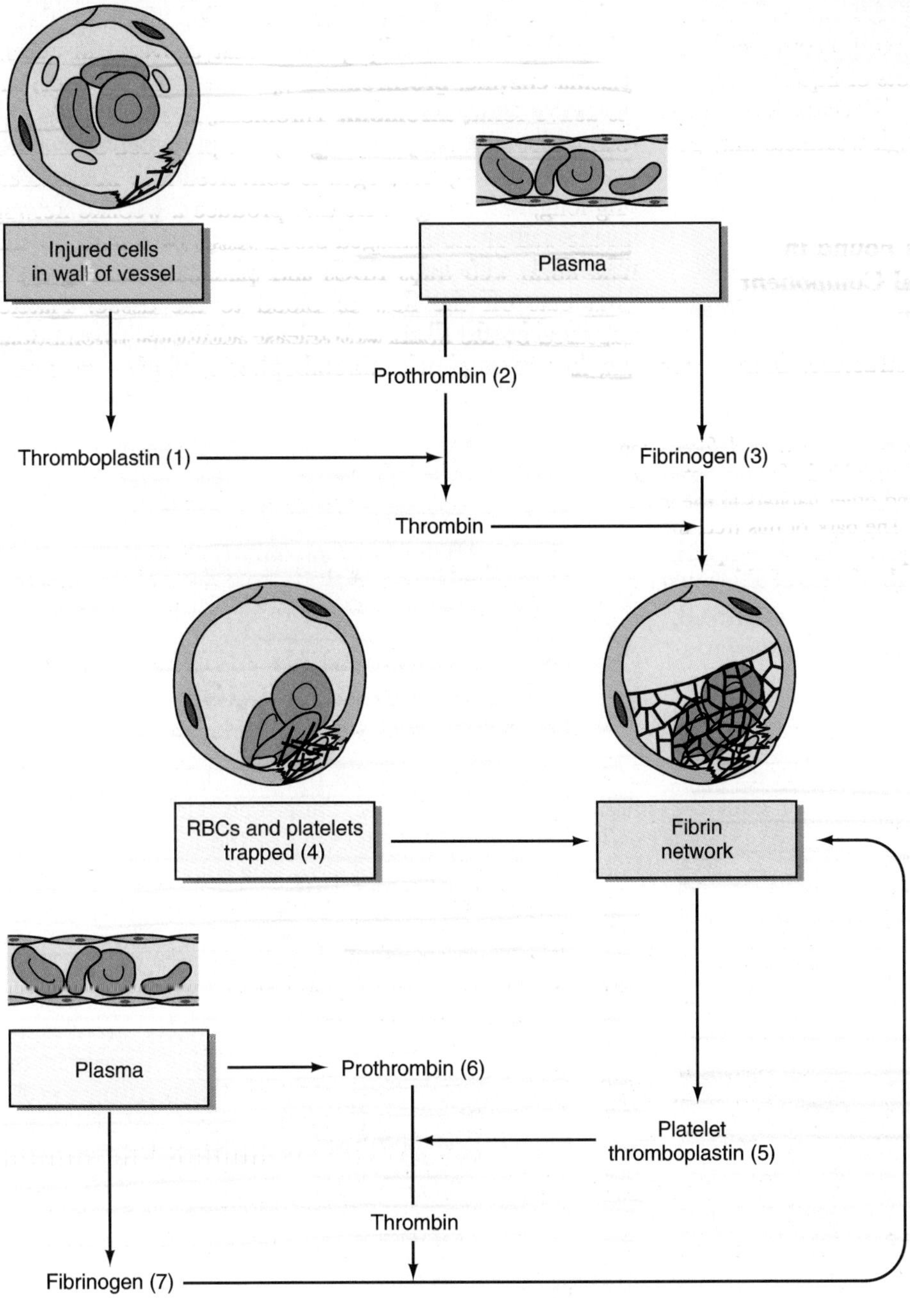

▻ **FIGURE 13–10 Blood Clotting Simplified** Injured cells in the wall of blood vessels release the chemical thromboplastin (1). Thromboplastin stimulates the conversion of prothrombin, found in the plasma, into thrombin (2). Thrombin, in turn, stimulates the conversion of the plasma protein fibrinogen into fibrin (3). The fibrin network captures RBCs and platelets (4). Platelets in the blood clot release platelet thromboplastin (5), which converts additional plasma prothrombin into thrombin (6). Thrombin, in turn, stimulates the production of additional fibrin (7).

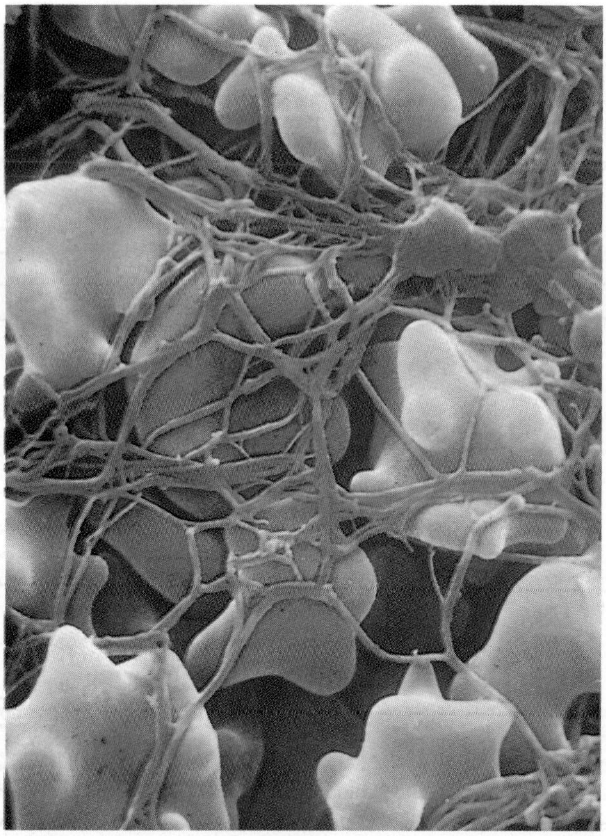

▻ **FIGURE 13–11 Blood Clot** A scanning electron micrograph of a fibrin clot that has already trapped platelets and RBCs, helping plug up a leak in a vessel. RBCs are yellow. Fibrin network is red.

cut or damaged blood vessel together. This closes the wound like a surgical stitch and accelerates healing.

Blood clots do not stay in place indefinitely. If they did, the circulatory system would eventually become clogged, and blood flow would come to a complete halt. Instead, clots are dissolved by a blood-borne enzyme known as **plasmin,** which is produced from an inactive form, **plasminogen.** Plasminogen is incorporated in the clot as it forms. It is then gradually converted to plasmin by an activating factor secreted by the endothelial cells of the blood vessel. This ensures that the plasmin dissolves the clot after the blood vessel damage has been repaired. TPA, a blood clot buster discussed briefly in the last chapter, is one such activator.

As important as blood clots are in protecting the body, normal clotting can also cause problems. As noted in previous chapters, blood clots can break loose from their site of formation, circulate in the blood, and become lodged in arteries narrowed by atherosclerosis, further restricting blood flow and causing considerable damage. Blood clots typically lodge in the narrow vessels serving the heart muscle, brain, and other vital organs.

Clotting Disorders Upset the Homeostatic Balance and Can Be Life Threatening. Tiny breaks occur in blood vessels with surprising regularity, but they are usually repaired without our ever knowing it. In some individuals, blood clotting is impaired because of: (1) an insufficient number of platelets; (2) liver damage, which hinders the production of clotting factors; or (3) genetic disorders that result in a lack of clotting factors.

A reduced platelet count may result from leukemia, as noted earlier, or from an exposure to excess radiation, which damages bone marrow where the megakaryocytes reside. Liver damage, which impairs the production of blood-clotting factors, can be caused by hepatitis, liver cancer, or excessive alcohol consumption. The most common genetic defect is called **hemophilia.** In people with the disease, the liver fails to produce necessary clotting factors. Problems begin early in life, and even tiny cuts or bruises can bleed uncontrollably, threatening the person's life. Because of repeated bleeding into the joints, victims suffer great pain and often become disabled; they often die at a young age.

Hemophiliacs can be treated by periodic transfusions of the missing blood-clotting factors. This therapy, however, is expensive and is required every few days. It has also put hemophiliacs at risk for **AIDS,** acquired immunodeficiency syndrome, which is caused by a virus that is transmitted primarily by sexual contact (Chapter 15). The AIDS virus invades certain white blood cells, known as helper T cells, resulting in a gradual deterioration of the immune function. The disease is considered 100% fatal; that is, everyone who gets it eventually dies, although some live much longer than others. Unfortunately, testing for the AIDS virus began late, and many transfusions of whole blood, blood plasma, and clotting factors were contaminated by the virus. Clotting factors produced by genetic engineering, however, have eliminated the need for transfusions of clotting agents taken from whole blood, reducing the risk to hemophiliacs of contracting AIDS. (For more on AIDS, see Chapter 15.)

ENVIRONMENT AND HEALTH: CARBON MONOXIDE

Modern society is powered by fossil fuels—coal, oil, and natural gas. When burned, these fuels release energy and a wide variety of pollutants. One of the most dangerous to human health is carbon monoxide, a colorless, odorless gas that results from incomplete combustion of organic fuels.[6]

Carbon monoxide (CO) emanates from stoves and furnaces in our homes and spews out of the tailpipes of our automobiles, causing increased levels along highways, in parking garages, and in tunnels. It also pours out of power plants and factory smokestacks, polluting the air in our cities. It is even a major pollutant in tobacco smoke. According to estimates by the Environmental Protection Agency, over 40 million Americans, or one of every six people in this country, are exposed to levels of CO in the outside air thought to be harmful to their health.

[6]When gasoline, coal, and other organic fuels burn completely, they produce carbon dioxide and water.

What makes CO so dangerous? Carbon monoxide, like oxygen, binds to hemoglobin. However, hemoglobin has a much greater affinity for CO than for oxygen (about 200 times greater). Consequently, CO "outcompetes" oxygen for the binding sites on the hemoglobin molecules and thus reduces the blood's ability to carry oxygen. For healthy people, CO does not create much of a problem, as long as levels are low. The body simply produces more RBCs or increases the heart rate to augment the flow of oxygen to tissues. CO does create a problem, however, when levels are so high that the body cannot compensate. At very high levels, CO becomes a deadly killer.

New research suggests that CO levels currently deemed acceptable by federal standards can, in some people, trigger chest pain (angina), which results from a lack of oxygen in the heart muscle. Levels such as those once thought safe in many workplaces or levels in the blood encountered after one hour in heavy traffic cause angina and abnormal ECGs in moderately active adults with coronary artery disease, according to a study published in the *New England Journal of Medicine.*

In levels commonly found in and around cities, CO is especially troublesome for the elderly or for individuals suffering from coronary artery and lung diseases. Carbon monoxide places additional strain on their heart, making the already weakened organ work harder. Thus, the elderly and infirm are advised to stay inside on high-pollution days.

Carbon monoxide is just one of many pollutants we breathe. No one knows what, if any, long-term effects it will have on our health. In the short term, however, it is clear that CO upsets oxygen delivery, which is so essential to homeostasis in the elderly and infirm. It has become another risk factor in an increasingly risky society, but it is not an insolvable problem. Much more efficient automobiles, widespread use of mass transit, overall reductions in driving, and alternative automobile fuels (such as hydrogen) can all reduce the level of CO in and around the cities where most Americans live and work (▻ Figure 13–12). More efficient factories, home furnaces, and power plants can also help reduce CO levels in our cities. Passive-solar homes and improvements in insulation can also cut our demand for fossil fuel energy (▻ Figure 13–13). Bans on smoking indoors could go a long way, too, in ridding our air of this transparent killer.

▻ **FIGURE 13–12 Air Saver** Commuter lines like this one in Portland, Oregon, significantly reduce air pollution, because light rail systems are far more efficient—per passenger mile traveled—than cars.

▻ **FIGURE 13–13 Passive-Solar Heating** The sun heats the author's home, located at 8000 feet above sea level in the Colorado Rockies. In the fall and winter, sunlight streams through south-facing windows, heating the interior of the house. Superinsulated walls and ceilings keep the heat from escaping. This house costs only about $100 per year to heat even though winter temperatures often remain well below zero for long stretches. The low use of heating fuel greatly reduces the output of pollution.

SUMMARY

BLOOD: ITS COMPOSITION AND FUNCTIONS

1. Blood is a watery tissue containing two basic parts: the plasma—which contains dissolved nutrients, proteins, gases, and wastes—and the formed elements—red blood cells, white blood cells, and platelets, which are suspended in the plasma.
2. The functions of the formed elements are summarized in Table 13–2.
3. The plasma constitutes about 55% of the volume of a person's blood, and the formed elements make up the remainder. The volume occupied by the blood cells and platelets is called the hematocrit.
4. Red blood cells (RBCs) are highly specialized cells that lack nuclei and organelles and are produced by the red bone marrow. The RBCs transport oxygen in the blood.
5. The concentration of RBCs in the blood is maintained by the hormone erythropoietin, produced by the kidneys in response to falling oxygen levels.
6. White blood cells (WBCs) are nucleated cells and are part of the body's protective mechanism to combat microorganisms.

WBCs are produced in the bone marrow and circulate in the bloodstream, but they do most of their work outside it, in the body tissues.

7. The most abundant WBCs are the neutrophils, which are attracted by chemicals released from infected tissue. Neutrophils leave the bloodstream and migrate to the site of infection by amoeboid movement.
8. Neutrophils are the first WBCs to arrive at an infection, where they phagocytize microorganisms, helping prevent the spread of bacteria and other organisms.
9. The second group of cells to arrive are the monocytes, which phagocytize microorganisms, dead cells, cellular debris, and dead neutrophils.
10. Lymphocytes are the second most numerous WBCs and play a vital role in immune protection.
11. Platelets are fragments of large bone marrow cells and are involved in blood clotting.
12. Platelets are coated by a layer of a sticky material, which causes them to adhere to irregular surfaces, such as tears in blood vessels. The process of blood clotting is summarized in Figure 13–10.

ENVIRONMENT AND HEALTH: CARBON MONOXIDE

13. Carbon monoxide is produced by the incomplete combustion of organic fuels in our homes, automobiles, factories, and power plants.
14. Carbon monoxide binds to hemoglobin and reduces the oxygen-carrying capacity of the blood.
15. At high concentrations, carbon monoxide can be lethal. At lower concentrations, it is harmful to people with cardiovascular disease and lung disease. For individuals with heart disease, it puts additional strain on the heart.

EXERCISING YOUR CRITICAL THINKING SKILLS

The stem cell in bone marrow is the subject of intense research, for scientists believe that by manipulating the cell they can find ways to better treat—perhaps even cure—a number of diseases, such as cancer, sickle-cell anemia, and AIDS. In order to manipulate the cell, however, scientists recognize that they need to know more about cell replication and differentiation. Through basic research, they hope to learn what, if anything, causes the cell to replicate and differentiate. Two schools of thought exist on the subject. The first says that the stem cell is subject to external influences, such as hormones, which direct its differentiation. The second school of thought contends that replication and differentiation occur randomly; that is, there are no outside controls.

At this point, can you see any problems in the way the debate is framed? You may want to review the critical thinking rules on page 18.

If after reviewing the debate and the critical thinking rules, you said the problem might be that researchers were over simplifying matters, you have pinpointed the problem. Put another way, researchers seem to have fallen into the dualistic thinking trap and may have set up a false dichotomy. It's quite possible that outside influences and random replication and differentiation both occur.

In an effort to settle this debate, Makio Ogawa of the Medical University of South Carolina, in Charleston, devised a method for growing human stem cells in a semisolid gel. He removed single cells from the stem cell colonies, then used these cells to start new colonies, which he could study. He found that the cells in the secondary colonies developed in a variety of cell types. That is, they appeared to differentiate randomly along several lines.

Think about the experiment for a few moments. Can you find any weaknesses with the design?

If you said that the researcher had isolated the cells from outside influences, such as hormones, that might be present in bone marrow and might influence differentiation, you're right. Cell culture studies—that is, studies of cells in culture dishes—although extremely useful, are often criticized because they do not expose cells to potentially important influences, such as hormones.

Can you think of any ways to solve this problem—that is, to study cell differentiation while preserving potential hormonal influences?

TEST OF TERMS

1. The liquid portion of blood is called ____________.
2. The proteins in the blood that destroy or inactivate viruses and bacteria are ____________ .
3. The ____________ is a biconcave disk that contains the protein ____________, which binds to oxygen.
4. Blood cells are produced in ______ ____________ marrow.
5. The hormone ____________ , produced by the kidney, stimulates the synthesis of ____________ when oxygen levels in the blood fall.
6. The heme group consists of a ____________ ring and an atom of iron.
7. ____________ is a reduction in the oxygen-carrying capacity of the blood.
8. The two white blood cells that phagocytize bacteria and viruses are ____________ and ____________ .
9. The ____________ is the white blood cell involved in immune protection.
10. ____________ is a cancer of the white blood cells.
11. The ____________ is a cell fragmentproduced from megakaryocytes.
12. ____________ , an inactive protein in the blood, is activated by thrombin and converted into long, branching fibers that form a weblike network in the wall of blood vessels that have been damaged.

13. ____________ is an enzyme in the blood that dissolves blood clots.
14. ____________ is a genetic defect that results in a lack of clotting factors and leads to uncontrollable bleeding.
15. ____________ ____________ is a colorless, odorless gas that binds to the hemoglobin molecule, reducing the oxygen-carrying capacity of the blood.

Answers to the Test of Terms are found in Appendix B.

TEST OF CONCEPTS

1. Describe the structure and function of each of the following: red blood cells, platelets, lymphocytes, monocytes, and neutrophils.
2. Define each of the following terms: leukemia, anemia, and infectious mononucleosis.
3. Explain how a blood clot forms and how it helps prevent bleeding.

CHAPTER

14

The Vital Exchange: Respiration

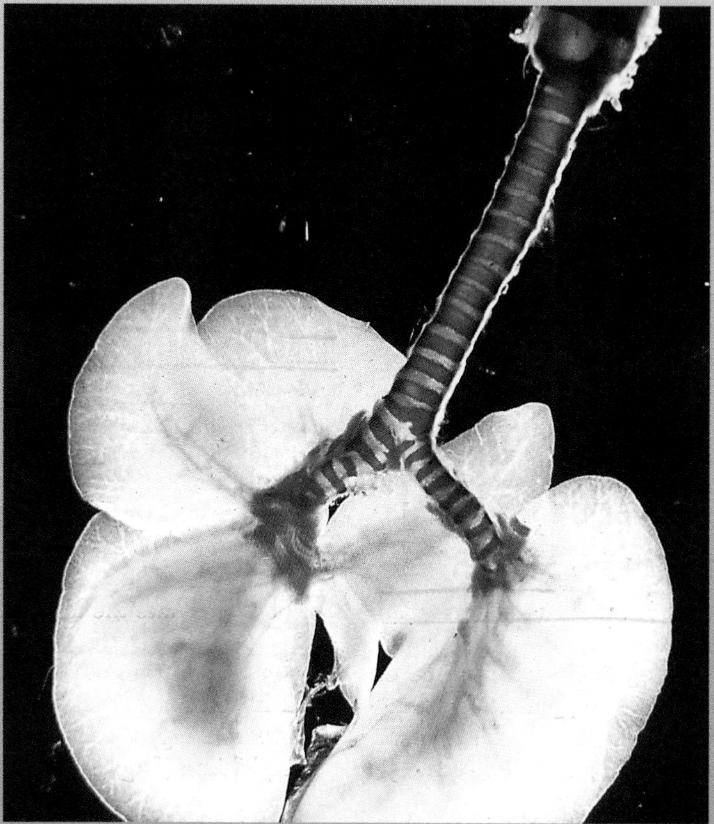

Respiratory system of a mammal showing larynx, trachea, bronchi, and lungs.

George F. Eaton was a robust and handsome Irishman who grew up in eastern Massachusetts, married, and raised three children. To support his family, Eaton worked as a fire fighter for 30 years. Fire fighting is dangerous work, in part because it exposes men and women to smoke containing numerous potentially harmful and sometimes lethal air pollutants.

Upon retirement, Eaton moved to Cape Cod, but his retirement years were cut short by **emphysema,** a debilitating respiratory disease resulting from the breakdown of the air sacs, or **alveoli,** in the lungs, where oxygen and carbon dioxide are exchanged between the air and the blood. As the walls of the alveoli degenerate, the surface area for the diffusion of oxygen and carbon dioxide greatly decreases. Emphysema worsens by the month and is an irreversible and incurable disease. Patients complain of shortness of breath; eventually, even mild exertion becomes trying, prompting one patient to describe the disease as a kind of "living hell."

The degeneration of the lungs in patients with emphysema creates a domino effect. One of the dominoes is the heart. Because oxygen absorption in the lungs declines substantially, the heart of an emphysemic patient must work harder. This effort puts additional strain on this already hard working organ and can lead to heart failure.

George Eaton died a slow and painful death as his lungs grew increasingly more inefficient. Relatives mourned his death and blamed it on the pollution to which he was exposed while working for the fire department. Fire fighting, however, was probably only part of the cause, for Eaton had smoked most of his adult life. Cigarette smoking is the leading cause of emphysema.

This chapter describes the respiratory system—its evolution, its function, and the diseases that affect it, like emphysema.

THE EVOLUTION OF RESPIRATORY SYSTEMS: AN OVERVIEW

What do the gills of a fish, the skin of a frog, and the lungs of mammals have in common? The answer is simple: all three are evolutionary adaptations to the same problem. That is, each one represents "evolution's answer" to an important biological challenge: getting oxygen into the body of a multicellular animal and getting rid of carbon dioxide.

Why is oxygen so important? If you think back on class discussions and your readings on cellular energy production, you may recall that all eukaryotic cells need oxygen for cellular respiration, the energy-yielding breakdown of glucose. That process produces abundant amounts of carbon dioxide that must be disposed of.

In the early history of life, single-celled eukaryotic organisms relied on diffusion to acquire oxygen and get rid of carbon dioxide. With the advent of multicellular animal life, though, diffusion was replaced by more efficient systems. Thus arose the respirable skin of the amphibian, the gill, the lung, and others. These specialized structures accommodated the needs of large, multicellular animals for oxygen.

Gills are structures found in aquatic vertebrates, mostly fishes, that increase the surface area for gas exchange. As shown in ▻ Figure 14–1, the gills of a fish consist of arched structures appropriately called **gill arches.** Each arch consists of many **gill filaments,** featherlike projections heavily endowed with blood vessels. Water containing oxygen enters a fish's mouth and is pumped backward over the gill filaments. Oxygen in the water is then absorbed across the relatively thin membrane of the gill filaments, moving into the bloodstream by diffusion. Carbon dioxide diffuses in the opposite direction.

If you have ever watched an aquarium fish floating idly in a tank, you may have noticed that it opens and closes its mouth, as if gulping water. This gulping action is vital to gaseous exchange, for it produces suction that draws water into the oral cavity. The water is then pushed out of the mouth and over the gills by movements of the gill covers, special plates that protect the gills from injury. When a fish is swimming, it ventilates its gills by opening its mouth and gill covers. This permits water to flow through the mouth and over the gill filaments. Some fish, such as mackerel, lack a pumping mechanism, and must therefore swim continuously with their mouth and gill covers open or die for lack of oxygen.

In terrestrial animals, several kinds of respiratory strategies have evolved. A few of them are presented here to highlight the diversity of ways in which animals meet the same requirement for gaseous exchange.

In the earthworm (a segmented worm, or annelid), gases are exchanged across the body wall. If you've ever picked up an earthworm, you know that its skin is slimy. That slime, or mucus, helps keep the earthworm's exterior moist so that gases can move in and out more easily.

Insects and other land-dwelling arthropods have tubular gas-exchange systems, as shown in ▻ Figure 14–2. Air diffuses into this system through tiny openings in the body wall called **spiracles** (spy-ruh-kuls). The spiracles empty into relatively large tubes, called **tracheae** (tray-key-eye). These tubes divide and subdivide, forming smaller and smaller tubules that course throughout the insect body. Gas exchange occurs in very tiny, dead-end tubules that terminate in body tissues. Moisture in these tubules, as on an earthworm's skin, is vital for the diffusion of oxygen and carbon dioxide.

In larger insects, the tracheal system is supplemented by air sacs, shown in Figure 14–2. These sacs are located near large muscles. As the insect moves, the muscles pump air out of the sacs into the tracheae. Air is drawn back into the sacs when the muscles relax. This interesting adaptation operates similarly to our lungs, but because the movement of air in and out of the sacs is incidental to movement, no additional energy is required for respiration.

Amphibians (frogs and salamanders) live on land and in

▻ **FIGURE 14–1 Gills** (*a*) Fish rely on specialized respiratory structures called gills to acquire oxygen from water and get rid of carbon dioxide. Water enters the mouth and is pumped across the gill arches, which are protected by the gill plates. (*b*) As shown here, the gill arches consist of featherlike gill filaments across which gas exchange occurs by diffusion.

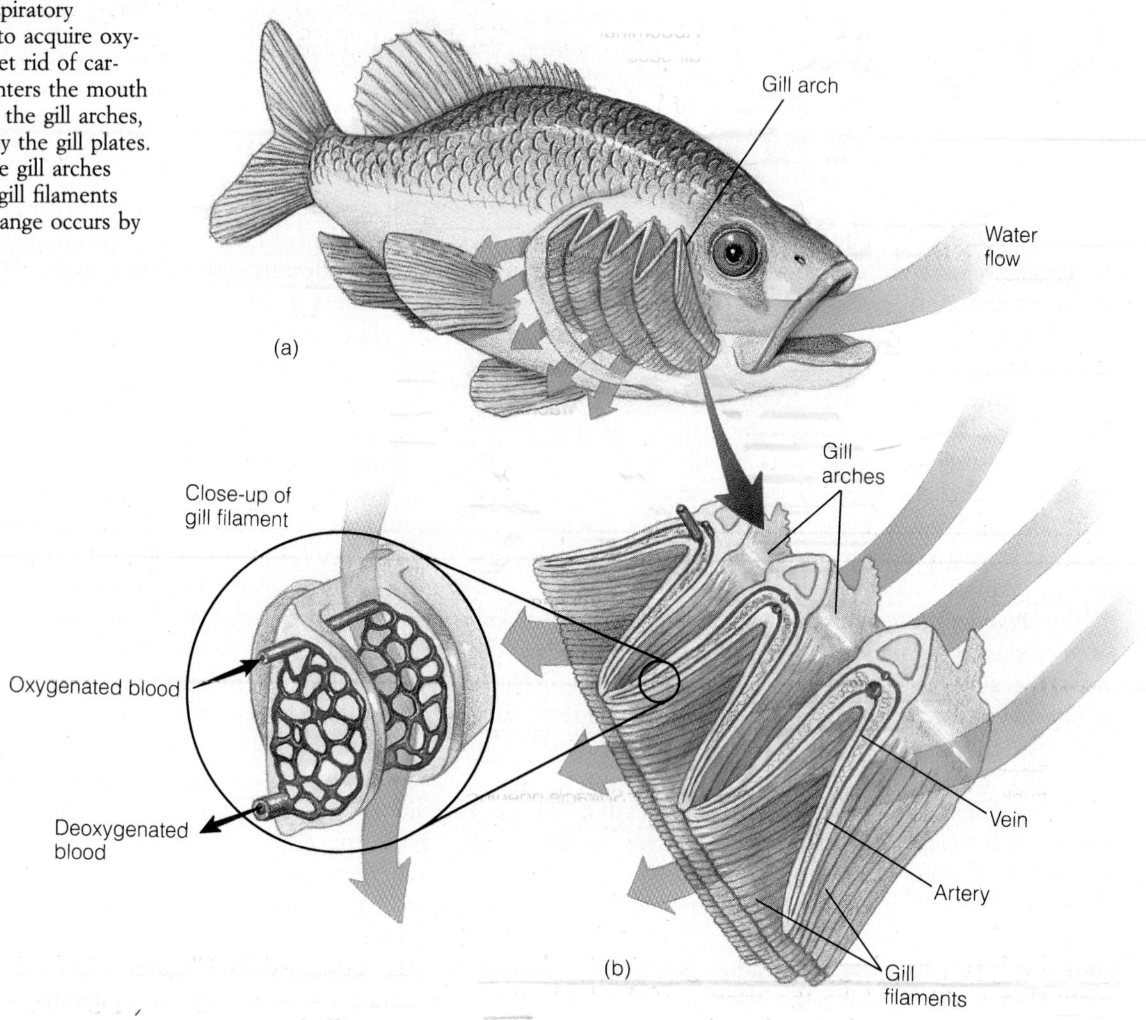

water. (The word amphibian comes from the Greek word *amphibios,* "living a double life.") Frogs have primitive, saclike structures called **lungs**, shown in ▻ Figure 14–3. As illustrated, the frog draws air into its mouth, closes it, then pushes the air into its lungs. However, lung respiration is less important than gas exchange through the animal's moist skin and across the moist surfaces of its mouth.

Mammals live on land, in the air, and in the oceans, lakes, and rivers, often moving from one medium to another in the course of their day. As a group, the mammals exhibit a higher rate of metabolism than reptiles, amphibians, and insects. This is due in part to the fact that mammals rely on internally generated heat to stay warm. That heat, of course, comes from glucose catabolism.

Endothermy, or internal heat production, no doubt led to the evolution of more efficient respiratory systems—systems that could acquire greater amounts of oxygen. Thus arose the mammalian lung and associated structures. We turn to the human respiratory system to study the mammalian lung.

STRUCTURE OF THE HUMAN RESPIRATORY SYSTEM

The human respiratory system, like those of other mammals, functions automatically, drawing air into the lungs, then letting it out, in a cycle that repeats itself about 16 times per minute at rest, or about 23,000 times per day. The respiratory system supplies oxygen to the body and gets rid of carbon dioxide. Oxygen is needed for cellular respiration; carbon dioxide is a waste product of this process.

In its role as a provider of oxygen and a disposer of carbon dioxide waste, the respiratory system helps maintain a constant internal environment necessary for normal cellular metabolism. Thus, like many other systems, it plays an important role in homeostasis.

▻ Figure 14–4 shows the structure of the human respiratory system. As illustrated, the respiratory system consists of two basic parts: an air-conducting portion and a gas-exchange portion (Table 14–1). The air-conducting portion is an elaborate set of passageways that transports air to and

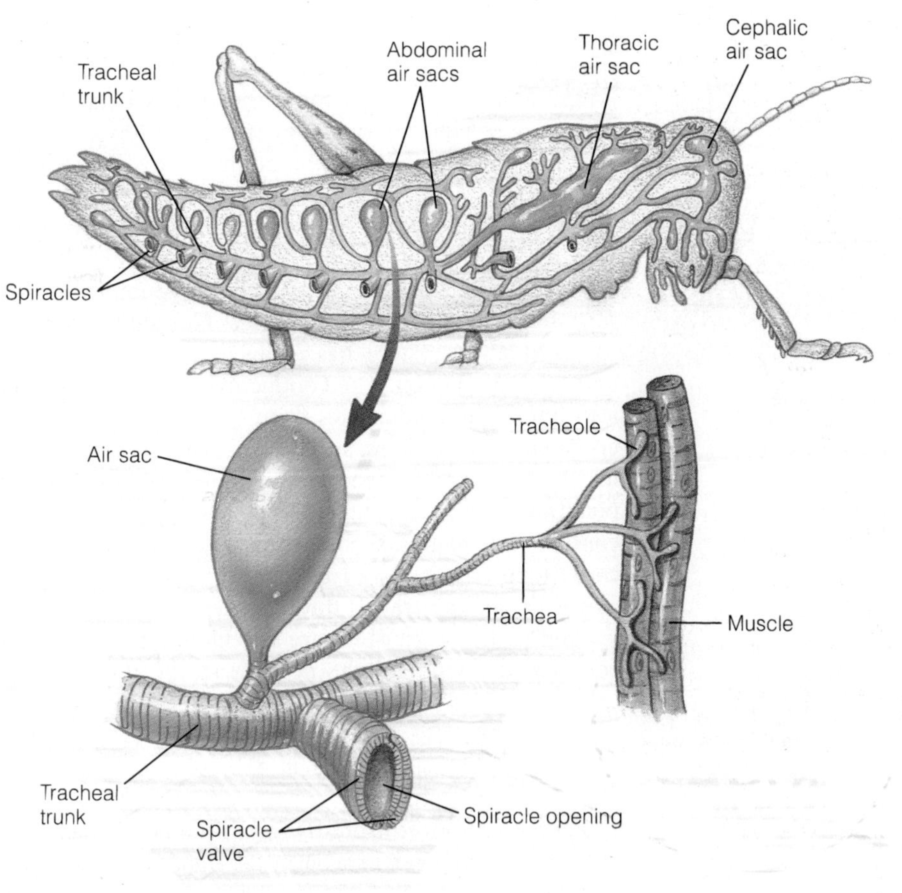

▻ FIGURE 14–2 Tracheal System of an Insect As shown here, insects contain an internal system of channels to transport oxygen to body cells. Air enters and leaves the system via tiny openings called spiracles. Air sacs are present in some insects and form reservoirs that store air.

from the lungs, two large, saclike organs in the thoracic cavity (Figure 14–4). Like the arteries of the body, these passageways start out large, then become progressively smaller and more numerous after entering the lung tissue, where the exchange of oxygen and carbon dioxide between the air and the blood takes place.

The lungs are the gas-exchange portion of the respiratory system. Each lung has millions of tiny, thin-walled alveoli (Figure 14–4). The walls of the alveoli contain numerous blood capillaries that absorb oxygen from the inhaled air and release carbon dioxide (▻ Figure 14–5, page 330).

The Conducting Portion of the Respiratory System Moves Air In and Out of the Body but also Filters the Air and Moistens the Incoming Air

Air enters the respiratory system through the nose and mouth, then is drawn backward into the pharynx (▻ Figure 14–6, page 330). The **pharynx** opens into the nose and mouth in the front and joins below with the larynx. The **larynx,** or voice box, is a rigid but hollow structure that houses the vocal cords (Figure 14–6). To feel it, gently put your fingers alongside your throat, then swallow. As shown in Figure 14–4, the larynx opens into the **trachea** below. You can feel the trachea below your Adam's apple, the protrusion of the laryngeal cartilage on your neck.

As explained in Chapter 11, food is prevented from entering the larynx by the epiglottis, a flap of tissue that closes off the opening to the larynx during swallowing. Occasionally, however, food goes the wrong way, accidentally entering the larynx and trachea. It may lead to violent coughing, a reflex that helps eject the food from the trachea. If the food cannot be dislodged by coughing, steps must be taken to remove it—and fast—or a person will suffocate. Health Note 14–1 (page 331) explains what to do when a person chokes.

As Figure 14–4 shows, the trachea is a short, wide duct. Starting in the neck below the larynx, it enters the thoracic cavity where it divides into two large branches, the right and left bronchi. The **bronchi** (singular, *bronchus*) enter the lungs alongside the arteries and veins. Inside the lungs, the bronchi branch extensively, forming progressively smaller tubes that carry air to the alveoli, mentioned earlier.

The trachea and bronchi are reinforced by hyaline cartilage in their walls (▻ Figure 14–7, page 332). The C-shaped cartilage "rings" of the trachea prevent the organ from collapsing during breathing, ensuring a steady flow of air in and out of the lungs. Plates of cartilage in the walls of the bronchi have the same effect. You can feel the cartilage rings in the trachea by gently rubbing your trachea just below the larynx.

The smallest bronchi in the lungs branch to form **bronchioles,** which lead to the alveoli. Like the arterioles of the

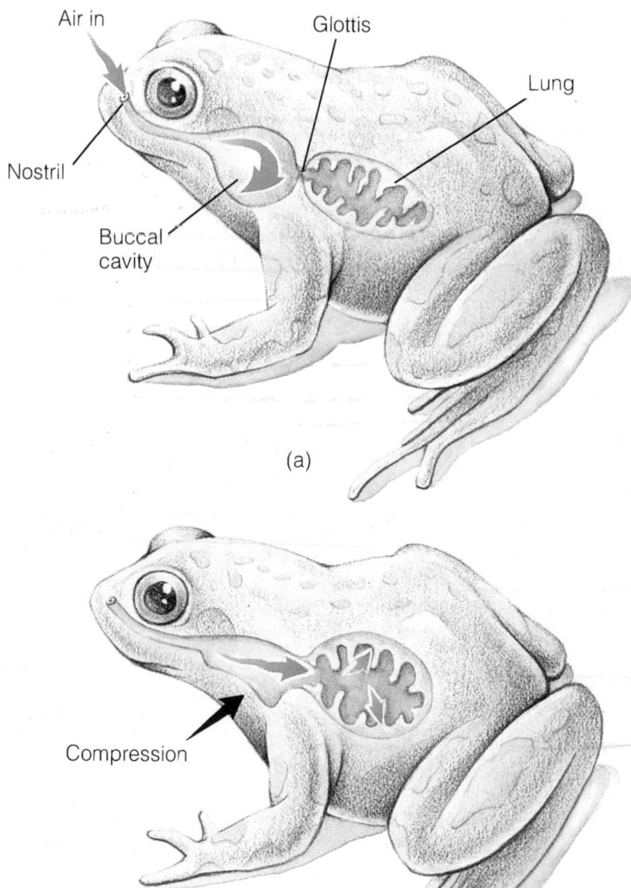

▻ **FIGURE 14–3 The Lungs of the Frog** Frogs rely principally on their skin for gas exchange but also have lungs. As illustrated here, air is drawn into the mouth cavity, then pushed back into the lungs.

circulatory system, the walls of the bronchioles consist largely of smooth muscle. As a result, the bronchioles can open and close, providing a means of controlling air flow in the lung. During exercise or during times of stress, for example, the smooth muscle in the bronchioles relaxes, opening the tubes and increasing the flow of air into the lungs. This action helps meet the body's need for more oxygen, in much the same way that the arterioles of capillary beds dilate to let more blood into body tissues in times of need.

The respiratory system is constantly in direct contact with the external environment and is therefore exposed to many infectious organisms as well as a number of potentially hazardous particles and gases. Not surprisingly, then, the respiratory system has evolved a number of mechanisms for protection. The conducting portion, for example, also filters many impurities from the air we breathe, especially airborne particles, such as dust and bacteria. Particles in the air exist in many sizes. Some are small and can penetrate deeply into the lung. Some of these particles may contain toxic metals, such as mercury and nickel, which can cause lung cancer. As a rule, larger particles are deposited as the inhaled air travels through the maze of passageways leading to the lungs. Many are also filtered in the nose. The convoluted interior of the nose slows the flow of air and causes the larger particles to drop out, in much the same way that sediment falls to the bottom in a slow-moving section of a stream. Hairs in the nasal cavity may also physically trap particles.

Particles removed from the air in the nose, trachea, and bronchi are trapped in a layer of **mucus,** a thick, slimy secretion deposited on the inside of much of the respiratory tract (▻ Figure 14–8, page 332). Mucus is produced by cells in the epithelium, called **mucous cells.**[1] The epithelium of the respiratory tract also contains numerous ciliated cells, an additional adaptation vital to keeping the respiratory system healthy. The cilia of these cells beat upward toward the mouth, slowly transporting mucus and its cargo of bacteria and dust particles. Operating day and night, they sweep the mucus toward the oral cavity, where it can be swallowed or expectorated (spit out). This process protects the respiratory tract and lungs from bacteria and potentially harmful particulates.

Like all homeostatic mechanisms, however, the respiratory mucous trap is not invulnerable. Bacteria and viruses do occasionally penetrate the lining, where they may proliferate, causing respiratory infections. Making matters worse, sulfur dioxide, a pollutant in cigarette smoke and urban air pollution, temporarily paralyzes, and may even destroy, cilia. Sulfur dioxide gas in the smoke of a single cigarette, for instance, can paralyze the cilia for an hour or more, permitting bacteria and toxic particulates to be deposited on the lining of the respiratory tract, and in some cases, even enter the lungs. Ironically, the cilia of a smoker are paralyzed when they are needed the most!

Because smoking impairs this natural protective mechanism, it should come as no surprise that smokers suffer

[1] You will notice that the adjective form is spelled *mucous* (as in mucous cell) and the noun form is spelled *mucus*.

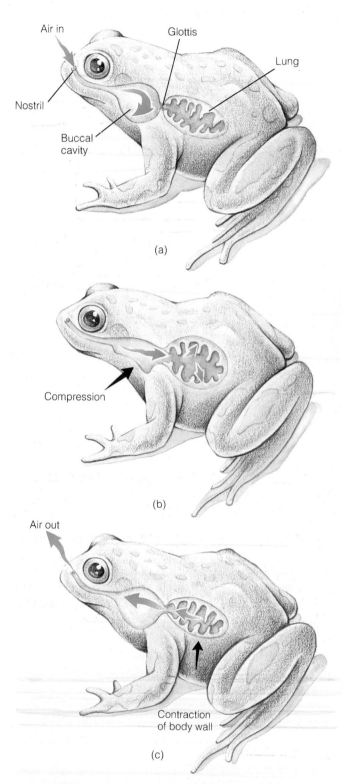

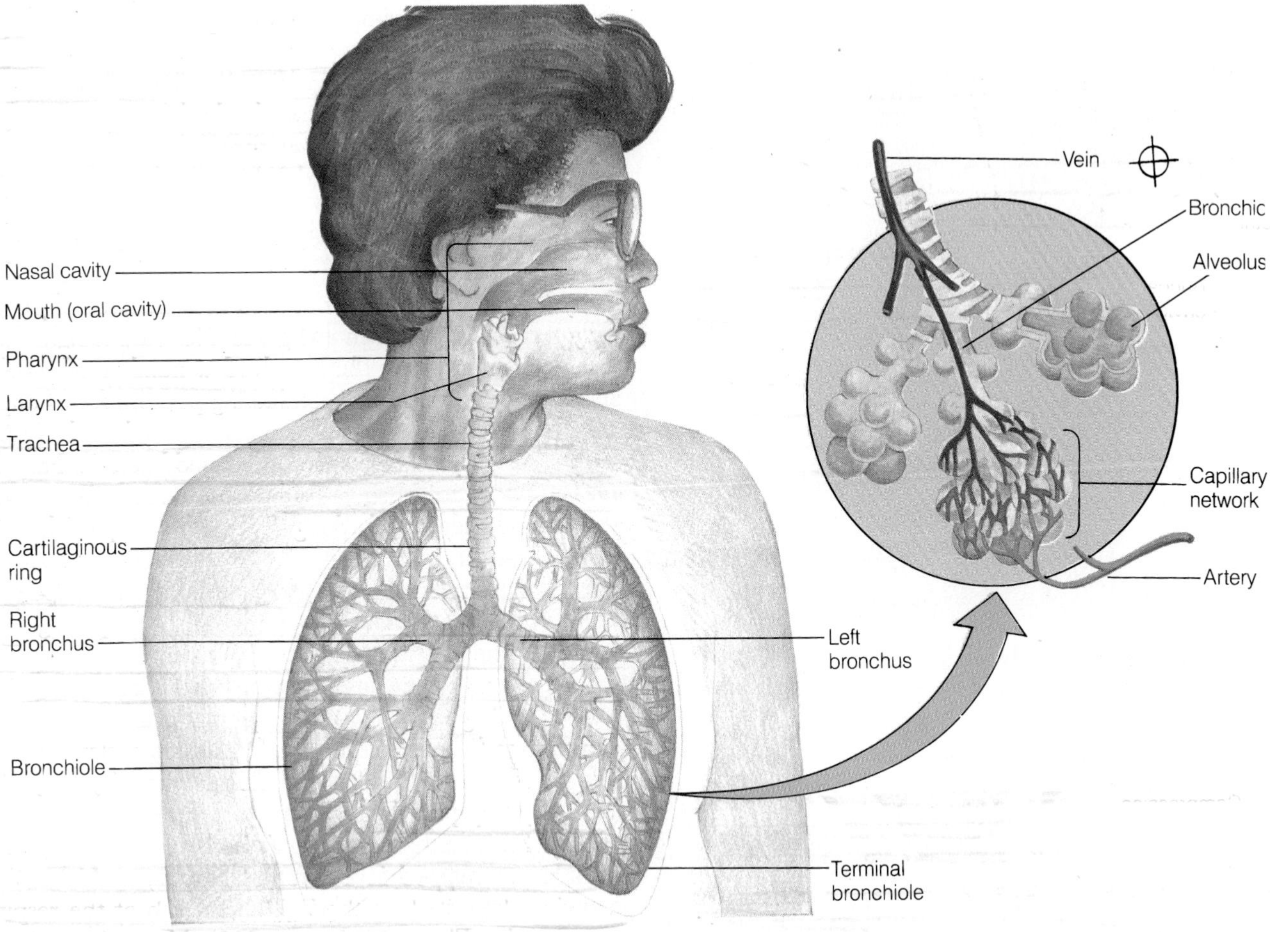

▻ **FIGURE 14–4 The Human Respiratory System** A cutaway showing the air-conducting portion and the gas-exchange portion of the human respiratory system. The insert shows a higher magnification of the alveoli, where oxygen and carbon dioxide exchange occurs.

more frequent respiratory infections than nonsmokers. Interestingly, alcohol also paralyzes the cilia of the respiratory system. As a result, alcoholics are much more prone to certain types of respiratory infections.

The conducting portion of the respiratory system also moistens and warms the incoming air. Beneath the epithelium of the respiratory tract is a rich network of capillaries that releases moisture and heat. Moisture protects the lungs from drying out, and heat protects them from cold temperatures. Except in extremely cold weather, by the time inhaled air reaches the lungs, it is nearly saturated with water and is warmed to body temperature.

TABLE14–1 Summary of the Respiratory System

ORGAN	FUNCTION
Air conducting	
Nasal cavity	Filters, warms, and moistens air; also transports air to pharynx
Oral cavity	Transports air to pharynx; warms and moistens air; helps produce sounds
Pharynx	Transports air to larynx
Epiglottis	Covers the opening to the trachea during swallowing
Larynx	Produces sounds; transports air to trachea; helps filter incoming air; warms and moistens incoming air
Trachea and bronchi	Warm and moisten air; transport air to lungs; filter incoming air
Bronchioles	Control air flow in the lungs; transport air to alveoli
Gas exchange	
Alveoli	Provide area for exchange of oxygen and carbon dioxide

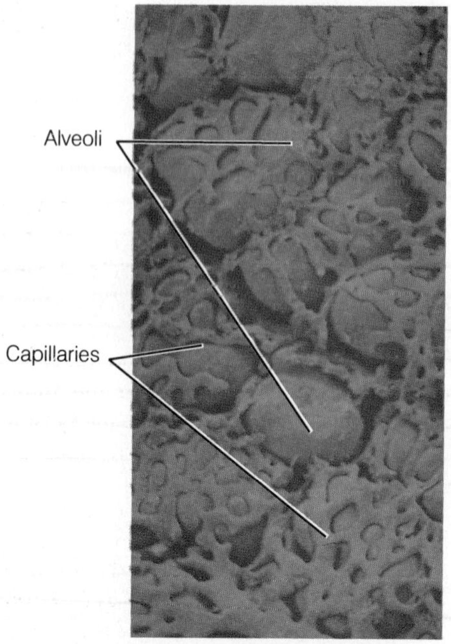

▷ **FIGURE 14–5 The Alveoli and Their Capillaries**
A scanning electron micrograph of the alveoli of the lung, showing the rich capillary network surrounding them.

On the way out, much of the water that was added to the air condenses on the lining of the nasal cavity which was cooled by the evaporation of water as the air was being drawn in. This mechanism is an adaptation that conserves water and also accounts for the reason our noses tend to drip in cold weather.

The Alveoli are the Site of Gaseous Exchange

The air we breathe consists principally of nitrogen and oxygen with small amounts of carbon dioxide and other gases (Table 14–2). Oxygen in the atmosphere is generated by the photosynthetic activity of plants and photosynthetic protists and is, of course, vital to humans and virtually all other living organisms. A constant supply must be delivered to human body cells in order for us to maintain cellular energy production.

As noted earlier, the conducting portion of the respiratory system delivers warmed, moistened, filtered air that is rich in oxygen to the bronchioles, which, in turn, deliver the air to the alveoli. Each lung contains an estimated 150 million alveoli, giving the lung the appearance of an angel food cake (▷ Figure 14–9a and 14–9b). Each alveolus is surrounded by an extensive capillary bed. The alveoli provide a surface area for absorption of oxygen in the lungs measuring 60 to 80 square meters, an area approximately the size of a tennis court.

As shown in Figures 14–9c and ▷ 14–10a (page 334), the alveoli are lined by a single layer of flattened cells, called **Type I alveolar cells.** These cells permit gases to move in and out the alveoli with ease. Thus, the large surface area of the lungs created by the alveoli and the relatively thin barrier between the blood and the alveolar air combine to produce a rather rapid and efficient diffusion of gases across the alveolar wall. The alveoli provide another example of the marriage of form and function produced by the process of evolution.

Another important cell in the alveoli is the **alveolar macrophage,** sometimes known as the **dust cell.** Alveolar macrophages remove dust and other particulates that reach the lungs (Figure 14–10a). Dust cells wander freely around and through the alveoli, engulfing foreign material that has escaped filtration. Once filled with particulates, the macrophages accumulate in the connective tissue surrounding the

▷ **FIGURE 14–6 Uppermost Portion of the Respiratory System**
Bony protrusions into the nasal cavity create turbulence that causes dust particles to settle out on the mucous coating. Notice that air passing from the pharynx enters the larynx. Food is kept from entering the respiratory system by the epiglottis, which covers the laryngeal opening during swallowing.

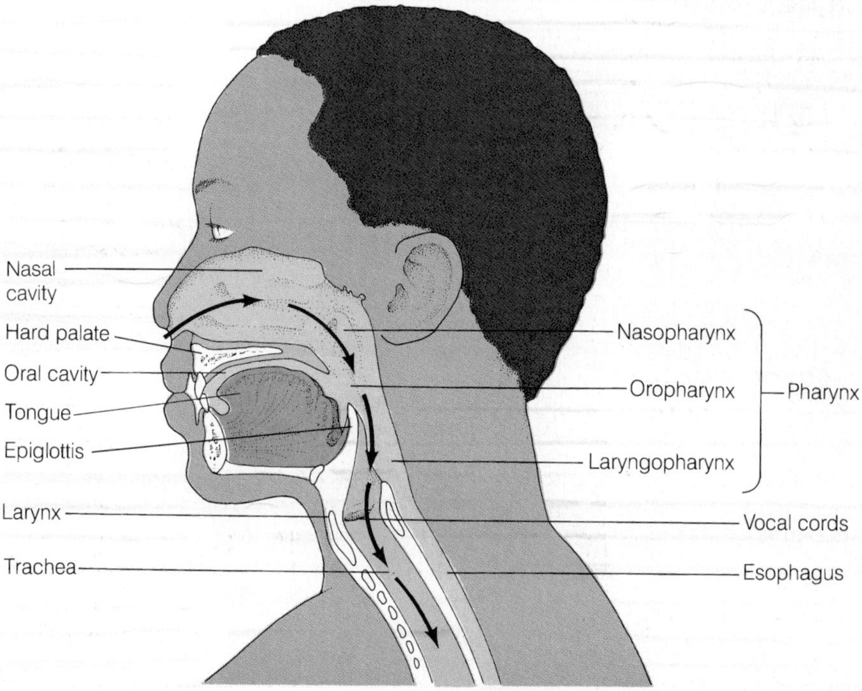

HEALTH NOTE 14–1

FIRST AID THAT MAY SAVE SOMEONE'S LIFE: THE HEIMLICH MANEUVER

During swallowing, food is normally excluded from the trachea by the epiglottis, as explained in the chapter. Each year, however, approximately 3000 Americans—on average, about 8 people a day—will choke to death on food that becomes lodged in the larynx or trachea. Many of these deaths could be prevented. Here's what should be done if you encounter a person who is choking on food. First, stand behind the victim, who may be either standing or sitting. Position yourself slightly to one side, as shown in ▻ Figure 1a. Place an arm across the victim's chest for support, and lean the person forward. With the heel of your other hand, give four hard thumps between the shoulder blades. This should clear the trachea. If it doesn't, don't give up. Try the Heimlich maneuver.

(a)

(b)

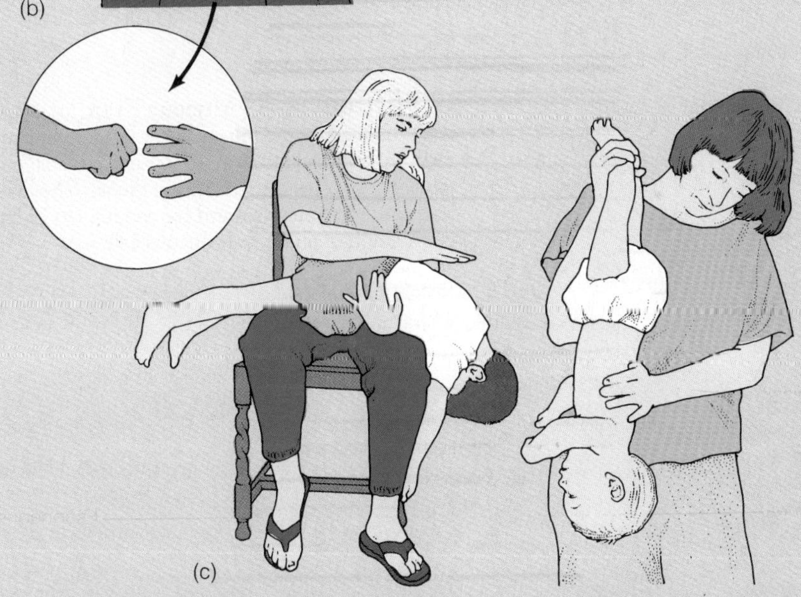

(c)

▻ **FIGURE 1 First Aid for Choking Victims** (*a*) As a first measure, stand behind and slightly to the side of the victim, and let him lean over one arm. Hit the victim's back between the shoulder blades with the heel of your hand. (*b*) If this does not dislodge the food, try the Heimlich maneuver. While standing behind the victim, wrap your arms around his waist. Press your fist against his waist with your thumb inward. Grasp your fist with your other hand, and pull sharply in and up. (*c*) A similar procedure can be performed on young children and babies in the positions shown here.

The **Heimlich maneuver** or **abdominal thrust,** is shown in Figure 1b. It is best performed with the individual standing. Position yourself behind the person, wrap your arms around the waist. Grasp your fist with the other hand, pressing it against the abdomen with the thumb pointing inward just above the person's navel. Now, give your fist a sharp pull inward and upward.

This should dislodge the food. If it doesn't, try it three more times. If that doesn't work, repeat the back blows. That should dislodge the food. If that doesn't work, call an ambulance.

Children are often victims of choking, and many parents make the mistake of trying to dislodge food by sticking their fingers down their child's throat to extract the food. Unfortunately, this may force the food deeper into the larynx or trachea. Figure 1c shows how to treat babies or young children who are choking on food or other objects. For a toddler, sit down and put the child across your knees with the head down. Give several thumps on the back between the shoulder blades with the heel of your hand (but softer than you would for an adult). Babies can be held face down with one arm supporting the body. Several light raps on the back should dislodge the food. Babies can also be held by the ankles while you rap on the back.

It is a good idea to practice these techniques on friends and family members, so that when the time comes you will be prepared. When you practice, however, be gentle, and be sure your arms are around the person's waist, not the chest.

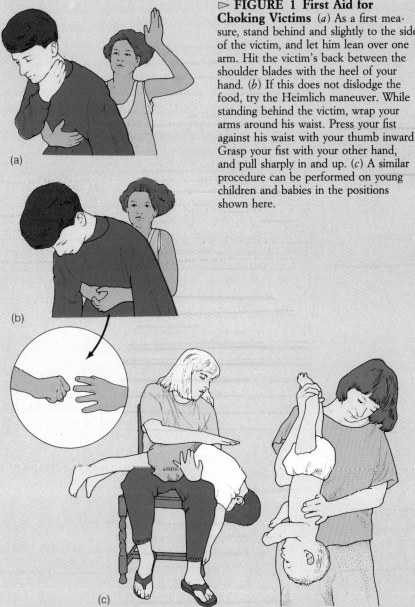

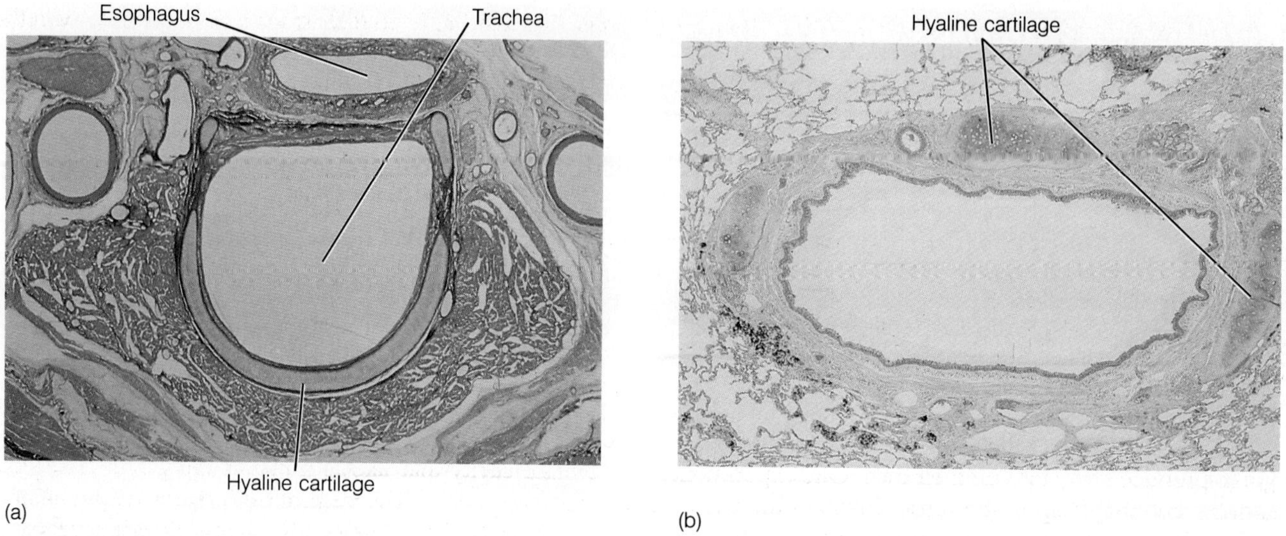

▻ **FIGURE 14–7 Cross Sections of the Trachea and Bronchus** (*a*) The trachea contains hyaline cartilage ribs, C-shaped segments of cartilage that give the organ internal support. (*b*) An intrapulmonary bronchus showing hyaline cartilage.

alveoli. At death, a smoker's lungs or the lungs of an urban resident are therefore often blackened by the accumulation of smoke and dust particles.

Another important cell is the **Type II alveolar cell,** shown in Figure 14–10a. Type II alveolar cells are large, round cells that produce **surfactant,** a detergentlike substance that dissolves in the thin layer of water lining the alveoli. The water covering the alveolar lining produces surface tension. **Surface tension** results from hydrogen bonds that form between water molecules (Chapter 2). In water, hydrogen bonds draw water molecules together. At the surface of a watery fluid, the hydrogen bonds draw water molecules together more tightly than elsewhere, creating a slightly denser region referred to as surface tension. Surface tension on a pond permits some insects, such as water striders, to walk on water and is the reason a drop of water beads up on your car windshield (▻ Figure 14–11).

In the alveoli, surface tension tends to draw the walls of the alveoli inward. Surfactant, however, reduces surface tension in the alveoli, thus decreasing forces that might otherwise cause the alveoli to collapse.

At birth, in some very small, premature infants, the Type II alveolar cells fail to produce enough surfactant. Consequently, surface tension causes the larger alveoli to collapse within a few hours of birth, reducing the surface area for absorption. This condition, known as **respiratory distress syndrome,** or **hyaline membrane disease,** is characterized by rapid, labored breathing, which may lead to exhaustion. If it is untreated, the lungs may collapse, causing death.

Currently, babies with respiratory distress syndrome are placed on respirators, which force air into their lungs, opening the alveoli. This treatment must continue until the lungs

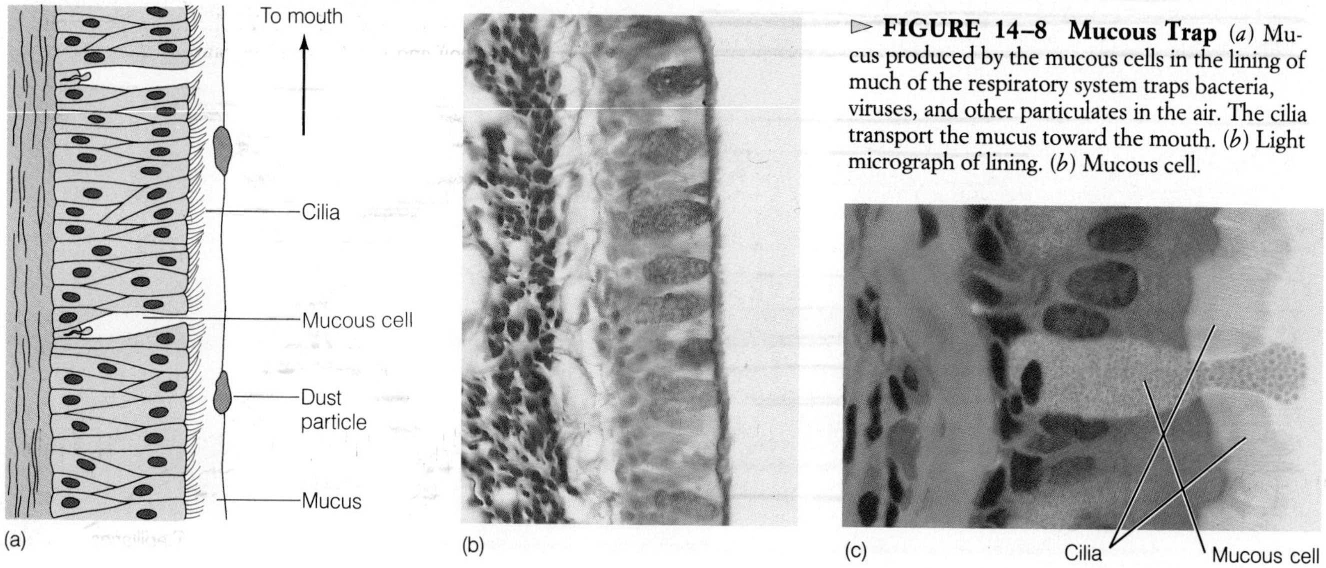

▻ **FIGURE 14–8 Mucous Trap** (*a*) Mucus produced by the mucous cells in the lining of much of the respiratory system traps bacteria, viruses, and other particulates in the air. The cilia transport the mucus toward the mouth. (*b*) Light micrograph of lining. (*b*) Mucous cell.

TABLE 14–2 Composition of Air

GAS	PERCENTAGE COMPOSITION
Nitrogen (N_2)	78
Oxygen (O_2)	21
Argon (Ar)	0.9
Carbon dioxide (CO_2)	0.03
Water vapor (H_2O)	Variable (0–4)
Pollutants	Variable

begin to produce surfactant on their own. One experimental treatment that may help in the future involves the use of a surfactantlike substance. Mixed with the air the baby breathes, this substance penetrates the lung, where it dissolves in the water lining the alveoli. This reduces surface tension and prevents the alveoli from collapsing. Treatment continues until a baby's own surfactant-producing cells mature and begin producing adequate amounts of surfactant. Although fairly successful, this procedure has not been approved by the Food and Drug Administration because of some unanswered ethical questions—notably, whether treating these babies, which often have multiple and often severe birth defects, is a prudent goal.

FUNCTIONS OF THE RESPIRATORY SYSTEM

The chief functions of the respiratory system are to (1) replenish the blood's oxygen supply, and (2) rid the blood of excess carbon dioxide. The respiratory system serves other functions as well. The vocal cords, described below, produce sounds that allow people and other mammals to communicate. The respiratory system houses the **olfactory membrane,** a specialized patch of epithelium in the roof of the nasal cavity that allows humans and other vertebrates to perceive odors. The respiratory system helps maintain pH balance by its influence on carbon dioxide levels.

In Humans, Sound Is Produced by the Vocal Cords and Is Influenced by the Tongue and Oral Cavity

Phonation, the production of sounds, is critical to many members of the animal kingdom. The eerie cry of the coyote, for example, signals to the pack a member's whereabouts and helps the members of the pack stay in contact (▻ Figure 14–12). The coyote's growl may signal to an intruder its intention to defend itself.

Humans exhibit a wider range of sounds for communication than other animals. These sounds are produced by two **vocal cords,** elastic ligaments inside the larynx. The vocal cords vibrate as air is expelled from the lungs (▻ Figure 14–13, page 336). The sounds generated by the vocal cords are further modified by changing the position of the tongue and the shape of the oral cavity.

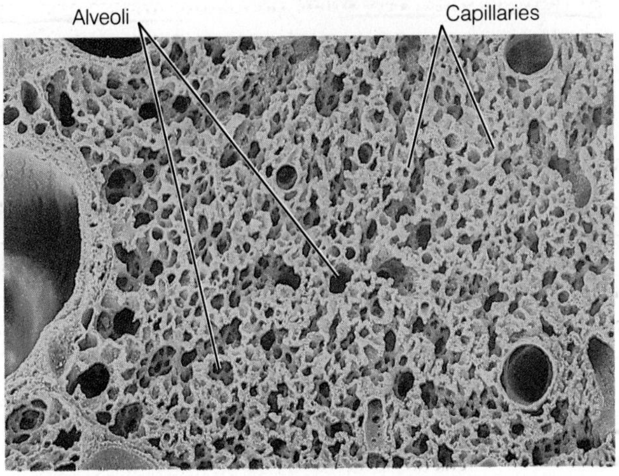

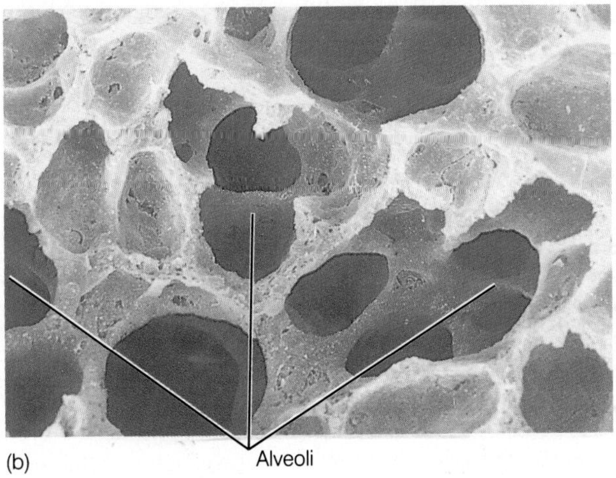

▻ **FIGURE 14–9 The Alveoli** (*a*) A scanning electron micrograph of the lung showing many alveoli. The smallest openings are capillaries in the alveolar walls. (*b*) A scanning electron micrograph of lung tissue showing alveoli. (*c*) An electron micrograph showing several alveoli and the close relationship of the capillaries.

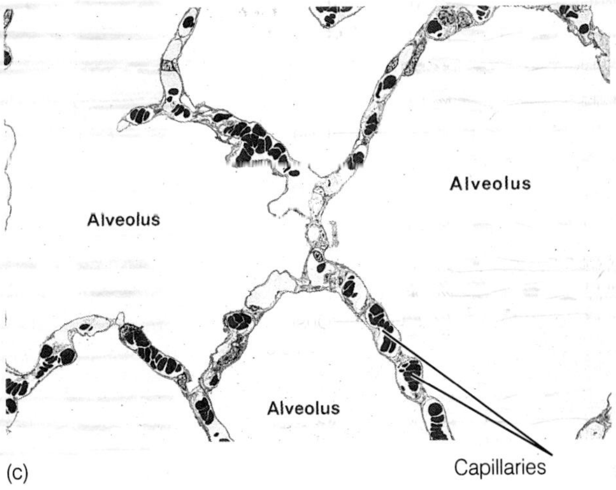

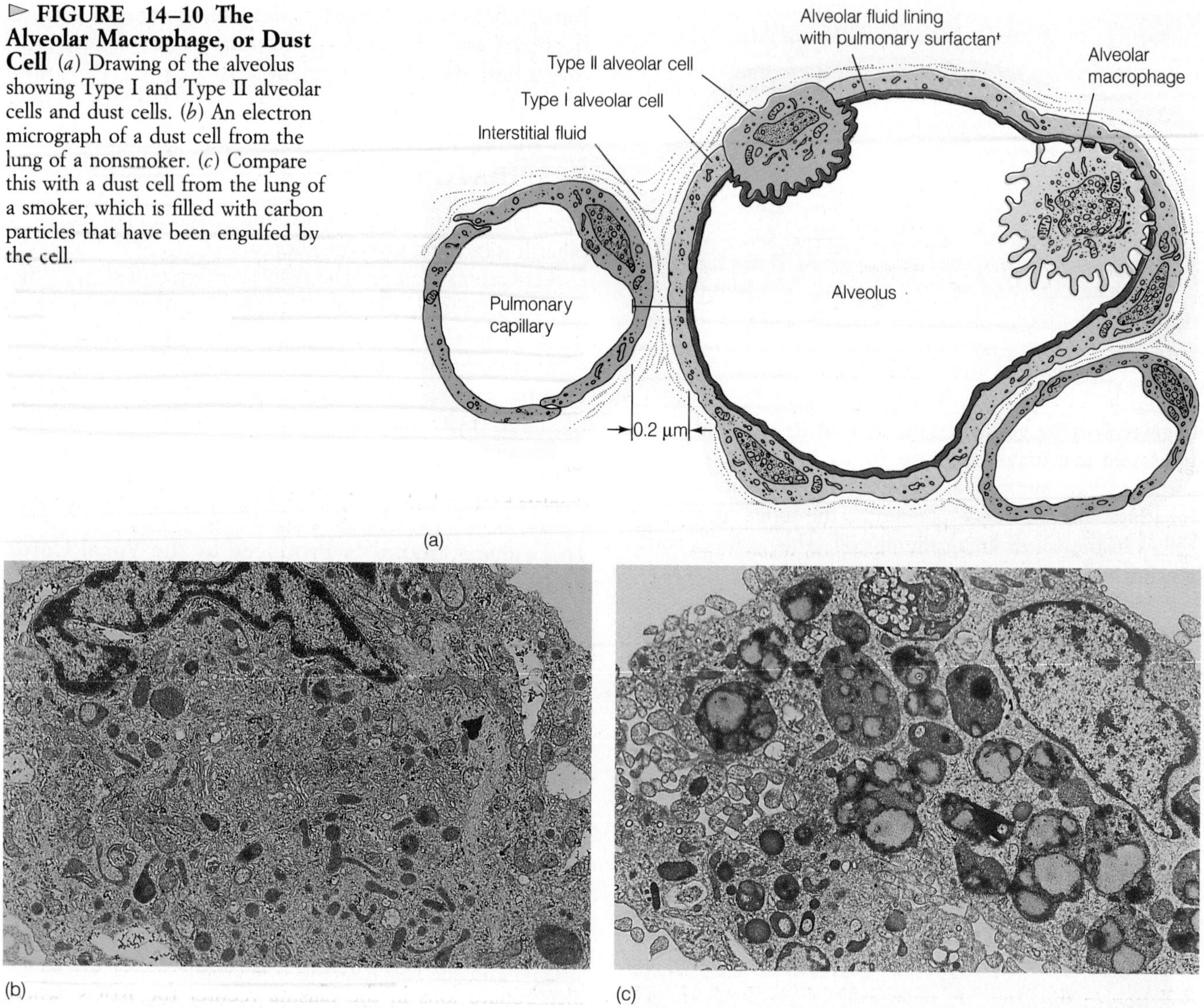

▻ **FIGURE 14–10 The Alveolar Macrophage, or Dust Cell** (*a*) Drawing of the alveolus showing Type I and Type II alveolar cells and dust cells. (*b*) An electron micrograph of a dust cell from the lung of a nonsmoker. (*c*) Compare this with a dust cell from the lung of a smoker, which is filled with carbon particles that have been engulfed by the cell.

The vocal cords vary in length and thickness from one person to the next. They also vary between men and women. Most men, for example, have longer, thicker vocal cords than women and, therefore, have deeper voices. Testosterone, a male sex hormone produced by the testes, is responsible for the growth of the larynx and the characteristic length and thickness of the male vocal cords.

In some respects, the vocal cords are like the strings of a guitar or piano. That is, they can be tightened or loosened, producing sounds of different pitch. The tighter the string on a guitar, the higher the note. In humans, muscles in the larynx that attach to the vocal cords make this adjustment. Relaxing the muscles lowers the tension on the cords, dropping the tone. Tightening the vocal cords has the opposite effect.

Bacterial and viral infections of the larynx result in a condition known as laryngitis. **Laryngitis** is characterized by inflammation of the lining of the larynx and the vocal cords. Inflammation thickens the cords, causing a person's voice to lower. Laryngitis may also be caused by irritation of the larynx from tobacco smoke, alcohol, excessive talking, shouting, coughing, or singing. In young children, inflammation results in a swelling of the lining that may impede the flow of air and impair breathing, resulting in a condition known as the croup.

Oxygen and Carbon Dioxide Move Rapidly Across the Alveolar and Capillary Walls, Flowing Down Concentration Gradients

Deoxygenated blood entering the lungs arrives from the right ventricle of the heart via the pulmonary arteries. This blood, you may recall from Chapter 12, laden with carbon dioxide picked up as the blood circulates through body tissues giving off oxygen. In the capillary beds of the lungs, carbon dioxide is released, and oxygen is added. These gases readily diffuse across the capillary and alveolar walls, driven by the concentration difference between the alveoli

(a)

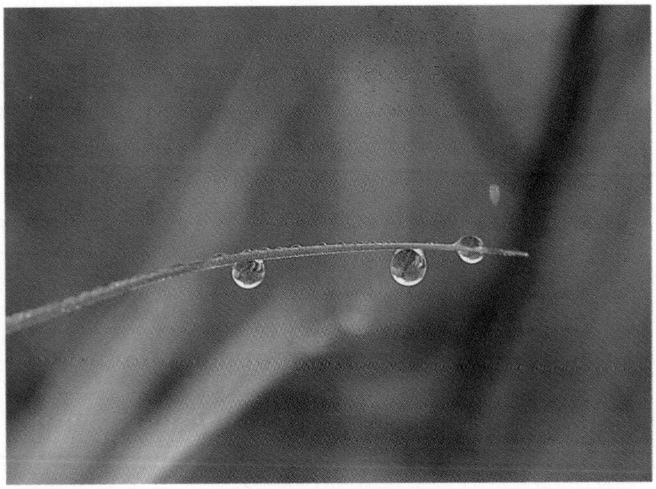

(b)

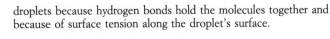

▻ **FIGURE 14–11 Surface Tension** (*a*) Some insects can walk on water because of surface tension, the tight packing of water molecules along the surface of a pond. (*b*) Water forms droplets because hydrogen bonds hold the molecules together and because of surface tension along the droplet's surface.

and capillaries.[2] ▻ Figure 14–14 illustrates the direction in which carbon dioxide and oxygen flow.

Oxygen in the alveolar air first diffuses through the alveolar epithelium, then into the extracellular fluid surrounding the capillaries. It then diffuses through the capillary wall and into the blood plasma (▻ Figure 14–15). From here, oxygen molecules cross the plasma membrane of the red blood cells (RBCs) and bind to hemoglobin molecules in their cytoplasm. About 98% of the oxygen in the blood is carried in the RBCs bound to hemoglobin; the rest is dissolved in the plasma and the cytoplasm of the RBCs.

In order to understand the details of carbon dioxide diffusion in the lung, we must go back to the body cells, where CO_2 is formed. In body tissues, carbon dioxide diffuses out of the cells and enters the blood plasma. Much of it then diffuses into the RBCs, where it is chemically converted to carbonic acid, H_2CO_3 (▻ Figure 14–16). This reaction is catalyzed by the enzyme **carbonic anhydrase** inside all RBCs. Carbonic acid molecules readily dissociate, however, forming bicarbonate ions and hydrogen ions. Many of the bicarbonate ions diffuse out of the RBCs into the plasma, where they are carried along with the blood. A small percentage (15% to 25%) of the carbon dioxide given off by body cells binds to hemoglobin (but not at the oxygen binding site), and an even smaller percentage (7%) dissolves in the plasma.

When blood rich in carbon dioxide reaches the lungs, bicarbonate ions in the plasma reenter the RBCs, where they combine with hydrogen ions to form carbonic acid. Carbonic acid, in turn, reforms carbon dioxide, as illustrated in ▻ Figure 14–17. Carbon dioxide then diffuses out of the RBC into the blood and finally into the alveoli down its concentration gradient. It is then expelled from the lungs during exhalation.

The uptake of oxygen and discharge of carbon dioxide occurring in the lungs "replenish" the blood in the alveolar capillaries. The oxygenated blood then flows back to the left atrium of the heart via the pulmonary veins. From here it empties into the left ventricle and is pumped to the body tissues via the aorta and its multitude of branches.

[2] Physiologists actually speak of differences in partial pressure. The partial pressure of a gas is caused by the collision of moving gas molecules with a surface. The partial pressure of oxygen is proportional to the impact of all the oxygen molecules striking the alveolar wall. Thus, the total pressure is directly proportional to the concentration of the gas molecules.

▻ **FIGURE 14–12 A Coyote Howls**

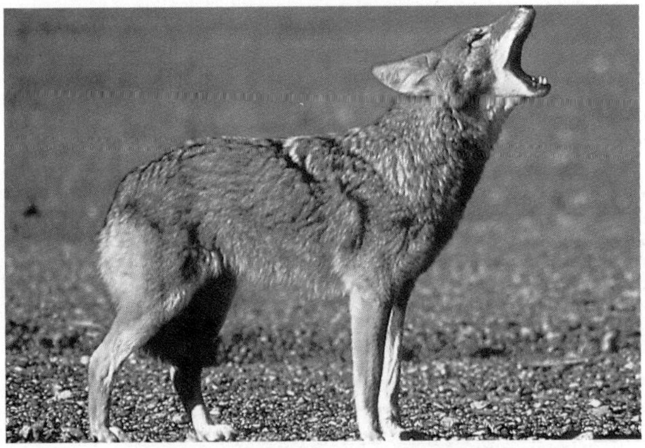

DISEASES OF THE RESPIRATORY SYSTEM

Given that the air we breathe is laden with bacteria, viruses, and other potentially dangerous pathogens and that the respiratory organs are in direct contact with the outside environment, the respiratory system is one of the main routes for infectious agents to enter the body. Protected by adaptations such as the mucous layer, the cilia along much

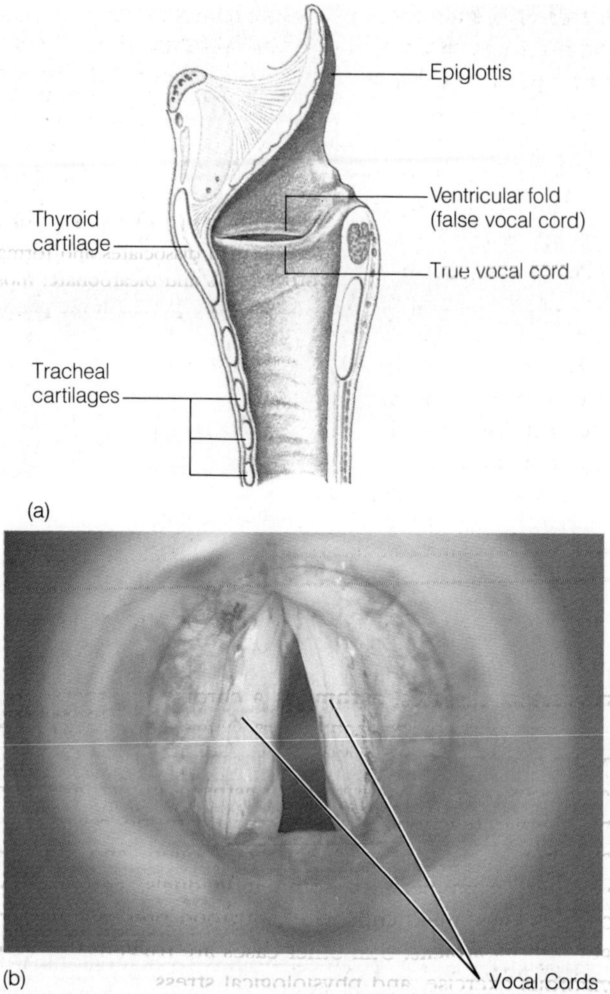

▻ **FIGURE 14–13 Vocal Cords** (*a*) Longitudinal section of the larynx showing the location of the vocal cords. Note the presence of two false vocal cords. They do not function in phonation. (*b*) View into the larynx of a patient also showing the vocal cords.

of the epithelium, the phagocytic cells in the lung, and others, the respiratory system nevertheless still falls victim to bacterial and viral invasion. These infectious agents may penetrate the epithelium, entering the underlying tissues where they proliferate madly. Some viruses, such as the influenza virus, multiply in the epithelial cells lining the respiratory tract.

Bacterial and viral infections of the respiratory tract can cause considerable discomfort, and some can be fatal. Infections may settle in many different locations in the respiratory system and are named by their site of residence. An infection in the bronchi is therefore known as **bronchitis.** An infection of the sinuses is known as **sinusitis.** (Bacterial and viral infections are discussed in more detail in the next chapter, but a few of the more common ones are listed in Table 14–3, on page 340.)

Once inside the respiratory tract, bacteria, viruses, and other microorganisms can spread to other organ systems. A good example is **meningitis,** a bacterial or viral infection of

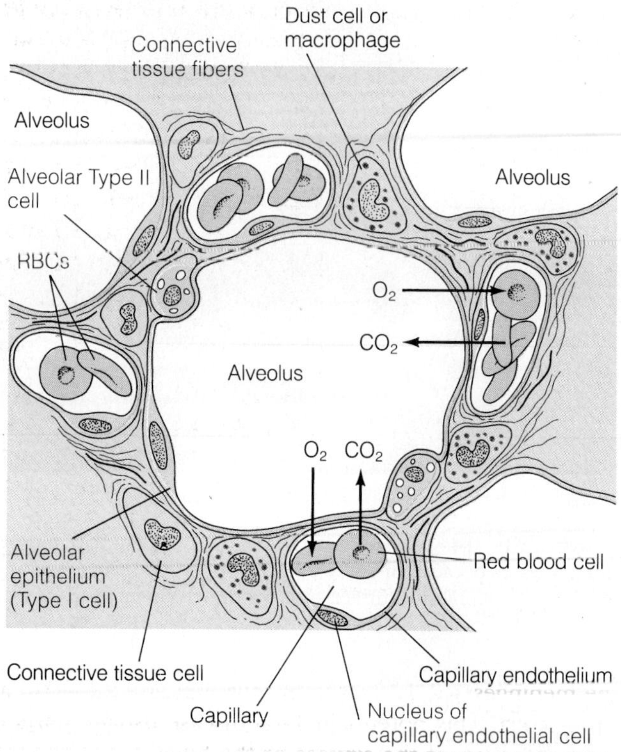

▻ **FIGURE 14–14 Close-Up of the Alveolus** Oxygen diffuses out of the alveolus into the capillary. Carbon dioxide diffuses in the opposite direction, entering the alveolar air that is expelled during exhalation.

▻ **FIGURE 14–15 Oxygen Diffusion** Oxygen travels from the alveoli into the blood plasma, then into the RBCs, where much of it binds to hemoglobin. When the oxygenated blood reaches the tissues, oxygen is released from the RBCs and diffuses into the plasma, then interstitial fluid and into body cells.

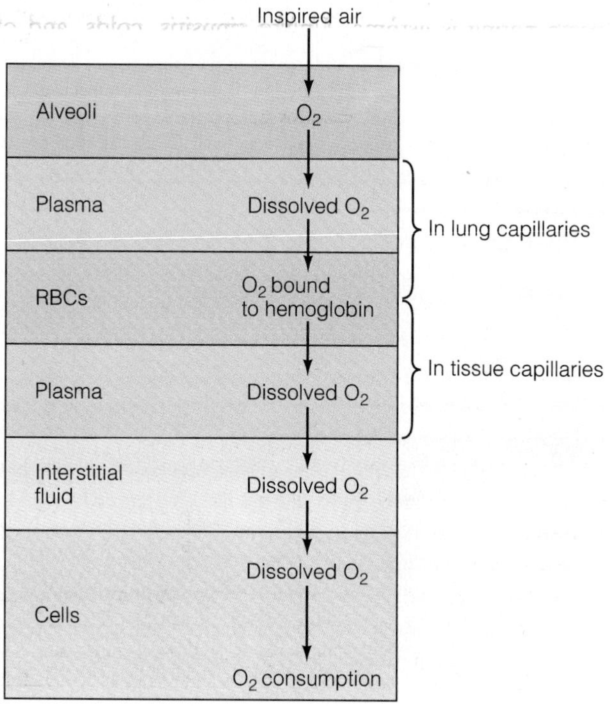

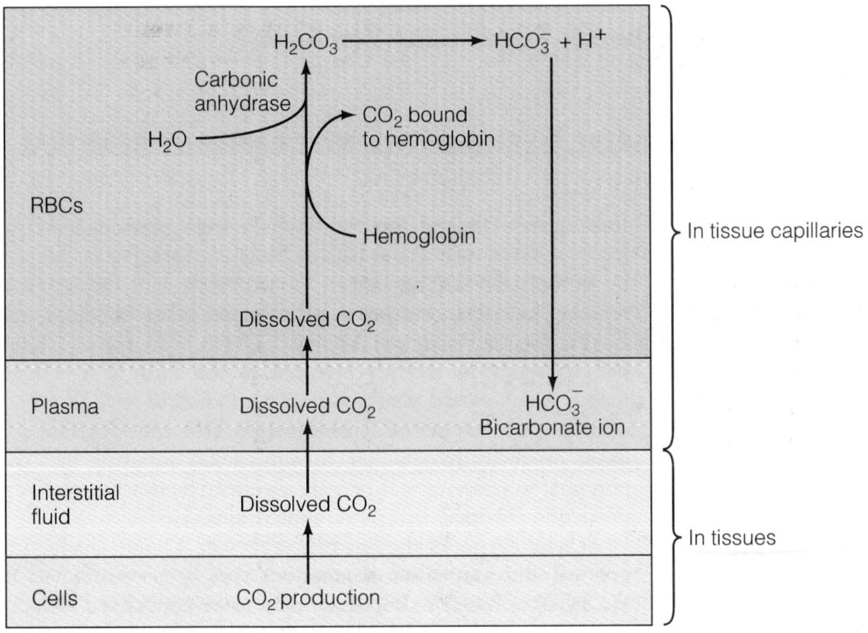

▻ **FIGURE 14–16 Bicarbonate Ion Production** Carbon dioxide (CO_2) diffuses out of body cells where it is produced and into the tissue fluid, then into the plasma. Although some carbon dioxide binds to hemoglobin and some is dissolved in the plasma, most is converted to carbonic acid (H_2CO_3) in the RBCs. Carbonic acid dissociates and forms hydrogen ions and bicarbonate, most of which is transported in the plasma.

the meninges, the fibrous layers surrounding the brain and spinal cord. This potentially fatal disease usually starts out as an infection of the sinuses or the lungs.

The lungs are also susceptible to airborne materials, among them asbestos fibers, which can cause two types of lung cancer and a debilitating disease known as **asbestosis**. Asbestosis is a build up of scar tissue that reduces the lung capacity. Because it is believed to be dangerous, many uses of asbestos have been banned in the United States, and asbestos used for insulation and decoration is being removed from buildings. For a discussion of the contro versy over removal, see the Point/Counterpoint in this chapter.

Another common disease of the respiratory system worth noting is asthma. Unlike sinusitis, colds, and other respiratory diseases, **asthma** is a chronic disorder—a disease that persists for many years. Marked by periodic episodes of wheezing and difficult breathing, asthma is not an infectious disease. Most cases of asthma, in fact, are caused by allergic reactions, or abnormal immune reactions, to common stimulants such as dust, pollen, and skin cells (dander) from pets. In some individuals, certain foods, such as eggs, milk, chocolate, and food preservatives, trigger asthma attacks. Still other cases are triggered by drugs, vigorous exercise, and physiological stress.

In asthmatics, irritants, such as pollen and dander, cause a rapid increase in the production of mucus by the bronchi and bronchioles, making breathing considerably more difficult. Irritants also stimulate the contraction of the smooth muscle cells in the walls of the bronchioles, constricting the

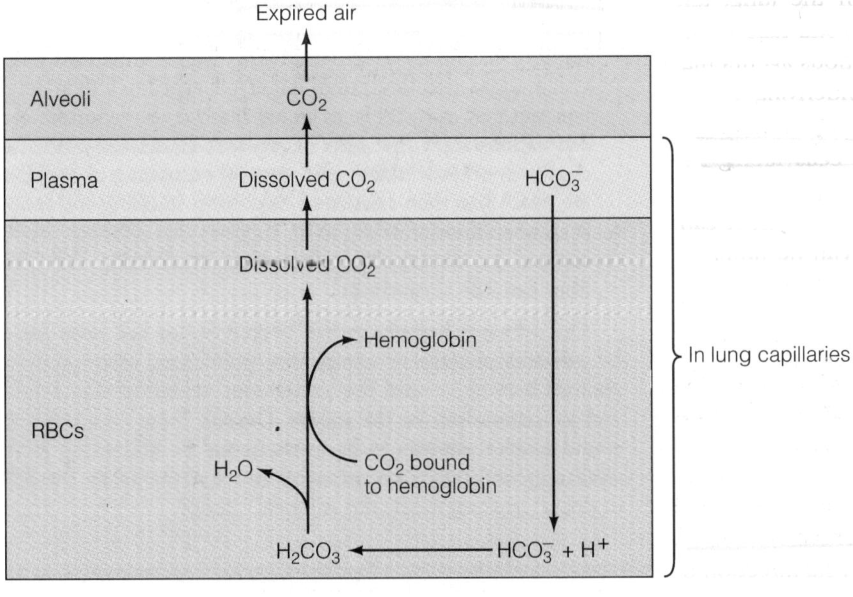

▻ **FIGURE 14–17 Carbon Dioxide Production from Bicarbonate** When the carbon-dioxide-laden blood reaches the lungs, bicarbonate ions combine with hydrogen ions in RBCs form carbonic acid, which dissociates, forming carbon dioxide gas. CO_2 diffuses out of the RBCs into the plasma, then into the alveoli.

MAINTENANCE OF ASBESTOS-CONTAINING MATERIAL IN BUILDINGS, NOT RAMPANT REMOVAL

Brooke T. Mossman

Brooke T. Mossman, Ph.D., is an associate professor of pathology at the University of Vermont College of Medicine. Her research, which is supported by the National Institutes of Health and the Environmental Protection Agency, concerns lung disease caused by asbestos fibers and implications for public policy on asbestos.

Despite the fact that scientific panels have concluded that the health risks of environmental (as opposed to worker) exposures to asbestos are low, the United States has been caught up in an asbestos removal fiasco, with projected expenditures of as much as $150 billion in the next 10 years. There is evidence that poorly performed asbestos removal may even increase air levels of fibers for extended periods and may increase the risk of disease in untrained individuals removing asbestos. The expenses incurred in removing asbestos over the past several years are approximately equivalent to the overall annual budgets of the National Institutes of Health. Thus, we are spending funds on asbestos removal desperately needed to remedy other significant health problems such as the AIDS epidemic, smoking, and drugs and alcohol.

Asbestos refers to a diverse group of naturally occurring fibers that are ubiquitous in the environment and mined for their use in insulation, friction materials, floor and ceiling tiles, and the like. Classic studies in the 1960s established that "blue" (crocidolite) asbestos, mined in parts of Africa and western Australia, caused a unique tumor called "mesothelioma" in asbestos miners and their family members. U.S. studies on insulation workers exposed to both amosite ("brown") asbestos and chrysotile ("white") asbestos and on shipyard workers (exposed largely to blue asbestos) demonstrated increases in numbers of mesotheliomas and lung cancers. Lung cancers were almost exclusively reported in heavy smokers, but occupational exposure to asbestos increased the risks of tumors in a multiplicative fashion.

Although a report by the National Academy of Sciences in 1984 suggested that environmental health risks of asbestos were low in comparison with cancer risks in workers, Congress in 1986 passed the Asbestos Hazard Emergency Response Act (AHERA), calling for inspection of asbestos-containing material (ACM) in private and public schools. If ACM is found, schools are required to notify parents and develop management plans for its control. The passage of AHERA, misconceptions about environmental cancer risks due to asbestos, and public panic encouraged the rampant removal of asbestos from schools and other public buildings. Costs of removal decimated school systems and engendered widespread litigation. More importantly, many poorly performed removals resulted in higher airborne concentrations of fibers than those observed before removal. Inadequately trained removal workers were often exposed to levels of asbestos fibers as high as those encountered in the past by workers developing cancer. Concurrently with the asbestos removal fiascoes occurring in schools and other buildings, the Environmental Protection Agency (EPA) in 1989 also initiated a ban on asbestos, recently reversed by the Fifth U.S. Circuit Court, which would phase out various industrial uses of asbestos over a 10-year period. Unfortunately, little consideration was given originally to whether substitutes for asbestos had the potential to cause cancer or were as effective as asbestos fibers in friction products and other applications.

Several events of the last two years will, I hope, change the political and journalistic perceptions that have contributed to the asbestos hysteria. Important data to be considered include:

1. Types of asbestos differ both chemically and physically. The types of asbestos (crocidolite, amosite) associated in the workplace with the development of mesothelioma are not the types most often encountered in buildings. Ninety percent to 95% of the bulk ACM in buildings is chrysotile (white) asbestos.
2. Asbestos fibers must be airborne and of a certain respirable size range to cause cancers in humans. Longer fibers (>5 microns), which are associated with the development of tumors in animals, are rarely found in building air samples.
3. In many situations, asbestos fibers are in a matrix or behind walls and are thus inaccessible. Unnecessary removal causes disruption of surfaces and may result in increased levels of airborne fibers and increased exposures to building occupants and removal workers.
4. The incidence of mesothelioma, a tumor associated with occupational exposure to asbestos, in U.S. women and men under 65 has been stable over the past several years, indicating that environmental exposure to asbestos is not resulting in increases in this disease.
5. Air monitoring of asbestos fibers in buildings, homes, and schools has indicated that levels are comparable to levels in outdoor air and thousands-fold less (even in situations where visual inspection indicated damaged asbestos) than levels encountered in workplaces where disease was reported. Risk estimations show that risks of asbestos-associated premature deaths in schoolchildren and general occupants of buildings are much less than calculated risks from smoking and radon.
6. A compendium of information suggests that asbestos-related cancers are dosage-dependent, thus invalidating the "one-fiber-can-kill" hypothesis.

The asbestos hysteria in the United States has been fueled by asbestos abatement companies, politicians, union activists, plaintiff lawyers . . . and the 'courtroom' scientists they employ, and sensationalism by the media. Despite these counterforces, sound science appears to be instrumental in influencing recent environmental policy on asbestos and statements by the EPA urging a "management, not removal" policy.

APPROPRIATE BALANCE BETWEEN REMOVAL AND IN-PLACE MANAGEMENT IS NEEDED

John M. Dement, Ph.D.

John M. Dement, Ph.D., is Director of Disease Prevention Research at the National Institute of Environmental Health Sciences. His research includes epidemiological studies of human populations exposed to asbestos and other fibers.

Extensive use of asbestos in building products in the United States has resulted in millions of tons of "in-place" material, much of which is deteriorating. The magnitude of asbestos exposure and the resulting risks to building occupants have been the subject of considerable debate. Although many would agree that some asbestos is being removed from schools and buildings that would pose less human health risk if properly managed in place, some authors go much further to imply that asbestos removal is *seldom* warranted and that "chrysotile asbestos, the type of fiber found predominantly in U.S. schools and buildings, is not a health risk in the nonoccupational environment." This conclusion is not supported by published literature concerning the carcinogenicity of chrysotile asbestos. Furthermore, most risk estimates for building occupants fail to consider exposure levels of airborne asbestos and exposure in buildings containing friable asbestos as a result of normal cleaning and maintenance operations.

The term *asbestos* refers to several fibrous mineral species and includes both serpentines (chrysotile) and amphiboles (amosite, crocidolite, tremolite, anthophyllite). Although these minerals have different crystal structures and chemical compositions, all produce respirable fibers that are capable of penetrating deeply into the lungs. Epidemiological studies of human populations exposed to asbestos have clearly demonstrated excess risks for a number of diseases, including asbestosis, lung cancer, mesothelioma, gastrointestinal cancers, and cancers at several other sites. Extensive reviews of the epidemiological literature clearly demonstrate the ability of both chrysotile and amphiboles to cause all of these diseases. The lung cancer exposure-response slope has been shown to vary widely; however, the steepest exposure-response relationship for lung cancer in any asbestos-exposed study has been observed among workers processing chrysotile. Furthermore, animal bioassay data clearly establish the carcinogenicity of chrysotile. The ability of chrysotile to produce asbestosis, lung tumors, and mesotheliomas in rats by inhalation has been amply demonstrated.

Although the incidence of all asbestos-related diseases increases with increasing dose, mesotheliomas have been observed among individuals with very brief periods of exposure. A recent study of mesothelioma cases in Wisconsin identified nine cases where the only identifiable source of asbestos exposure was living or working in a building containing asbestos. These data are consistent with a detailed case report of mesothelioma due to building materials containing amosite asbestos.

Assessing airborne exposures in buildings for purposes of risk assessment is problematic due to the variable nature of asbestos fiber release. Although *average* exposures are very low, *peak* exposures can occur during normal building maintenance and cleaning. Significantly elevated exposure levels have been demonstrated during routine cleaning operations such as sweeping and dusting in asbestos-contaminated buildings. Elevated levels of airborne fibers have been clearly demonstrated in schools containing damaged, friable asbestos. The significance of these peak exposures becomes more clear upon review of recent data that demonstrate asbestos-related diseases among school maintenance and custodial personnel. In these studies, the prevalence of asbestos-related changes in chest X-rays ranged from 11% to 27% for workers whose only known asbestos exposure had occurred in schools.

The public health implications for asbestos-containing materials in place are not insignificant. Risk assessments must consider risks to cleaning and maintenance personnel as well as to building occupants where damaged materials create elevated exposure for general occupants. Although the individual risk to building occupants may be small, the impact on the overall population is very significant given the number of people exposed. EPA regulations promulgated under the Asbestos Hazard Emergency Response Act (AHERA) *do not* require that asbestos be removed but require assessments and management plans. The solution is appropriate triage to ensure that the least-risk alternative is chosen. When asbestos is removed, it must be done using established methods that will minimize exposures to building occupants and removal workers. Sweeping conclusions concerning the lack of a health risk for chrysotile asbestos in the nonoccupational environment are not supported by the available literature and may lead to inappropriate asbestos-control strategies.

SHARPENING YOUR CRITICAL THINKING SKILLS

1. Summarize the key points and supporting arguments of each author.
2. Do you see any errors in the reasoning of either author? If so, where? What critical thinking rules were violated?
3. Given the extremely technical nature of this debate, can you decide whether asbestos in public schools should be removed?

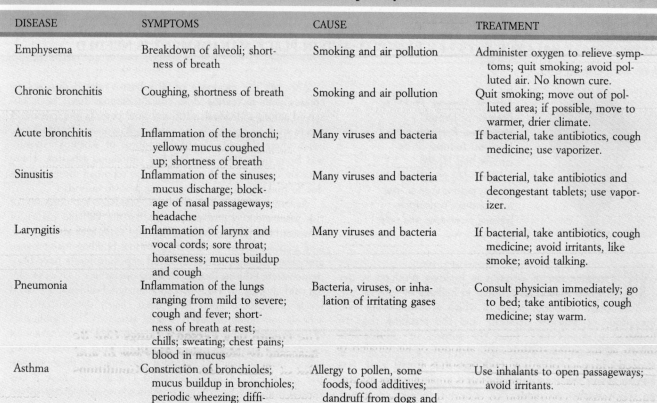

TABLE14–3 Common Respiratory Diseases

DISEASE	SYMPTOMS	CAUSE	TREATMENT
Emphysema	Breakdown of alveoli; shortness of breath	Smoking and air pollution	Administer oxygen to relieve symptoms; quit smoking; avoid polluted air. No known cure.
Chronic bronchitis	Coughing, shortness of breath	Smoking and air pollution	Quit smoking; move out of polluted area; if possible, move to warmer, drier climate.
Acute bronchitis	Inflammation of the bronchi; yellowy mucus coughed up; shortness of breath	Many viruses and bacteria	If bacterial, take antibiotics, cough medicine; use vaporizer.
Sinusitis	Inflammation of the sinuses; mucus discharge; blockage of nasal passageways; headache	Many viruses and bacteria	If bacterial, take antibiotics and decongestant tablets; use vaporizer.
Laryngitis	Inflammation of larynx and vocal cords; sore throat; hoarseness; mucus buildup and cough	Many viruses and bacteria	If bacterial, take antibiotics, cough medicine; avoid irritants, like smoke; avoid talking.
Pneumonia	Inflammation of the lungs ranging from mild to severe; cough and fever; shortness of breath at rest; chills; sweating; chest pains; blood in mucus	Bacteria, viruses, or inhalation of irritating gases	Consult physician immediately; go to bed; take antibiotics, cough medicine; stay warm.
Asthma	Constriction of bronchioles; mucus buildup in bronchioles; periodic wheezing; difficulty breathing	Allergy to pollen, some foods, food additives; dandruff from dogs and cats; exercise	Use inhalants to open passageways; avoid irritants.

openings and making it even more difficult for asthmatics to move air in and out of their lungs.

Asthma is fairly common in school children but often disappears as they grow older. As a result, about 2% of the adult population suffers from asthma. Nevertheless, asthma is a serious disease. Periodic attacks can be quite disabling; some may even lead to death. By one estimate, several thousand Americans die each year from severe asthma attacks. Victims are generally elderly individuals who are suffering from other diseases.

Although asthma is incurable, the severity of an attack can be greatly lessened by proper medical treatment. One of the most common treatments is an oral spray (inhalant) containing the hormone epinephrine, which stimulates the bronchioles to open (▻ Figure 14–18). Screening tests can help a patient find out what substances trigger an asthmatic attack so they can be avoided.

BREATHING AND THE CONTROL OF RESPIRATION

Air moves in and out of the lungs in much the same way that it moves in and out of the bellows that blacksmiths use to fan their fires. Breathing, however, is largely an involuntary action, controlled by the nervous system.

Air Is Moved In and Out of the Lungs by Changes in the Intrapulmonary Pressure

During breathing, air must first be drawn into the lungs. This process is known as **inspiration, or inhalation** (Table 14–4). Following inspiration, air must be expelled. This is known as **expiration, or exhalation.** For a discussion of an

▻ **FIGURE 14–18 Asthma Relief** Constriction of the bronchioles can be released by epinephrine inhalant spray.

interesting evolutionary variation on this theme, see Spotlight on Evolution 14-1.

Inhalation is an active process that is controlled by the brain. Nerve impulses traveling from the brain stimulate the **diaphragm,** a dome-shaped muscle (unique to mammals) that separates the abdominal and thoracic cavities (▷ Figure 14–19a). These impulses cause the diaphragm to contract. When it contracts, the diaphragm flattens and lowers. Much in the same way that pulling the plunger of a syringe draws in air, the contraction of the diaphragm draws air into the lungs.

Inhalation also requires the **intercostal muscles,** short, powerful muscles that lie between the ribs. (They are the meat on barbecued ribs.) Nerve impulses traveling to these muscles cause them to contract as the diaphragm is lowered. When the intercostal muscles contract, the rib cage lifts up and out (Figure 14–19a). Together, the contractions of the intercostal muscles and the diaphragm increase the volume of the thoracic cavity (Figure 14–19b). This, in turn, decreases the **intrapulmonary pressure,** the pressure in the alveoli, which draws air in through the mouth or nose downward into the trachea, bronchi, and lungs. At rest, each breath delivers about 500 milliliters of air to the lung. This is known as the **tidal volume,** the amount of air inhaled or exhaled with each breath when a person is at rest.

As you have just seen, inhalation is an active process that requires muscle contraction to occur. In a person at rest, exhalation is a decidedly passive process (not requiring muscle contraction) beginning after the lungs fill with air. At this point in the cycle, the diaphragm and intercostal muscles relax. The relaxed diaphragm rises and resumes its domed shape, and the chest wall falls slightly inward. These changes reduce the volume of the thoracic cavity, raising the pressure and forcing the air out. The lungs also participate in passive exhalation. Containing numerous elastic connective tissue fibers, they fill like balloons during inspiration. When inhalation ceases, the lungs simply recoil (shrink), forcing air out.

Although exhalation is a passive process in an individual at rest, it can be made active by enlisting the aid of the muscles of the wall of the chest and abdomen. The forceful expulsion of air is known as **forced exhalation.** Try this by taking in a breath of air, then actively forcing it out. During forced exhalation, contraction of the abdominal muscles increases the intra-abdominal pressure, forcing the abdominal organs upward against the diaphragm. Contraction of the muscles in the wall of the chest also reduces the volume of the chest, helping force the air out of your lungs.

Inhalation can also be consciously augmented by a forceful contraction of the muscles of inspiration. (You can test this by taking a deep breath.) Forced inhalation increases the amount of air entering your lungs. Athletes often actively inhale and exhale just before an event to increase oxygen levels in their blood. A competitive swimmer, for example, may take several deep breaths before diving into the pool for a race. Deep breathing, while effective, can be dangerous, for reasons explained shortly.

TABLE 14–4 Summary of Inhalation and Exhalation

Inhalation
- Nerve impulses from the breathing center stimulate the muscles of inspiration – the diaphragm and intercostal muscles.
- Contraction of the intercostal muscles causes the rib cage to move up and out.
- Contraction of the diaphragm causes it to flatten.
- Volume of the thoracic cavity increases.
- Intrapulmonary pressure decreases.
- Air flows into the lungs through the nose and mouth.

Exhalation
- Nerve impulses from the breathing center feed back on it, shutting off stimuli to muscles of inspiration.
- The intercostal muscles relax and the rib cage falls.
- The diaphragm relaxes and rises.
- The lungs recoil.
- Air is pushed out of the lungs.

The Health of a Person's Lungs Can Be Assessed by Measuring Air Flow In and Out of Them Under Various Conditions

Studies show that children who are exposed to tobacco smoke at home suffer a decrease in lung capacity—that is, a decrease in their ability to move air in and out of their lungs. Several measurements of lung capacity are routinely used by physicians to determine the "health" of a person's lungs.

Measurements of lung function are taken under controlled conditions. As shown in ▷ Figure 14–20a (page 344), patients breathe into a machine that measures the amount of air moving in and out of the lung at various times.

Figure 14–20b shows a graph of some of the common measurements. The first is the tidal volume, which, as noted earlier, is the amount of air that moves in and out during passive breathing. Another important measurement is the **inspiratory reserve volume**—the amount of air that can be drawn into the lungs during active inspiration. Deep inhalation pulls in four to six times more air than the tidal volume, or 2000 to 3000 milliliters, depending on the size of the individual.

After exhalation, under resting conditions, the lungs still contain a considerable amount of air—about 2400 milliliters. Forced exhalation will expel about half of that air. The amount that can be exhaled after a normal exhalation is called the **expiratory reserve volume.** The remaining 1200 milliliters is known as the **residual volume.**

Lung disease often results in changes in the amount of air that can be moved in and out of the lung or changes in the residual volume. Asthma, for example, reduces the inspiratory reserve volume—the amount of air that can be drawn into the lungs by forced inspiration—because the constricted bronchioles limit air flow.

SPOTLIGHT ON EVOLUTION 14–1

AVIAN ADAPTATIONS TO HIGH-ALTITUDE LIVING

Radar observers scan the night skies monitoring airline traffic. Time and again they track flocks of birds flying as high as 6500 meters (21,000 feet) above sea level, where oxygen is low and the air bitterly cold. But birds are capable of surviving at even higher altitudes. One species, in fact, is routinely found on the slopes of Mount Everest, 8200 meters (27,000 feet) above sea level. Even more remarkably, bar-headed geese pass directly over the summit of the Himalayas at altitudes of 9200 meters (30,000 feet) during the annual migration. How do birds manage to breathe at such altitudes?

The answer lies in the bird's unique respiratory system, unmatched among vertebrates. Shown in ▻ Figure 1a, the respiratory system of the bird consists of a trachea, eight or nine supplemental air sacs, and two lungs. The air sacs hold nine times more air than the bird's lungs, but are poorly vascularized and therefore are not involved in gas exchange.

Unlike the human respiratory system, in which air flows into the lungs, then back out, retracing its steps like the tides, the bird's respiratory system is characterized by a unidirectional air flow, an evolutionary adaptation that greatly increases efficiency. In order for air to pass through the respiratory system of the bird, however, two cycles of inhalation and exhalation are required. During the first inhalation, the volume of the thoracic cavity increases, drawing air into the trachea. As illustrated in Figure 1b, however, most of the air bypasses the lungs and enters the posterior air sacs. During exhalation, the volume of the thoracic cavity decreases, and the air in the posterior air sacs passes into the lungs.

During the next inhalation, while new air is being drawn into the posterior air sacs, the air in the lungs is pushed into the anterior air sacs. During the second exhalation, the air in the anterior air sacs is expelled, while the air in the posterior air sacs is pushed into the lungs. As a result, the lung always contains a relatively fresh supply of air.

A peek inside the bird's lungs reveals another interesting adaptation that improves the efficiency of respiration. Unlike mammalian lungs, which contain a series of branching tubules that end in dead-end sacs (alveoli) where oxygen exchange occurs, bird lungs contain channels through which air flows continuously in one direction. Tiny blind-ended tubules branch off from these channels, coming in intimate contact with blood capillaries that draw off the oxygen and dump carbon dioxide. In the bird lung, moreover, air and blood flow in opposite directions, which is more efficient than the mammalian system. Combined with the one-way flow of air, this highly efficient system helps birds acquire the oxygen they need at very high altitudes.

Besides providing birds an efficient means of acquiring oxygen and ridding the body of carbon dioxide, this unique arrangement of lungs and air sacs helps lighten the bird's body, reducing energy use during flight. Hollow, air-filled bones found in many species may also assist the bird in flight. Ornithologists (biologists who study birds) believe that the air sacs probably also serve as a repository for heat generated by metabolism in the bird's flight muscles, thus helping the bird stabilize its body temperature and maintain homeostasis.

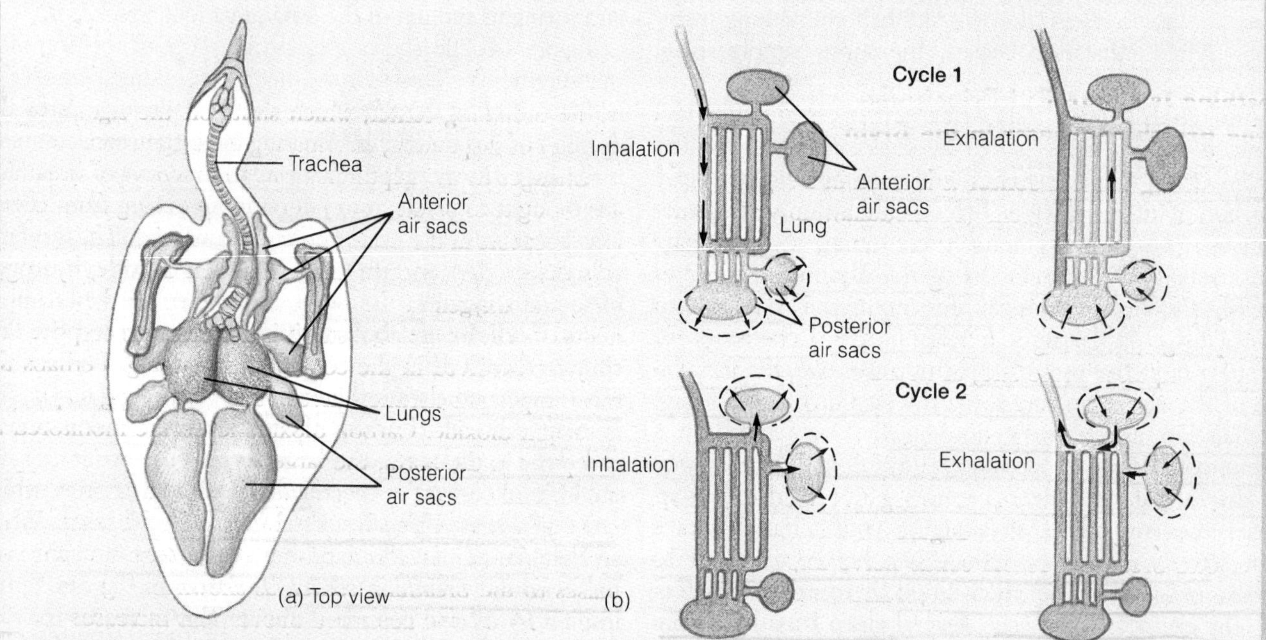

▻ **FIGURE 1 Respiratory System of the Bird** (*a*) Note the air sacs on both sides of the lungs. (*b*) Schematic illustrating flow of air in the respiratory system of the bird.

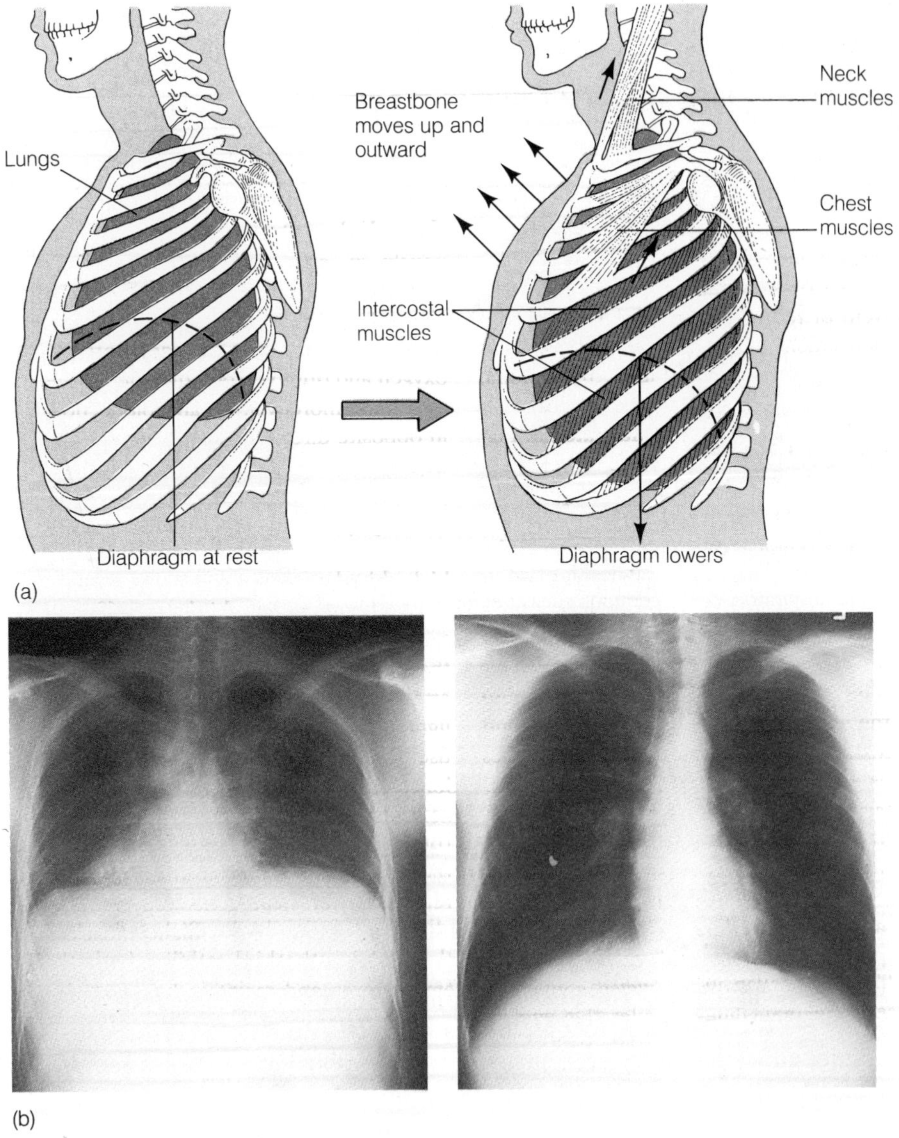

▻ **FIGURE 14–19 The Bellows Effect** (*a*) The rising and falling of the chest wall caused by contraction of the intercostal muscles is shown in this diagram, illustrating the bellows effect. Inspiration is assisted by the diaphragm, which lowers, like the plunger on a syringe, drawing air into the lungs. (*b*) X-rays showing the size of the lungs in full exhalation (*left*) and full inspiration (*right*).

Breathing Is Controlled Principally by the Breathing Center in the Brain

The **breathing control center** is located in a region of the brain called the brain stem (or medulla). In some ways similar to the sinoatrial node of the heart, the breathing center contains nerve cells that give off periodic impulses that stimulate contraction of the intercostal muscles and the diaphragm, resulting in inhalation. When the lungs fill, the nerve impulses cease and, as noted earlier, the muscles relax. This, in turn, decreases the volume of the thoracic cavity and forces air out of the lungs.

Several mechanisms are responsible for the termination of these impulses. The first is a negative feedback loop, shown on the right side of ▻ Figure 14–21. Here's how it works. The breathing center sends nerve impulses to the diaphragm and intercostal muscles; it also sends impulses to a nearby region of the brain stem, which is a kind of "relay center" that transmits nerve impulses back to the breathing center. When these impulses arrive, they inhibit the neurons in the breathing center, which shuts off the signals to the muscles of inspiration, terminating the inspiration.

Changes in the depth and rate (frequency) of breathing are thought to result from neural input arising from chemical receptors in the brain and certain arteries. These receptors detect the concentration of carbon dioxide, hydrogen ions, and oxygen in the body. Since they are key components of cellular metabolism, it is therefore no surprise that they are involved in the control of breathing. Perhaps the most important chemical involved in controlling respiration is carbon dioxide. Carbon dioxide levels are monitored by receptors in the aorta, the large artery that carries oxygenated blood out of the heart, and the carotid arteries, which carry oxygenated blood to the brain from the aorta. When carbon dioxide levels rise, these receptors transmit impulses to the breathing center, as shown in Figures 14–21 and ▻ 14–22 and described above. This increases the rate of breathing. A decline in carbon dioxide levels has the opposite effect.

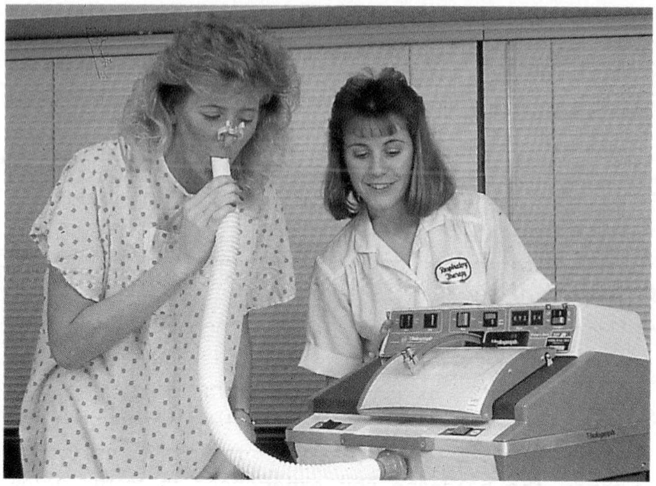

(a)

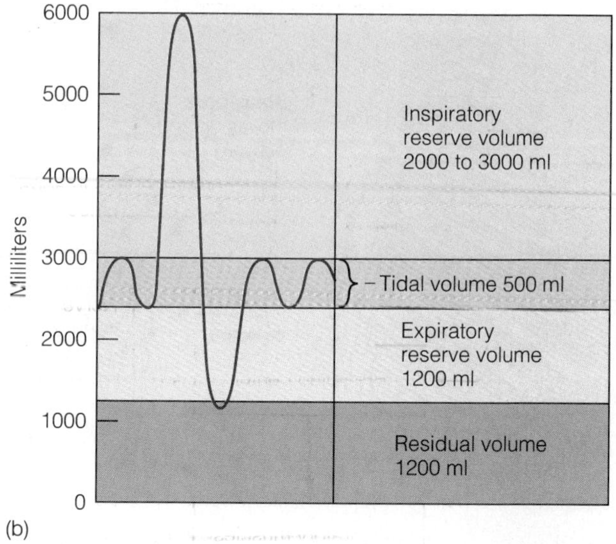

(b)

▻ **FIGURE 14–20 Measuring Lung Capacity** *(a)* This machine allows physicians to determine tidal volume, inspiratory reserve volume, and other lung-capacity measurements to determine the health of an individual's lung. *(b)* This graph shows several common measurements.

In the brain, carbon dioxide diffuses into the **cerebrospinal fluid (CSF),** a clear liquid found within cavities in the brain known as the **ventricles.** In the CSF, carbon dioxide is converted into carbonic acid, which then dissociates to form bicarbonate and hydrogen ions. A rise in carbon dioxide in the blood, therefore, results in an increase in the H^+ concentration of the CSF. The increase in H^+, in turn, is detected by chemical receptors, or **chemoreceptors,** in the brain. These receptors send impulses to the breathing center, triggering an increase in the rate and depth of breathing (Figure 14–22).

The chemoreceptors in the brain and arteries allow the body to align respiration with cellular demands. During exercise, for example, cellular energy production increases to meet body demands. As noted in Chapter 6, energy production requires oxygen and generates carbon dioxide waste. The carbon dioxide produced during exercise increases the depth and rate of breathing. This has two effects. First, it helps lower the concentration of carbon dioxide in the blood. (Breathing slows when levels return to normal.) Second, increased ventilation also makes more oxygen available for energy production.

The body also contains a set of oxygen receptors. These receptors are not as sensitive as the H^+ receptors so oxygen levels must fall considerably before the oxygen receptors begin generating impulses. This fact can have pro-

▻ **FIGURE 14–21 Breathing Center** The breathing center controls respiration. It sends periodic impulses along the nerves to the muscles of inspiration, causing them to contract. The center also sends impulses along another route to a relay center in the brain stem. Impulses from here feed back to the breathing center, shutting off the impulses that stimulate inspiration. Chemical receptors in the brain and certain arteries and stretch receptors in the lung also alter the activity of the breathing center.

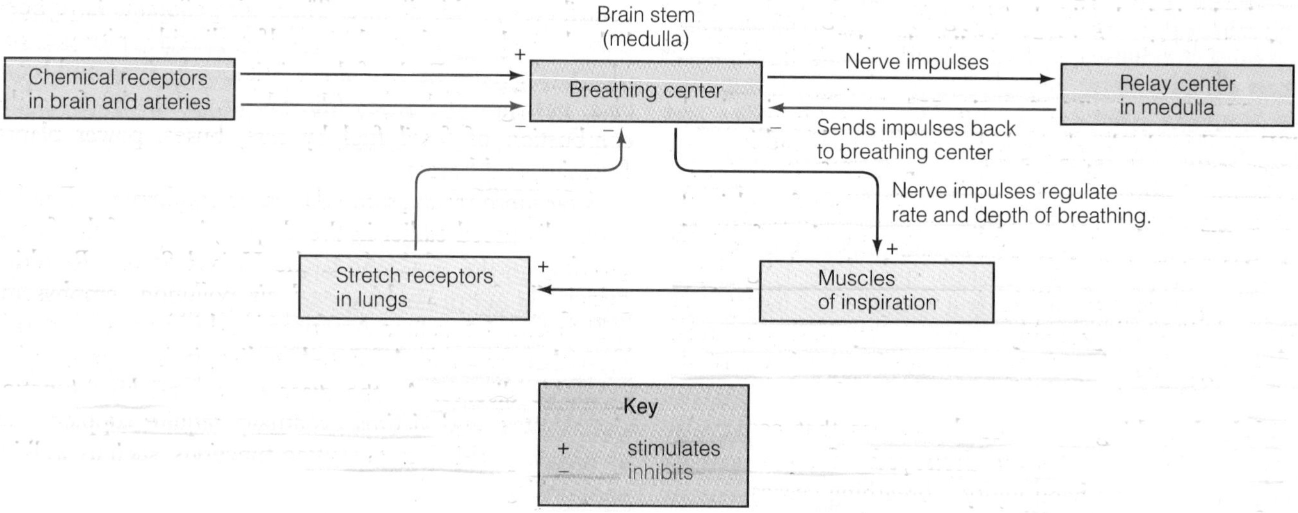

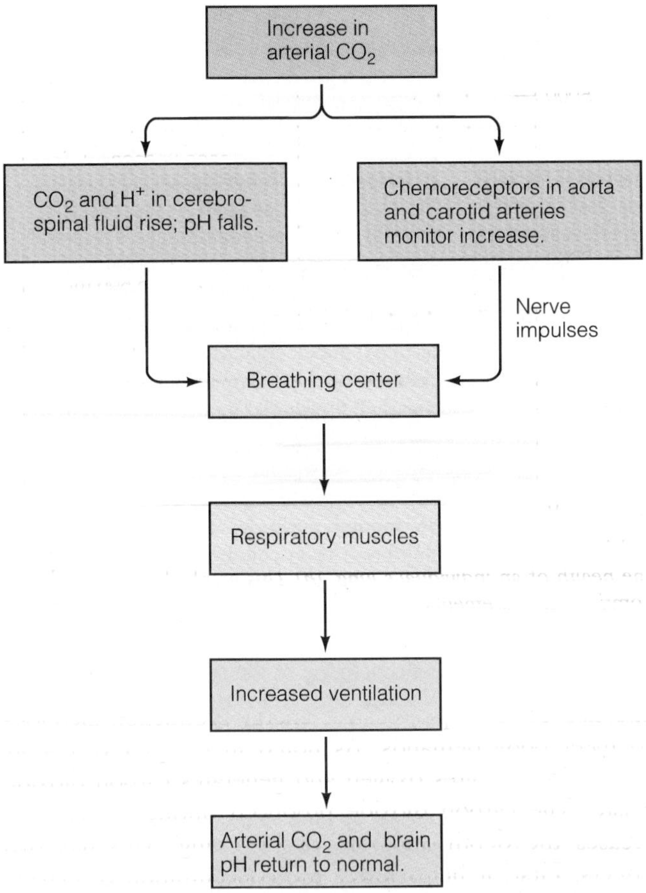

▻ **FIGURE 14–22 Chemical Control of Breathing**
CO_2 and hydrogen ions are the chief chemical controls of breathing.

found consequences for divers and swimmers. Repeated deep and rapid breathing, or **hyperventilation,** for example, makes it possible for divers to hold their breath underwater longer. Hyperventilation decreases carbon dioxide levels in the blood and H^+ concentrations in the cerebrospinal fluid, reducing the urge to breathe. When the diver enters the water, oxygen levels in the blood may fall so low that the brain is deprived of oxygen, causing the individual to lose consciousness. Ironically, the decrease in blood oxygen levels is not enough to stimulate breathing; the diver blacks out well before the H^+ concentration in the CSF reaches the level needed to stimulate breathing.

The third control mechanism consists of sensory nerve fibers known as stretch receptors, which are found in the lungs. When the lung is full, nerve impulses from the stretch receptors are transmitted to the breathing center. These impulses turn off the breathing center. Stretch receptors, however, probably function only during exercise, when large volumes of air are moved in and out of the lungs.

If the breathing center or the nerves that convey impulses to these muscles are destroyed—for example, by the polio virus or a head injury—breathing ceases.

ENVIRONMENT AND HEALTH: AIR POLLUTION

The air of the industrialized world contains a multitude of potentially harmful chemicals, or pollutants, that are damaging to human health. Air pollutants generated in our homes, factories, and cities in fact now claim the lives of thousands of Americans each year.

Air pollution upsets the normal homeostatic balance of our bodies, affecting millions of us in often subtle, but sometimes pronounced, ways. Unfortunately, most people are unaware of the dangers of air pollution because the line between cause and effect is not always clear. Consider, for instance, the headache you experienced in traffic going home from school or work. Was it caused by tension, or could it have been caused by carbon monoxide emissions from cars, buses, and trucks? And what about the runny nose and sinus condition you experienced last winter? Were they caused by a virus or bacterium, or by pollution?

One classic study of air pollution on the East Coast illustrates the relationship between air pollution and upper respiratory problems. The researchers found that the level of sulfur dioxide, a pollutant produced by automobiles, power plants, and factories, increased during the winter months in New York City. In one episode, certain weather conditions trapped the pollutants, raising ground-level sulfur dioxide concentrations. During this episode, upper respiratory illness in New York residents skyrocketed (▻ Figure 14–23). Colds, coughs, nasal irritation, and other symptoms increased fivefold in a few days. Soon after the pollution levels returned to normal, the symptoms subsided.

Air pollution is partly responsible for other long-term diseases. One of these is **chronic bronchitis,** a persistent irritation of the bronchi characterized by excess mucus production, coughing, and difficulty breathing. Today, one out of every five American men between the ages of 40 and 60 suffers from chronic bronchitis.

Although the leading cause of chronic bronchitis is cigarette smoking, studies show that urban air pollution also contributes to this disease. Three air pollutants have been linked to chronic bronchitis: sulfur dioxide, nitrogen oxides, and ozone. Each of these irritates the lung and bronchial passages and arises (directly or indirectly) from the combustion of fossil fuel by cars, buses, power plants, factories, and homes.

A far more troublesome disease is emphysema. Emphysema, discussed earlier in the chapter, is one of the fastest growing causes of death in the United States. Resulting principally from smoking and air pollution, emphysema afflicts over 1.5 million Americans and is more common in men than women. As noted earlier, it is a progressive, incurable disease. As the disease worsens, lung function deteriorates, and victims eventually require supplemental oxygen to perform even routine functions, such as walking or speaking.

HEALTH NOTE 14–2

SMOKING AND HEALTH: THE DEADLY CONNECTION

Urban air pollution worries many Americans, and for good reason. However, the city air that many of us breathe is benign compared with the "air" that 65 million Americans voluntarily inhale from cigarettes. Loaded with dangerous pollutants in concentrations far greater than those of our cities, cigarette smoke takes a huge toll on citizens of the world. In the United States, for example, an estimated 390,000 people—about 1000 every day—die from the many adverse health effects of tobacco smoke, including heart attacks, lung cancer, and emphysema.

Smoking also costs society a great deal in medical bills and lost productivity. According to the Worldwatch Institute, every pack of cigarettes sold in the United States costs our society $1.25 to $3.17 in medical costs, lost wages, and reduced productivity—that's about $125 billion to $400 billion a year!

Smoking is a principal cause of lung cancer, claiming the lives of an estimated 130,000 men and women in the United States each year. Depending on how many packs they smoke each day, smokers are 11 to 25 times more likely to develop lung cancer than nonsmokers.

Unfortunately, nonsmokers are also affected by the smoke of others. Nonsmokers inhale tobacco smoke in meetings, in restaurants, at work, and at home. For years, these "passive smokers" have had little to say about their exposure to other people's smoke. Today, however, as a result of a growing awareness of the dangers, new regulations are banning smoking in many public places and in the workplace, except in designated areas.

This trend has been spurred, in part, by research showing that passive smokers are more likely to develop lung cancer than people who manage to steer clear of smokers. In a study of Japanese women married to men who smoked, researchers found that the wives were as likely to develop lung cancer as people who smoked half a pack of cigarettes a day! A recent report by the U.S. Environmental Protection Agency estimates that passive smoking causes 500 to 5000 cases of lung cancer a year in the United States. Passive smokers who are exposed to tobacco smoke for long periods also suffer from impaired lung function equal to that seen in light smokers (people who smoke under a pack a day).

Cigarette smoke in closed quarters may also cause angina (chest pains) in individuals who are suffering from atherosclerosis of the coronary arteries. Carbon monoxide in cigarette smoke is responsible for angina attacks. Research also shows that smokers are more susceptible to colds and other respiratory infections. And smoking affects children. One study showed that children from families in which both parents smoked suffered twice as many upper respiratory infections as children from nonsmoking families. Recently, researchers from the Harvard Medical School reported finding a 7% decrease in lung capacity in children raised by mothers who smoked. The researchers believe that this change may lead to other pulmonary problems later in life.

Smoking has been shown to affect fertility in women as well. Studies show, for example, that women who smoke over a pack of cigarettes a day are half as fertile as nonsmokers. Smoking may also affect the outcome of pregnancy. According to the 1985 U.S. Surgeon General's Report, women who smoke several packs a day during pregnancy are much more likely to miscarry and are also more likely to give birth to smaller children. On average, the children of these women are 200 grams (nearly 0.5 pounds) lighter than children born to nonsmoking mothers.[1] Finally, children of women who smoke heavily during pregnancy generally score lower on mental aptitude tests during early childhood than children whose mothers do not smoke.

Tobacco smoke contains numerous hazardous substances that damage the delicate lining of the respiratory system. Nicotine and sulfur dioxide, for example, paralyze the cilia lining the respiratory tract. As noted in the chapter, one cigarette can knock the cilia out of action for an hour or more, eliminating the natural cleansing mechanism.

Tobacco smoke is also laden with microscopic carbon particles. Many toxic chemicals adhere to these particles and are transported into the lungs. A dozen or so of these chemicals are known to cause cancer. Carbon particles penetrate deeply into the lungs, where they accumulate in the alveoli and alveolar walls, turning healthy tissue into a blackened mass that often becomes cancerous (▶ Figure 1). Tobacco smoke may also paralyze the alveolar macrophages, making a bad situation even worse.

Toxin-carrying particles adhere to the lungs, larynx, trachea, and bronchi. Virtually any place they stick, they can cause cancer. That is why smokers are five times more likely than nonsmokers to develop laryngeal cancer and four

The leading cause of emphysema is smoking, a habit of 65 million Americans. But emphysema is also caused by urban air pollution. Not surprisingly, smokers who live in polluted urban settings have the highest incidence of the disease. Like chronic bronchitis, emphysema is caused by ozone, sulfur dioxide, and nitrogen oxides.

Researchers believe that 85% of the nearly 150,000 cases of lung cancer in the United States are caused by smoking (for more on the effects of smoking and the controversy surrounding tobacco advertising, see Health Note 14–2). The remaining 15,000 or so cases are thought to be caused by a variety of other factors, including urban and workplace air pollutants as well as natural pollutants, such as radioactive radon gas. Radon is emitted from radium, which occurs naturally in the soil in many parts of the country.

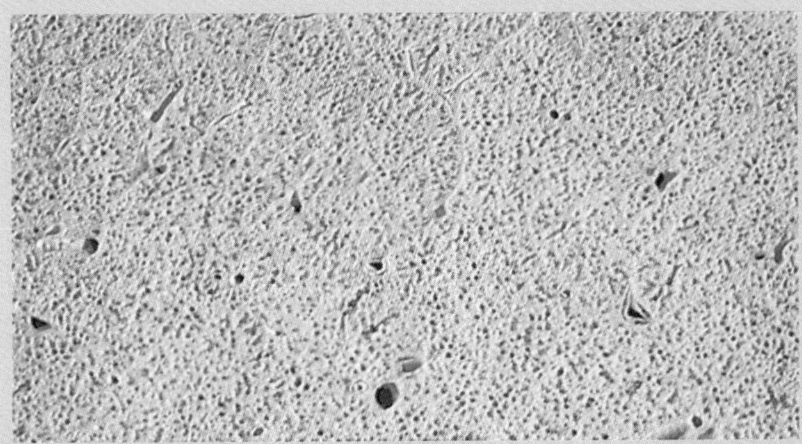

(a)

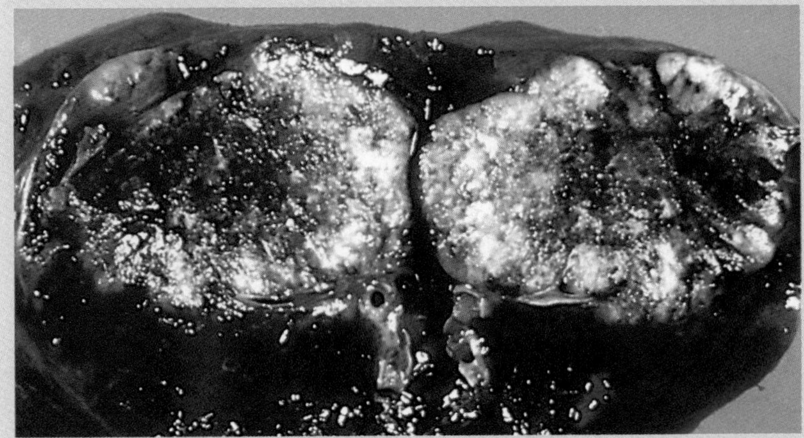

(b)

▻ **FIGURE 1 The Normal and Cancerous Lung** (*a*) The normal lung appears spongy. (*b*) The cancerous lung from a smoker is filled with particulates and tumor tissue.

times more likely to develop cancer of the oral cavity.

Nitrogen dioxide and sulfur dioxide in tobacco smoke penetrate deep into the lungs, where they dissolve in the watery layer inside the alveoli. Nitrogen dioxide is converted to nitric acid; sulfur dioxide is converted to sulfuric acid. Both acids erode the alveolar walls, sometimes leading to emphysema.

If tobacco smoke is so dangerous, you may ask, why don't we ban smoking or discontinue generous government subsidies to tobacco growers? Surely, an air pollutant or food contaminant that killed hundreds of thousands of Americans each year would be banned.

Part of the answer lies in the fact that tobacco use has a long history in the United States, and many people think that because smoking is a voluntary act, individuals should have the right to make their own decision. Furthermore, smoking supports a $30-billion-a-year tobacco industry that employs about 2 million people, including tobacco farmers, advertising people, retailers, and so on. The tobacco industry lobbies diligently to protect the rights of smokers.

The dangers of smoking are becoming well known. As a result of widespread publicity and public pressure, smoking has dropped substantially in the United States. In 1990, for example, only 26% of American adults smoked, down from 34% in 1985 and down substantially from the 1950s and 1960s, when well over half of all men and over one-third of all women engaged in this potentially lethal habit.

Despite the downturn in smoking, an estimated 65 million Americans still smoke. Literally millions of people around the world will continue to die from smoking over the coming decade. And while cigarette smoking is on a decline in the United States, tobacco manufacturers have set their sights on the Third World, hoping to capture a rapidly growing market.

[1] The Surgeon General's office publishes an annual report that summarizes new findings on the effects of smoking on reproduction and health.

No one knows the exact toll of urban air pollution. It is not something that can be determined easily, if at all, because people are exposed to so many different pollutants over their lifetimes. However, a recent report issued by the federal government estimates that approximately 51,000 Americans die each year from lung disease caused by urban air pollution. The authors of the study predict that, by the year 2000, the number of victims will climb to nearly 60,000 per year, illustrating once again that human health is clearly dependent on a clean environment. This statistic also illustrates that the respiratory system of humans is not well adapted to the physical environment we have made for ourselves. Cultural evolution has progressed in ways that overwhelm past biological evolution.

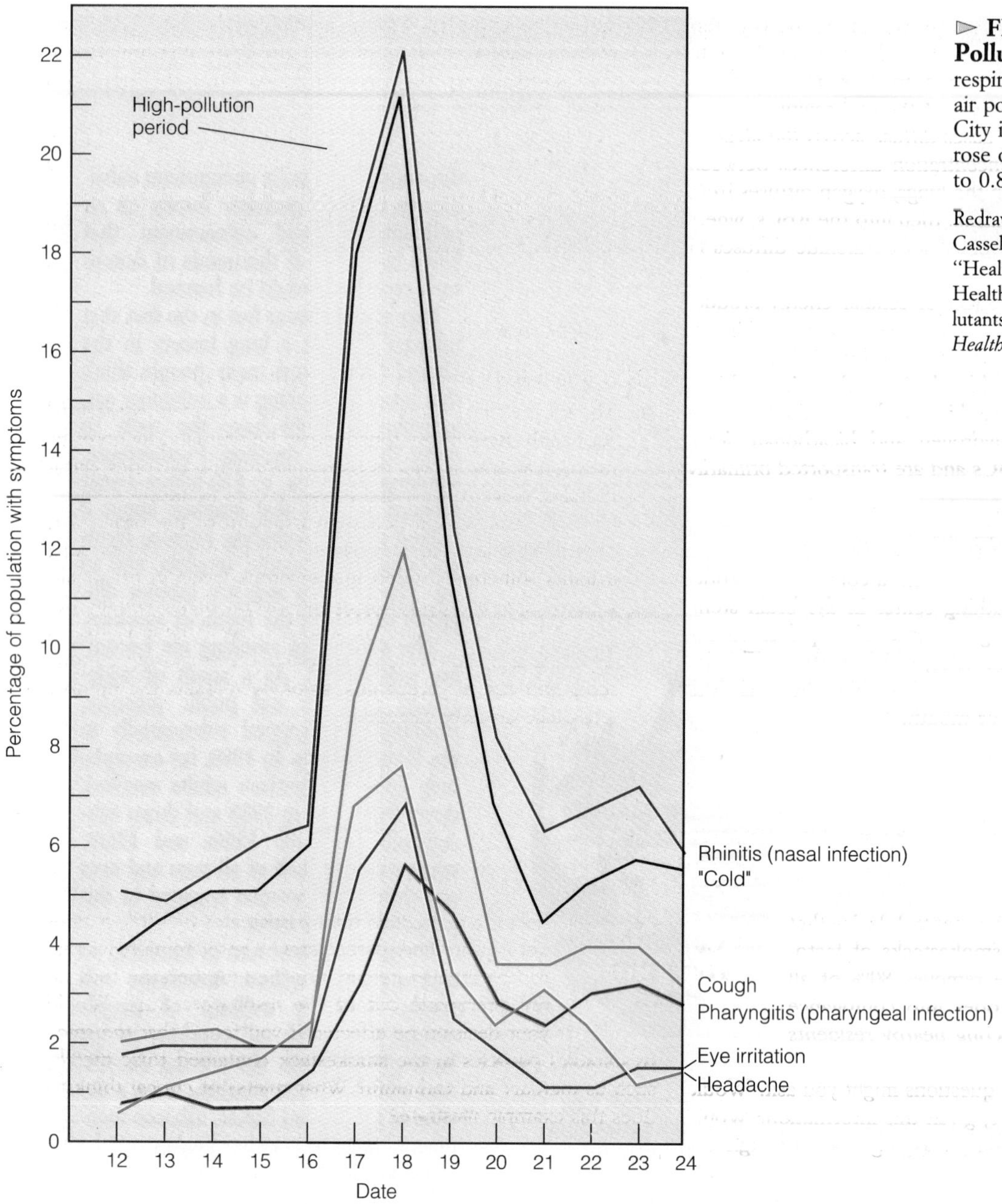

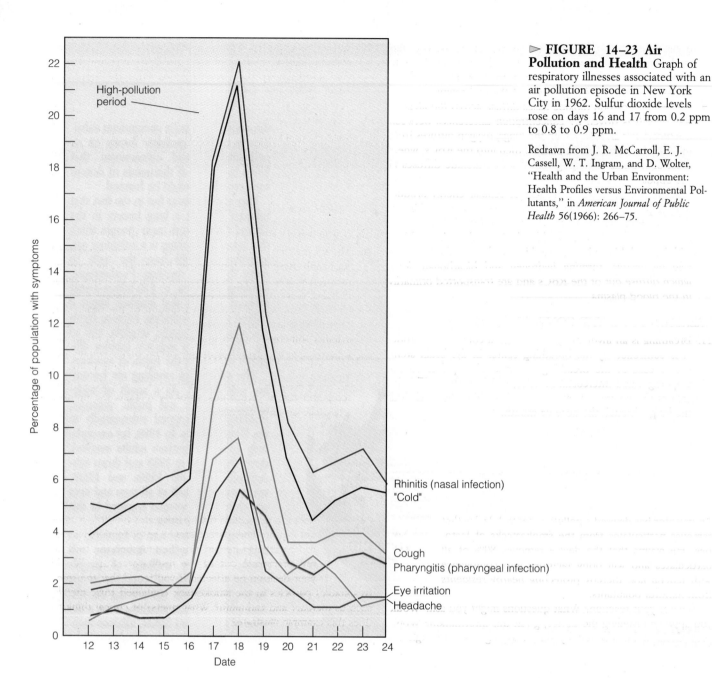

▷ **FIGURE 14–23 Air Pollution and Health** Graph of respiratory illnesses associated with an air pollution episode in New York City in 1962. Sulfur dioxide levels rose on days 16 and 17 from 0.2 ppm to 0.8 to 0.9 ppm.

Redrawn from J. R. McCarroll, E. J. Cassell, W. T. Ingram, and D. Wolter, "Health and the Urban Environment: Health Profiles versus Environmental Pollutants," in *American Journal of Public Health* 56(1966): 266–75.

SUMMARY

EVOLUTION OF RESPIRATORY SYSTEMS: AN OVERVIEW

1. Animals have evolved a wide array of respiratory systems to acquire oxygen and get rid of carbon dioxide.
2. Single-celled eukaryotic organisms relied on diffusion to acquire oxygen and get rid of carbon dioxide.
3. With the advent of multicellular animal life, though, more efficient life-supporting systems evolved, among them the respirable skin of the amphibian, the gill, and the lung.

STRUCTURE OF THE HUMAN RESPIRATORY SYSTEM

4. The respiratory system consists of an air-conducting portion and a gas-exchange portion.
5. The air-conducting portion transports air from outside the body to the alveoli in the lungs, the site of gaseous exchange.
6. The alveoli are tiny, thin-walled sacs formed by a single layer of flattened epithelial cells that facilitate diffusion. Surrounding the alveoli are capillary beds that pick up oxygen and expel carbon dioxide.
7. The lining of the alveoli is kept moist by water. Surfactant, a phospholipid produced in the lung, reduces the surface tension inside the alveoli and prevents their collapse.

FUNCTIONS OF THE RESPIRATORY SYSTEM

8. The respiratory system conducts air to and from the lungs,

exchanges gases, and helps produce sounds. Sound is generated as air rushes past the vocal cords, causing them to vibrate. The sounds are modified by movements of the tongue and changes in the shape of the oral cavity.

9. Oxygen and carbon dioxide gases diffuse across the alveolar wall, which is driven by concentration differences between the blood and alveolar air. In the lungs, oxygen diffuses from the alveoli into the blood plasma, then into the RBCs, where most of it binds to hemoglobin. Carbon dioxide diffuses in the opposite direction.
10. Carbon dioxide, a waste product of cellular energy production, is picked up by the blood flowing through capillaries. Some carbon dioxide is dissolved in the blood. Most, however, enters the RBCs in the bloodstream where it combines with hemoglobin or is converted into carbonic acid. Carbonic acid dissociates, forming hydrogen and bicarbonate ions, which diffuse out of the RBCs and are transported primarily in the blood plasma.

BREATHING AND THE CONTROL OF RESPIRATION

11. Breathing is an involuntary action with a conscious override. It is controlled by the breathing center in the brain stem. Nerve cells in the breathing center send impulses to the diaphragm and intercostal muscles, which contract. This increases the volume of the chest cavity, which draws air into the lungs through the nose or mouth.
12. When the impulses stop, inspiration ends. Air is then expelled passively as the chest wall returns to the normal position, and the diaphragm rises. The recoil of the lungs also assists in expelling the air.
13. The breathing center is regulated by a negative feedback loop that it generates itself. It is also regulated by outside influences—notably, levels of carbon dioxide in the blood.
14. Expiration can be augmented by enlisting the aid of abdominal and chest muscles, as can inspiration.
15. The rate of respiration can be increased by rising blood carbon dioxide levels or falling oxygen levels and by an increase in physical exercise.

ENVIRONMENT AND HEALTH: AIR POLLUTION

16. The proper functioning of the respiratory system is essential for health. Respiratory function can be dramatically upset by microorganisms as well as by pollution from factories, automobiles, power plants, and even our own homes.
17. Chronic bronchitis, a persistent irritation of the bronchi, is caused by sulfur dioxide, ozone, and nitrogen oxides, lung irritants sometimes found in dangerous levels in urban air.
18. Emphysema, a breakdown of the alveoli that gradually destroys the lung's ability to absorb oxygen, is similarly induced. Despite the role of air pollution in causing emphysema and chronic bronchitis, smoking remains the number one cause of these diseases.

EXERCISING YOUR CRITICAL THINKING SKILLS

An inventor has devised a pollution-control device that removes particulates from the smokestacks of factories. He claims that the device removes 80% of all particulates and will bring factories into compliance with federal law, thereby protecting nearby residents from harmful pollutants.

What is your reaction? What questions might you ask? Would you approve installing the device, given this information? Would your decision be affected by data showing that although the device does reduce total particulates by 80%, it does not capture finer particulates? These are particles that can be inhaled deeply into the lung because they do not precipitate out of the respiratory tract. Would your decision be affected if you found that the small (respirable) particles in the smokestack contained toxic metals, such as mercury and cadmium? What rule(s) of critical thinking does this example illustrate?

TEST OF TERMS

1. ____________ is a disease of the lung characterized by the progressive breakdown of the alveoli.
2. The ____________ is that part of the respiratory system that conducts air from the nose and mouth to the larynx.
3. Air travels from the larynx to the, ____________ a ribbed duct that leads to the lungs, then splits into right and left ____________, which penetrate the lungs.
4. The ____________, the muscular ducts that conduct air to the alveoli, can contract and relax like arterioles, thus providing a way to control the flow of air inside the lung.
5. Much of the lining of the conducting portion of the respiratory system contains ____________ cells, which produce a secretion that traps dust particles and bacteria.
6. A chemical, ____________, is produced by certain cells in the alveoli and helps reduce surface tension, preventing the alveoli from collapsing.
7. The phagocytic cell that wanders in and out of the alveoli is called a ____________ cell.
8. Inside the larynx are two elastic cords that vibrate when air breezes past. These are the ____________ ____________.
9. The patch of epithelium in the roof of the nasal cavity that perceives smell is called the ____________.
10. Most oxygen is carried in the blood bound to ____________ in the RBCs. Most carbon dioxide is transported as ____________.
11. The region of the ____________ that controls inspiration is called the ____________ ____________.

12. The active process in which air is drawn into the lungs is called ____________. It is caused by an enlargement of the chest cavity, which results from the contraction of the ____________ and the ____________ muscles.
13. The passive expulsion of air is called ____________.
14. Persistent irritation of the bronchi, leading to buildup of mucus and coughing, is called ____________ ____________ and is often caused by pollution in cigarette smoke and urban air.
15. The ____________ volume is the amount of air moved in and out of the lungs at rest. The ____________ ____________ volume is the amount of air that can be expelled from the lungs after a normal, passive exhalation.

Answers to the Test of Terms are found in Appendix B.

TEST OF CONCEPTS

1. Trace the flow of air from the mouth and nose to the alveoli, and describe what happens to the air as it travels along the various passageways.
2. Draw an alveolus, including all cell types found there. Be sure to show the relationship of the surrounding capillaries. Show the path that oxygen and carbon dioxide must take.
3. Trace the movement of oxygen from alveolar air to the blood in alveolar capillaries. Describe the forces that cause oxygen to move in this direction. Do the same for the reverse flow of carbon dioxide.
4. Why would a breakdown of alveoli in emphysemic patients make it more difficult for them to receive adequate oxygen?
5. Describe how sounds are generated and refined.
6. A baby is born prematurely and is having difficulty breathing. As the attending physician, explain to the parents what the problem is and how it could be corrected.
7. Smoking irritates the trachea and bronchi, causing mucus to build up and paralyze cilia. How do these changes affect the lung?
8. Describe inspiration and expiration, being sure to include discussions of what triggers them and the role of muscles in bringing about these actions.
9. How does the breathing center regulate itself to control the frequency of breathing?
10. Exercise increases the rate of breathing. How?
11. Turn to Figure 14–20. Explain each of the terms.
12. What is asthma? What are its symptoms? How does it affect one's inspiratory reserve volume?
13. Debate this statement: urban air pollution has very little overall impact on human health.
14. Do you agree or disagree that the single most effective way of reducing deaths in the United States would be to ban smoking. Explain.

CHAPTER

15

The Immune System

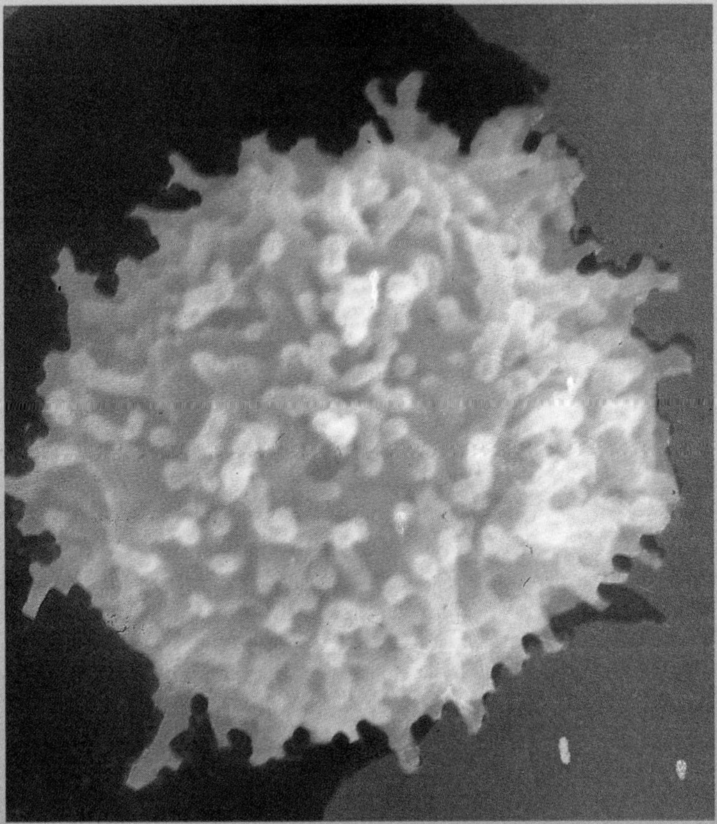

Scanning electron micrograph of a lymphocyte, one of the key players in immune protection.

Multicellular plants and animals evolved in a world teeming with viruses and single-celled organisms—bacteria and protistas. These organisms were the evolutionary ancestors of multicellular life. The vast majority of these microorganisms do not interact with multicellular animals directly. Therefore, even though they may be part of important environmental functions, essential for the web of life, most of them live their lives seemingly apart from the animals all around them. Anyone who has suffered from a cold or the flu will attest, however, that some microorganisms do indeed cross our paths, with sometimes dismal consequences. Fortunately, multicellular animals have evolved a series of highly effective defenses against bacteria, viruses, and other infectious agents.

This chapter examines the human body's defense mechanisms against disease-causing, or pathogenic, microorganisms. It looks briefly at inherent defenses against cancer and also discusses the devastating disease known as AIDS. The discussion of AIDS examines the causes of the disease, how it can be prevented, and some of the more recent efforts to find a cure. We begin with a brief overview of infectious agents.

VIRUSES AND BACTERIA: AN INTRODUCTION

Two of the most important infectious agents are viruses and bacteria. A brief discussion of these agents will help you understand the information that follows on the immune system. (Chapter 24 provides a much more detailed coverage of bacteria and viruses.)

A **virus** is a submicroscopic intracellular parasite. It consists of a nucleic acid core, consisting of either DNA or RNA, and an outer protein coat, the **capsid** (▻ Figure 15–1). Some viruses have an outer envelope that lies outside the capsid. The envelope is structurally similar to the plasma membranes of eukaryotic cells.

▻ **FIGURE 15–1 General Structure of a Virus** (*a*) The virus consists of a nucleic acid core of either RNA or DNA. Surrounding the viral core is a layer of protein known as the capsid. Each protein molecule in the capsid is known as a capsomere. (*b*) Some viruses have an additional protective coat known as the envelope.

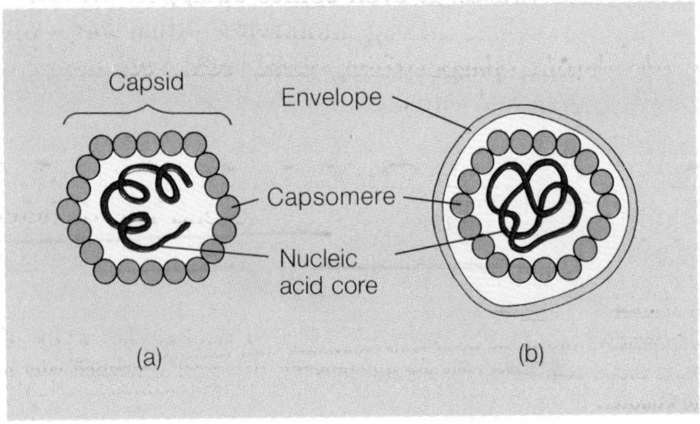

Viruses invade cells and commandeer their metabolic machinery. In the process, they convert host cells into miniature virus factories, often killing their hosts.

Viruses most often enter the body through the respiratory and digestive systems and spread from cell to cell in the bloodstream and lymphatic system. However, other avenues of entry are also possible—for example, sexual contact.

The immune system kills many viruses, but some may take refuge in cells, reemerging under stress or some other influence. One example is the virus responsible for genital herpes, tiny sores that periodically emerge on the genitals, thighs, and buttocks of men and women.

Bacteria (singular, **bacterium**) are microorganisms that do not require host cells to replicate. Bacteria are prokaryotic cells containing cytoplasm enveloped by a plasma membrane (▻ Figure 15–2). Outside the plasma membrane is a thick, rigid cell wall. Bacteria contain a single circular "strand" of DNA. Many bacteria also contain tiny circular pieces of extrachromosomal DNA, known as **plasmids** (described in Chapter 24).

Although they are best known for their role in causing sickness and death, most bacteria perform useful functions. Soil bacteria, for example, help recycle nutrients, thus helping ensure the continuation of life. This chapter concerns itself with potentially harmful bacteria that invade the human body.

THE FIRST AND SECOND LINES OF DEFENSE

Thousands of different bacteria and viruses and other microorganisms (for example, single-celled protistas) are found in the air we breathe and the food and water we consume. The protective mechanisms that evolved to protect multicellular organisms like humans from these hordes of potentially harmful microorganisms can be divided into three groups.

In Humans, the First Line of Defense Consists of the Skin, Epithelial Linings of the Respiratory, Digestive, and Urinary Systems, and Body Secretions that Destroy Harmful Microorganisms

Human skin forms an outer protective layer, which repels many potentially harmful microorganisms. As noted in Chapter 10, the skin consists of a relatively thick layer of epidermal cells overlying the rich vascular layer known as the dermis. Epidermal cells are produced by cell division in the base of the epidermis. As the basal cells proliferate, they move outward, become flattened, and die. The dead cells contain a protein called **keratin,** which forms a fairly waterproof protective layer that not only reduces moisture loss but also protects underlying tissues from microorganisms. The cells of the epidermis are tightly joined by special structures referred to as tight junctions. These structures also help impede water loss and microbial penetration.

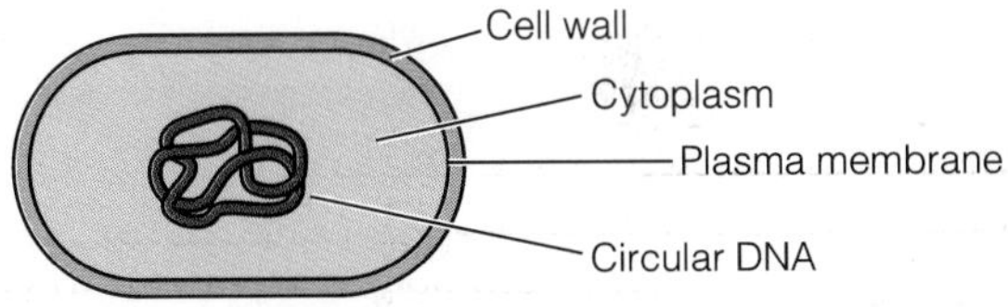

▻ **FIGURE 15–2 General Structure of a Bacterium** Bacteria come in many shapes and sizes, but all have a circular strand of DNA, cytoplasm, and a plasma membrane. Surrounding the membrane of many bacteria is a cell wall.

The epithelial linings of the respiratory, digestive, and urinary systems are also part of the body's wall of protection. They keep potentially harmful microorganisms from invading the underlying tissues. A break in these linings, like an opening in a fortress wall, permits microorganisms to enter. Chapter 11, for example, pointed out how a break in the lining of the digestive tract, caused by diverticulosis, can spawn a dangerous bacterial infection that spreads through the body.

The first line of defense also includes several protective chemicals. The skin, for example, produces slightly acidic secretions that impair bacterial growth. The stomach lining produces hydrochloric acid, which destroys many ingested bacteria. Tears and saliva contain an enzyme called **lysozyme,** which dissolves the cell wall of bacteria, killing them. Cells in the lining of the trachea and bronchi produce mucus, which not only traps bacteria but also has antimicrobial properties. Cilia in the lining of the respiratory system sweep the mucus toward the mouth, where it can be expectorated or swallowed (Chapter 14).

The protective mechanisms described above are all nonspecific; that is, they operate indiscriminately against all microorganisms.

The Second Line of Defense Combats Infectious Agents that Penetrate the First Line and Consists of Cellular and Chemical Responses

The first line of defense is not impenetrable. Even tiny breaks in the skin or in the lining of the respiratory, digestive, and urinary tracts permit viruses, bacteria, and other microorganisms to enter the body. Fortunately, a second line of defense exists. It involves a whole host of chemicals and cellular agents that work together to combat the invaders. Like the first line of defense, these mechanisms are nonspecific.

The Inflammatory Response Is a Major Part of the Second Line of Defense. Damage to body tissues triggers a series of reactions, that is part of the **inflammatory response.** The word *inflammatory* comes from the Latin word *inflammare* meaning "in flame" and refers to the heat given off by a wound. A protective measure, the inflammatory response is also characterized by redness, swelling, and pain—symptoms not unknown to anyone who has ever been cut or had a splinter.

The inflammatory response is a kind of chemical and biological warfare waged against bacteria, viruses, and other microorganisms. It begins with the release of a variety of chemical substances by injured tissue. Some chemicals attract resident macrophages that reside in body tissues and neutrophils found in the blood (Chapter 13). These cells quickly begin phagocytizing bacteria that enter the wound. Soon after these cells begin to work, a yellowish fluid begins to exude from the wound. Called pus, it contains dead white blood cells (mostly neutrophils), microorganisms, and cellular debris, which accumulate at the site of inflammation.

Tissue injury also stimulates the release of chemical substances that cause blood vessels to dilate and leak (▻ Figure 15–3). One such substance is **histamine.** Histamine stimulates the arterioles to dilate, causing the capillary networks to swell with blood.[1] The increase in the flow of blood through an injured tissue is responsible for the heat and redness around a cut or abrasion. Heat increases the metabolic rate of cells in the injured area and therefore accelerates the healing process.

Still other substances released by injured tissues increase the permeability of capillaries, augmenting the flow of plasma into a wounded region. Plasma carries with it oxygen and nutrients that facilitate healing. It also carries the molecules necessary for blood clotting. As described in Chapter 13, the clotting mechanism walls off injured vessels and helps reduce blood loss.

Plasma leaking into injured tissues causes swelling, which stimulates pain receptors in the area. Pain receptors send nerve impulses to the brain. Pain also results from chemical toxins released by bacteria and from chemicals released by injured cells themselves. One important pain-causing chemical is prostaglandin. Aspirin and other mild painkillers inhibit the synthesis and release of prostaglandins, reducing pain.

Although it evokes pain, the flow of fluid into body tissues is helpful in certain circumstances. Injury to joints, for example, results in local swelling that helps immobilize joints. Swelling is nature's way of protecting joints and allowing tissues to mend.

Inflammation occurs in virtually all tissues invaded by bacteria and viruses. It even comes equipped with its own cleanup crew—late-arriving monocytes—that mops up after the battle, phagocytizing dead cells, cell fragments, dead bacteria, and viruses.

Three Additional Chemicals Are Part of the Second Line of Defense. The second line of defense includes three additional substances: pyrogens, interferons, and complement.

[1] Histamine is produced by mast cells, platelets, and basophils, a type of white blood cell. Mast cells are a connective tissue cell described later in the chapter.

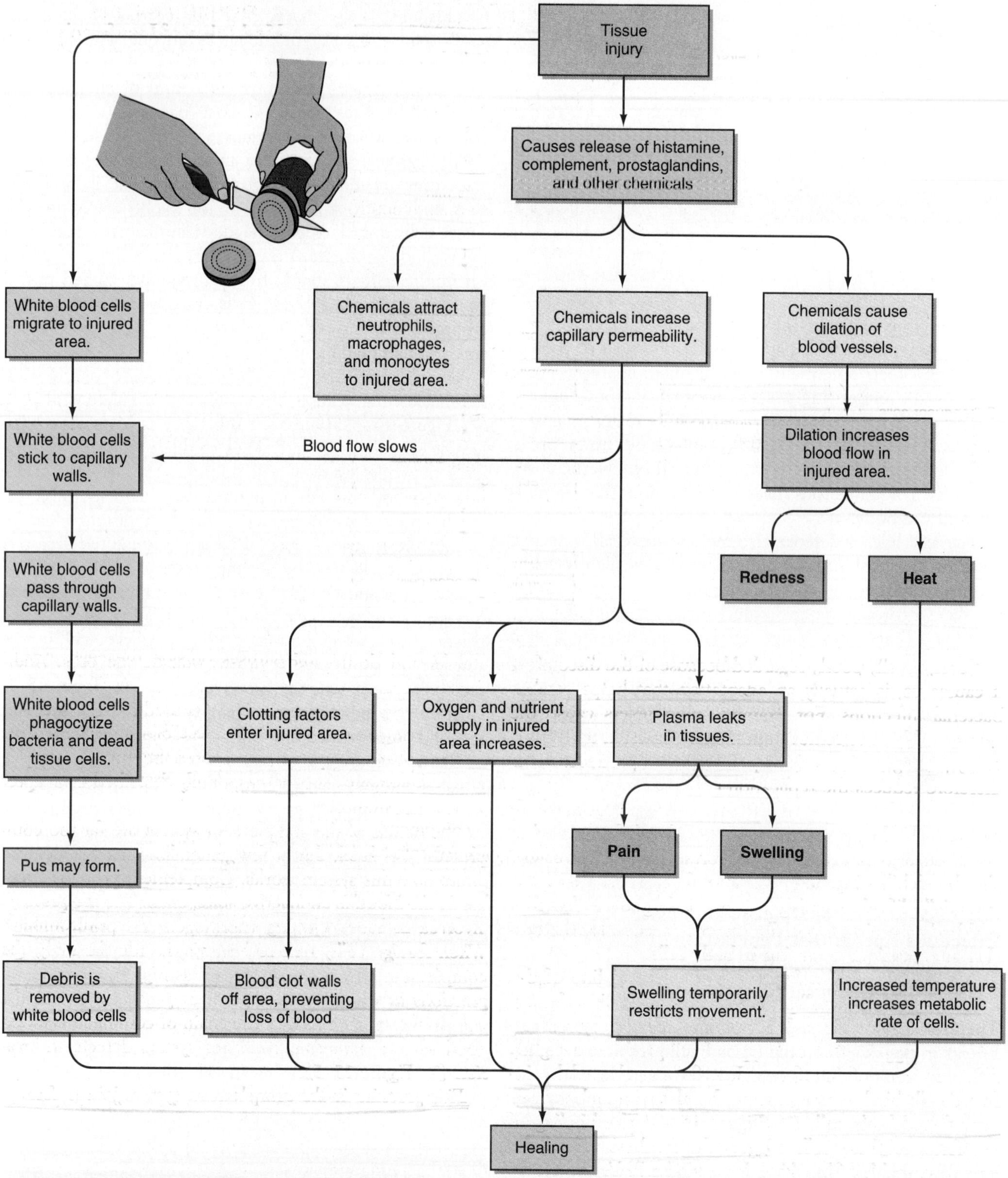

FIGURE 15–3 The Inflammatory Response

Pyrogens are molecules released primarily from macrophages that have been exposed to bacteria and other foreign substances. Pyrogens travel to a region of the brain called the hypothalamus. In the hypothalamus is a group of nerve cells that controls the body's temperature, in much the same way that a thermostat regulates the temperature of a room. Pyrogens turn the thermostat up, increasing body temperature and causing a fever.

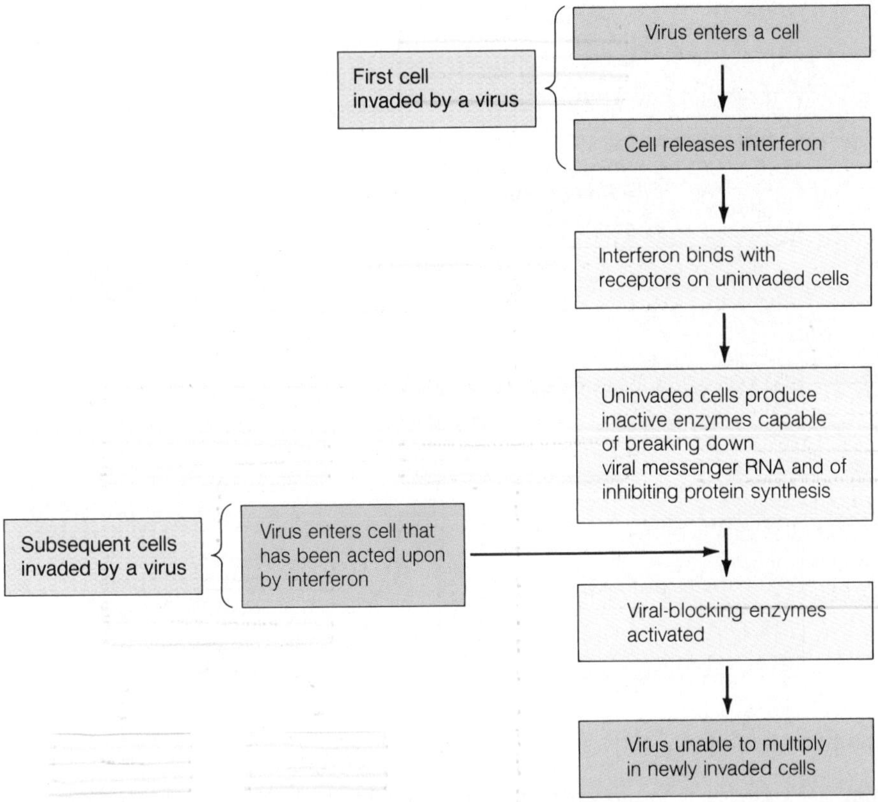

FIGURE 15–4 The Mechanism of Interferon's Action Interferon protects cells from viral infection.

Fever, usually poorly regarded because of the discomfort it causes us, is actually an adaptation that helps combat bacterial infections. For example, mild fevers cause the spleen and liver to remove additional iron from the blood. Many pathogenic bacteria require iron to reproduce. Fever therefore reduces the replication of bacteria and helps the body battle them. Fever also increases metabolism, which facilitates healing and accelerates cellular defense mechanisms, such as phagocytosis. Important as it is, fever can also be debilitating, and a severe fever (over 105°F) is potentially life-threatening because it begins to denature vital body proteins, especially enzymes needed for the biochemical reactions occurring in body cells.

Another chemical safeguard in the second line of defense is a group of small proteins known as the interferons. **Interferons** are released from cells infected by viruses. Research suggests that each type of cell produces a slightly different form of interferon. Interferons released by infected cells bind to receptors on the plasma membranes of noninfected body cells (Figure 15–4). The binding of interferon to these cells triggers the synthesis of cellular enzymes capable of breaking down viral mRNA and blocking viral protein synthesis. These enzymes, however, remain inactive until a virus enters the cell.

Interferons do not protect cells already infected by a virus, they simply stop the spread of viruses from one cell to another. In essence, the production and release of interferon are the dying cell's last acts to protect other cells of the body. Interferons are a remarkable chemical adaptation that stops the spread of viruses while the immune response attacks and destroys the viruses outside the cells. Additional effects of the interferons are listed in Table 15–1.

Another group of chemical agents used to fight infection are the **complement proteins.** These blood proteins form the **complement system,** so named because it complements the action of antibodies, briefly described in Chapter 13 and elsewhere.

The details of the complement system are far too complex for this book, but a few generalizations will demonstrate how this system works. Complement proteins circulate in the blood in an inactive state, much like the proteins involved in blood clotting (fibrinogen and prothrombin). When foreign cells, such as bacteria, invade the body, the complement system is activated. This triggers a cascade of reactions in which one complement protein activates the next in the series, much as the chain of command is awakened when a nation's surveillance system detects an invasion (Figure 15–5).

Five proteins in the complement system join to form a

TABLE15–1 Functions of Interferons

Protect cells against viruses by destroying viral mRNA and inhibiting protein synthesis
Enhance the phagocytic activity of macrophages
Stimulate the production of antibodies
Stimulate the activity of cytotoxic T cells
Suppress tumor growth

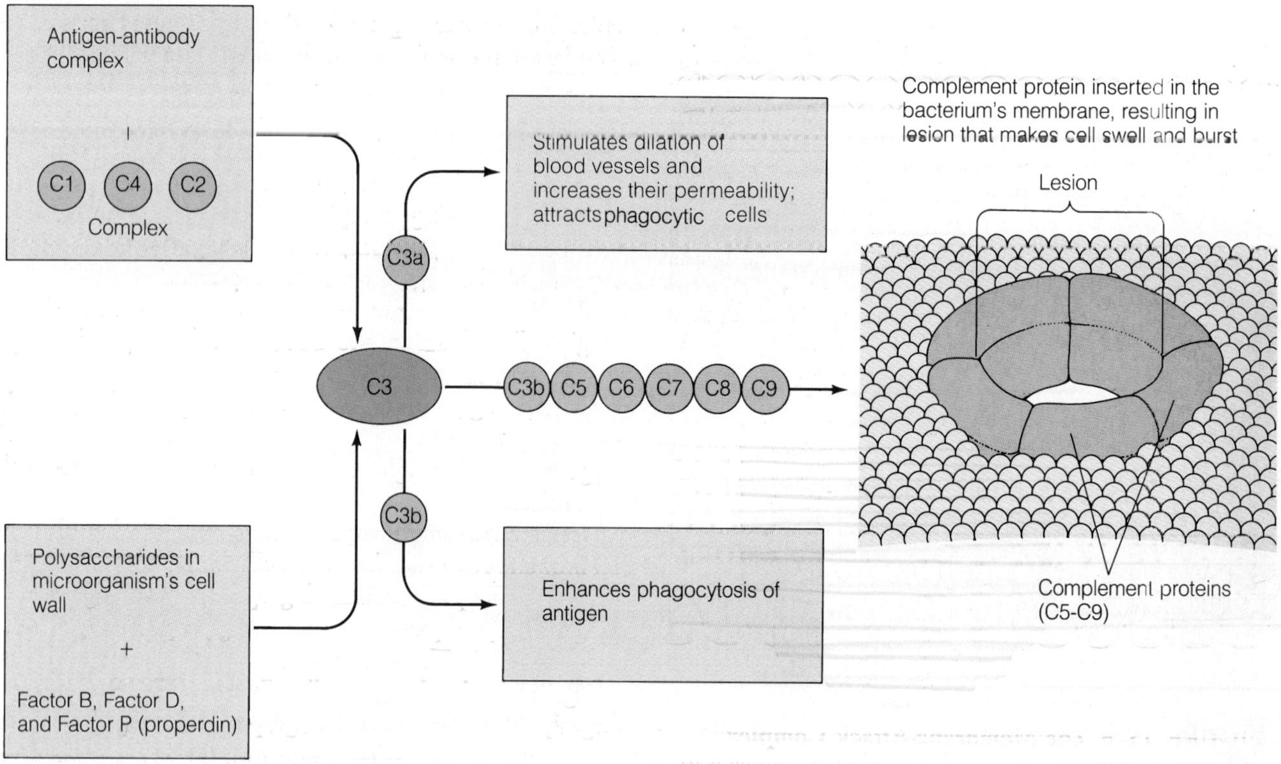

▻ **FIGURE 15–5 The Complement System** The complement system is activated by the presence of antibodies bound to their antigens and by the polysaccharides in the cell wall of some bacteria and fungi. These triggers activate C3 protein, which splits, forming two fragments. The C3a fragment stimulates the inflammatory response. The C3b fragment binds to bacteria, making them easier to phagocytize. C3b also activates additional complement proteins, which are inserted into the membrane of bacteria. They create an opening, or lesion, in the membrane, which, if large enough, can kill the cell.

large protein complex, known as the **membrane-attack complex** (▻ Figure 15–6). The membrane-attack complex embeds in the plasma membrane of bacteria, creating an opening into which water flows. The influx of water causes bacterial cells to swell, burst, and die.

Several of the activated complement proteins also function on their own and are part of the inflammatory response. Some of them, for example, stimulate the dilation of blood vessels in an infected area, described earlier. Others increase the permeability of the blood vessels, allowing white blood cells and nutrient-rich plasma to pass more readily into an infected zone. Certain complement proteins may also act as chemical attractants, drawing macrophages, monocytes, and neutrophils to the site of infection, where they phagocytize foreign cells. Another complement protein (C3b) binds to microorganisms, forming a rough coat on the intruders that facilitates their phagocytosis.

In summary, the first and second lines of defense are composed of physical barriers, chemical weapons, and a cellular defense mechanism. These mechanisms are nonspecific; that is, they do not target specific infectious agents. The skin, for example, repels most bacteria and fungi. Macrophages and neutrophils devour whatever foreign substances enter the body tissues. Fever helps combat dividing bacteria by reducing iron levels. In so doing, the first and second lines of defense lighten the work load of the third and final line of defense, the immune system.

THE THIRD LINE OF DEFENSE: THE IMMUNE SYSTEM

The **immune system** is not a distinct organ system like the digestive or respiratory system. Rather, it is a functional system consisting of many millions of lymphocytes, a type of white blood cell. Humans, for instance, contain an estimated 2 trillion lymphocytes—that's 2000 billion. These cells circulate in the blood and lymph but also take up residence in the lymphoid organs, such as the spleen, thymus, lymph nodes, and tonsils, as well as other body tissues (Figure 12-21). The cells of the immune system selectively target foreign substances and foreign organisms. As a result, the immune system is said to be specific.

The immune system is an important homeostatic mechanism that eliminates foreign organisms—including bacteria, viruses, single-celled protists, and even many parasites—that penetrate the outer defenses of the body. It also helps reduce cellular dissent from within—that is, the emergence of cancer. Thus, in a world filled with infectious agents and natural mutagens (agents that cause mutation, some of

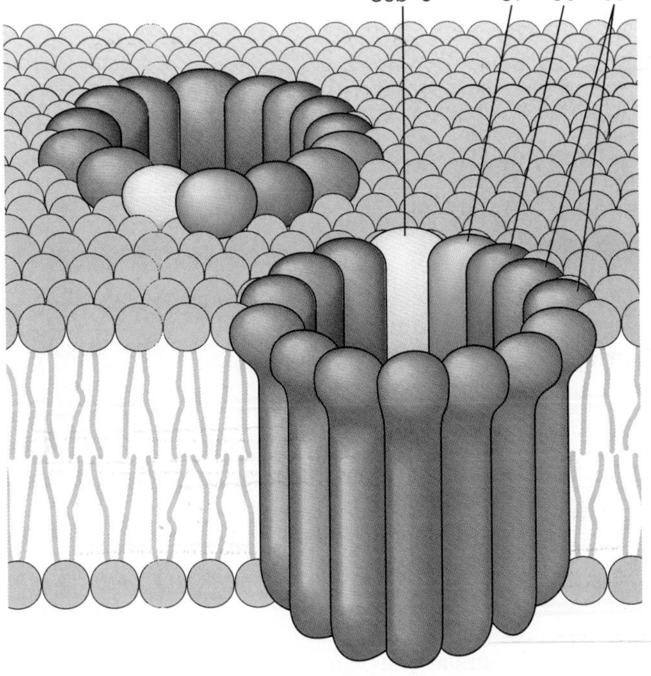

▻ **FIGURE 15–6 The Membrane-Attack Complex**
Complement proteins embed in a cell's membrane, causing it to leak, swell, and burst. By Dana Burns from John Ding-E Young and Zanvil A. Kohn, "How Killer Cells Kill." Copyright © January 1988 by Scientific American, Inc. All rights reserved.

which might lead to cancer), the immune system is an important evolutionary advance.

Lymphocytes Detect Foreign Substances in the Body and Mount an Attack on Them

One of the chief functions of the immune system is to identify what belongs in the body and what does not. The ability to recognize foreign materials, while essential to survival, is a rather difficult task. It is a little like having to sort through the thousands of items in your family's home, including the contents of every drawer in the kitchen and every jar of nails in the garage, each and every day of the year to determine if someone has brought something as tiny as a pin into the house.

Once a foreign substance has been detected, the immune system mounts an attack to eliminate it. Therefore, like all homeostatic systems, the immune system requires receptors to detect a change and effectors to bring about a response. In the immune system, the lymphocytes serve both of these functions.

Foreign Substances that Trigger an Immune Response are Proteins and Polysaccharides with Large Molecular Weights

The immune response is triggered by large foreign molecules, notably proteins and polysaccharides. These molecules are called **antigens** (*anti*body-*gen*erating substances). The larger the molecule is, the greater its antigenicity.

As a rule, small molecules generally do not elicit an immune reaction. In some individuals, however, small, nonantigenic molecules like formaldehyde, penicillin, and the poison ivy toxin bind to naturally occurring proteins in the body, forming complexes. The large complexes so formed are unique compounds foreign to the body and capable of eliciting an immune response.

The immune system responds to viruses, bacteria, and single-celled fungi in the body. It also responds to parasites, such as the protozoan that causes malaria. Viruses, bacteria, fungi, and parasites elicit a response because they are enclosed by a membrane, or coat, that contains large-molecular-weight proteins or polysaccharides—that is, antigens.

Cells transplanted from one person to another also elicit an immune response, because cells from another individual contain a unique "cellular fingerprint," resulting from the unique array of plasma membrane glycoproteins, as noted in Chapter 3. The immune system is activated by these antigens on the foreign cells. Cancer cells also present a slightly different chemical fingerprint, making them essentially foreign cells within our own bodies to which the immune system responds. Although cancer cells evoke an immune response, it is often not sufficient to stop the disease.

Antigens stimulate the proliferation and differentiation of two types of lymphocytes: the **T lymphocytes,** commonly called **T cells,** and the **B lymphocytes,** also known as **B cells** (Chapter 13). The **immune reaction,** therefore, is the response of B and T cells. As you will see in later sections, B and T cells react differently and respond to different types of antigens. As a rule of thumb, B cells recognize and react to microorganisms such as bacteria. They also respond to a few viruses and bacterial toxins, chemical substances released by bacteria. When activated, B cells produce antibodies to these antigens. In contrast, T cells recognize and respond to body cells that have gone awry, such as cancer cells, or cells that have been invaded by viruses. T cells also respond to transplanted tissue cells and larger disease-causing agents, such as single-celled fungi and parasites. Unlike B cells, T cells attack their targets directly.

Immature B and T Cells Are Incapable of Responding to Antigens But Soon Mature and Gain that Ability

Lymphocytes are produced in the red bone marrow and released into the bloodstream. These immature cells circulate through the blood and lymph (▻ Figure 15–7).

T Cells. Cells destined to become T lymphocytes take up temporary residence in the thymus, a lymphoid organ located above the heart. Inside the thymus, these lymphocytes mature in two to three days. During this time, lym-

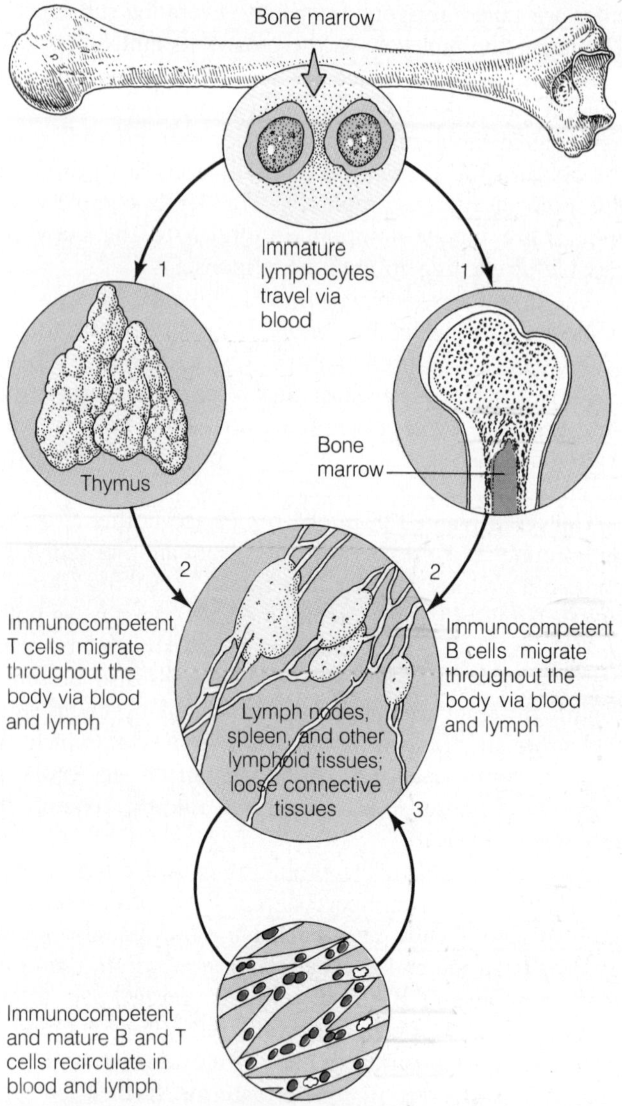
Bone marrow
Immature lymphocytes travel via blood
1
1
Thymus
Bone marrow
2
2
Immunocompetent T cells migrate throughout the body via blood and lymph
Immunocompetent B cells migrate throughout the body via blood and lymph
Lymph nodes, spleen, and other lymphoid tissues; loose connective tissues
3
Immunocompetent and mature B and T cells recirculate in blood and lymph

▻ **FIGURE 15–7 B- and T-Cell Immunocompetence**
Immunocompetence, the ability to respond to specific antigens, is conferred in the bone marrow, in the case of B cells, or the thymus, in the case of T cells. The cells then migrate in the blood and lymph to lymphoid organs, such as the lymph nodes, spleen, and loose connective tissue underlying many epithelia.

phocytes are said to develop **immunocompetence,** because they gain the capacity to respond to specific antigens. During this process, each differentiated T cell (so named because it becomes immunocompetent in the thymus) produces a unique type of membrane receptor that will bind to one—and only one—type of antigen. Over an individual's lifetime, millions of antigens will be encountered. Thanks to immunocompetence developed during fetal development, each of us is equipped with millions of different programmed T cells to respond to the onslaught of antigens long before we encounter them.

B Cells. B cells mature and differentiate in the bone marrow.[2] Afterwards, immunologically competent B cells circulate in the blood and take up residence in connective and lymphoid tissues. They therefore become part of the body's vast cellular reserve, stationed at distant outposts, awaiting the arrival of the microbial invaders.

By various estimates, several million immunologically distinct B and T cells are produced in the body early in life. Over a lifetime, only a small fraction of these cells will be called into duty.

B Cells Provide Humoral Immunity Through the Production of Antibodies

The immune response consists of two separate but related reactions: humoral immunity, provided by the B cells, and cell-mediated immunity, involving T cells (Table 15–2). Let's consider humoral immunity and the B cells first.

When an antigen first enters the body, it binds to B cells programmed during their residence in the bone marrow to respond to that particular antigen (▻ Figure 15–8).[3] These cells soon begin to divide, producing a population of immunologically similar cells, also known as a **clone.** As the clone expands, some of the B cells begin to differentiate, forming another kind of cell, the plasma cell. **Plasma cells** are differentiated B cells and contain a prominent rough endoplasmic reticulum on which they produce prodigious amounts of antibody. Antibodies released from plasma cells circulate in the blood and lymph, where they bind to the antigens that triggered the response. Because the blood and lymph were once referred to as body "humors," this arm of the protective immune response is called **humoral immunity.**

[2] B cells are so named because they develop immunocompetence in a part of the chicken's digestive system known as the bursa. Humans lack this organ.
[3] As you will soon see, this process is a bit more complex and involves the macrophage.

TABLE 15-2 Comparison of Humoral and Cell-Mediated Immunity

HUMORAL	CELL-MEDIATED
Principal cellular agent is the B cell.	Principal cellular agent is the T cell.
B cell responds to bacteria, bacterial toxins, and some viruses.	T cells responds to cancer cells, virus-infected cells, single-celled fungi, parasites, and foreign cells in an organ transplant.
When activated, B cells form memory cells and plasma cells, which produce antibodies to these antigens.	When activated, T cells differentiate into memory cells, cytotoxic cells, suppressor cells, and helper cells; cytotoxic T cells attack the antigen directly.

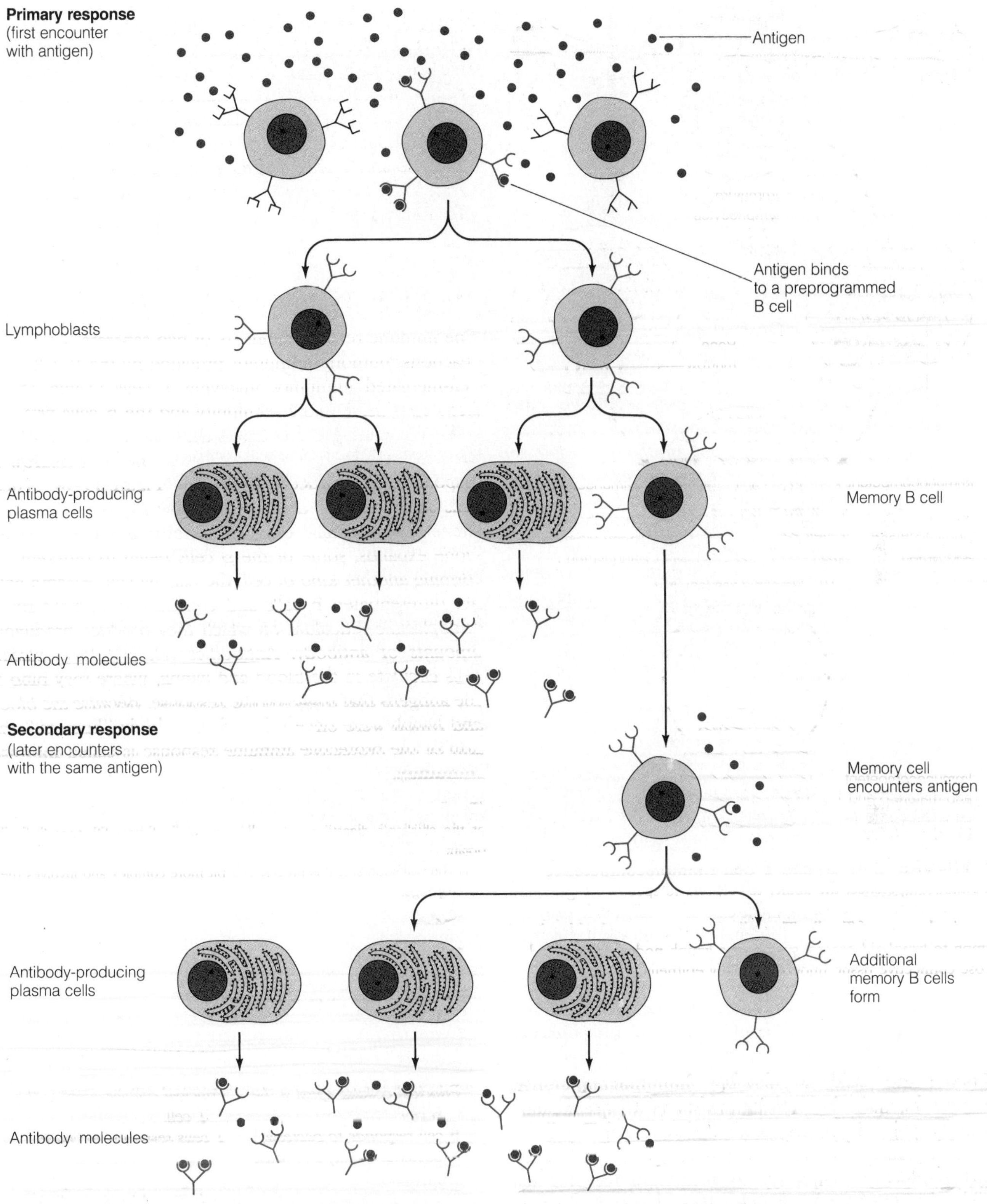

▻ **FIGURE 15–8 B-Cell Activation** Immunocompetent B cells are stimulated by the presence of an antigen, producing an intermediate cell, the lymphoblast. The lymphoblasts divide, producing plasma cells and some memory cells. Memory cells respond to subsequent antigen encroachment, yielding a rapid, secondary response.

▻ **FIGURE 15–9 Primary and Secondary Responses** (*a*) The primary (initial) immune response is slow. It takes about 10 days for antibody levels to peak. Almost no antibody is produced during the first week as plasma cells are being formed. (*b*) The secondary response is much more rapid. Antibody levels rise almost immediately after the antigen invades. T cells show a similar response pattern.

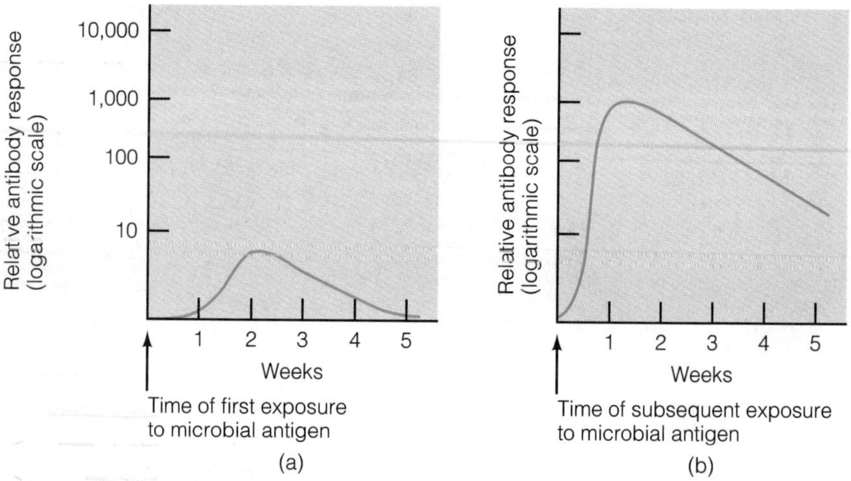

The Initial Reaction to an Antigen Is Slower and Weaker Than Subsequent Responses. The first time an antigen enters the body, it elicits an immune response, but the initial reaction—or **primary response**—is relatively slow and of small magnitude (▻ Figure 15–9). During the primary response, antibody levels in the blood do not begin to rise until approximately the beginning of the second week *after* the intruder was detected. This delay occurs because it takes time for B cells to multiply and form a sufficient number of plasma cells. Antibody levels usually peak about the end of the second week, then decline over the next three weeks, partly explaining why it takes most people about a week to 10 days to combat a cold or the flu.

If the same antigen enters the body at a later date, however, the immune system acts much more quickly and more forcefully (Figure 15–9). This greatly fortified reaction constitutes the **secondary response.** As Figure 15–9 illustrates, during a secondary response antibody levels increase rather quickly, only a few days after the antigen has entered the body. The amount of antibody produced also greatly exceeds quantities generated during the primary response. Consequently, the antigen is quickly destroyed, and a recurrence of the illness is prevented.

The Rapidity of the Secondary Response Is the Result of the Production of Memory Cells During the Primary Response. As shown in Figure 15–8, during the primary response, some lymphocytes divide to produce memory cells. **Memory cells** are immunologically competent B cells that do not transform into plasma cells. Instead, they remain in the body awaiting the antigen's reentry. These cells therefore create a relatively large reserve force of antigen-specific B-cells. When the antigen reappears, memory cells proliferate rapidly, producing numerous plasma cells that quickly crank out antibodies to combat the foreign invaders. As illustrated in Figure 15–8, during the secondary response, the memory cells also generate additional memory cells that remain in the body in case the antigen should reappear at some later date.

Immune protection afforded by memory cells can last 20 years or longer and explains why once a person has had a childhood disease, such as the mumps or chicken pox, it is unlikely that he or she will contract it again. Resistance to disease that is provided by the immune system is known as **immunity.** From an evolutionary standpoint, this adaptation serves us extremely well, greatly reducing the incidence of infectious disease. Without it, humans would probably not be able to survive.

Antibodies Act in Four Ways to Destroy Antigens. Antibodies belong to a class of blood proteins called the globulins, introduced in Chapter 13. Antibodies are specifically called **immunoglobulins.** Each antibody consists of four peptide chains (▻ Figure 15–10). The chains are joined by disulfide bonds—that is, covalent bonds that form between sulfur atoms of different amino acid side groups. Two small chains intertwine with two larger chains, forming T-shaped molecules. It is the arms of the T that bind to antigens and confer specificity, in much the same way that the active sites of enzymes result in enzyme specificity. Immunologists have discovered five classes of antibodies, each with a slightly different role (Table 15–3).

Antibodies destroy foreign organisms and antigens through four mechanisms: (1) neutralization, (2) agglutination, (3) precipitation, and (4) complement activation. We will consider each one very briefly.

Neutralization. During **neutralization,** antibodies bind to viruses, forming a complete coating around them. This action prevents viruses from binding to plasma membrane receptors of body cells. If a virus cannot bind to a plasma membrane receptor, it cannot get inside most cells (▻ Figure 15–11, far right).

Neutralization also helps destroy bacterial toxins. A toxic protein, for instance, may be so heavily coated with antibody that it is rendered ineffective. Toxins and viruses neutralized by their antibody coating are eventually engulfed by macrophages and other phagocytic cells, as shown in Figure 15–11.

Agglutination. Antibodies deactivate foreign cells (bacteria and red blood cells transfused into another person) by **agglutination**—a clumping of the foreign cells (Figure

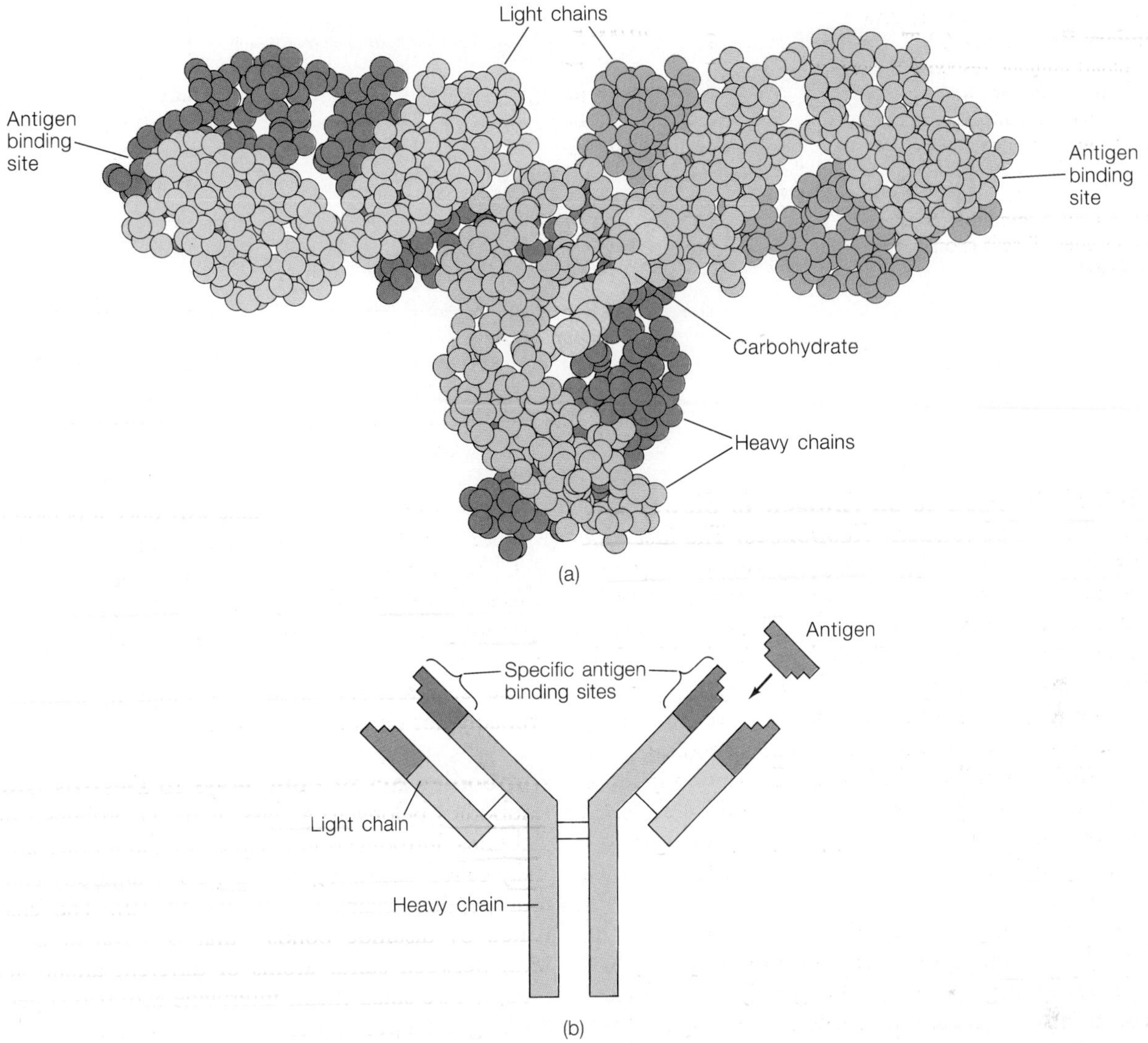

▻ **FIGURE 15–10 Antibody Structure** (*a*) A three-dimensional model of an antibody showing the four chains. The molecule is T shaped before binding to an antigen. After it binds, it becomes Y shaped; as shown in (*b*), a diagrammatic representation of the structure of an antibody molecule. It shows the four protein chains, two large (heavy chains) and two small (light chains). Note that the antigens bind to the arms of the molecule.

15–11). During agglutination, a single antibody may bind to several antigens, causing them to clump together. These antigen-antibody complexes are then phagocytized and removed from blood and body fluids by macrophages and other phagocytic cells.

Precipitation. Antibodies also bind to soluble antigens (for example, a protein), forming much larger, water-insoluble complexes that precipitate (fall out) out of solution, where they are engulfed by phagocytic cells.

Activation of the complement system. The final mechanism by which antibodies help rid the body of bacteria is through the activation of the complement system. As noted earlier in this chapter, the complement system is a family of blood-borne proteins that is part of the nonspecific immune response to antigens. The complement system is activated by the presence of antigen-antibody complexes (antibodies bound to antigens). When activated, the complement system produces a membrane-attack complex that embeds in the plasma membrane of bacterial cells, causing them to leak, swell, and burst. Some proteins of this system stimulate the inflammatory process, and some coat microorganisms, facilitating their phagocytosis by macrophages.

Macrophages in the Body's Tissues Play a Key Role in Activating B Cells. Macrophages are phagocytic cells found in connective tissue, lymphoid tissue, and the organs of the lymphatic system (for example, lymph nodes); as you may recall from Chapter 13, these cells arise from monocytes, which escape from the bloodstream and set up residence in body tissues.

Macrophages play several important roles in the immune response. First, they phagocytize bacteria and other antigens at the site of infection. They also phagocytize antigen-antibody complexes, dead cells, and dead microorganisms,

TABLE 15–3 Types and Functions of the Immunoglobulins

CLASS		LOCATION AND FUNCTION
IgD	 Monomer	Present on surface of many B cells, but function uncertain; may be a surface receptor for B cells; plays a role in activation of B cell
IgM	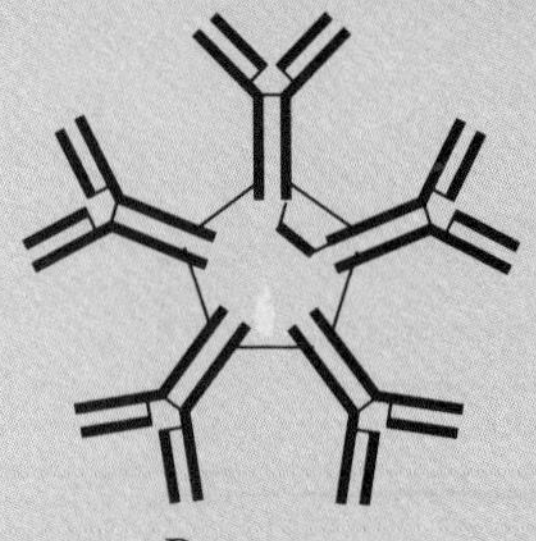 Pentamer	Found on surface of B cells and in plasma; acts as a B-cell surface receptor for antigen, secreted early in primary response; powerful agglutinating agent
IgG	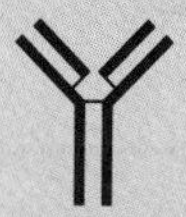 Monomer	Most abundant immunoglobulin in the blood plasma; produced during primary and secondary response; can pass through the placenta, entering fetal bloodstream, thus providing protection to fetus
IgA	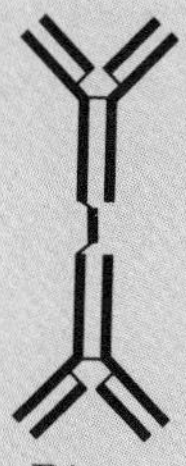 Dimer	Produced by plasma cells in the digestive, respiratory, and urinary systems, where it protects the surface linings by preventing attachment of bacteria to surfaces of epithelial cells; also present in tears and breast milk; protects lining of digestive, respiratory, and urinary systems
IgE	 Monomer	Produced by plasma cells in skin, tonsils, and the digestive and respiratory systems; responsible for allergic reactions, including hay fever and asthma

helping mop up the debris of the battle. Third, macrophages also help to activate T- and B-cell differentiation. B cells, in fact, cannot differentiate into plasma cells and produce antibodies without macrophages.

▻ Figure 15–12 is a simplified illustration showing the role of the macrophage in B-cell activation. As illustrated, macrophages first engulf invading bacteria. The macrophages then transfer antigens from the surface of the bacterium to their own plasma membrane. The macrophages then cluster around B cells, "presenting" the bacterial antigen to them. B cells that are programmed to respond to that antigen are activated when they encounter the concentrated bacterial antigen on the macrophage plasma membrane. As a result, the B cells begin to divide and differentiate, forming antibody-producing plasma cells and memory cells. Macrophages also secrete a chemical called **interleukin 1,** which enhances the proliferation and differentiation of activated B cells.

Macrophages also present antigen to certain T cells, called helper T cells (described in more detail shortly and shown in Figure 15–12). When activated, the helper T cells produce a chemical substance known as **B-cell growth factor.** B-cell growth factor enhances the proliferation and differentiation of B cells much as does interleukin 1. It also enhances antibody production by the plasma cells (Figure 15–12).

T Cells Differentiate Into At Least Four Cell Types, Each with a Separate Function in Cell-Mediated Immunity

T cells provide a much more complex form of protection

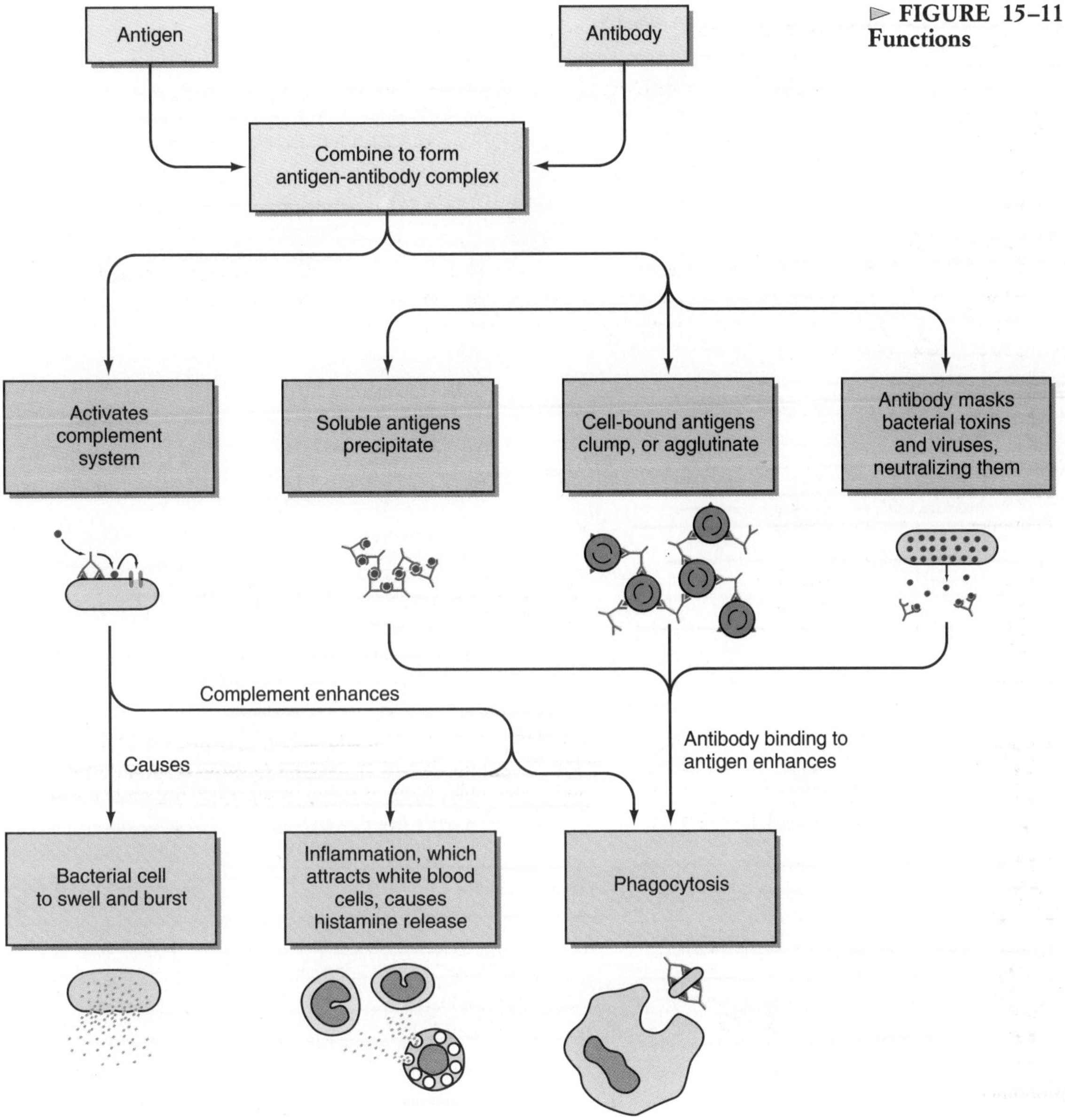

▻ **FIGURE 15–11 Antibody Functions**

than B cells. Like B cells, they respond to the presence of antigens by undergoing rapid proliferation. T cells, however, differentiate into at least four cell types: (1) memory T cells, (2) cytotoxic T cells, (3) helper T cells, and (4) suppressor T cells (Table 15–4).

Memory T cells form a cellular reserve force that plays a crucial role in the secondary response. **Cytotoxic T cells** perform two essential roles (Table 15–4). Some cytotoxic T cells attack and kill body cells that have been infected by viruses. When a virus infects a cell, antigenic proteins in the virus's envelope become incorporated in the plasma membrane of the host cell. Cytotoxic T cells bind to that antigen and destroy the host cell. They also attack and kill bacteria, parasites, single-celled fungi, cancer cells, and foreign cells introduced during blood transfusions or tissue or organ transplants. Cytotoxic T cells bind to antigenic molecules in the membranes of these cells and release a chemical called **perforin-1** (▻ Figure 15–13). Perforin-1 molecules become embedded in the plasma membrane of the target cell. These molecules join to form pores, similar to those produced by the membrane-attack complex of the complement system. These pores cause the plasma membrane to leak, destroying the target cell within a few hours. After it has delivered its lethal payload, the cytotoxic cell detaches and is free to hunt down other antigens.

Helper T cells enhance the immune response and are activated by the presence of certain antigens. As noted earlier, helper T cells produce a B-cell growth factor, enhancing humoral immunity. They also enhance cell-mediated immunity by releasing a molecule known as **interleukin 2.** This substance increases the activity of cytotoxic T, suppressor T, and even helper T cells.

▻ **FIGURE 15–12 The Role of the Macrophage in Immunity** Macrophages "present" antigen to B cells and helper cells. They also phagocytize antibody-coated bacteria.

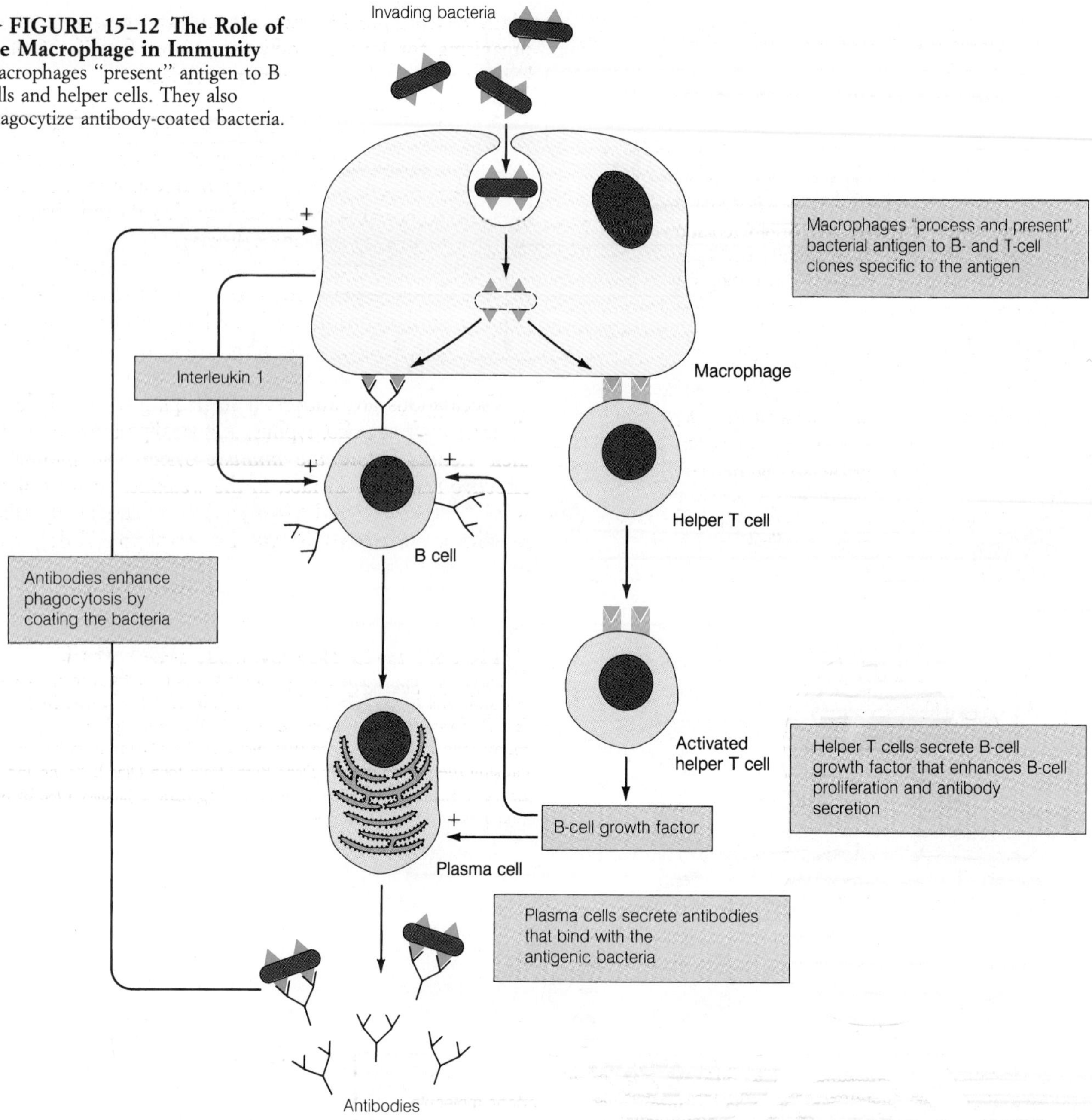

Invading bacteria
Macrophages "process and present" bacterial antigen to B- and T-cell clones specific to the antigen
Macrophage
Interleukin 1
Helper T cell
B cell
Antibodies enhance phagocytosis by coating the bacteria
Activated helper T cell
Helper T cells secrete B-cell growth factor that enhances B-cell proliferation and antibody secretion
B-cell growth factor
Plasma cell
Plasma cells secrete antibodies that bind with the antigenic bacteria
Antibodies

Helper T cells are the most abundant of all the T cells (making up 60% to 70% of the circulating T cells). They have been likened to the immune system's master switch. Without them, antibody production and T-cell activity would be greatly reduced. In fact, the immune response without helper T cells would be almost nonexistent. In their absence, antigens would stimulate a few B and T cells, then the process would come to a halt. Because the AIDS virus preferentially infects helper T cells, people suffering from the disease are unable to mount an effective immune response and eventually die from bacterial infections or cancer.

The role of **suppressor T cells** is less well understood. Research suggests that they "turn off" the immune reaction as the antigen begins to disappear—that is, as the antigen is phagocytized. The activity of suppressor T cells, therefore, increases as the immune system finishes its job. Suppressor cells release chemicals that reduce B- and T-cell division.

Two Types of Immunity Are Possible: Active and Passive

One of the major medical advances of the last century was the discovery of **vaccines,** which help prevent bacterial and viral infections. Vaccines contain inactivated or weakened viruses, bacteria, or bacterial toxins. When injected into the body, the antigens in vaccines elicit an immune response.

TABLE15–4 Summary of T Cells	
CELL TYPE	ACTION
Cytotoxic T cells	Destroy body cells infected by viruses, and attack and kill bacteria, fungi, parasites, and cancer cells
Helper T cells	Produce a growth factor that stimulates B-cell proliferation and differentiation and also stimulates antibody production by plasma cells; enhance activity of cytotoxic T cells
Suppressor T cells	May inhibit immune reaction by decreasing B- and T-cell activity and B- and T-cell division
Memory T cells	Remain in body awaiting reintroduction of antigen, at which time they proliferate and differentiate into cytotoxic T cells, helper T cells, suppressor T cells, and additional memory cells

Many vaccines provide immunity or protection from microorganisms for long periods, sometimes for life. Others, however, give only short-term protection.

Vaccines stimulate the immune reaction because the weakened or deactivated organisms (or toxins) they contain still possess the antigenic proteins or carbohydrates that trigger B- and T-cell activation. Because their antigens have been seriously weakened or deactivated, however, vaccines usually do not cause disease.

Vaccination provides a form of protection that immunologists call **active immunity**—so named because the body actively produces memory T and B cells that protect a person against future infections. Viral or bacterial infections also produce active immunity.

Vaccinations are important in helping us avoid deadly diseases such as polio, typhus, and smallpox that often kill their victims before the immune system can mount an effective response. In fact, in the wealthier nations of the world, like the United States, vaccines have nearly eliminated many infectious diseases like small pox. Today, many

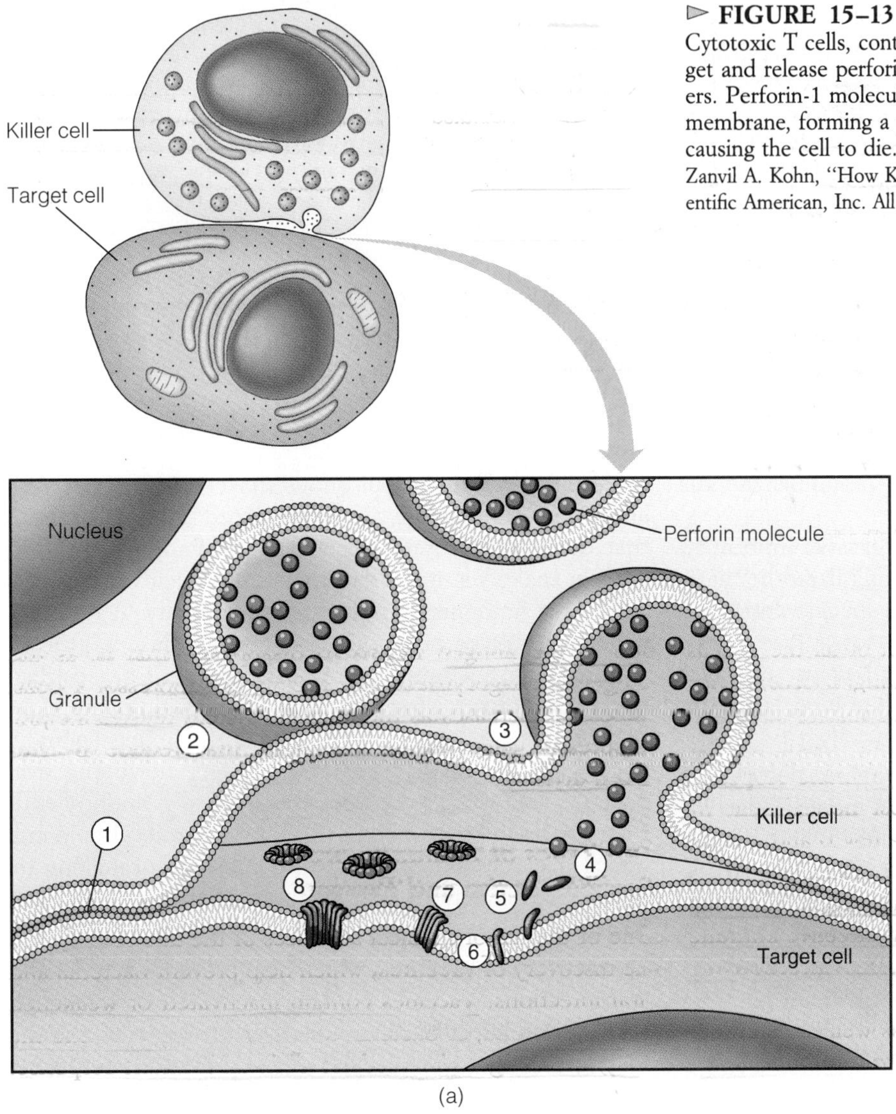

FIGURE 15–13 How Cytotoxic T Cells Work
Cytotoxic T cells, containing perforin-1 granules, bind to their target and release perforin-1, then detach in search of other invaders. Perforin-1 molecules congregate in the target plasma membrane, forming a pore that disrupts the plasma membrane, causing the cell to die. By Dana Burns from John Ding-E Young and Zanvil A. Kohn, "How Killer Cells Kill." Copyright © January 1988 by Scientific American, Inc. All rights reserved.

HEALTH NOTE 15–1

BRINGING BABY UP RIGHT: THE IMMUNOLOGICAL AND NUTRITIONAL BENEFITS OF BREAST MILK

A baby is born into a dangerous world in which bacteria and viruses abound. Complicating matters, the immune system of a newborn child is poorly developed. Fortunately, newborns are protected by passive immunity—antibodies that have traveled from their mothers' blood. Antibodies also travel to the infant in breast milk (▻ Figure 1).

Several immunoglobulins are present in breast milk. One of these is called **secretory IgA.** It is present in very high quantities in **colostrum,** a thick fluid produced by the breast immediately after delivery—before the breast begins full-scale milk production. Colostrum is so important, in fact, that some hospitals give "colostrum cocktails" to newborns who are not going to be breast fed by their mothers. Nurses remove the colostrum from the mother's breast with a breast pump and feed it to the baby in a bottle.

Colostrum, says Sarah McCamman, a nutritionist at the University of Kansas Medical School, coats the lining of the intestines. The IgA antibodies in colostrum prevent bacteria ingested by the infant from adhering to the epithelium and gaining entrance. Breast milk also contains lysozyme, an enzyme that breaks down the cell walls of bacteria, destroying them.

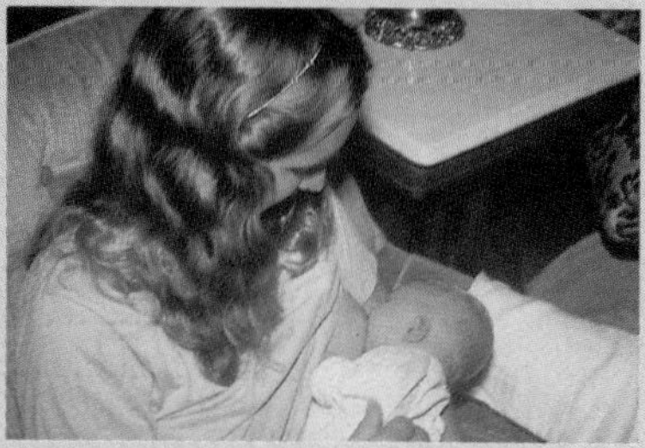

▻ **FIGURE 1 Breast Feeding**
Mother breast feeding her newborn infant.

Unfortunately, not all medical personnel agree on the benefits of breast milk. One problem, they say, is that breast milk has unusually low levels of iron. This fact has led many physicians to recommend iron supplements for newborns. A more careful analysis, however, shows that breast-fed infants generally do not suffer from iron deficiency because the percentage of iron absorbed from breast milk is extraordinarily high. Thus, low levels of iron in breast milk are offset by the high absorption.

The wisdom of iron supplements has also been questioned on other grounds. Researchers, for example, have found that iron supplements increase the incidence of harmful bacterial infection in newborns. As noted in the chapter, iron is a limiting factor in many pathogenic bacteria. Low levels of iron in breast milk, therefore, may reduce bacterial replication in an infant's intestinal tract, aiding in protecting the newborn.

In general, breast-fed babies are healthier than bottle-fed babies. The incidence of gastroenteritis (inflammation of the intestine), otitis (ear infections), and upper respiratory infections is lower in breast-fed babies. Studies also show that children breast fed for at least six months contract fewer childhood cancers than their bottle-fed counterparts. The incidence of childhood lymphoma, a cancer of the lymph glands, in bottle-

---

physicians recommend vaccines against the influenza virus for young children and the elderly.

The second type of immunity, called **passive immunity,** is a temporary form of protection, resulting from the injection of immunoglobulins (antibodies to specific antigens). Immunoglobulins remain in the blood for a few weeks, protecting an individual from infection. Because the liver slowly removes these molecules from the blood, a person gradually loses protection.

Immunoglobulins are given to prevent or counteract certain infections already under way. Travelers to Third World countries are often given immunoglobulins to viral hepatitis (liver infection) as a preventive measure.

In addition, immunoglobulins are used to treat individuals who have been bitten by poisonous snakes (▻ Figure 15–14). The venom in poisonous snakes is a mixture of proteins, enzymes, and polypeptides that may attack body cells, especially nerve cells and cardiac cells. In the United States, the most common poisonous snakebite comes from rattlesnakes.[4] Bites of poisonous snakes can be treated by antivenom, immunoglobulins produced in other animals that quickly destroy or deactivate the immunogenic proteins in snake venom before they can have adverse effects.

Passive immunity can also occur naturally. A fetus, for instance, receives antibodies through the placenta, the organ that transfers nutrients from the mother's bloodstream to the fetal blood. Maternal antibodies remain in the blood of an infant for several months while the infant's immune system is still developing. They protect newborns from bacteria and viruses. Mothers also transfer antibodies to their babies in breast milk. The maternal antibodies in milk attack bacteria and viruses in the intestine, protecting the infant from infection. (For more on this topic, see Health Note 15–1.)

[4]In the United States, 15% of untreated rattlesnake bites and only about 1% of treated bites are fatal.

fed babies is nearly double the rate in breast-fed children for reasons not yet understood.

New research also suggests that certain proteins in breast milk may stimulate the development of a newborn's immune system. In laboratory experiments, the still-unidentified proteins speed up the maturation of B cells and prime them for antibody production. These soluble proteins may also activate macrophages, which play a key role in the immune system.

Breast milk is also more digestible and more easily absorbed by infants than formula. Formula is a mixture of cow's milk, proteins, vegetable oils, and carbohydrates. It is only an approximation of mother's milk and is not broken down and absorbed as completely as breast milk.

Because of a growing awareness of the benefits of breast feeding, virtually every national and international organization involved with maternal and child health supports breast feeding, says McCamman.

Breast feeding can have a major health impact in this country. Unfortunately, the benefits are not as widely known as many people would like, even among health-care professionals. Fortunately, more and more health workers, including physicians, are beginning to understand the benefits of breast feeding and are promoting this option. As a result, many middle-class American women are now choosing to breast-feed.

Unfortunately, says McCamman, there is "a huge population of low-income, poorly educated . . . women who choose not to nurse." The federal government may be playing an unwitting role in their decision. A national program aimed at improving child nutrition provides free formula to needy women, perhaps discouraging mothers from breast feeding.

Another reason is economic. Low-income women often work at jobs that do not provide maternity leave. Thus, these women must return to work soon after giving birth.

Still another reason for the low rate of breast feeding in low-income women is that women need a lot of support to nurse. "People think nursing is innate, natural, and easy," says McCamman. That is not always the case, however. In some instances, getting started requires guidance and education. Without that education and support, breast feeding can be a difficult and painful experience that discourages many women. Breast feeding among all women, rich and poor, may also be discouraged by attitudes and fear of embarrassment. In fact, many open-minded people find breast feeding in public or even semipublic settings embarrassing.

Given the many benefits of breast feeding, McCamman recommends it to all mothers who can. Physicians can help by educating their patients on the benefits of breast feeding. "Doctors should present the information on breast and bottle feeding," says McCamman, "outlining the pros and cons of both methods. Then, let the woman choose. Too few doctors do that today, so women aren't making informed decisions."

(a) (b)

▻ **FIGURE 15–14 Poison and Antidote** (*a*) Poisonous snakes like this rattler inject venom into their victims. (*b*) Venom can be milked from the snake and is used to produce antivenin, a serum containing immunoglobulins that neutralize the venom.

TABLE15–5 Summary of Blood Types

↑ BLOOD TYPE	↑ ANTIGENS ON PLASMA MEMBRANES OF RBCs	↑ ANTIBODIES IN BLOOD	SAFE TO TRANSFUSE To	SAFE TO TRANSFUSE From
A	A	b*	A, AB	A, O
B	B	a	B, AB	B, O
AB	A + B	—	AB	A, B, AB, O
O	—	a + b	A, B, AB, O	

*Lowercase *b* indicates antibody to B antigen.

Vaccination Fears in the United States. Vaccines have helped lower the incidence of many infectious diseases in the United States and other Western countries. Vaccines for diphtheria, tetanus, whooping cough, polio, measles, mumps, and congenital rubella (German measles), for example, have reduced the occurrence of these often-lethal diseases in the United States by more than 99%.

Despite the successes of vaccines, publicity concerning their rare, but sometimes devastating, side effects has created something of a medical dilemma in the United States, Japan, and Great Britain. In 1976 and 1977, for example, a mass-immunization program in the United States for the swine flu, one type of influenza, resulted in the paralysis of a number of people. As a result of public concern over this and other incidents, many parents have chosen not to have their children vaccinated. Excessive media attention given to the rare but serious complications, say some critics, has harmed efforts to promote vaccination. Because of this publicity, public fear of vaccination, and lawsuits, pharmaceutical companies have become increasingly reluctant to invest the huge sums of money needed to develop and market vaccines.

Researchers also believe that another cause for the decline in vaccination stems from the success of previous immunization programs, which have greatly reduced the incidence of most infectious diseases. Parents reared in an environment free of such diseases, researchers say, are often unaware of the dangers of infectious disease. Having their children immunized seems unimportant. Public health officials are concerned that the incidence of infectious diseases such as polio and measles may increase as a consequence.

Harmful side effects from conventional vaccines are often caused by reactions to certain "nonessential" antigens on the injected microorganism. These antigens frequently play little or no role in immunity. Therefore, by eliminating these antigens from vaccines, researchers hope to develop even safer alternatives to vaccines in use today.

PRACTICAL APPLICATIONS: BLOOD TRANSFUSIONS AND TISSUE TRANSPLANTATION

Although important in protecting us from microorganisms, the immune system presents something of a dilemma during blood transfusions and tissue transplants—biological interventions unwitnessed in evolution.

Blood Transfusions Require Careful Cross-Matching of Donors and Recipients

The surface of the red blood cell (RBC) membrane contains many inherited antigens (glycoproteins). One group of antigens is used to determine blood type. As you probably know, four blood types exist: A, B, AB, and O. The letters refer to one type of antigen present on the plasma membrane of RBCs of an individual.

As illustrated in Table 15–5, individuals with type A blood contain RBCs whose plasma membranes contain the A antigen. RBCs in individuals with type B blood contain type B antigens. People with AB blood have both A and B antigens, and people with type O have neither antigen.

To perform a successful blood transfusion, physicians must match donors and recipients. Individuals with type A blood, for example, can receive blood from people with the same blood type but cannot receive blood from people with type B. Individuals with type B blood can generally receive blood from people having type B blood but not from type A donors.

Cross-matching the blood is essential to prevent an immune reaction. Immune reactions during improper transfusions result from antibodies found in the bloodstream of most people. As shown in Table 15–5, each blood type carries a specific type of antibody. People with type A blood, for example, naturally contain antibodies to the B antigen. (As a result, people with type A blood cannot receive type B blood.) People with type B blood contain antibodies to the A antigen. For reasons not well understood, these antibodies appear in the blood during the first year of life.

Problems arise when incompatible blood types are mixed. For example, imagine that an individual with type A blood is accidentally given type B blood (▻ Figure 15–15) The antibodies to type B blood found in the recipient bind to the transfused RBCs (containing type B antigens), causing them to agglutinate (clump) and hemolyze (burst). Hemolysis and agglutination constitute the **transfusion reaction.**

RBC clumping restricts blood flow through capillaries, reducing oxygen and nutrient flow to cells and tissues. Massive hemolysis results in the release of large amounts of

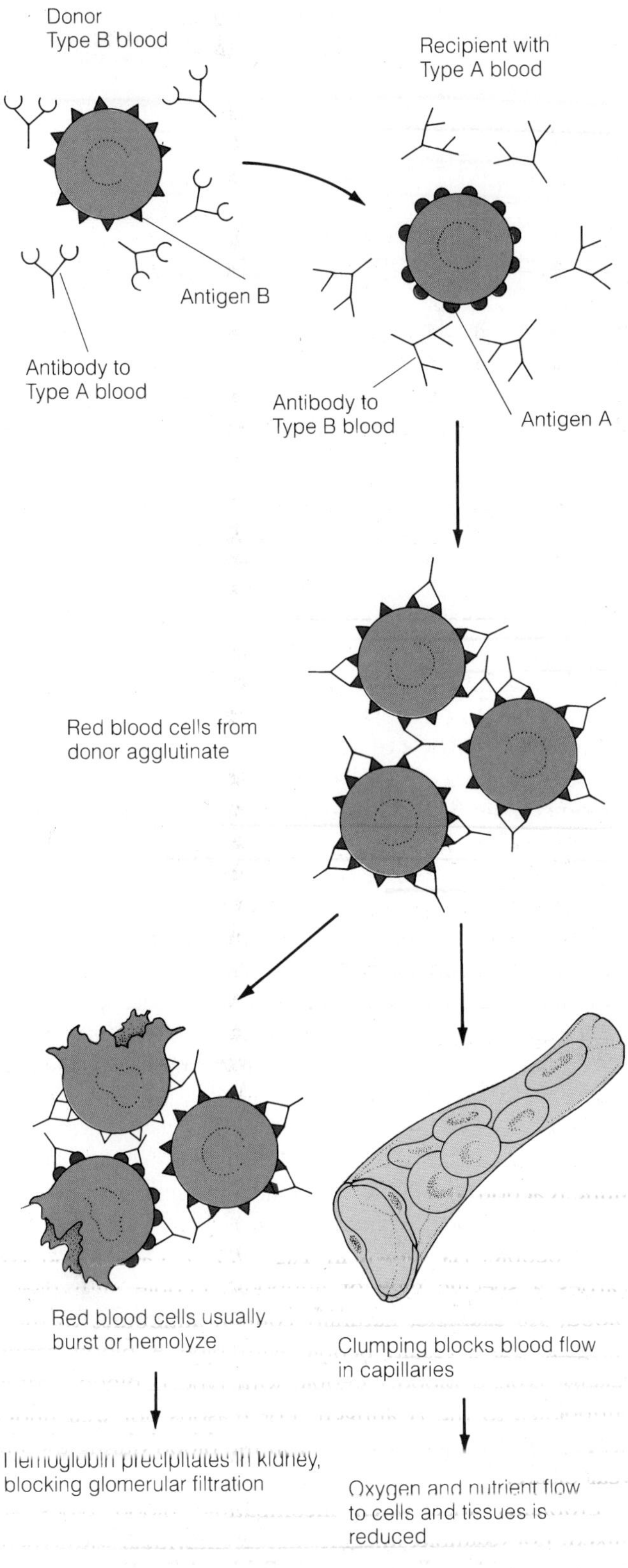

▻ **FIGURE 15–15 Transfusion Reaction** Type B blood transfused into an individual with type A blood results in a transfusion reaction, characterized by agglutination and hemolysis.

hemoglobin into the blood plasma. Hemoglobin precipitates in the kidney, blocking the tiny tubules that produce urine, which often results in acute kidney failure.

Because of the possibility of a transfusion reaction, successful transfusions require careful matching of the blood types of the donor and recipient. As Table 15–5 shows, RBCs from individuals with type O blood have neither A nor B antigens. Therefore, type O blood can be transfused into individuals with all four types: A, B, AB, and O. Type O individuals are said to be **universal donors.**

Type O blood, while free of antigens, contains antibodies to both A and B antigens. Therefore, individuals with type O blood can receive only type O blood. Any other type of blood would cause a transfusion reaction.

As shown in Table 15–5, individuals with type AB blood contain RBCs with both A and B antigens but no antibodies related to the ABO system. These people can therefore receive blood from all others and are consequently referred to as **universal recipients.** AB blood can be safely transfused only into individuals with AB blood.

The terms *universal donor* and *universal recipient* are somewhat misleading, however, because RBCs also contain other antigens that can cause transfusion reactions. The most important of these is called the Rh factor. This antigen was first identified in rhesus monkeys, hence the designation. People whose cells contain the Rh antigen, or **Rh factor,** are said to be **Rh positive.** Those without it are **Rh negative.**

Unlike the ABO system, in the Rh system antibodies are produced *only* when Rh-positive blood is transfused into the bloodstream of a person with Rh-negative blood. The first transfusion of Rh-positive blood into an Rh-negative person generally does not result in a transfusion reaction, but a second transfusion does. To reduce the likelihood of a transfusion reaction, Rh-negative people should receive only Rh-negative blood, and Rh-positive people should receive only Rh-positive blood.

The Rh factor becomes particularly important during pregnancy. Problems can arise if an Rh-negative mother has an Rh-positive baby (▻ Figure 15–16). Even though the maternal and fetal bloodstreams are separate, small amounts of fetal blood usually enter the maternal bloodstream at birth, causing the immune reaction. Rh antibodies will form in the maternal bloodstream, and the woman will be sensitized to the Rh factor.

If the woman is not treated at the time and becomes pregnant again with an Rh-positive baby, maternal antibodies to the Rh factor will cross the placenta. These antibodies cause fetal RBCs to agglutinate, then break down, eventually resulting in anemia and hypoxia (lack of oxygen to tissues). Unless the baby receives a blood transfusion (of Rh-negative blood) *before* birth and several after birth, it is likely to have brain damage and may even die.

To prevent antibody production in Rh-negative women who give birth to Rh-positive babies, physicians inject antibodies to fetal Rh-positive RBCs into the mother soon after she has given birth. (The antibody containing serum is called RhoGAM.) These antibodies bind to Rh-positive RBCs from the fetus before a woman's immune system can respond to them. This prevents a woman from being sen-

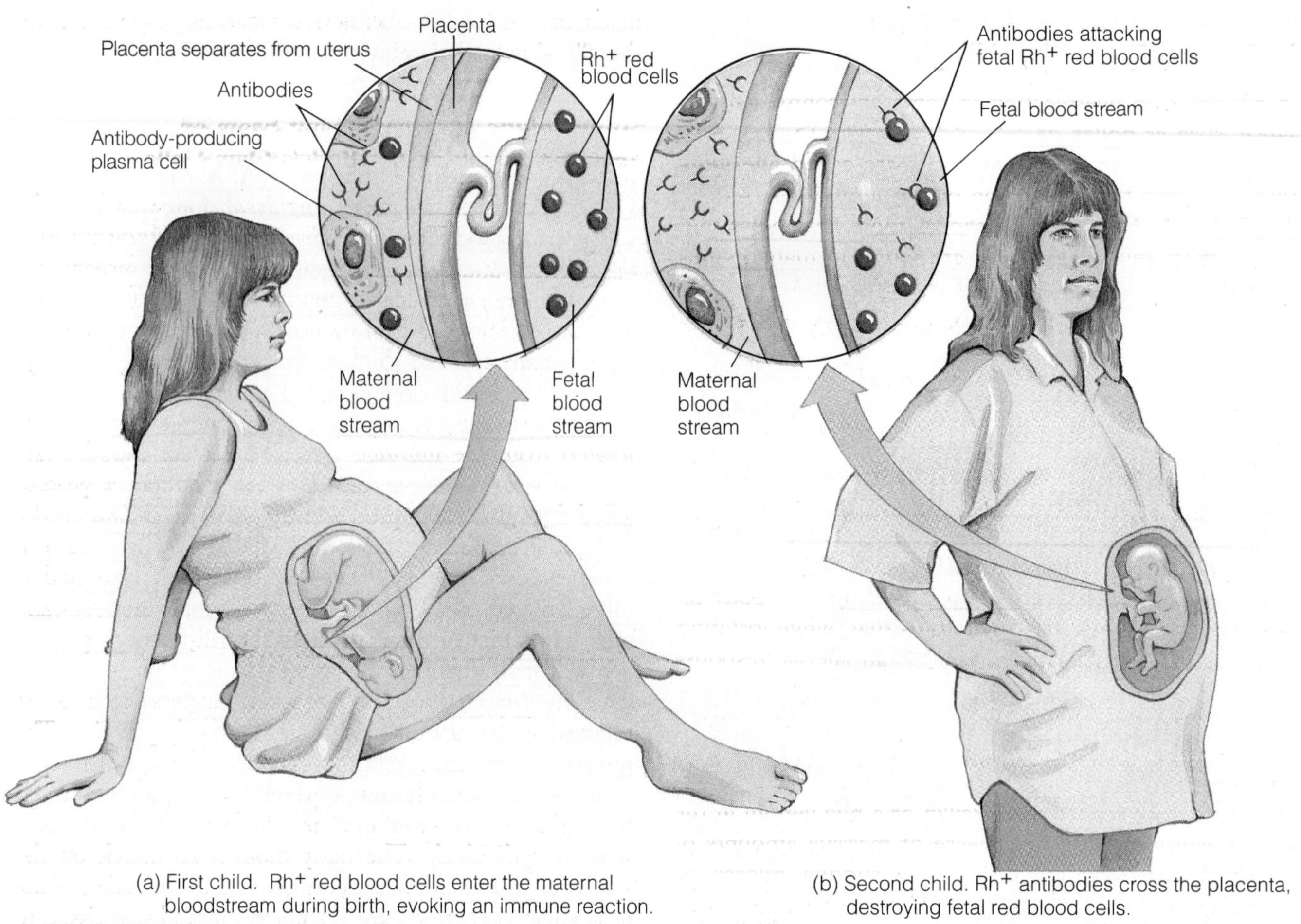

FIGURE 15–16 The Rh Factor and Pregnancy Rh-positive cells from the fetus enter the mother's blood at birth. If the mother is Rh negative, her immune system responds, producing antibodies to the Rh-positive RBCs and destroying them. Problems arise if the mother becomes pregnant again and has another Rh-positive baby. If the mother was not treated the first time, antibodies to Rh-positive RBCs cross the placenta and destroy fetal RBCs.

sitized. In other words, RhoGAM prevents B cells from being stimulated and obviously blocks the production of memory cells to Rh factor. To be effective, however, the treatment must be given within 72 hours after the baby is born.

Tissue Transplantation Often Evokes Cell-Mediated Immunity, Which Can Be Blocked by Certain Drugs

Tissue transplantation is a much more complex matter. Only three conditions exist in which a person can receive a transplant and not reject it. One is if the tissue comes from the individual himself or herself. For burn victims, surgeons might use healthy skin from one part of the body to cover a badly damaged region. The second instance is when a tissue is transplanted between identical twins—individuals from a single fertilized ovum that split and formed two embryos. These individuals are genetically identical and therefore have identical cellular antigens.

A third instance occurs when tissue rejection is inhibited by specific drugs. For example, heart, liver, and kidney transplants are successful only when recipients are treated with immunosuppressive drugs—that is, drugs that suppress the immune system. This treatment must be continued throughout the life of the patient. Unfortunately, most immune suppressants have numerous side effects and often leave the patient vulnerable to bacterial and viral infections. In the 1980s, a new drug known as **cyclosporin** was introduced. This drug suppresses the formation of interleukin 2 by helper T cells, thus greatly reducing cell-mediated immunity without affecting B cells. Patients who receive the drug are therefore able to combat many bacterial infections with antibodies.

DISEASES OF THE IMMUNE SYSTEM

The immune system, like all other body systems, can malfunction resulting in a wide range of symptoms. This section will look at two disorders: allergies and autoimmune diseases.

Making matters worse, HIV may be taking refuge in the body. In 1988, a research team announced that out of 100 homosexual men studied, 4 initially showed antibodies to the AIDS virus, but slowly lost them. This process usually occurs only in the late stages of AIDS, when the immune system is too weak to produce antibodies. These 4 men, however, had no overt symptoms of AIDS. Had the virus been conquered by the immune system, or had it simply entered a latent phase, "hiding out" as does the herpes virus?

Research suggests that HIV may take up residence in bone marrow stem cells that give rise to lymphocytes. The virus, therefore, may remain in these cells and be passed on to new white blood cells produced by cell division. Thus, once the virus is in the body, it may be there forever. As a result, completely eliminating the virus from the body may be impossible. One of the upshots is that if HIV goes into hiding, AIDS-infected blood donors may escape detection, even with the new genetic tests. AIDS-infected blood cells could unknowingly be passed to thousands of patients over the coming years.

The Good News about AIDS. Despite the discouraging news, researchers and much of the general public remain determined to find a cure for AIDS and a way to prevent it.

One encouraging development was revealed in 1988. Two California research groups announced that they had successfully transplanted parts of the human immune system into mice. Human immune cells in mice may provide scientists with a tool to study AIDS and to test potential cures for it, thus accelerating the pace of research.

On another front, researchers have developed a genetically engineered weapon that could kill cells infected with the AIDS virus, possibly curing the disease once it has developed. As noted above, cells infected with the AIDS virus produce a protein known as gp120, which is part of the capsid (▻ Figure 15–20). This protein also ends up in the plasma membrane of infected cells.

Researchers have genetically engineered a bacterium that binds to the gp120 protein. The bacterium carries with it a toxin that kills the HIV-infected cells. Preliminary studies indicate that noninfected cells are unharmed by this treatment. Although initial studies suggest that this treatment will not work, researchers hope that the technique can be improved or modified, making it possible to kill enough infected cells in AIDS patients to halt the disease. One question that must be answered before this procedure can be tried in people is whether AIDS-infected cells killed by this technique will degenerate and release active AIDS viruses that could spread to other body cells.

Researchers are also testing an HIV vaccine that could be administered as a preventive measure. The rapid rate at which the virus mutates, however, has stymied these efforts. Nevertheless, many potential vaccines are in clinical trials.

▻ FIGURE 15–20 Duping the AIDS Virus The AIDS virus has a protein, gp120, in its capsid. When it infects cells, it produces more gp120 to make new capsids. However, some gp120 ends up in the infected cell's plasma membrane, marking it. By genetically engineering a bacterium that can locate the infected cells by finding gp120, medical researchers may be able to hunt and kill infected cells, stopping the spread of the virus.

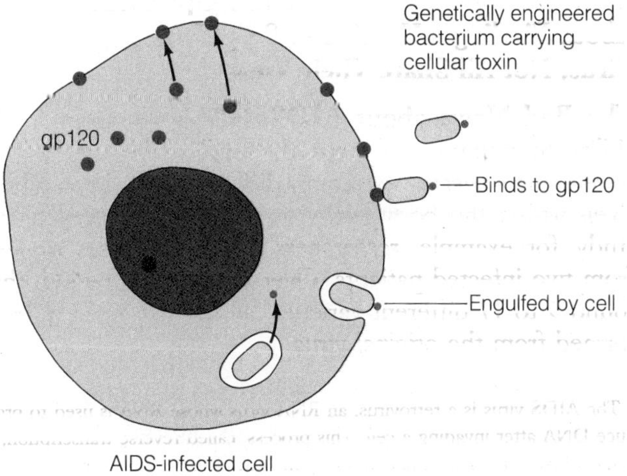

ENVIRONMENT AND HEALTH: MULTIPLE CHEMICAL SENSITIVITY

Richard Sharp was a physicist for a major aviation company in California. Today, he is confined to two stripped-down rooms equipped with special filters that remove all air contaminants.

Sharp is one of many Americans who has developed a condition known as multiple chemical sensitivity (MCS). He has become a prisoner in his own home, unable to venture forth into modern society without suffering extreme discomfort, even debilitation. MCS is thought to be caused by exposure to a number of common household and industrial chemicals, including formaldehyde, solvents, acrylic resins, mercury compounds, and pesticides. Chemicals such as formaldehyde, which is commonly found in carpeting, plywood, furniture, and many other household products, are thought to bind to naturally occurring proteins in the body. This process produces foreign substances against which the immune system mounts an attack. In other words, common household and industrial substances turn the immune system against the body.

Individuals become sensitive to low levels of chemicals over long periods. As a result, this condition is typically referred to as **hypersensitivity.** Victims also exhibit cross-reactions; that is, they react to other chemically similar substances. Massive exposures to certain chemicals may also elicit a hypersensitivity reaction.

The symptoms of MCS vary, ranging from life-threatening to mild. The most common symptoms are tension, memory loss, fatigue, sleepiness, headaches, confusion, and depression. Many victims of MCS suffer from

POINT/COUNTERPOINT Tracking Victims of AIDS

ANONYMOUS TESTING IS THE ANSWER

Earl F. Thomas

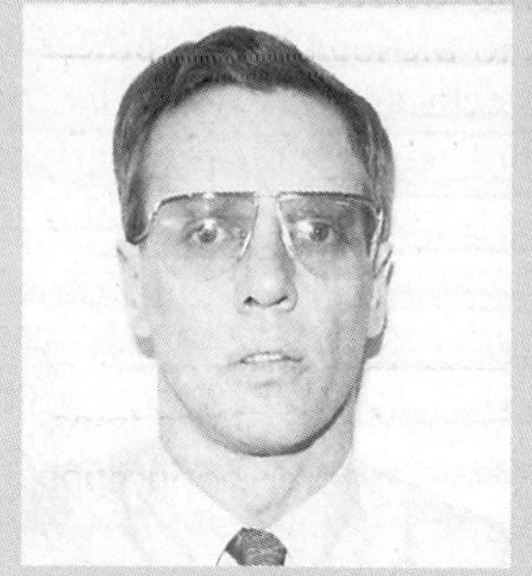

Earl F. Thomas is on the boards of directors for the Colorado AIDS Project and the People With AIDS Coalition of Colorado. Thomas was diagnosed as having AIDS in 1986.

Confidential testing for HIV (in which the names and addresses of individuals who test positive are reported to health officials) versus anonymous testing continues to be a highly controversial topic. In today's society, the word *AIDS* still breeds fear, especially the fear of discrimination. These fears, quite simply, are keeping people from being tested.

Health departments and AIDS organizations stress that persons who have reason to believe that they may have been infected with the HIV virus should be tested, for several reasons. First, the earlier a person knows he or she is HIV positive, the earlier treatment can be started to slow the progression of the disease or simply to buy time while researchers explore better treatments. Second, knowledge of one's HIV status is crucial in determining what behavioral changes need to be made.

If the testing system discourages individuals from obtaining knowledge about their HIV status, however, no one benefits. Unfortunately, name reporting tends to create an atmosphere of distrust between health officials and those who wish to be tested. These people already feel disenfranchised from society and cannot believe that the public health system is concerned about their personal rights and welfare. They believe that information concerning their HIV status goes beyond the health department to others who may have a need to know, and they fear that the information will eventually find its way to persons who have no need to know. For instance, numerous links in the information chain (nurses, laboratory workers, therapists, and secretaries) all have access to a person's medical information. Confidential medical information is not so confidential after all. As a result of these concerns, many persons, even in high-risk groups, refuse to be tested at all.

In a survey conducted by the Educational Department of the Colorado AIDS Project in 1989, 32% of 1,112 respondents (homosexual and bisexual men) cited confidentiality concerns as their reason for not being tested. Despite the health department's insistence that fears of information leaks are poorly founded, they are very real to many individuals. More importantly, these fears prevent individuals who have engaged in high-risk behaviors from being tested.

Further information that supports anonymous testing comes from the state of Oregon. When anonymous HIV testing was offered along with confidential testing, there was a 50% increase in the demand to be tested during the first four months of the program. People from all segments of society sought out the anonymous test sites. Confidential testing sites reported *no* increase during the same period. These data strongly suggest that reporting by name is not the solution to HIV testing.

Fear of testing is justified. There is documented evidence of discrimination against people who are HIV-positive. Individuals have been denied housing and, in certain cases, have been evicted from their homes, have lost their jobs, have been denied access to public education, and have been shunned by families, friends, and co-workers.

Partner notification (contact tracing) may have a place when trying to control the spread of AIDS, but at what cost? Some feel that contact tracing is not a viable option due to the monetary cost. In six months in 1988 the state of Colorado spent $450,000 on partner notification. The result of this enormous expenditure was that 52 people were found.

The experiences of other health departments clearly show that partner notification can be carried out with reasonable success when testing is anonymous. If lists are being maintained, people shy away from testing and forfeit the possibility of early intervention treatment and partner notification information. When health officials act as contact tracers, they are perceived as police. In our society, police-state tactics will never work, and no one benefits.

Anonymous testing would reduce a serious impediment to a powerful, collective anti-AIDS effort. And most important, we could get an answer to whether people are truly avoiding testing because of reportability.

gastrointestinal problems as well—nausea, indigestion, and cramps. Some exhibit respiratory symptoms as well, including frequent colds, bronchitis, and shortness of breath. Skin rashes are not uncommon. Many people report allergy-like symptoms such as nasal stuffiness and sinus infections.

MCS remains one of the most mysterious of all human diseases. Victims often witness a sudden deterioration of their health. Physicians, family members, friends, and even the victims themselves are baffled by the disease. As a result, patients are often likely to be labeled "psychiatric cases." People suffering from MCS must often get rid of

NOTIFICATION WORKS

John Potterat

John Potterat is an authority on AIDS and sexually transmitted disease (STD) control. He has published numerous articles in medical journals dealing with STD and AIDS control and is currently director of the STD and AIDS programs in Colorado Springs.

The process of reporting people with the AIDS virus (HIV) by name to public health officers and in turn tracing their contacts should not be controversial. Such procedures have been standard public health practice for serious communicable diseases for nearly a century. Formal notification allows society to define the disease burden (surveillance) and to counterattack (control). You cannot control a communicable disease if you do not know who has it and who might be next to have it; moreover, you need to find those directly affected.

Notifying partners of people infected with sexually transmitted disease has been an effective control tool for 50 years. The fundamental reason that health officers are involved in this notification process is that STD patients are not good at referring their own sexual partners. Such "self-referral" fails more often than it succeeds: less than a third of STD partners are successfully referred for medical evaluation. Partner referral by HIV patients is even less successful (despite frequent assurances by patients that they "will take care of it!"). Part of this failure is due to the reluctance of HIV patients to face their partners (fear of anger or reprisal); part is due to selective notification (denial that "nice" partners can be infected); and part to failure to convince partners (partner denial). Trusting the notification process to infected people alone is a luxury that society can ill afford.

Those exposed to HIV have a right to know. Important sexual (safer practices) and reproductive (postponing pregnancy) decisions depend on knowledge of exposure and its outcome. Many persons are unaware that their partners have histories of needle exposures or of bisexuality. The duty to warn people has compelling moral, legal, and historical foundations. In free societies, notification is a straightforward, confidential process. Medical workers who detect HIV infection report the case by name and address to the local health officer who then discreetly contacts the patient to counsel him or her and to obtain identifying information on sexual and needle partners. People are persuaded, not coerced, into voluntary cooperation. Although counseling is "mandatory," blood testing is optional.

Even if "treatment" for partners were to consist solely of personal counseling to discourage behaviors that facilitate transmission or accelerate disease progression, partner notification would be worthwhile.

A disease control procedure should be acceptable to people. Partner notification by health officers has been well received by the affected populations. The majority (70% to 80%) of notified partners accept blood testing, and almost all who decline testing accept counseling. Although organized gay advocacy groups have generally opposed both HIV reporting by name and partner notification by health officers, when approached individually and sympathetically, gay men have generally cooperated.

Health officers are responsible for maintaining the physical security of HIV records; such records are also immune from any discovery process. They cannot be subpoenaed or released to potentially adversarial agents like insurance, police, or employer investigators. Whatever discrimination is suffered by infected people, none of it stems from disease notification to or by public health officers.

Notification initiatives are affordable, acceptable to patients, and effective in reaching high-risk people. It is well known to health officers that those at highest risk are least inclined to appear for counseling and least likely to use safer practices. While notification is not a panacea, it is one of the most useful measures for containing this tragic epidemic.

SHARPENING YOUR CRITICAL THINKING SKILLS

1. Summarize Thomas's reasons for keeping AIDS testing anonymous.
2. Summarize Potterat's views in support of name reporting.
3. Do you agree or disagree with the following statement: both writers believe that their approach will provide the greatest protection to the public health, but they differ in their approach. Explain.
4. Of the two basic approaches, which do you think would be most effective in reducing the spread of AIDS?

all cleaning agents, pesticides, perfumes, deodorants, and other household chemicals.

At first, many physicians and scientists dismissed this disease, but a growing body of evidence shows that many common chemicals can indeed alter the immune system. The National Research Council estimates that 15% of the U.S. population experiences some degree of chemical hypersensitivity. Studies show that 5% of the workers exposed to an agent used in the manufacture of plastics, TDI (toluene diisocyanate), develop asthmalike symptoms. TDI apparently binds to proteins in the respiratory tract, creating foreign substances that stimulate a hypersensitivity reaction. Individuals who have been hypersensitized have difficulty breathing when exposed to TDI, tobacco smoke, and air pollutants. In Japan, 15% of all cases of asthma in men can be directly attributed to industrial exposure to chemicals.

Other chemicals bind to proteins in the skin, creating foreign substances to which the immune system reacts. Formaldehyde, for example, results in a condition called **contact dermatitis,** characterized by a skin rash. T cells attack the cells of the skin, destroying them. Even low levels of formaldehyde found in newsprint dyes, some cosmetics, and photographic papers are sufficient to induce rashes.

Other chemicals apparently act by suppressing immune function, making individuals more susceptible to infectious agents. Chronic workplace exposure to benzene, for example, results in a reduction in the number of circulating lymphocytes. Benzene probably depresses lymphocyte production in the bone marrow. In rabbits, benzene exposure results in an increased susceptibility to various infectious agents.

Dioxins, PCBs, ozone, certain pesticides, and a variety of other chemicals suppress the immune response in laboratory animals and humans. However, the overall significance of immune suppression and hypersensitivity in human populations remains unknown. Nonetheless, this section illustrates the subtle but potentially far-reaching effects of toxic chemicals and underscores the importance of a healthy environment for maintaining a healthy population.

SUMMARY

THE FIRST AND SECOND LINES OF DEFENSE

1. The first line of defense against viruses, bacteria, and other infectious agents is the skin and the epithelia of the respiratory, digestive, and urinary systems. Some epithelia also produce protective chemical substances that kill microorganisms.
2. The second line of defense consists of cells and chemicals that the body produces to combat infectious agents that penetrate the epithelia.
3. One of the chief combatants in the second line of defense is the macrophage, a cell derived from the monocyte. Macrophages are found in connective tissue lying beneath the epithelia, where they phagocytize infectious agents, preventing their spread. Neutrophils and monocytes also invade infected areas from the bloodstream and destroy bacteria and viruses.
4. Another portion of the second line of defense consists of the chemicals released by damaged tissue, which stimulate arterioles in the infected tissue to dilate. The increase in blood flow raises the temperature of the wound. Heat stimulates macrophage metabolism, accelerating the rate of the destruction of infectious agents. Heat also speeds up the healing process by accelerating the metabolic rate of the cells involved in the recovery.
5. Still other chemicals increase the permeability of the capillaries, causing plasma to flow into the wound and increasing the supply of nutrients for macrophages and other cells fighting the invader.
6. The increase in blood flow, the release of chemical attractants, and the flow of plasma into the wound constitute the inflammatory response.
7. Another part of the secondary line of defense is pyrogens, chemicals released primarily by macrophages exposed to bacteria, which raise body temperature and lower iron availability, thus decreasing bacterial replication.
8. Interferons, a group of proteins released by cells infected by viruses, are also part of the second line of defense. Interferons travel to other virus-infected cells, where they inhibit viral replication.
9. The blood also contains the complement proteins, which circulate in an inactive state, becoming activated when the body is invaded by bacteria. They too are part of the second line of defense.
10. Some of the complement proteins stimulate the inflammatory response. Others embed in the plasma membrane of bacteria. There, they combine to form membrane-attack complexes, which create holes in the bacterial plasma membranes, killing these pathogens. Another complement protein binds to the invader, making it more easily phagocytized by macrophages.

THE THIRD LINE OF DEFENSE: THE IMMUNE SYSTEM

11. The immune system consists of billions of lymphocytes that circulate in the blood and lymph and take up residence in the lymphoid organs.
12. The lymphocytes recognize antigens—foreign cells and foreign molecules, mostly proteins and large-molecular-weight polysaccharides.
13. T and B cells are two types of lymphocytes produced in red bone marrow. The T cells become immunocompetent—able to respond to a particular antigen—in the thymus. B cells gain this ability in the bone marrow. During this process, the T and B cells gain specific membrane receptors that respond to specific antigens.
14. Immunocompetent B cells encounter antigens (often presented to them by macrophages) to which they are programmed to respond, then begin to divide, forming plasma cells and memory cells. Plasma cells produce huge amounts of antibodies. The memory cells enable the body to respond more quickly to future invasions by the same antigen.
15. Antibodies are small protein molecules produced by plasma cells. They bind to specific antigens, destroying them either by precipitation, agglutination, or neutralization, or by activation of the compliment system.
16. When T and B cells first encounter an antigen, they react slowly in what is called the primary response. As a result, there is often a period of illness before the pathogen is removed by the immune system. Because numerous memory cells are produced during the first assault, a reappearance of the antigen elicits a much faster and more powerful reaction,

the secondary response. Consequently, the pathogen is usually vanquished before symptoms of illness occur.
17. Resistance created by a response to an antigen is called immunity.
18. When activated by antigen, T cells multiply and differentiate, forming memory T cells, cytotoxic T cells, helper T cells, and suppressor T cells whose functions are summarized in Table 15–4.
19. A solution containing a dead or weakened virus, bacterium, or bacterial toxin that is injected into people to provide immunity is called a vaccine. Vaccines produce an active immunity, which may last for years.
20. Passive immunity can be conferred by injecting antibodies into a patient or by the transfer of antibodies from a mother to her baby through the bloodstream or milk. Passive immunity is short-lived, lasting at most only a few months.

PRACTICAL APPLICATIONS: BLOOD TRANSFUSIONS AND TISSUE TRANSPLANTATION

21. Blood transfusions require careful matching of donor and recipient blood types. The antigens on the membranes of body cells are more complex. Only cells from the same individual or an identical twin will be accepted. All others are rejected by the T cells, unless the system is suppressed with drugs, which often leave the patient vulnerable to bacterial infections.

DISEASES OF THE IMMUNE SYSTEM

22. The most common malfunctions of the immune system are allergies, extreme overreactions to some antigens.
23. Allergies are caused by IgE antibodies, produced by plasma cells. IgE antibodies bind to mast cells, which causes the cells to release histamine and other chemical substances that induce the symptoms of an allergy—production of mucus, sneezing, and itching.
24. Autoimmune diseases, another type of immune system disorder, result from an immune attack on the body's own cells. Autoimmune diseases may result when normal proteins are modified by chemicals or genetic mutations so that they are no longer recognizable as self. Another possible cause is the sudden presence of proteins that are normally isolated from the immune system. Still another cause of autoimmune reaction is the exposure to antigens that are nearly identical to body proteins.

AIDS: FIGHTING A DEADLY VIRUS

25. AIDS is a disease of the immune system caused by HIV, a virus that attacks helper T cells, severely impairing a person's immune system.
26. Patients grow progressively weaker and may develop cancer and bacterial infections because of their diminished immune response.
27. AIDS is spread through body fluids during sexual contact and blood transfusions or through needles shared by drug users.
28. Stopping the virus has proved difficult, in large part because it mutates so rapidly and appears to take refuge in body cells. Recent accomplishments in research, however, offer some promise.

ENVIRONMENT AND HEALTH: MULTIPLE CHEMICAL SENSITIVITY

29. Many individuals suffer from multiple chemical sensitivity. Chronic exposure to low levels of pollutants or short-term exposure to high levels may alter the immune system, causing a wide range of symptoms.
30. Research shows that some toxic chemicals cause immune hypersensitivity, evoking allergylike symptoms. Others stimulate autoimmune responses. Still others cause immune suppression.

EXERCISING YOUR CRITICAL THINKING SKILLS

You have been selected as a juror for a trial involving a physician who refused to perform surgery on an AIDS patient. The patient had been in an automobile accident and had suffered a ruptured spleen, causing internal bleeding. The physician refused to perform an operation that could have saved the patient's life, because she was afraid of contracting AIDS.

Consider the following facts presented by the attorneys for the plaintiffs (the AIDS patient's family). The plaintiffs argue that the physician violated her code of ethics, which obligates her to treat all patients. They also argue that she knowingly allowed a patient to die and that she should be punished by having to pay damages as compensation for the lost life. The AIDS patient had received a transfusion of HIV-contaminated blood several years earlier and had only three to six months to live. Nevertheless, these months were valuable to him and to his family.

The defense attorneys admit that the physician refused treatment and caused the premature death of her patient. They say, however, that she acted rightfully. During surgery, sharp instruments frequently pierce the protective gloves of surgeons, exposing them to the AIDS virus. Refusing to operate protected not only the physician but also her entire surgical team. That team could, over the course of years, save hundreds of lives. In refusing to operate, the physician was considering the greater good—the benefit of her services to the public. The surgeon also had a husband and two children. By protecting herself, she was also taking into account the good of her family.

How would you decide such a case? Should the physician be forced to pay damages? Why? Or was she correct in her decision? Why? What would you have done if you were the physician?

As a side note, hospitals often assemble teams of doctors and other medical personnel who are willing to treat AIDS patients to avoid problems such as the one described above.

TEST OF TERMS

1. A virus contains a nucleic acid core and a coat of protein called the ____________ .
2. The redness and swelling around a cut or abrasion are part of the ____________ ____________.
3. ____________ are chemical substances, produced primarily by macrophages, that increase body temperature during a bacterial or viral infection.
4. ____________ are a group of chemicals produced by virus-infected cells that impair viral replication in other virus-infected cells.
5. A(n) ____________ is a large-molecular-weight carbohydrate or protein that evokes an immune response.
6. The ability of B and T cells to respond to specific foreign invaders is called ____________ .
7. Antibodies provide ____________ immunity, protection against bacteria and viruses occurring primarily in the blood and lymph.
8. The secondary response to an antigen occurs more rapidly than the primary response because of the formation of ____________ cells.
9. Antibodies belong to a class of proteins called ____________ .
10. Coating a virus or bacterial toxin with antibody is called ____________ .
11. Blood cells from an incompatible source clump together, because the recipient's blood contains antibodies to the donor's RBCs. This process is called ____________ .
12. T cells differentiate into a cell, the ____________ ____________ cell, that travels throughout the body, destroying bacteria and virus-infected cells and tumor cells directly.
13. A ____________ contains dead or weakened bacteria or viruses that elicit an immune response without causing disease. It provides a form of immunity.

Answers to the Test of Terms are found in Appendix B.

TEST OF CONCEPTS

1. Describe the structure of a typical virus.
2. Based on your previous studies (Chapter 3), in what ways are bacteria similar to human body cells, and in what ways are they different?
3. The human body consists of three lines of defense. Describe what they are and how they operate.
4. Describe the inflammatory response, and explain how it protects the body.
5. Define each of the following terms, and explain how they help protect the body: pyrogen, interferon, and complement.
6. The first and second lines of defense differ substantially from the third line of defense. Describe the major differences.
7. How does the immune system detect foreign substances?
8. Describe how the B cell operates. Be sure to include the following terms in your discussion: bone marrow, immunocompetence, plasma membrane receptors, primary response, plasma cell, antigen, antibody, and secondary response.
9. Describe the four mechanisms by which antibodies "destroy" antigens.
10. Describe the events that occur after a T cell encounters its antigen.
11. What is the difference between active and passive immunity?
12. A child is stung by a bee, swells up, and collapses, having great difficulty breathing. What has happened? What can be done to save the child's life?
13. What is an autoimmune disease? Explain the reasons it forms.
14. What is AIDS? What are the symptoms? What causes it?

CHAPTER

16

The Urinary System: Ridding the Body of Wastes and Maintaining Homeostasis

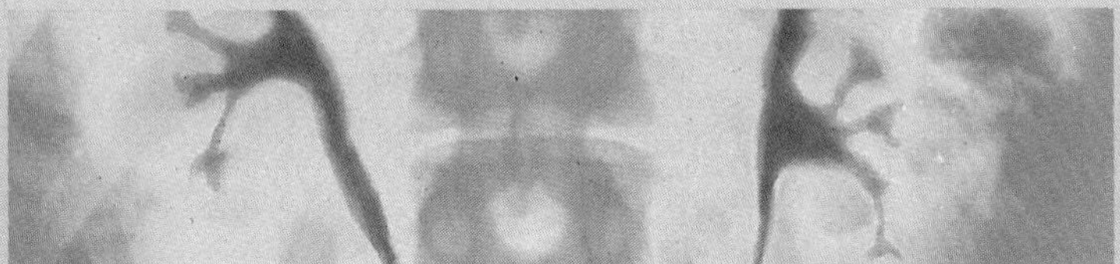

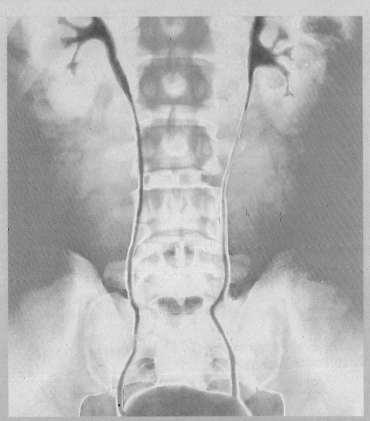

Colorized x-ray of human urinary system, showing kidneys, ureters, and urinary bladder.

Leon Markowitz is lowered into a large pool of warm water in a special room in the hospital where he will spend the next few hours (▻ Figure 16–1). Physicians position a large, cylindrical device in the water in front of one of his kidneys, the organs that filter the blood and help maintain normal blood concentrations of nutrients and wastes. Over the next few hours, as Markowitz listens to tapes of Paul Simon, ultrasound waves, undetectable by the human ear, will silently smash a mass of calcium phosphate and other chemicals that has formed in his kidney and is obstructing the flow of urine and causing excruciating pain.

A decade earlier, surgeons would have had to cut an incision 15 to 20 centimeters (6 to 8 inches) long in Markowitz's side to remove the stone. He would have spent 7 to 10 days recovering in the hospital and would have had to recuperate at home eight more weeks before returning to work. With this new technique, known as ultrasound lithotripsy ("stone crushing"), patients usually return home within a day.

This chapter describes the ways in which animals rid themselves of potentially harmful wastes derived from metabolism. It focuses principally on the structure and function of the vertebrate urinary system and presents information that will be useful to you in your life. It also points out how the urinary system contributes to homeostasis and describes some common diseases, such as kidney stones, that disturb the function of this important organ system.

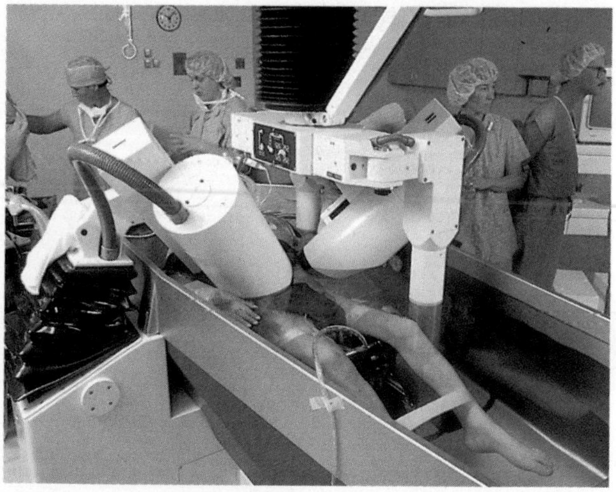

▻ **FIGURE 16–1 Lithotripsy** Ultrasound waves, undetectable to the human ear, bombard the kidney stones in this man, smashing them into sandlike particles that can be passed relatively painlessly in the urine.

GETTING RID OF WASTES: EVOLUTION MEETS THE CHALLENGE

Ralph Waldo Emerson once said that as soon as there is life, there is risk. A biologist might look at things differently, noting that as soon as there is life, there is waste. Put more bluntly, all living things produce waste. Fortunately, elaborate systems of recycling have evolved in the biosphere so that the waste of one organism is a resource for another. The most obvious example is carbon dioxide, a waste product of glucose catabolism. As noted earlier in the text, carbon dioxide is one of the main raw materials of photosynthesis.

All organisms face the same challenge—how to get rid of wastes that are produced internally. For single-celled organisms, like the protists, the answer is simple: waste diffuses or is actively transported out of their bodies and is released into the medium (air, water, soil) they inhabit.

Moving up the evolutionary tree, we encounter some of the simple multicellular animals, like sponges and sea stars. Although multicellular, these organisms do not have specific waste-removing organs. The cells on their body surface eliminate wastes directly into the surrounding environment and into the extracellular fluid.

In truth, though, very few multicellular animals rely on body cells to carry out waste-management functions. Most of them have special organs that rid the body of waste and also help regulate internal concentrations of ions and water, a function vital to homeostasis. Consider two examples: earthworms and insects.

In earthworms and other members of the group (phylum) they belong to (the annelids), fluid from the body cavity enters two small, tubular structures, the **metanephridia** (pronounced, met-uh-nef-rid-e-uh) in each segment of the body. The metanephridia collectively form a primitive kidney (▻ Figure 16–2). Fluid enters through an opening, then travels through the tubule to the **bladder,** an expanded region that stores the liquid containing wastes and other substances. Useful materials in the fluid are reabsorbed into the blood stream—that is, pass back into the blood. Wastes are eliminated through another opening that empties the waste onto the earthworm's external body surface. The nephridia rid the earthworm of water and salts; other wastes, like ammonia, diffuse across the body surface directly.

Another interesting waste-disposal system is found in terrestrial insects (▻ Figure 16–3). As illustrated, small, dead-end tubules attach to the gut. These tubules are bathed in body fluids and extract ions and water from them, thus helping get rid of excess water and wastes. The mechanism of transport is explained in the figure. Interestingly, cells that line the hindgut reabsorb useful materials, thus conserving essential ions and other substances necessary for life.

Waste disposal reaches its zenith in the vertebrates, endowed as they are with several mechanisms to rid the body of potentially harmful substances. The next two sections discuss the human organs of excretion and the anatomy of the urinary system.

ORGANS OF EXCRETION

Metabolism in human body cells produces enormous amounts of waste, such as carbon dioxide, ammonia, and

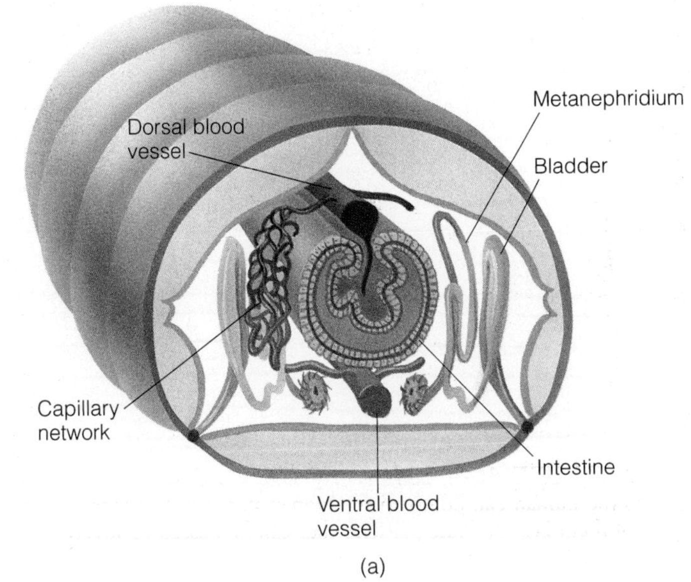

(a)

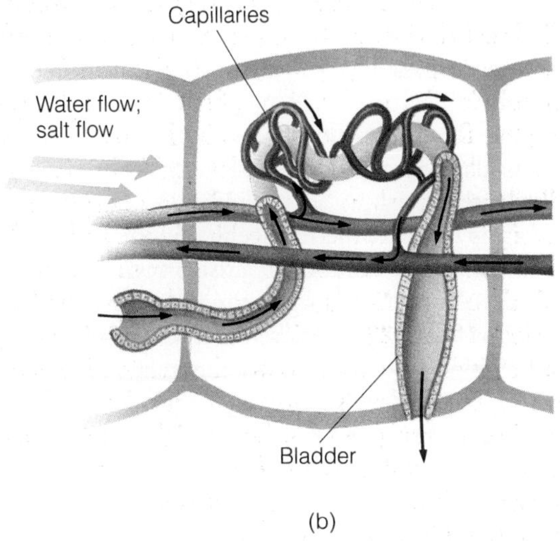

(b)

FIGURE 16–2 Waste Removal in the Earthworm (*a*) The earthworm's body consists of many connected segments, each of which contains a pair of tubules called metanephridia. (*b*) Waste-containing fluids in the body cavity enter the tubules and are excreted through openings to the outside after useful materials have been reabsorbed into the blood stream.

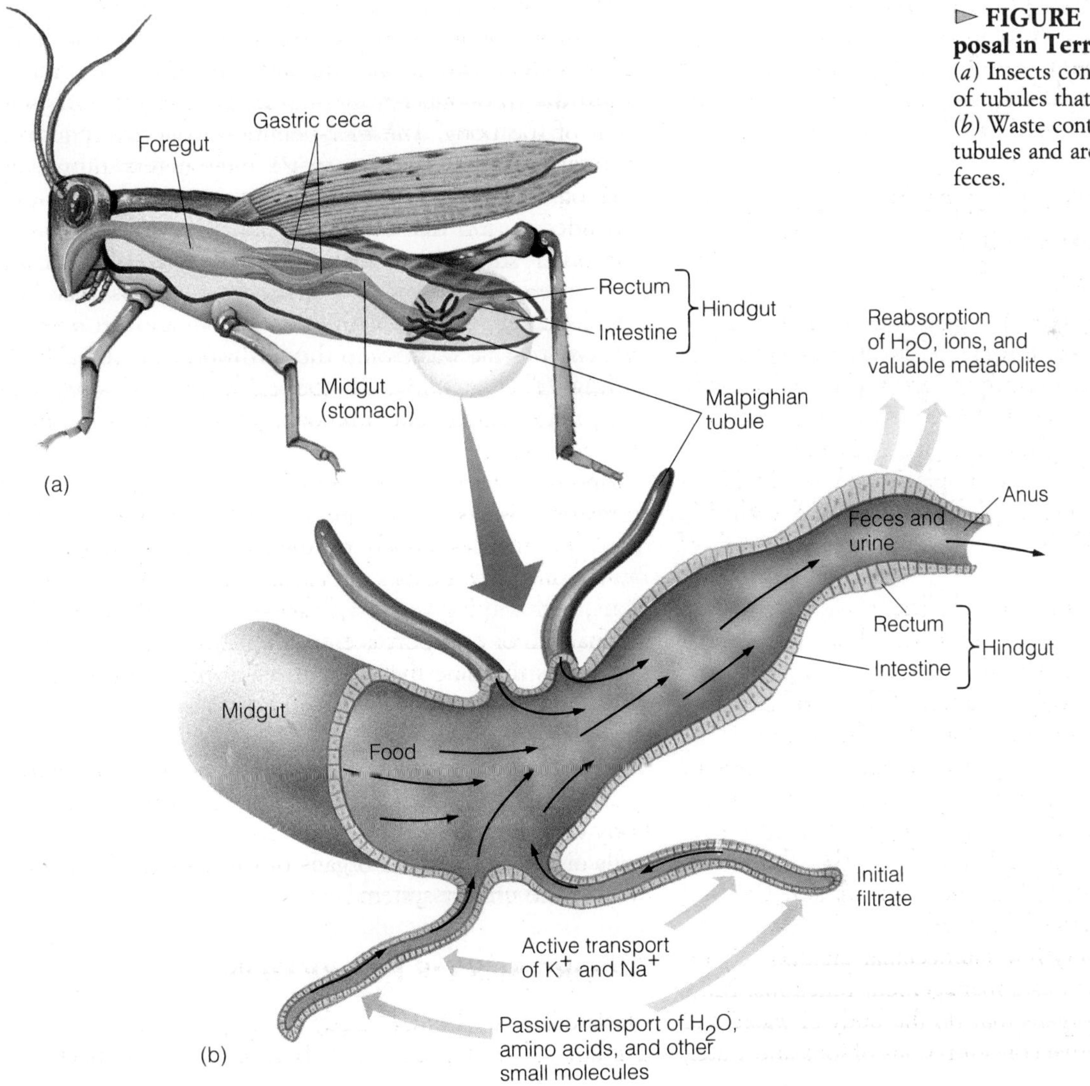

FIGURE 16–3 Waste Disposal in Terrestrial Insects (*a*) Insects contain two to hundreds of tubules that connect to the gut. (*b*) Waste containing fluids enter the tubules and are expelled with the feces.

urea. Like the toxic wastes produced by factories in an industrial society, these potentially harmful substances must go somewhere. Just as in an industrial society, however, they cannot be dumped carelessly. If these hazardous substances were discarded in one of the body's storage depots, they would surely poison body cells and kill us in a few days.

Table 16–1 lists the major metabolic wastes and other chemicals excreted from the body. Take a moment to study the list.

As you can see, the nitrogen-containing wastes like urea, uric acid, and ammonia arise from several different sources. Ammonia (NH_3), for instance, comes from the breakdown of amino acids, which occurs principally in the liver. Amino acid breakdown generally results from an excess of protein in the diet or a shortage of carbohydrate, causing the body to break down protein to acquire amino acids for energy production. The amino groups (NH_2) removed from the amino acids in a reaction called **deamination** are converted into ammonia.

Ammonia is a highly toxic chemical. A small amount of the ammonia produced by the liver is excreted in the urine, but the liver converts most ammonia to urea. The chemical structure of urea is shown in ▻ Figure 16–4.

Another by-product of metabolism in humans is **uric acid,** which the liver produces as it breaks down nucleotides. Uric acid is generally produced in very small quantities and excreted in the urine. In adults, excess production may result in the deposition of uric acid crystals in the bloodstream and in joints, where they cause considerable pain. This condition is known as **gout.** Uric acid may also

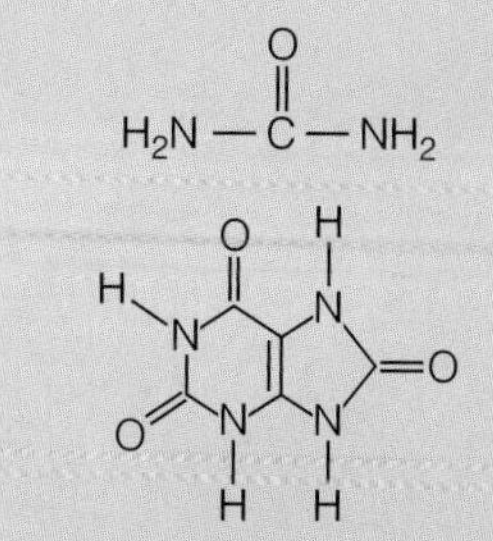

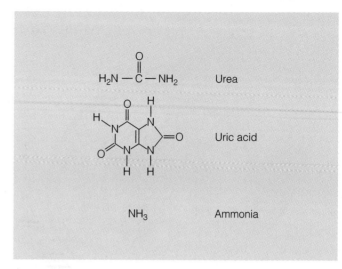

▻ **FIGURE 16–4 Structure of Urea, Uric Acid, and Ammonia** These three waste products are released by cells.

TABLE 16–1 Important Metabolic Wastes and Substances Excreted from the Body

CHEMICAL	SOURCE	ORGAN OF EXCRETION
Ammonia	Deamination of amino acids in liver	Kidneys
Urea	Derived from ammonia	Kidneys, skin
Uric acid	Nucleotide catabolism	Kidneys
Bile pigments	Hemoglobin breakdown in liver	Liver (into small intestine)
Carbon dioxide	Breakdown of glucose in cells	Lungs
Water	Food and water; breakdown of glucose	Kidneys, skin, and lungs
Inorganic ions*	Food and water	Kidneys and sweat glands

*Ions are not a metabolic waste product like the other substances shown in this table. Nonetheless, ions are excreted to help maintain constant levels in the body.

appear in the urine of some babies as orange crystals. Although alarming to a parent, their presence is generally not a sign of trouble.

The **bile pigments,** derived from the breakdown of RBCs in the liver, are another well-known waste product. Hemoglobin of the RBCs contains four protein subunits, each with an iron-containing heme group. Bile pigments are derived from the heme groups and transferred from the liver to the gallbladder, where they are stored along with bile salts, the emulsifying agents that aid in the digestion of fats. Bile pigments are released with the bile salts and are passed along the digestive tract. Some are reabsorbed, and the rest are eliminated with the feces.

The liver also produces another water-soluble pigment during the breakdown of hemoglobin. Known as **urochrome,** this yellow pigment is dissolved in the blood and passes to the kidneys, where it is excreted with the urine. Urochrome gives urine its yellowish color.

Table 16–1 also lists ions. Even though they are not end products of metabolism, ions are excreted from the body by various organs. This excretion is essential to maintain constant levels in the blood, the tissue fluid, and the cytoplasm of cells and is therefore essential to homeostasis.

THE URINARY SYSTEM

Removing wastes is obviously a very important function. Not surprisingly, evolution has "provided" several avenues by which humans (and other animals) get rid of, or excrete, cellular wastes. In fact, if the number of organs involved in a body function is an indication of the importance of a process, excretion would have to rate among the most important of all. In humans, excretion of wastes occurs in the lungs, the skin, the liver, the kidneys, and even the intestines.

Of all the organs that participate in removing waste,

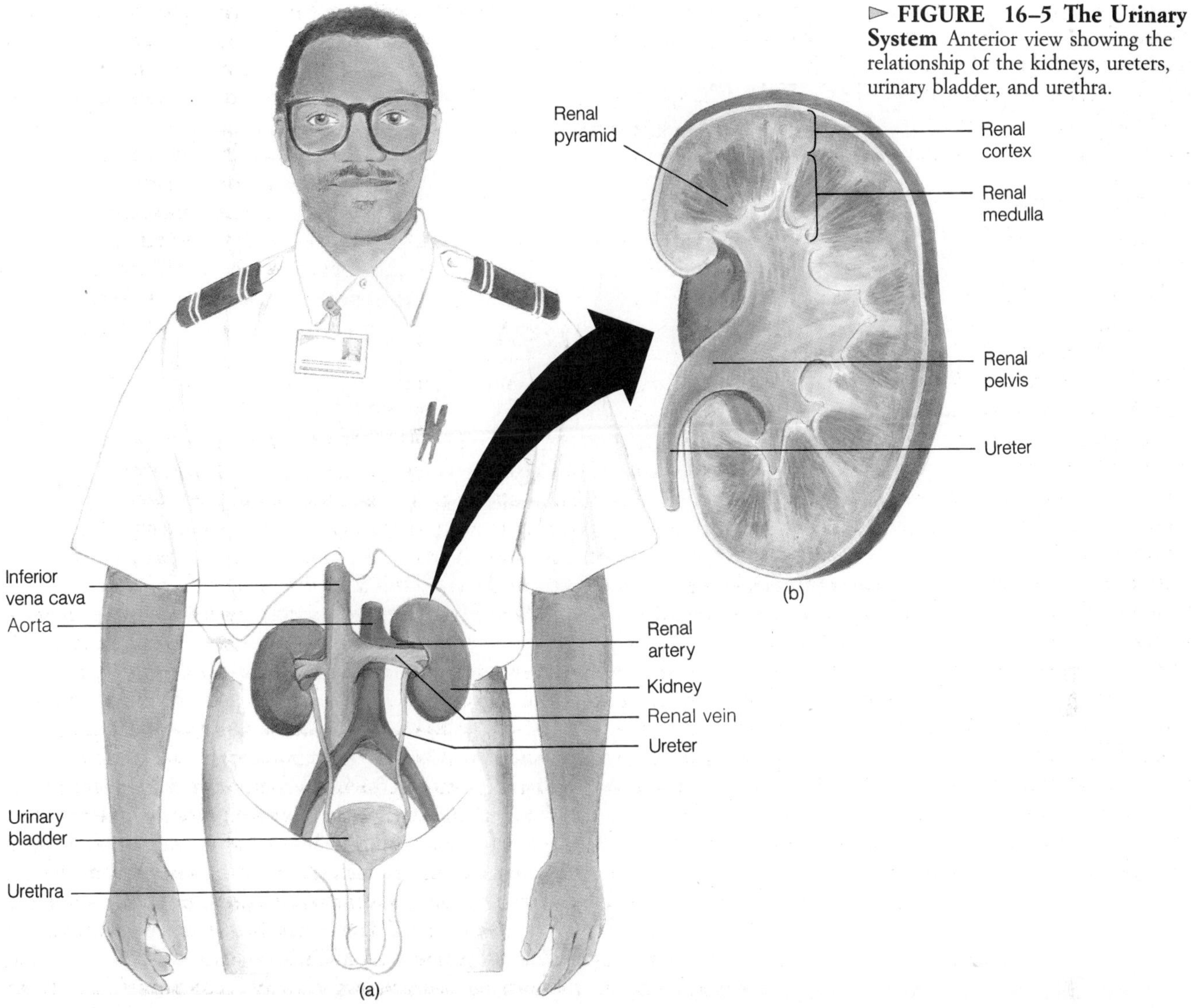

▻ **FIGURE 16–5 The Urinary System** Anterior view showing the relationship of the kidneys, ureters, urinary bladder, and urethra.

however, the kidneys rank as one of the most important, for they rid the body of the greatest variety of dissolved wastes. The kidneys also play a key role in regulating the chemical constancy of the blood.

As illustrated in ▻ Figure 16–5, humans come equipped with two kidneys, which are a part of the **urinary system.** The urinary system also includes the (1) ureters, which drain the kidneys, (2) the urinary bladder, which stores urine, and (3) the urethra, which transports the urine to the outside. The functions of these organs are described in more detail below and are summarized in Table 16–2.

The Urinary System Consists of the Kidney, Ureters, Bladder, and Urethra

The kidneys lie on either side of the vertebral column and are about the size of a person's fist. Surrounded by a layer of fat and located high in the posterior abdominal wall beneath the diaphragm, the human kidneys are oval structures, slightly indented on one side like kidney beans (Figure 16–5). Arterial blood flows into the kidneys through the renal arteries, which enter at each indented region. The **renal arteries** are major branches of the abdominal aorta (Figure 16–5). Inside the kidney, much of the blood-borne wastes are filtered out and eliminated in the urine. **Urine** is

TABLE 16–2 Components of the Urinary System and Their Functions

COMPONENT	FUNCTION
Kidneys	Eliminate wastes from the blood; help regulate body water concentration; help regulate blood pressure; help maintain a constant blood pH
Ureters	Transport urine to the urinary bladder
Urinary bladder	Stores urine; contracts to eliminate stored urine
Urethra	Transports urine to the outside of the body

▻ **FIGURE 16–6 The Urinary Bladder and Urethra** These drawings show the differences in the urethras of men (*a*) and women (*b*). The smooth muscle at the juncture of the urinary bladder and urethra forms the internal sphincter. The pelvic diaphragm is a flat sheet of muscle covering the lower boundary of the pelvic cavity. It forms the external sphincter and is under voluntary control.

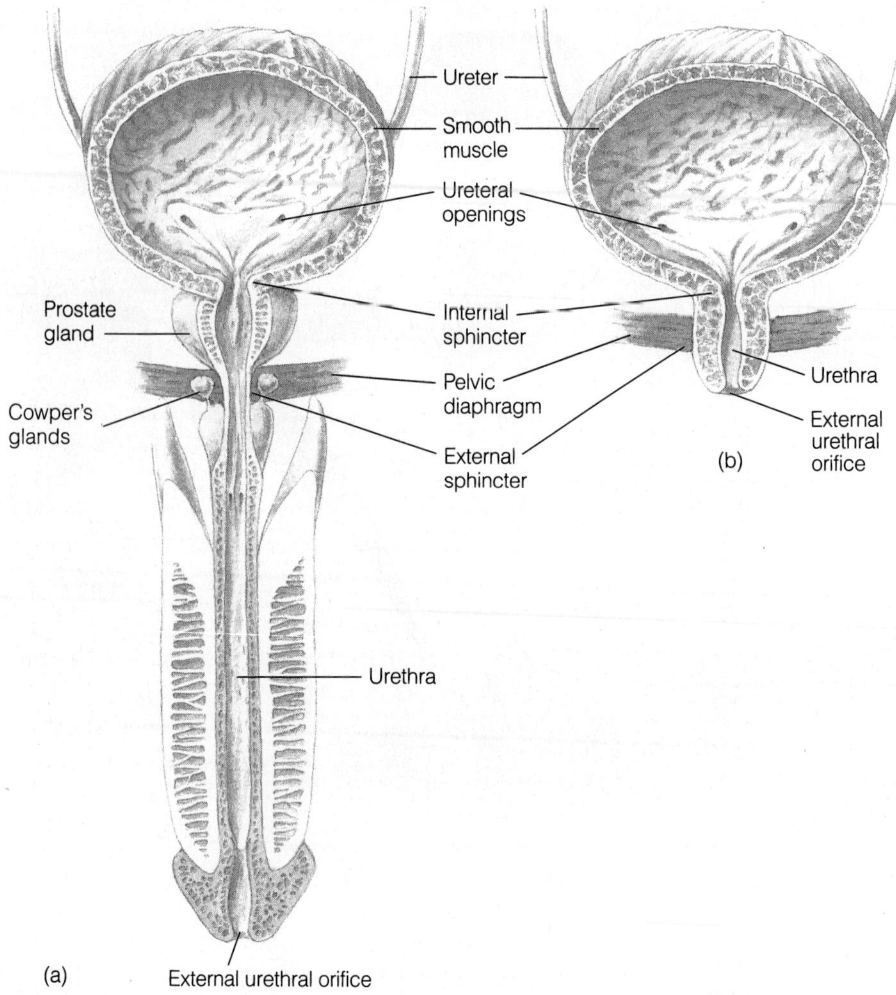

a yellowish fluid containing water, inorganic ions, nitrogenous wastes (such as urea), very small amounts of hormones, ions, and other chemical substances. After the blood has been filtered, it leaves the kidneys via the **renal veins,** which drain into the inferior vena cava, the vessel that transports venous blood to the heart.

Dissolved wastes are removed by numerous microscopic filtering units in the kidney known as **nephrons.** The nephrons produce urine, which is drained from the kidney via the ureters. The **ureters** are muscular tubes that transport urine to the urinary bladder (Figure 16–5). Waves of peristaltic contractions, involuntary smooth muscle contractions like those in the digestive tract, help propel the urine in the ureter to the urinary bladder. The **urinary bladder,** which lies in the pelvic cavity, is a temporary receptacle for urine and is located just behind the pubic bone. The bladder's distensible walls contain a relatively thick layer of smooth muscle that stretches as the bladder fills. When the bladder is full, its walls contract, forcing the urine out through the urethra.

The **urethra** is a narrow tube, measuring approximately 4 centimeters (1.5 inches) in women and 15 to 20 centimeters (6 to 8 inches) in men. The additional length in men largely results from the fact that the urethra travels through the penis (▻ Figure 16–6). The difference in the length of the urethra between men and women has important medical implications. The shorter urethra in women, for example, makes women more susceptible to bacterial infections of the urinary bladder. Bacteria can travel up the urethra of women, invading the bladder, far more easily than they can in men.[1] Bladder infections may result in an itching or burning sensation and an increase in the frequency of urination. They may also cause blood to appear in the urine. Urinary tract infections can be treated with antibiotics. If untreated, however, infections may spread up the ureters to the kidneys, where they can seriously damage the nephrons and impair kidney function.

The Human Kidney Consists of Two Zones, an Outer Cortex and Inner Medulla

Each kidney is surrounded by a connective tissue capsule, the **renal capsule.** The internal structure of the kidney is

[1] Sexual intercourse increases the frequency of urinary bladder infections in many women. To help avoid the problem, physicians advise women to empty their bladders soon after intercourse.

▻ **FIGURE 16–7 Cross Section of the Kidney** (*a*) Human kidney cut in two showing the cortex, medulla, and renal pelvis. (*b*) A drawing of the kidney showing the course of the arteries and veins.

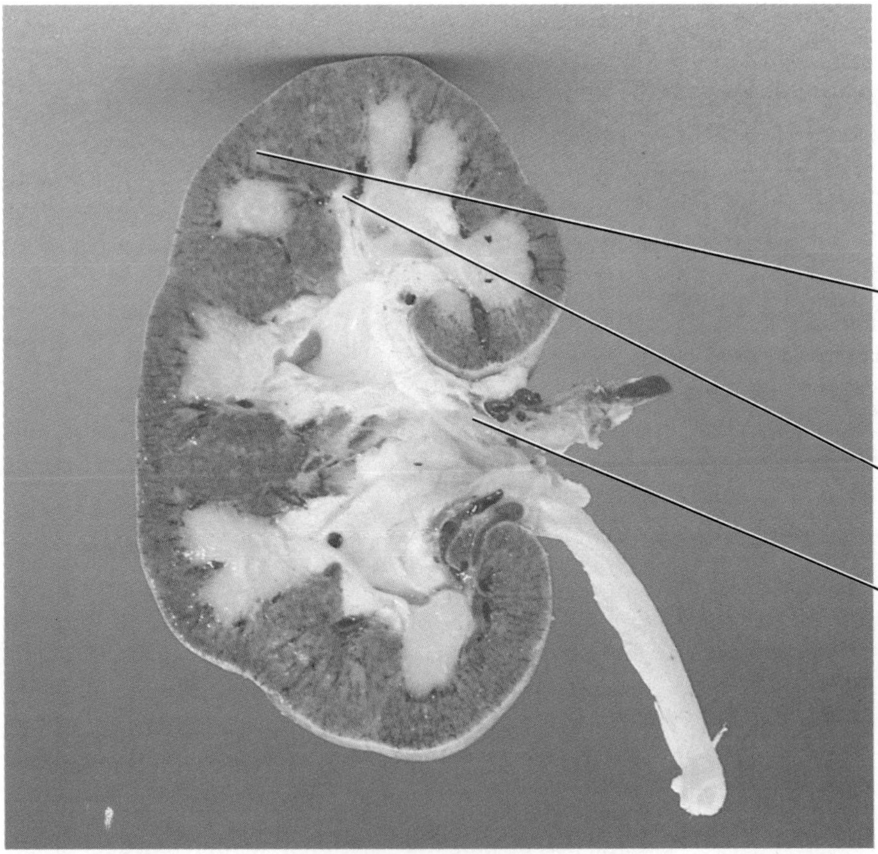

(a)

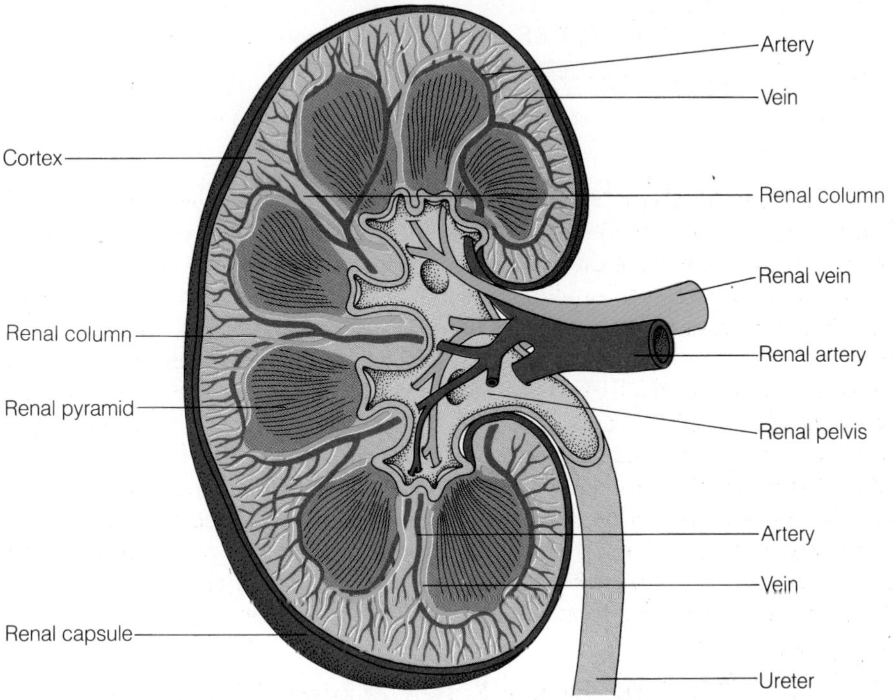

(b)

shown in ▻ Figure 16–7. As illustrated, the kidney is divided into two distinct zones. The outer zone is the **renal cortex,** containing many small filtering units, or nephrons. The inner zone, the **renal medulla,** consists of cone-shaped structures known as **renal pyramids** and intervening tissue called **renal columns.** The renal pyramids contain small ducts that transport urine from the nephrons into a central receiving chamber known as the **renal pelvis.**

Nephrons Consist of Two Parts, a Glomerulus and a Renal Tubule

Each kidney contains 1 million to 2 million nephrons, visible only through a microscope (▻ Figure 16–8a; Table 16–3). Each nephron consists of a tuft of capillaries, the **glomerulus** (Latin for "ball of yarn"), and a long, twisted tube, the **renal tubule** (Figures 16–8a and 16–8c). The renal tubule, in turn, consists of four segments: (1) Bowman's capsule, (2) the proximal convoluted tubule, (3) the loop of Henle, and (4) the distal convoluted tubule.

As illustrated in Figure 16–8c, **Bowman's capsule** is a double-walled structure that surrounds the glomerulus. The inner wall of the capsule fits closely over the glomerular capillaries and is separated from the outer wall by a small space, **Bowman's space.** To understand the relationship between the glomerulus and Bowman's capsule, imagine that your fist is a glomerulus. Then imagine that you are holding a partially inflated balloon in your other hand. If you push your fist (glomerulus) into the balloon, the layer immediately surrounding your fist would be equivalent to the inner layer of Bowman's capsule. It is separated from the outer layer of the capsule by Bowman's space.

The outer wall of Bowman's capsule is continuous with the **proximal convoluted tubule,** a sinuous, or winding, section of the renal tubule. The tubule soon straightens, then descends and reascends, forming a thin, U-shaped segment of the renal tubule known as the **loop of Henle.** The loops of Henle of some nephrons extend into the medulla.

The loop of Henle drains into the fourth and final portion of the renal tubule, the **distal convoluted tubule,** another winding segment. Each distal convoluted tubule drains into a straight duct called a **collecting tubule.** The collecting tubules merge to form larger ducts that course through the renal pyramids and empty into the renal pelvis.

The nephrons filter large amounts of blood and produce from 1 to 3 liters of urine per day. The urinary output depends, in large part, on the intake of fluids. In general, the more fluid you drink, the more urine you produce.

≈ FUNCTION OF THE URINARY SYSTEM

With the basic anatomy of the kidney and urinary system in mind, let us turn our attention to the function of the urinary system, focusing first on the nephron.

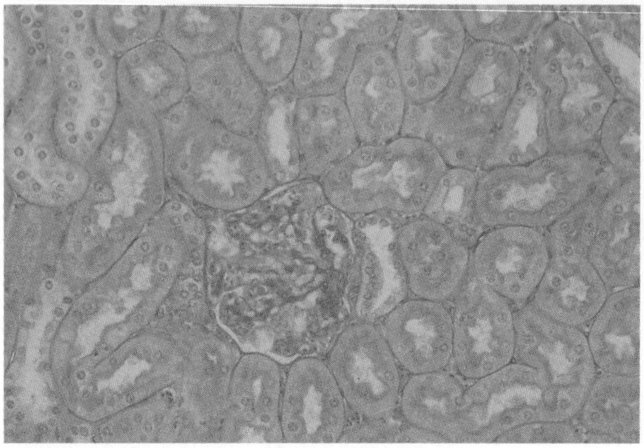

(a)

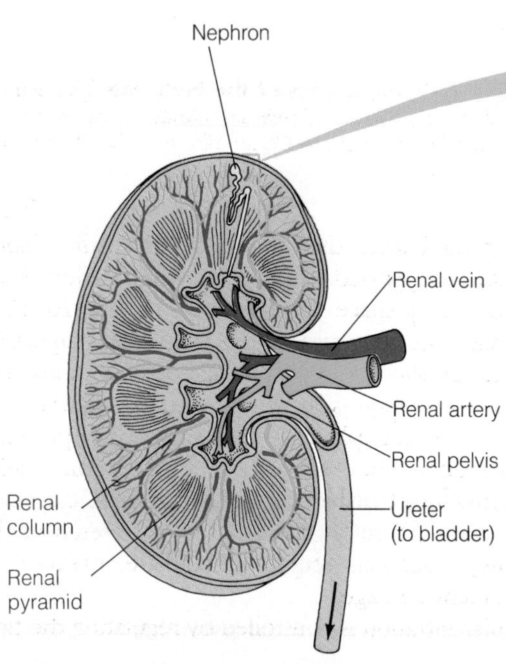

(b)

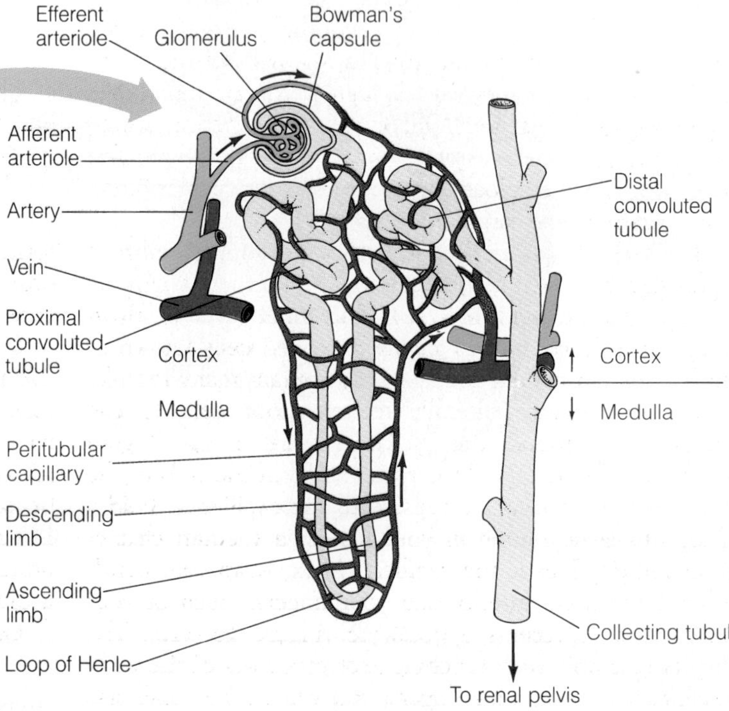

(c)

▻ **FIGURE 16–8 The Nephron** (*a*) A microscopic view of the cortex of the kidney showing the many tubules packed together and a single glomerulus. (*b*) A cross section of the kidney showing the location of the nephrons. (*c*) A drawing of a nephron.

TABLE 16–3 Components of the Nephron and Their Function

COMPONENT	FUNCTION
Glomerulus	Mechanically filters the blood
Bowman's capsule	Mechanically filters the blood
Proximal convoluted tubule	Reabsorbs 75% of the water, salts, glucose, and amino acids
Loop of Henle	Participates in countercurrent exchange, which maintains the concentration gradient
Distal convoluted tubule	Site of tubular secretion of H^+, potassium, and certain drugs

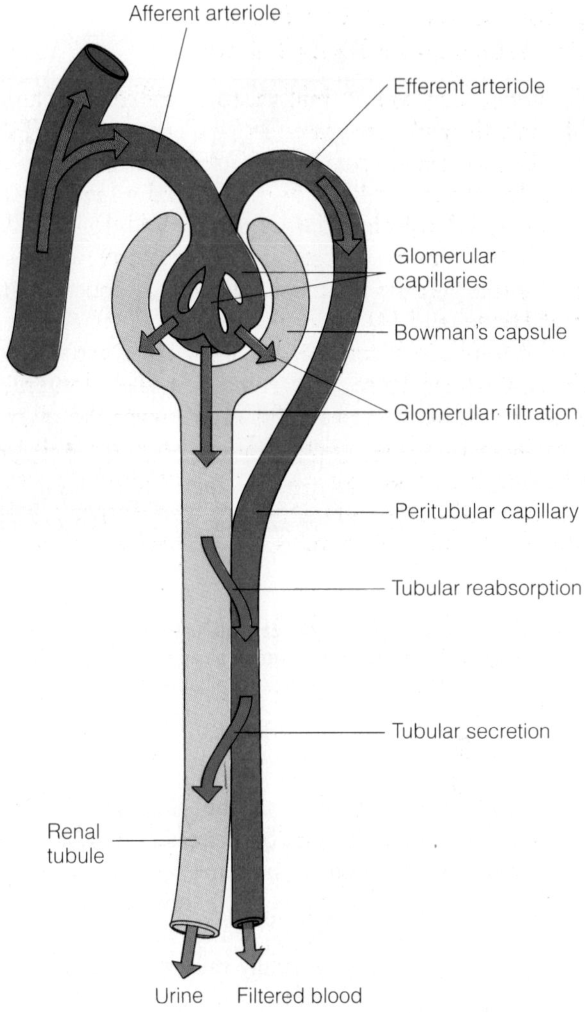

▻ **FIGURE 16–9 Physiology of the Nephron** The nephron carries out three processes: glomerular filtration, tubular reabsorption, and tubular secretion. All contribute to the filtering of the blood.

Blood Filtration in Nephrons Involves Three Processes: Glomerular Filtration, Tubular Reabsorption, and Tubular Secretion

Glomerular Filtration. The first step in the purification of the blood is **glomerular filtration** (▻ Figure 16–9). As noted earlier, the glomeruli are tufts of capillaries (▻ Figure 16–10). Blood flows into the glomeruli through the **afferent arterioles.** As blood flows through the glomerular capillaries, water and dissolved materials are forced through the endothelium of the capillaries by blood pressure. The resulting liquid, called the **glomerular filtrate,** travels through the inner layer of Bowman's capsule and enters Bowman's space (▻ Figure 16–11a).

Glomerular filtration is akin to the filtration that takes place in a sieve. In the kidney, the "sieve" consists of two cell layers. The innermost is the endothelium of the capillaries. Shown in the Figure 16–11a, the cells of the endothelium of the glomerular capillaries contain numerous openings, or **fenestrae** ("windows"), that permit water, ions, glucose, and even small proteins to pass through. The fenestrae, however, block larger components, such as blood cells, platelets, and large blood proteins.

The second layer of the glomerular filtration membrane is formed by the inner membrane of Bowman's capsule (Figures 16–11a and 16–11b). The inner layer of Bowman's capsule consists of highly branched cells known as **podocytes** (so named because they contain many footlike processes; *podocytes* literally means "foot cells"). The podocytes surround the glomerular capillaries (Figure 16–11c). To understand the relationship of the cells of the inner layer of Bowman's capsule to the capillaries, hold a piece of plastic tubing in your hands; a vacuum cleaner hose will do. This is the capillary. Next, wrap your hands around the tube, interlocking your fingers. Each of your hands now represents a podocyte (Figure 16–11c). The fingers resemble the branching foot processes of the cells. Looking down on your fingers, you will notice small slits between them. These openings, known as **filtration slits,** form a physical barrier that prevents the smaller blood proteins that have passed through the endothelium from entering Bowman's space (Figures 16–11b and 16–11c).

In summary, the fenestrae of the glomerular capillaries and the filtration slits allow the passage of water, ions, and many small- to medium-sized molecules but prevent the passage of blood cells, platelets, and most blood proteins. Kidney infections may damage the filtration membrane—so much so that blood cells and proteins can enter Bowman's space. In such cases, blood is excreted in the urine. Kidney infections require immediate attention to avoid permanent damage.[2]

Glomerular filtration is controlled by regulating the flow

[2] Blood in the urine is usually the result of a urinary bladder infection or a kidney stone, both of which require immediate attention as well.

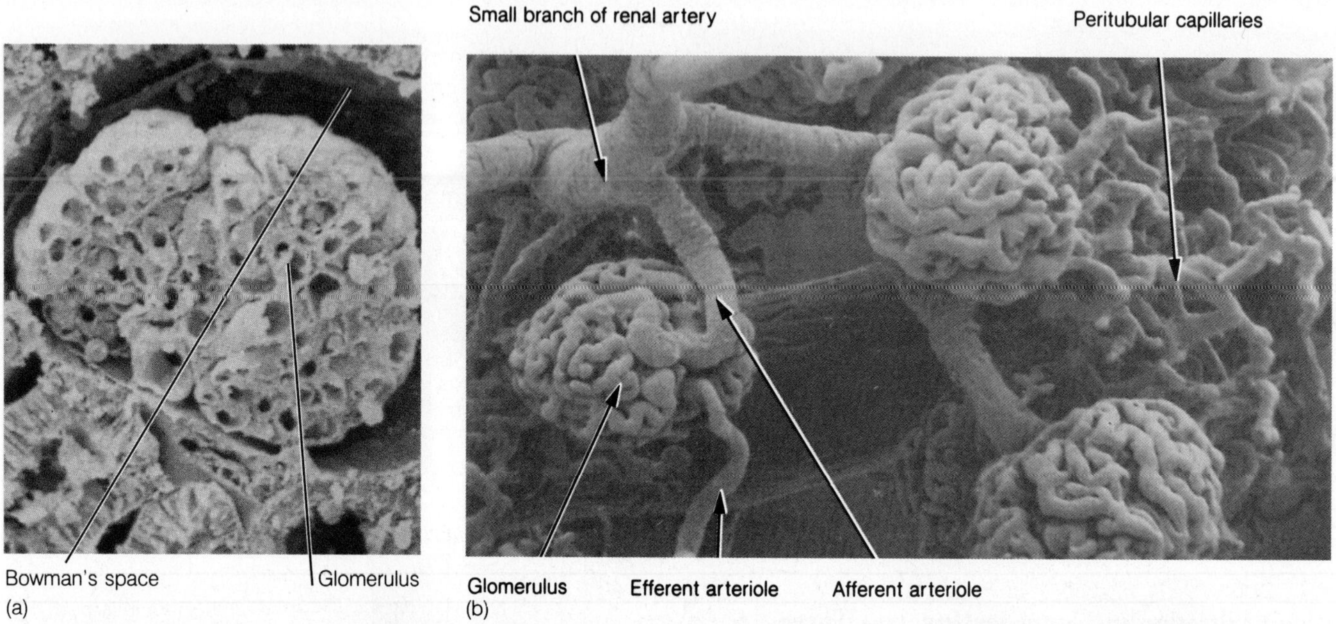

▻ **FIGURE 16–10 The Glomerulus** (*a*) A scanning electron micrograph of a glomerulus and Bowman's capsule. (*b*) A scanning electron micrograph of glomerular capillaries. Part (*b*) from Richard G. Kessel and Randy H. Kardon, *Tissues and Organs: A Text-Atlas of Scanning Electron Microscopy*. Copyright © 1979 W.H. Freeman and Company. Reprinted with permission.

of blood and blood pressure in the glomerulus. Blood pressure in the glomerulus can be increased by dilation of the afferent arterioles in much the same way that opening a faucet increases water pressure in a hose fitted with a nozzle. The **efferent arterioles,** which drain blood from the glomerular capillaries, naturally raise glomerular blood pressure, because they are slightly smaller in diameter than the afferent arterioles. Slight constriction of the efferent arterioles further increases blood pressure within the glomerular capillaries in much the same way that putting a clamp on a garden hose increases pressure upstream from the point of constriction. In the capillaries, an increase in blood pressure inside the glomerular capillaries increases the rate of filtration.

Tubular Reabsorption. Each day, approximately 180 liters (45 gallons) of filtrate is formed by glomerular filtration. Despite this massive outpouring of filtrate, the kidneys produce only about 1 to 3 liters of urine each day. In other words, only about 1% of the filtrate actually leaves the kidneys as urine. What happens to the rest of the fluid filtered by the glomerulus?

Most of the fluid filtered by the glomeruli is reabsorbed, passing from the renal tubule back into the bloodstream. The movement of water, ions, and molecules from the renal tubule to the bloodstream is called **tubular reabsorption** (see Figure 16–9). During tubular reabsorption, water containing nutrients and ions passes from the renal tubule into the **peritubular capillaries,** networks of capillaries surrounding the nephrons. The peritubular capillaries are branches of the efferent arterioles.

The peritubular capillaries reabsorb water, nutrients such as glucose, and various ions such as sodium that are filtered out of the blood in the glomerulus. Tubular reabsorption therefore helps conserve water and various dissolved materials. Table 16–4 shows that 99% of the water molecules, 99.5% of the sodium ions, and 100% of the glucose molecules filtered out of the blood during glomerular filtration are eventually reabsorbed. Without tubular reabsorption, glomerular filtration would require massive water and nutrient intake to offset the loses.

The value of tubular reabsorption lies in its selectivity–that is, its ability to selectively reabsorb valuable molecules

TABLE 16–4 Fate of Various Substances Filtered by Kidneys

SUBSTANCE	AVERAGE PERCENTAGE OF FILTERED SUBSTANCE REABSORBED	AVERAGE PERCENTAGE OF FILTERED SUBSTANCE EXCRETED
Water	99	1
Sodium	99.5	0.5
Glucose	100	0
Urea (a waste product)	50	50
Phenol (a waste product)	0	100

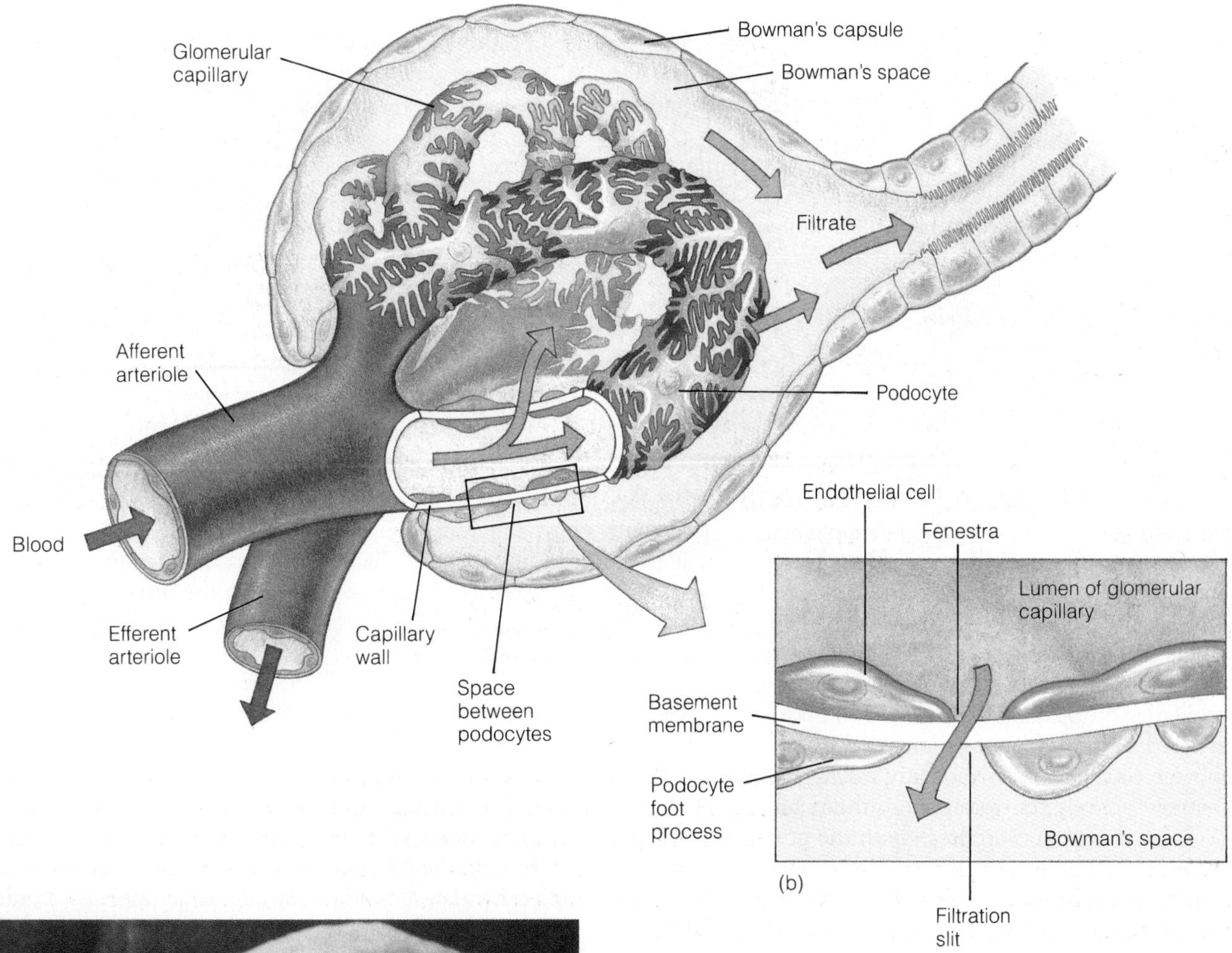

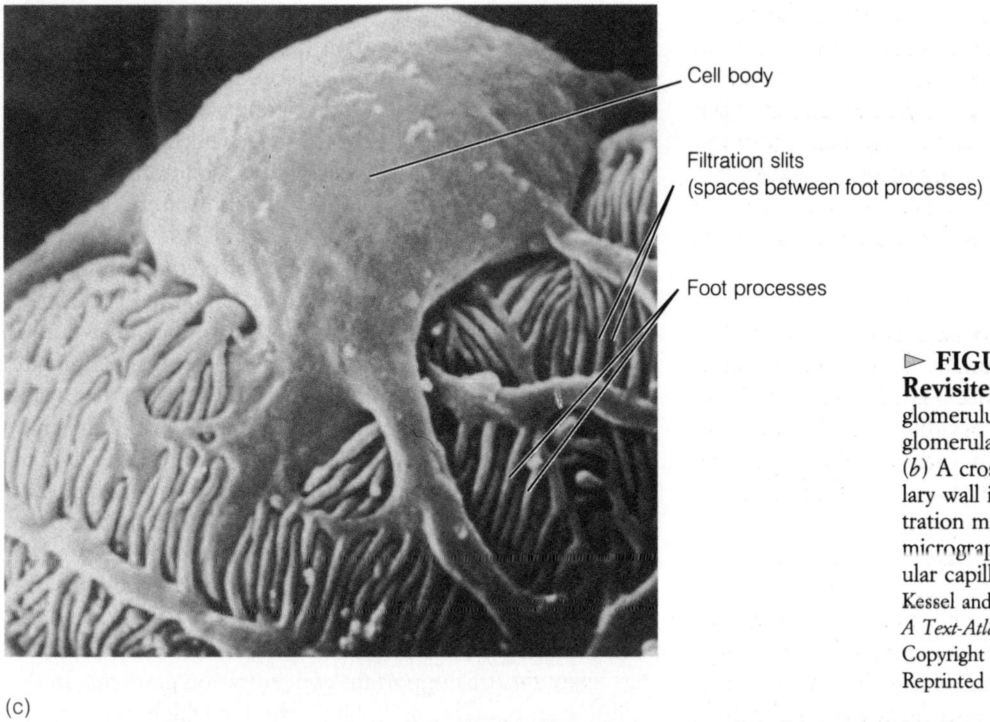

▻ **FIGURE 16–11 The Glomerulus Revisited** (*a*) A simplified drawing of the glomerulus showing the relationship of the glomerular capillaries to Bowman's capsule. (*b*) A cross section of a glomerular capillary wall illustrating the structure of the filtration membrane. (*c*) A scanning electron micrograph of the podocytes on the glomerular capillaries. Part (*b*) from Richard G. Kessel and Randy H. Kardon, *Tissues and Organs: A Text-Atlas of Scanning Electron Microscopy*. Copyright © 1979 W.H. Freeman and Company. Reprinted with permission.

like water and glucose while letting wastes pass through. As a result, important chemicals are conserved, and wastes are excreted. The one exception is urea, a waste product of amino acid catabolism. As shown in Table 16–4, only 50% of the urea that enters the nephron in the filtrate is reabsorbed, and the rest is eliminated in the urine.

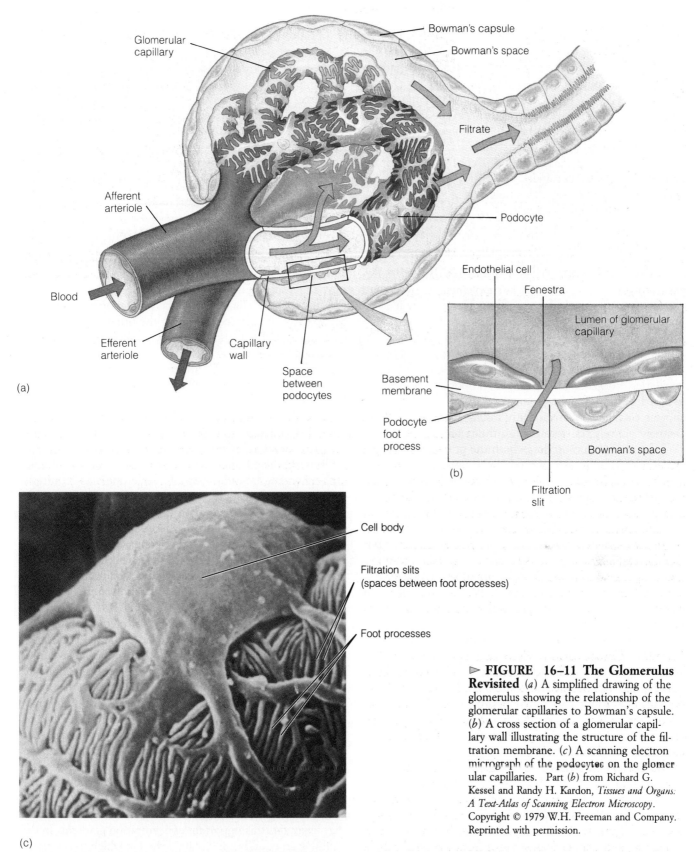

Tubular reabsorption occurs in all parts of the renal tubule, but the bulk of it occurs in the proximal convoluted tubule. In the proximal convoluted tubule, about 70% of all sodium ions and water are reabsorbed. The positively charged sodium ions are actively transported out of the tubule, whereas chloride ions, urea, and other negatively charged ions follow passively, moving down an electrochemical gradient.[3] The increase in the concentration of sodium and the negatively charged ions in the extracellular fluid, in turn, causes water molecules to move outward by osmosis. As noted in Chapter 3, osmosis is the movement of water from a region of high water concentration to one of lower water concentration.

The proximal convoluted tubule is also responsible for the reabsorption of calcium ions, glucose, and vitamins from the filtrate.

Tubular Secretion Is the Transport of Waste Products From the Peritubular Capillaries Into the Renal Tubule. Waste disposal is supplemented by a third process, known as **tubular secretion.** Here is how it operates. Wastes not filtered from the blood as it passes through the glomerular capillaries remain in the blood that enters the peritubular capillaries. Some of these wastes are then transported *into* the renal tubule from the peritubular capillaries. For example, hydrogen and potassium ions that escaped filtration in the glomeruli diffuse out of the peritubular capillaries into the surrounding extracellular fluid, then are actively transported into the renal tubule.

Tubular secretion occurs in both the proximal and distal convoluted tubules, but mostly in the latter. This process helps rid the blood of wastes and also helps regulate the H^+ concentration of the blood. The latter aids in maintaining a constant pH and supplements buffers in the blood and extracellular fluid, especially bicarbonate ions.[4]

The urine leaving the nephron consists mostly of water and a variety of dissolved waste products. Blood draining from the peritubular capillaries is purified, or cleared of most wastes, and contains most of the ions, water, and nutrients that entered the kidney.

The blood draining from the peritubular capillaries empties into small veins. These veins converge to form the renal vein, which transports the filtered blood out of the kidney and into the inferior vena cava and then on to the heart.

Summary of Renal Filtration

Let me take a few moments to summarize the key points of the filtration process.

1. Glomerular filtration is the first phase in the purification of the blood and the production of urine. Blood plasma is forced out of the glomerular capillaries through the filtration membrane into Bowman's capsule.
2. The next phase is tubular reabsorption. Valuable nutrients and ions that might otherwise be lost in the urine are reabsorbed via the renal tubule. Except for urea, waste products pass through the renal tubule untouched, eventually leaving the kidney via the ureter.
3. Some additional purification occurs via tubular secretion—the transport of unfiltered wastes from the peritubular capillaries to the renal tubule.

Countercurrent Exchange Helps Concentrate the Urine

Urine leaves the distal convoluted tubule and enters the collecting tubules. The collecting tubules descend through the medulla and converge to form larger ducts that drain into the renal pelvis. As the collecting tubules descend through the medulla, much of the remaining water escapes by osmosis, further concentrating the urine and conserving body water. What causes this outward movement of water?

As illustrated in ▻Figure 16–12, the concentration of sodium chloride (NaCl) in the extracellular fluid of the medulla increases toward the renal pelvis. As the urine flows down through the medulla, then, water moves out by osmosis through the highly permeable collecting tubules in an attempt to equalize the concentration of the two solutions.

The sodium chloride concentration gradient in the medulla is produced and maintained by the loop of Henle. Understanding how the loop of Henle operates requires a look at its permeability to various substances—notably, water and salt. We begin with the ascending limb.

As the urine moves up the ascending limb toward the distal convoluted tubule, sodium ions are actively transported out. Chloride ions follow passively, moving by diffusion (Figure 16-12).[5] Water cannot follow because the ascending limb is rather impermeable to water. The transport of sodium ions and diffusion of chloride ions is greatest in the lowest portion of the loop of Henle, as indicated in Figure 16–12 by the thick arrows. Thus, the concentration of sodium and chloride ions in the extracellular fluid is greatest at the bottom of the loop. Because the outward movement of sodium and chloride decreases as the urine moves up, the extracellular concentration of sodium chloride decreases.

Now let us turn to the descending limb. As indicated by the solid arrows in Figure 16–12, sodium chloride ions pumped out of the ascending limb enter the descending limb. These ions will soon be transported back up the ascending limb and will be pumped out as the urine ascends once again, creating a self-perpetuating closed circuit that maintains this important concentration gradient. In the medulla, an equilibrium is established in which the outward movement of sodium chloride in the ascending limb equals the inward movement of sodium chloride in the descend-

[3] As you recall from Chapter 3, the term *electrochemical gradient* refers to a charge and concentration difference across a membrane.

[4] To prevent the buildup of hydrogen ions inside the renal tubule, some cells of the tubule produce ammonia (NH_3). Ammonia diffuses into the renal tubule, where it combines with hydrogen ions, forming ammonium ions (NH_{4+}).

[5] Some researchers believe that the chloride ions are actively transported out and the sodium ions follow passively down the resulting electrical gradient.

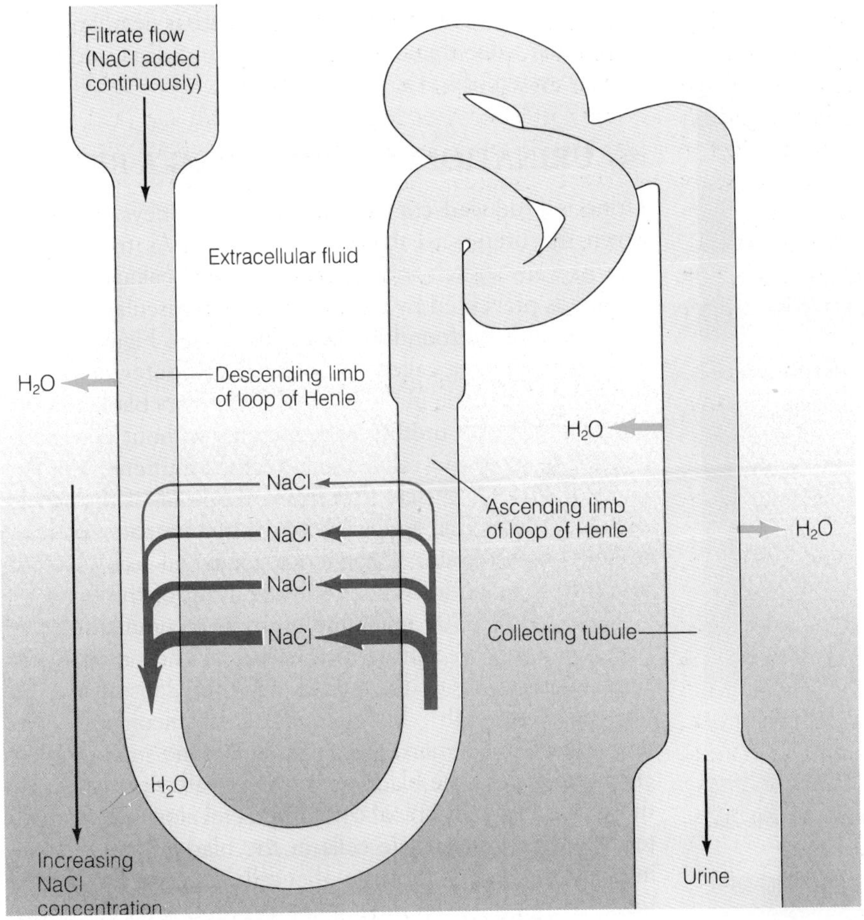

▻ **FIGURE 16–12 The Concentration of Urine by the Nephron** A countercurrent-exchange mechanism ensures the concentration gradient in the extracellular fluid surrounding the nephron, which facilitates the movement of water out of the collecting tubule and helps conserve body water and concentrate urine. (See the text for an explanation of countercurrent-exchange mechanism.)

ing limb. The extracellular concentration therefore remains the same and is always greatest at the bottom of the loop, for the reasons described above.

This mechanism is called **countercurrent exchange.** The term *countercurrent exchange* applies to any system in which fluids, flowing in opposite directions, exchange chemical substances or heat. Arteries and veins, for example, contain blood flowing in opposite directions and often lie side by side. In the arm, warm blood traveling in the artery loses its heat to the slightly cooled venous blood returning to the heart. This mechanism helps conserve body heat and is the reason your hands are slightly colder than your torso. It is also one reason ducks can stand on ice and not be bothered by the cold.

In the kidney, countercurrent created by the descending and ascending limbs of the loops of Henle exchanges sodium and chloride ions. This exchange maintains the concentration gradient in the medulla, which, in turn, ensures the outward movement of water (by osmosis) from the descending limbs. As the urine flows through the collecting tubules and through the concentrated medulla, water also escapes via osmosis, further increasing the concentration of the urine. The water lost from the descending limb and collecting tubules is eventually reabsorbed into the bloodstream.

Practical Applications: Kidney Stones

Urine contains various dissolved wastes. Approximately 90% of the dissolved waste consists of three substances: urea, sodium ions, and chloride ions. Varying amounts of other chemical substances are also present. Increases or decreases in the level of these substances may signal underlying health problems. As a result, physicians routinely analyze the dissolved components in their patients' urine as a way of checking for underlying metabolic problems. For example, diabetes mellitus, or sugar diabetes, a defect in glucose uptake by cells in the body, results in the presence of glucose in the urine.

Excess calcium, magnesium, and uric acid in the urine, caused in many cases by inadequate fluid intake, may crystallize in the renal pelvis of the kidney, forming deposits known as **kidney stones** (▻ Figure 16–13a). These small deposits enlarge by **accretion**—the deposition of materials on the outside of the stones—causing them to grow in much the same way that a snowball enlarges as it is rolled through wet snow (Figure 16–13b).

Many kidney stones enter the ureter on their own and are passed to the bladder. As they are propelled along the ureter, however, their sharp edges can dig into its wall, stimulating pain fibers. Small stones that pass to the urinary

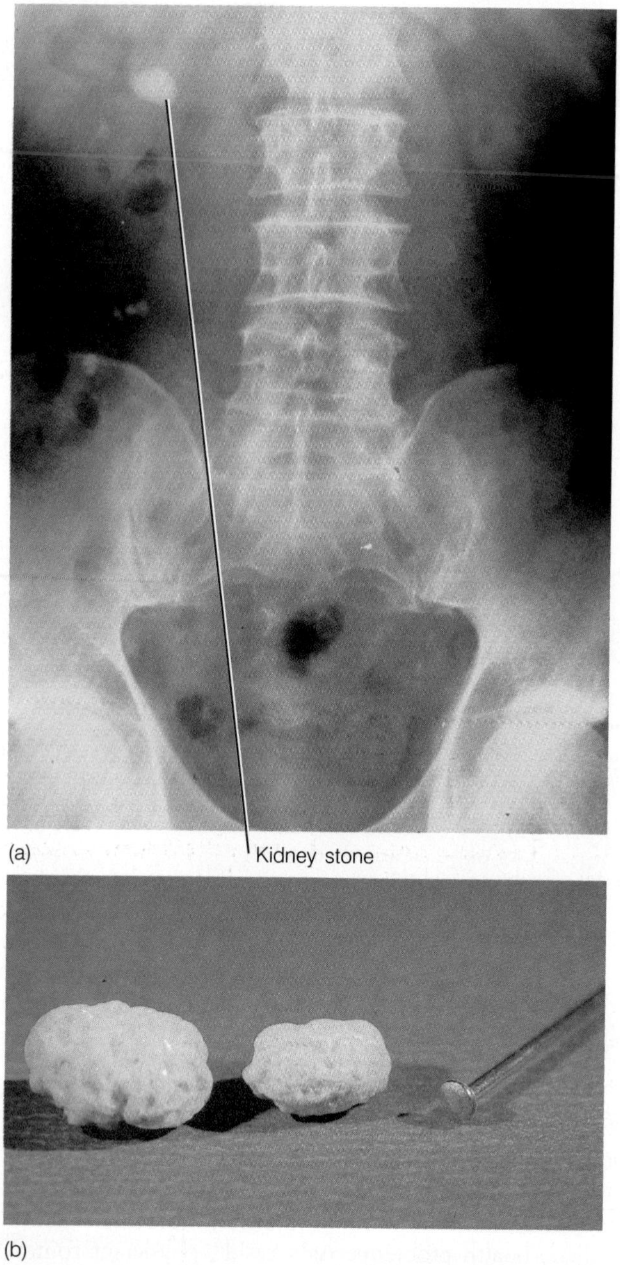

▻ **FIGURE 16–13 Kidney Stones** (*a*) An X-ray of a kidney stone. (*b*) Kidney stones removed by surgery. (See nail for size.)

bladder are often excreted in the urine via the urethra. This may also be accompanied by considerable pain. Larger stones, however, often lodge in the ureters or in the renal pelvis, where they obstruct the flow of urine. Pressure increases inside the kidney, causing considerable damage to the nephrons if untreated.

For years, kidney stones were removed surgically. Today, however, the relatively new medical technique of ultrasound lithotripsy is generally used. In this technique, mentioned in the opening vignette of this chapter, physicians bombard kidneys with ultrasound waves, which shatter the stones, producing fine, sandlike grains that can be passed in the urine without incident. The procedure is nearly painless and much safer than surgery. Table 16–5 lists additional urinary system disorders.

URINATION: CONTROLLING A REFLEX

Urine is produced continuously by the kidneys and flows down the ureters to the urinary bladder. As the bladder fills, its walls stretch and become thinner. Leakage into the urethra is prevented by two sphincters—muscular "valves" not unlike those found in the stomach (see Figure 16–6). The first sphincter, called the **internal sphincter,** is formed by a smooth muscle in the neck of the urinary bladder at its junction with the urethra and operates without conscious control. The second valve, the **external sphincter,** is a flat band of skeletal muscle that forms the floor of the pelvic cavity. The external sphincter is a voluntary gateway controlled by the brain. When both sphincters are relaxed, urine is propelled into the urethra and out of the body.

When 200 to 300 milliliters of urine accumulate in the bladder, stretch receptors in the wall of the organ begin sending impulses to the spinal cord via sensory nerves (▻ Figure 16–14b). In the spinal cord, incoming nerve impulses stimulate nerve cells that supply the smooth muscle in the wall of the bladder. Nerve impulses generated in these cells leave the spinal cord and travel along nerves that terminate on the muscle cells in the bladder wall. These impulses, in turn, stimulate the cells to contract, which forces the ends of the ureters shut, preventing the backflow of urine. Contraction of the wall also forces the internal sphincter to open, letting urine escape into the urethra.

Urine does not escape, however, until the external sphincter is relaxed. The external sphincter is supplied by a continuous barrage of nerve impulses from the brain and spinal cord, which keep this muscle in a constant state of contraction. In order for the sphincter to open, these impulses must cease. As shown in Figure 16–14, in babies and very young children nerve impulses arriving in the spinal cord from the stretch fibers also inhibit the nerve cells supplying the external sphincter, permitting it to open. In babies and very young children, then, urination is a reflex; that is, there is no conscious override. Not until children grow older (2 to 3 years) can they begin to control urination.

Conscious control occurs at the external sphincter and is exerted by the brain. Thus, the brain can override the urination reflex. Here's how it works: As the bladder fills, nerve impulses from stretch receptors travel to the spinal cord; some of these impulses then travel to the brain, creating a conscious awareness of the situation. If the time and place are inappropriate, a person consciously overrides the reflex. The brain simply sends nerve impulses back down the spinal cord to the skeletal muscles of the external sphincter. These impulses keep the sphincter muscles contracted. While awaiting the signal to release this control, the bladder can expand to hold up to 800 milliliters of urine. At this stage, waiting may become quite painful.

Adults sometimes lose control over urination, resulting

TABLE16–5 Common Urinary Disorders

DISEASE	SYMPTOMS	CAUSE
Bladder infections	Especially prevalent in women; pain in lower abdomen; frequent urge to urinate; blood in urine; strong smell to urine	Nearly always bacteria
Kidney stones	Large stones lodged in the kidney often create no symptoms at all; pain occurs if stones are being passed to the bladder; pains come in waves a few minutes apart.	Deposition of calcium, phosphate, magnesium, and uric acid crystals in the kidney, possibly resulting from inadequate water intake
Kidney failure	Symptoms often occur gradually: more frequent urination, lethargy, and fatigue; should the kidney fail completely, patient may develop nausea, headaches, vomiting, diarrhea, water buildup, especially in the lungs and skin, and pain in the chest and bones.	Immune reaction to some drugs, especially antibiotics; toxic chemicals; kidney infections; sudden decreases in blood flow to the kidney—for example, resulting from trauma
Pyelonephritis	Infection of the kidney's nephrons; sudden, intense pain in the lower back immediately above the waist, high temperature, and chills	Bacterial infection

in a condition referred to as **urinary incontinence.** Urinary incontinence has several causes. In some instances, it results from traumatic injury to the spinal cord, which disrupts descending nerve fibers carrying the impulses from the brain that override the urination reflex. In such cases, the urinary reflex remains intact, and the bladder empties as soon as it reaches a certain size, much as it does in a young child.

Mild urinary incontinence is much more common. It is characterized by the escape of urine when a person sneezes or coughs and, in women, usually results from damage to the external sphincter during childbirth. Childbirth stretches the

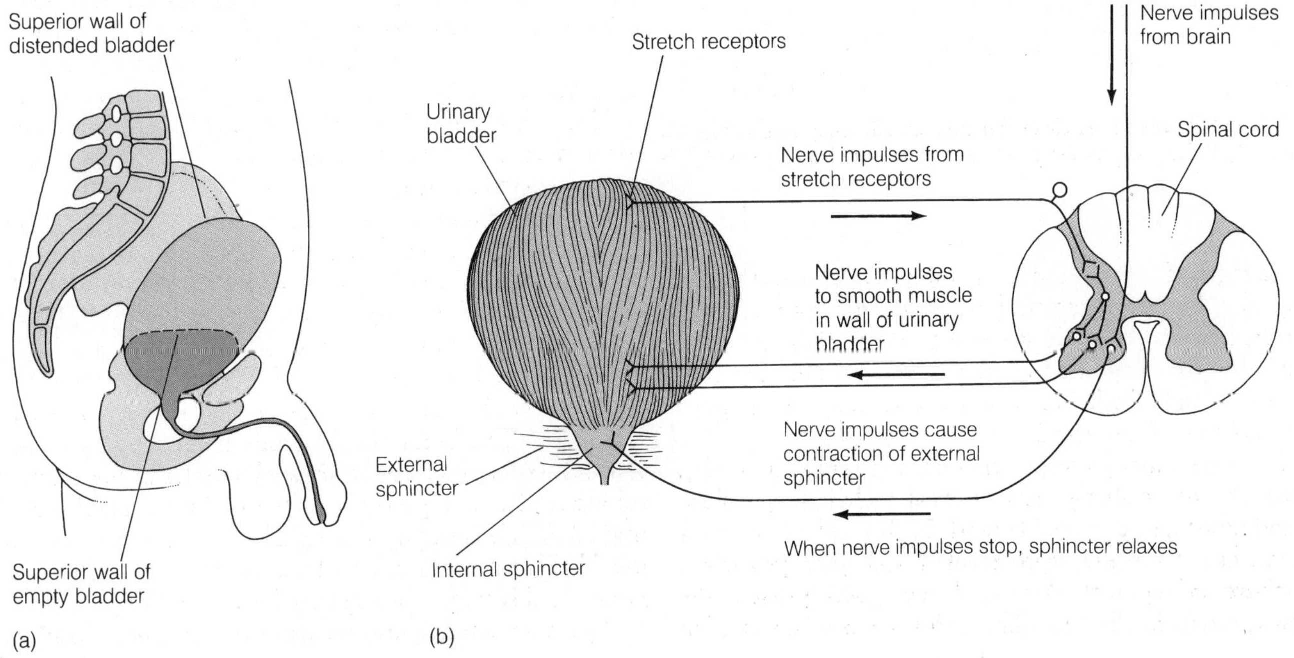

▻ **FIGURE 16–14 Urination in Babies** (*a*) The bladder before and after it fills, showing how much this organ can expand to accommodate urine. (*b*) Stretch receptors signal the distension of the urinary bladder. Nerve impulses travel to the spinal cord. In the spinal cord, they stimulate nerve cells that send impulses back to the bladder, causing muscle contraction in the wall which forces the internal sphincter open. Nerve impulses also travel up the spinal cord to the brain signaling the need to void (not shown here). In adults the brain sends signals back to the external sphincter, causing it to relax if the time and place for urination are appropriate.

skeletal muscles that compose the external sphincter, reducing their effectiveness and making such accidents embarrassingly common. For this reason, many women undertake exercise programs to strengthen the muscles before and after childbirth. Urinary incontinence may also occur in men whose external sphincters have been injured in surgery on the prostate gland, which surrounds the neck of the urinary bladder.

THE KIDNEYS AS ORGANS OF HOMEOSTASIS

The massive flow of blood through the kidneys and the high rate of glomerular filtration combine to ensure a very thorough filtering of blood borne impurities. This filtering, in turn, helps the body control the composition of the blood to maintain homeostasis.

Besides Ridding the Body of Wastes, the Kidneys Also Help Regulate the Body's Water Content

As noted earlier, much of the water filtered by the glomeruli is passively reabsorbed by the renal tubules and returned to the bloodstream. However, the rate of water reabsorption, can be increased or decreased to alter urine production. The ability to adjust water reabsorption and urine output allows the body to rid itself of excess water when a person has consumed too much and also helps conserve body water when an individual is becoming dehydrated.

The Hormone ADH Increases the Permeability of the Distal Convoluted Tubules and Collecting Tubules, Increasing Water Reabsorption and Conserving Body Water. Water reabsorption is controlled in large part by **antidiuretic hormone (ADH)** in a negative feedback mechanism. ADH is released by an endocrine gland at the base of the brain known as the pituitary (Chapter 20).[6] ADH release is regulated by two receptors (▻ Figure 16–15). The first is a group of nerve cells in a region of the brain just above the pituitary gland known as the hypothalamus. These cells monitor the osmotic concentration of the blood. The production and release of ADH are also controlled by receptors in the heart, which detect changes in blood volume (which reflect water levels).

Consider an example. If you were to play soccer on a hot summer afternoon, you would undoubtedly perspire heavily and lose a significant amount of body water and salts. If you did not replace the water you were losing, you would become dehydrated. As a result, your blood volume

[6] ADH is manufactured by the hypothalamus and transported to the posterior lobe of the pituitary gland via modified nerve cells, called neurosecretory cells, which are described in Chapter 20.

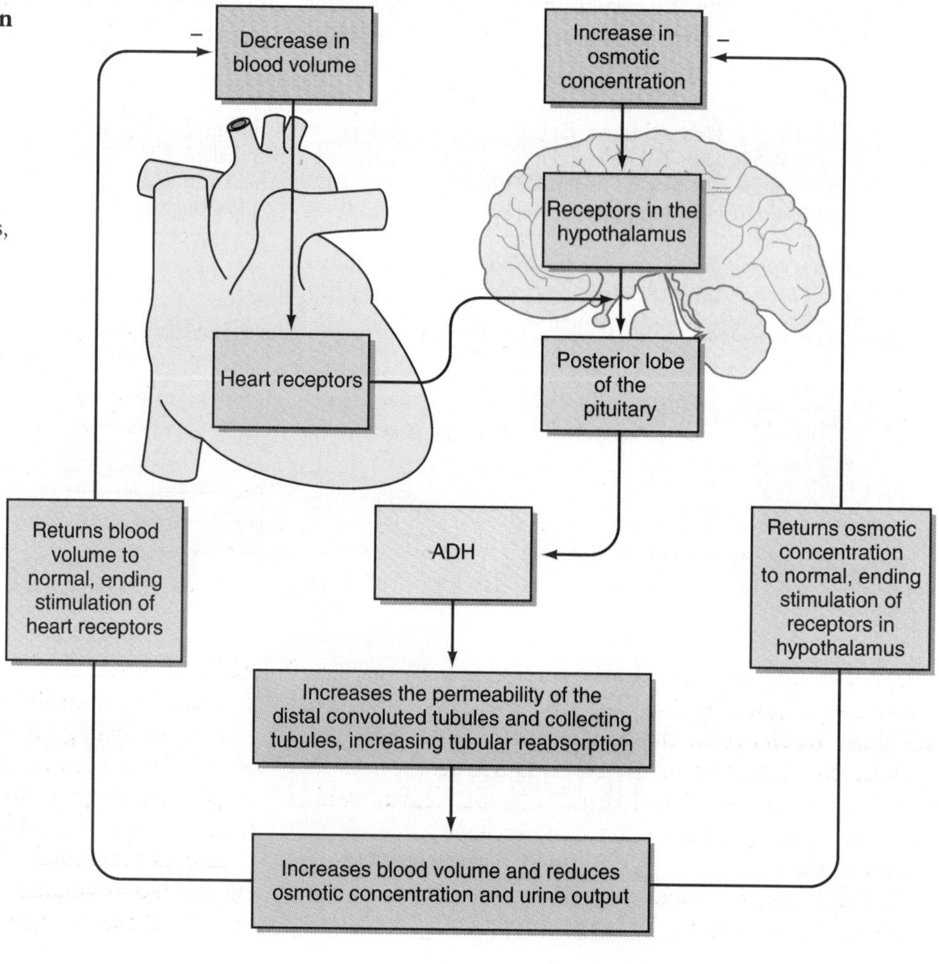

▻ FIGURE 16–15 ADH Secretion ADH secretion is under the control of the hypothalamus. When the osmotic concentration of the blood rises, receptors in the hypothalamus detect the change and trigger the release of ADH from the posterior lobe of the pituitary. Detectors in the heart respond to changes in blood volume. When it drops, they send signals to the brain, causing the release of ADH.

SPOTLIGHT ON EVOLUTION 16–1

EVOLUTION'S WATER CONSERVATION CHAMPION

It is high summer in Organ Pipe Cactus National Monument in southern Arizona. The midday sun beats down. The temperature on the desert floor is nearly 120°F. Few animals stir.

Deep in its burrow, the kangaroo rat sleeps (see ▻ Figure 1). Like many other desert species, the kangaroo rat is behaviorally adapted to the desert, sleeping underground during the day to escape the fierce heat and coming out at night to feed. This behavioral adaptation is just one of many adaptations aimed at conserving water in the desert. Like other mammals, the kangaroo rat has a built-in water recycler in its nasal passages. As dry desert air is inhaled, it passes over moist tissues in the nasal cavity, picking up moisture. This hydrated air protects the lungs from desiccation. Evaporation of water caused by the incoming dry air cools the nasal passageway. When the moist air is later exhaled, water vapor condenses on the cool surfaces of the nasal passages, thus conserving body water.

It is important to note, however, that the countercurrent exchange of heat and moisture is not an adaptation unique to desert life. It is present in birds and lizards as well as mammals. Nevertheless, research suggests that although this evolutionary strategy is widespread, some desert rodents like the kangaroo rat are particularly good at recovering water when breathing dry air. Overall, the nasal countercurrent system of this creature reduces evaporative losses by 70% to 74%.

Desert rodents also produce relatively dry feces, due to the efficient uptake of water by the large intestine. This adaptation conserves water and, like the nasal countercurrent system, is found in many other rodents. In the desert, it serves the kangaroo rat extremely well.

Rodents also have the ability to concentrate their urine. The urine of the laboratory white rat, for instance, is twice as concentrated as human urine. The kangaroo rat and other desert rodents produce urine about two to five times more concentrated than that of humans.

The ability to concentrate urine can be traced to the nephrons of the kidney. As a rule, the longer the loops of Henle, the greater an animal's ability to concentrate urine. The beaver, for example, which lives in aquatic environments where water conservation is not a priority, has relatively short loops of Henle. Humans have long and short loops, but kangaroo rats have uniformly long loops, giving them a remarkable ability to concentrate urine.

Collectively, the water conservation adaptations of the kangaroo rat result in such small water losses that this animal is able to get by without drinking water or consuming live plants. The kangaroo rat can, in fact, get all of the water it needs from seeds and dry plant matter. How?

As you may recall from Chapter 6, energy catabolism produces water, known as metabolic water. Metabolic water and small quantities of water in the food the kangaroo rat eats, combined with behavioral and physiological water-conservation measures, liberate the kangaroo rat from free water, making this animal ideally suited to its desert home.

▻ **FIGURE 1 Kangaroo Rat** This creature is well adapted to desert conditions and is an excellent model of water conservation.

would fall, and the osmotic concentration of your blood would rise (Figure 16–15). The decrease in blood volume obviously results from the loss of water. The increase in osmotic concentration of the blood is due to the fact that perspiration carries off water and salts but leaves behind many osmotically active chemicals, including blood proteins and glucose.

The decrease in blood volume and the rise in osmotic concentration trigger the release of ADH (Figure 16–15). ADH circulates in the blood to the kidney and there increases the permeability of the distal convoluted tubules and the collecting tubules to water. This increases the rate of tubular reabsorption of water, which reduces urinary output and helps restore the volume and proper osmotic concentration of the blood.

Excess water intake has just the opposite effect, increasing the blood volume and decreasing its osmotic concentration. These changes reduce ADH secretion. As ADH levels in the blood fall, the permeability of the distal convoluted tubules and the collecting tubules decreases. As a result, tubular reabsorption declines, and more water is lost in the urine. This helps decrease blood volume and also helps restore the proper osmotic concentration of the blood.

Caffeine Increases Urine Output Without Affecting ADH Secretion. Water in the fluids we drink affects urinary output through ADH secretion, but certain chemical substances in common beverages act on the kidneys directly and have a dramatic effect on urine production. Coffee and (most nonherbal) teas, for example, increase urine production because they contain caffeine.[7] Caffeine in teas, coffee, and certain soft drinks is a **diuretic,** a chemical that increases urination. Caffeine has two major effects on the kidney. First, it increases glomerular blood pressure, which, in turn, increases glomerular filtration. (As a result, more filtrate is formed.) Second, caffeine decreases the tubular reabsorption of sodium ions. As noted earlier, water follows sodium ions out of the renal tubule during tubular reabsorption. A decrease in sodium reabsorption therefore results in a decline in the amount of water leaving the renal tubule and an increase in urine output.

Ethanol Inhibits the Secretion of ADH by the Pituitary. Alcoholic beverages also increase urination. Ethanol in alcoholic beverages is a diuretic, but works by inhibiting the secretion of ADH. This reduces water reabsorption and increases water loss, giving credence to the quip that you don't buy wine or beer, you rent it!

Diabetes Insipidus Is a Rare Disease Characterized by Excessive Drinking and Excessive Urine Output. Severe head injuries can halt the production of ADH, leading to a disease known as **diabetes insipidus.** This condition is not to be confused with diabetes mellitus (commonly called sugar diabetes), a disorder involving the hormone insulin. **Diabetes insipidus** is characterized by frequent urination (polyuria) and excessive liquid intake (polydipsia). It results from insufficient ADH output. Patients with the disease produce up to 20 liters (5 gallons) of colorless, dilute urine per day. Diabetes insipidus gets its name from the fact that the urine is dilute and tasteless (insipid).[8] Sleeping through the night is impossible, for patients are continually awakened by thirst or the urge to urinate.

Diabetes insipidus can be treated in a variety of ways, depending on the severity of the disorder. In patients whose urine output is only slightly elevated, dietary salt restrictions and antidiuretics (drugs that reduce urine output) will work. In severe cases, patients must receive synthetic ADH. ADH can be administered by injections or nose drops. If the pituitary damage has resulted from a head injury, treatment may be required only for a year or so, depending on the rate of recovery. If the damage is permanent, however, treatment will be required for the rest of a person's life.

The Hormone Aldosterone Also Controls Water Reabsorption in the Kidney. ADH controls the amount of water in the body, and that, in turn, helps regulate the concentration of dissolved substances. ADH is therefore, a key component of the homeostatic mechanism that maintains the body's water and chemical balance. ADH is aided by another hormone, aldosterone. **Aldosterone** is a steroid hormone produced by the two **adrenal glands,** which sit atop the kidneys like loose-fitting stocking caps (see Figure 20–1a).

Aldosterone levels in the blood rise and fall in response to three factors: (1) blood pressure, (2) blood volume, and (3) osmotic concentration. A decrease in the blood pressure, for example, triggers the release of this hormone from the outer portion of the adrenal gland, the **adrenal cortex** (▻ Figure 16–16).

Aldosterone increases the reabsorption of sodium ions by the distal convoluted tubules and collecting tubules—that is, the passage of sodium ions from the renal tubule to the blood. Sodium enters the peritubular capillaries, thus increasing the sodium concentration of the blood. Water follows sodium, moving by osmosis into the bloodstream. The outflow of water, therefore, increases the blood volume and blood pressure and lowers the osmotic concentration.

As illustrated in Figure 16–16, the release of aldosterone is controlled by a fairly complex sequence of events. A reduction in blood pressure or a reduction in the volume of filtrate in the renal tubule causes certain cells in the kidney to produce an enzyme called **renin.** In the blood, renin cleaves a segment off a large plasma protein called **angiotensinogen.** The result is a small peptide molecule called **angiotensin I.** Angiotensin I is inactive but is converted into the active form, **angiotensin II,** by further enzymatic action. Angiotensin II stimulates aldosterone secretion.

Kidney Failure May Occur Suddenly or Gradually and Can Be Treated by Dialysis, a Mechanical Filtering of the Blood

The importance of the kidneys is most obvious when they stop working, a condition known as **renal,** or **kidney, failure.** Renal failure generally results from one of four causes: the presence of certain toxic chemicals in the blood, immune reactions to certain antibiotics, severe kidney infections, and sudden decreases in blood flow (for example, after an injury).

Renal failure may occur suddenly, over a period of a few hours or a few days, a condition referred to as **acute renal failure.** Renal function may also deteriorate slowly over many years, resulting in **chronic renal impairment** (so named because the kidneys never really quit). Chronic

[7] Part of the increase is due to the fluid contained in these drinks.

[8] Diabetes mellitus is so named because the urine is sweet. The Latin word for "honey" is *mellifer.*

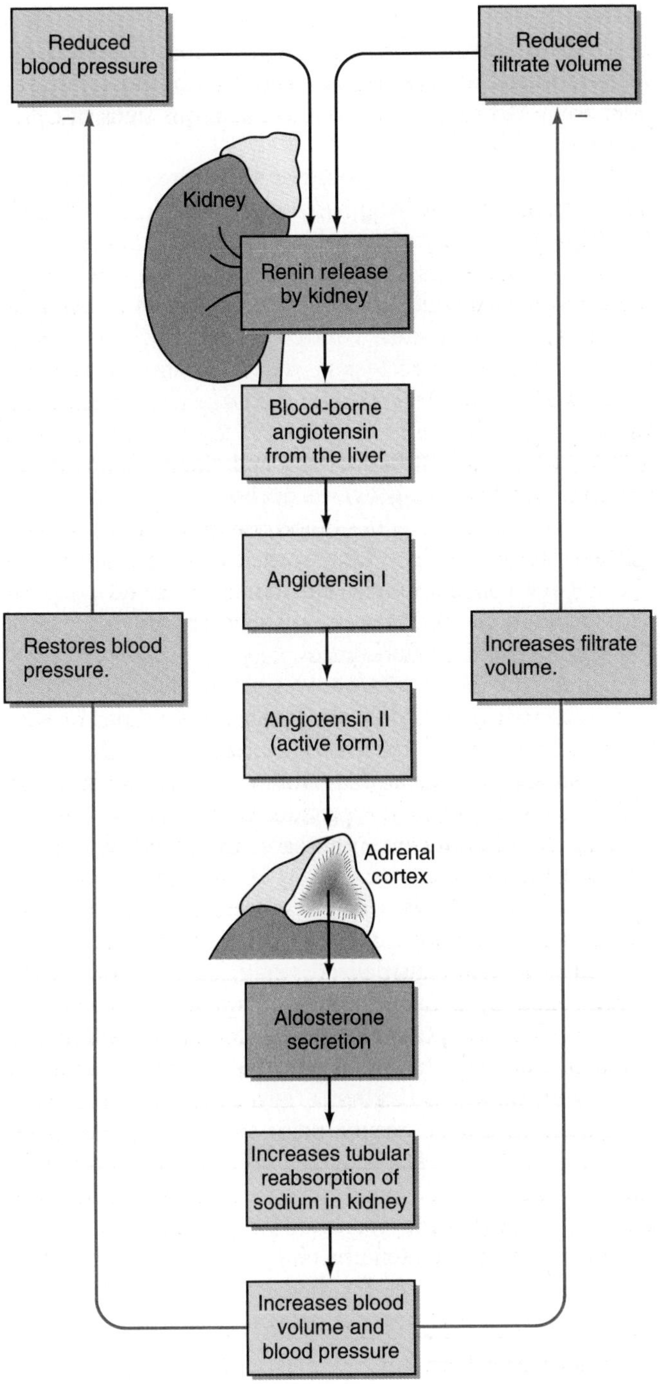

▻ **FIGURE 16–16 Aldosterone Secretion** Aldosterone is released by the adrenal cortex. Its release, however, is stimulated by a chain of events that begins in the kidney.

renal impairment may lead to a complete or nearly complete shutdown, known as **end-stage failure.**

Kidney failure (whether acute or end stage) is generally fatal unless treated quickly. When the kidneys stop working, water begins to accumulate in the body. Toxic wastes build up in the blood. Homeostasis is disrupted by an imbalance of chemicals that are normally regulated by the kidneys. Death generally occurs in two to three days if renal failure is untreated. Death usually results from an increase in the concentration of potassium ions in the blood and tissue fluids. Potassium is essential for the normal function of heart muscle, but excess potassium destroys the rhythmic contraction of the heart, causing fibrillation (Chapter 12).

Treating renal failure depends on the underlying cause. If the problem is caused by an acute loss of blood, transfusions may be required. Patients whose kidneys have shut down, even temporarily, may require **renal dialysis,** a procedure in which toxic materials in the patient's blood are filtered using a machine called a **dialysis unit.** Blood is drawn out of a vein and passed through a piece of tubing that transports the blood to an osmotic filter. After filtration, the blood is pumped back into the patient's bloodstream (▻ Figure 16–17). Dialysis requires several hours and must be repeated every two or three days. Some patients have dialysis units at home and simply hook themselves up each night before they go to bed.

Complete kidney failure can be treated by kidney transplants. Transplants are generally most successful when they come from closely related family members, for reasons explained in the previous chapter.

For years, the complete or nearly complete destruction of kidney function was almost always fatal. Thanks to renal dialysis and kidney transplantation, many patients today

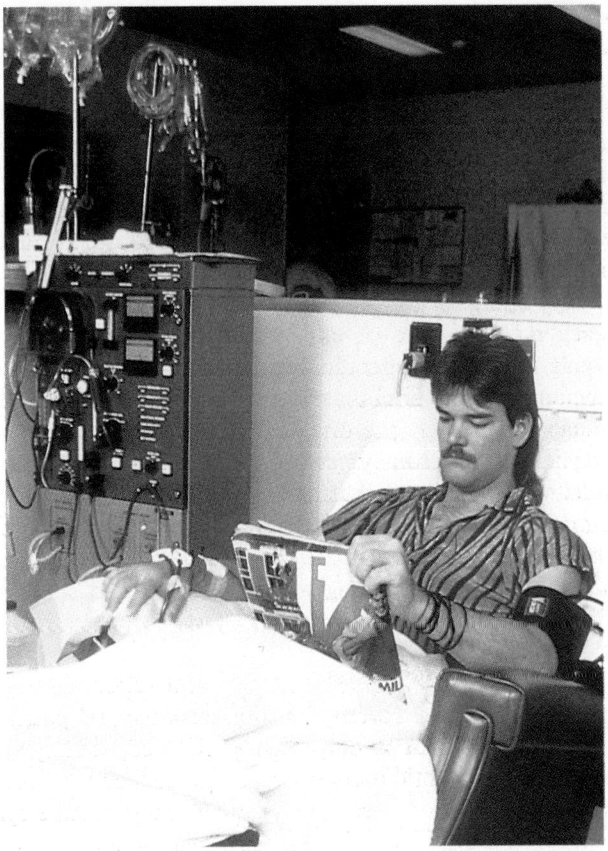

▻ **FIGURE 16–17 Renal Dialysis** When the kidneys fail, the blood must be filtered by a dialysis machine, which runs the blood through filters that remove the impurities.

can live normal, healthy lives. These procedures, especially transplants, are costly, however.

ENVIRONMENT AND HEALTH: MERCURY POISONING

The kidneys are elaborate filters that help maintain the proper levels of nutrients, such as glucose, amino acids, vitamins, and a number of important ions, such as potassium and chloride. The kidneys also help dispose of various endogenous (internally generated) wastes, such as urea, and eliminate numerous potentially harmful exogenous (externally produced) substances, including some drugs, food additives, pesticides, and toxic chemicals that enter the bloodstream from cigarette smoke.

Unfortunately, the kidneys can be seriously damaged by a variety of environmental contaminants. For example, most heavy metals, such as mercury and cadmium, are potent nephrotoxins–toxic chemicals that affect the nephrons. Even relatively low doses of heavy metals can damage the kidneys.

Nephrons possess several protective mechanisms to reduce the impact of heavy metals. Lysosomes inside the cells of the renal tubule, for example, bind to and engulf these toxins, helping decrease the cytoplasmic concentration of these harmful substances. As blood levels increase, however, the protective mechanisms are overwhelmed, and kidney cells begin to die.

Exposure to low levels of various heavy metals increases urinary output and increases the concentrations of amino acids and glucose in the urine—good indicators that tubular reabsorption is not working well. At higher levels, heavy metals damage the nephrons and may result in renal failure and death.

One of the portions of the nephron most sensitive to heavy metals is the proximal convoluted tubule. These metals disrupt its operation and seriously impair kidney function. Other portions of the nephron can also be affected.

One common heavy metal found in the water we drink and occasionally even the food we eat is mercury. In high concentrations in the blood, mercury produces acute renal failure. This condition results from vasoconstriction of the afferent arterioles, which reduces the flow of blood into the glomeruli. Mercury also produces numerous signs of cellular deterioration.

In the 1950s, Japan announced an outbreak of mercury poisoning in residents who had consumed fish and shellfish taken from Minimata Bay, which had been contaminated by a nearby plastics factory. Over 100 people were affected. Victims developed numbness of the limbs, lips, and tongue and lost muscular control, becoming clumsy. Most suffered from blurred vision, deafness, and mental derangement. All told, 17 people died and 23 were permanently disabled, in large part due to the inability of their kidneys to cope with the mercury and subsequent nervous system effects. The tragedy was deepened by the discovery of birth defects in 19 babies born to women who had eaten contaminated seafood. Many of their mothers showed no signs of mercury poisoning, illustrating a general principle of toxicology: the younger an organism, the more sensitive it is to harmful toxic substances.

The Minimata Bay incident was not an isolated event. A similar tragedy occurred on the Japanese island of Honshu. In Sweden in the early 1960s, mercury poisoning killed large numbers of birds that had been feeding on seeds treated with a mercury fungicide, a coating that retards mildew. Swedes who ate pheasant and other birds were also poisoned by mercury.

Fortunately, incidents of overt mercury poisoning are rare. However, mercury is one of the more common water pollutants in the industrialized world, and people are exposed to low levels from a variety of sources. For example, mercury is a by-product of the production of the plastic polyvinyl chloride (commonly called PVC plastic), which is used to manufacture children's toys, beach balls, car seats, and other products. Mercury is also emitted into waterways by a variety of chemical manufacturers and is released by coal-fired power plants and garbage incinerators that burn batteries thrown out in our trash.

Many sources of mercury production are on the rise. Without tighter controls on their emissions, it is possible that low-level mercury poisoning may become more and more common in the years to come.

SUMMARY

GETTING RID OF WASTES: MEETING THE EVOLUTIONARY CHALLENGE

1. All organisms produce waste and therefore all organisms face the same challenge: getting rid of wastes that are produced internally.
2. For single-celled organisms, waste simply diffuses or is actively transported out and is released into the medium (air, water, soil) they inhabit.
3. In some simple multicellular animals, such as sponges and sea stars, the cells on the body surface eliminate wastes directly into the surrounding environment and into the extracellular fluid.
4. Most multicellular animals have excretory organs that facilitate the removal of waste and also help regulate internal concentrations of ions and water.
5. In earthworms and other annelids, fluid from the body cavity enters two small tubular structures, the metanephridia, present in each segment of the body. They transport the waste to the earthworm's body surface.
6. In terrestrial insects dead-end small tubules attach to the gut

and are bathed in body fluids from which they extract ions and water.
7. Waste disposal reaches its zenith in the vertebrates.

ORGANS OF EXCRETION

8. Table 16–1 lists the major metabolic wastes and other molecules excreted from the body.
9. Metabolic wastes are removed by the organs of excretion, including the skin, lungs, liver, and kidneys.

THE URINARY SYSTEM

10. One of the most important organs of excretion in vertebrates is the kidney. Kidneys remove impurities from the blood but also help regulate the water levels and ionic concentrations of the blood.
11. Blood enters the kidneys in the renal arteries and is delivered to millions of nephrons.
12. The nephrons produce urine, which drains from the kidneys into the ureters, slender muscular tubes that lead to the urinary bladder. Urine is stored in the bladder, then voided through the urethra.
13. Each nephron consists of a glomerulus, a tuft of highly porous capillaries, and a renal tubule, where urine is produced.
14. The renal tubule consists of four parts: (a) Bowman's capsule, (b) the proximal convoluted tubule, (c) the loop of Henle, and (d) the distal convoluted tubule. The distal convoluted tubules of nephrons drain into collecting tubules, which converge and empty urine into the renal pelvis.

THE FUNCTION OF THE URINARY SYSTEM

15. Blood filtration is accomplished by three processes: glomerular filtration, tubular reabsorption, and tubular secretion.
16. Glomerular filtration occurs in the glomerulus, producing a liquid called the filtrate.
17. The filtrate is processed as it flows along the renal tubule. Water, ions, and nutrients are largely reabsorbed as they travel along the tubule. This process is called tubular reabsorption. Water and reabsorbed nutrients and ions pass into a network of capillaries, the peritubular capillaries, surrounding each nephron. What is left is a concentrated liquid, the urine.
18. Not all wastes are filtered from the blood in the glomerulus. Those that remain pass into the peritubular capillaries with the blood. These substances may be transported out of the peritubular capillaries into the renal tubule in a process known as tubular secretion. Hydrogen and potassium ions, for example, are secreted into the renal tubule.
19. Calcium, magnesium, and other materials can precipitate out of the urine in the renal pelvis, forming kidney stones, which can block the outflow of urine. Smaller stones may be passed along the ureters to the bladder and are often eliminated during urination. Larger stones that remain in the kidney must be removed surgically or via ultrasound lithotripsy.

URINATION: CONTROLLING A REFLEX

20. Urination is a reflex in babies and very young children. In older children and adults, the urination reflex still operates, but it is overridden by a conscious control mechanism.

THE KIDNEYS AS ORGANS OF HOMEOSTASIS

21. Each drop of blood in your body flows through the kidneys many times in a single day. This flow allows for a thorough filtering of the blood and also helps the body control the blood's chemical composition.
22. The concentration of water and dissolved substances in the blood is controlled by two hormones: ADH and aldosterone.
23. ADH is secreted by the posterior lobe of the pituitary gland. It is released when the osmotic concentration of the blood increases or when blood volume decreases.
24. ADH increases the permeability of the distal convoluted tubules and the collecting tubules to water. When ADH is present, water reabsorption increases. A lack of ADH causes diabetes insipidus, an overexcretion of urine.
25. Aldosterone is produced by the adrenal cortex. Aldosterone stimulates the reabsorption of sodium ions by the nephron. Water follows the sodium out of the renal tubule, increasing blood pressure and blood volume.

ENVIRONMENT AND HEALTH: MERCURY POISONING

26. The kidneys help regulate the levels of harmful toxins produced by the body and help eliminate toxins taken into the body from air, water, and food. But they are not immune to many potentially harmful substances. Heavy metals, for example, destroy some of the cells of the renal tubule and can restrict blood flow to the glomeruli. As a result, heavy metals can damage the kidney, impairing the function of this important homeostatic organ.

EXERCISING YOUR CRITICAL THINKING SKILLS

1. Each year, 24,000 new cases of kidney cancer are diagnosed in the United States. Ten thousand people die from the disease annually. Chemotherapy is virtually the only treatment available for patients whose cancers have spread, but it is beneficial in fewer than 10 percent of the cases. In fact, on average, patients receiving chemotherapy survive only about nine months after they have been diagnosed with the disease.

Recently, a new treatment was introduced that holds some promise for those with kidney cancer. This treatment, known as autolymphocyte therapy, or ALT for short, is marketed by a company in Newton, Massachusetts.

Costing about $22,000 per patient, ALT is designed to strengthen the immune system of patients with malignant kidney tumors. Here's how it works: Blood is withdrawn from patients with kidney cancer. Lymphocytes are extracted from the blood, then treated with monoclonal antibodies (antibodies to the kidney tumor). This process activates the lymphocytes, causing them to produce a chemical substance known as cytokine. Cytokines are natural compounds found in the body that boost the activity of lymphocytes in the immune system. The cytokines produced from this procedure are divided into six batches and stored for later use.

Once a month, for the next six months, patients donate more of their own lymphocytes. The cells are treated with cytokines, then reinjected into the patient, where they apparently go to work on the tumors.

Results of one study on the effectiveness of ALT were published in April 1990 in the British medical journal the *Lancet.* This study of 90 individuals showed that patients who had undergone the procedure survived 22 months after diagnosis, two and a half times longer than patients treated with chemotherapeutic agents.

Less than a year after the publication of this clinical trial, a treatment center opened in Boston. It offers ALT to patients who fit the profile of those who were helped by the drug in the trial.

Assume that you are considering investing in the company that markets ALT. Using your critical thinking skills and your knowledge of good scientific method, what concerns do you have, and what questions would you ask before investing in this company? Write your list on a separate page.

After preparing your list, compare your concerns and questions with mine to see how we match.

1. Concern: A clinical trial on 90 patients is extremely small. With so few patients, the reliability of the results is in question. Question: Are more clinical tests under way?
2. Question: Does the treatment have any adverse impacts?
3. Questions: Is the marketing company unbiased? Is it promoting a product that may turn out to be ineffective, opening the company to lawsuits for fraud? In other words, is it letting financial concerns outweigh the need for good scientific research and carefully controlled experiments?

Now suppose that you have kidney cancer and have the $22,000 to pay for ALT. Would you do it? Are any of the questions or concerns above still relevant? How would you go about determining whether you should try the procedure?

To be fair to all parties concerned, let me point out that approximately two-thirds of the insurance companies in the United States pay for the treatment. As a rule, most companies do not pay for experimental procedures. In other words, they must be satisfied that ALT is a useful and effective treatment before reimbursing clients. In your view, has the insurance industry decided wisely about this treatment? Why or why not?

2. After reading the following hypothetical scenario, you will be asked to make a decision. After you have made your decision, you will be asked some questions that may help you begin to clarify your values. Clarifying your values is essential to critical thinking because it helps you understand your own biases.

Here's the scenario: You are a state legislator considering legislation on prioritizing medical expenditures. Your subcommittee will make a recommendation to the legislature to adopt or reject a plan that would shift state funding from an organ transplant program to a prenatal care program. The state currently pays for organ transplants for needy families, spending over $5 million per year. A group of legislators, however, is proposing that this money be used to fund a program that would offer free checkups for pregnant women as well as advice on drug and alcohol use during pregnancy. The program would also offer information on maternal and infant nutrition to expectant mothers.

The proponents of medical care prioritization say that the money now required for one organ transplant could fund prenatal care for about a thousand mothers. By spending the money on prenatal care, the state could reach thousands of pregnant women who are too poor to see a doctor. Proponents also estimate that 20% of all newborns in your state are born addicted to cocaine, which affects mental and physical development.

Opponents of the bill point out that if needy families are denied money for organ transplants, dozens of children will die. They present the case of Jason Lowry to illustrate what will happen. Jason is 12 years old. His family lives on welfare. Jason needs a liver transplant, which will cost $100,000. Without it, the boy is certain to die. If the state chooses to fund prenatal care instead of organ transplants, dozens of children needing liver transplants will die each year.

You have a choice. Would you recommend the bill that transfers funding to prenatal care or continue funding organ transplants? Why? Make a list of reasons why you supported or opposed the bill.

Now take a moment to ponder your reasons. Was your decision based on economics? Was it based on relative benefits—that is, the benefits for a few versus the benefits for many? Was your decision based on benefits for future generations? Or were you mostly concerned with immediate effects—for example, saving a few lives now?

Imagine, if you will, that Jason was your son. How does that affect your decision? Does the issue take on a different meaning? What general observations can you make about your objectivity? Did it change as the issue came "closer to home"? To what extent do personal interests affect your decisions about other questions, such as environmental issues? Give some examples.

TEST OF TERMS

1. The kidneys are the filtering organs of the ____________ system. Blood enters each of the kidneys through the ____________ arteries.
2. Urine drains from the kidneys into the urinary bladder through two slender muscular tubes, the __________.
3. Urine drains from the urinary bladder through the ____________ to the outside of the body.
4. The outermost region of the kidney is called the ____________ .
5. Urine produced in the nephrons empties into the ____________ ____________, a hollow chamber inside the kidney.
6. The nephron is the filtering unit of the kidney. It consists of two parts, the ____________, a tuft of capillaries, and the ____________ ____________, a long, twisted tube.
7. The nephron's tuft of capillaries is supplied by the ____________ arteriole. These capillaries are highly porous and allow much of the water and dissolved substances in the blood to enter ____________ ____________, in Bowman's capsule.
8. The highly branched cells surrounding the capillaries of the glomerulus are called ____________ .
9. The capillary network surrounding much of the nephron, into which water and dissolved substances are reab-

sorbed, is called the ____________ ____________. The movement of materials from these capillaries into the nephron is called ____________ ____________.

10. The band of smooth muscle at the neck of the urinary bladder that is under reflex control is called the ____________ ____________.
11. Periodic filtering of the blood by an artificial filter is called ____________.
12. Alcohol inhibits the release of a hormone, ____________, from the posterior lobe of the pituitary. This hormone ____________ the permeability of the distal convoluted tubules and collecting tubules.
13. ____________, found in coffee and other beverages, increases urine output and is known as a ____________.
14. The adrenal glands produce another hormone, ____________, which affects the reabsorption of sodium ions and therefore helps control ionic balance and water levels in the body.
15. Mercury is a heavy metal that when present in high levels can destroy cells of the ____________.

Answers to the Test of Terms are found in Appendix B.

TEST OF CONCEPTS

1. Draw the various parts of the urinary system, and describe what each one does.
2. Draw a nephron, then label its parts. Explain what happens to the filtrate in each section of the nephron.
3. Trace the flow of blood into and out of the kidney. Be sure to include details of the pathway once it reaches the afferent arteriole.
4. Describe the three ways in which the kidney filters the blood.
5. A drug inhibits the uptake, or reabsorption, of water by the distal convoluted tubules and collecting tubules. What effect would this drug have on urine output, urine concentration, blood pressure, blood volume, and the concentration of the blood?
6. Describe how ADH controls blood pressure and the water content of the body. Describe the hormonal and physiological changes in the body that take place when excess liquid is ingested. Do the same for dehydration.
7. How does urination differ between newborns and adults? Explain what is meant by this statement: in older children and adults, urination is a reflex with a conscious override.
8. Aldosterone helps regulate blood pressure and water content. In what ways is this hormone different from ADH?
9. You have just finished your residency in family medicine. A patient comes to your office complaining that he drinks water all day long and spends much of the rest of the day in the bathroom urinating. What tests would you order? What diagnosis would you suspect?

CHAPTER

17

The Nervous System: Integration, Coordination, and Control

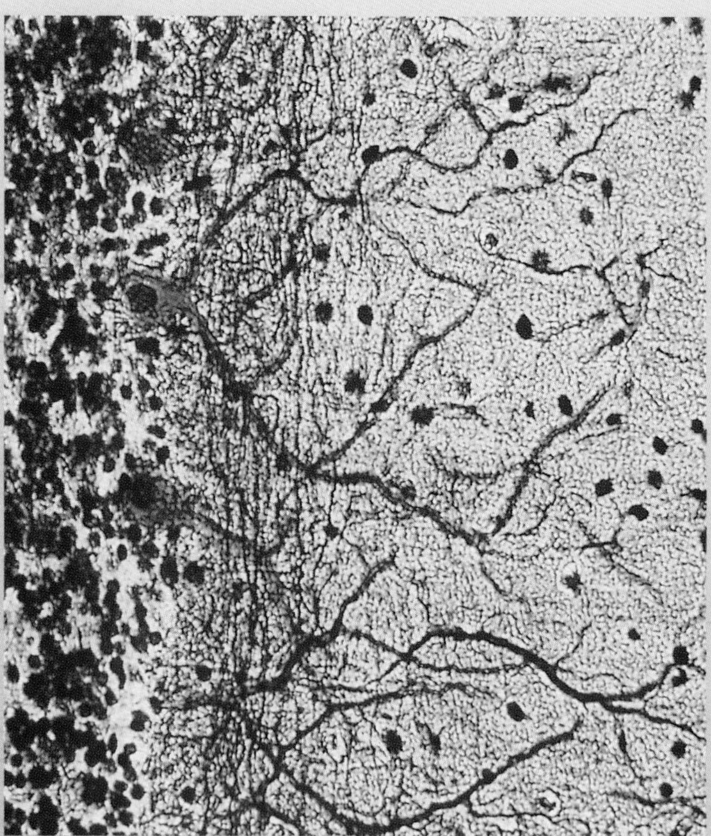

Light micrograph of neurons of the human cerebellum.

Ernest Hemingway's novels and short stories won him the Nobel Prize in literature (▻ Figure 17–1). Despite success and widespread popularity, however, Hemingway was a troubled man, and he eventually committed suicide. What ultimately caused him to take his life no one can know, but some believe that he was suffering from a rare and painful nervous system disorder known as **trigeminal neuralgia.** This disease results in periodic and unexplained flashes of intense pain along the course of the **trigeminal nerve,** which supplies the face. A slight breeze or the pressure of a razor can set off this pain, which lasts a minute or more. In some patients, the pain recurs for no apparent reason every few minutes for weeks on end. Some observers believe that the pain Hemingway felt may have become unbearable. Combined with personal conflict, it may have caused the writer to take his life.

Today, physicians treat mild cases of trigeminal neuralgia with drugs. In extreme cases, however, they may elect to sever the trigeminal nerve as it leaves the brain. This procedure ends the pain, but because it cuts off the inflow of other sensory information, it leaves half of the victim's face, tongue, and oral cavity numb—a little like the feeling you get when a dentist injects novocaine into your gums.

This chapter describes the anatomy and physiology of the nervous system. You will study the nerve that may have caused Hemingway so much trouble and examine the ways in which the nervous system helps regulate homeostasis.

▻ **FIGURE 17–1 Was the Agony Too Much?** Ernest Hemingway may have suffered from trigeminal neuralgia, a painful disorder of the nervous system.

≈ OVERVIEW OF THE NERVOUS SYSTEM

The nervous system of vertebrates governs the functions of the body, exerting its control over muscles, glands, and organs. The nervous system also controls heartbeat, breathing, digestion, and urination. It helps regulate blood flow as well as the osmotic concentration of the blood.

In its governing capacity, the nervous system receives input from a large number of sources. Input from the body helps the nervous system "manage" body functions in much the same way that citizen letters and phone calls help elected officials govern society.

The human nervous system provides additional functions not seen in other animal species. For example, the brain is the site of **ideation**—the formation of ideas. It also allows us to think about and plan for the future. The brain also permits humans to reason—that is, to judge right from wrong, logical from illogical. Although some species display a rudimentary ability to reason, the function is best expressed in humankind.

Lest we forget, the nervous system also allows us to manipulate our environment for our own purposes. More than any other species alive today, humans are reshaping the planet's surface. We topple tropical forests to make room for cattle, drain and fill swamps to build homes, split atoms to generate energy, and catapult men and women into outer space. Joan McIntyre, an author and critic, once wrote, "The ability of our minds to imagine, coupled with the ability of our hands to devise our images, brings us a power almost beyond control." Today, however, humankind has begun to realize that many of the alterations we have made in the planet actually threaten our own long-term future (Chapter 33). Our brains, then, are something of a double-edged sword. They give us a power to create but also an incredible power to destroy. In some cases, the products of our cultural evolution threaten biological evolution. Clearly, the noxious by-products of many technologies overwhelm homeostatic systems that evolved in a much cleaner world.

The Nervous System Consists of Two Main Anatomical Subdivisions, the Central and Peripheral Nervous Systems

The brain, spinal cord, and nerves make up the nervous system (▻ Figure 17–2). The brain and spinal cord constitute the **central nervous system (CNS)** and are housed in the skull and vertebral canal, respectively. Three layers of connective tissue, known as the **meninges,** surround the brain and spinal cord (▻ Figure 17–3). The outer layer consists of fibrous connective tissue and is known as the **dura mater** ("hard mother"). The middle layer is the **arachnoid layer,** so named for its spider-weblike appearance. The innermost layer of the meninges is the **pia mater** (literally, "tender mother"). The pia mater is a delicate, vascular layer that adheres closely to the brain and spinal cord. The space between the arachnoid layer and pia mater is filled with a liquid called **cerebrospinal fluid.**

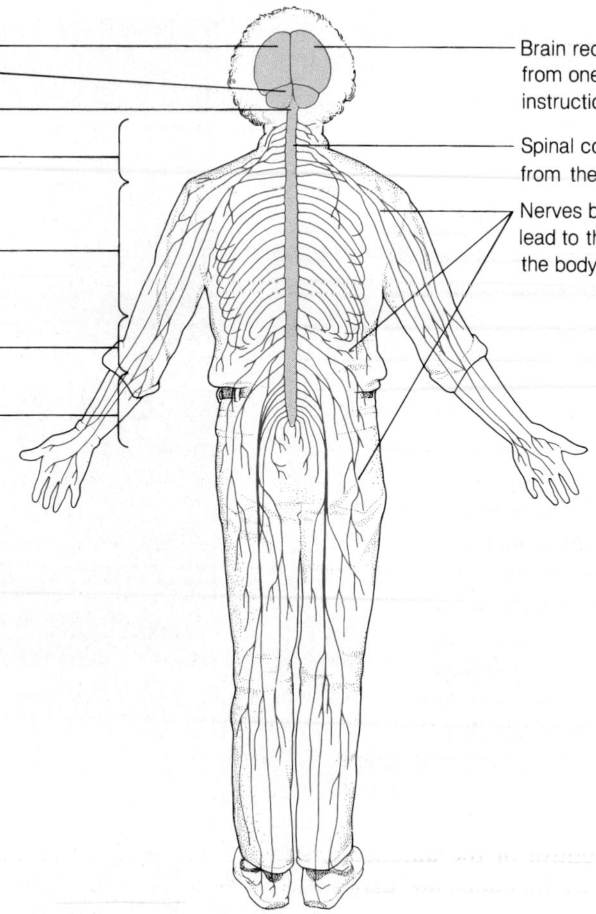
Cerebrum, the brain's higher control center
Cerebellum coordinates movement
Brain-stem controls many autonomic functions
Cervical nerves (8 pairs) serve mainly the arms
Thoracic nerves (12 pairs) lead to the skin, internal organs, and muscles of the chest
Lumbar nerves (5 pairs) serve the abdominal wall and legs
Sacral and coccygeal nerves (6 pairs) lead mainly to the legs
Brain receives and processes information from one set of nerves and issues instructions to the body along another set
Spinal cord, the main nerve trunk to and from the brain
Nerves branching from the spinal cord lead to the arms, legs, and all parts of the body

▻ **FIGURE 17–2 The Nervous System** The human nervous system is a network of nerves connected to the brain and spinal cord. Nerves compose the peripheral nervous system. The spinal cord and brain make up the central nervous system.

Nerves are bundles of nerve fibers (axons and dendrites) of nerve cells, or **neurons** (discussed below). Nerves transport messages to and from the CNS. Nerves and various receptors, which respond to a variety of internal and external stimuli, constitute the **peripheral nervous system (PNS).**

The CNS receives all sensory information from the body. Right now, for instance, your CNS is literally being bombarded with sensory impulses, which travel by nerves from receptors in your body. These receptors alert you to the room temperature, the presence of wind, traffic sounds, and the touch of the page. They transmit visual images of words and figures on the pages. This information is processed in the CNS. Some information is stored in memory. Some of it is blocked. Some incoming information is ignored, and some elicits responses. A particularly exciting section you read, for example, might accelerate your heart rate. A frightening thought might cause you to cringe.

The brain processes incoming stimuli and often responds by sending nerve impulses to muscles and glands (effectors). These impulses also travel along the nerves, but in the opposite direction. In summary, nerves carry two types of information: (1) **sensory impulses** traveling *to* the CNS from sensory receptors in the body, and (2) **motor impulses** traveling *away from* the CNS to effector muscles and glands.

Sensory information pouring into the CNS is integrated with information stored in memory. Thus, a new fact may trigger memories of previous knowledge, causing you to think about a problem in a new way. Memory also influences the way we respond to stimuli. A pet cat brushing against your leg, for example, may elicit a smile. The sensation is not startling, because your memory reminds you of the cat's presence. If you didn't own a cat, the brush of fur along your leg would probably elicit an entirely different response.

The Peripheral Nervous System Can Be Divided into Two Subdivisions, the Somatic and Autonomic Nervous Systems

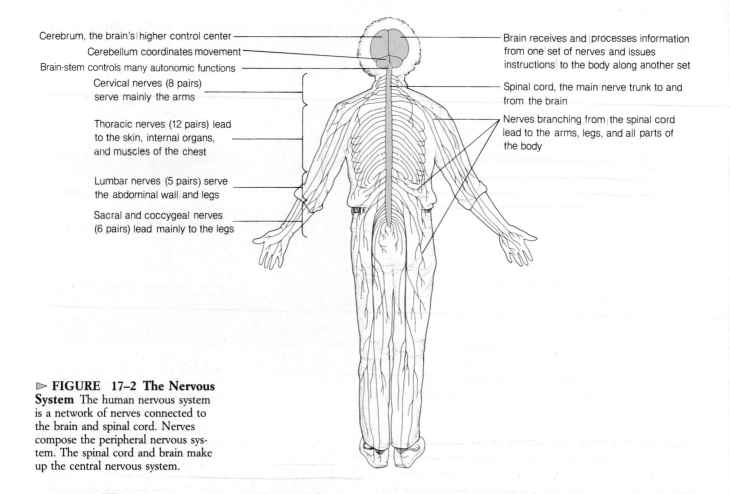

As shown in ▻ Figure 17–4, the peripheral nervous system can be divided into two parts: the somatic nervous system and the autonomic nervous system. The **somatic nervous system** is that portion of the PNS that controls voluntary functions, such as muscle contractions that lead to the movement of the limbs. It also controls certain involuntary reflex actions, like the knee-jerk response. That part of the PNS that controls involuntary functions other than reflexes, such as heart rate, is known as the **autonomic nervous system (ANS)** (Figure 17–4). Breathing and di-

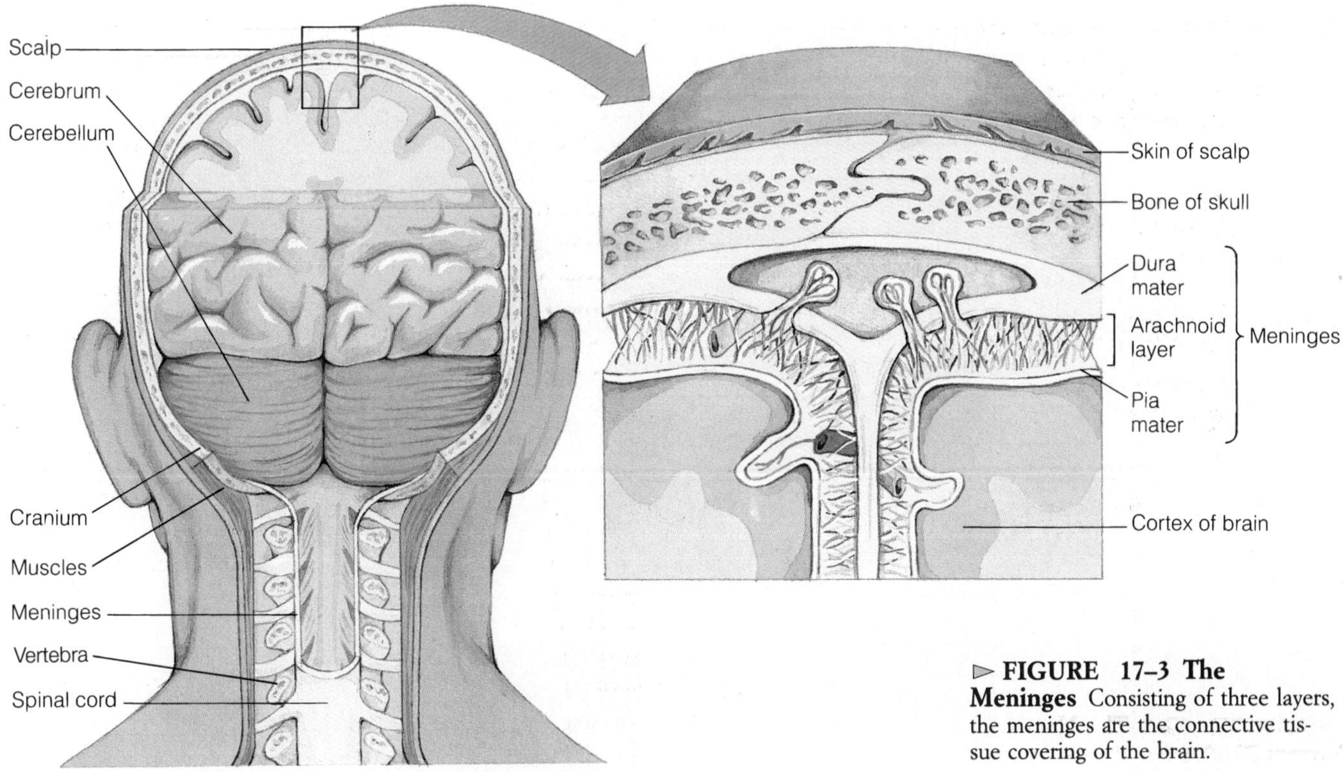

▻ **FIGURE 17–3 The Meninges** Consisting of three layers, the meninges are the connective tissue covering of the brain.

gestion are also under the control of the autonomic nervous system.[1] Many other body functions are under autonomic control and are regulated by negative feedback loops. From an evolutionary perspective, the ANS is vital to survival. Imagine, if you will, how much more difficult our lives would be if we had to consciously control our breathing and other automatic functions.

Figure 17–4 shows that each branch of the peripheral nervous system consists of two types of neurons—**sensory, or afferent, neurons,** which transmit information to the central nervous system, and **motor,** or **efferent, neurons,** which transmit information to the effectors.

STRUCTURE AND FUNCTION OF THE NEURON

Before we look in more detail at the structure and function of the nervous system, let us look at the neuron, or nerve cell. The **neuron** is the fundamental structural unit of the nervous system. This highly specialized cell generates bioelectric impulses and transmits them from one part of the body to another—alerting us to a variety of internal and external stimuli and permitting us to respond.

All Neurons Consist of a Cell Body Containing the Nucleus

Neurons come in several shapes and sizes. Despite these differences, nerve cells share several common characteristics. All neurons, for example, consist of a more or less spherical central portion, called the cell body (Figure 17–5). The **cell body** houses the nucleus, most of the cell's cytoplasm, and numerous organelles. Metabolic activities in the cell body sustain the entire neuron, providing energy and synthesizing materials necessary for proper cell function. Two organelles of particular interest are microtubules and microfilaments. These structures form the cytoskeleton of the neuron and are responsible for the neuron's characteristic shape.

All Nerve Cells Contain Processes that Transmit Bioelectric Impulses. Neuronal processes that transmit impulses to the cell body are referred to as **dendrites.** Those that transmit impulses away from the cell body are called **axons.** Neurobiologists generally classify neurons by their anatomy—notably, the type of processes they contain. Accordingly, three distinct types of neuron are found in the human nervous system. Shown in ▻ Figure 17–6 (page 410), they are the unipolar, the bipolar, and the multipolar. The **unipolar neuron** has a single cellular process that splits into an axon and a dendrite. The **bipolar neuron** has two cellular processes, one axon and one dendrite, on opposite sides of the cell body. The **multipolar neuron** contains a single long process, the axon, and numerous short, branching dendrites.

Multipolar neurons are the most abundant and will be the focus of our discussion (see Figure 17–5).[2] In multipo-

[1] Note that breathing can be controlled voluntarily, but for the most part it is an involuntary action.

[2] Multipolar neurons are involved in the efferent pathways of the PNS and carry motor information to effector organs, such as glands and muscles.

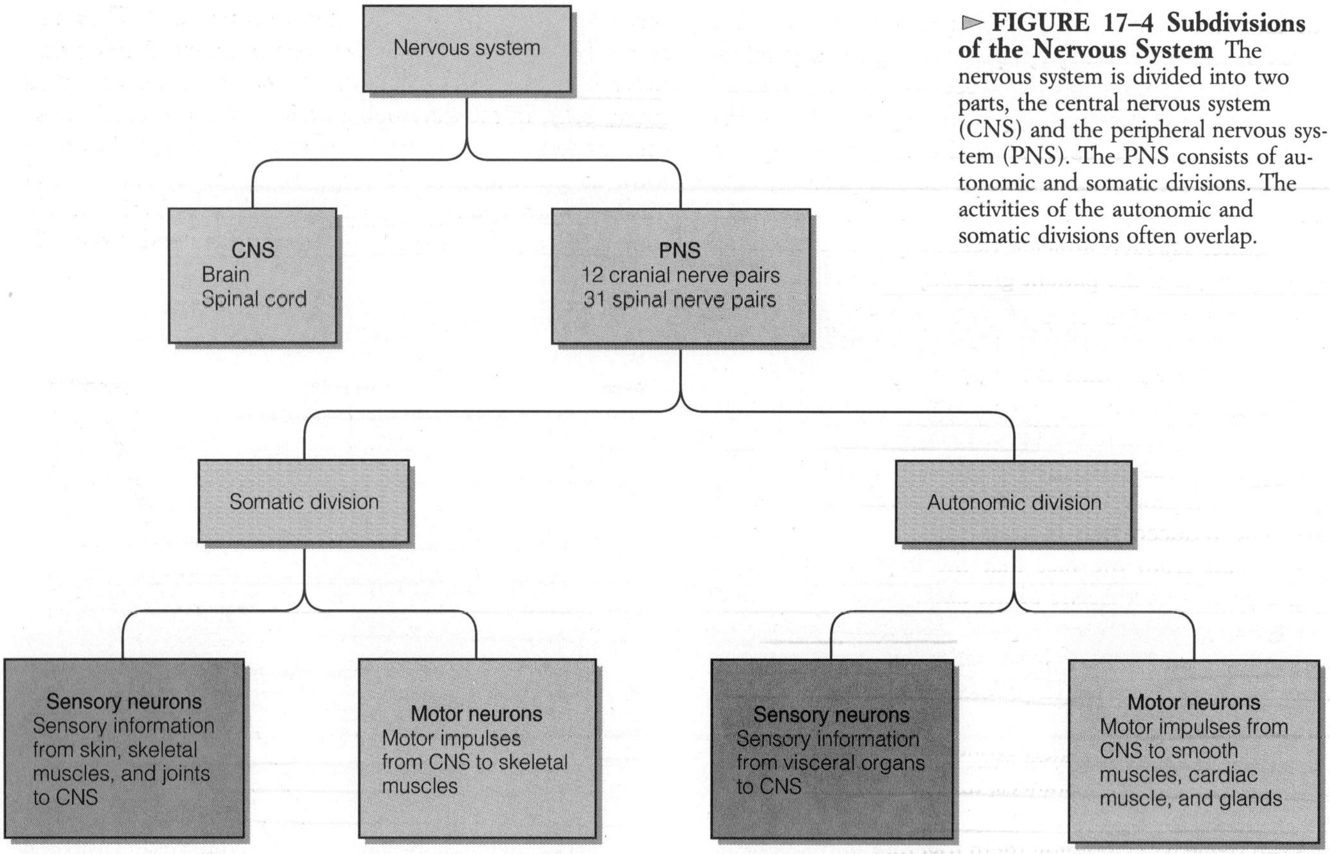

▻ **FIGURE 17–4 Subdivisions of the Nervous System** The nervous system is divided into two parts, the central nervous system (CNS) and the peripheral nervous system (PNS). The PNS consists of autonomic and somatic divisions. The activities of the autonomic and somatic divisions often overlap.

lar neurons, the short, branching dendrites transmit impulses to the cell body. (The arrows in Figure 17–5b indicate the direction in which an impulse travels.) The cell bodies of multipolar neurons are located in the spinal cord and brain. After reaching the cell body, impulses travel down the long, unbranched axon. Axons of multipolar neurons often leave the CNS with the nerves of the PNS or, alternatively, may connect one part of the CNS to another (thus remaining inside the CNS). Axons occasionally also give off side branches, known as **axon collaterals** (Figure 17–5b). When an axon reaches its destination, it often branches profusely, giving off many small fibers. These fibers terminate in tiny swellings known as **terminal boutons** (end buttons), or **terminal bulbs.** Nerve impulses reaching the terminal ends of axons may be transmitted to other neurons, muscle fibers, or glands.

The Myelin Sheath Increases the Rate of Transmission of Nerve Impulses and Is Found on Axons in Both the CNS and PNS. The axons of many multipolar neurons in both the central and peripheral nervous systems are coated with a protective layer called the **myelin sheath** (▻ Figure 17–7a). The myelin sheath is formed by nonconducting cells of the nervous system known as **glial cells.** In the PNS, the glial cells that form the myelin sheath are referred to as **Schwann cells,** after the early German cytologist Theodore Schwann. In the CNS, the myelin sheath is laid down by glial cells known as **oligodendrocytes.**

During embryonic development, Schwann cells in the PNS attach to the growing axon, then begin to encircle it (Figure 17–7d). As they do, they leave behind a trail of plasma membrane, which wraps around the axon, forming many concentric layers—like an elastic bandage wrapped around your wrist (Figures 17–7b and 17–7c).

The entire myelin sheath of an axon is formed by numerous Schwann cells, which align themselves along the length of the axon. Each lays down a separate patch of myelin. Because the plasma membrane of the Schwann cell is about 80% lipid, the myelin sheath is mostly lipid (mostly triglyceride) and appears glistening white when viewed with the naked eye. As shown in Figures 17–7a and 17–7b, each segment of myelin is separated by a small unmyelinated segment known as a **node of Ranvier.** In the CNS, a single oligodendrocyte produces myelin for several axons (Figure 17–7e).

The myelin sheath permits nerve impulses to travel with great speed down axons. As illustrated in Figure 17–7a, impulses "jump" from node to node like a stone skipping along the surface of the water. (More on this process later.)

Destruction of the myelin sheath of nerve cells in the central nervous system results in a condition known as **multiple sclerosis.** The destroyed myelin is replaced by plaque that disrupts the transmission of impulses. Thought to be an autoimmune disease, multiple sclerosis can affect any part of the CNS. Early symptoms are generally mild—weakness or a tingling or numbing feeling in one part of the body. Temporary weakness may cause a person to stumble and fall. Some people report blurred vision, slurred speech, and difficulty controlling urination. In many cases, these symptoms disappear, never to return. In other cases, individuals suffer repeated attacks. Recovery after each attack is incomplete, and the patient gradually deteriorates, losing vision and becoming progressively weaker. Fortunately, many treatments are available, and only a small number of multiple sclerosis patients are crippled by the disease.

Although most axons are covered with myelin, some axons are unmyelinated. Found in both the central and peripheral nervous systems, unmyelinated axons conduct impulses much more slowly than their myelinated counterparts. The reduced rate of transmission in unmyelinated axons results from the fact that the impulse must travel along the entire membrane of the axon—more like a wave moving along the surface of a pond. As a general rule, then, the most urgent types of information are transmitted via myelinated fibers; less urgent information is transmitted via unmyelinated fibers.

Interestingly, unmyelinated axons are also associated with Schwann cells. In such instances, though, the Schwann cells merely encase the unmyelinated axons, as shown in ▻ Figure 17–8 (page 412), holding them together in a bundle.

Microtubules Inside Axons Function in Axonal Development and also Transport Materials from the Cell Body to the Axon Terminal. Inside all axons are bundles of microtubules (▻Figure 17–9, page 412). These microtubules facilitate the transport of materials produced in the cell body to the axon terminal. They also play a key role in the development of axons. During embryonic development, for example, nerve cells start out as round cells. In the developing brain and spinal cord, these cells undergo a dramatic transformation. Axons begin to form as microtubules develop inside the cytoplasm and push outward against the plasma membrane. As the microtubules grow, the axons elongate and extend from the

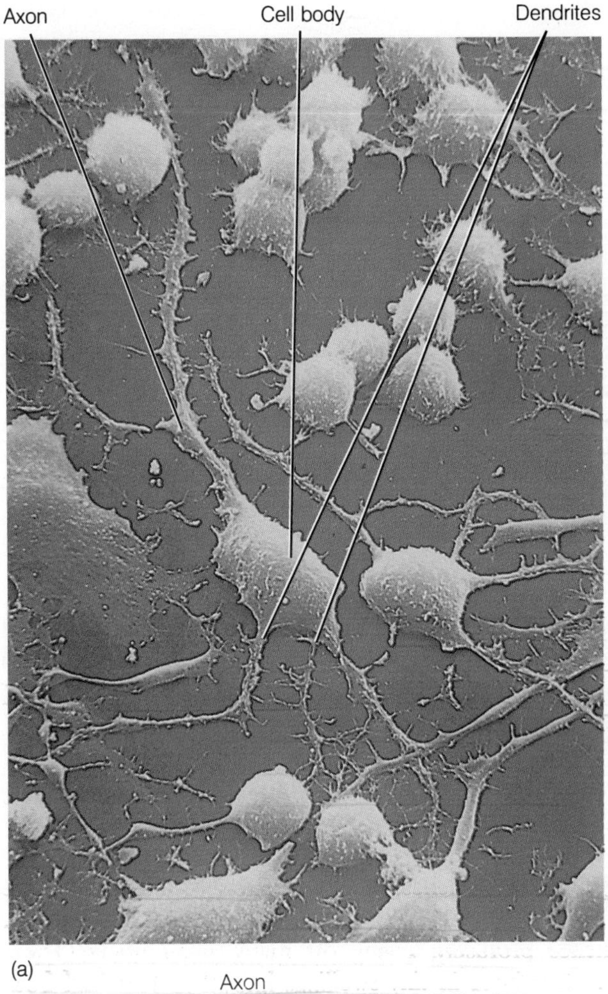

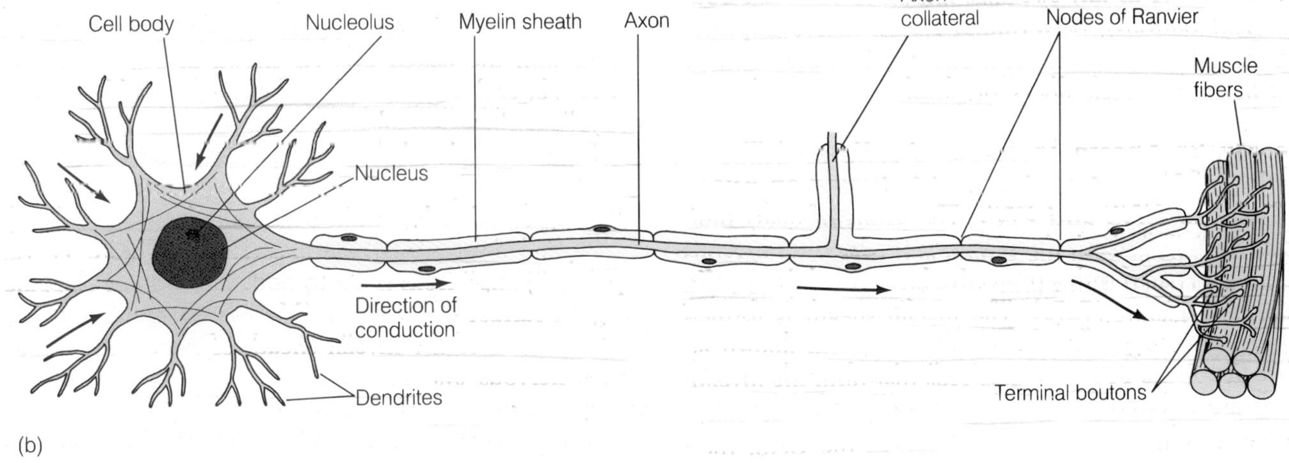

▻ **FIGURE 17–5 A Neuron** (*a*) A scanning electron micrograph of the cell body and dendrites of a multipolar neuron. The multipolar neuron resides within the central nervous system. Its multiangular cell body has several highly branched dendrites and one long axon. (*b*) Collateral branches may occur along the length of the axon. When the axon terminates, it branches many times, ending on individual muscle fibers.

▷ **FIGURE 17–6 Three Types of Neurons** (*a*) Unipolar. (*b*) Bipolar. (*c*) Multipolar.

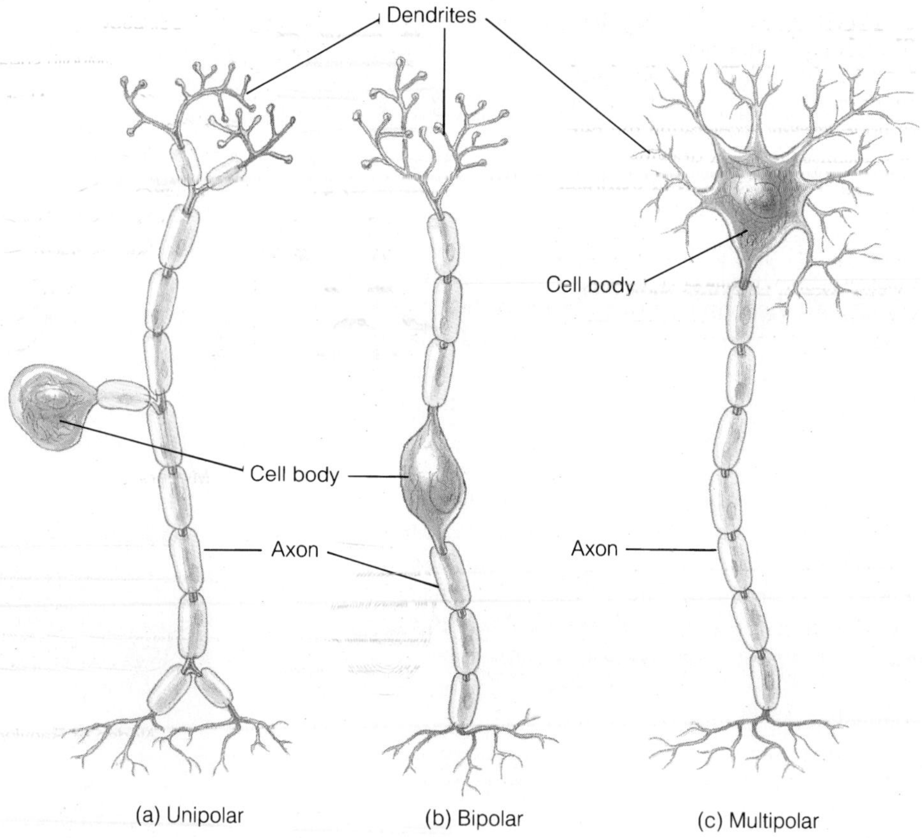

embryonic brain and spinal cord to the muscles and glands they will supply. Some axons within the central nervous system (brain and spinal cord) extend to other regions of the CNS, thus helping connect its parts.

Neurons Are Highly Specialized Cells That Have Lost the Ability to Divide. Like a few other cells that undergo cellular differentiation during embryonic development, nerve cells lose the ability to divide. Therefore, nerve cells that die generally cannot be replaced by cell division of other nerve cells.

Fortunately, there are mechanisms to offset nerve cell damage. Consider what happens in a stroke. A **stroke** occurs when brain cells are damaged as a result of a sharp reduction in blood supply. Strokes result from one of three causes: (1) cerebral hemorrhage, a break in an artery of the brain; (2) cerebral thrombosis, a blood clot that forms in a brain artery narrowed by atherosclerosis; and (3) cerebral embolism, a blood clot from another source that lodges in an artery of the brain (▷ Figure 17–10). All three problems have the same end result: they terminate blood flow to vital brain cells, killing them.

If a person survives a stroke, undamaged neurons in the brain may take over the function of damaged brain cells, thus permitting partial neurological recovery. Recovery also occurs when cells that were injured but not killed by the loss of blood regain their function. Recovery generally takes a long time, which explains why people who have suffered from strokes require long-term rehabilitation.

The fate of a nerve cell after injury depends on its location in the nervous system. In the central nervous system, for example, damaged axons cannot be repaired. In the peripheral nervous system, axonal regeneration is possible. Thus, a severed nerve in the arm may be able to regenerate new axons. As illustrated in ▷ Figure 17–11, a severed axon in the PNS generally degenerates from the point of injury to the muscle or gland it supplied. The segment of the axon still attached to the cell body may grow back to replace the degenerated section. During regeneration, the segment of the axon attached to the cell body elongates and expands along the hollow tunnel left in the myelin sheath by the degenerated section of axon. Eventually, the axon reestablishes connections with the muscles or glands it once supplied, providing partial or even nearly complete recovery of control.

Research suggests that regeneration of axons in the PNS is possible because Schwann cells release a neuronal growth-promoting factor. Oligodendrocytes, which are responsible for myelin formation in the CNS, do not produce a similar substance. Experiments show that Schwann cell transplants in the CNS promote axonal regeneration. Researchers hope some day to be able to isolate, purify, and

▻ **FIGURE 17–7 The Myelin Sheath and Saltatory Conduction** (*a*) The myelin sheath allows impulses to "jump" from node to node, greatly accelerating the rate of transmission. (*b*) A drawing showing the arrangement of Schwann cell membrane in the myelin sheath. (*c*) A transmission electron micrograph of an axon in cross section showing a myelin sheath. Drawings show how the myelin sheath is formed in the PNS (*d*) and CNS (*e*).

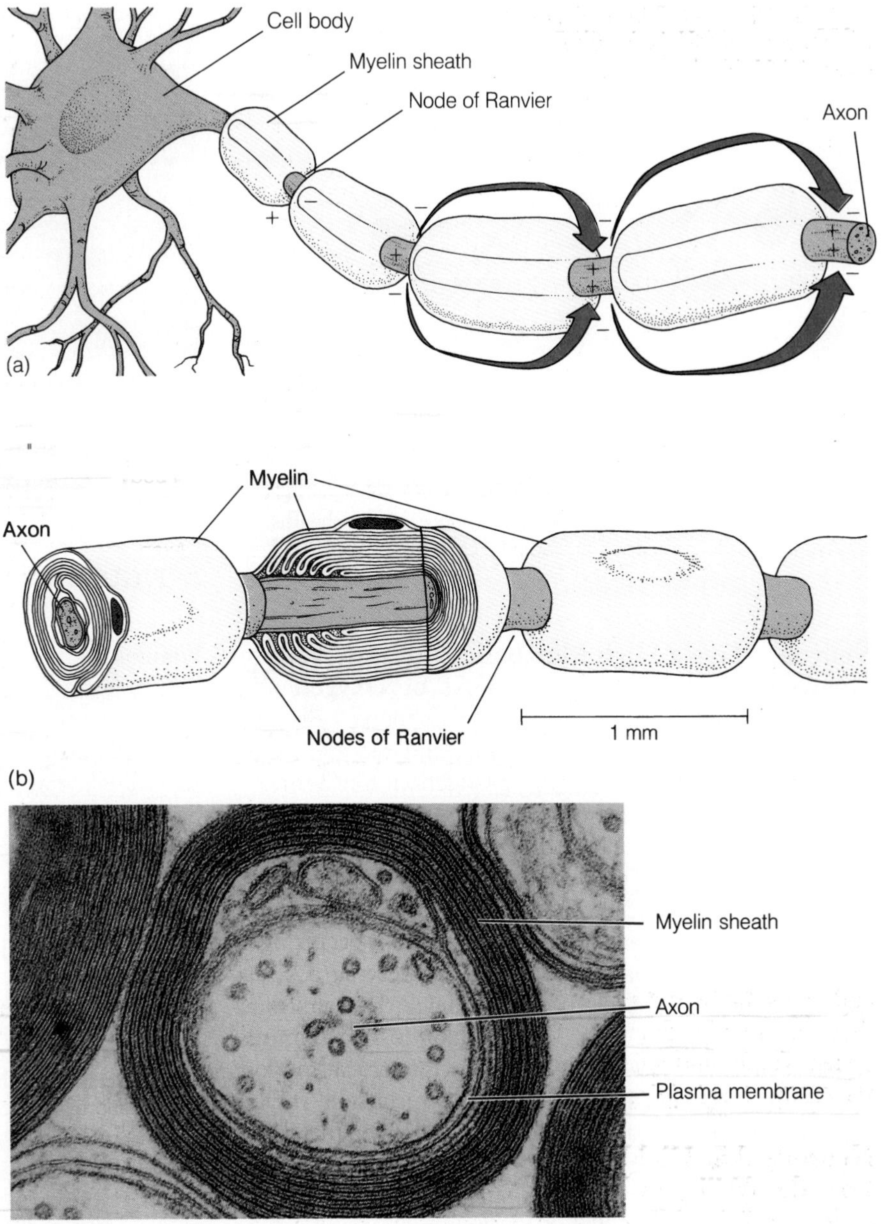

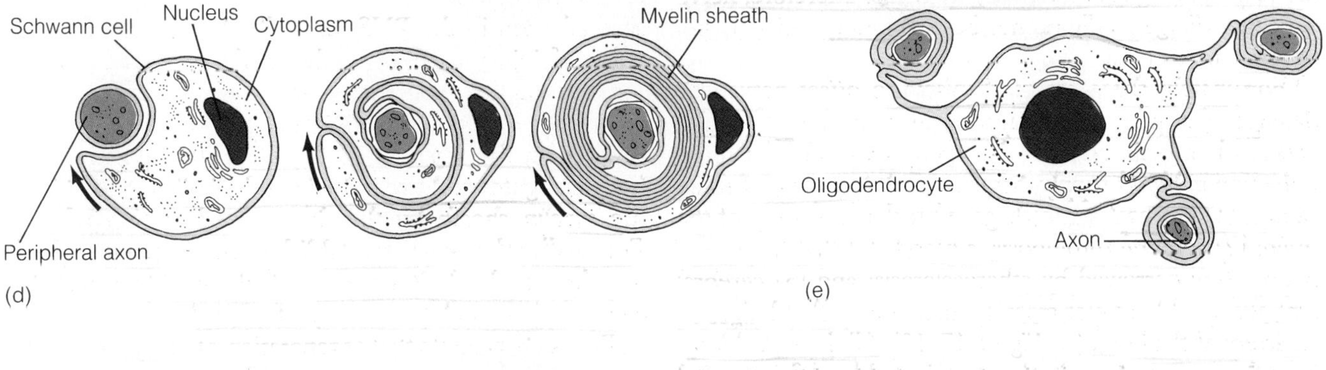

manufacture the neuronal growth-promoting factor to stimulate axonal regeneration in victims of strokes and accidents.

One final note: neurosurgeons can facilitate axonal regeneration in the PNS by microsurgery performed under a dissecting microscope more elaborate than the one you may have used in your high school biology class. Surgeons sew the severed ends of nerves together after lining up the empty myelin sheaths with the regenerating nerve fibers to facilitate axonal regrowth.

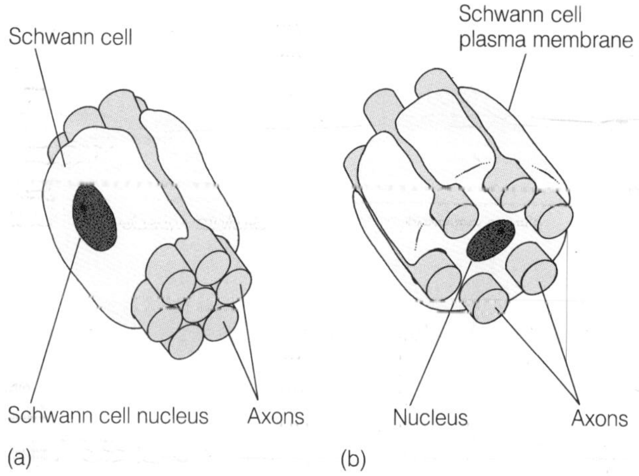

▻ **FIGURE 17–8 Unmyelinated Axons** (*a*) Schwann cells encompass groups of axons, but do not produce myelin sheaths. (*b*) Axons may be embedded individually as well.

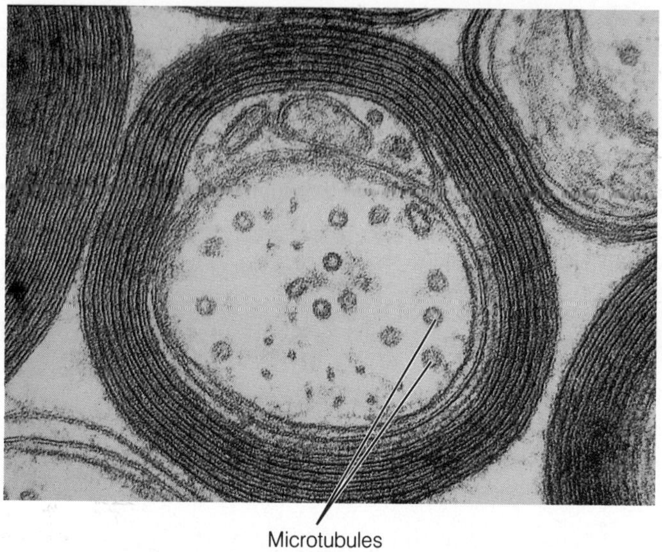

▻ **FIGURE 17–9 Microtubules in Axon**

Neurons Have a High Metabolic Demand, Making Them Highly Susceptible to Loss of Oxygen and Glucose. Besides being unable to divide, nerve cells have an extraordinarily high metabolic rate and require a constant supply of oxygen for energy production. Furthermore, neurons cannot generate ATP in the absence of oxygen via fermentation (glucose breakdown to lactic acid) the way most other cells can. Thus, if the amount of oxygen flowing to the brain is drastically reduced, neurons begin to die within minutes. To prevent brain damage from occurring in someone who has collapsed with a severe heart attack, has drowned, or has suffered an electric shock, rescuers must start resuscitating the victim within four to five minutes. Although victims may be revived after this crucial period, the lack of oxygen in the brain often results in varying degrees of brain damage. Generally, the longer the deprivation, the greater the damage.

Interestingly, if a person is submerged in icy water, resuscitation may be successful if begun within an hour. In most cases, victims recover without any detectable brain damage. In one exceptional case a young boy from Fargo, North Dakota, was underwater for five hours before he was recovered and resuscitated. Much to his parents' delight, the boy not only survived the incident but also had no apparent ill effects. Why is it that people who are submerged in cold water can be revived without suffering brain damage? Cold water greatly slows metabolism in the brain, dramatically reducing oxygen demand. As a result, brain cells are preserved, and brain damage is minimized or avoided.

Nerve cells are also highly dependent on glucose for energy production. Unlike most other body cells, they cannot use fatty acids to generate energy. They also cannot store glucose as can liver and muscle cells. As a result, when blood glucose levels fall, nerve cells are first to "feel" the ill effects.

A decline in blood glucose is normally prevented by homeostatic mechanisms described in Chapter 10. In diabetics, however, blood glucose levels may fall dangerously low if too much insulin is taken or if not enough glucose is ingested. Deprived of glucose, brain cells begin to falter. Individuals may become dizzy and weak. Vision may blur. Speech may become awkward. The person may be mistaken for a drunk and may undergo seizures or become unconscious.

▻ **FIGURE 17–10 Stroke** PET (positron emission tomography) scan of brain revealing damaged region of the cerebral cortex following a stroke. This color image was generated by a computer that converts readings of the rate of emission from radioactive glucose molecules injected into the patient. Damaged (dark region) areas show the lowest glucose uptake. Highest glucose uptake and highest emissions are in red.

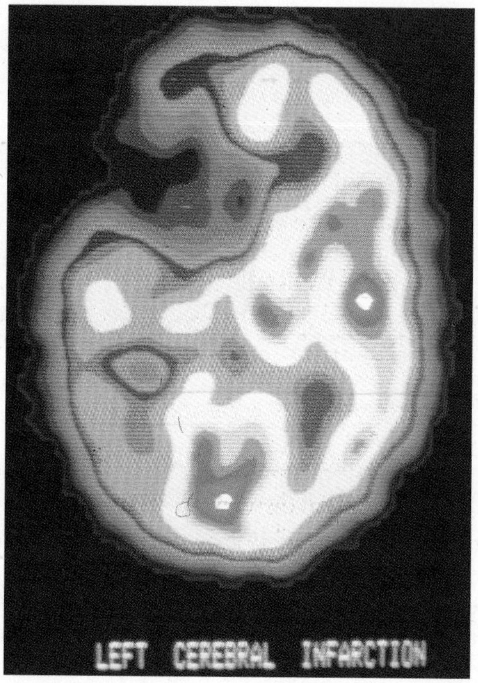

▷ **FIGURE 17–11 Axonal Regeneration** (*a*) A severed axon can regenerate in the peripheral nervous system. The segment from the cut to the effector organ degenerates. (*b*) The myelin sheath remains, providing a tunnel (*c*) through which the axonal stub can regrow, often reestablishing previous contacts and restoring motor function.

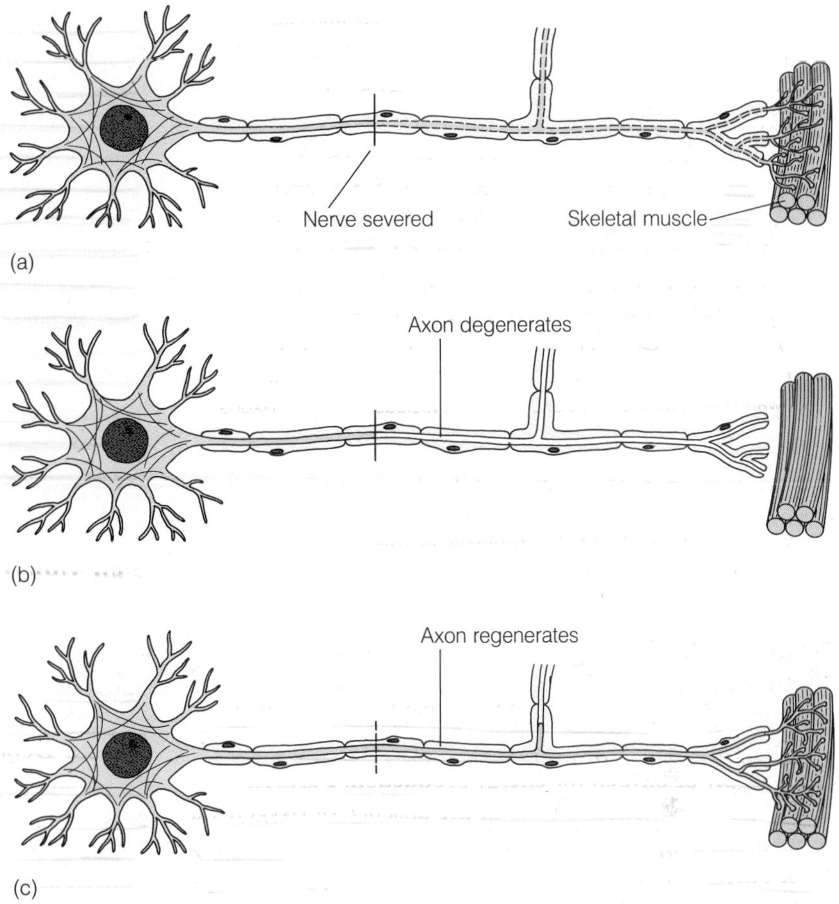

Chronic low blood glucose (hypoglycemia) often results in severe headaches. People may become aggressive when blood sugar levels fall. In extreme cases, low blood sugar triggers convulsions and death. Other body cells are not adversely affected by a decline in blood glucose because they switch to alternative fuels, fats and proteins.

Bioelectric Impulses Traveling in Nerve Cells Result From the Flow of Ions Across the Plasma Membrane of these Cells The evolution of the nervous system is important to the survival and propagation of animals because it keeps them attuned to changes in the environment. It also helps them make many thousands of adjustments needed to survive in an ever-changing environment.

These functions are possible because of nerve impulses transmitted through the PNS and CNS. Nerve impulses are not like the electric current that powers computers and light bulbs, which is formed by the flow of electrons. Rather, nerve impulses are small ionic changes in the membrane of the neuron that move along the plasma membrane of a nerve cell.

To understand the nerve cell impulse, or **bioelectric impulse,** so named to distinguish it from electricity, we begin by examining the plasma membrane of a neuron. If you placed a tiny electrode on the outside and inserted another on the inside of the plasma membrane of a neuron and hooked them up to a voltmeter, you would be able to measure a small voltage, much like that produced in a battery. In the simplest terms, **voltage** is a measure of the tendency of charged particles to flow from one pole of the battery to the other. The higher the voltage, the greater the tendency for electrons to flow through a wire connected to the poles. In the nerve cell, however, electrons do not flow from one side of the membrane to another, sodium and potassium ions do.

In neurons, the potential difference, or voltage, is a measure of the force that can drive sodium ions from one side of the membrane to the other. For now, however, it is important just to remember that a small voltage exists across the plasma membrane of the neuron. It is so small, in fact, that it is measured in millivolts. A millivolt is 1/1000 of a volt.

As shown in ▷ Figure 17–12a, the potential difference in a typical neuron is about −60 millivolts. This is known as the **membrane potential,** or **resting potential,** for it is the membrane potential of a nerve cell at rest. The minus sign is added because the plasma membrane is positively charged on the outside and negatively charged on the inside, for reasons beyond the scope of this book (Figure 17–12b). It is important to note, however, that sodium ions are found in greater concentration outside the neuron, whereas potassium ions are found in greater concentration inside the cell. In fact, neurons expend a great deal of energy to maintain this concentration imbalance, which, as you will soon see, is essential to the production of nerve impulses. Cellular energy is used to power active transport

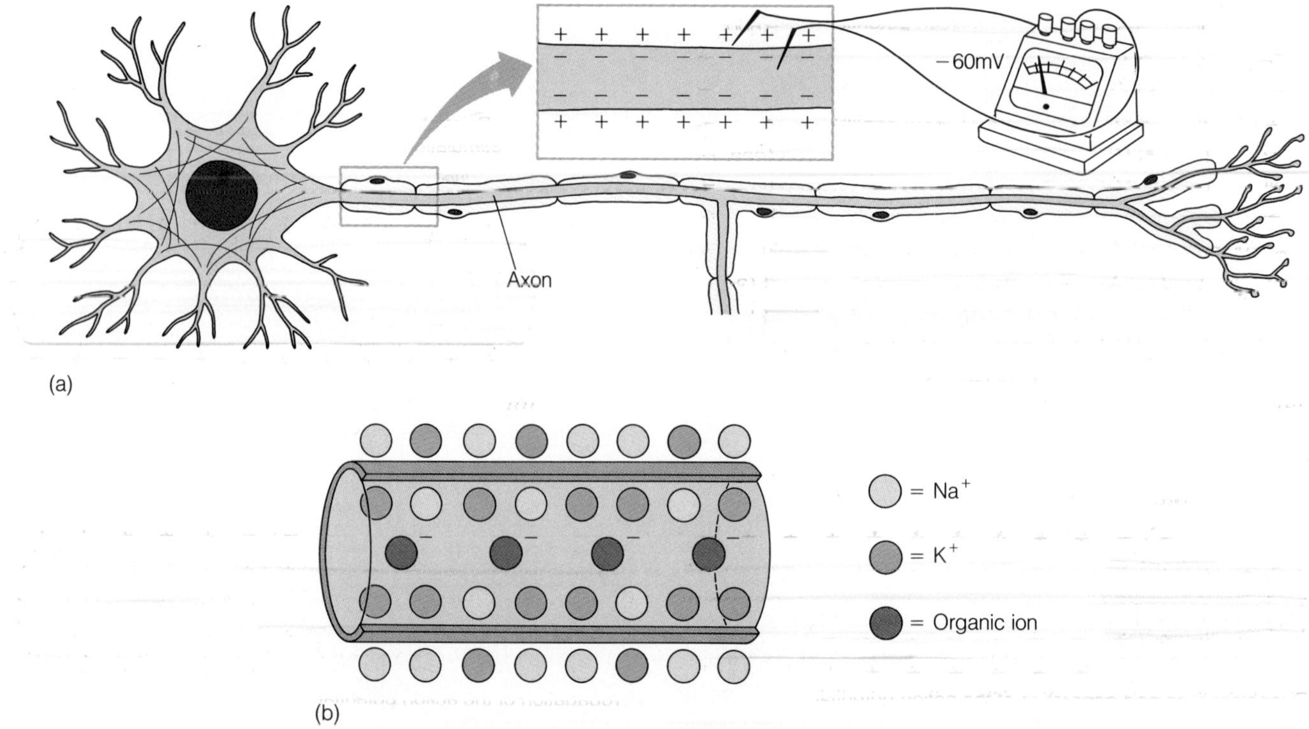

▻ **FIGURE 17–12 Resting Potential** (*a*) Electrodes placed on both sides of the plasma membrane of a neuron measure a tiny potential difference, roughly −60 millivolts (mV). This is the resting potential. (*b*) It results from a chemical disequilibrium caused by the active transport of sodium (Na^+) ions out of the neuron and potassium (K^+) ions into the cytoplasm and the presence of negatively charged organic ions in the cytoplasm.

pumps located in the plasma membrane of neurons. These pumps transport sodium ions that leak into the cytoplasm of the neuron back out into the surrounding fluid, helping maintain the external concentration. They also transport potassium ions that have leaked into the extracellular fluid back into the cell.

After a Nerve Cell Is Stimulated, the Membrane Undergoes Dramatic Changes in Permeability to Various Ions, Resulting in Sudden Shifts in Membrane Potential. The plasma membrane of a nerve cell is a lot like a loaded gun. That is, it has a built-up charge. The charge consists of sodium ions concentrated on the outside of the cell. When the neuron is stimulated, the membrane undergoes a rapid change, "discharging" the load. The first change occurring in the membrane is a rapid increase in its permeability to sodium ions. Neurophysiologists believe that stimulating the nerve cell causes protein pores in the plasma membrane to open. Sodium flows into the cell through these pores.

Electrodes implanted in a nerve cell detect the sudden influx of positively charged sodium ions and register a shift in the resting potential from −60 millivolts to +40 millivolts. The change in voltage occurs at the site of stimulation and is called **depolarization.**[3] Immediately after depolarization, the membrane returns to its previous state, which is referred to as **repolarization.**

The electrical charge across the membrane is known as an **action potential** and is graphically represented in ▻ Figure 17–13a. This graph shows (1) a brief upswing, depolarization, as the voltage goes from −60 millivolts to +40 millivolts, and (2) a rapid downswing, repolarization, the return to the resting potential.

The action potential occurs so rapidly and the membrane returns to the resting state so quickly (about 3/1000 of a second) that neurons can be stimulated in rapid succession. Such brisk recovery allows us to respond swiftly and forcefully to danger and to perform rapid muscle movements. A nerve cell can also transmit many impulses in sequence, because only a small number of sodium ions are exchanged with each impulse.

As noted above, depolarization results in the rapid influx of sodium ions. During repolarization, the membrane shifts from +40 millivolts back to −60 millivolts, the resting potential. Repolarization results from two factors: a sudden decrease in the membrane's permeability to sodium ions, which stops the influx of sodium ions, and a rapid efflux of positively charged potassium ions (Figure 17–13d). The outflow of potassium ions results from concentration and electrical forces. That is to say, potassium ions flow out of the axon down a concentration gradient. Because the outside of the membrane becomes negatively charged during depolarization, an electrical gradient also comes into play.

[3] Depolarization means that the membrane loses its previous polarization.

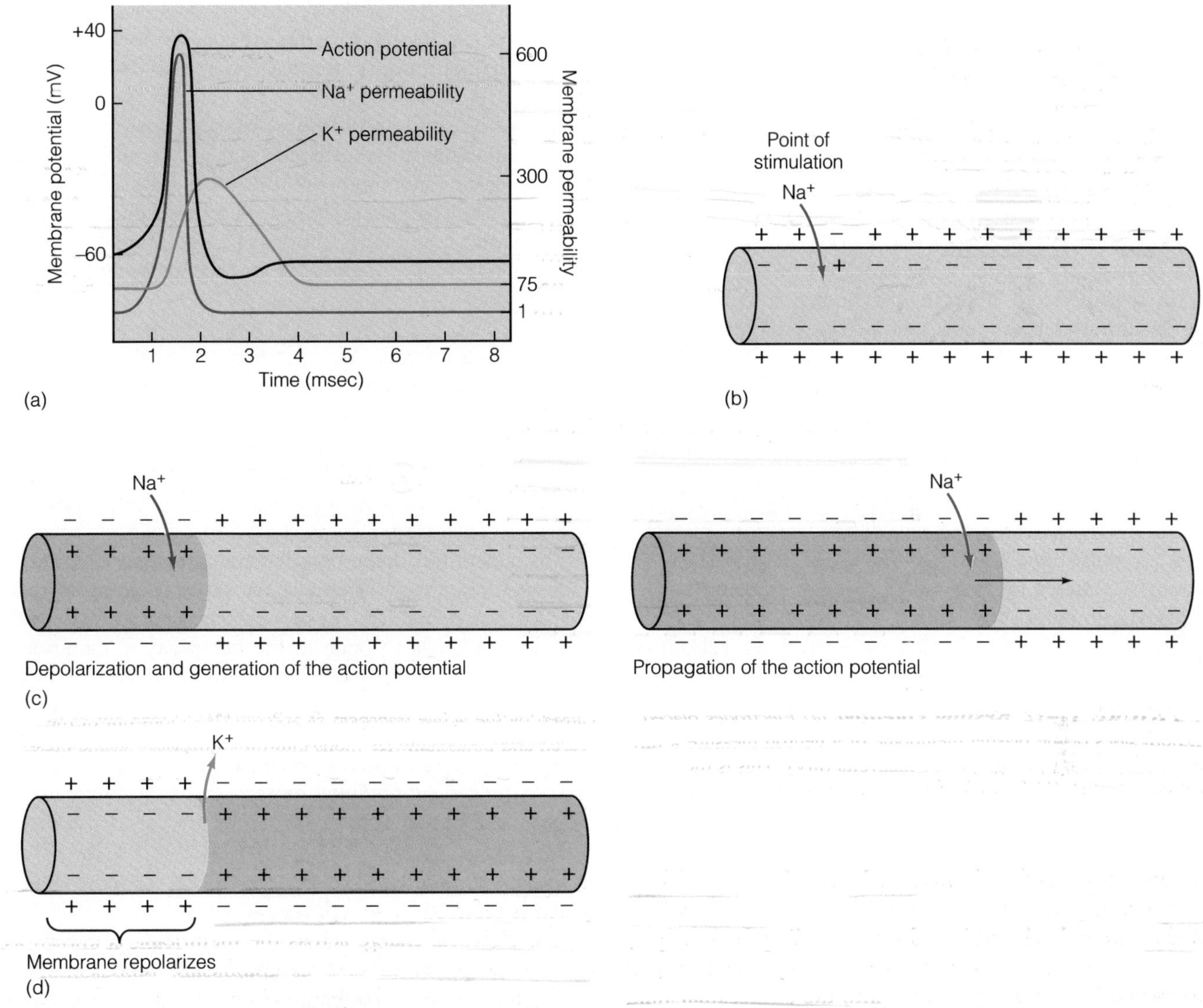

FIGURE 17–13 Action Potential (*a*) Stimulating the neuron creates a bioelectric impulse, which is recorded as an action potential. The resting potential shifts from −60 millivolts to +40 millivolts. The membrane is said to be depolarized. This graph shows the shift in potential and the change in sodium (Na^+) and potassium (K^+) ion permeability, which is largely responsible for the action potential. (*b*) This drawing shows the influx of sodium ions and the depolarization that occur at the point of stimulation. (*c*) The impulse travels along the membrane as a wave of depolarization. (*d*) The efflux of potassium ions restores the resting potential, allowing the neuron to transmit additional impulses almost immediately.

The net outward movement of potassium ions helps reestablish the resting potential.

When a neuron is no longer being stimulated, it quickly reestablishes the proper sodium and potassium ion concentrations inside and outside the cell. It does this by pumping sodium ions that flowed into the cell during activation out of the axon and by pumping potassium ions that flowed out of the neuron during repolarization back in. The plasma membranes of nerve cells contain numerous "sodium-potassium active transport pumps" that transport sodium out of and potassium into the cell. (You may recall from Chapter 3 that active transport pumps require energy in the form of ATP.) The sodium-potassium pumps, therefore, help neurons reestablish the chemical disequilibrium so necessary for normal nerve cell function. Interestingly, about 30% of the energy you burn at rest is used to operate this pump.[4]

The Nerve Impulse Moves Along the Membrane of the Neuron Because Depolarization in One Region Stimulates Depolarization in Neighboring Regions. Stimulating a nerve cell artificially at one point creates an inward rush of sodium ions, which dramatically shifts the resting potential. The influx of sodium ions is followed by an outward rush of potassium ions that returns the membrane potential to normal. This constitutes the bioelectric impulse. How does the bioelectric impulse travel along the nerve cell?

[4] The sodium-potassium pump is found in all cells, not just neurons.

Many studies to explore the conduction of impulses by neurons have been made using the giant unmyelinated axons of the squid. These studies show that a change in membrane permeability in the stimulated region, which results in depolarization, causes a change in the sodium permeability of neighboring regions. Thus, depolarization in one region of the plasma membrane of an axon stimulates depolarization in adjacent regions. This process continues along the length of the axon.[5]

In unmyelinated fibers in the human nervous system, nerve impulses travel like waves from one region to the next as they do in the giant squid axon. In myelinated fibers, however, the depolarization "jumps" from one node of Ranvier (the section between adjacent Schwann cells) to another. This movement is called **saltatory conduction** (from the Latin word *saltare,* meaning "to jump"). Shown in Figure 17–7a, this "jumping" of the impulse from node to node greatly increases the rate of transmission. In fact, a nerve impulse travels along an unmyelinated fiber at a rate of about 0.5 meter per second (1.5 feet per second). In a myelinated neuron, the impulse travels 400 times faster—that is, about 200 meters (650 feet) per second (or about 400 miles per hour). The difference in the rate of transmission is largely due to a difference in the total amount of axonal membrane that must be depolarized and repolarized.[6] Saltatory conduction also conserves energy, because it reduces the amount of energy needed to pump sodium and potassium ions. Like so many adaptations, it contributed significantly to the evolution of complex multicellular animals.

Nerve Impulses Travel from One Neuron to Another Across Synapses. Nerve impulses travel from one neuron to another across a small space that separates them, as shown in ▻ Figure 17–14. The juncture of two neurons is called a synapse. A **synapse** consists of (1) a terminal bouton (or some other kind of axon terminus), (2) a gap between the adjoining neurons, called the **synaptic cleft,** and (3) the membrane of the dendrite or postsynaptic cell (Figure 17–14b). The neuron that transmits the impulse is called the **presynaptic neuron;** the one that receives the impulse is called the **postsynaptic neuron.**

When a nerve impulse reaches a terminal bouton, depolarization of the plasma membrane causes a rapid influx of calcium ions into the bouton. Calcium ions stimulate the release (by exocytosis) of a chemical substance known as a **neurotransmitter,** which is stored in small vesicles in the terminal bouton.

At least 30 chemicals serve as neurotransmitters. Produced and packaged in vesicles in the cell body of the neuron, neurotransmitters are transported down the axon along the microtubules to the terminal bouton, where they are stored until needed. When the bioelectric impulse arrives, the vesicles bind to the presynaptic membrane and release the neurotransmitter (by exocytosis) into the synaptic cleft.

Neurotransmitters diffuse across the synaptic cleft between adjoining nerve cells and bind to protein receptors in the plasma membrane of the postsynaptic (receiving) neuron. The binding of most neurotransmitters to the postsynaptic membrane stimulates a rapid increase in the permeability of the membrane of the postsynaptic cell to sodium ions.

Neurotransmitters May Excite or Inhibit the Post-Synaptic Membrane. In the brain, a single nerve cell may have as many as 50,000 synapses. In some of these synapses, neurotransmitters stimulate the uptake of sodium ions, which slightly depolarizes the postsynaptic neuron. Synapses that depolarize the postsynaptic neuron are known as **excitatory synapses.** In other synapses, however, neurotransmitters hyperpolarize the membrane—that is, increase the voltage difference across the membrane, making it less excitable. These neurotransmitters stimulate chloride channels to open in the postsynaptic membrane. Because the chloride ion concentration outside the cell is higher than it is inside, the opening of the chloride channels results in an influx of chloride ions, making the interior of the postsynaptic cell more negative. This influx, in turn, makes the resting potential more negative and renders the neuron less excitable. Such synapses are called **inhibitory synapses.**[7]

Whether a neuron fires (generates an action potential) depends on the summation of excitatory and inhibitory impulses. If the number of excitatory impulses exceeds the inhibitory impulses, a nerve impulse will be generated. If not, the neuron will not fire. This phenomenon provides the nervous system with a way of integrating incoming information—that is, determining a response by the kinds of input it receives.

Transmission across the synapse can occur in only one direction because only terminal boutons contain neurotransmitter substances and because receptors for these substances are found only in the postsynaptic membrane.

Neurotransmitters Are Quickly Removed from the Synaptic Cleft. Transmission across the synapse is remarkably fast, requiring only about 1/1000 of a second, or one millisecond. Synaptic transmission is also a transitory event. A short burst of neurotransmitter is released each time an impulse reaches the terminal bouton. It binds to the postsynaptic membrane, elicits a change, and is then removed from the synaptic cleft by three routes: (1) enzymatic destruction, (2) reabsorption by the terminal bouton or absorption by glial cells in the brain, and (3) diffusion away from the synapse.

[5]The rate of transmission actually varies depending on the diameter of the axon.

[6]In normal nerve cells, stimulation usually does not occur on the axon but, rather, on the dendrites.

[7]Inhibitory neurotransmitters may also stimulate the opening of potassium channels. Potassium ions flow out of the postsynaptic cell, making the interior more negative.

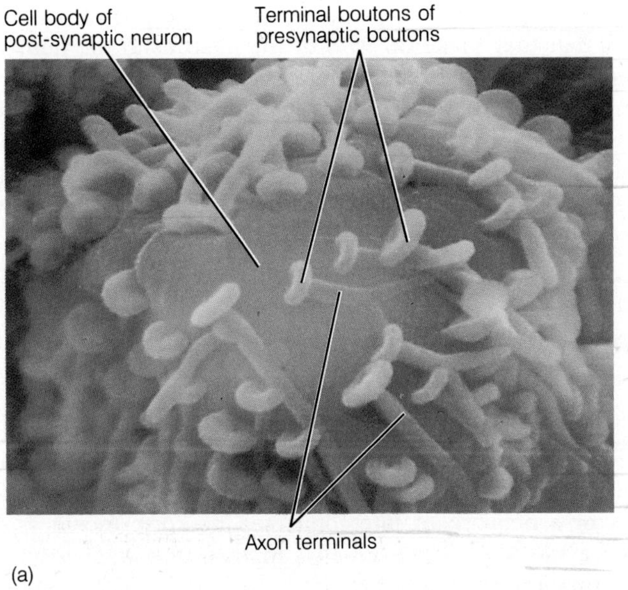

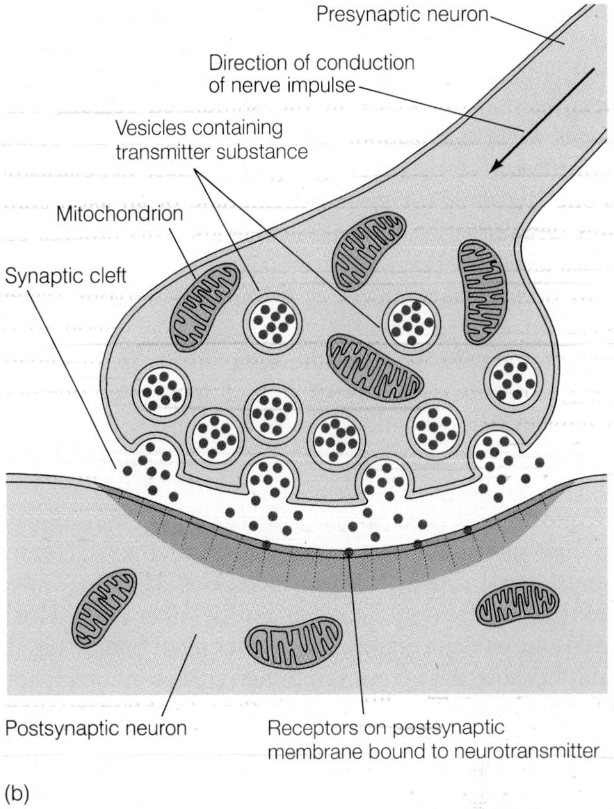

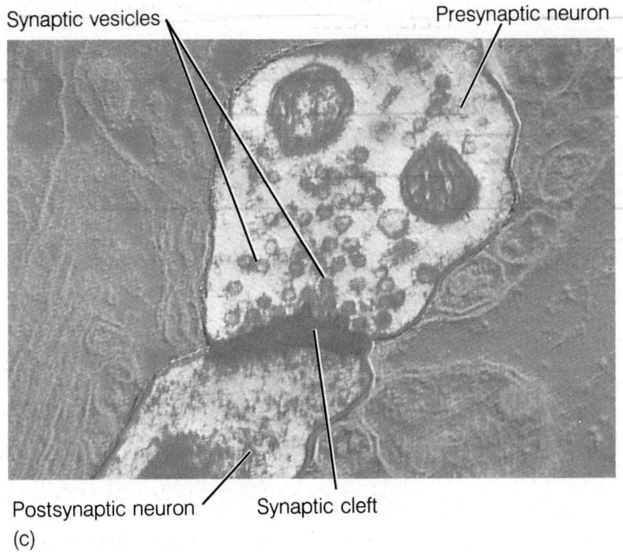

▻ **FIGURE 17–14 The Terminal Bouton and Synaptic Transmission** (*a*) A scanning electron micrograph showing the terminal boutons of an axon ending on the cell body of another neuron. (*b*) The arrival of the impulse stimulates the release of neurotransmitter held in membrane-bound vesicles in the axon terminals. Neurotransmitter diffuses across the synaptic cleft and binds to the postsynaptic membrane, where it elicits another action potential that travels down the dendrite to the cell body. (*c*) A transmission electron micrograph showing the details of the synapse.

Many Chemical Substances, Including Drugs and Insecticides, Exert their Effect by Altering Synaptic Transmission. Certain drugs and common chemicals impair the removal of a neurotransmitter from the synaptic cleft. As long as a neurotransmitter remains bound to the receptors on the postsynaptic membrane, the postsynaptic neuron remains activated. Some insecticides, for instance, act on the neurotransmitter acetylcholine (a-seat-'l-ko-leen). Acetylcholine is the main excitatory neurotransmitter in the neurons that innervate (supply) skeletal muscle cells. Many commonly used insecticides, however, block the action of **acetylcholinesterase,** an enzyme found in many synapses of insects and people that destroys acetylcholine after it has elicited a reaction in the muscle cell. Organophosphates, such as malathion and parathion, are two examples. Thus, insecticides that block the enzyme kill insects by disrupting nerve transmission. Insects become hyperstimulated and are so incapacitated they eventually die. Unfortunately, insecticides have the same effect on people. These nerve poisons are therefore quite harmful to farm workers and pesticide applicators, who are often exposed to high levels of the pesticides at work. In humans, blocking acetylcholinesterase greatly decreases the removal of acetylcholine from synapses, causing the postsynaptic nerve to be stimulated repeatedly. Muscles go into spasms. At low levels, these insecticides cause blurred vision, headaches, rapid pulse, and profuse sweating. At higher doses, victims may begin to writhe uncontrollably and can die in a short time.

Each year, an estimated 100,000 to 300,000 Americans (mostly farm workers) are poisoned by pesticides. By various estimates 200 to 1000 of them die. Worldwide, an estimated 500,000 people are poisoned and 5000 to 14,000 people die each year from pesticides. These problems and the widespread contamination of the environment have led some farmers to reduce pesticide use and to rely on other, nonpolluting methods of pest control. Crop rotation, insect-resistant crops, natural predators (ladybugs and praying mantises), and a variety of alternatives are practical and cost-effective. As a rule, these strategies also benefit the soil and surrounding environment and are therefore essential to developing a more sustainable system of agriculture.

Some anesthetics may also alter synaptic transmission, decreasing the transmission of pain impulses. Others, however, seem to operate on the nerve cell itself. They apparently alter protein pores in the plasma membrane of neurons that regulate the flow of sodium ions into and out of the nerve cells. By blocking the flow of sodium, these anesthetics "paralyze" sensory nerves carrying pain messages to the brain.

Caffeine and cocaine also affect synaptic transmission. Caffeine increases synaptic transmission, thus increasing overall neural activity. It is no wonder that coffee makes some people so jittery. Cocaine, on the other hand, blocks the uptake of neurotransmitters by terminal boutons in the brain. Because the neurotransmitter remains in the synaptic cleft for a longer time, neural activity is greatly increased. Increased neural transmission in the brain results in a heightened state of alertness and euphoria, commonly known as a "high," which lasts for 20 to 40 minutes. Euphoria, however, is followed by a period of depression and anxiety, which causes many people to seek another high. Excessive cocaine use can result in serious mental derangement—in particular, delusions that others are out to get the user. In this state, heavy users may become violent.

Nerve Cells Can Be Grouped into Three Functional Categories

Nerve cells can be categorized by structure or function. For our purposes, a functional classification is most useful. According to this system, nerve cells fall into three distinct groupings: (1) sensory neurons, (2) motor neurons, and (3) interneurons.

Sensory neurons carry impulses from **sensory receptors** in the body to the central nervous system. Sensory receptors come in many shapes and sizes and respond to a variety of stimuli, such as pressure, pain, heat, and movement (Chapter 18).

Motor neurons carry impulses from the brain and spinal cord to effectors, such as the muscles and glands of the body. Sensory information entering the brain and spinal cord via sensory neurons often stimulates motor neurons, creating a desired response.

In some cases, intervening neurons called **interneurons,** or **association neurons** are required to transmit impulses from the sensory neurons to motor neurons. Interneurons may also transmit impulses from sensory neurons to various parts of the CNS.

The importance of interneurons is underscored by the fact that they make up 99% of the neurons in the human central nervous system. As noted below, these neurons play an important role in coordinating complex activities. They are the neural communication network that transmits impulses from one part of the CNS to another to help bring about coordinated actions. And they are essential to many body reflexes.

EVOLUTION OF THE NERVOUS SYSTEM

Like many other systems, the nervous systems of the animal kingdom exist on a continuum from the simplest to the most advanced. Those classified as primitive generally consist of a network of interconnected neurons known as a **nerve net** (▻ Figure 17–15). In these systems, sensory information is conveyed to motor neurons, which transmit impulses to effectors. Primitive nervous systems lack a central nervous system.

Advanced nervous systems are characterized by the presence of central nervous systems that consist of aggregations of nerves that process and integrate information. In these organisms, a network of nerves akin to the nerve net conveys sensory and motor information to and from the CNS. In between the two extremes on the evolutionary continuum are a multitude of organisms with nervous systems representing the evolutionary progression from the simplest to the most complex.

As one might surmise, the more advanced a nervous system is, the more sophisticated an animal's behavior will be. In organisms with primitive nervous systems, for example, behavioral responses to light and heat are usually limited to increases or decreases in the rate of movement. In those animals with more advanced nervous systems, like humans, the responses are generally more complex and may even involve reasoning.

Interestingly, the evolution of the nervous system corresponds with the evolution of complexity in body structure, as with other systems discussed in earlier chapters. That is to say, as multicellular animal life developed, the nervous system became more complex.

Perhaps one of the most significant developments in the evolution of the nervous system was the emergence of adaptations that accelerate nerve impulse conduction. In invertebrates, the presence of fibers of very large diameter permitted rapid transmission of nerve impulses vital to survival in the environment. The giant nerve axons in invertebrates, for instance, permit the sudden movement of earthworms, crayfish, and squid vital to escape from predators. In vertebrates, saltatory conduction improved the ability to respond quickly to environmental stimuli. With this overview in mind, we next take a brief look at four nervous systems along the continuum, starting with the simplest.

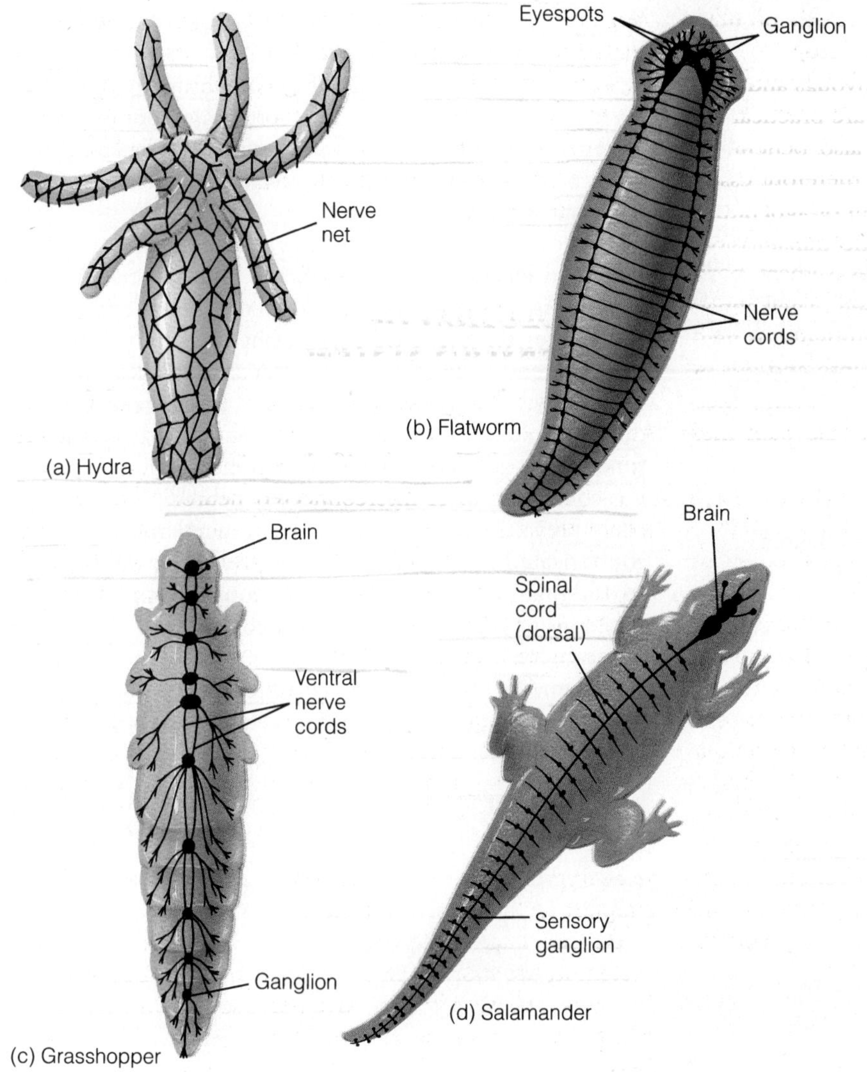

FIGURE 17–15 Overview of Evolution of the Nervous System (*a*) Jellyfish and other cnidarians like this hydra contain a network of nerves that respond to stimuli and evoke responses. (*b*) In flatworms, the simple nerve net is modified to consist of two nerve cords that run the length of the body, giving off fibers to muscles. An aggregation of nerves in the head region is called the ganglion. (*c*) In arthropods, like the grasshopper, further central nervous system development is evident. (*d*) Vertebrate nervous systems have distinct brains and spinal cords that give off nerves to body structures.

The Cnidarians Contain the Simplest Nervous Systems of all Animals

Figure 17–15a shows the structure of the nervous system of a cnidarian—a group including jellyfish and hydras. As illustrated, many cnidarians contain nerve nets, loosely organized networks of nerve cells. Lacking central control, the nervous system of cnidarians is largely based on reflex pathways between epithelial receptor cells and contractile cells in the body wall. In one pathway, which is concerned with feeding, nerve cells transmit impulses from receptor cells in the tentacles to contractile cells around the mouth.

Even in the relatively simple cnidarians, some evidence of centralization occurs. In jellyfish, for instance, special clusters of nerve cells are found near sensory structures. These nerve cell clusters represent very early signs of the formation of a central nervous system and are involved in tasks such as swimming.

Some Flatworms Contain Nerve Nets, but Others Contain Specialized Aggregations of Nerve Cells

Biologists believe that more complex nervous systems evolved from the nerve nets. In the simplest flatworms, nerve nets like those described above form the entire nervous system. In more highly developed flatworms, however, sensory structures such as the paired eyespots are concentrated near the anterior (front) end of the organism. This adaptation gives a crawling animal such as planaria an advantage because the forward end is the first to encounter stimuli. Not surprisingly, the head end also contains aggregations of nerve cells forming a brainlike structure known as a **ganglion.** Ganglia (plural) help coordinate responses to the stimuli flatworms encounter. Two nerve cords also run the length of the body. Attached to the ganglia, the nerve cords give off branches that supply muscles needed for swimming and crawling.

The Nervous Systems of Annelids and Arthropods Exhibit More Advanced Forms of Centralization

Annelids (such as the earthworm) and arthropods (such as the grasshopper) have aggregations of nerve cells in their front ends, forming rudimentary brains, and single nerve cords that run the length of their bodies, giving off branches to body parts. Both animals are segmented, and each segment is served by a pair of nerves and an aggregation of nerve cells (ganglion) located in the nerve cord. Several giant axons are found in the nerve cord as well. These axons transmit nerve impulses rapidly to muscles, permitting the otherwise slow-moving worm to retract its body quickly, as anyone who has ever tried to catch an earthworm at night knows.

The Vertebrate Nervous System Is the Most Advanced of all, but Even Within Vertebrates Considerable Differences are Observed

The nervous system of vertebrates represents the culmination of a long and successful biological progression. Consisting of well-defined central and peripheral portions, the vertebrate nervous system has undergone many changes in evolution. A few general comments are appropriate before we move on to the human nervous system, the most complex of all vertebrate systems.

Located on the back side of the organism and protected by the backbone and skull, the vertebrate central nervous system consists of a well-defined brain and spinal cord. The brain is divided into distinct regions with specialized structures and functions.

Three trends are evident in vertebrate evolution. The first is an increase in the size of the brain relative to body weight. As a general rule, the more complex an organism is, the greater the ratio of brain weight to body weight. The second trend is an increase in compartmentalization. That is, although vertebrate brains have three general parts, during evolution these parts have become structurally and functionally specialized. The third trend is toward increasing complexity of a region of the forebrain. This is especially prevalent in mammals. As a general rule, the more sophisticated an animal's behavior, the better developed is its forebrain, or cerebrum, which houses such specialized functions as hearing, vision, motor control, and ideation. In humans, for instance, 80% of the total brain mass is in the cerebrum. ▻ Figure 17–16 illustrates the evolution of the forebrain (indicated by color) in several vertebrates.

▻ FIGURE 17–16 Evolution of the Vertebrate Brain
One of the trends in the evolution of the vertebrate brain is the increase in the size of the forebrain. The forebrain, or cerebrum, houses many important sensory and motor functions. This trend is especially evident in mammals, represented here by the cat and human.

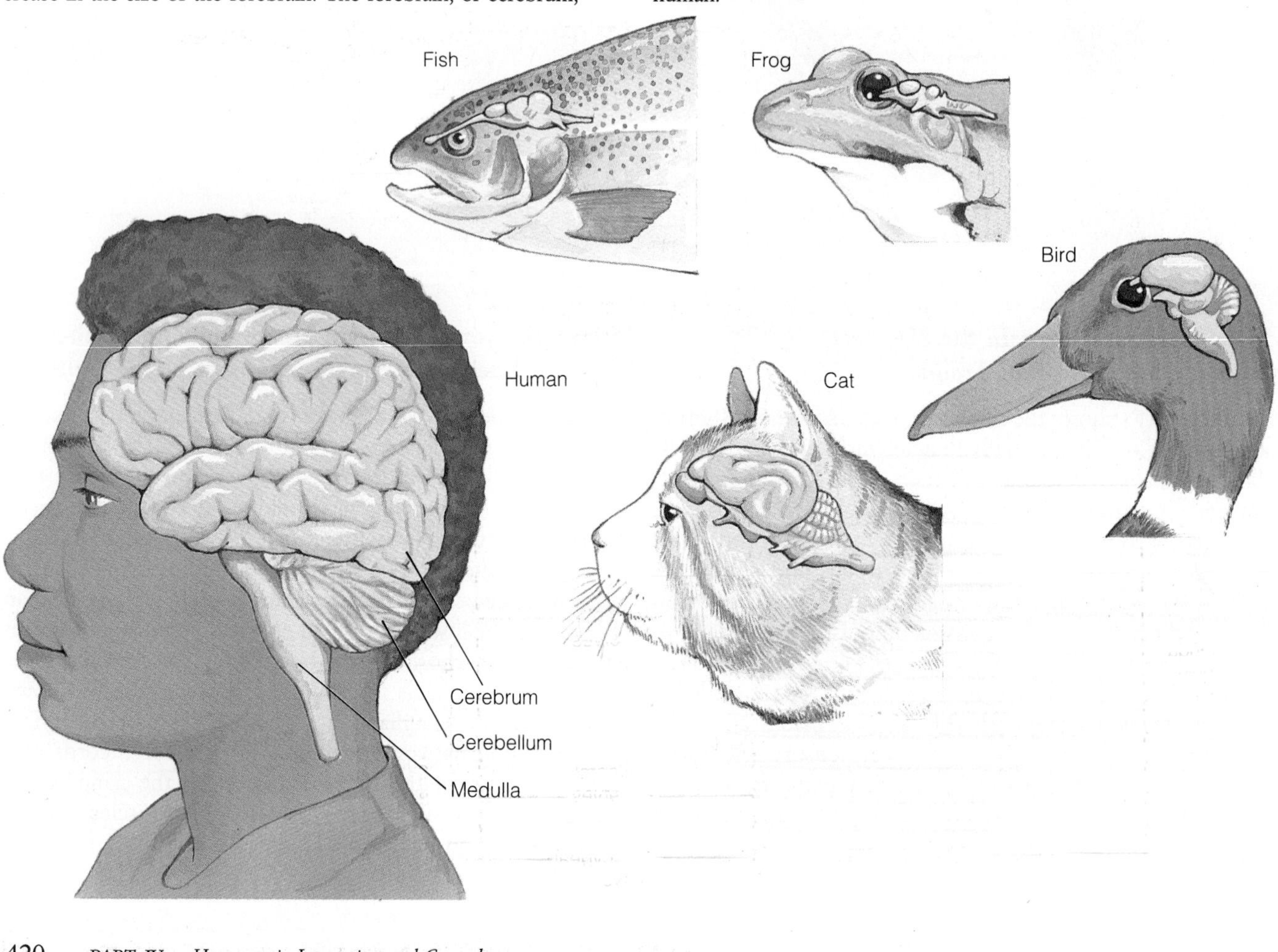

THE SPINAL CORD AND NERVES

With your understanding of the neuron and this brief overview of the evolution of nervous systems, we can now turn our attention to the nervous system of humans, beginning with the spinal cord and nerves.

The Spinal Cord Is Part of the CNS, Transmits Information To and From the Brain, and also Houses Many Reflexes

The spinal cord is a long, ropelike structure about the diameter of a person's little finger. The spinal cord connects to the brain above and courses downward through the vertebral canal formed by the vertebrae (▻ Figure 17–17). The spinal cord gives off nerves along its course that innervate the skin, muscles, bones, and organs of the body. The spinal cord ends into the lower back (at about the level of the second lumbar vertebra), at which point it gives off a series of nerves called the **cauda equina** ("horse's tail"), which supply the lower sections of the body.

As shown in ▻ Figure 17–18, the central portion of the spinal cord is an H-shaped zone of gray matter. **Gray matter** consists of nerve cell bodies of interneurons and motor neurons and is so named because it appears gray to the naked eye. Surrounding the gray matter are fiber tracts consisting of axons and a much smaller number of den-

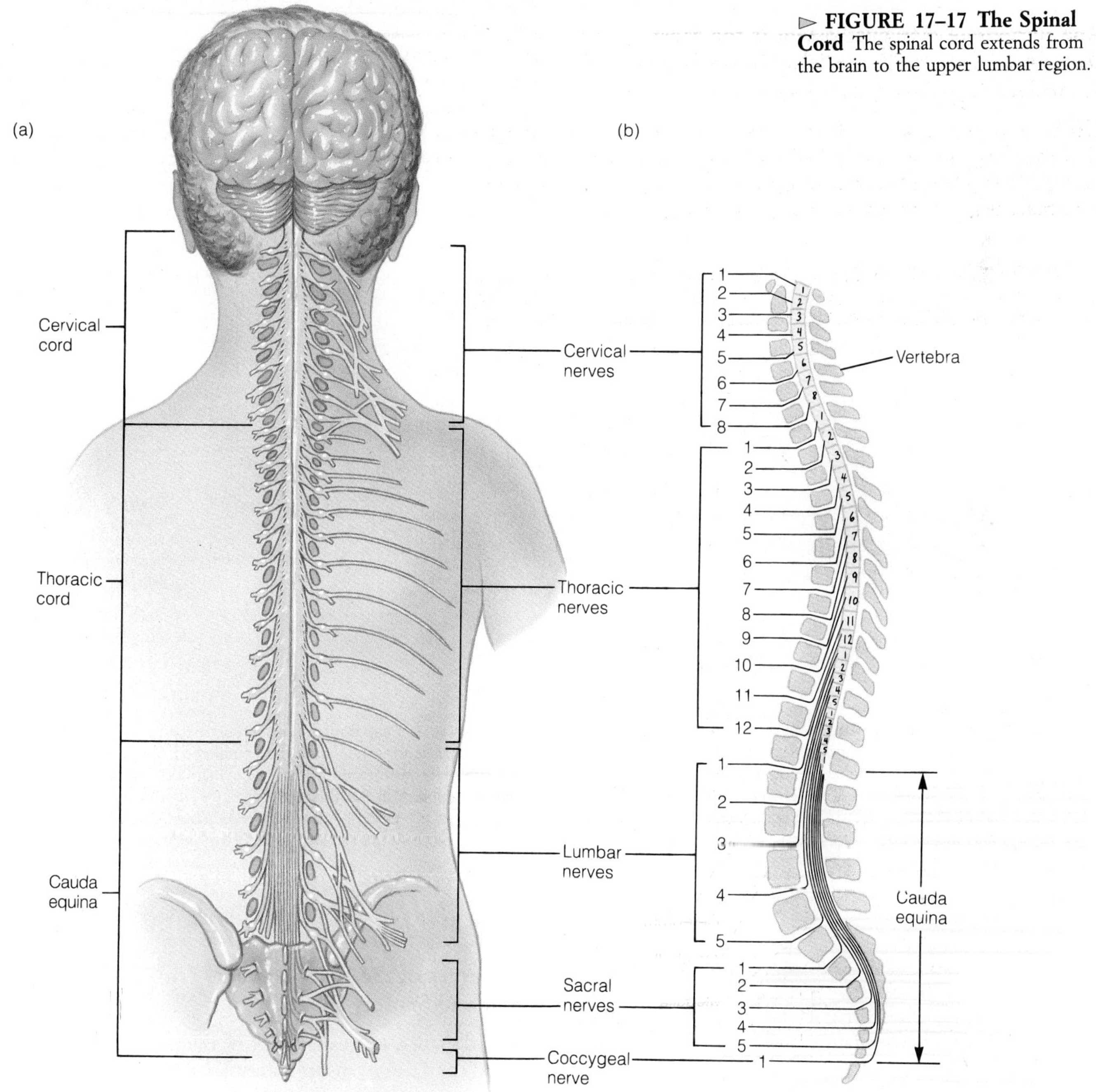

▻ **FIGURE 17–17 The Spinal Cord** The spinal cord extends from the brain to the upper lumbar region.

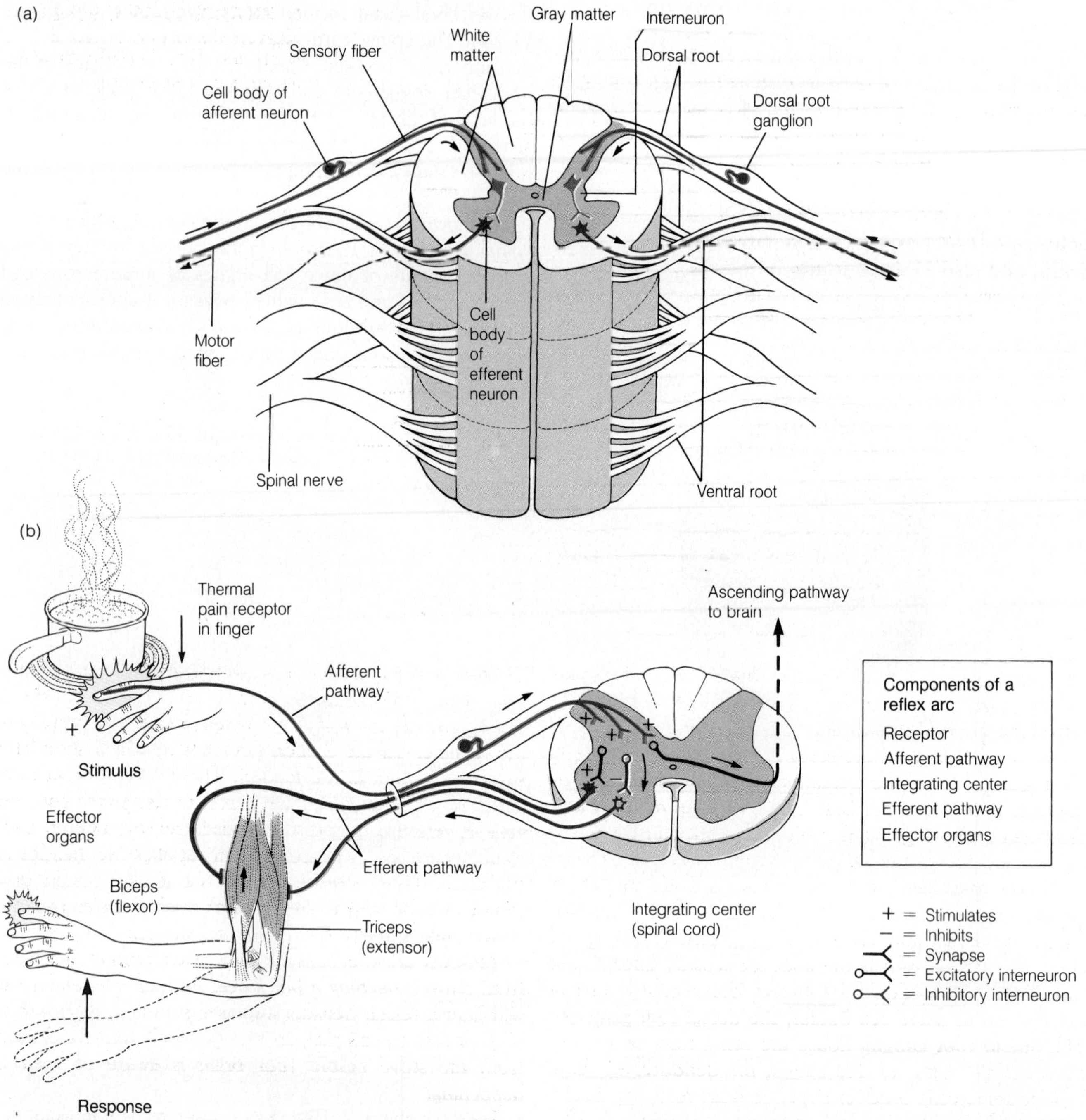

FIGURE 17–18 The Spinal Cord and Dorsal Root Ganglia (*a*) Spinal nerves are attached to the spinal cord by two roots, the dorsal and ventral roots. The dorsal root carries sensory information into the spinal cord. The ventral root carries motor information out of the spinal cord. The spinal nerve often contains both sensory and motor fibers. (*b*) When you accidentally touch a hot pan on the stove, you withdraw your hand before the brain even knows what's happening. This occurs because of a reflex arc. Sensory fibers send impulses to the spinal cord. The sensory impulses stimulate motor neurons in the spinal cord. This causes muscle contraction in the flexor muscles (+) and inhibits muscle contraction in the extensor muscles (−), allowing you to withdraw your hand. Nerve impulses also ascend to the brain to let it know what is happening.

drites that travel up and down the spinal cord, carrying information to and from the brain. The fiber tracts form the **white matter,** whose characteristic color comes from the myelin sheaths of the many axons coursing through this portion of the CNS.

The Nerves of the PNS Contain Motor and Sensory Fibers

As noted earlier, nerves are part of the peripheral nervous system. They carry sensory information to the spinal cord

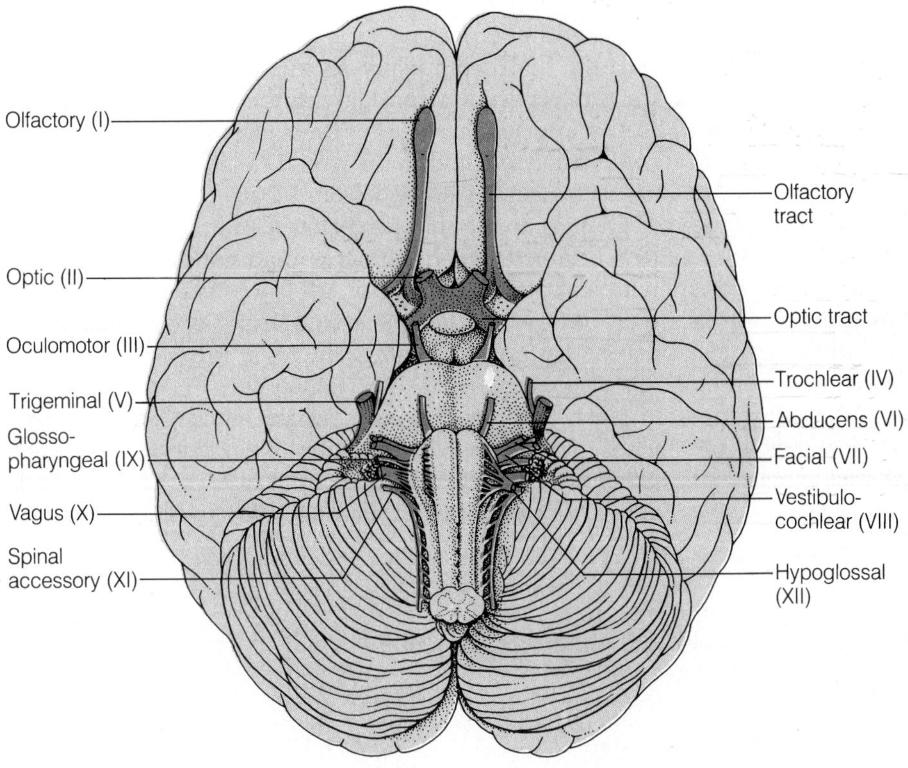

▶ **FIGURE 17–19 Cranial Nerves** The 12 pairs of cranial nerves arise from the underside of the brain and brain stem.

and brain and motor information out. Some nerves are strictly motor and some are strictly sensory, but many are mixed, having both motor and sensory fibers.

Nerves arising from the brain are called **cranial nerves** (▶ Figure 17–19). The cranial nerves supply structures of the head and several key body parts, such as the heart and diaphragm. The trigeminal nerves mentioned in the introduction are one of the 12 pairs of cranial nerves.

Nerves associated with the spinal cord are known as **spinal nerves** (Figure 17–18a). Each spinal nerve has two roots, the dorsal and ventral, which attach to the spinal cord. The **dorsal root** is the inlet for sensory information traveling to the spinal cord. On each dorsal root is a small aggregation of nerve cell bodies, the dorsal root ganglion. The **dorsal root ganglia** house the cell bodies of sensory neurons. As Figure 17–18a shows, the dendrites of these bipolar nerve cells conduct impulses from receptors in the body to the dorsal root ganglia; the axons, in turn, carry the impulse into the spinal cord along the dorsal root.

As noted earlier, sensory fibers entering the spinal cord often end on interneurons (Figure 17–18a). Interneurons receive input from many sensory neurons and process this information, acting like a receptionist in a busy corporate office. Interneurons transmit the impulses to nearby multipolar motor neurons, whose axons leave in the ventral root of the spinal nerve, carrying impulses to muscles and glands. This anatomical arrangement of neurons allows information to enter and leave the spinal cord quickly and forms the basis of the **reflex arc,** a neuronal pathway by which sensory impulses from receptors reach effectors without traveling to the brain (Figure 17–18b). Some reflex arcs contain interneurons, and some do not.

When a physician taps a rubber hammer on the tendon just below your kneecap (patellar tendon), he or she is testing one of your body's many reflex arcs. The tapping of the hammer on the tendon stretches it, which stimulates stretch receptors in the tendon. These receptors, in turn, generate nerve impulses that travel to the spinal cord via sensory neurons. In this reflex, each sensory neuron ends directly on a motor neuron, which supplies the muscles of the thigh. Thus, a quick tap on the tendon results in a motor impulse sent to the anterior thigh muscles (quadriceps), causing them to contract and the knee to jerk.

Reflexes are mechanisms that often protect the body from harm. Touching a hot stove, for example, elicits the withdrawal reflex. Sensory impulses stimulate the muscles of your arm to contract, so that you pull your hand away from the stove before your brain is aware of what is happening.

Babies come equipped with a number of important reflexes. Rub your finger on the cheek of a newborn, and it immediately turns its head toward your finger. This reflex helps babies find the mother's nipple. Crying is also a reflex. When a baby is hungry, thirsty, wet, or uncomfortable, it cries, a reflex sure to get attention.

The spinal cord also serves as a route by which sensory information is transmitted to the brain. Sensory impulses travel along special tracts lying outside the central H-shaped zone of the spinal cord. Because of this arrangement, sensory information that enters via the spinal nerves can also reach vital brain centers. Although the incoming sensory information may elicit a reflex, the brain is still informed of the problem, allowing for appropriate follow-up action.

The Severity of a Spinal Cord Injury Depends on the Location of the Injury and the Extent of Damage

An automobile crash, a bullet wound, or even a bad fall can damage or sever the fiber tracts of the spinal cord. Injury to the cord usually results in permanent damage, because, as noted previously, axons of the central nervous system do not usually regenerate.

The amount of damage depends on where the cord is injured and the severity of the injury. Sensory fibers traveling to the brain and motor fibers traveling from the brain to the spinal cord run in separate tracts in the white matter. If both tracts are severed—for example, by a severe vertebral fracture—all sensory and motor functions below the level of the injury are lost. As a result, muscles supplied by nerves below the injury become paralyzed and are unable to contract voluntarily. Segments of the body below the injury lose all sensation.

If the spinal cord injury occurs high in the neck (above the fifth cervical vertebra), it cuts the nerve fibers traveling to the muscles that control breathing. These muscles become paralyzed, and the person dies fairly quickly. This is how a hangman's noose kills its victim.

Damage to the cord just below the fifth cervical vertebra (the fifth vertebra in the neck) does not affect breathing, but paralyzes the legs and arms. The condition is known as **quadriplegia.** If the spinal cord injury occurs below the nerves that supply the arms, the result is **paraplegia,** paralysis of the legs.

New studies at the Craig Rehabilitation Center in Denver show that the administration of large doses of an anti-inflammatory steroid drug shortly after a head or neck injury stops swelling of the spinal cord. Swelling is thought

TABLE 17–1 Summary of Brain Structures and Functions

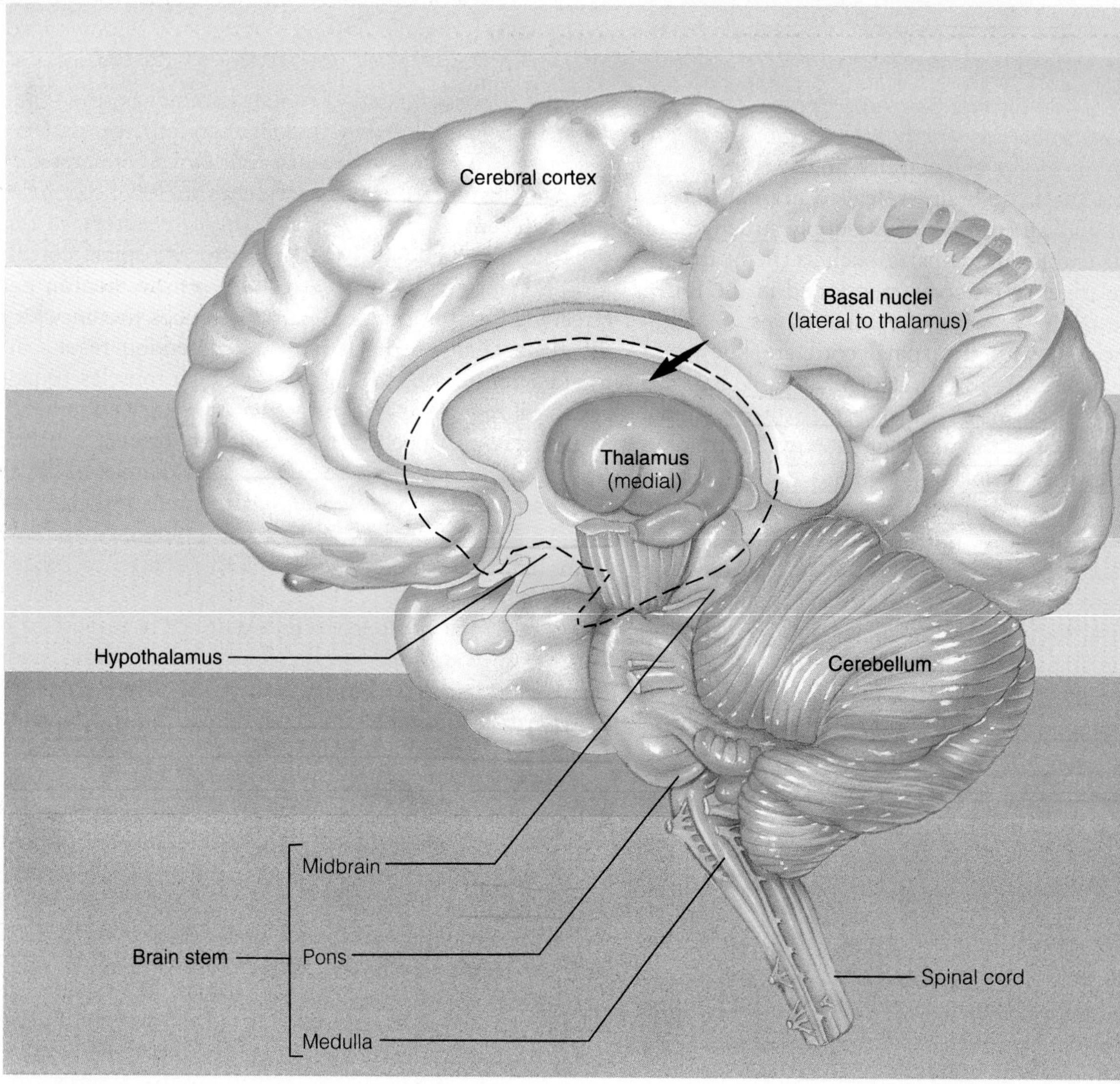

BRAIN COMPONENT
Cerebral cortex
Basal nuclei
Thalamus
Hypothalamus
Cerebellum
Brain stem *(midbrain, pons, and medulla)*

to greatly increase the amount of damage resulting from an injury to the spinal cord. This treatment, therefore, may greatly reduce paralysis following an injury to the spinal cord.

Research is also under way to find ways to stimulate the regeneration of axons in the brain and spinal cord. Some early results are promising, suggesting that one day physicians may be able to offset spinal cord damage.

THE BRAIN

About the size of a cantaloupe, the human brain is an extraordinary organ, a product of many millions of years of evolutionary trial and error. Responsible for producing art and music and remarkable feats of engineering and abstract reasoning, it also allows us to ponder right and wrong and remember a vast amount of information. Table 17–1 provides an overview of the structures and functions of the major components of the brain.

TABLE 17–1 (*continued*)

MAJOR FUNCTIONS
1. Sensory perception 2. Voluntary control of movement 3. Language 4. Personality traits 5. Sophisticated mental events, such as thinking, memory, decision making, creativity, and self-consciousness
1. Inhibition of muscle tone 2. Coordination of slow, sustained movements 3. Suppression of useless patterns of movement
1. Relay station for all synaptic input 2. Crude awareness of sensation 3. Some degree of consciousness 4. Role in motor control
1. Regulation of many homeostatic functions, such as temperature control, thirst, urine output, and food intake 2. Important link between nervous and endocrine systems 3. Extensive involvement with emotion and basic behavioral patterns
1. Maintenance of balance 2. Enhancement of muscle tone 3. Coordination and planning of skilled voluntary muscle activity
1. Origin of majority of peripheral cranial nerves 2. Cardiovascular, respiratory, and digestive control centers 3. Regulation of muscle reflexes involved with equilibrium and posture 4. Reception and integration of all synaptic input from spinal cord; arousal and activation of cerebral cortex 5. Sleep centers

The Cerebral Hemispheres Function in Integration, Sensory Reception, and Motor Action

The brain is housed in the skull, a bony shell that protects it from injury. ▻ Figure 17–20a shows some of the externally visible parts of the human brain. The largest and most conspicuous is the **cerebrum,** a convoluted mass of nervous tissue that, as noted earlier, constitutes about 80% of the total brain mass.

The cerebrum is divided into two halves, the right and left **cerebral hemispheres.** Each hemisphere has a thin outer layer of gray matter, the **cerebral cortex** (Figure 17-20b). The cerebral cortex contains glial cells and the cell bodies of numerous multipolar neurons. Lying beneath the cerebral cortex is a thick central core of white matter. It contains bundles of myelinated axons that give it a white appearance. The axons transmit impulses from one part of the cerebral cortex to another, permitting the integration of the activities of various parts of the cortex. The axons of the white matter also carry impulses from the cerebral cortex to the spinal cord, permitting information to flow from the brain to motor neurons that control many of the body muscles.

As shown in Figure 17–20a, the cerebral cortex is thrown into numerous folds, called **gyri** (singular, *gyrus*). The gyri are separated by numerous valleys, known as **sulci** (singular, *sulcus*).

Deep within the white matter of the cerebral hemispheres are masses of gray matter known as the **basal ganglia.** The nerve cells of the basal ganglia inhibit motor function, helping fine-tune muscular control.

The Cerebrum Is Divided into Four Major Lobes, and Each Lobe Contains Areas that House Specific Functions. The cerebral hemispheres can be divided into four major **lobes,** shown in Figure 17–20a. They are the frontal lobe, the parietal lobe, the occipital lobe, and the temporal lobe. Within each of these lobes are areas that house specific functions, illustrating that the brain did not escape the evolutionary forces that resulted in compartmentalization and specialization (▻ Figure 17–21). The functional areas within the lobes can be broadly classified into three types: motor cortex, sensory cortex, and association cortex. The **motor cortex** stimulates muscle activity. The **sensory cortex** receives sensory stimuli, and the **association cortex** integrates information, bringing about coordinated responses.

Figure 17–21 shows the major cortical areas and lists some of their functions. As you can see, the cortex contains areas for receiving sound (primary auditory cortex) and generating speech (Broca's area), areas that receive input from the eyes (primary visual cortex), and regions that receive sensory information from the body (primary sensory cortex). It also contains regions that control muscle

▻ **FIGURE 17–20 The Brain** (*a*) The cerebral cortex consists of lobes as shown here. The lobes, in turn, can be divided into sensory, association, and motor areas (not shown). (*b*) A section through the brain showing the gray and white matter of the cortex and deeper structures.

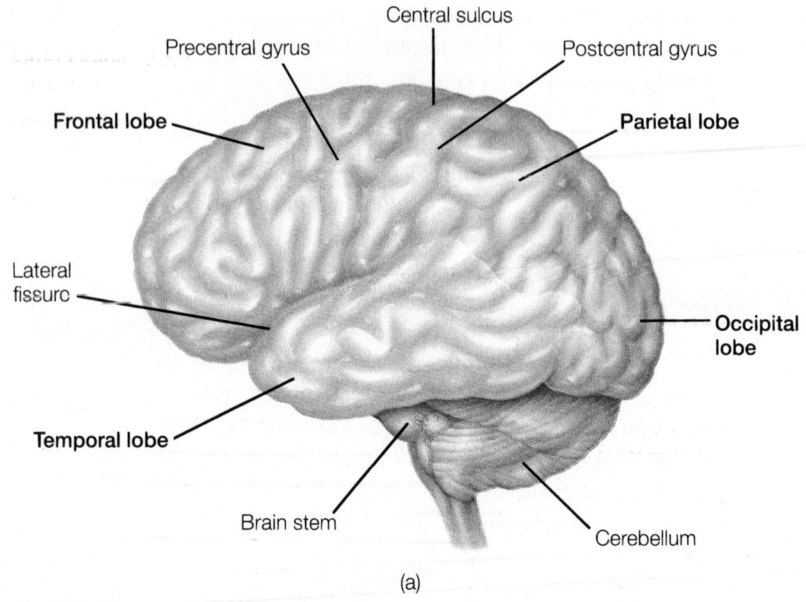

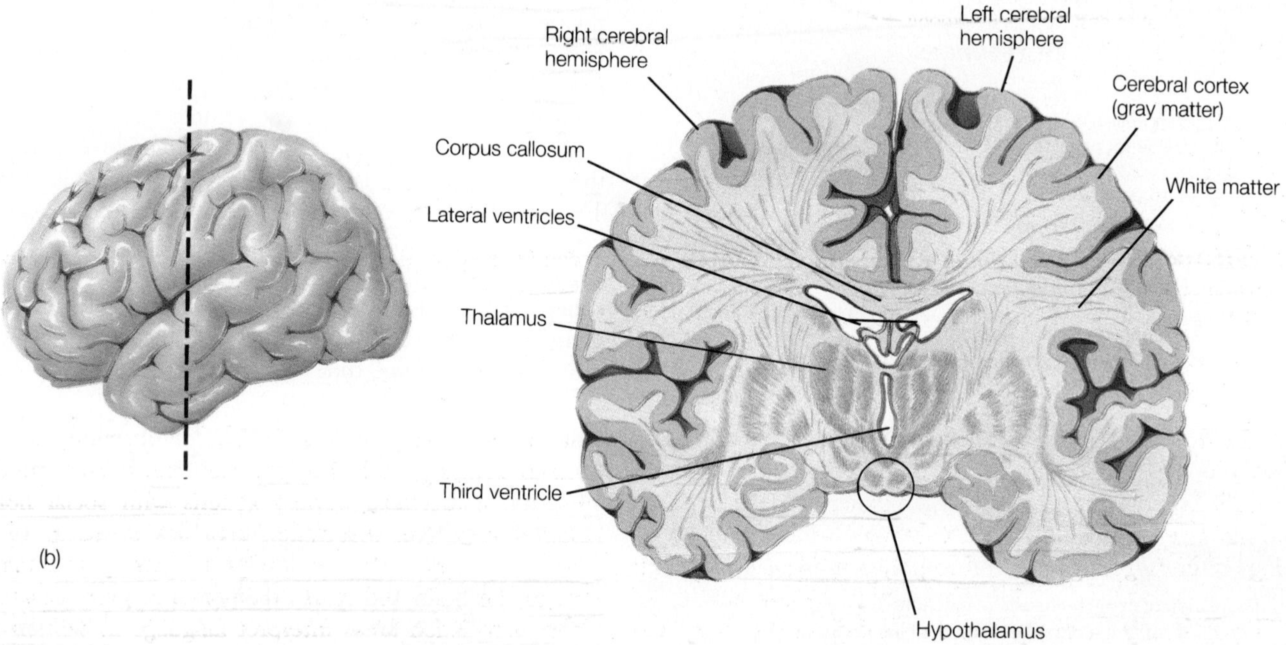

movement and areas that allow for planning (prefrontal association cortex).

The following section discusses some of the major areas of the cortex, broadening your understanding of how the brain works. We begin with the primary motor cortex.

The Primary Motor Cortex Controls Voluntary Movement. The **primary motor cortex** occupies a single gyrus (ridge) on each hemisphere, just in front of the central sulcus (Figure 17–21). The primary motor cortex controls voluntary motor activity—for example, the muscles in your hand that are turning the pages of your book.

The neurons in the primary motor cortex are arranged in a very specific order. Thus, each region of the motor cortex controls a specific body part. As illustrated in ▻ Figure 17–22a, the neurons that control the muscles of the knee are located in the uppermost region of the primary motor cortex. Hip muscle control occurs below that. Muscles of the hand are controlled by neurons located even lower.

To bring about a voluntary movement, a conscious thought stimulates the generation of nerve impulses in the primary motor cortex. The impulse then travels from the brain down the spinal cord to the motor neurons in the spinal cord. The axons of these neurons leave the spinal cord, traveling in the nerves, and terminate on the muscles of the body.

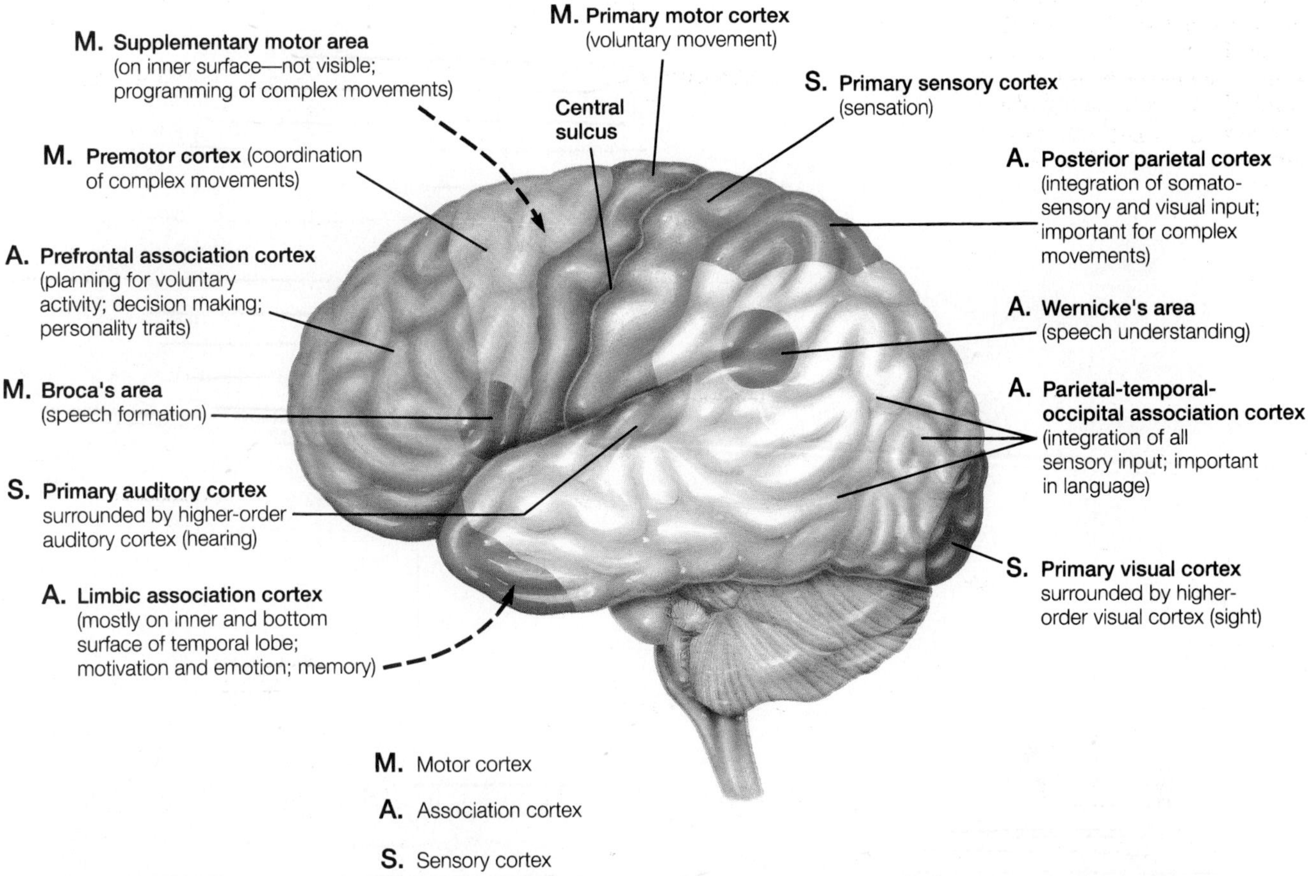

▻ **FIGURE 17–21 Functional Regions of the Cortex** The cerebral cortex has three principal functions: receiving sensory input, integrating sensory information, and generating motor responses. Special sensory areas handle vision, smell, taste, and hearing.

In front of the primary motor area is the **premotor cortex.** The premotor area is also involved in controlling muscle contraction. However, the movements that the premotor cortex controls are less voluntary—for example, typing or the fingering required to play a musical instrument.

The Primary Sensory Cortex Receives Sensory Information From the Body. Just behind the central sulcus is another long ridge of tissue running parallel to the primary motor area. Known as the **primary sensory cortex,** it is the destination of many sensory impulses traveling to the brain. As in the primary motor area, different parts of this ridge correspond to different parts of the body (Figure 17–22b). Electrical stimulation of the primary sensory area will elicit sensations that appear to be coming from specific body parts.

The Association Cortex Is the Site of Integration and Sometimes Houses Complex Intellectual Activities. The **association cortex** consists of large expanses of cerebral cortex where integration occurs. In the frontal lobe is a region of the association cortex (prefrontal association cortex) that houses complex intellectual activities, such as planning and ideation. This region also modifies behavior, conforming human actions with social norms. Another important association area lies posterior to the primary sensory cortex. It interprets sensory information sent to the brain and stores memories of past sensations. Other association areas interpret language in written and spoken form.

Hearing, Vision, Taste, and Smell Are also Housed in Specific Cortical Regions. Figure 17–21 also illustrates the presence of patches of sensory cortex for hearing (primary auditory cortex) and vision (primary visual cortex). Although they are not shown, separate regions also exist for taste and smell.

Unconscious Functions Are Housed in the Cerebellum, Hypothalamus, and Brain Stem

Consciousness resides in the cerebral cortex. However, many body functions occur at an unconscious level. Heartbeat, breathing, and many homeostatic functions, for ex-

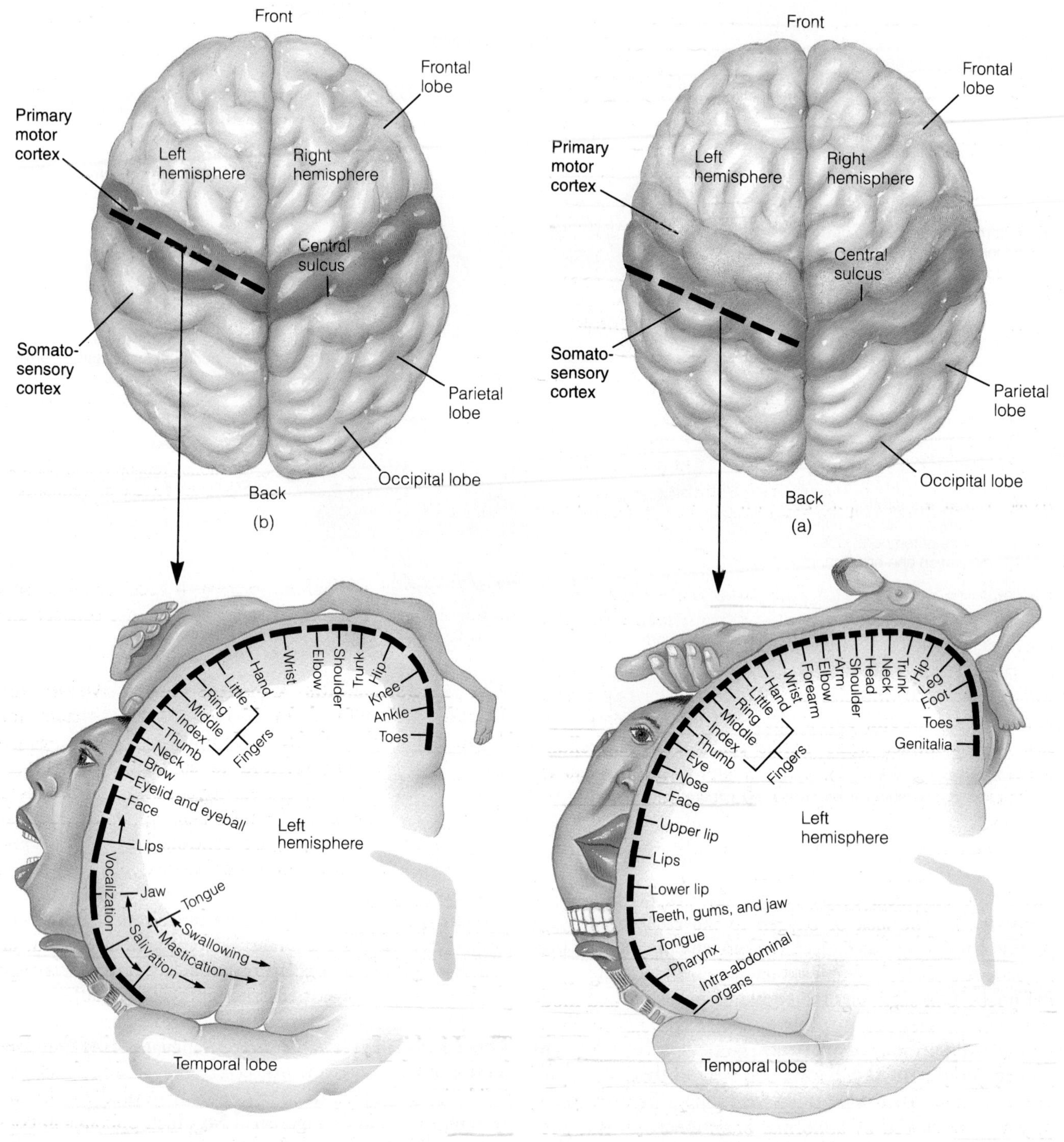

▻ **FIGURE 17–22 Primary Motor and Sensory Areas** (*a*) Top view of the brain showing the primary motor cortex. Below is a map of the location of motor functions within the primary motor cortex. (*b*) Cross section through the primary sensory cortex (somatosensory cortex). Below is a map of the location of regions within the primary sensory cortex.

ample, all take place without conscious control.[8] One region of the brain that controls unconscious actions is the cerebellum. The **cerebellum** is the second largest structure of the brain. As Figure 17–21 shows, the cerebellum sits below the cerebrum on the brain stem.

[8] Conscious control is possible for many automatic functions, but for the most part, breathing and swallowing are controlled automatically in lower brain centers.

The Cerebellum Controls Muscle Synergy and Helps Maintain Posture. The cerebellum plays several key roles. One of them is synergy. Neurophysiologists define **synergy** as the coordination of skeletal muscle contrac-

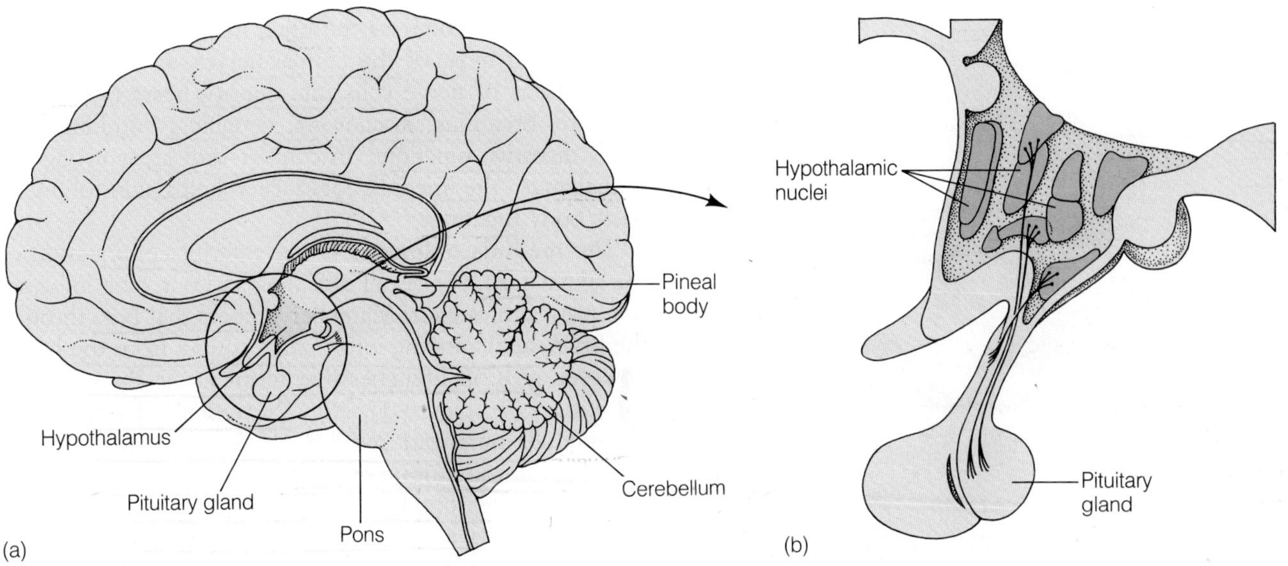

▻ **FIGURE 17–23 Cross Section of the Brain** (*a*) The hypothalamus is clearly visible in a cross section of the brain. (*b*) The hypothalamus is at the base of the brain, just above the pituitary gland. It regulates many autonomic functions and plays a particularly important role in controlling the release of hormones by the pituitary.

tion and the movement of body parts. To understand the concept, first hold your arm straight out, then bring your hand to your chest. To perform this simple action, the biceps (muscles in the front of the upper arm) contracted while the triceps (muscles in the back of the upper arm) relaxed. For smooth, coordinated movement, some muscles must contract while others relax. The cerebellum, when operating normally, ensures that opposing sets of muscles work together to bring about smooth motion, or synergy.

Damage to the cerebellum may occur during childbirth if the blood supply to the baby is interrupted—for example, if the umbilical cord accidentally wraps around the baby's neck. The lack of oxygen to the brain may cause permanent damage to the cerebellum, resulting in a loss of a synergistic control of skeletal muscles. Mild damage generally results in slight rigidness (called spasticity) and moderately jerky motions. More severe damage can result in serious impairment, with body motions becoming extremely jerky and simple tasks requiring several attempts. This condition is known as **cerebral palsy.** Cerebral palsy may also be caused by abnormal brain development, possibly due to exposure to harmful chemicals (such as alcohol) during embryonic development. Many children with this disorder have some degree of mental retardation, although some others are highly intelligent.

The cerebellum also helps maintain posture. It receives impulses from sense organs in the ear that detect body position. The cerebellum sends impulses to the muscles to maintain or correct posture.

The Thalamus Is a Relay Center. Just beneath the cerebrum is a region of the brain called the thalamus (see Figure 17–20). The **thalamus** is a relay center, like a switchboard in a telephone system. It receives all sensory input, except for smell, then relays it to the sensory and association cortex.

The Hypothalamus Controls many Autonomic Functions Involved in Homeostasis. Beneath the thalamus is the hypothalamus. It consists of many aggregations of nerve cells, referred to as **nuclei**[9] (▻ Figure 17–23). The nuclei control a variety of autonomic functions all aimed at maintaining homeostasis. Appetite and body temperature, for example, are controlled by some of the hypothalamic nuclei, as are water balance, blood pressure, and sexual activity.

The hypothalamus is a primitive brain center. Perhaps its best known function is control of the pituitary gland, an endocrine gland that regulates many body functions through the hormones it releases (described in Chapter 20).

The Limbic System Is the Site of Instinctive Behavior and Emotion. Instincts, among other functions, reside in a complex array of structures called the **limbic system,** shown in ▻ Figure 17–24, which operates in conjuction with the hypothalamus. Instincts are among the most fundamental responses of organisms. They include the protective urge of a mother, the territorial assertions of a male, and the fight-or-flight response an animal experiences in the face of adversity.

The limbic system also plays a role in emotions—fear, anger, and so on. Electrodes in some areas of the limbic system may elicit primitive rage. In other areas, electrical stimulation elicits placidity. Stimulation of specific regions within the limbic system of humans may elicit sensations

[9] These are not to be confused with the nuclei of cells.

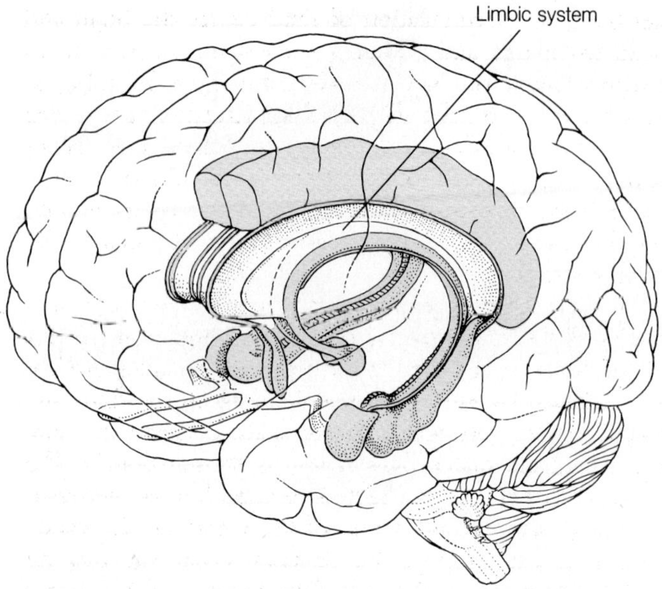

▻ **FIGURE 17–24 The Limbic System** This odd assortment of structures is the seat of emotions and instincts, among other functions.

patients describe as joy, pleasure, fear, or anxiety, depending on the site of stimulation.

Many Basic Body Functions Are Controlled by the Brain Stem. The **brain stem** controls additional primitive functions and consists of three parts, the medulla oblongata, the pons, and the midbrain (▻ Figure 17–25). The brain stem connects the brain to the spinal cord and gives rise to most of the cranial nerves. It also contains aggregations of nerve cells that control heart rate, blood pressure, breathing, swallowing, coughing, vomiting, and many digestive functions. In concert with areas in the hypothalamus, the medulla oblongata helps control many basic body functions.

The **medulla oblongata** is the anterior continuation of the spinal cord. Nearly all incoming and outgoing information passes through it. Many nerve fibers that pass through the brain stem carrying information to and from the brain, however, give off branches that terminate in the **reticular formation** (▻ Figure 17–26). The reticular formation of the medulla oblongata receives all incoming and outgoing information, monitoring activity like a security guard at a doorway. This information is then projected to the cerebral cortex via special nerve fibers that activate cortical neurons, in much the same way that a security guard might keep the "boss" informed of the comings and goings at the door. These nerve fibers are referred to as the **reticular activating system (RAS).**

Nerve fibers of the RAS help maintain wakefulness or alertness. When we are asleep, the flow of information through the RAS is greatly reduced, and the cortex "sleeps." Sleep may be disturbed by a biting insect, which stimulates sensory nerves in the skin. These nerves transmit impulses to the spinal cord, which then transmits impulses to the brain. Impulses reaching the RAS travel to the cortex and may awaken the sleeper.

The RAS awakens us and keeps us alert, but it may also

▻ **FIGURE 17–25 Brain Stem** The brain stem is a primitive portion of the brain and consists of the midbrain, pons, and medulla oblongata.

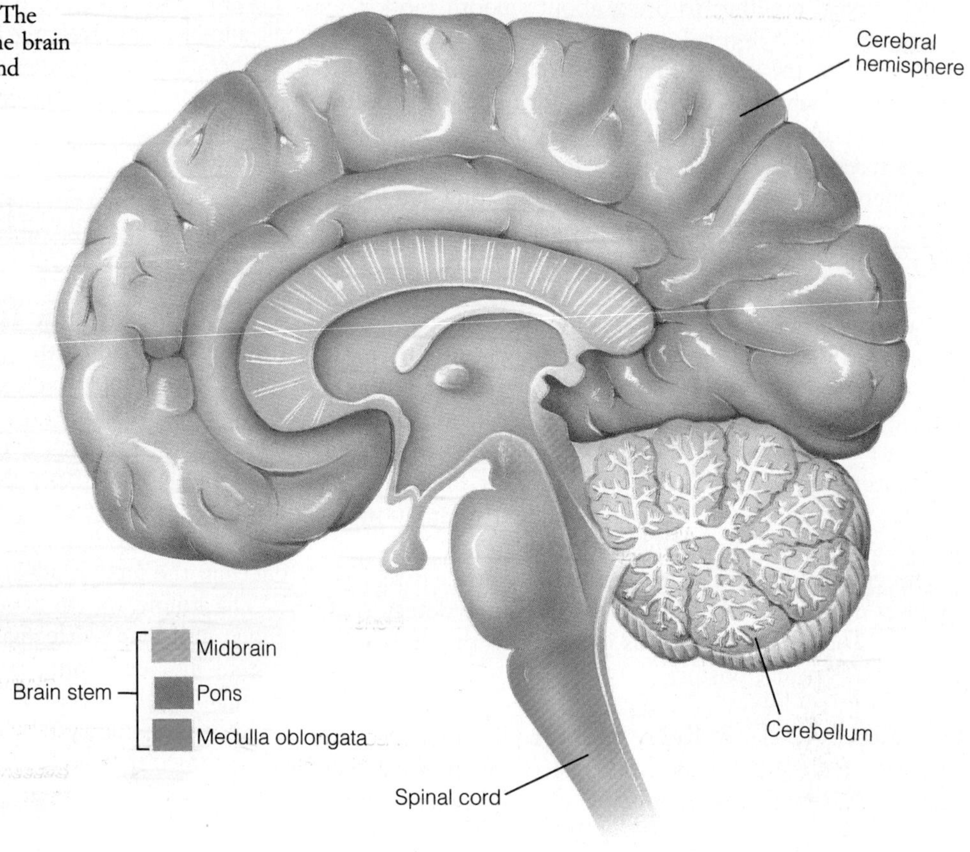

prevent us from falling asleep from time to time. Pain from a bad sunburn, for example, prevents many people from falling asleep. That is because pain impulses traveling to the brain from sensors in the skin enter the RAS, generating impulses that travel to the cortex and keep the cells active despite one's fatigue.

Cerebrospinal Fluid Cushions the Central Nervous System

A watery fluid, similar to blood plasma and interstitial fluid, surrounds the brain and spinal cord, filling the arachnoid space between the inner and outer layers of the meninges and also filling internal cavities in these organs (see Figure 17–3). This fluid, called **cerebrospinal fluid (CSF),** serves as a cushion that protects the brain and spinal cord from traumatic injury. Cerebrospinal fluid also fills special cavities within the brain known as **ventricles** (▻ Figure 17–27).

Cerebrospinal fluid is produced by the highly vascularized lining of the lateral ventricles and is slowly drained out of the brain and into the bloodstream. Generally, the amount produced equals the amount removed. If the flow of CSF is blocked, however, fluid builds up in the brain, causing damage. If the circulation of CSF is blocked in a newborn, the accumulation of fluid causes the brain and head to enlarge and produces a condition referred to as **hydrocephalus** (literally, "water on the brain") (▻ Figure 17–28). The ventricles fill with fluid, stretching the cortex and permanently damaging brain cells. If the damage is not severe, surgeons can treat hydrocephalus by implanting a plastic tube that drains the fluid from the ventricles into the thoracic cavity. This draining permits the brain to develop normally.

Because CSF bathes the central nervous system, examination of the fluid provides a means of detecting infections of the brain, spinal cord, and meninges. Samples of CSF are obtained by inserting a needle between the third and fourth lumbar vertebrae into the space between the dura mater and pia mater. This region is chosen because the spinal cord ends at the first or second lumbar vertebra. Should a needle contact one of the nerves of the cauda equina, which travel in the vertebral canal, the damage would probably be much less than if it pierced the spinal cord. CSF is then withdrawn and examined.

One dangerous condition detected by this method is **meningitis,** an infection of the meninges by certain viruses or bacteria, which, as noted in Chapter 14, usually begins in the respiratory system. Meningitis requires immediate

▻ **FIGURE 17–26 The Brain Stem and Reticular Activating System** The brain stem is composed of the pons and medulla (not labeled). The reticular formation resides in the brain stem, where it receives input from incoming and outgoing neurons. Fibers projecting from the reticular formation to the cortex constitute the reticular activating system.

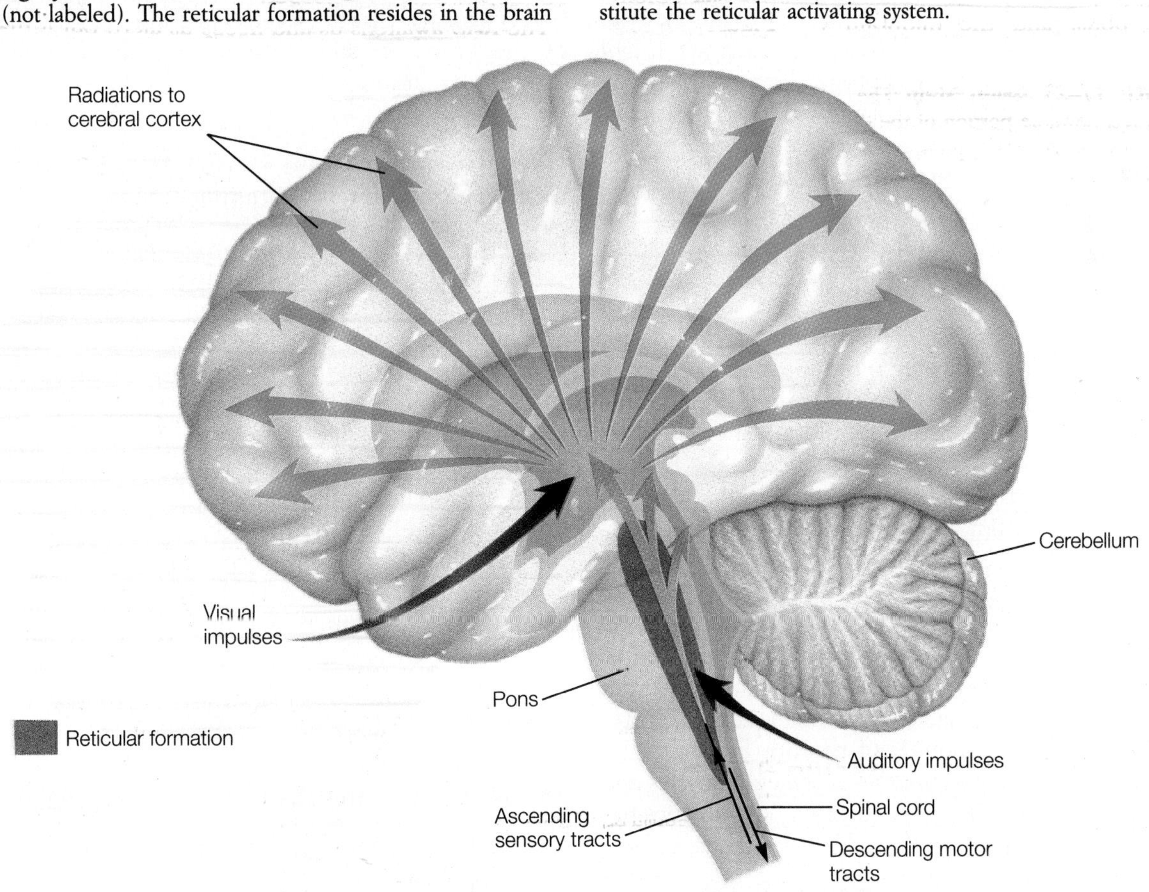

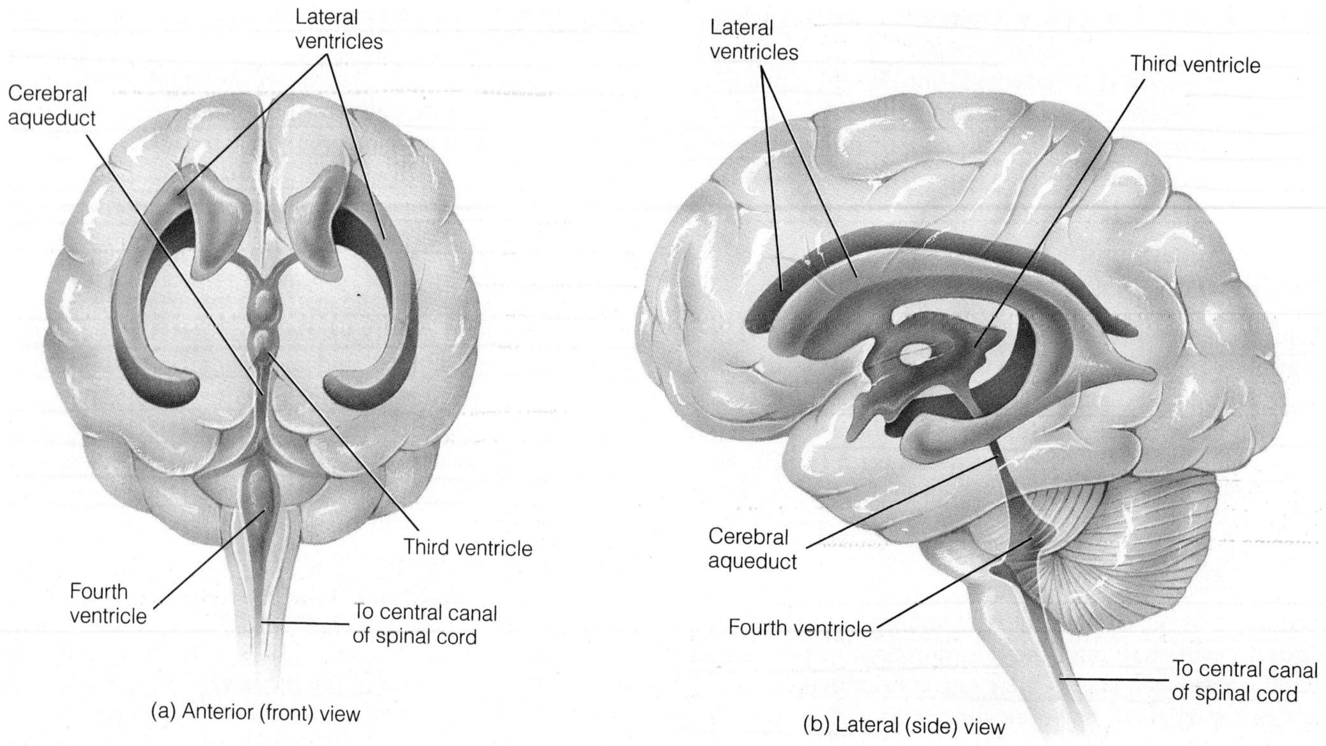

FIGURE 17–27 Ventricles
The brain contains four internal cavities, called ventricles, which are filled with a plasmalike fluid known as cerebrospinal fluid.

attention because it can result in sudden death. In adults, meningitis is characterized by fever, headache, nausea, vomiting, a stiff neck, and an inability to tolerate bright light (photophobia). These symptoms develop over the course of a few hours. In babies and children, the symptoms are often less obvious, causing parents to overlook a potentially fatal condition until it is too late.

Electrical Activity of the Brain Varies Depending on Activity Level or Level of Sleep

Electrodes applied to different parts of the scalp detect underlying electrical activity in the brain and produce a tracing known as an **electroencephalogram (EEG).** Some sample tracings of the electrical activity at different times are shown in Figure 17–29. As illustrated, the type of brain wave recorded depends on one's level of cortical activity. The waves often appear irregular, but distinct patterns can sometimes be observed, especially during the different phases of sleep.

EEGs are used to diagnosis brain dysfunction, because diseases or injuries of the cerebral cortex often give rise to altered EEG patterns. Perhaps the most common disorder is **epilepsy.** Epilepsy is characterized by periodic seizures that occur when a large number of neurons fire spontaneously, producing involuntary spasms of skeletal muscles and alterations in behavior. Intensive research on epileptics has shown that in about two-thirds of the cases, patients have no identifiable structural abnormality in the brain. In the remaining group, epileptic seizures can usually be traced to brain damage at birth, severe head injury, or inflammation of the brain. In some instances, epileptic seizures are caused by brain tumors.

Headaches Have Many Causes But Are Rarely the Result of Life-Threatening Anomalies

No discussion of the brain would be complete without considering headaches. **Headaches** are the most common form of pain and may result from one of several causes. For example, many headaches result from tension—sustained tightening of the muscles of the head and neck when a person is nervous, stressed, or tired. Headaches also commonly result from swelling of the membranes lining the sinuses—either from infections or allergies. Eyestrain is also a common cause of headaches, as is the dilation of cerebral blood vessels, which may be associated with high blood pressure or excessive alcohol consumption.

More serious causes of headache are increased intracranial pressure, resulting from a brain tumor or bleeding into the skull. Inflammation caused by an infection of the meninges (meningitis) or the brain itself (encephalitis), although rare, requires immediate attention.

THE AUTONOMIC NERVOUS SYSTEM

As noted earlier, the nervous system consists of two parts: the peripheral nervous system (the spinal and cranial

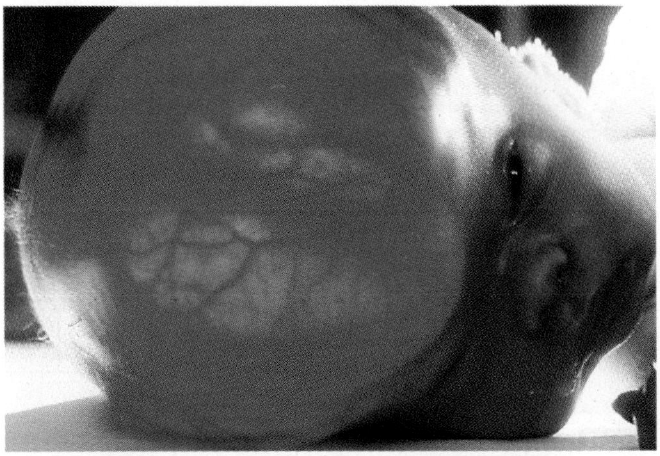

▻ **FIGURE 17–28 Hydrocephalus** This birth defect results from a blockage in the ventricles, which causes CSF to build up, thinning the cortex and causing severe brain damage.

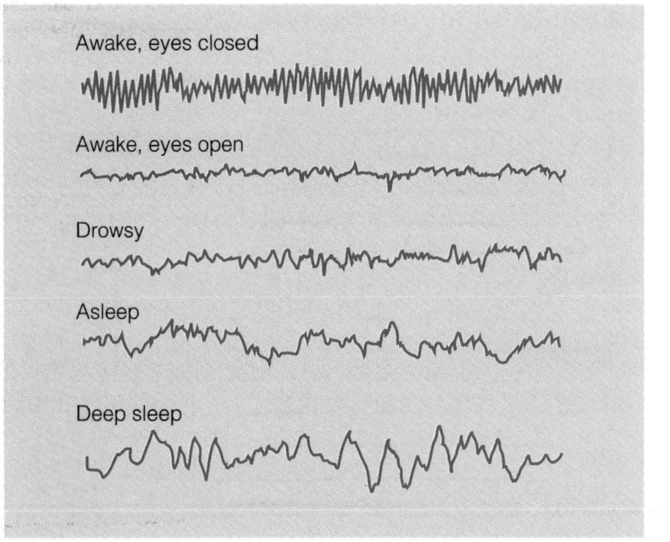

▻ **FIGURE 17–29 Electroencephalogram** Several sample tracings of brain waves during different activities and stages of sleep.

nerves) and the central nervous system (brain and spinal cord). The PNS consists of a somatic division and an autonomic division, both of which contain motor and sensory neurons (see Figure 17–4).

The **somatic division** receives sensory information from the skin, muscles, and joints, which it transmits to the CNS. Motor impulses from the CNS, in turn, are sent to skeletal muscles. In contrast, the **autonomic division** transmits sensory information from organs to the CNS, which delivers motor impulses to smooth muscle, cardiac muscle, and glands.

The autonomic nervous system (ANS) functions automatically, usually at a subconscious level. It innervates all internal organs and has two subdivisions: the sympathetic and the parasympathetic.

The **sympathetic division** of the autonomic nervous system functions in emergencies and is largely responsible for the **fight-or-flight response.** This response occurs when an individual is startled or faced with danger. A sudden scare results in sympathetic nerve impulses that bring about a whole host of responses, preparing an animal to fight or flee the scene. One change is a rapid increase in heart rate, which many of us have experienced when startled. The fight-or-flight response also results in pupil dilation and an increase in breathing rate. Blood flow to the intestines decreases, and blood flow to skeletal muscles increases, providing them with additional oxygen.

In contrast, the **parasympathetic division** of the ANS brings about internal responses associated with the relaxed state. It therefore reduces heart rate, contracts the pupils, and promotes digestion.

Most organs are supplied with both parasympathetic and sympathetic fibers. As a general rule, the parasympathetic and sympathetic fibers have antagonistic (or opposite) effects. One stimulates activity, and the other reduces activity. This provides the body with a means of fine-tuning organ function.

≈ LEARNING AND MEMORY

Few questions intrigue humans more than how the brain functions in learning and memory. **Learning** is a process in which an individual acquires knowledge and skills, usually from experience, instruction, or some combination of the two. Learning depends on one's ability to store information in the brain, a phenomenon called **memory.**

As most of us can attest, newly acquired knowledge is first stored in **short-term memory.** Short-term memory holds information for periods of seconds to hours. The phone number you just looked up or the name of a new acquaintance, for example, fall into short-term memory. If you don't do something to "fix" these memories, they will soon vanish (▻ Figure 17–30). Cramming for tests places a lot of information into short-term memory. Unfortunately, soon after the test is over, the information seems to fade into oblivion. The reason people often cannot recall details of a traumatic accident is because short-term memory has been wiped out.

Long-term memory, on the other hand, holds information for much longer periods—from days to years—and has a much greater storage capacity than short-term memory "banks."

Transferring information from short-term to long-term memory is called **consolidation.** In the *Study Skills* section at the front of the book, I described several ways to consolidate information, including repetition, mnemonics, and rhymes. All of these tools, and others, help shift information to your long-term memory.

The process of recalling information stored in either short-term or long-term memory is called **remembering.** As illustrated in Figure 17–30, short-term memory recall is generally faster than long-term memory recall. That is,

FIGURE 17–30 A Model of Memory

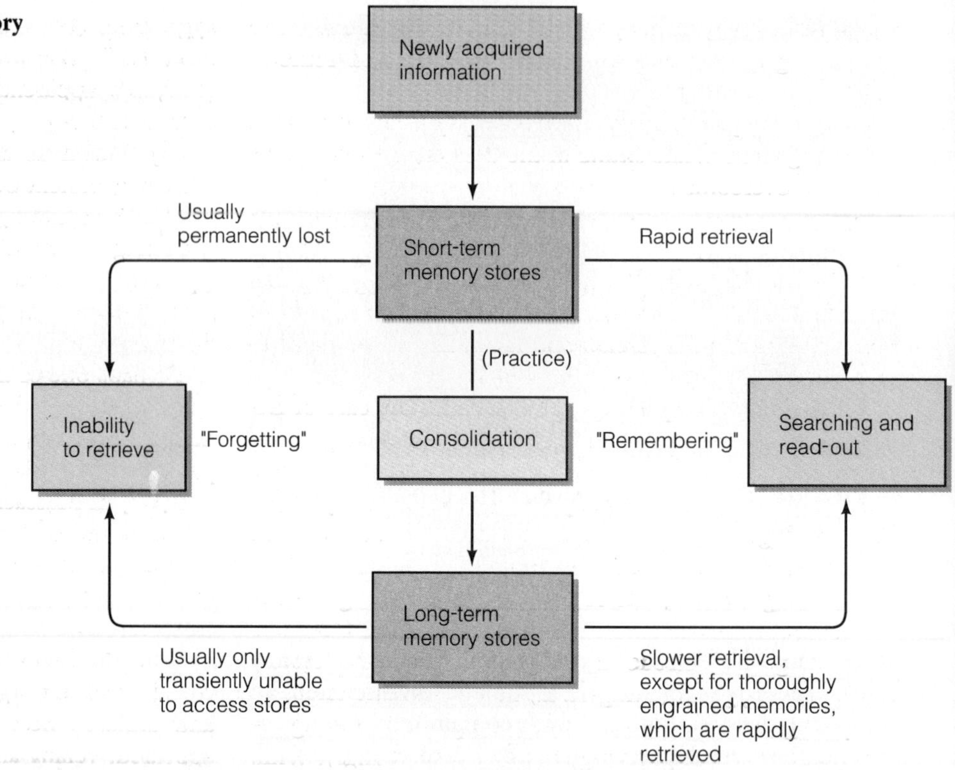

unless the long-term memory you are retrieving is one that is thoroughly ingrained (for example, your name and place of birth), it usually takes longer to recall than something in short-term memory. By the same token, information lost from short-term memory is generally gone forever. Information you cannot recall from long-term memory is often still there; it just takes awhile to remember it or some special stimulus to extract it from the cobwebs.

Memory Is Stored in Multiple Regions of the Brain

Interestingly, neurons involved in storing memories are widely distributed in the cerebral cortex (especially the temporal lobe) as well as other regions (the cerebellum and limbic system). The temporal lobes of the cerebrum and one part of the limbic system (hippocampus) appear essential for transferring short-term memories into long-term memory.

Short-Term and Long-Term Memory Appear to Involve Structural and Functional Changes of the Neurons

How are memories stored? Although no one knows how memories are stored in the brain, some evidence suggests that certain anatomical and physiological changes in the neuronal circuitry may be the basis of our memory.

Research suggests, however, that short-term and long-term memories may involve different mechanisms. Studies on slugs and snails suggest that short-term memory may involve transient modifications in the function of existing synapses. That is, synapses may become altered by experiences or knowledge. Whether this mechanism holds true for humans and other animals remains to be seen.

In contrast, long-term memory involves relatively permanent structural and functional changes in the neurons of the brain. For example, studies that compared the brains of experimental animals reared in a sensory-deprived environment with those reared in a sensory-rich environment reveal marked differences in the microscopic structures of the brains of the two groups. The brains of those animals exposed to a sensory-rich environment displayed greater branching and elongation of dendrites in regions of the brain thought to serve as memory storage.

Some studies suggest that protein synthesis may be involved in long-term memory, because drugs that block protein synthesis interfere with long-term memory. What proteins are involved remains unknown.

ENVIRONMENT AND HEALTH: THE VIOLENT BRAIN

Dr. Vernon H. Mark, a neurosurgeon at the Harvard Medical School, has a special interest in the violent mind. He has studied and treated a large number of violent patients in his years of medical practice. One of the most memorable was a woman named Julie. She came to Dr. Mark at age 21, suffering from epileptic seizures and sudden, unpredictable outbursts of violence.

Julie's problems began early in life when a mumps infection spread to her brain, causing severe inflammation. At the age of 10, she began suffering from epileptic seizures. After many of her seizures, the girl would burst into fits of anger, attacking people. Depression and remorse followed her outbursts.

At age 18, Julie and her father, a physician, went to a movie together. While sitting in the theater, she felt a wave of terror overcoming her. She went to the restroom and stared at herself in the mirror, clutching a dinner knife that she carried to protect herself. Another woman in the restroom accidentally bumped into Julie, triggering her rage. Julie attacked her, plunging the knife into her chest. Her father heard the scream and raced in. He was able to administer first aid to save the woman's life.

Three years later, Julie was placed under the care of Dr. Mark. To find out where the problems were arising in her brain, he implanted a number of electrodes in her brain that would allow him to study the electrical activity of her brain during seizures. He was also able to stimulate the electrodes by remote control to study the behavioral effects of activating various parts of her brain. Stimulating most of the electrodes had little effect on her behavior. However, stimulating an electrode in a region of the limbic system called the amygdala (a-mig-da-la) had a dramatic effect on the young woman. She became unresponsive, staring vacantly in front of her, although her face turned angry. While this was occurring, the brain cells in her amygdala and hippocampus fired erratically. Dr. Mark had stimulated a full-fledged epileptic seizure. All of a sudden, Julie exploded violently.

Dr. Mark repeated the experiment later and got a similar response, confirming that he had found the focal point of Julie's seizures and rages. Over the next several months, he set out to destroy the tissue, sending radio waves along the electrode. This raised the temperature of the electrode and destroyed the cells little by little. Today, Julie's seizures and violent outbursts have ended. She lives a normal life.

This account is a remarkable testimony to the dramatic effects of brain trauma. Damage in crucial areas, no matter how slight, can upset the normal operations of the brain, resulting in erratic behavior or throwing patients into a violent rage. Dr. Richard Restak, a neurologist and author of numerous popular books on the brain, notes that alcohol and other drugs can also upset the delicate balance of the brain, creating violent outbursts. Such drugs can dismantle (at least temporarily) the brain's intricate system of checks and balances, eliciting unusual, and even violent, behavior. "It is humbling and frightening," says Restak, "to consider that our rationality is dependent on the normal function of tissue within our skulls." It is even more frightening to realize that the balance can be so easily upset. "We have the capacity, if everything is operating correctly within our brains, of composing a Bill of Rights or the Constitution," says Restak. "But in the presence of a barely measurable electrical impulse within the limbic system, our much vaunted rationality can be replaced by savage attacks and seemingly inexplicable violence."

Alcohol dampens the inhibitions that hold aggression in check in some people. Many other drugs affect levels of aggression. (For more on drugs and the brain, see Health Note 17-1). Barbiturates, for example, increase aggression. Marijuana apparently reduces aggression. Marijuana, however, is sometimes laced with PCP (phencyclidine), commonly known as angel dust. This street drug unleashes extremely violent behavior.

PCP acts through the limbic system, shutting off the inhibiting effects of the cortex on primitive rage centers. Episodes of sudden, unprovoked violence due to PCP have become common in many of our major cities, says Restak.

Perhaps one of the most shocking stories of the impacts of chemicals on the brain is that of David Garabedian, a man described by his family and friends as mild-mannered, passive, even docile. Garabedian worked for a lawn-care company in Massachusetts. On March 29, 1983, he arrived at the home of Eileen Muldoon to estimate the cost of treating her lawn with chemical insecticides. He knocked on the door but found no one at home. Garabedian decided to make an estimate and leave it for her.

When he had completed his work, however, he experienced a sudden urge to urinate. He went to the backyard and urinated near the house. Just then, Mrs. Muldoon appeared, yelling angrily at him. Garabedian became confused, apologized, and tried to explain his plight. According to his testimony, she turned away from him, refusing to listen. Garabedian tapped her on the shoulder to say he was sorry, but the woman turned on him and clawed his face with her fingernails. Garabedian exploded, grabbing the woman by the neck and strangling her. As she lay motionless on the ground, he hurled large rocks at her head, smashing her skull.

A month before the murder, Garabedian had undergone a remarkable personality change. Usually amicable, he became easily angered and abusive toward his family. He complained of tension, nervousness, impatience, and terrible nightmares. He also complained of numerous physical symptoms such as nausea, diarrhea, headaches, and frequent urination. These symptoms suggested to some that the young man was suffering from pesticide poisoning.

Dr. Peter Spencer, a toxicologist at the Albert Einstein College of Medicine in New York, testified at the trial that "David Garabedian was involuntarily intoxicated with a chemical in the lawn products that he was exposed to on a daily basis." He not only sprayed chemicals on lawns and trees but also mixed them, pouring large amounts into trucks each night in preparation for the next day's spraying.

One chemical, in particular, has been singled out as the possible culprit. It is carbaryl, a powerful inhibitor of acetylcholinesterase. Physicians believe that the carbaryl, and possibly other pesticides that Garabedian routinely handled, inhibited acetylcholinesterase in his brain. Acetylcholine then accumulated in the synaptic clefts in neurons in his limbic system, triggering uncontrollable rage.

A psychiatrist who testified at the trial said, "In my view, David did not have the capacity to control his behavior because his brain was poisoned." Despite this testimony,

HEALTH NOTE 17–1

THE CAUSES AND CURES OF ADDICTION

Marleen Whitehead lives a lie. Each morning, she kisses her husband good-bye, then drives off to work. On her way to the office, she pulls off the highway for a moment, opens a flask she keeps under the seat, and takes a drink. Throughout the day, she laces her coffee with scotch, so no one will suspect that she is drinking.

Like millions of other Americans, Whitehead's life is on the road to disaster. Her boss is aware of her drinking, and her job is at risk. What makes her, and millions of others like her who cannot get through a day without a drink or a fix from a needle, so dependent on a drug?

Many biologists think that genetics plays a key role in alcoholism. Behavioralists, however, argue that a person's upbringing and psychology create alcohol dependency. Most likely, biological and behavioral factors are both involved in alcoholism and other addictive behavior.

Let's review some of the facts. A recent study in mice showed that addictive drugs, such as alcohol and cocaine, stimulate the brain's pleasure center. The pleasure center is a region of the brain called the nucleus accumbens that, when stimulated, brings gratification.

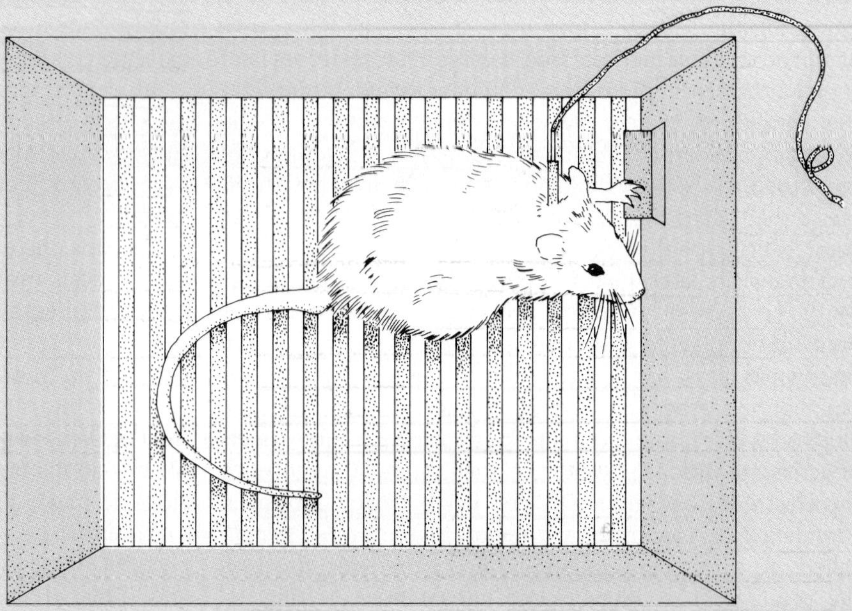

▻ **FIGURE 1 The Pleasure Center** An electrode implanted in the pleasure center of a rat's brain can be activated when the rat presses the bar. To obtain this stimulation, the rat will forsake sex, food, and water. Addictive drugs may stimulate the pleasure center.

Stimulation of the pleasure center by drugs, researchers believe, may underlie all forms of drug addiction.

Experiments in which researchers implanted electrodes in the pleasure centers of rats showed that the rodents could be trained to press a bar to stimulate the center (▻ Figure 1). Some rats, in fact, pressed the bar hundreds of times per hour, ignoring food, water, and sex. After 15 to 20 hours of continual pressing, the rats collapsed from exhaustion. But when they awoke, they began pressing again.

Experiments with rats and mice suggest that the pleasure center is activated

Garabedian was found guilty of first-degree murder and is serving a life sentence.

Drugs, alcohol, pesticides, and industrial chemicals enter our bodies. In small amounts, they may be harmless, but in higher concentrations and in combination, they may disrupt chemical balance and cellular structure, upsetting homeostasis. Nowhere is the effect more pronounced than in the brain.

SUMMARY

OVERVIEW OF THE NERVOUS SYSTEM

1. The nervous system controls a wide range of functions in many animals and helps regulate homeostasis. The human nervous system also performs many other functions, such as ideation, planning for the future, learning, and remembering.
2. The nervous system consists of two anatomical subdivisions: the central nervous system (CNS), made up of the brain and spinal cord, and the peripheral nervous system (PNS), comprising the spinal and cranial nerves.
3. Receptors in the skin, skeletal muscles, joints, and organs, transmit sensory input to the CNS via sensory neurons. The CNS integrates all sensory input and generates appropriate responses. Motor output leaves the CNS in motor neurons.
4. The PNS has two functional divisions: the autonomic nervous system, which controls involuntary actions such as heart rate, and the somatic nervous system, which largely controls voluntary actions such as skeletal muscle contractions and provides the neural connections needed for many reflex arcs.

by cocaine and amphetamines. These drugs stimulate the production of large amounts of a neurotransmitter known as dopamine. In a recent study, researchers implanted small tubes into rats' pleasure centers and areas involved in muscle movement. The researchers administered various doses of addictive and nonaddictive drugs while the rats moved freely in their cages. The researchers then extracted fluid from the tubes to measure dopamine levels.

Drugs that are addictive to humans and rewarding to rats (amphetamine, cocaine, morphine, methadone, ethanol, and nicotine) increased dopamine concentrations in both brain areas. Levels, however, were much higher in the pleasure center. Although these results cannot be extrapolated to humans, some researchers believe that they help build a case for the theory that dopamine release in the pleasure center is a common denominator in all forms of drug addiction. Other researchers think that other neurotransmitters may also be involved in addiction.

This research helps explain the neural and biochemical basis of addiction, but what about the underlying cause? Why do some people become addicted to drugs whereas others can take them or leave them?

Research into the genetics of alcoholism has built a case for the assertion that alcoholism is largely the result of defective genes. Donald W. Goodwind of the University of Kansas Medical Center in Kansas City found that children of alcoholics had an increased risk of becoming alcoholic even when reared by adoptive (nonalcoholic) parents. This and a number of other similar studies suggest that the environment is less important than a person's genetic heritage. If one or both of your biological parents is an alcoholic, say researchers, you are much more likely to be one yourself.

Nevertheless, some researchers think that the scientific community has accepted the genetic findings too readily and uncritically. They say that the genetic theory of alcoholism may be a simplified view of the causes of the disease. Alcoholism probably results from a complex interaction of environmental factors and genetics.

New research also shows that alcoholism results from physical, personal, and social characteristics that predispose a person to drink excessively. Herbert Fingarette, a professor at the University of California at Santa Barbara who has studied alcohol addiction for years, contends that alcoholism has psychological not biological roots. Fingarette's arguments reflect the findings of many psychological studies. This research indicates that alcoholism and other addictions are more habits than diseases. Addictive behavior, the studies suggest, typically revolves around immediate gratification.

Alcohol enhances social and physical pleasure, increases sexual responsiveness and assertiveness, and reduces tension up to a point. Unfortunately, the initial physical stimulation, brought on by low doses of alcohol, can lead some people into an addictive cycle. The expectation of improved feelings drives people to drink. But higher doses of alcohol dampen arousal, sap energy, and cause hangovers. This, in turn, say psychologists, leads to a craving for alcohol's stimulating effects—that is, a craving to feel good again. The repetitive cycle of pleasure and displeasure is addiction.

The controversy over the roots of alcoholism and other addictive behaviors will undoubtedly continue for years, pitting biologists against psychologists. Although it is impossible to predict the outcome of future research, it is possible that the intermediate position will hold sway: that addiction may be explained by both genetics and psychology.

STRUCTURE AND FUNCTION OF THE NEURON

5. The fundamental unit of the nervous system is the neuron, a highly specialized cell that generates and transmits bioelectric impulses from one part of the body to another.
6. Three types of neurons are found in the body: sensory neurons, interneurons, and motor neurons.
7. All neurons have more or less spherical cell bodies. Extending from the cell body are two types of processes: dendrites, which conduct impulses to the cell body, and axons, which conduct impulses away from the cell body.
8. Many axons are covered by a layer of myelin, which increases the rate of impulse transmission. In the PNS, myelin is laid down by Schwann cells during embryonic development. In the CNS, myelin is produced by oligodendrocytes.
9. The terminal ends of axons branch profusely, forming numerous fibers that end in small knobs called terminal boutons.
10. During cellular differentiation, nerve cells lose their ability to divide. Because nerve cells cannot divide, neurons that die cannot be replaced by existing cells. Axons in the PNS may regenerate, but axonal regeneration in the CNS is rare.
11. Nerve cells have exceptionally high metabolic rates and require a constant supply of oxygen. Because nerve cells rely exclusively on glucose for energy, decreases in blood glucose levels can have deleterious effects on the nervous system.
12. The small electrical potential across the membrane of nerve cells is known as the membrane, or resting, potential.
13. When a nerve cell is stimulated, its plasma membrane increases its permeability to sodium ions. Sodium ions rush in, causing depolarization, which spreads down the membrane.
14. Depolarization is followed by repolarization, a recovery of the resting potential stemming largely from the efflux of potassium ions. The depolarization and repolarization of the neuron's plasma membrane constitute a bioelectric impulse or action potential.
15. Nerve impulses travel along the plasma membranes of dendrites and unmyelinated axons, but in myelinated axons, impulses "jump" from node to node.
16. When a bioelectric impulse reaches the terminal bouton, it

stimulates the release of neurotransmitters contained in membrane-bound vesicles.

17. Released into synaptic clefts, neurotransmitters diffuse across the cleft, where they bind to receptors in the postsynaptic membrane. They may excite or inhibit the postsynaptic membrane. Whether a neuron fires depends on the sum of excitatory and inhibitory impulses it receives.
18. After stimulating the postsynaptic membrane, neurotransmitters may diffuse out of the synaptic cleft, be reabsorbed by the axon terminal, or be removed by enzymes.
19. Some insecticides inhibit the activity of the enzymes that deactivate neurotransmitter substance, creating a wide range of nervous system effects.

EVOLUTION OF THE NERVOUS SYSTEM

20. The nervous systems of animals exist on a continuum from the simplest to the most advanced. The simplest systems generally consist of a network of interconnected neurons, or nerve net.
21. Advanced nervous systems contain a network of nerves and a central nervous system, an aggregation of nerves where information processing and integration occur.
22. The evolution of the nervous system corresponds with the evolution of complexity in body structure.
23. Perhaps one of the most significant developments in the evolution of the nervous system was the emergence of adaptations that accelerate nerve impulse conduction—in vertebrates, saltatory conduction and in invertebrates, the presence of axons of very large diameter.

THE SPINAL CORD AND NERVES

24. The spinal cord descends from the brain through the vertebral canal to the lower back. It carries information to and from the brain, and its neurons participate in many reflexes.
25. Two types of nerves emanate from the CNS: spinal and cranial. Spinal nerves arise from the spinal cord and may be sensory, motor, or mixed. Cranial nerves attach to the brain and supply the structures of the head and several key body parts.
26. Spinal nerves are attached to the spinal cord via two roots, a dorsal root, which brings sensory information into the cord, and a ventral root, which carries motor information out. Sensory fibers entering the cord often end on interneurons, which frequently end on motor neurons in the spinal cord, forming reflex arcs. Interneurons may also send axons to the brain.

THE BRAIN

27. The brain is housed in the skull. The cerebrum with its two cerebral hemispheres is the largest part of the brain.
28. The outer layer of each hemisphere is the cortex. It contains gray matter, which houses nerve cell bodies, and underlying white matter, which contains myelinated nerve fibers that transmit nerve impulses to and from the gray matter.
29. The cerebral cortex consists of many discrete functional regions, including motor, sensory, and association areas.
30. Consciousness resides in the cerebral cortex, but a great many functions occur at the unconscious level in parts of the brain beneath the cortex.
31. The cerebellum coordinates muscle movement and controls posture. The hypothalamus regulates many homeostatic functions. The limbic system houses instincts and emotions. The brain stem, like the hypothalamus, helps regulate basic body functions.
32. A watery fluid surrounds the brain and spinal cord. Known as cerebrospinal fluid, it serves as a cushion that helps protect the brain and spinal cord from traumatic injury. CSF also fills the central canal of the spinal cord and cavities within the brain, the ventricles.
33. Electrodes applied to different parts of the scalp detect electrical activity in the brain and produce a tracing known as an electroencephalogram. The type of brain wave recorded depends on one's level of cortical activity. EEGs are used to diagnosis some brain dysfunctions.
34. Headaches are the most common form of pain and generally result from tension, swelling of the membranes lining the sinuses, eyestrain, or dilation of cerebral blood vessels.

THE AUTONOMIC NERVOUS SYSTEM

35. The autonomic nervous system (ANS) is a subdivision of the PNS and transmits sensory information from organs to the CNS and delivers motor impulses to smooth muscle, cardiac muscle, and glands.
36. The ANS innervates all internal organs and has two subdivisions: the sympathetic and the parasympathetic.
37. The sympathetic division of the ANS functions in emergencies and is responsible in large part for the fight-or-flight response, accelerating heart rate and breathing, dilating the pupils, and shunting blood to the skeletal muscles, which provides additional oxygen and nutrients needed to combat an adversary or to escape.
38. The parasympathetic division of the ANS reduces heart rate, contracts the pupils, and promotes digestion—responses associated with the relaxed state.

LEARNING AND MEMORY

39. Learning is a process in which an individual acquires knowledge and skills from experience, instruction, or some combination of the two. Learning depends on one's ability to store information in the brain—that is, memory.
40. Newly acquired knowledge is first stored in short-term memory, which retains information for periods of seconds to hours. Short-term memories may be transferred to long-term memory, which holds information for periods of days to years.
41. Memory is stored in multiple regions of the brain—in the cortex, especially the temporal lobe, as well as the cerebellum and limbic system.
42. Short-term and long-term memory appear to involve structural and functional changes of the neurons.

ENVIRONMENT AND HEALTH: THE VIOLENT BRAIN

43. Abnormal brain activity can result in bizarre behavior in humans. This is especially true if damage occurs in parts of the limbic system. Violent rage may result.
44. Damage may result from infections or harmful chemicals, such as pesticides and drugs. Alcohol can also bring on rage by eliminating normal inhibitions.

EXERCISING YOUR CRITICAL THINKING SKILLS

Elevated blood glucose levels in diabetics damage the kidneys and nervous system, in some instances causing renal failure and blindness. Diabetics also suffer from intestinal disorders, which researchers think may be caused by damage to the nerves that innervate the intestines.

One researcher, intent on pinpointing the cause of the intestinal problems in diabetics, decided to examine the microscopic anatomy of the paravertebral ganglia. These ganglia are clusters of nerve cell bodies along the spine that are involved in the autonomic nervous system, which as you learned in this chapter controls the function of many organs in the body, including the intestines and urinary bladder.

The researcher first examined ganglia in older diabetic rats. In his studies, he found that nerve endings (terminal boutons) in the ganglia were quite swollen, some of them 30 times their normal size. This observation led him to hypothesize that diabetes was causing the swelling, which was due to the accumulation of vesicles containing neurotransmitter.

To test this hypothesis, the researcher examined control groups composed of nondiabetic rats of the same age that had been kept under similar conditions. To his surprise, many of the nerve terminals in the nondiabetic rats' ganglia were also swollen. (This example illustrates how important a control group is to good science.)

The researcher concluded that some other factor was responsible for the swelling. From the little bit of detail given so far, can you come up with any ideas? (Read back over the material for a hint . . . it's there.)

If you said age, you are right. After thinking about the commonalities of the two groups, the researcher realized that all of his animals were old. He hypothesized that age might have been responsible for the swelling. If you were involved in the study, how would you test this new hypothesis?

If you said that you would compare the ganglia of young and old nondiabetic rats, you're right again. That's exactly what he did. When the researcher made the comparison, he found no evidence of nerve terminal swelling in the young rats.

Like other good researchers, intent on solving one problem, this one found a partial answer to another: aging. As the body ages, the autonomic nervous system can malfunction, causing a wide range of problems in elderly people, including loss of bladder control and fainting when standing too quickly.

The researcher in this example decided to see if this age-related swelling occurred in humans as well. He collected ganglia from 56 people aged 15 to 93 who had died from a variety of causes. In people under 60, there were few swollen nerve terminals, but after age 60 the number of swollen nerve terminals increased dramatically.

Although it is true that the researcher was unsuccessful in finding answers to the original problem and that much research is still needed to determine why the swollen nerve terminals exist in older animals, this case study is a good example of how using critical thinking skills and good experimental methods can lead to important results.

TEST OF TERMS

1. The brain and spinal cord are part of the ____________ nervous system. The nerves belong to the ________ nervous system.
2. Heartbeat and breathing are largely controlled by the ____________ nervous system.
3. Many neurons contain short, branching fibers called ____________, which conduct impulses to the cell body, and a long, unbranched fiber, the ____________, which carries impulses away from the cell body.
4. Many axons of the PNS have a coating, the __________ __________, which is laid down by Schwann cells and accelerates nerve cell transmission.
5. The swellings at the ends of axons are called __________ __________. They release chemical substances called ____________, which stimulate a bioelectric impulse in adjoining nerve cells.
6. The voltage measured across the plasma membrane of an inactive neuron is called the ____________ ____________ and is about ________ millivolts.
7. Stimulating the plasma membrane of a nerve cell results in a drastic change in the membrane's permeability to ____________ ____________. The change in the resting potential that occurs when the membrane is stimulated is called a(n) ____________ ____________.
8. When a nerve impulse reaches a terminal bouton, depolarization causes the influx of calcium ions, which stimulates the release of a neurotransmitter, which is stored in vesicles, into the ____________ ____________. Neurotransmitter substances bind to receptors on the ____________ ____________, stimulating either the uptake of sodium or chloride ions, depending on the substance.
9. ____________ is the main excitatory neurotransmitter in many synapses. Certain insecticides block ____________, an enzyme that removes this transmitter from the synapse.
10. Nerves attached to the spinal cord are called ________ ________. Each of these has two roots. Motor fibers leave via the ____________ root.
11. The neuron that transmits the impulse from the sensory neuron to the motor neuron in the spinal cord is called a(n) ____________.
12. The portion of the cerebral cortex that controls voluntary muscle movement is the ____________ ____________ cortex.
13. The ridges in the cerebral cortex are called ____________. The grooves between them are known as ____________.

14. Sensory fibers carrying information from all over the body terminate in a part of the cerebral cortex called the ____________ ____________ cortex, which is located just behind the ____________ ____________.
15. The ____________ are three fibrous layers surrounding the brain, the ____________ ____________, arachnoid layer, and ____________ ____________.
16. The regions of the cortex that integrate incoming information are called the ____________ cortex.
17. The ____________ system has many functions. It is the site of the primitive rage response.
18. The ____________ is in charge of coordinating skeletal muscle activity. Problems in this region of the brain cause spasticity.
19. Just above the pituitary and attached to it is the region of the brain called the ____________. It controls eating behavior and monitors the composition of the blood. Aggregations of nerve cell bodies in this region are called ____________.
20. The ____________ ____________ system is found in the medulla oblongata. It receives all incoming and outgoing information and sends impulses to the cerebral cortex, keeping it active.
21. The brain and spinal cord are bathed in a fluid, known as ____________ fluid, which also fills the central canal of the spinal cord and the hollow cavities inside the brain, called the ____________.
22. Electrical activity of the neurons of the cerebral cortex can be picked up by electrodes placed on the scalp. A tracing of the electrical activity is known as an ____________, or ____________.
23. The ____________ division of the autonomic nervous system increases the heart rate and breathing.
24. The ____________ division of the autonomic nervous system brings about internal responses generally associated with the relaxed state.
25. Newly acquired facts are first stored in ____________-____________ memory, which holds information for periods of ____________ to ____________.

Answers to the Test of Terms are found in Appendix B.

TEST OF CONCEPTS

1. The nervous system performs a great many functions. Describe them. What functions do you think are unique to humans?
2. Describe how the resting and action potentials are generated. Describe how the plasma membrane of a neuron is repolarized.
3. Draw a typical multipolar neuron, and label its parts. Show the direction in which an action potential travels.
4. Draw a typical synapse, label the parts, and explain how a nerve impulse is transmitted from one nerve cell to another.
5. Describe how each of the following substances affects synaptic transmission: caffeine, cocaine, and malathion.
6. In general, how do anesthetics function?
7. Describe the progression of the evolution of the nervous systems of animals.
8. Name and describe the various divisions of the nervous system.
9. Draw a cross section through the spinal cord showing the spinal nerves. Label the parts, and explain a reflex arc and how nerve impulses entering a spinal nerve also travel to the brain.
10. A physician can stimulate various parts of the brain and get different responses. What effects would you expect if the electrodes were placed in the premotor area? The primary motor cortex? The sensory motor cortex?
11. The brain is a delicate organ, and slight shifts in electrical activity can create bizarre behavior. Do you agree with this statement? Give examples.

CHAPTER

18

The Senses

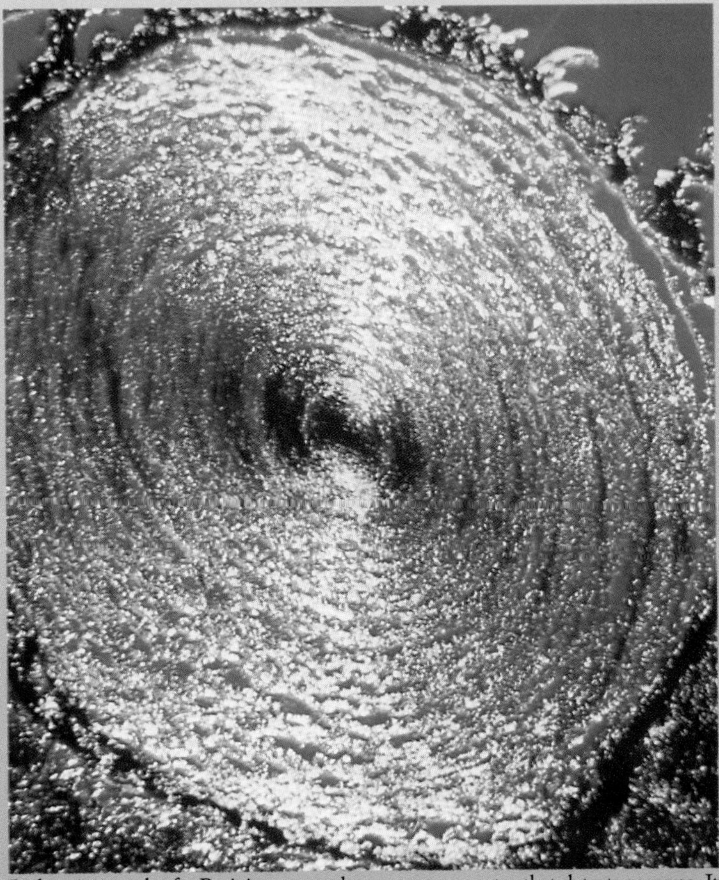

Light micrograph of a Pacinian corpuscle, a sensory receptor that detects pressure. It is found in the skin and other locations.

Debra Cartwright noticed something strange one day while she was eating dinner. For no apparent reason, the college sophomore had lost her senses of smell and taste. Doctors were puzzled at first by the finding, for Cartwright seemed to be in fine health. She was not suffering from a cold, sinus infection, or even any allergies that might have blocked her nasal passages and impaired her ability to smell and taste. When she had a blood test, though, they found the cause of her problem: low blood levels of the micronutrient zinc. Her physician prescribed a zinc supplement, and in a few days her senses of smell and taste returned.

This story illustrates the importance of a nutritionally balanced diet and the impact of a dietary deficiency on two important body functions—the senses of taste and smell.[1] It is just one of many examples presented in this book that illustrates how body functions can be altered by external factors such as stress, pollution, and diet.

This chapter examines animal senses, dividing the discussion into two broad categories: the general senses and the special senses. The chapter focuses primarily on human sensory structures. As a result, your studies will yield some information useful throughout your life and, incidentally, will explain the reason why a zinc deficiency eliminates the sense of taste. Finally, this chapter also describes sensory adaptations of other animal species that permit them to hear, to see, to smell, and to taste.

THE GENERAL AND SPECIAL SENSES

Staying aware of one's environment and responding to changes in that environment require a system of surveillance not unlike that found at high-security weapons facilities, where both inside and outside activities are routinely monitored to detect intruders. The surveillance system of the human body, like that of many other animals, consists of numerous receptors that detect internal and external changes. In humans, these receptors are found in the skin, internal organs, bones, joints, and muscles. They detect stimuli that give rise to the **general senses:** pain, temperature, light touch, pressure, and a sense of body and limb position (Table 18–1).

The human body is endowed with five additional senses, known as the **special senses.** They are taste, smell, vision, hearing, and balance and are generally made possible by special sensory organs (the eyes, for example). These sophisticated detectors greatly increase our ability to perceive stimuli in the environment and are one of the most remarkable developments in the evolution of complex multicellular organisms.

Receptors in the animal kingdom involved in the general and special senses fall into five functional categories: (1) mechanoreceptors, (2) chemoreceptors, (3) thermoreceptors, (4) photoreceptors, and (5) nociceptors (pain receptors). **Mechanoreceptors** are those that are activated by mechanical stimulation—for example, touch or pressure. **Chemoreceptors** are activated by chemicals in the food we eat, the air we breathe, or levels of certain chemicals (for example, hydrogen ions and oxygen) in our blood. **Thermoreceptors** are activated by heat and cold, and **photoreceptors** are sensitive to light. **Nociceptors** are stimulated by tissue damage, such as that caused by pinching, tearing, or burning. Interestingly, intense stimulation of all the other receptors also gives rise to the sensation of pain. A much rarer sense receptor, which detects minute electrical signals, is discussed in Spotlight on Evolution 18-1.

THE GENERAL SENSES

Sit back in your chair for a moment, close your eyes, and concentrate on what you feel. A cold draft may be stirring

[1] Zinc deficiencies are quite rare, and dietary supplements containing zinc are generally not needed if you are eating a well-balanced diet. Furthermore, dietary supplements can be dangerous. Just a few times the recommended daily allowance of zinc can cause serious problems.

TABLE 18–1 Summary of General and Special Senses

SENSE	STIMULUS	RECEPTOR
General senses	Pain	Naked nerve endings
	Light touch	Merkel's discs; naked nerve endings around hair follicles; Meissner's corpuscles; Ruffini's corpuscles*; Krause's end-bulbs*
	Pressure	Pacinian corpuscles
	Temperature	Naked nerve endings
	Proprioception	Golgi tendon organs; muscle spindles; receptors similar to Meissner's corpuscles in joints
Special senses	Taste	Taste buds
	Smell	Olfactory epithelium
	Sight	Retina
	Hearing	Organ of Corti
	Balance	Crista ampularis in the semicircular canals; maculae in utricle and saccule

*These may both be alternate forms of the Meissner's corpuscle and may be stimulated by light touch.

SPOTLIGHT ON EVOLUTION 18–1

THE PLATYPUS: NATURE'S JOKE OR MARVEL OF EVOLUTION?

In 1798, scientists at the British Museum in London received an odd-looking specimen from Australia. About the size of a cat, with the bill of a duck and the tail and fur of a beaver, this web-footed creature had to be a fake, a clever joke sewn together from the parts of other animals (▷ Figure 1). Certain that their irreverent Australian colonials had foisted a joke on them, the British scientists proceeded to hunt for the sutures that held this bizarre animal together. After a painstaking search, however, they found none. Today that same specimen remains in the museum, sporting scalpel marks made by the skeptics searching for elusive stitches.

The odd-looking animal is the duck-billed platypus, an egg-laying mammal that has only recently begun to be studied, a process made difficult by the creature's lifestyle. Although it is fairly common in eastern Australia and Tasmania (an island south of Australia), the duck-billed platypus feeds underwater and only at night, spending most of its life curled up inside its underground burrow. Wary of humans, it suffers extreme stress in captivity. Too much stress and the animal dies. Accordingly, getting the animal to reproduce in captivity has been nearly impossible.

Despite its elusive nature, researchers are beginning to discover the unique adaptations that promote its survival and reproduction in the wild. One of them is a dual sensory system—one that functions on land and another that functions in water. On land, the animal depends on its eyes, ears, and nose, but when the creature dives underwater, these sense organs clamp shut. How does the platypus steer its course underwater?

In 1986, researchers found that its snout is an electronic receiver that picks up minute electrical signals. In laboratory studies, the animal responded to extremely small electrical fields as weak as those created by a shrimp flicking its tail.

In 1988, another team of researchers found minute pores on the bill's side containing nerve endings, which are the actual detectors of electrical fields. Small glands release a fluid that keeps the pores from drying out. Underwater, the animal scans back and forth with its bill, searching for prey. The bill is also extremely sensitive to touch and may help the animal feel its way around its aquatic home.

Besides these remarkable sensory adaptations, the platypus exhibits other characteristics helpful to its survival. Unlike most aquatic mammals, which propel themselves with webbed hind feet and tails, the platypus relies principally on its front legs and feet for propulsion. With its oversized front feet pulling it along, the platypus uses its tail as a rudder. Unlike the tail of a beaver, which is also used for propulsion, the platypus tail is filled with fat. Thus, the tail also serves as a fat depot for reserve energy, and at night it serves as a blanket, for when the creature curls up to sleep, it wraps its tail around its body.

▷ FIGURE 1 The Duck-Billed Platypus

Despite its shyness, a trapped platypus is a force to be reckoned with. Its sharp hind claws deliver a painful scratch, worsened by the release of a strong venom. Hunters who have tangled with the creature report suffering severe pain, swelling, and weeks of partial paralysis.

After laying their eggs, female platypuses remain sheltered for several months in their burrows, leaving only occasionally to feed. Taking refuge in their burrows, which they seal with mud, is thought to be an evolved behavioral strategy, which protects their offspring from predators. It does, however, pose a problem—notably, a lack of oxygen. Here too the platypus exhibits an adaptation vital to its survival. Like prairie dogs, which spend much of their time in low-oxygen burrows, the platypus has a high concentration of red blood cells.

The duck-billed platypus has confused more than British scientists over the years. The aborigines of Australia, in fact, believe that the first platypus resulted from the rape of a duck by a water rat. Early taxonomists thought that it might be a primitive mammal linking humans to reptiles. Today, the platypus is assigned to an order known as *Monotremata* which means "one hole"—referring to the single orifice used for excreting waste and reproduction. This order consists of only two other members, both spiny anteaters (one found in Australia and the other in New Guinea).

Writer Eric Hoffman notes that the platypus seems to fare well because of its adaptations. "Ultimately," he writes, "the animal that once seemed to be nature's joke could, with its extraordinary mix of specialized adaptations, prove to be nature's last laugh."

SOURCE. Adapted from Eric Hoffman, "Paradoxes of the Platypus," *International Wildlife* (1990) 20(1): 18–21.

at your ankles. You may feel the pressure of the chair on your buttocks and warmth emanating from a nearby reading lamp. You may feel your cat brushing against the hairs on your arm. You may feel a slight pain from a bruise or gas pressure in your intestines from the burritos you ate for lunch. Now move your arm. Even though your eyes are closed, you can feel it moving.

The various sensations you have just experienced fall

▻ **FIGURE 18–1 Receptors** General sense receptors are either (*a*) naked nerve endings or (*b*) encapsulated nerve endings.

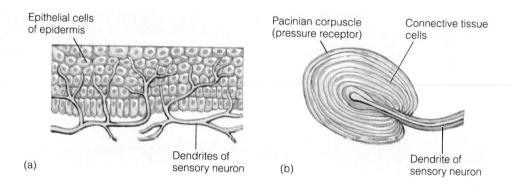

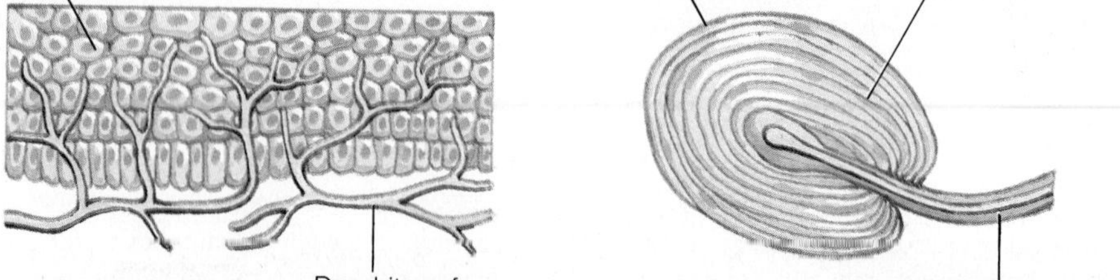

into the group of general senses. Receptors for the general senses detect internal and external stimuli and relay messages to the spinal cord and brain via sensory nerves (Chapter 17). Sensory input to the central nervous system may elicit a conscious response; for example, the touch of a cat may cause you to reach down and pat your furry friend. Some stimuli will cause unconscious responses; for example, heat may cause you to perspire, or cold may cause you to shiver. Others may simply be registered in the cerebral cortex, making you aware of the stimulus. Still other stimuli may elicit no response at all.

Receptors for the general body senses come in many shapes and sizes, but they generally fit into one of two groups based on structure: naked nerve endings or encapsulated receptors, nerve endings associated with cells or groups of cells (▻ Figure 18–1). As you shall soon see, the naked nerve endings and encapsulated receptors in the body serve very specific roles. Some are thermoreceptors; others are mechanoreceptors, and so on.

Naked Nerve Endings in Body Tissues Detect Pain, Temperature, and Light Touch

Naked nerve endings are the terminal ends of the dendrites of sensory neurons. Located in the skin, bones, and internal organs and in and around joints, they are responsible for at least three sensations: pain, temperature, and light touch. Therefore, naked nerve endings may be nociceptors, thermoreceptors, or mechanoreceptors.

Pain. Two types of pain are experienced: somatic pain and visceral pain. **Somatic pain** results when receptors in the skin, joints, muscles, and tendons are stimulated. These receptors are activated when tissues are damaged and are part of a protective mechanism of considerable adaptive value in the evolution of complex multicellular organisms. They signal that something is awry and alert the brain to elicit some action.

Somatic pain receptors respond to several types of stimuli when tissue is injured. Some respond to cutting, crushing, and pinching. Others respond to temperature extremes. Still others respond to irritating chemicals released from injured tissues.

Visceral pain results from the stimulation of naked nerve endings in body organs (the viscera). Pain receptors in body organs are stimulated by distension in some and by **anoxia,** a lack of oxygen, in others. For example, intestinal pain you feel when you have a gas buildup results from the stretching of naked nerve fibers in the wall of the intestine. These nerve endings are mechanoreceptors. The pain felt

▻ **FIGURE 18–2 Referred Pain** Visceral pain is often felt on the body surface at the points indicated by the colored areas.

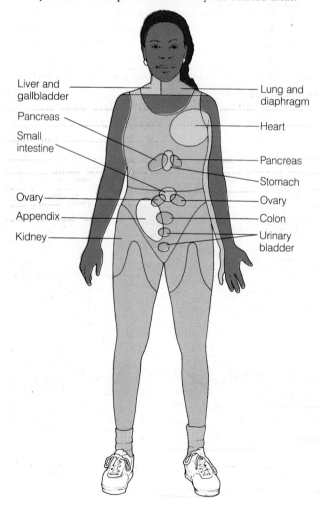

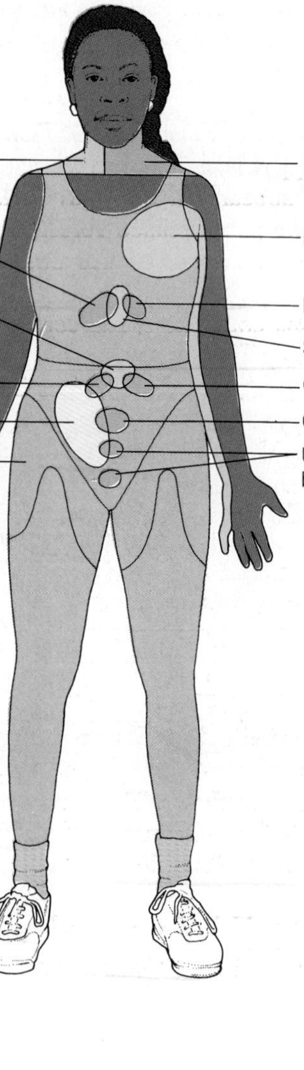

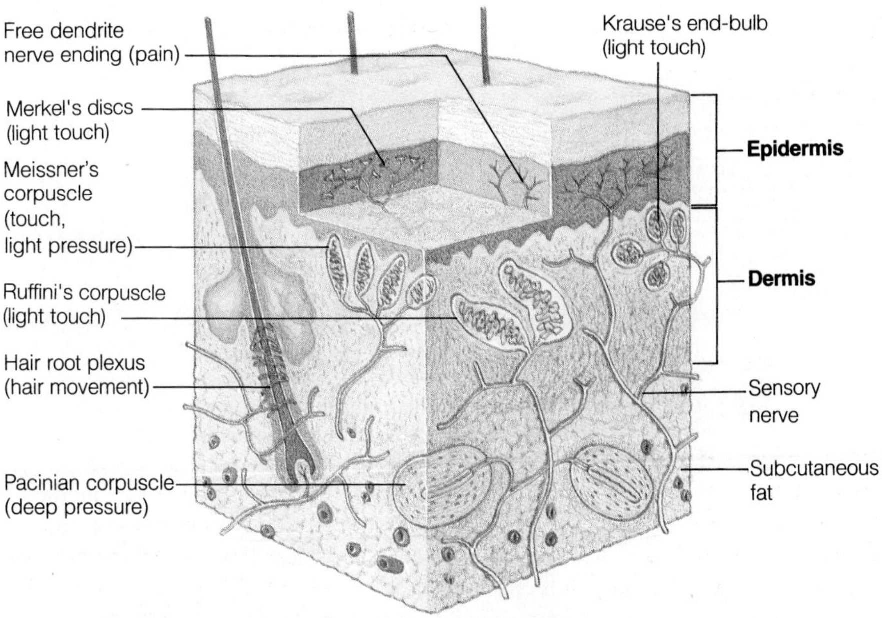

▻ **FIGURE 18–3 General Sense Receptors** The skin houses many of the receptors for general senses. Receptors fall into two categories: naked nerve endings and encapsulated receptors, shown here.

during a heart attack results from a lack of oxygen flowing to the heart muscle (Chapter 12).

Visceral pain and somatic pain result from quite different stimuli and are perceived very differently as well. Somatic pain is easily identified, whereas visceral pain is vague and difficult to localize. Visceral pain is generally felt on the body surface at a site some distance from its origination. For example, anginal pain, caused by a lack of oxygen to the heart muscle, appears in the chest and along the inside of the left arm (▻ Figure 18–2). Pain from the lung and diaphragm appears in the neck.

Pain that appears on the body surface away from the location of the pain is called **referred pain.** Physiologists do not know the cause of this phenomenon, but many think that it results from the fact that pain fibers from internal organs enter the spinal cord at the same location that the sensory fibers from the skin enter. The brain, they hypothesize, interprets the impulses from pain fibers supplying the organs as pain from a somatic source. (See Health Note 18–1 for some techniques to relieve pain.)

Light Touch. Light touch is perceived by two anatomically distinct mechanoreceptors. The first receptor is located at the base of the hairs in our skin. As shown in ▻ Figure 18–3, naked nerve endings (dendrites) wrap around the base of the hair follicles. When a hair is moved—for example, by a gentle touch—these nerve fibers are stimulated. The second light-touch mechanoreceptor is the **Merkel's disc.** Shown in Figure 18–3, Merkel's discs consist of small cup-shaped cells on which naked nerve endings terminate. These receptors, located in the outer layer of the epidermis of the skin, are activated by gentle pressure on the skin.

Temperature. Naked nerve endings in the skin detect heat and cold. Heat receptors respond to temperatures from 25°C (77°F) to 45°C (113°F). If the temperature of the skin rises above this range, pain receptors are activated, creating a burning sensation. In contrast, cold receptors respond to temperatures from 10°C (50°F) to 20°C (68°F). If the temperature drops below 10°C, pain receptors respond.

Encapsulated Receptors Contain a Naked Nerve Ending Surrounded by One or More Layers of Cells that Form the Capsule

Shown in Figure 18–3, the largest encapsulated nerve ending is the **Pacinian corpuscle.** It consists of a naked nerve ending surrounded by numerous concentric cell layers. The Pacinian corpuscle (▻ Figure 18–4), resembles a small

▻ **FIGURE 18–4 Pacinian Corpuscle** This receptor, often located in the dermis of the skin, detects pressure.

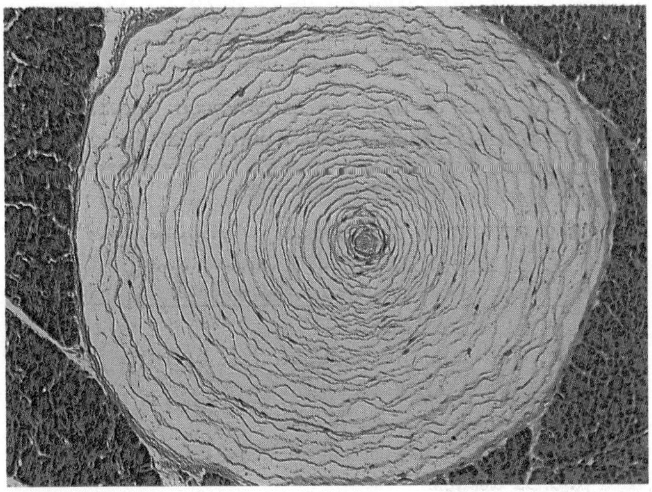

HEALTH NOTE 18–1

OLD AND NEW TREATMENTS FOR PAIN

Millions of people suffer from chronic or persistent pain. In the United States, for example, an estimated 70 million people are tormented by back pain. Another 20 million suffer from migraine headaches.

Despite its prevalence, pain is one of the least understood medical problems in the world. The study of pain, in fact, is so poorly funded and so widely ignored by the medical community that some have called it an orphan science. Making matters worse, physicians are often poorly trained in dealing with pain.

For many decades, pain was treated with painkillers such as morphine and codeine. Their addictive nature led researchers to look for other techniques. One of the more promising was acupuncture, a technique used by the Chinese for thousands of years. Acupuncture relies on thread-thin needles inserted in the skin near nerves and rotated very quickly (▻ Figure 1).

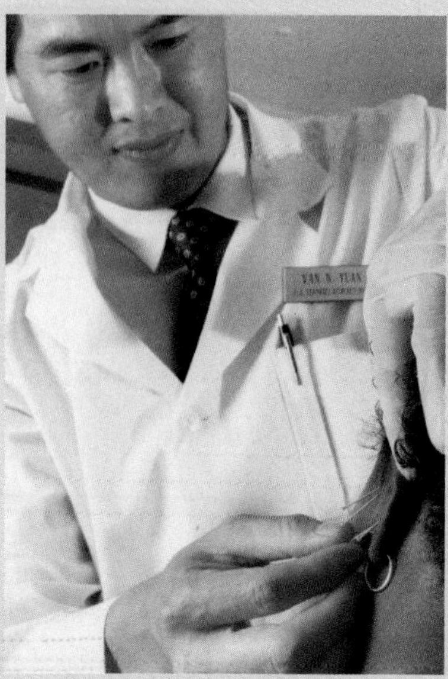

▻ **FIGURE 1 Patient Undergoing Acupuncture** Acupucture is generally used to alleviate pain, but it has other applications as well. This patient is being treated in an experimental program to combat drug addiction.

Neurologists are not certain how acupuncture works, but many think that it blocks pain by overloading the neuronal circuitry. They note that two types of nerve fibers transmit sensory information from the body to the central nervous system: small- and large-diameter fibers. Small-diameter fibers carry pain messages. Large-diameter fibers carry many other forms of sensory information from receptors in the skin—for example, pressure and light touch. The dendrites of both small- and large-diameter sensory nerve cells often terminate in the same location in the spinal cord. From here they send impulses to the brain, signaling pain or some other sense.

Neurobiologists believe that acupuncture needles stimulate the large-diameter nerve fibers. This stimulation blocks nerve impulses carried by the smaller nerve fibers, thus blocking pain messages to the brain. Physicians who are trained to use drugs and surgery to solve most pain remain skeptical about the usefulness of acupuncture. Over the past 10 years, however, a small but steady stream of research has confirmed the painkilling effect of this treatment.

Joseph Helms, a physician with the American Academy of Acupuncture in Berkeley, California, performed acupuncture on 40 women with menstrual pain. Some women received real acupuncture treatment. Others received placebo treatments (shallow needle treatments that did not reach the acupuncture points). In the group of women receiving acupuncture, 10 out of 11 showed a marked decrease in pain.

---

onion pierced by a thin wire. Located in the deeper layers of the skin, in the loose connective tissue of the body, and elsewhere, Pacinian corpuscles are stimulated by pressure, such as the pressure you feel sitting in your chair.

Another common encapsulated sensory receptor is the Meissner's corpuscle. **Meissner's corpuscles** are smaller than Pacinian corpuscles. These oval receptors contain two or three spiraling dendritic ends surrounded by a thin cellular capsule (Figures 18–3 and ▻ 18–5). Like the naked nerve endings surrounding the base of the hair follicle and Merkel's discs, Meissner's corpuscles are thought to be mechanoreceptors that respond to light touch. Located just beneath the epithelium in the outermost layer of the dermis, Meissner's corpuscles are most abundant in the sensitive parts of the body, such as the lips and the tips of the fingers.

Two of the more controversial encapsulated receptors are Krause's end-bulbs and Ruffini's corpuscles (see Figure 18–3). Most scientists believe that these receptors are ac-

▻ **FIGURE 18–5 Meissner's Corpuscle** This receptor, found just beneath the epidermis, detects light touch.

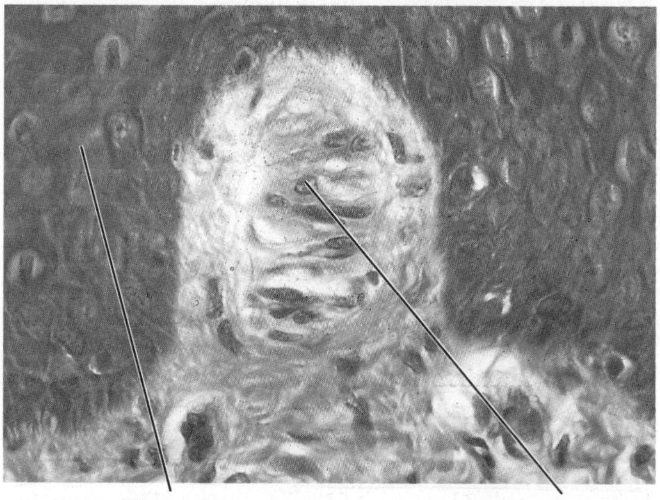

Patients reported an approximately 50% decrease in pain. In the placebo group only 4 out of 11 reported a lessening of pain. Only 1 of 10 people given no treatment showed improvement. Acupuncture also reduced the need for painkilling drugs during treatment by over half. Remember, however, that the Critical Thinking section in Chapter 1 suggested that experiments using small numbers of subjects are themselves subject to question. Further studies are needed to confirm these results.

While acupuncture is slowly earning a respected place in the treatment of pain, researchers are also experimenting with a technique called transcutaneous electrical nerve stimulation, or TENS. Patients are fitted with electrodes that attach to the skin above the nerves that transmit pain signals to the central nervous system. A small battery supplies energy that stimulates the electrodes. When the pain begins, patients press a button on the battery pack that sends a tiny current to the electrode. The current is conducted through the skin and blocks the pain impulses.

TENS can be used to reduce pain after surgery. One study showed that this technique reduced the amount of painkillers that doctors had to administer by two-thirds and cut hospital stays by one or two days. Someday, dentists may use TENS instead of novocaine. TENS may also be used to reduce the pain of childbirth and could help treat the pain that athletes suffer.

TENS works like acupuncture, although part of its success may be psychological. In a study of 93 patients with chronic pain, researchers found that over one-third (36%) of them reported no pain or greatly reduced pain when they thought they were being stimulated but really were not. How effective was it when the battery really worked? About half of the patients reported no pain or greatly reduced pain. The slight difference suggests to some physicians that TENS may be overrated. To those suffering from chronic pain, it can be a godsend.

Severe pain can also be treated by surgery. Doctors may cut nerves or destroy small parts of the brain to get rid of chronic pain. Unfortunately, pain recurs in 9 of 10 patients who have undergone surgery, usually within a year or so. Recurring pain results from the partial regrowth of axons. Even after another operation, the pain frequently returns, often with much greater intensity. Consequently, physicians are now looking for alternative measures to eliminate chronic pain.

One promising measure is deep brain stimulation. Electrodes can be implanted in parts of the brain and stimulated to block pain impulses before they reach the sensory cortex, where pain is perceived. The electrodes are connected to a portable battery worn on the belt or implanted under the skin. When the pain begins, the patient turns on the current, blocking the pain impulses.

Research shows that deep brain stimulation is an effective blocker of even the most powerful pain stimuli. Yet it does not upset other brain functions. Unfortunately, this technique requires surgery, and implanting electrodes may cause hemorrhaging in some patients, which can result in permanent paralysis or a loss of feeling in parts of the body. Infections may also develop where the electrodes enter the skull. The 75% success rate and the reduced suffering make most patients more than willing to accept the risks.

tually structural variations of the Meissner's corpuscle and are therefore stimulated by light touch.

Proprioception (sense of position) is provided by special encapsulated receptors located in the joints of the body. Resembling Meissner's corpuscles, these receptors inform us of the position of our limbs and alert us to movements of the body. Proprioception is also served by the muscle spindle and the Golgi tendon organ (▻ Figure 18–6). **Muscle spindles,** or **neuromuscular spindles,** are found in the skeletal muscles of the body. A muscle spindle consists of several modified muscle fibers with sensory nerve endings wrapped around them; a thin capsule of connective tissue surrounds the entire structure. When a muscle is extended or stretched, spindle fibers are stimulated. Nerve impulses generated by the spindle are then transmitted to the spinal cord via sensory nerves. Impulses reaching the spinal cord may ascend to the cerebral cortex, helping us remain aware of body position. The nerve impulses generated by muscle spindles (caused by stretching the muscle) may also stimulate motor neurons in the spinal cord, eliciting a reflex contraction of the muscle that is being stretched. Contraction of the muscle counteracts the stretching.

Golgi tendon organs are mechanoreceptors that are functionally similar to the muscle spindle but are located in **tendons**—the structures that connect muscles to bones. Also known as **neurotendinous organs,** Golgi tendon organs are composed of connective tissue fibers surrounded by dendrites and encased in a capsule (Figure 18–6). When a muscle contracts, the tendon stretches and stimulates the receptor. Like the muscle spindle, this receptor alerts the brain to movement and body position. Impulses from the Golgi tendon organ can also stimulate reflex contraction of muscles. The knee-jerk reflex described in the last chapter is a good example.

FIGURE 18–6 **Stretch Receptors** The muscle spindle and Golgi tendon organ both detect muscle stretch. The dendrites of the sensory neurons from these receptors send impulses to the spinal cord. These impulses may trigger motor neurons to fire; returning signals cause the muscle to contract. Sensory signals reaching the spinal cord also ascend to the brain, alerting it to the muscle movement.

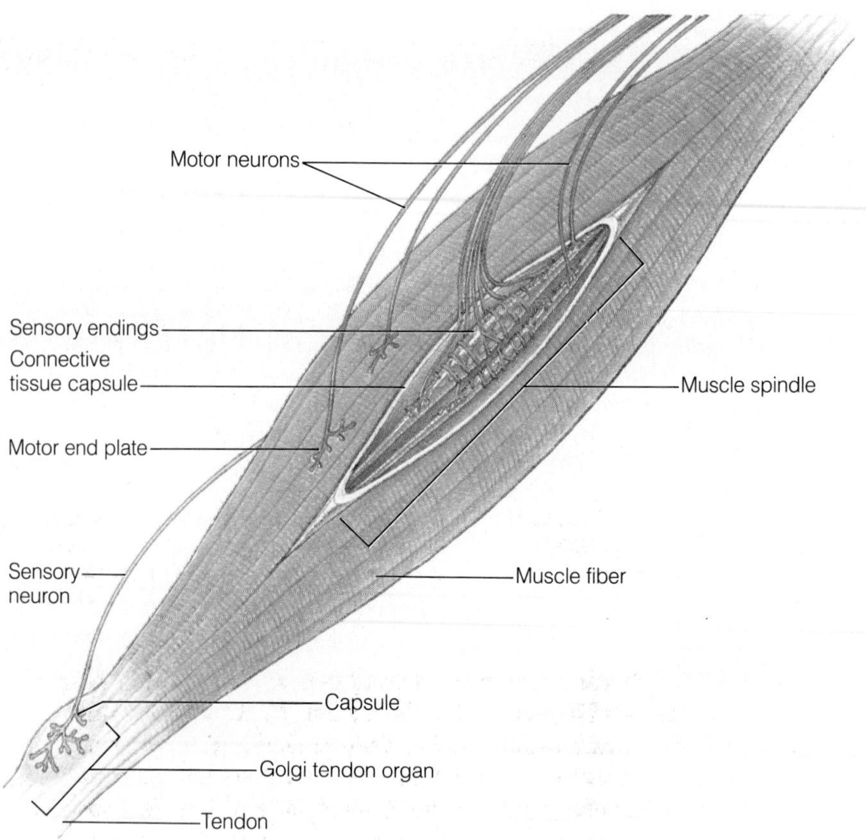

Many Receptors Stop Generating Impulses After Exposure to a Stimulus For Some Length of Time

Pain, temperature, and pressure receptors are all subject to a phenomenon known as sensory adaptation.[2] **Adaptation** occurs when sensory receptors stop generating impulses even though the stimulus is still present. You have probably witnessed the phenomenon dozens of times in your lifetime. Recall, for example, the first time you wore a ring or contact lenses. At first, the sensation may have nearly driven you mad, but after a few days or perhaps a few hours, the discomfort waned, and the stimulus seemed to have disappeared. Pressure receptors that were originally alerting the brain stopped generating impulses, relieving you of what would have otherwise been unrelenting discomfort.

Interestingly, not all receptors adapt. Muscle stretch receptors and joint proprioceptors are two examples. Because the CNS must be continuously apprised of muscle length and joint position to maintain posture, adaptation of these receptors would be counterproductive.

Receptors Play an Important Role in Homeostasis

Many general sense receptors play an important role in homeostasis and no doubt evolved because of the selective advantage they conferred on organisms. Mechanoreceptors that detect changes in blood pressure and chemoreceptors that respond to the ionic concentration of the blood are important in maintaining proper water balance and blood pressure. These receptors help regulate blood volume and blood concentration through mechanisms involving the kidney. Chemoreceptors that detect levels of carbon dioxide and hydrogen ions in the blood and cerebrospinal fluid also help regulate respiration, a function vital to normal body function. In Chapter 20, you will study additional chemoreceptors that detect levels of various hormones, nutrients, and ions in the blood and body fluids. These detectors may stimulate hormonal responses that correct potentially disruptive chemical imbalances.

[2] Not to be confused with the adaptation occurring in evolution.

TASTE AND SMELL: THE CHEMICAL SENSES

The special senses are taste, smell, vision, hearing, and balance. The following sections discuss the special senses, beginning with taste.

Taste Buds Are the Receptors for Taste and Respond to Chemicals Dissolved in the Food We Eat

In humans and other mammals, the tongue contains receptors for taste. Known as the **taste buds,** these microscopic,

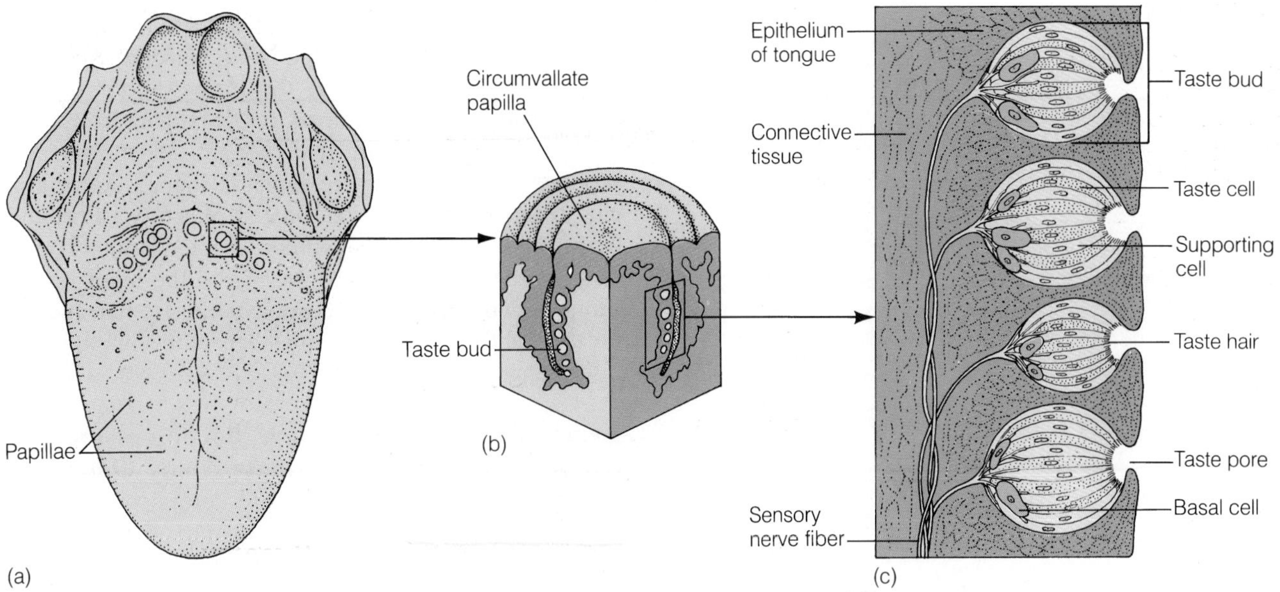

▻ **FIGURE 18–7 Taste Receptors** (*a*) Taste buds are located on the upper surface of the tongue and are concentrated on the (*b*) papillae. (*c*) The structure of the taste bud.

onion-shaped structures are located in the surface epithelium of the tongue and on small protrusions, **papillae,** on the upper surface of the tongue (▻ Figure 18–7a). Taste buds are also found on the roof of the oral cavity (on the hard and soft palates), the pharynx, and the larynx, but in smaller numbers.

Taste buds are chemoreceptors and are stimulated by ions and molecules in the food we eat. These substances dissolve in the saliva and enter the **taste pores,** small openings that lead to the interior of the taste bud. As Figure 18–7c illustrates, taste buds consist of two cell types, receptor cells and supporting cells (which recent evidence suggests may be the same cell type, just in a different phase of its life cycle). The **supporting cells** lie on the outside of the taste bud and resemble the staves of a wooden barrel. The **receptor cells** lie inside, like so many pickles jammed lengthwise into a jar. The ends of the receptor cells possess large microvilli, which are known as **taste hairs.** Taste hairs project into the taste pore. The membrane of the taste hairs contains receptors that bind to food molecules dissolved in water. This stimulates the receptor cells, which then stimulate the dendrites of the sensory nerves that are wrapped around the receptor cells.[3] Impulses from the taste buds are then transmitted to the taste centers in the cerebral cortex.

In the opening paragraph of this chapter, you were introduced to an unfortunate student who lost her sense of taste because of a dietary zinc deficiency. Research has shown that zinc, which is normally found in low concentrations in the saliva, stimulates division of the cells in taste buds. Because cells of the taste buds are lost from normal wear and tear, a zinc deficiency reduces cellular replacement, and the taste buds eventually cease operation.

The Taste Buds Respond to all Four Primary Flavors but Are Generally Preferentially Responsive To One. Humans can discriminate among thousands of taste sensations. The taste sensations are a combination of four basic flavors: sweet, sour, bitter, and salty. Sweet flavors result from sugars and some amino acids. Sour flavors result from acidic substances. Salty tastes result from metal ions (like sodium in table salt). Bitter flavors result from chemical substances belonging to a group called alkaloids, among them caffeine, but also some non-alkaloid substances, such as aspirin.

All taste buds respond, in varying degree, to all four taste sensations, but they respond preferentially to one taste. Taste buds are also distributed unevenly on the surface of the tongue. A simple experiment in which drops of various substances are placed on the tongue shows that the tip of the tongue is most sensitive to sweet flavors because it contains a higher proportion of taste buds that respond preferentially to sweet flavors (▻ Figure 18–8). The sides of the tongue are most sensitive to sour flavors, and the back of the tongue is most sensitive to bitter flavors. Salty taste is more evenly distributed, with slightly increased sensitivity on the sides of the tongue near the front.

Food contains many different flavors. What we taste, therefore, depends on the relative proportion of the four basic flavors. For example, grapefruit juice tastes sour because of the predominance of acidic substances. Adding sugar to grapefruit juice, however, gives it a sweet-sour

[3] Note: unlike the receptors in the previous section, the receptor cells of taste buds are not part of nerve cells but are independent cells.

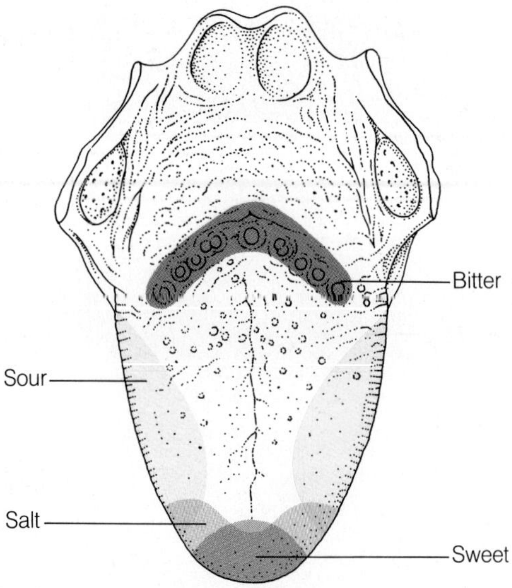

▻ **FIGURE 18–8 The Distribution of Taste on the Upper Surface of the Tongue**

flavor. If you add enough sugar, you can mask the sour taste almost entirely. As you will soon see, the sense of smell also plays a role in determining the taste of food.

The Olfactory Epithelium Is a Patch of Receptor Cells that Detects Odors

Smell is also a chemical sense. The receptors for smell are located in the roof of each nasal cavity in a patch of cells called the **olfactory epithelium,** or **membrane** (▻ Figure 18–9). The olfactory membrane contains receptor cells and supporting cells. **Receptor cells** are neurons whose cell bodies lie in the olfactory membrane. The dendrites of these neurons extend to the surface of the olfactory membrane, terminating in six to eight long projections, **olfactory hairs,** or **olfactory cilia** (Figure 18–9). The membranes of the olfactory hairs, like those in the receptor cells of the taste bud, contain receptors for molecules. When airborne molecules are trapped on the thin, watery layer on the surface of the cell, they bind to the receptors, activating the neurons. This causes the neurons to generate impulses that are sent to the **olfactory bulb,** a complex neural structure that contains neurons that synapse with the dendrites of the olfactory receptor cells. Axons of the neurons in the olfactory bulb cells then travel to the brain via the **olfactory nerve.** Supporting cells are interspersed among receptor cells.

Humans can distinguish tens of thousands of odors, many at very low levels. The olfactory receptors are so sensitive, in fact, that even a single molecule binding to the olfactory hairs can produce a bioelectric impulse. But the sense of smell in humans is not as sensitive as that of other animals such as dogs, wolves, and coyotes. A dog's keen sense of smell is due to the presence of an olfactory membrane about 20 times larger than that of humans.

Like taste, odor discrimination is thought to depend on combinations of primary odors. Unfortunately, neuroscientists do not agree on what the primary odors are. One system suggests seven primary odors ranging from pepperminty to floral to putrid. It is thought that molecules that produce a similar odor are similarly shaped and that specific receptor binding sites on the olfactory hairs bind to those molecules with a common shape. Various combinations of the primary odors give rise to the many odors we can perceive. Some researchers hypothesize that there may be thousands of kinds of smell receptors, providing odor discrimination.

Even Though the Olfactory Receptors are Sensitive and Able to Discriminate Many Odors, They Quickly Become Adapted. Receptors for smell adapt within a short period—about a minute. If you have ever worked on a dairy farm or lived near one, or if you have been around a baby in diapers on a long car trip, you understand (and are grateful for) olfactory adaptation.

Smell Influences the Sense of Taste, and Vice Versa. Hold a piece of hot apple pie to your nose, and take a deep breath. It smells so good you can almost taste it. In fact, you *are* tasting it. Molecules given off by the pie enter the nose, reach the mouth through the pharynx, and dissolve in the saliva, where they stimulate taste receptors. Just as odors stimulate taste receptors, food in our mouths also stimulates olfactory receptors. Molecules released by food enter the nasal cavities, dissolving in the water on the surface of the olfactory membrane and stimulating the receptor cells.

The complementary nature of taste and smell is abundantly evident when a person suffers from nasal congestion. As you have probably noticed, a viral infection in the nose often makes food seem bland. Cold sufferers complain that they cannot taste their food. This phenomenon results from the buildup of mucus in the nasal cavities, which blocks the flow of air. Mucus may also block the olfactory hairs. Therefore, food loses its "taste" when you have a stuffy nose because your sense of smell is impaired. (You can test this by holding your nose while you eat.)

The Ability to Detect Various Chemicals in the Environment—that is, Chemoreception—Is a Universal Trait of Animals and Even Protists

When it comes to finding a mate, recognizing their territory or the territory of others, or moving about in the environment, most animals depend heavily on chemoreceptors. In insects, for example, females give off small quantities of chemical substances (known as pheromones) that attract males. This adaptation greatly increases the likelihood of breeding and depends on special chemoreceptors in the males that can detect pheromones in parts-per-billion concentrations. Interestingly, many fishes have taste buds all over their body. Chemical receptors are even found in

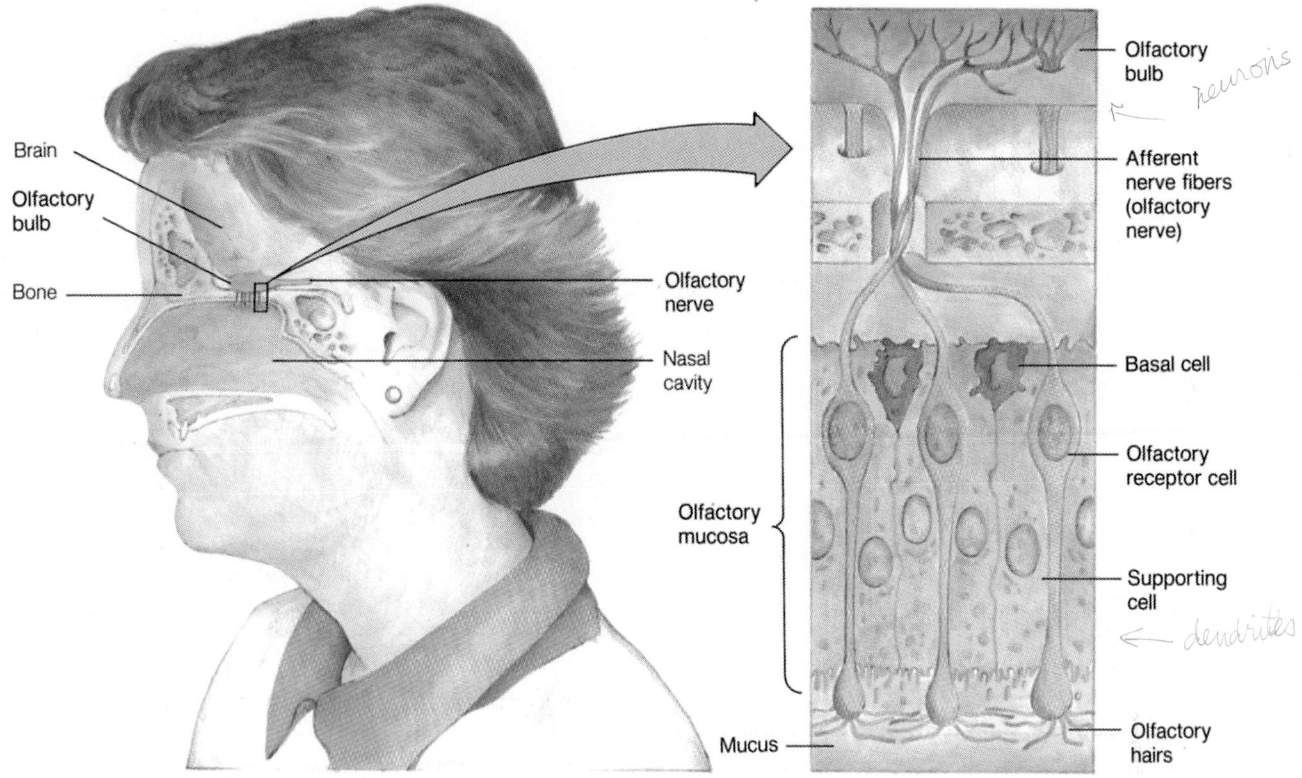

▻ **FIGURE 18–9 Location and Structure of the Olfactory Epithelium** Olfactory receptors are located in the olfactory epithelium in the roof of the nasal cavity. Chemicals in the air dissolve in the watery fluid bathing the surface of the cells, then bind to receptors on the plasma membranes of the olfactory hairs. The olfactory receptors terminate in the olfactory bulb. From here, nerve fibers travel to the brain.

some of the most primitive members of the animal kingdom, such as hydra and planaria. In hydra, for example, special chemoreceptors pick up the scent of wounded prey and induce swallowing movements that cause the prey to be ingested.

Most of what is known about chemoreception in animals other than vertebrates comes from studies of insects. As shown in ▻ Figure 18–10, insects have sensory hairs on their mouthparts and feet. To taste food, all an insect has to do is land on it. The sensory hairs therefore help insects determine what is food and what is not. As a rule, the sensory hairs contain more than one receptor cell, each of which responds maximally to a different chemical. Just as in humans, the pattern of responses probably helps insects discriminate among many tastes. Besides alerting an insect to the presence of food and helping it distinguish the type of food, the sensory hairs activate or inhibit the feeding behavior.

Some sensory hairs detect airborne chemicals, such as pheromones. Flying insects, such as moths, use their olfactory sensory hairs to detect mates but also to locate food plants.

THE VISUAL SENSE: THE EYE

The vertebrate eye is one of the most extraordinary products of evolution. It contains a patch of photoreceptors that permits us to perceive the remarkably diverse and colorful environment we live in.

The Vertebrate Eye Houses the Light-Sensitive Layer Known as the Retina

In humans, the eyes are roughly spherical organs located in the eye sockets, or **orbits,** cavities formed by the bones of the skull. The eye is attached to the orbit by six muscles, the **extrinsic eye muscles,** which control eye movement. Small tendons connect these muscles to the outermost layer of the eye. The location of the eye of a vertebrate—whether on the side of the head or in front—tells you a great deal about its lifestyle. If the eyes are in front, it is most likely a predator (▻ Figure 18–11a). If the eyes are on the side of the skull, the animal is most likely a prey species—that is, an animal hunted and killed by predators (Figure 18–11b). The location of the eyes on the side of the head gives a prey species a wider field of vision and increases its ability to spot a potential predator.

The Sclera and Cornea. As shown in ▻ Figure 18–12, the wall of the human eye consists of three layers (Table 18–2). The outermost is a durable fibrous layer, which consists of the **sclera,** the white of the eye, and the **cornea,** the clear part in front, which lets light into the interior of

⊳ **FIGURE 18–10 Sensory Hairs** Insects have sensory hairs on their mouthparts and feet that help them determine what is food and what is not. Sensory hairs usually contain more than one receptor cell, each of which responds maximally to one chemical.

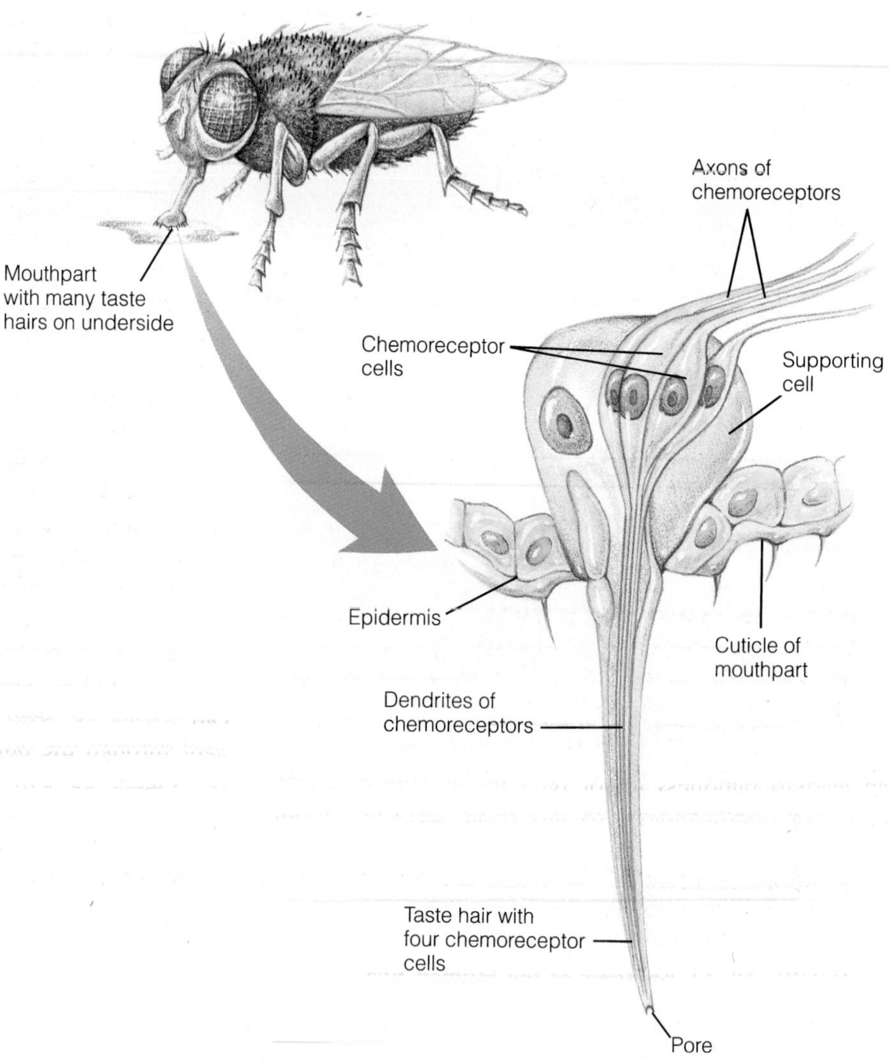

the eye (Figure 18–12). Tendons of the extrinsic eye muscles attach to the sclera.

The Choroid, Ciliary Body, and Iris. The middle layer is a heavily pigmented and highly vascularized region. As Figure 18–12 shows, the middle layer consists of three parts: the choroid, the ciliary body, and the iris. The **choroid** is the largest portion of the middle layer. It contains a larger amount of melanin, a pigment that absorbs stray light the way the black interior of a camera does. The blood vessels of the choroid supply nutrients to the eye. Anteriorly, the choroid forms the ciliary body. The **ciliary body** contains smooth muscle fibers, which control the shape of the lens, permitting us to focus incoming light given off by objects or reflected from their surface. The third part of the middle layer is the iris. The **iris** is the colored portion of the eye visible through the cornea. Looking in a mirror, you can see a dark opening in the iris called the pupil. The **pupil** allows light to penetrate the eye. The blackness you see through the pupil is the choroid layer and the pigmented section of the retina, discussed below. Like the ciliary body, the iris contains smooth muscle cells. The smooth muscle of the iris regulates the diameter of the pupil. Opening the pupil lets more light in, and narrowing it reduces the amount of light that can enter. The pupils open and close reflexively in response to light intensity. This reflex is an adaptation that protects the light-sensitive inner layer, the retina. The pupils also constrict when the eye focuses on a near object, a process discussed later.

The Retina. The innermost layer of the eye is the **retina.** The retina consists of an outer, pigmented layer that complements the light-absorbing function of the choroid layer, and an inner layer, the neural layer, consisting of photoreceptors and associated nerve cells. The retina is weakly attached to the choroid and can become separated from it as a result of trauma to the head. A detached retina

▻ **FIGURE 18–11 Eye Location** You can tell a lot about an animal by the location of its eye. (*a*) Predators have forward-facing eyes, which give them three-dimensional vision necessary for capturing prey. (*b*) Prey species have eyes on the side of their heads for better viewing the approach of predators.

can lead to blindness if not repaired by surgery. Today, doctors of ophthalmology usually repair detached retinas with lasers.

The **photoreceptors** of the retina are modified nerve cells located in the outermost portion of the neural layer, adjacent to the pigmented layer. Two types of photoreceptors are present in the retina: rods and cones (Table 18–3). The **rods,** so named because of their shape, are sensitive to low light (▻ Figures 18–13b and 18–13c). Thus, the rods function on moonlit evenings, yielding grayish, somewhat vague images. The **cones,** also named because of their shape, are photoreceptors that sense colors and operate only in brighter light. They are responsible for visual acuity—sharp vision.

In order for light to stimulate the photoreceptors, it must pass through the ganglion and bipolar cell layers, two layers of neurons that lie in front of the photoreceptors in the retina (Figure 18–13). As Figure 18–13b shows, the rods and cones synapse with the bipolar neurons. These, in turn, synapse with ganglion cells. Nerve impulses from the photoreceptors travel from the bipolar neurons to the ganglion cells. The axons of the ganglion cells course along the inner surface of the retina and unite at the back of the eye to form the **optic nerve.** The optic nerve leaves at the **optic disc,** or **blind spot,** so named because it contains no photoreceptors and is therefore insensitive to light. The blood vessels that enter and leave the eye do so with the optic nerve. These blood vessels and their branches can readily be seen by an ophthalmologist by shining a light through the pupil onto the posterior wall of the eye (▻ Figure 18–14).

Rods and cones are found throughout the retina, but the cones are most abundant in a tiny region of each eye lateral to the optic disc. This spot is called the **macula lutea,**

▻ **FIGURE 18–12 Anatomy of the Human Eye**

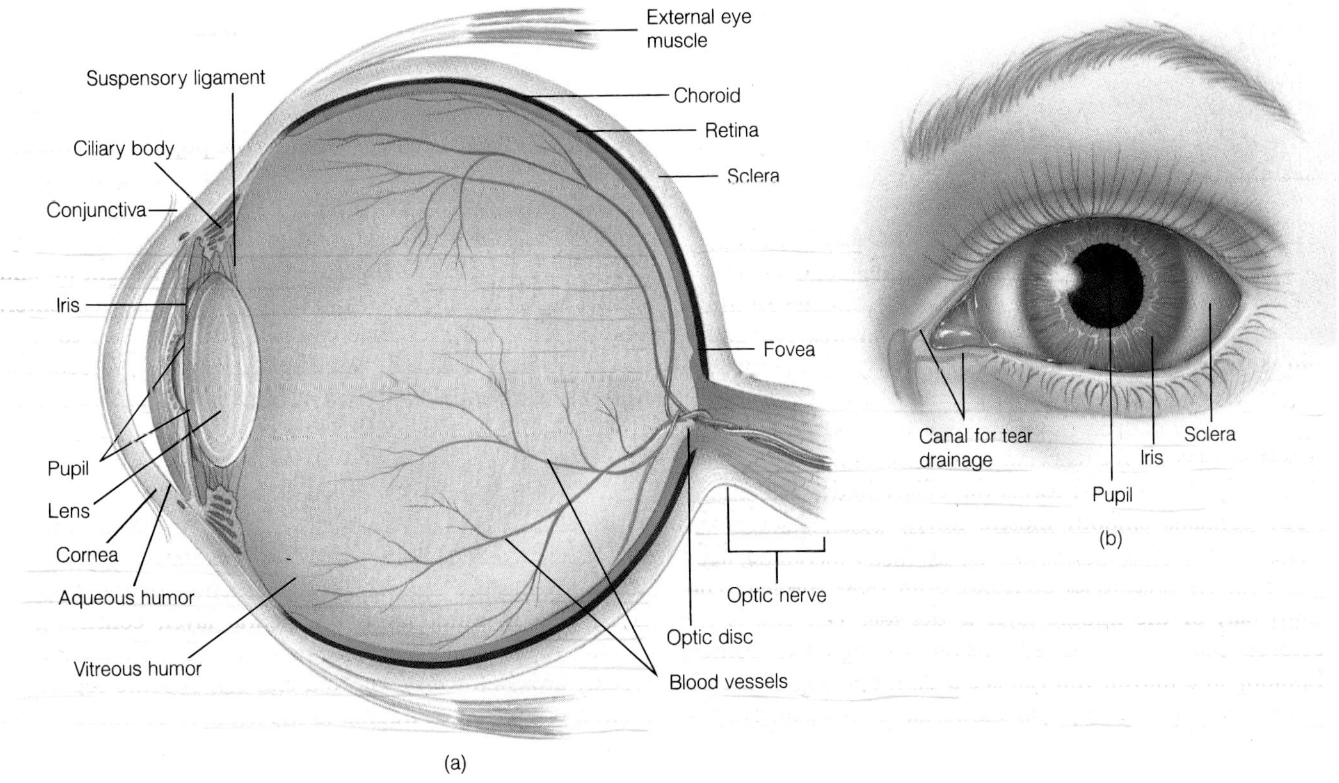

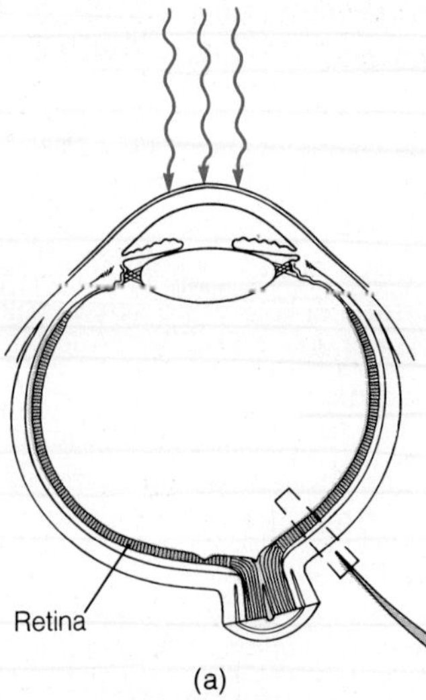

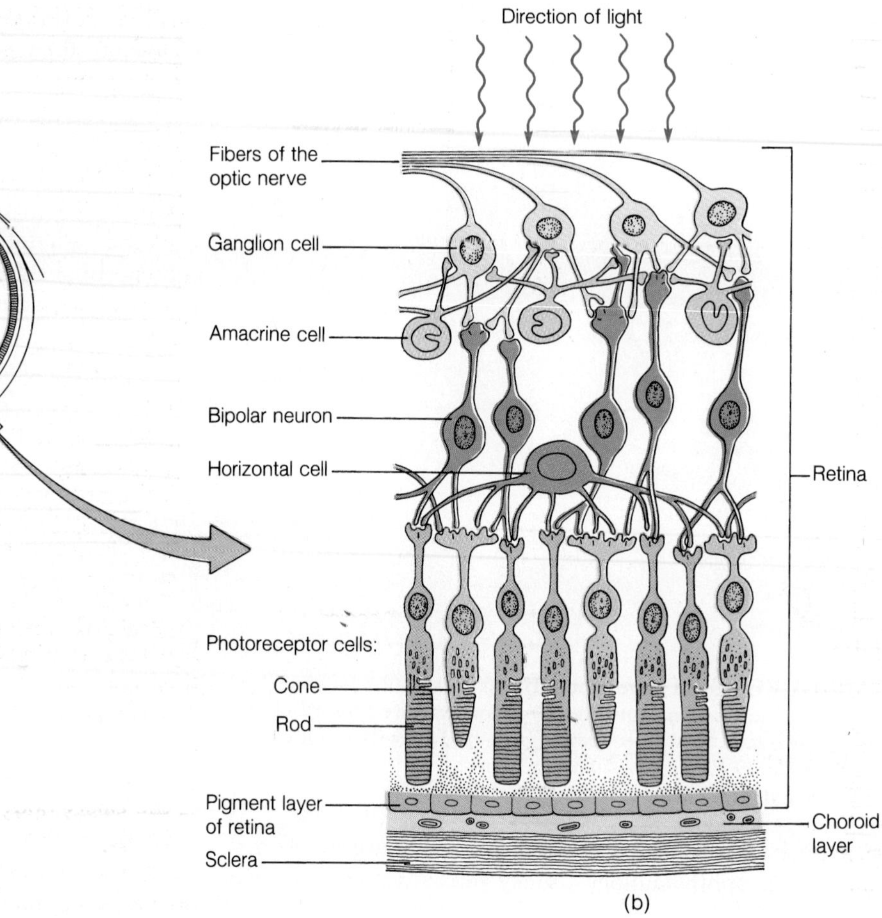

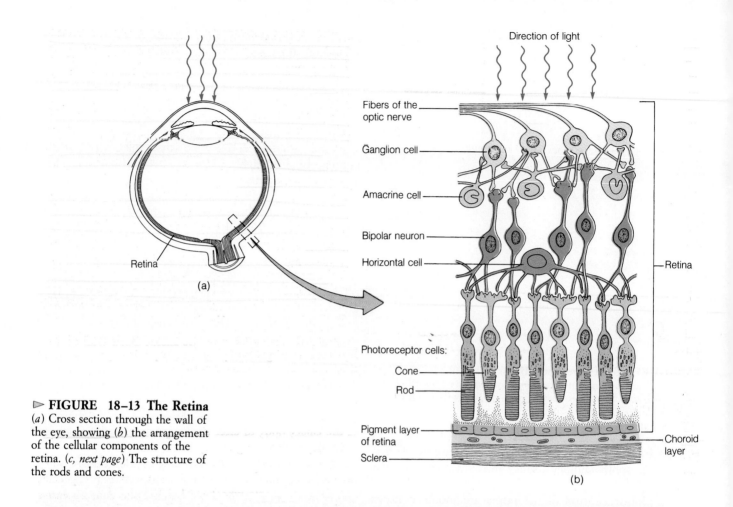

▻ **FIGURE 18–13 The Retina** (*a*) Cross section through the wall of the eye, showing (*b*) the arrangement of the cellular components of the retina. (*c, next page*) The structure of the rods and cones.

TABLE 18–2 Structures and Functions of the Eye

STRUCTURE		FUNCTION
Wall		
Outer layer	Sclera	Provides insertion for extrinsic eye muscles
	Cornea	Allows light to enter; bends incoming light
Middle layer	Choroid	Absorbs stray light; provides nutrients to eye structures
	Ciliary body	Regulates lens, allowing it to focus images
	Iris	Regulates amount of light entering the eye
Inner layer	Retina	Responds to light, converting light to nerve impulses
Accessory structures and components		
	Lens	Focuses images on the retina
	Vitreous humor	Holds retina and lens in place
	Aqueous humor	Supplies nutrients to structures in contact with the anterior cavity of the eye
	Optic nerve	Transmits impulses from the retina to the brain

TABLE 18–3 Summary of Rods and Cones

PHOTORECEPTOR	DAY OR NIGHT	COLOR VISION	LOCATION
Rods	Night vision	No	Highest concentration in the periphery of the retina
Cones	Day vision	Yes	Highest concentration in the macula and fovea

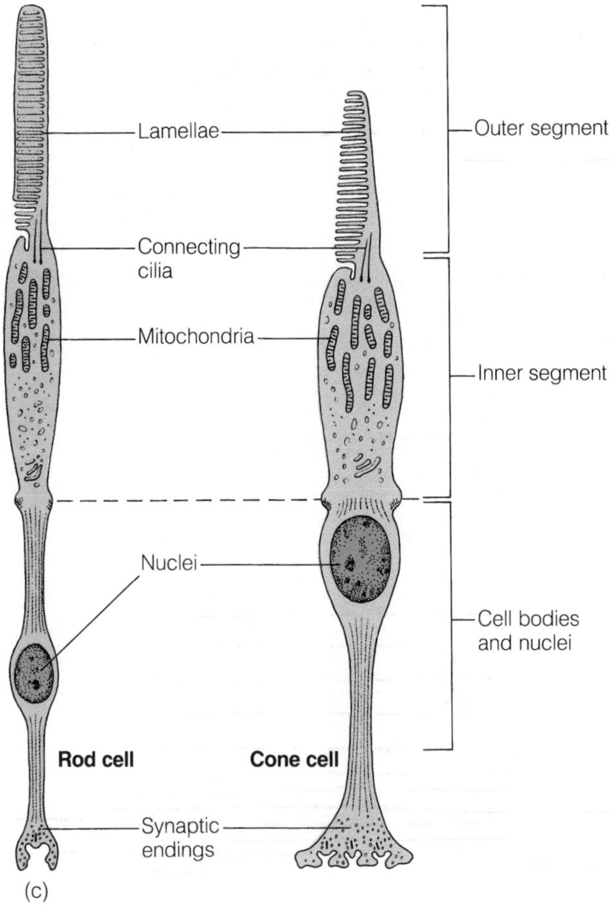

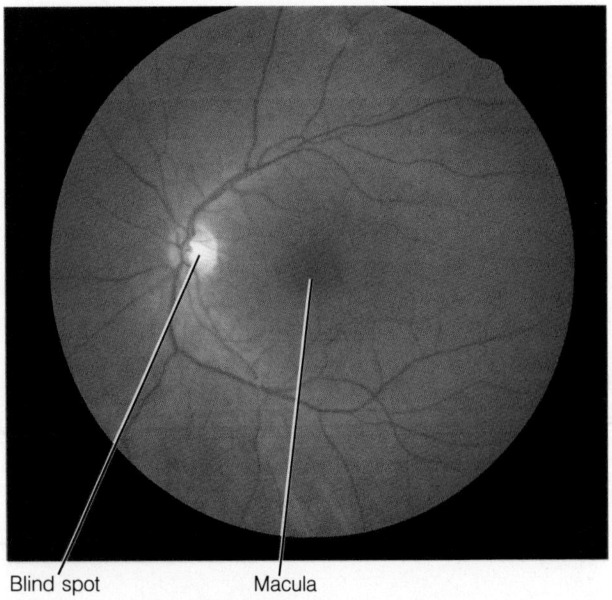

▻ **FIGURE 18–14 View of the Inside Back Wall of the Eye Seen through an Ophthalmoscope** The optic disk (blind spot) and macula are both indicated.

literally "yellow spot." In the center of the macula is a minute depression, about the size of the head of a pin, known as the **fovea centralis** (central depression). The fovea contains only cones. The number of cones in the retina decreases progressively from this point outward, whereas the number of rods increases. The greatest concentration of rods is found in the periphery of the retina.

Images from our visual field are cast onto the retina, and impulses are transmitted to the visual cortex of the brain's occipital lobe, where they are interpreted. When we focus on an object, the image is projected upside down onto the fovea. The sharpest vision occurs at the fovea, because it contains the highest concentration of cones and because the bipolar neurons and ganglion cells do not cover the cones in this region as they do throughout the rest of the retina. Even though the image is upside down, the brain processes the information and gives us a right-side-up image; that is, it allows us to perceive objects in their true position.

The Lens. Light is focused on the retina by the lens. The **lens** is a transparent structure that lies behind the iris (▻ Figure 18–15). This flexible, clear structure is attached to the ciliary body by thin fibers composing the **suspensory ligament.** This connection allows the smooth muscle of the ciliary body to alter the shape of the lens, an action necessary for focusing the eye.

In older individuals, the lens may develop cloudy spots, known as **cataracts.** The loss of transparency is especially prevalent in people who have been exposed to excessive sunlight or excessive ultraviolet light at work or elsewhere. Patients with this disease complain of cloudy vision. Looking out on the world to them is a little like looking through frosted glass. Interestingly, cataract risk may increase as the Earth's ozone layer continues to be eroded by pollutants such as chlorofluorocarbons from refrigerators, air conditioning units, and other sources, as discussed in Chapter 33. As noted earlier in the book, the ozone layer blocks harmful ultraviolet light.

Recent research suggests that the color of an individual's eyes affects the risk of developing a cataract. Dark-eyed people run the highest risk of cataracts. Brown- and hazel-eyed subjects had more cataracts than did blue-, gray-, and green-eyed patients. Researchers suggest that melanin in the irises of dark-eyed people may absorb solar radiation, causing more damage to the nearby lens. To lower your risk of cataracts, many eye doctors suggest wearing sunglasses with a coating that reduces ultraviolet penetration. But beware, there is evidence that some manufacturers' claims about ultraviolet protection may be inaccurate.

Ophthalmologists once treated cataracts surgically by removing the afflicted lens. Patients were then fitted with a thick pair of glasses or a pair of contact lenses, which compensated for the missing lens. Today, however, lenses are routinely replaced by artificial plastic lenses that provide nearly normal vision.

Research has shown that cataracts may result when a

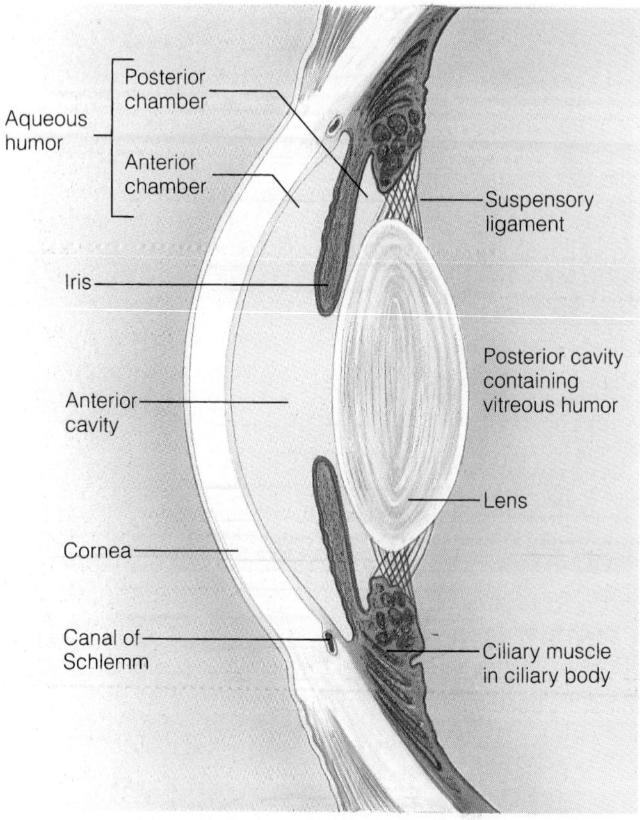

▻ **FIGURE 18–15 Detailed Cross Section of the Anterior Cavity.** Arrows show the flow of aqueous humor.

lens protein called **crystallin** denatures. Researchers are hoping that new drugs may be able to reverse the process, helping restore vision without surgery.

The lens separates the interior of the eye into two cavities of unequal sizes. Everything in front of the lens is the **anterior cavity;** everything behind it is the **posterior cavity.** The posterior cavity is filled with a clear, gelatinous material, the **vitreous humor** ("glassy liquid"). Formed during embryonic development, the vitreous humor remains throughout life, holding the lens and retina in place. Being clear, the vitreous humor transmits light faithfully to the retina.

The anterior cavity is divided into two portions: the **anterior chamber** and the **posterior chamber** (Figure 18–15). The anterior chamber lies between the iris and the cornea; the posterior chamber lies between the iris and the lens. A thin liquid, chemically similar to blood plasma, fills the anterior and posterior chambers of the eye and is called the **aqueous humor** ("watery fluid").

Unlike the vitreous humor, the aqueous humor is constantly replaced. New fluid is produced by capillaries in the ciliary body. The fluid enters the posterior chamber, then flows forward into the anterior chamber, where it drains into a canal known as the **canal of Schlemm,** which is located at the junction of the sclera and cornea. From here, the plasmalike fluid flows into the bloodstream.

Aqueous humor provides nutrients to the cornea and lens and carries away cellular wastes. In normal, healthy individuals, aqueous humor production is balanced by absorption. If the outflow is blocked, however, aqueous humor builds up inside the anterior chamber, creating internal pressure. This disease, called **glaucoma,** occurs gradually and imperceptibly. If untreated, the pressure inside the eye can damage the retina and optic nerve, causing blindness. Because the incidence of glaucoma increases after age 40, doctors recommend an annual eye examination for men and women over 40. If diagnosed early, glaucoma can be treated with eye drops that increase the rate of drainage, thus reducing intraocular pressure. In severe cases, surgery may be required to increase the outflow.

The Lens Focuses Light on the Retina

To understand how the lens operates, you have to understand a little bit about light.

Refraction. First of all, visible light travels in waves. Light waves travel at a constant rate in any given medium such as air or water. When light passes from one medium to another, however, its velocity changes. When light passes from a less dense medium, such as air, to a denser medium, such as the cornea, it slows down. Anytime light changes speed in passing from one substance to another, it bends. The bending of light is called **refraction** (▻ Figure 18–16).

Focusing the Image. Light traveling through a camera lens is bent. The lens of the camera is designed to bend light enough to focus the image on the film. The lens of the eye also bends incoming light rays, focusing images on the photoreceptors of the retina. Lying in front of the lens is the cornea; it also bends incoming light rays (▻ Figure 18–17). Although we usually think of the lens as the structure that allows us to focus, most of the bending of incoming rays takes place in the cornea. However, the cornea is a fixed structure. Like the lens on a fixed-focus camera, the cornea cannot be adjusted to focus on nearby objects. Without the adjustable lens, the eye would be unable to focus on objects close at hand.

The lens is resilient like a rubber ball. Its shape is controlled by the muscles in the ciliary body that are attached via the suspensory ligament. When the muscles of the ciliary body are relaxed, the suspensory ligament is taut, and the lens is somewhat flattened. As Figure 18–17a shows, light from distant objects comes to the eye as nearly parallel rays. The fixed refractive power of the cornea and the refractive power of the lens in its relaxed state are sufficient to bend these beams to bring them into focus on the retina.

As Figure 18–17b shows, light from nearby objects is divergent. To focus on nearby objects, the lens must become more curved. When the eye focuses on nearby objects, the ciliary smooth muscle contracts, causing the suspensory ligament to relax (▻ Figure 18–18b). The lens thickens and shortens, becoming more curved, in much the

▻ **FIGURE 18–16 Refraction** The pencil in the glass of water appears to bend. What is happening, though, is that the light rays coming to our eyes are bent as they pass through the water and glass, making the pencil appear bent.

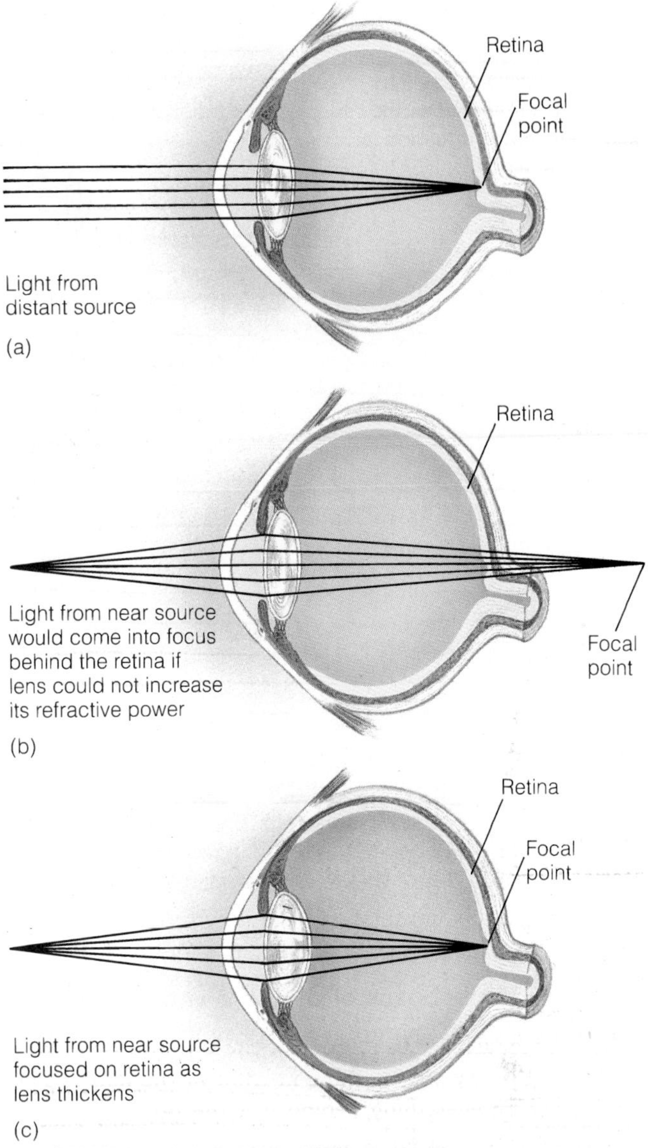

▻ **FIGURE 18–17 Refraction of Light by the Cornea and Lens** (*a*) Light rays from distant objects are parallel when they strike the eye. The refractive power of the cornea and resting lens are sufficient to bring them into focus on the retina. (*b*) Light rays from nearby objects are divergent. The cornea has a fixed refractive power and cannot change. The rays would focus behind the retina if the lens could not alter its refractive power. (*c*) To focus the image, the ciliary muscles contract. This lessens pressure on the suspensory ligaments, which allows the lens to thicken and shorten, becoming more curved and more refractive.

same way that a rubber ball flattened between your hands will return to normal when you reduce the pressure on it. The automatic adjustment in the curvature of the lens required to focus on a nearby object is called **accommodation** (Figure 18–18). Accommodation increases the refractive (bending) power of the lens.

Accommodation is enhanced by pupillary constriction. As the eyes focus on a nearby object, the pupils constrict. Pupillary constriction is a reflex that eliminates divergent rays of light that would otherwise strike the periphery of the lens. The lens would be unable to bend these rays sufficiently to bring them into focus on the retina. Thus, without pupillary constriction, images of nearby objects would be quite blurred.

Convergence. The human eyes are movable. Six muscles located outside the eye, the extrinsic eye muscles, are responsible for movement (▻ Figure 18–19). As noted earlier, these muscles attach to the orbit (the bony eye socket) and to the sclera (the white of the eye) and allow for a wide range of movements.

The eyes generally move in unison like a pair of synchronized swimmers. Synchronized movement is an evolutionary adaptation that ensures that images are focused on the foveas of both eyes at the same time. To test the synchronized movement, close one of your eyes. Place an index finger gently over the closed lid. Then hold the other hand in front of your face and move it back and forth, then up and down, following it with your opened eye. You should feel your closed eye moving in sync, even though the lid is shut.

▷ **FIGURE 18–18**
Accommodation (*a*) The lens is flattened when the ciliary muscles are relaxed. (*b*) When the ciliary muscles contract, tension on the suspensory ligaments is reduced and the lens shortens and thickens.

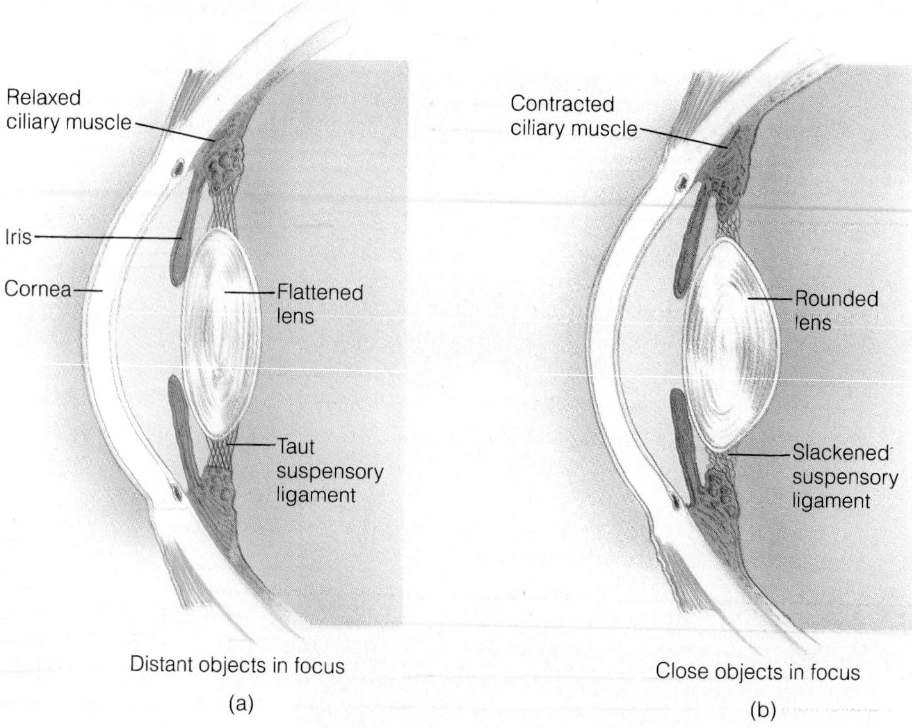

When a nearby object is viewed, the eyes turn inward, or converge. Convergence ensures that the image is focused on each fovea. Convergence occurs during all near-point work—reading, writing, sewing, knitting—and puts strain on the extrinsic eye muscles, contributing to eyestrain.

Alterations in the Shape of the Lens and Eyeball Cause the Most Common Visual Problems

In the normal eye, objects farther than 6 meters (20 feet) away fall into perfect focus on the back of the relaxed eye ▷ Figure 18–20a). Many individuals, however, have imperfectly shaped eyeballs or defective lenses. These imperfections result in two visual problems: nearsightedness and farsightedness.

▷ **FIGURE 18–19 Extrinsic Eye Muscles** These muscles move the eye in all directions. They attach to the bony orbit and the sclera.

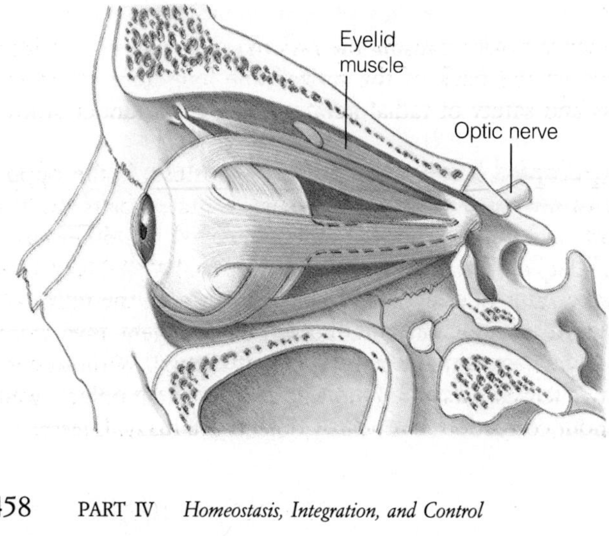

Myopia. Nearsightedness or **myopia,** results when the eyeball is slightly elongated (Figure 18–20b). Without corrective lenses, the parallel light rays arising from distant images fall into focus in front of the retina. In contrast, nearby images with much more divergent light rays tend to be in focus in the uncorrected eye. People with myopia, therefore, can see near objects without corrective lenses—hence the name nearsightedness. Myopia may also result when the lens is too strong—that is, too concave. This lens bends the light too much, causing the image to come into focus in front of the retina.

Myopia is quite common. Approximately one of every five Americans needs glasses to correct for it. Myopia is caused by many factors and tends to run in families; it generally appears around age 12, often worsening until a person reaches 20.

Myopia can be corrected by contact lenses or prescription glasses that cause incoming light rays to diverge (Figure 18–20b). Contact lenses fit on the surface of the cornea and bend the incoming light rays outward, compensating for the shape of the eye or a defective ocular lens.

Although useful in correcting vision, contact lenses can cause serious problems. The largest study of their use in the United States suggests that extended-wear contact lenses, those worn overnight and for up to a week at a time, carry a greater risk of serious complications than daily-wear lenses. A study of over 22,000 lens wearers showed that sight-threatening complications, such as corneal abrasion, growth of blood vessels into the cornea, corneal ulcers, and severe corneal scarring, were two to

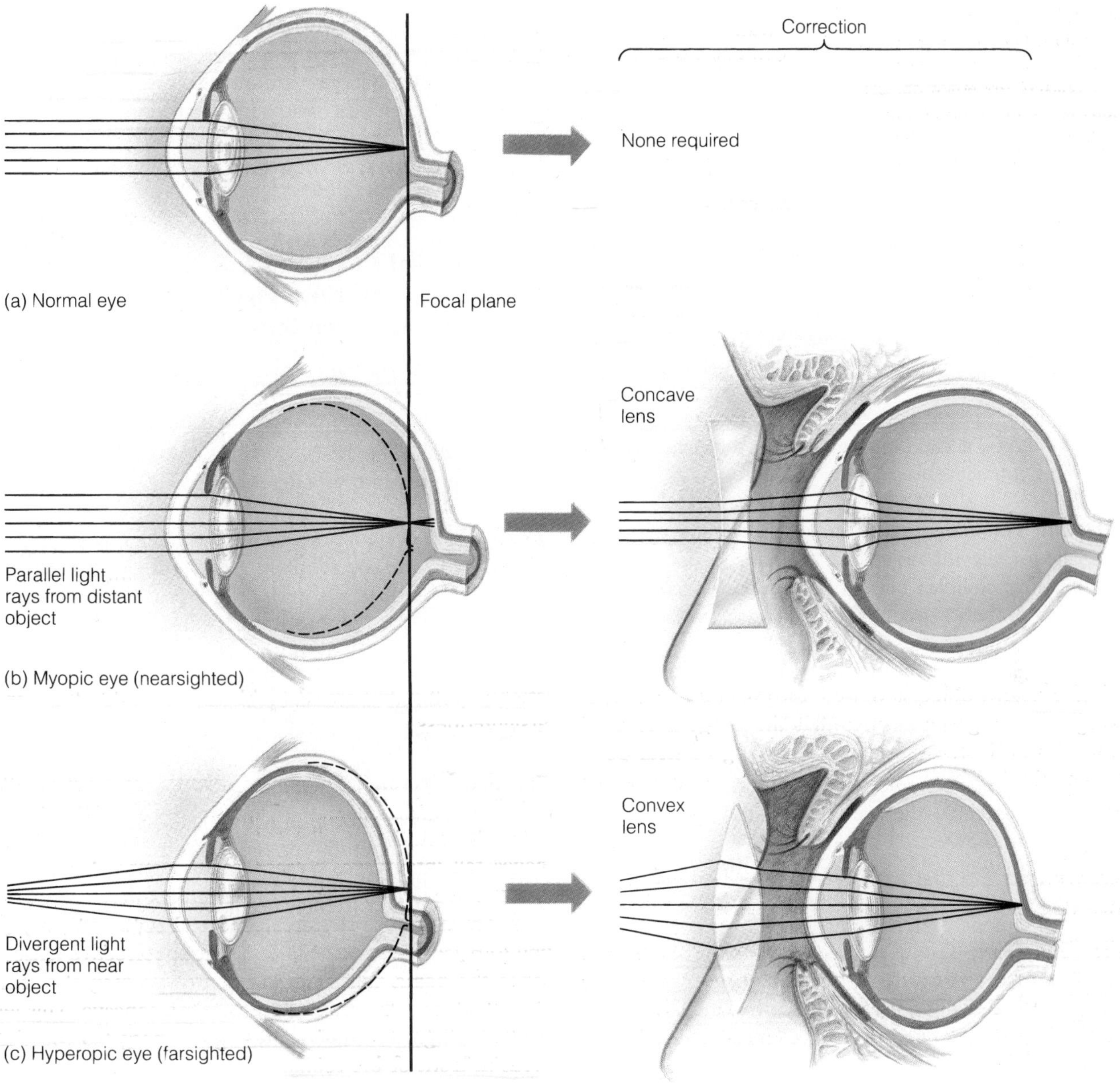

▻ **FIGURE 18–20 Common Visual Problems** (*a*) The normal eye, (*b*) the myopic (nearsighted) eye, and (*c*) the hyperopic (farsighted) eye.

four times more prevalent in users of extended-wear contacts than in those using daily-wear lenses.

Researchers found that 1 in 2000 daily-wear lens users contracted corneal ulcers; 1 in 300 suffered from other serious complications. In extended-wear users, however, 1 in 500 suffered from corneal ulcers, and 1 in 100 had other serious reactions. The increase in risk suggests one should think carefully before choosing extended-wear contacts.

Surgeons have also developed a method to decrease the refractive power of the cornea, called **radial keratotomy.** Numerous small, superficial incisions are made in the cornea, radiating from the center like the spokes of a bicycle wheel. This procedure flattens the cornea and reduces its refractive power, causing the rays to diverge and come into focus on the back of the retina. The long term effectiveness and safety of radial keratotomy are still under study.

Hyperopia. Hyperopia, or **farsightedness,** is the opposite of myopia. That is, it results when the eyeballs are too short or the lens is too weak (too convex). In either case, the light falls into focus behind the retina. Parallel light rays from distant objects usually fall into focus on the retina—hence the name farsightedness. But divergent rays from nearby objects cannot be focused sufficiently. Without corrective lenses, farsighted individuals see distant objects well without correction, but nearby objects are fuzzy. Glasses or contact lenses that bend the light inward help bring near objects into sharp focus on the retina (Figure 18–20c).

Hyperopia is generally present from birth and is usually diagnosed during childhood. Like myopia, it tends to run in families.

Astigmatism. In the normal eye, the cornea and lens have uniformly curved surfaces. Either of these structures may be slightly disfigured, however. The surface of the cornea, for example, may have a slightly different curvature in the vertical plane than it does in the horizontal plane. This unequal curvature is called an **astigmatism.** It creates fuzzy images because light rays are bent differently by the different parts of the lens or cornea. Astigmatism is usually present from birth and does not grow worse with age. Like other conditions, it can be corrected with specially ground glasses and contact lenses.

Eyestrain. For many years, humans have used their eyes principally for viewing objects at a distance—for example, watching their children play or wild animals roam. Near-point vision probably occupied little of their time. It should come as no surprise, then, to learn that the human eye is best suited for distance vision, as are most other vertebrate eyes. Intensive near-point vision so common in today's world strains the eyes and can result in a progressive deterioration of eyesight. Frequent readers often become more nearsighted as they become older. No one knows why, but research suggests that the eye may elongate as a result of constant near-point use.

To reduce eyestrain and deterioration of eyesight, ophthalmologists advise that computer operators look away from their screens and that readers look up from their materials regularly, letting their eyes focus on distant objects. This action relaxes the ciliary muscles, reducing eyestrain. Some ophthalmologists suggest that computer users "blink at every period and look up after every paragraph."

Presbyopia. The aging process brings with it many joys: hair loss, hearing loss, and arthritis. Aging also results in a loss in the resiliency of the lens. Thus, when the ciliary muscles contract to allow one to focus on a nearby object, the lens responds slowly or only partially, making it difficult to focus. This condition, known as **presbyopia,** (prez-bee-ope-ee-a) usually begins around the age of 40 and can be corrected by glasses worn for near-point work such as reading.[4] With this understanding of how the eye focuses light on the retina, we now turn our attention to the retina itself.

The Rods and Cones of the Retina Contain Photosensitive Pigments that Dissociate When Struck by Light

As you have seen, the cornea and lens focus images on the retina. The image cast on the rods and cones of the retina is converted into bioelectric impulses that are transmitted to the brain. The brain, in turn, translates this flood of nerve impulses into a coherent image of our environment.

Rods. Each eye contains about 100 million rods. Sensitive to low light, the rods contain a pigment called **rhodopsin.** When light strikes the rods, rhodopsin molecules in the cytoplasm of the photoreceptors split into two component molecules, retinal and opsin. **Retinal** is a derivative of vitamin A. **Opsin** is an enzyme. In the dark, the rods release a steady stream of a neurotransmitter. The neurotransmitter released by the rods inhibits bipolar neurons from firing (▻ Figure 18–21a). When light stimulates the rods, however, the breakdown of rhodopsin inhibits the release of the neurotransmitter (Figure 18–21b). This removes the inhibition on the bipolar neurons, allowing them to send impulses to the brain.

Rhodopsin is extremely sensitive to light and therefore permits the rods to function in dim light. Because rhodopsin is so easily dissociated, the molecules break down rapidly during daylight hours, dissociating as quickly as they are regenerated. Consequently, bright light reduces the amount of rhodopsin in the rods, making them ineffective in daylight. In dim light, however, rhodopsin is regen-

[4] You may have seen your parents or grandparents holding a phone number at arm's length to read it. They may have argued that their eyes weren't going bad, their arms were just too short!

▻ **FIGURE 18–21 Effect of Light on Rods**

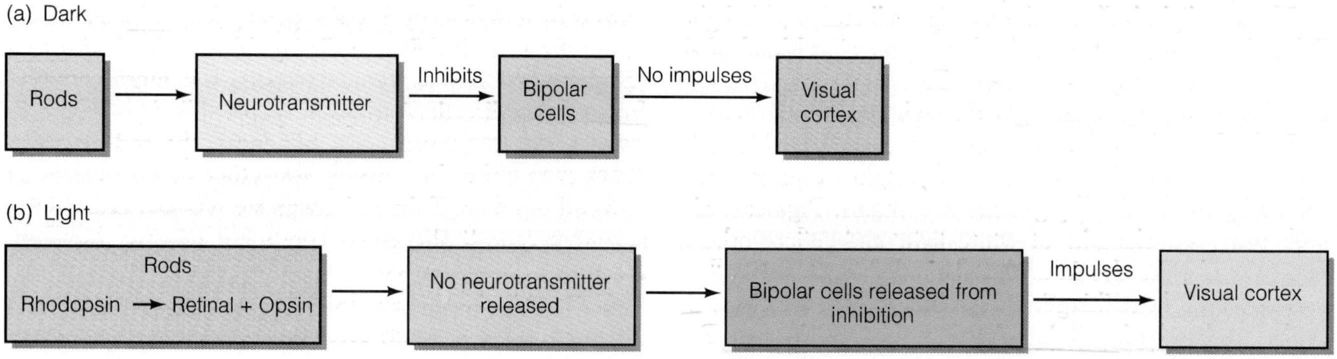

erated, and the rods become functional. Stepping into a dark movie theater on a bright day, you have probably noticed that, at first, your vision is greatly impaired. That is because your rods, which have been bleached by bright outdoor light, require some time to recover and become fully functional. As rhodopsin molecules are regenerated and the rods begin functioning, your eyesight returns.

Rhodopsin molecules in the rods are broken down and regenerated in a continuous cycle. Although retinal and opsin are recycled to replenish rhodopsin molecules, some retinal is lost or destroyed and must be replaced. Retinal is produced in the body from vitamin A, which is found in a variety of foods, such as carrots, spinach, fortified milk, and peaches. Rhodopsin concentrations decline when vitamin A intake is reduced, decreasing the sensitivity of the rods so that they are unable to respond to dim light. Therefore, a dietary deficiency of vitamin A often leads to **night blindness,** a marked reduction in night vision that is easily corrected by eating foods rich in vitamin A.

Besides replenishing retinal supplies, vitamin A is important in maintaining the cornea. As noted in Chapter 11, a deficiency of this fat-soluble vitamin results in corneal dryness, which may, if severe, lead to ulceration. If the deficiency persists, corneal ulcers may result in permanent blindness. In the Third World, an estimated 100,000 people lose their eyesight each year because of severe vitamin A deficiency.

Cones. Each eye contains about 3 million cones. Like the rods, the cones contain photosensitive pigments. However, the cones operate under bright light and are also sensitive to different colors of light, thus making color vision possible.

Three types of cones are found in the human retina: blue cones, green cones, and red cones—so named because of their sensitivity to a particular color of light. Each type contains a unique type of pigment, which responds optimally to one particular color of light (▷ Figure 18–22). The pigments inside the cones dissociate when struck by colored light. As in the rods, the dissociation of the photopigments in the cones decreases the release of neurotransmitter, "unleashing" bipolar neurons. Nerve impulses traveling to the brain convey information the visual cortex requires to construct a visual image. To understand how color vision works, we will first look at the properties of visible light and examine why an object is a certain color.

The sun, light bulbs, and neon signs all emit visible light. White light from the sun and other sources is a blend of all of the colors of the rainbow (▷ Figure 18–23). You can test this by shining white light through a prism.

Light from the sun and other sources strikes objects in the environment. The color of an object, however, is determined by the kinds of pigments it contains. Pigments absorb some wavelengths of white light and reflect others. For example, the dye in a blue flannel shirt absorbs all of the colored light striking it except blue, which it reflects. It is the unabsorbed wavelengths that the eye detects. Put

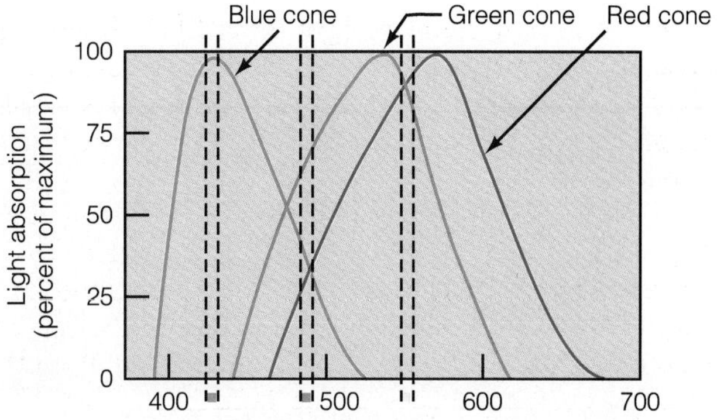

Color perceived	Percent of maximum stimulation		
	Red cones	Green cones	Blue cones
■	0	0	100
■	31	67	36
■	83	83	0

▷ **FIGURE 18–22 Sensitivity of the Three Types of Cones to Different Colors of Light** The color of light perceived is determined by the type or ratio of cones stimulated. The ratios of stimulation of the three cone types are shown for three sample colors.

another way, it is the unabsorbed wavelengths that give an object its characteristic color (▷ Figure 18–24). Thus, a leaf appears green because the pigments in the leaf absorb all of the colored light striking it except green, which they reflect.

Color vision occurs because each type of cone responds optimally to one color. Thus, a purely blue object stimulates "blue" cones (see Figure 18–22). Red light reflected from an object stimulates "red" cones. Green light stimulates "green" cones as well as some red and blue cones. How do we see so many in-between shades, such as yellow? Yellow light has a wavelength between red and green. Yellow light stimulates both red and green cones. Blue cones are not stimulated at all. The brain interprets the signal it receives from the red and green cones as yellow light. Thus, the relative proportion of different cones stimulated determines the color we perceive.

Color Blindness. About 5% of the human population suffers from color blindness. **Color blindness** is a hereditary disorder, more prevalent in men than women. Characterized by a deficiency in color perception, color blindness ranges from an inability to distinguish certain shades to a complete inability to perceive color. The most common form of this disorder is red-green color blindness. In individuals with red-green color blindness, the red or green cones may either be missing altogether or be present in reduced number. If the red cones are missing, red objects appear green. If the green cones are missing, however, green objects appear red.

Color blindness can be detected by simple tests (▷ Figure 18–25). Many color-blind people are unaware of

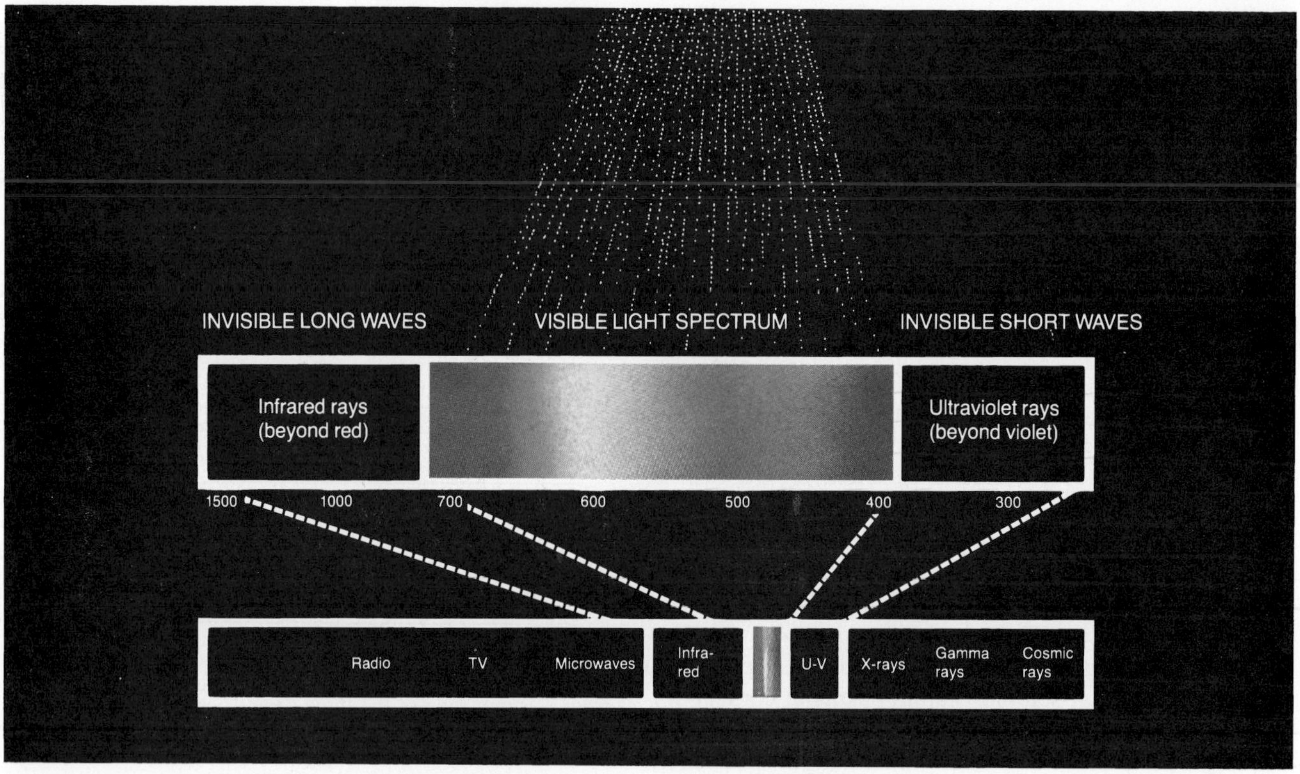

▻ **FIGURE 18–23 The Electromagnetic Spectrum** The sun produces a wide range of electromagnetic radiation, a small portion of which is visible to the eye. Called visible light, it consists of a variety of colors. (Numbers on the spectrum are wavelength measured in nanometers.)

their condition or untroubled by it. They rely on a variety of visual cues, such as differences in intensity, to distinguish red and green objects. They also rely on position cues. In traffic lights, for example, the red light is always at the top of the signal; green is on the bottom. Although the colors may appear more or less the same, the position of the light helps color-blind drivers determine whether to hit the brakes or step on the gas.

▻ **FIGURE 18–24 Color** The color of an object results from the reflection of certain wavelengths of light.

Overlapping Visual Fields Give Us Depth Perception

Human eyes are located in the front of the skull, looking forward, much like those of other predatory animals. Each eye has a visual field of about 170 degrees, but the visual

▻ **FIGURE 18–25 Color Blindness Chart** People with red-green color blindness cannot detect the number 29 in this chart.

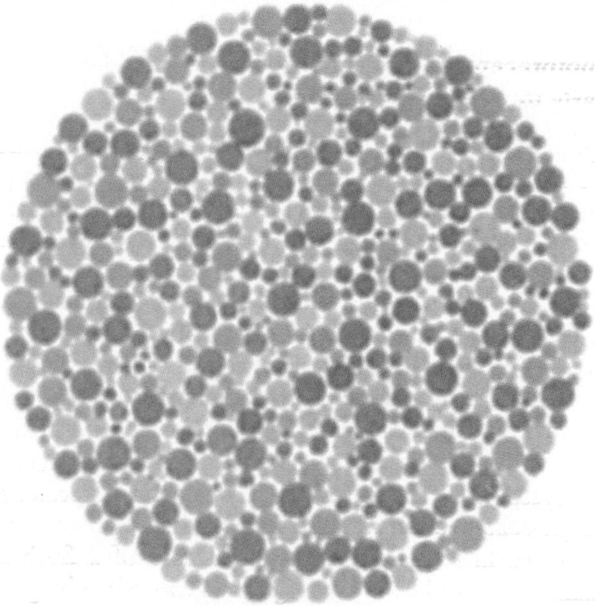

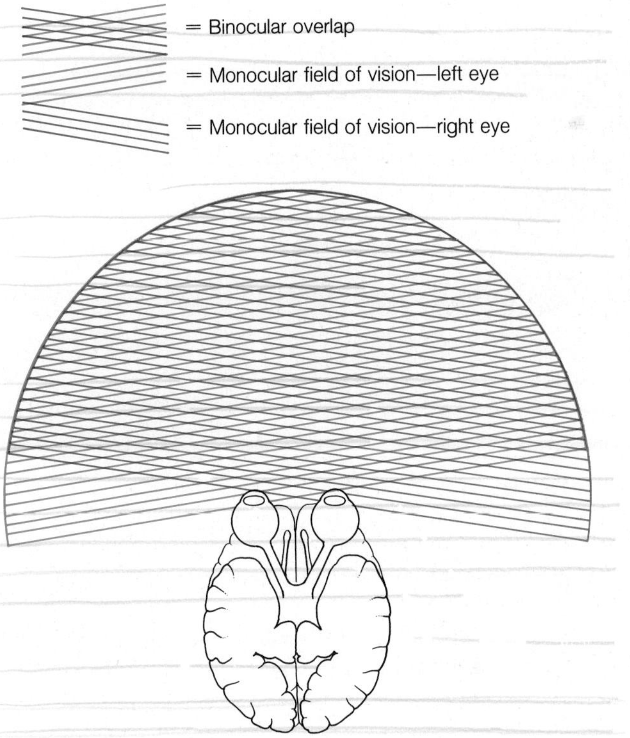

▻ **FIGURE 18–26 Overlapping Visual Fields** The overlap helps us perceive depth—three-dimensional relationships.

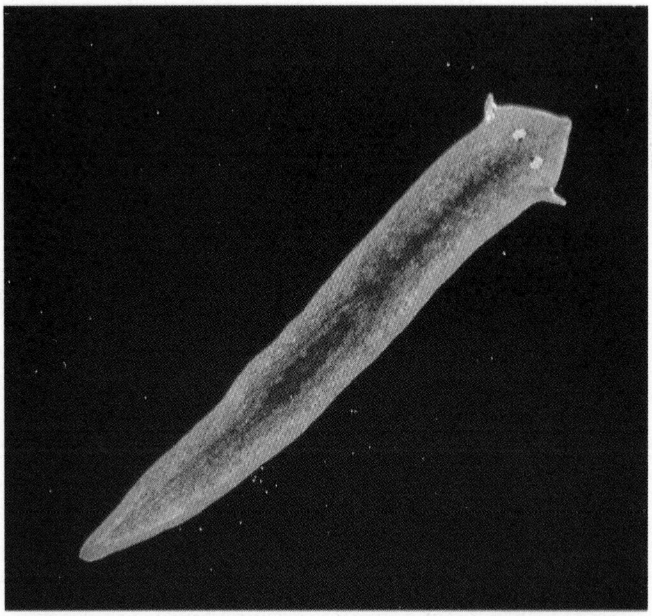

▻ **FIGURE 18–27 Planaria Eyespots** Eyespots like those found in the flatworm planaria perceive light but do not provide vision.

fields overlap considerably (▻ Figure 18–26). The overlapping of visual fields gives us the ability to judge the relative position of objects in our visual field; that is, it gives us **depth perception,** also known as **stereoscopic vision.** This ability is also important for predators that need to judge the location of prey with great accuracy, especially ones like the peregrine falcon, which dives at its bird prey at speeds over 150 miles per hour. A lack of depth perception in such an animal would be devastating!

Photoreceptors Appear Very Early in Animal Evolution

Specialized structures that detect light appear in some of the most primitive animals known to science. These photoreceptors range in complexity from simple cellular organelles to clusters of light-sensitive cells to complex structures like the vertebrate eye. Despite the variety, the fundamental process of light detection is quite similar. In most instances, that is, light detection depends on photochemical reactions that involve visual pigments.

A brief look at the different types of photoreceptors gives an overview of the evolutionary progression that led to the formation of the complex human eye. We begin with the eyespots of flatworms and photosynthetic protozoans (single-celled eukaryotes like *Euglena*).

Shown in ▻ Figure 18–27, **eyespots** of a freshwater planarian worm known as *Dugesia* consist of nothing more than two cup-shaped groups of cells in the head region. These structures permit the organism to detect light, but they are not sophisticated enough to provide vision as we know it.

In the animal kingdom, primitive vision is first found in certain molluscs—a group that includes shellfish, snails, and octopi (▻ Figure 18–28). Even among molluscs, there is a wide range of eye structures, from simple eyespots to sophisticated eyes like ours with a lens, iris, and retina. The most sophisticated eyes belong to squid and octopi and are capable of forming clear images.

Most insects and crustaceans, such as crabs, have **compound eyes** consisting of many photoreceptive units (▻ Figure 18–29). Each photosynthetic unit consists of a small crystalline lens that focuses light on a photoreceptor cell beneath it. Scientists believe that each photoreceptive unit in a compound eye samples a small portion of the visual field. Information then travels to the insect brain, where it is assembled into one coherent image.

The eyes of almost all vertebrates are capable of forming extremely clear images of the world. The human eye, described earlier, is an excellent representative. Despite this commonality in structure, vertebrate eyes do contain modifications that reflect natural selection. In birds of prey (raptors), for instance, the eyes face forward to provide the depth of field vital to diving at prey, but they also contain a higher concentration of cones than those of any other vertebrate. Cones are responsible for visual acuity as well as color vision. Raptor eyes are also large in relation to head size. This is especially evident in owls and eagles, whose eyes are about the same size as the human eye. Large eye size permits a greater accumulation of light,

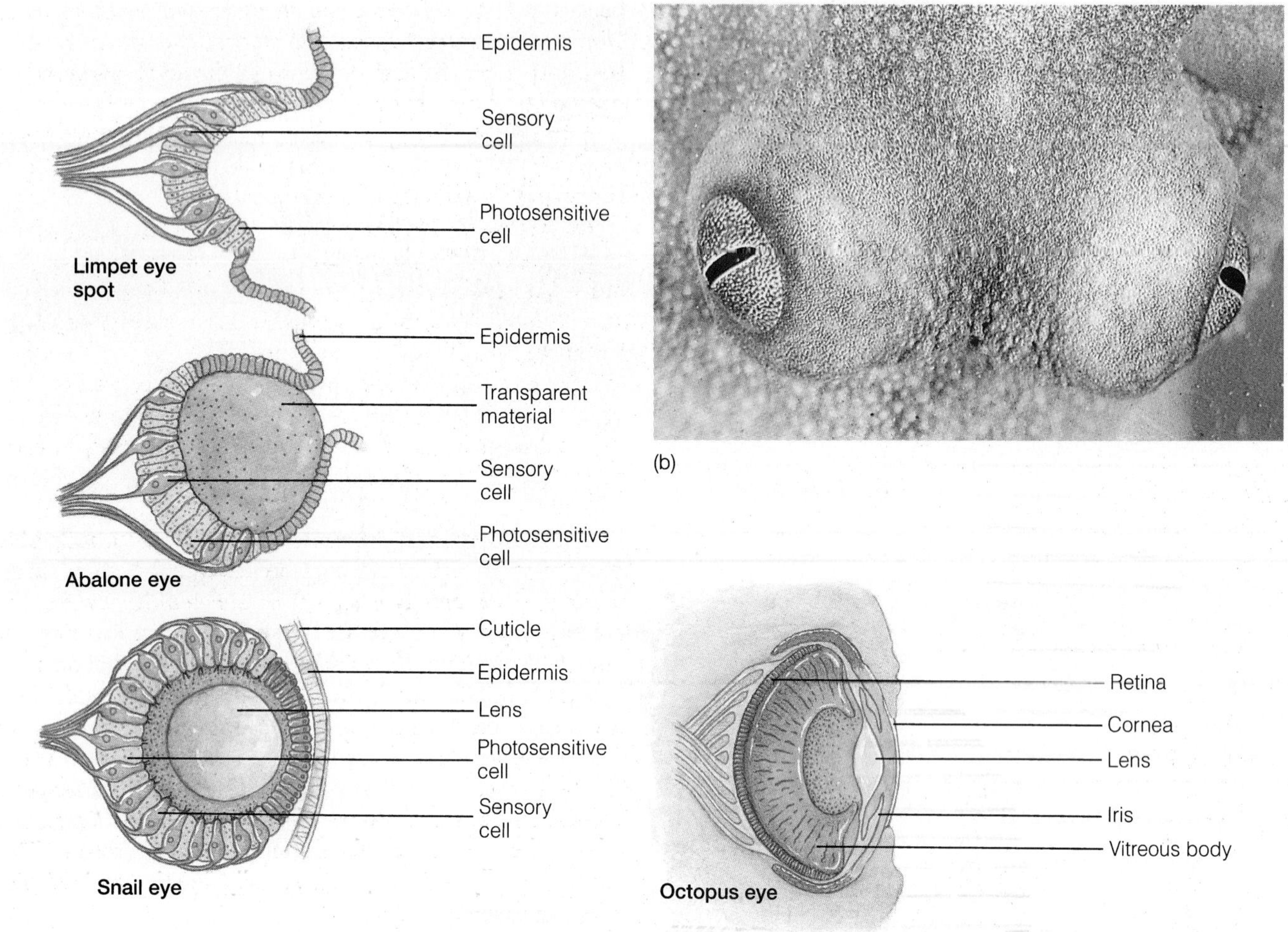

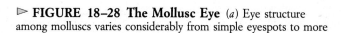

▻ **FIGURE 18–28 The Mollusc Eye** (*a*) Eye structure among molluscs varies considerably from simple eyespots to more complex eyes with lenses, retinas, and other structures. (*b*) The octopus eye is similar to human eye.

necessary for seeing in dim light or at night, as in the case of the owl. The eye's large size also separates the lens from the retina, contributing to telescopic vision. (Telescopic lenses for cameras are longer than regular lenses.) Because of telescopic imaging and the higher density of cones, an eagle can see three times better than a human. An eagle can see a rabbit a mile away. You or I would have to be 500 meters, or less than a third of a mile, away to see the rabbit.

HEARING AND BALANCE: THE COCHLEA AND MIDDLE EAR

The human ear is an organ of special sense. It serves two functions: it detects sound, and it detects body position, helping us maintain balance (Table 18–4).

The Ear Consists of Three Anatomically Separate Portions: the Outer, Middle, and Inner Ears

The Outer Ear. The **outer ear** consists of an irregularly shaped piece of cartilage covered by skin, the **auricle,** and the earlobe, a flap of skin that hangs down from the ear (▻ Figure 18–30a). The outer ear also consists of a short tube, the **external auditory canal,** which transmits airborne sound waves to the middle ear (Figure 18–30a). The external auditory canal is lined by skin containing modified sweat glands that produce earwax, or **cerumen.** Earwax traps foreign particles, such as bacteria, and contains a natural antibiotic substance that may help reduce ear infections.

The Middle Ear. The **middle ear** lies entirely within the temporal bone of the skull (Figure 18–30b). The eardrum, or **tympanic membrane,** separates the middle ear cavity from the external auditory canal. The tympanic membrane oscillates when struck by sound waves in much the same way that a guitar string vibrates when a note is sounded by another nearby instrument.

Inside the middle ear are three minuscule bones, the **ossicles.** Starting from the outside, they are the **malleus** (hammer), **incus** (anvil), and **stapes** (stirrup). As illustrated in Figure 18–30b, the hammer-shaped malleus abuts the tympanic membrane. When the membrane is struck by sound waves and vibrates, it causes the malleus to rock

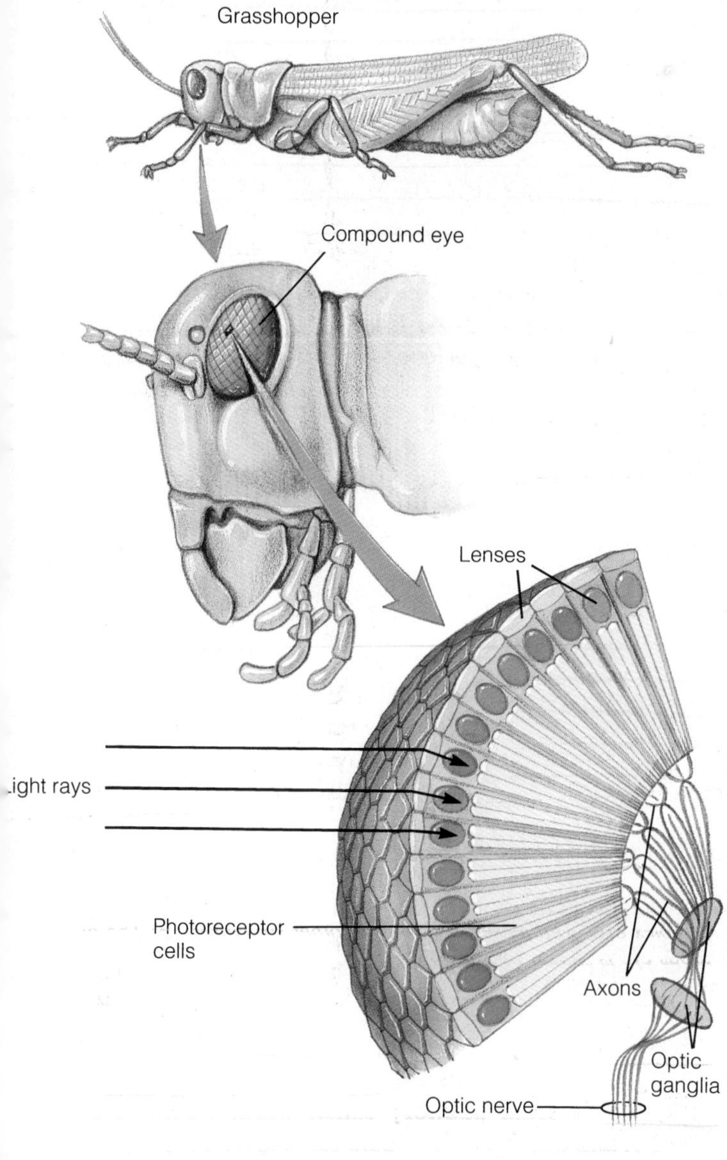

▻ **FIGURE 18–29 Compound Eye** The insect eye consists of many small units, each with a lens that focuses light on an underlying photoreceptor.

back and forth. The malleus, in turn, causes the incus to vibrate. The incus causes the stapes, the stirrup-shaped bone, to move in and out against the **oval window,** an opening to the inner ear covered with a membrane like the skin on a drum. Thus, vibrations created in the eardrum are amplified as they are transmitted to the inner ear, where the sound receptors are located.

As Figure 18–30 illustrates, the middle ear cavity opens to the pharynx via the **auditory,** or **eustachian, tube.** The eustachian tube extends downward at an angle and opens into the nasopharynx. It serves as a pressure valve. Normally, the eustachian tube is closed. Yawning and swallowing, however, cause it to open. This allows air to flow into or out of the middle ear cavity, equalizing the internal and external pressure on the eardrum, as you will notice when taking off in an airplane or ascending a mountain highway in your car.

Scuba divers and swimmers can sustain considerable damage to an eardrum if they are not careful. As a swimmer descends, pressure from the water builds, pushing the eardrum inward. To prevent the eardrum from tearing, air must be forced into the middle ear cavity. This can be done by simply holding one's nose, clamping one's mouth shut, and gently blowing. Air is forced through the eustachian tube into the middle ear cavity, equalizing the pressure. When a diver ascends, just the opposite happens: air pressure increases inside the middle ear cavity. The diver must release the pressure or else suffer a broken eardrum. This release usually occurs quite naturally, but it can be facilitated by yawning.

The Inner Ear. The inner ear occupies a much larger cavity in the temporal bone than the middle ear and contains two sensory organs, the cochlea and the vestibular apparatus. The **cochlea** is shaped like a snail shell and houses the receptors for hearing. The **vestibular apparatus** consists of two parts: the semicircular canals and the vestibule (Figure 18–30b). The **semicircular canals** are three ringlike structures set at right angles to one another. They house receptors for body position and movement. The **vestibule** is a bony chamber lying between the cochlea and

TABLE 18–4 Structures and Functions of the Ear

PART	STRUCTURE	FUNCTION
Outer ear	Auricle	Funnels sound waves into external auditory canal
	Ear lobe	
	External auditory canal	Directs sound waves to the eardrum
Middle ear	Tympanic membrane, or eardrum	Vibrates when struck by sound waves
	Ossicles	Transmit sound to the cochlea in the inner ear
Inner ear	Cochlea	Converts fluid waves to nerve impulses
	Semicircular canals	Detect head movement
	Saccule and utricle	Detect head movement and linear acceleration

▻ **FIGURE 18–30 Cross Section Showing the Structure of the Ear** (*a*) Notice the structures of the outer, middle, and inner ears. (*b*) Note also that the receptors for balance and sound are located in the inner ear.

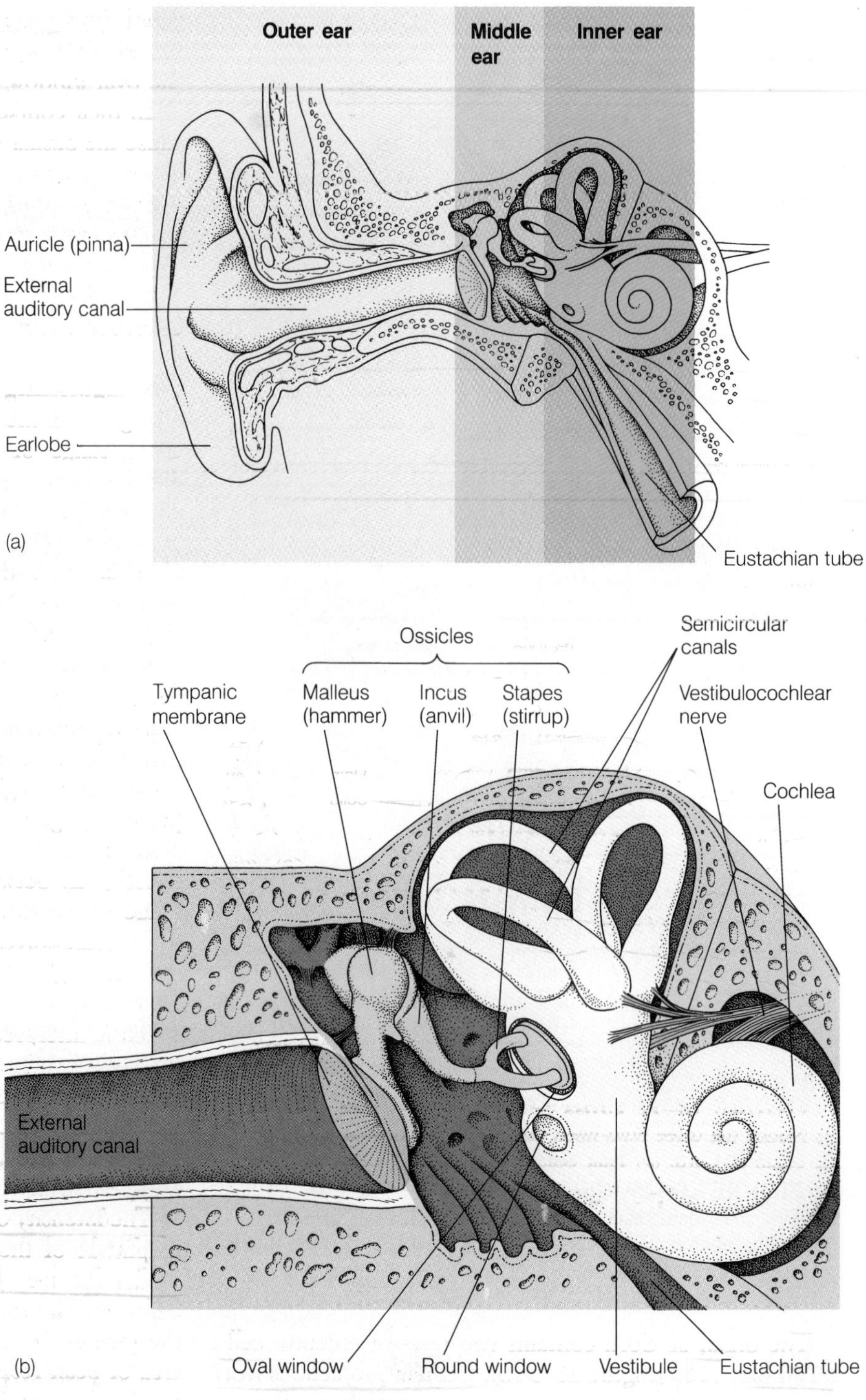

semicircular canals. It also houses receptors that respond to body position and movement.

Hearing Requires the Participation of Several Structures

The detection and perception of sound require the action of the eardrum, the ossicles, and the cochlea. As noted previously, sound waves enter the external auditory canal, where they strike the tympanic membrane, or eardrum. The eardrum vibrates back and forth, causing the ossicles of the middle ear cavity to vibrate. The ossicles, in turn, transmit sound waves to the cochlea, which houses the receptor for sound.

Structure and Function of the Cochlea. The cochlea is a hollow, bony spiral. A cross section through this remarkable structure reveals three fluid-filled canals (▻ Figure 18–31a). Separating the middle canal from the lowermost one is a flexible membrane called the **basilar membrane.** It supports the **organ of Corti,** the receptor organ for sound.

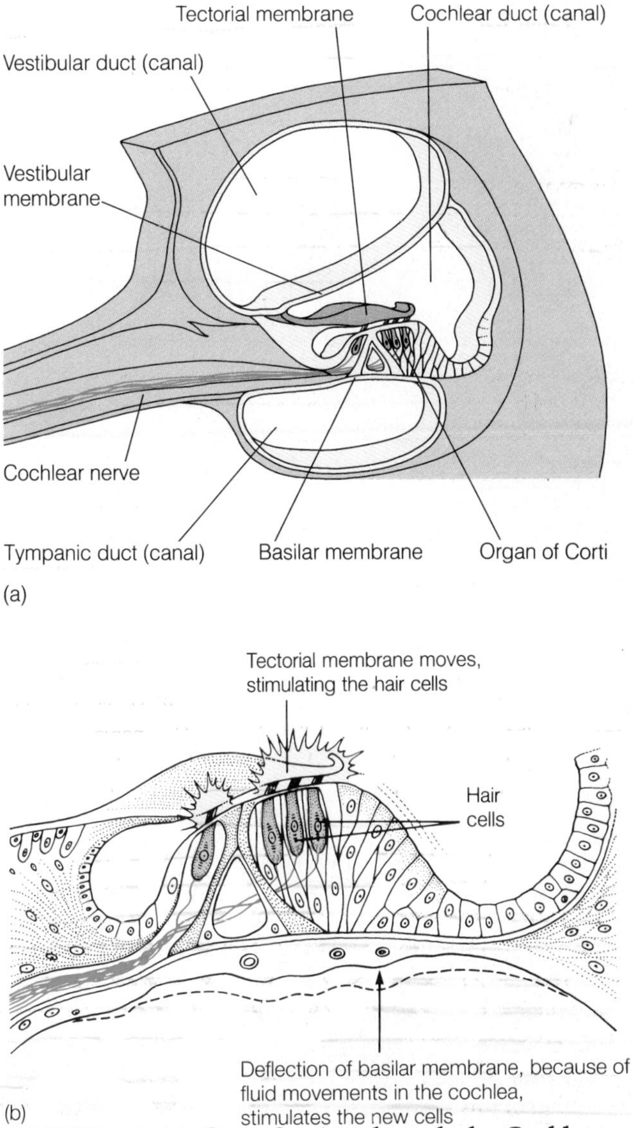

FIGURE 18–31 Cross Section through the Cochlea (*a*) Notice the three fluid-filled canals and the central position of the organ of Corti. (*b*) Hair cells of the organ of Corti are embedded in the overlying tectorial membrane. When the basilar membrane vibrates, the hair cells are stimulated.

The organ of Corti contains two rows of receptor cells, called **hair cells** (Figure 18–31b). Cellular projections from the hair cells contact the overlying **tectorial membrane.** When sound waves are transmitted from the middle ear to the inner ear, they create waves of compression and decompression in the fluid in the uppermost canal of the cochlea, the **vestibular canal.** As shown in Figure 18–32a, in which the cochlea is unwound, the stirrup transmits vibrations to the drumlike **oval window,** the opening in the bony cochlea. Fluid pressure waves are created in the vestibular canal; they then travel through the thin vestibular membrane into the middle canal, the **cochlear duct (canal).** From here, the pressure waves pass through the basilar membrane into the lowermost canal, the **tympanic canal.** Pressure is relieved by the outward bulging of the **round window,** an opening in the bony cochlea, which, like the oval window, is spanned by a flexible membrane.

In their course through the cochlea, the pressure waves cause the basilar membrane to vibrate. This vibration stimulates the hair cells. The hair cells respond by releasing a neurotransmitter, which stimulates the dendrites that wrap around the bases of the hair cells. In the cochlea, sound waves are converted to pressure waves, then to nerve impulses. The nerve impulses leave the cochlea via the **vestibulocochlear nerve,** one of the 12 cranial nerves.

Distinguishing Pitch and Intensity. A pitch pipe helps a singer find the note to begin a song. On the pitch pipe are a range of notes from high to low. The ear can distinguish between these various pitches, or frequencies, in large part because of the structure of the basilar membrane. The basilar membrane underlying the organ of Corti is stiff and narrow at the oval window, where fluid pressure waves are first established inside the cochlea (Figure 18–32b). As the basilar membrane proceeds to the apex of the spiral, however, it becomes wider and more flexible. The change in width and stiffness results in marked differences in its ability to vibrate. The narrow, stiff end, for example, vibrates maximally when pressure waves from high-frequency sounds ("high notes") are present (Figure 18–32c). The far end of the membrane vibrates maximally with low-frequency sounds ("low notes"). In between, the membrane responds to a wide range of intermediate frequencies.

Pressure waves caused by any given sound stimulate one specific region of the organ of Corti. The hair cells stimulated in that region send impulses to the brain, which it interprets as a specific pitch. Each region of the organ of Corti sends impulses to a specific region of the auditory cortex in the temporal lobe of the brain. Thus, the auditory cortex can be mapped according to tone in much the same way the motor and sensory cortex can be mapped (Chapter 17).

The intensity of a sound, or its loudness, depends on the amplitude of the vibration in the basilar membrane. The louder the sound, the more vigorous the vibration of the eardrum. The more vigorous the vibration of the eardrum, the greater the deflection of the basilar membrane in the area of peak responsiveness. The greater the deflection of the basilar membrane, the more hair cells are stimulated.

The auditory system is remarkably sensitive, so much so that it can detect sounds that deflect the membrane only a fraction of the diameter of a hydrogen atom. It is no wonder, then, that loud noises can cause so much damage to hearing. Loud rock music or sirens, for example, cause extreme vibrations in the basilar membrane, destroying hair cells over time and causing partial deafness.

Hearing Loss. As people grow older, many lose their hearing. Hearing loss usually occurs so slowly that most people are unaware of it. In some cases, though, people

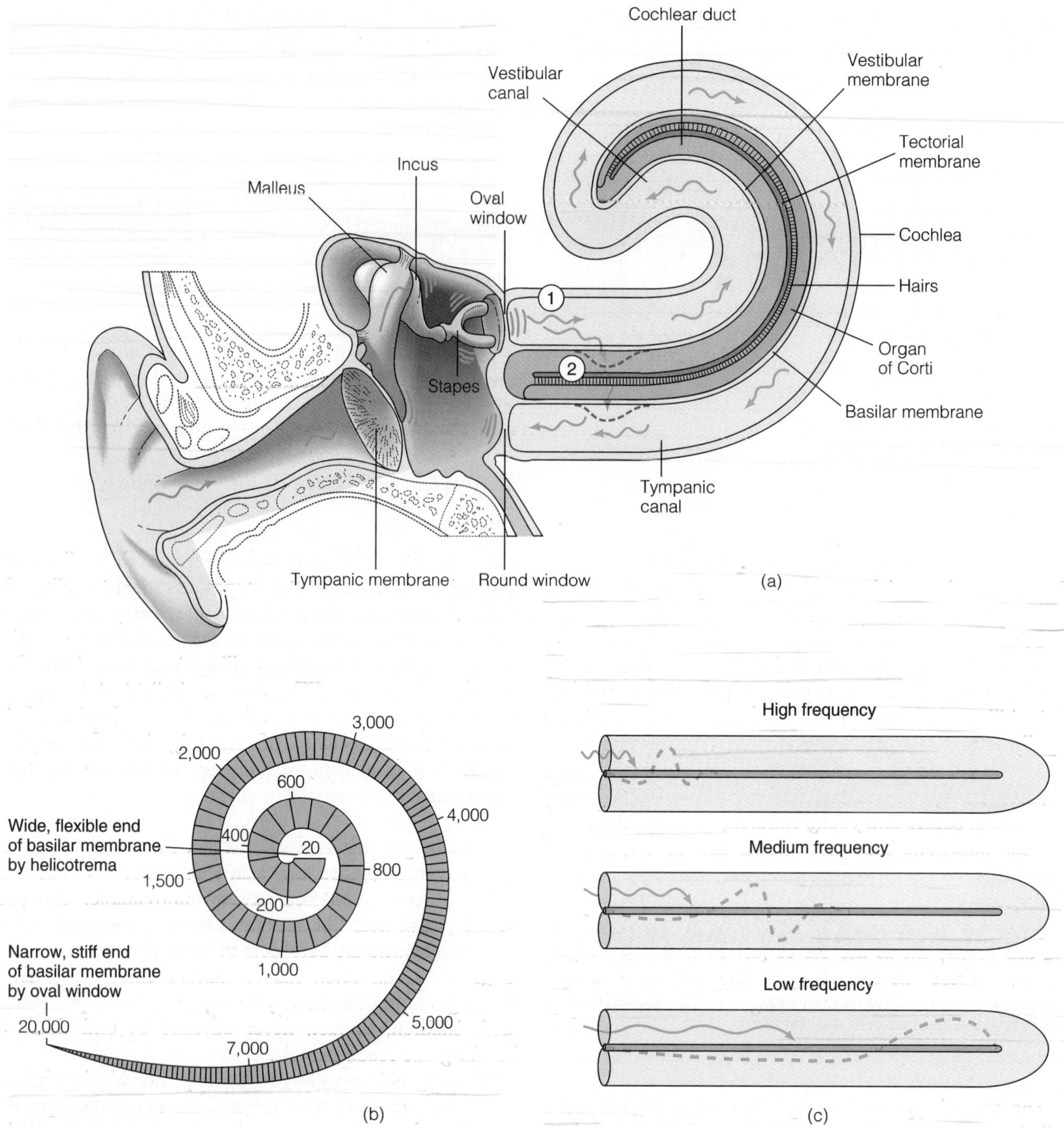

The numbers indicate the frequencies with which different regions of the basilar membrane maximally vibrate.

▻ **FIGURE 18–32 The Transmission of Sound Waves through the Cochlea** The cochlea is unwound here to simplify matters. (*a*) Vibrations are transmitted from the stirrup (stapes) to the oval window. Fluid pressure waves are established in the vestibular canal and pass to the tympanic canal, causing the basilar membrane to vibrate. (*b*) A representation of the basilar membrane, showing the points along its length where the various wavelengths of soundare perceived. Notice that the basilar membrane is narrowest at the base of the cochlea at the oval window end and widest at the apex. (*c*) High-frequency sounds set the basilar membrane near the base of the cochlea into motion. Hair cells send impulses to the brain, which interprets the signals as a high-pitch sound. Low-frequency sounds stimulate the basilar membrane where it is widest and most flexible.

lose their hearing suddenly. A loud explosion, for example, can damage the hair cells or even break the ossicles.

Hearing losses may be temporary or permanent, partial or complete. Basically, hearing loss falls into one of two categories, depending on the part of the system that is affected. The first is **conduction deafness,** which occurs when the conduction of sound waves to the inner ear is impaired. Conduction deafness may result from excessive

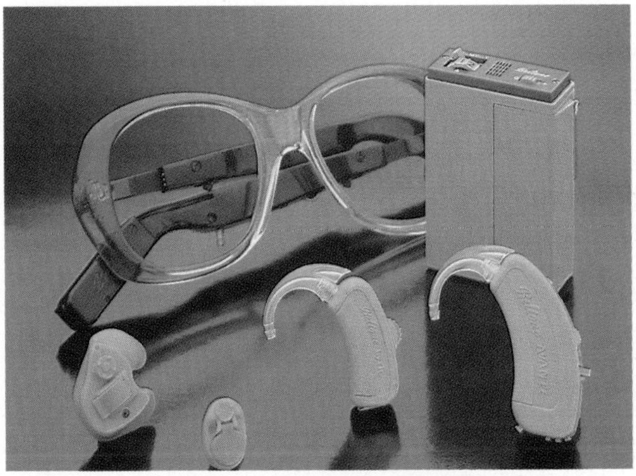

▻ **FIGURE 18–33 Hearing Aids** Worn by people with conduction deafness, hearing aids send sound impulses through the bone of the skull to the cochlea.

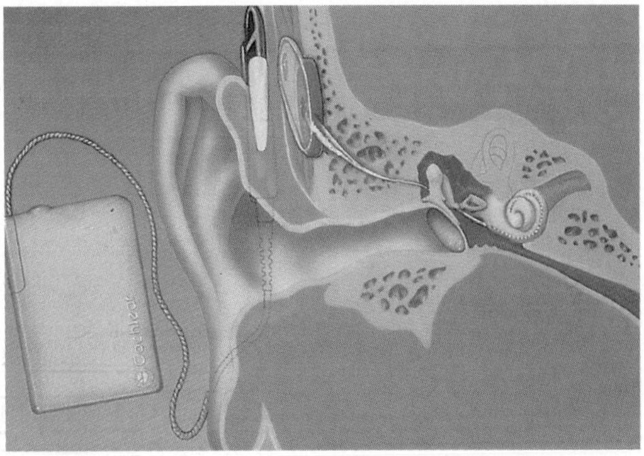

▻ **FIGURE 18–34 Cochlear Implant** The cochlear implant can correct for nerve deafness. Electrodes convey electrical impulses from a small microphone mounted in the ear to the auditory nerve.

earwax in the auditory canal or a rupture of the eardrum. Damage to the ossicles, and even a rupture of the oval window are additional causes of conduction deafness.

Conduction deafness most often results from infections in the middle ear. Bacterial infections may result in the buildup of scar tissue, causing the ossicles to fuse and lose their ability to transmit sound. Infections of the middle ear usually enter through the eustachian tube. Thus, a sore throat or a cold can easily spread to the ear, where it requires prompt treatment. Ear infections are especially common in babies and young children. If undetected, middle ear infections may slow down the development of speech. Untreated, such infections may lead to permanent deafness.

Conduction deafness is treated by hearing aids. A **hearing aid** usually fits in the ear or just behind it (▻ Figure 18–33). These devices bypass the defective sound-conduction system by transmitting sound waves through the bone of the skull to the inner ear. These cause fluid pressure waves to form in the cochlea and stimulate the basilar membrane.

The second type of hearing loss is neurological and is called **nerve,** or **sensineural, deafness.** Sensineural deafness results from physical damage to the hair cells, the vestibulocochlear nerve, or the auditory cortex. Explosions, extremely loud noises, and some antibiotics, for example, can all damage the hair cells in the organ of Corti, creating partial or even complete deafness. The auditory nerve, which conducts impulses from the organ of Corti to the cortex, may degenerate, thus ending the flow of information to the cortex. Tumors in the brain or strokes may destroy the cells of the auditory cortex.

Although the ear is quite vulnerable to loud noise and other problems, it contains a mechanism to protect itself from damage. This mechanism consists of two tiny skeletal muscles (the smallest in the body) located in the middle ear cavity. One of these muscles inserts on the malleus, and the other attaches to the stapes. As noted earlier, the malleus attaches to the eardrum, and the stapes attaches to the oval window. Loud noises stimulate a reflex contraction of the middle ear muscles, pulling them away from their membrane contacts. Whenever an individual is exposed to loud noise, this reflex reduces the conduction of the noise to the inner ear. Unfortunately, the reflex requires about 40 milliseconds, so it cannot protect the ear from explosions.

Correcting Profound Deafness. More than 2 million Americans are profoundly deaf, a condition that has until recently been considered virtually untreatable. The profoundly deaf hear nothing, not even the sound of sirens.

Children who are born deaf or are deafened before they begin to speak often fail to mature emotionally. Even reading comprehension may be impaired. Some profoundly deaf children, in fact, never advance beyond third- or fourth-grade reading levels.

Hearing aids usually cannot help individuals who are born deaf or those who suffer from nerve damage. Researchers, however, have developed a new device called a **cochlear implant,** which simulates the function of the inner ear (▻ Figure 18-34). This device picks up sound and transmits it to a receiver implanted inside the skull. The signal then travels to an electrode implanted in the vestibulocochlear nerve in the cochlea. Electrical impulses in the electrode stimulate the nerve, creating impulses that travel to the auditory cortex.

Today, hundreds of adults and children are equipped with cochlear implants. The single-electrode model, which provides the deaf with only a rudimentary hearing capacity, is quickly becoming obsolete thanks to the advent of newer multiple-electrode models. These models detect and transmit a wider range of sounds. Recipients of the new models

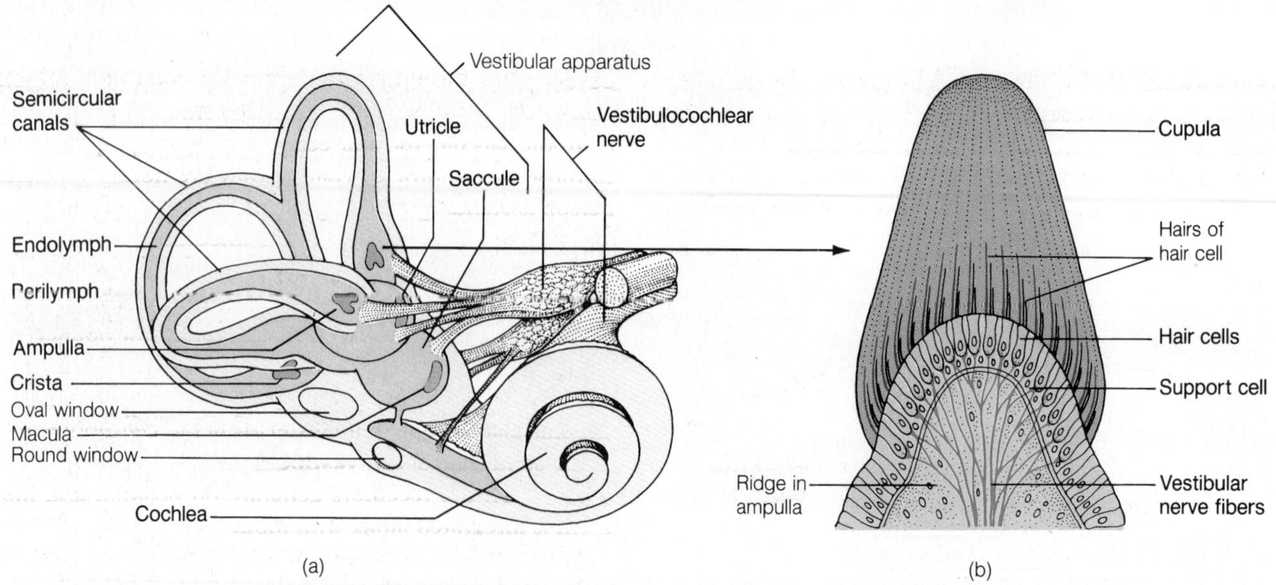

▻ **FIGURE 18–35 Location and Structure of the Cristae** (*a*) This illustration shows the location of the cristae in the ampullae of the semicircular canals. The semicircular canals are filled with endolymph. (*b*) When the head spins, the endolymph is set into motion, deflecting the gelatinous cupula of the crista, thus stimulating the receptor cells.

can perceive many distinct words, not just the sound of a telephone or an automobile horn.

Although it is doubtful that "normal" hearing will ever be fully restored by such devices, people who have lived in the silent world point out that any sound is better than no sound at all. For them, the cochlear implant provides valuable outside stimuli, such as sirens and horns, the importance of which many of us take for granted. The cochlear implant also helps the deaf monitor and regulate their own voices, and it makes lip reading easier. For deaf children, a cochlear implant could mean the difference between learning to speak or a lifetime of silence.

The Vestibular Apparatus Houses Receptors that Detect Body Position and Movement

The cochlea lies alongside the **vestibular apparatus,** which as noted earlier, consists of the semicircular canals and the vestibule, a bony chamber between the cochlea and semicircular canals. The vestibular apparatus houses receptors that detect body position and movement (▻ Figure 18–35).

The Semicircular Canals. The three semicircular canals are arranged at right angles to one another. Each canal is filled with a fluid called **endolymph.** As illustrated in Figure 18–35a, the base of each semicircular canal expands to form the **ampulla.** On the inside wall of each ampulla is a small ridge of tissue, or **crista.** Each crista consists of a patch of receptor cells, each of which contains numerous microvilli and a single cilium embedded in a cap of gelatinous material, the **cupula,** as illustrated in Figure 18–35b. Dendrites of sensory nerves wrap around the base of the receptor cells, and the cupula extends into the cavity of the ampulla, where it is bathed in endolymph.

Rotation of the head causes the endolymph in the semicircular canals to move. The movement of the fluid deflects the cupula, which stimulates the hair cells to release a neurotransmitter that excites the sensory neurons. These neurons send impulses to the brain, alerting it to the rotational movement of the head and body.

Because the semicircular canals are set in all three planes of space, movement in any direction can be detected. By alerting the brain to rotation and movement, the semicircular canals contribute to our sense of balance. They are therefore important to **dynamic balance**—helping us stay balanced when in movement.

The Utricle and Saccule. Two additional receptors play a role in balance and detection of movement, the **utricle** and **saccule** (Figure 18–35). These membranous compartments inside the vestibule provide input under two distinctly different conditions: at rest and during acceleration in a straight line. As a result, these receptors contribute to **static balance**—helping us stay balanced when we are not moving—and dynamic balance.

The utricle and saccule contain small receptor organs known as **maculae** ("spots"), which are akin to the cristae of the semicircular canals. Each macula consists of a patch of receptor cells, which is structurally similar to those found in the cristae. The cilia and microvilli of the receptor cells in the maculae are also embedded in a gelatinous cap. One notable difference with the cristae, however, is the presence of numerous small crystals of calcium carbonate, called **otoliths** (literally, "ear rocks"), which are embedded in the gelatinous material. The otoliths make the gelatin heavier than the surrounding fluid (▻ Figure 18–36a).

Although similar in structure, the maculae in the utricle and saccule are oriented in different directions. Thus, when a person is standing, the hair cells of the utricle are oriented vertically. Bending the head forward, as illustrated in Figure 18–36b, causes the otoliths and gelatinous cap to droop forward. Pulled downward by gravity, the gelatinous cap stimulates the receptor cells. Nerve impulses are transmitted to the brain, alerting it to the head movement. Moving forward—say, by walking or running—causes the gelatinous mass to slide backward, as shown in Figure 18–36c, and the utricle therefore also transmits information to the brain on linear acceleration.

In the saccule, the hair cells are oriented perpendicularly to those of the utricle. That is, they are horizontal when a person is standing or sitting. These receptor cells are therefore stimulated when the head is tilted back—for instance, when a person lies down. The saccule also responds to acceleration and deceleration but in a vertical direction—for example, when one rides on an elevator or bounces on a trampoline.

The information provided by the semicircular canals and the maculae is sent to a cluster of nerve cell bodies in the brain stem called the **vestibular nuclei.** Here all of the information the receptors generate on position and movement is integrated along with input from the eyes and from receptors in the skin, joints, and muscles, as discussed earlier in the chapter.

From the vestibular nuclei, information flows in many directions. One major pathway leads to the cerebral cortex. This path makes us conscious of our position and movement. Another path leads to the muscles of the limbs and torso. Signals to the muscles help maintain our balance and, if necessary, correct body position.

In some people, activation of the vestibular apparatus results in motion sickness, characterized by dizziness and nausea. The exact cause of motion sickness is not known.

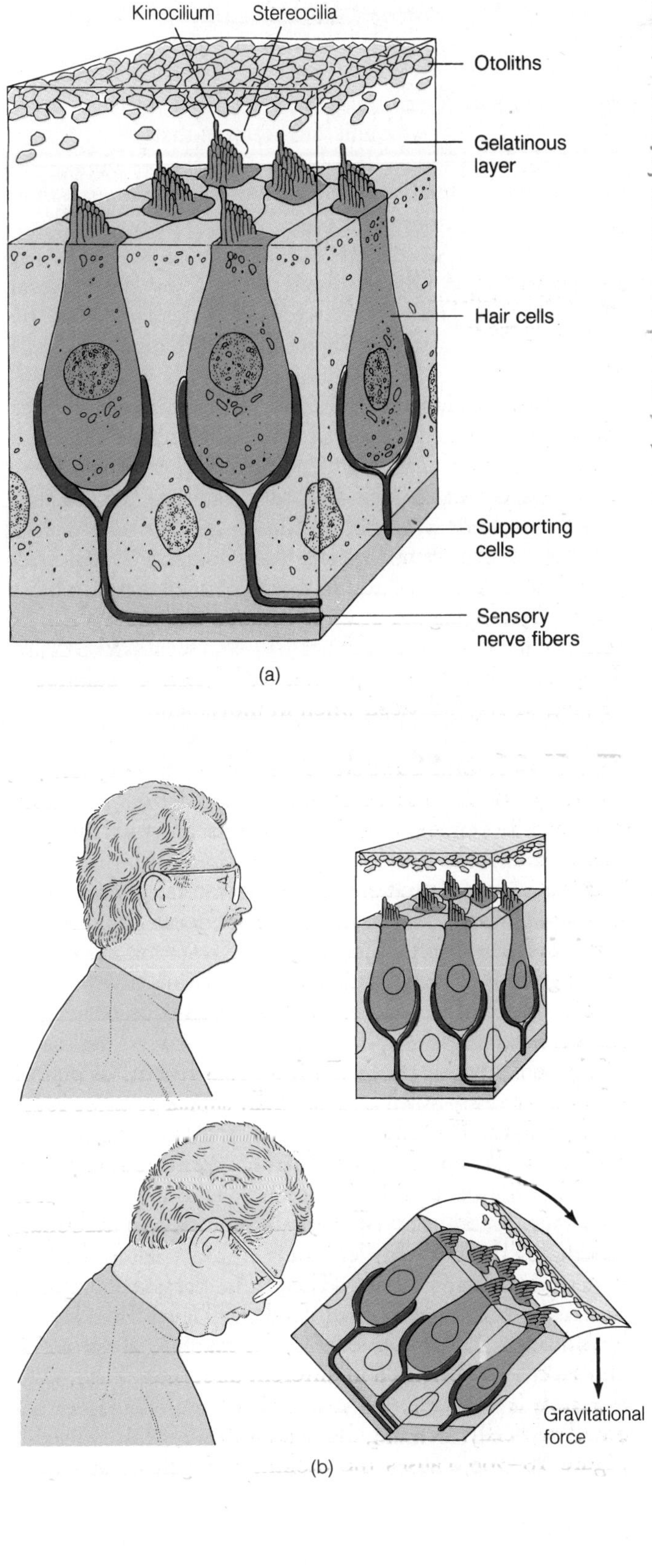

▶ **FIGURE 18–36 The Macula** (*a*) Receptor cells in the saccule and utricle are surrounded by supporting cells. Otoliths embedded in the gelatinous cap make the layer heavier than the surrounding fluid. (*b*) Position of the macula of the utricle in an upright position and when head is tilted forward. (*c*) Deflection of the otoliths during forward motion.

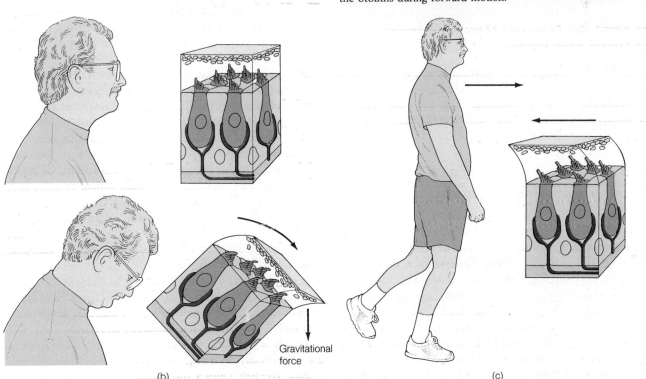

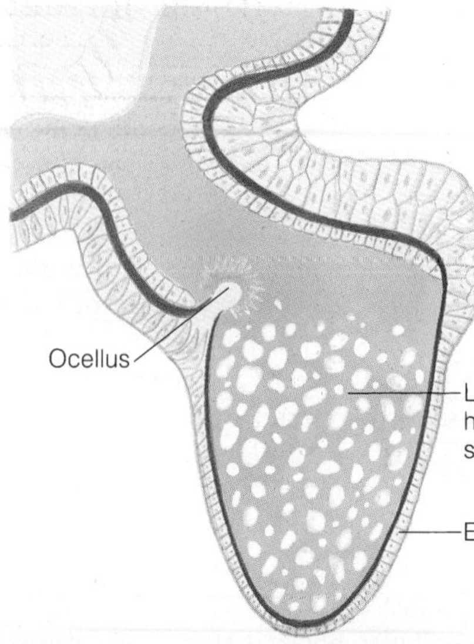

(a)

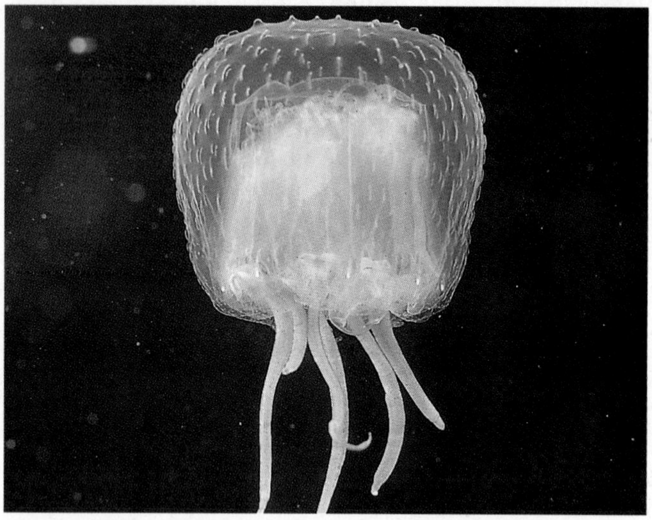

(b)

▻ **FIGURE 18–37 Statocysts in Jellyfish** Small flaps of tissue along the lower margin of the jellyfish contain sensory receptors that help the organism sense and correct body position. When the organism tilts, small crystals in the statocyst press against neighboring hair cells, alerting the nervous system of a change in body position.

The Ability to Detect Sound and Body Position Are Prevalent in the Animal Kingdom

Turn most animals upside down, and they will right themselves. Why? Because many animals possess mechanoreceptors that detect body position. These receptors contain hair cells akin to those found in the semicircular canals and utricle and saccule. Hair cells contain one or more surface processes, or hairs, that bend in response to gravity or other forces. This bending produces bioelectric signals that are sent to the brain, alerting the organism of a change in position.

A good example is the position receptor of the jellyfish, shown in ▻ Figure 18–37. As illustrated, the **statocyst** consists of a flap of tissue containing hair cells and small crystals. When the body moves, the crystals press against the hair cells, stimulating them. They send information to the nervous system that is used to maintain body position.

Sound detection is also widespread in the animal kingdom. For example, many insects have body hairs that vibrate at the same frequency as biologically important sounds. The body hairs of certain caterpillars, for instance, vibrate in synch with the wingbeat of wasps that prey on them, thus permitting the caterpillar to take evasive action.

Some insects also have pressure-detecting organs on various parts of their bodies, such as the front legs. These organs contain a tympanic membrane that responds to vibrations. Attached to the underside of the membrane are numerous sensory cells that generate nerve impulses that, in turn, travel to the insect brain.

In many fishes, cells containing minute hair cells are located in longitudinal canals known as the **lateral lines**—that is, grooves that run along the length of the body on both sides (▻ Figure 18–38). These hair cells detect low-frequency vibrations of water flow and turbulence, allowing the fish to locate underwater obstructions and monitor the movement of other fishes swimming close by when in murky water.

Some vertebrates, including bats, dolphins, and whales, come equipped with radar. These animals emit high-frequency sound waves that echo off objects or potential predators or prey. The sound waves bounce off the object or animal and are picked up by the animals' ears.

ENVIRONMENT AND HEALTH: NOISE POLLUTION

Noise may be one of the most widespread environmental pollutants in modern industrial societies. Unknown to many, noise may be turning our nation into a country of the hearing-impaired. Traffic noise, airport noise, loud music, crowd noise, and other loud sounds so common in modern society may be slowly destroying our hearing. By the time many New York City residents reach age 20, their hearing has been so impaired that they hear only as well as a 70-year-old African Bushman who has lived a life free of city noises.

For many years, scientists attributed the decline in hearing to middle ear infections, certain antibiotics, and the natural deterioration of the hair cells. New research, however, suggests that noise is probably the principal cause of hearing loss in modern societies.

Like that of any other pollutant, the damage caused by noise is related to two factors: exposure level (how loud a noise is) and the duration of exposure (how long one is exposed to it). In general, the louder the noise, the more damaging it is. In addition, the longer you are exposed to a damaging level of noise, the more hearing loss you will suffer. Extremely loud noise can result in immediate, per-

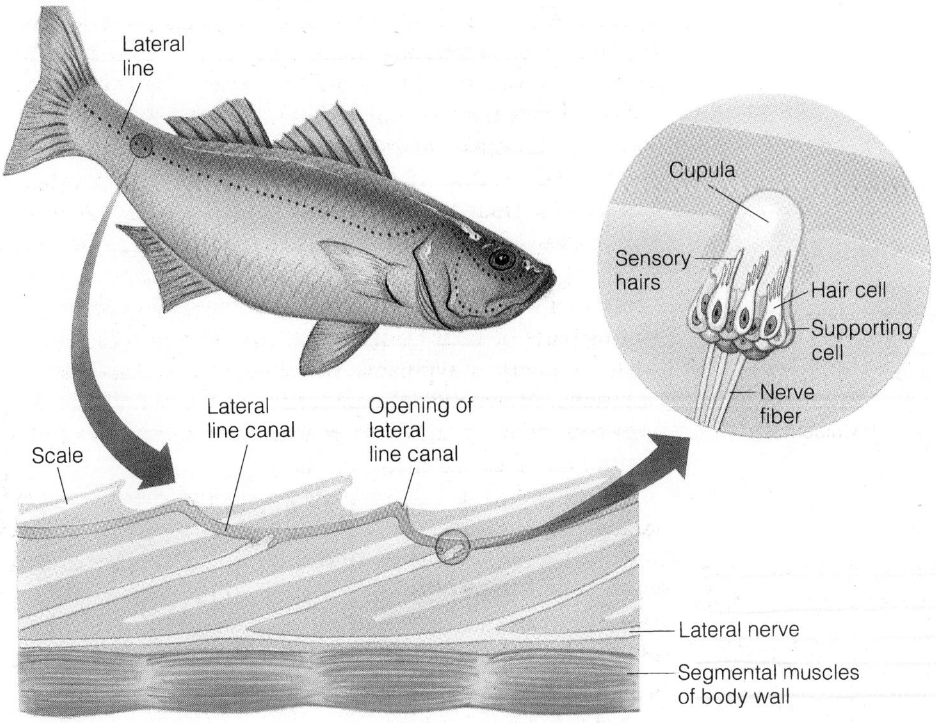

▻ **FIGURE 18–38 Lateral Line** The lateral line of fishes contains groups of sensory cells that detect pressure waves in water, alerting a fish to the presence of other fishes and obstacles.

manent damage. An explosion, for instance, can rupture the eardrum or fracture the bony ossicles, resulting in conduction deafness.

The noise to which people are exposed in factories and at construction sites is sufficient to cause gradual hearing loss. A worker may notice a dulled sense of hearing after working in a noisy environment; this is called a **temporary threshold shift.** Over time, the continued assault leads to a **permanent threshold shift,** complete hearing loss in certain frequencies. In most cases, hearing loss occurs so gradually that workers do not notice it until it is too late.

The intensity of sound is measured in **decibels (db).** Table 18–5 shows the scale and lists some common sounds. Surprising new research published by the EPA shows that continuous exposure to sounds over 55 decibels can result in hearing loss. Light traffic and an air conditioner operating 6 meters (18 feet) from you are 60-decibel sounds.

Besides deafening us and cutting us off from the important sounds of modern life, noise disturbs communications, rest, and sleep. It raises our level of stress, which, in turn, shortens our lives. When hearing is impaired, communication falters, and tensions often rise. People who are losing their hearing complain that they feel inadequate in social situations.

Hearing loss is not inevitable. You can take steps to protect your hearing. Keep your stereo at a reasonable level. Avoid noisy events. Cover your ears when an ambulance or fire engine approaches. Wear ear plugs or ear guards when operating noisy equipment like vacuum cleaners, chain saws, or construction equipment. Get treatment if you develop an ear infection. Prevention is the best medicine, because once you lose your hearing, it is gone forever, and so is an important part of your life.

SUMMARY

THE GENERAL SENSES

1. The general senses are pain, light touch, pressure, temperature, and position sense. Receptors for these senses may be naked or encapsulated nerve endings.
2. Receptors fit into five functional categories: (a) mechanoreceptors, (b) thermoreceptors, (c) photoreceptors, (d) chemoreceptors, and (e) pain receptors (nocireceptors).
3. Sensory stimuli may be internal or external. Some stimuli cause reflex actions. Still others may stimulate physiological changes.
4. The naked nerve ending receptors in the body include those stimulated by pain, light touch, and temperature. The encapsulated receptors include those that detect pressure (Pacinian corpuscles), light touch (Meissner's corpuscles, Krause's end-bulbs, and Ruffini's corpuscles), and muscle extension (muscle spindles and Golgi tendon organs).
5. Many sensory receptors cease responding to prolonged stimuli, a phenomenon called adaptation. Pain, temperature, pressure, and olfactory receptors all adapt.

TABLE 18–5 The Decibel Scale

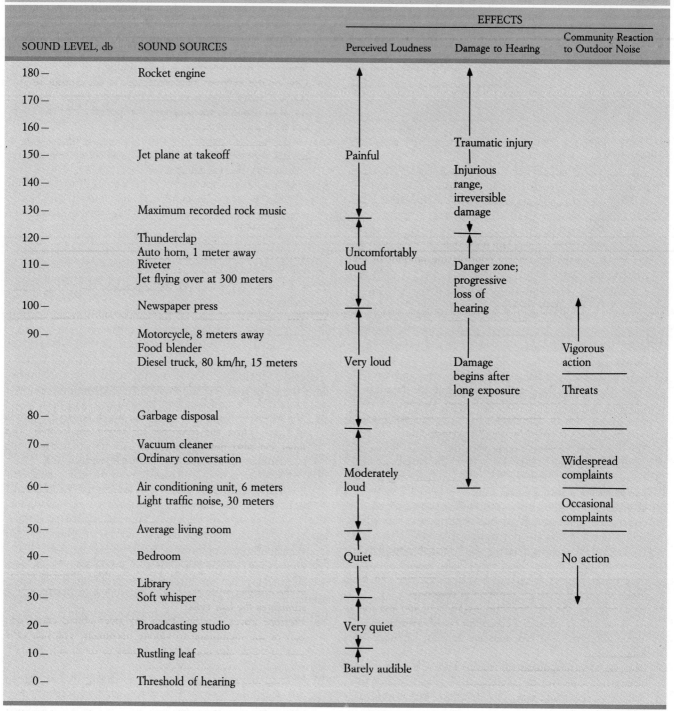

SOUND LEVEL, db	SOUND SOURCES	EFFECTS: Perceived Loudness	EFFECTS: Damage to Hearing	EFFECTS: Community Reaction to Outdoor Noise
180–	Rocket engine			
170–				
160–				
150–	Jet plane at takeoff	Painful	Traumatic injury	
140–			Injurious range, irreversible damage	
130–	Maximum recorded rock music			
120–	Thunderclap Auto horn, 1 meter away	Uncomfortably loud		
110–	Riveter Jet flying over at 300 meters		Danger zone; progressive loss of hearing	
100–	Newspaper press			
90–	Motorcycle, 8 meters away Food blender Diesel truck, 80 km/hr, 15 meters	Very loud	Damage begins after long exposure	Vigorous action Threats
80–	Garbage disposal			
70–	Vacuum cleaner Ordinary conversation			Widespread complaints
60–	Air conditioning unit, 6 meters Light traffic noise, 30 meters	Moderately loud		Occasional complaints
50–	Average living room			
40–	Bedroom	Quiet		No action
30–	Library Soft whisper			
20–	Broadcasting studio	Very quiet		
10–	Rustling leaf	Barely audible		
0–	Threshold of hearing			

SOURCE: From Jonathan Turk et al., *Environmental Science*, copyright © 1984 by Saunders College Publishing, a division of Holt, Rinehart and Winston, Inc. Reprinted by permission of the publisher.

TASTE AND SMELL: THE CHEMICAL SENSES

6. The special senses are served by more elaborate receptor organs, providing for taste, smell, vision, hearing, and balance.
7. Taste receptors called taste buds are located principally on the upper surface of the tongue. Taste buds contain receptor cells. Food molecules dissolve in the saliva and bind to the membranes of the microvilli of the receptor cells.
8. Taste buds respond to four flavors: salty, bitter, sweet, and sour.
9. The receptors for smell are located in the olfactory membrane in the roof of the nasal cavities. The receptor cells in the olfactory membrane respond to thousands of different molecules, which bind to membrane receptors on the olfactory hairs, stimulating nerve impulses that are transmitted to the brain via the olfactory nerve.

10. The ability to detect various chemicals in the environment is a universal trait of animals and even protists.
11. Chemoreception is used to find food and mates, recognize territories, and move about in the environment.
12. Insects have sensory hairs on their mouthparts and feet. The sensory hairs contain more than one receptor cell, each of which responds maximally to a different chemical. Some sensory hairs detect airborne chemicals, such as pheromones.

THE VISUAL SENSE: THE EYE

13. The eye is the receptor for visual stimuli and is located in the orbit, a bony socket.
14. The wall of the human eye contains three coats. The outermost coat is the fibrous layer and consists of the sclera (the white of the eye) and the cornea (the clear anterior structure that lets light shine in).
15. The middle layer consists of the choroid (the pigmented and vascularized section), the ciliary body (whose muscles control the lens), and the iris (which controls the amount of light entering the eye).
16. The innermost layer is the retina, which contains two types of photoreceptors. Rods function in dim light and provide black-and-white vision, and cones operate in bright light and provide color vision. Cones are also responsible for visual acuity. The highest concentration of cones is found in the fovea centralis.
17. Light is focused on the retina by the cornea and lens. The cornea has a fixed refractive power, but the lens can be adjusted to bend light according to need. The muscles of the ciliary body play an important role in this process. Focusing on near objects is aided by pupillary constriction.
18. The lens may become cloudy with old age or because of exposure to excess ultraviolet light. This condition, called cataracts, can be corrected by surgically removing the lens and replacing it with a plastic one.
19. The eye is divided into two cavities. The posterior cavity lies behind the lens and is filled with a gelatinous material, the vitreous humor. The anterior cavity is filled with a plasmalike fluid called the aqueous humor, which nourishes the lens and other eye structures in the vicinity. If the rate of absorption of aqueous humor decreases, however, pressure can build inside the anterior cavity, resulting in glaucoma.
20. As people age, the lens becomes less resilient and less able to focus on nearby objects, a condition called presbyopia.
21. Three additional eye problems are myopia, hyperopia, and astigmatism.
22. Myopia, or nearsightedness, results from a lens that is too strong (too concave) or an elongated eyeball. In the uncorrected eye, the image from distant objects comes in focus in front of the retina.
23. Hyperopia results from a weak lens (too convex) or a shortened eyeball. In the uncorrected eye, light rays from an image nearby would come into focus behind the retina.
24. Astigmatism is an irregularly curved lens or cornea that creates fuzzy images.
25. Three types of cones are present in the eye: red, green, and blue. Each type responds maximally to one specific color of light. Intermediate colors activate two or more types of cones.
26. Color blindness is a genetic disorder, more common in men than women. It results from a deficiency or an absence of one or more types of cones. Red-green color blindness is the most common type.
27. Structures that detect light appear in some of the most primitive animals on Earth and range in complexity from simple cellular organelles to clusters of light-sensitive cells to the vertebrate eye. Despite this variety, the fundamental process of light detection is quite similar.
28. One of the simplest photoreceptors is the eyespot of the planaria, which consists of two cup-shaped groups of cells in the head region of the organism that permit it to detect light but do not permit vision as we know it.
29. In the animal kingdom, primitive vision is first found in molluscs. Some molluscs like squid and octopi have sophisticated eyes that permit clear vision.
30. Most insects and crustaceans, such as crabs, have compound eyes that consist of many photoreceptive units, each with a small crystalline lens that focuses light on a photoreceptor cell beneath it.

HEARING AND BALANCE: THE COCHLEA AND MIDDLE EAR

31. The human ear consists of three portions: the outer, middle, and inner ears.
32. The outer ear consists of the auricle and external auditory canal, both of which direct sound to the eardrum.
33. The middle ear consists of the eardrum and three small bones, the ossicles, which transmit vibrations to the inner ear.
34. The auditory, or eustachian, tube helps equilibrate the pressure inside the middle ear cavity.
35. The inner ear contains the cochlea, which houses the organ of Corti, where the receptors for sound are located. The inner ear also houses receptors for movement and head position: the semicircular canals, utricle, and saccule.
36. The cochlea is a spiral-shaped, bony structure that contains three fluid-filled canals. Separating the middle canal from the lower one is the flexible basilar membrane, which supports the organ of Corti. Hair cells in the organ of Corti are embedded in the relatively rigid tectorial membrane.
37. Sound waves create vibrations in the eardrum and ossicles, which are transmitted to fluid in the cochlea. Pressure waves in the cochlea cause the basilar membrane to vibrate, which stimulates the hair cells.
38. Pressure waves resulting from any given sound cause one part of the membrane to vibrate maximally. The hair cells stimulated in that region send signals to the brain, which it interprets as a specific frequency.
39. Hearing loss may occur as a result of damage or blockage to the conducting system: the external auditory canal, the eardrum, and the ossicles. Damage to the hair cells, the auditory nerve, or the auditory cortex are forms of nerve or sensineural deafness.
40. The semicircular canals are three hollow rings filled with a fluid called endolymph. The receptors for head movement are located in an enlarged portion at the base of each canal, the ampulla.
41. Fluid movement inside the semicircular canals deflects the gelatinous cap (cupula) lying over the receptor cells, stimulating them and alerting the brain to head movements.
42. The semicircular canals are set in all three planes of space, so movement in any direction can be detected.

43. Two membranous sacs in the vestibule, the utricle and saccule, contain receptors called maculae that respond to linear acceleration and tilting of the head.
44. Mechanoreceptors that detect body position are common in the animal kingdom. They contain hair cells akin to those found in the vertebrates. Hair cells respond to gravity or other forces, producing bioelectric impulses that are sent to the brain, alerting the organism of a change in position.
45. Sound detection is also widespread in the animal kingdom. Many insects have body hairs that vibrate at the same frequency as biologically important sounds. Some insects also have pressure-detecting organs on various parts of their bodies, such as the front legs.
46. In many fishes, cells containing minute hair cells are located in longitudinal grooves that run along the length of the body on both sides. These receptor cells detect low-frequency vibrations of water flow and turbulence.

ENVIRONMENT AND HEALTH: NOISE POLLUTION

47. Noise damages the ears. Extremely loud noises can rupture the eardrum or break the ossicles. Less intense noises, however, generally destroy hearing gradually by damaging hair cells.
48. In most people, hearing loss occurs so gradually as to be undetected. Individuals can take steps to avoid hearing loss.

EXERCISING YOUR CRITICAL THINKING SKILLS

In an experiment to determine how chemicals affect vision, two researchers expose rats to varying levels of a toxin, chemical A. They find that at low doses, the chemical has no effect, but higher doses result in severe vision loss and blindness. The researchers immediately ask the Food and Drug Administration to ban the chemical from production for fear it might similarly affect humans. You are head of the FDA. Using your critical thinking skills, how would you go about considering the request? What factors would help you determine whether the request was valid? What studies might you want to see done?

TEST OF TERMS

1. The general senses are ________, temperature, light touch, pressure, and proprioception, or position sense.
2. Receptors fall into two broad groups: naked nerve endings and ________ ________.
3. Light touch is perceived by two sensors: naked nerve endings surrounding hair follicles and ________ ________.
4. Pressure is perceived by ________ corpuscles located in the deeper layers of the skin and around various organs.
5. The receptor situated in the superficial layer of the dermis that responds to touch is the ________ ________.
6. The ________ ________ is a stretch receptor found in skeletal muscles.
7. Stretch receptors in tendons are called ________ ________ ________.
8. ________ occurs when a sensory receptor stops sending impulses, even though the stimulus is still present.
9. Taste, hearing, and vision are three of the ________ ________.
10. The taste receptors in the oral cavity and on the dorsal (upper) surface of the tongue are called ________ ________. They are especially abundant on the ________, small protrusions on the surface of the tongue.
11. Receptors for the sense of smell are located in the ________ membrane found in the roof of each nasal cavity. Receptor cells in the membrane are modified ________ neurons.
12. The eye consists of three layers. The outermost, fibrous layer consists of the ________, the white of the eye, and the ________, the clear, anterior portion that allows light to enter.
13. The middle layer of the eye is heavily ________ and vascularized. It consists of the ________, the ________ ________, and the iris.
14. The inner layer of the eye is the light-sensitive portion of the eye and is called the ________. It contains two types of receptors, the ________, which confer color vision, and the rods, which operate best in ________ light.
15. Nerve impulses leave the retina via the optic nerve, which is formed from the axons of the ________ cells.
16. Sharpest vision occurs when an image is cast on the ________ ________, a small depression in the retina lying lateral to the blind spot, or ________.
17. The ________ is a flexible structure used to focus light coming from nearby objects on the retina. It is attached to the ________ ________ by the suspensory ligaments. It may become cloudy with age, a condition known as ________.
18. The ________ ________ is a gelatinous mass that occupies the posterior cavity of the eye.
19. ________ results from the excess buildup of ________ ________ in the anterior cavity of the eye.
20. The bending of light is called ________. It occurs anytime light waves ________ ________.

21. Eye movements are caused by the ______________ ______________ muscles.
22. Myopia is also called __________. It results from a(n) ____________ or a(n) ____________ ____________.
23. The surgical technique that corrects for myopia is called ______________ ____________.
24. An irregularly curved lens or cornea results in a condition known as ____________.
25. ____________ is the pigment of the rods. It breaks down into two molecules when struck by light.
26. Color blindness is a ____________-____________ trait.
27. Sound waves are directed into the ______________ ______________ canal to the ________, which separates the outer ear from the middle ear.
28. Extremely loud noises may damage the ____________, the bones in the middle ear cavity, which transmit sound to the ________ ________ of the cochlea.
29. The ________ ________ leads from the middle ear cavity to the nasopharynx and serves as a pressure release valve.
30. Movement of the head is detected by sensors in the ____________ of the ____________ canals in the inner ear. These canals are filled with a fluid called ____________.
31. Static balance is provided in part by two receptor patches, the ________, located in the utricle and saccule.
32. The ____________ ____________ ____________ is the receptor organ for sound in the cochlea. It consists of three canals: the ____________ canal, the cochlear duct, and the tympanic canal.
33. A rupture of the eardrum or a fusion of the bones in the middle ear results in ____________ deafness.
34. A transitory loss of hearing is called a ____________ ____________ shift.

Answers to the Test of Terms are found in Appendix B.

TEST OF CONCEPTS

1. Define the terms *general senses* and *special senses.*
2. Using your knowledge of the senses and of other organ systems gained from previous chapters, describe the role that sensory receptors play in homeostasis. Give specific examples to illustrate your main points.
3. Make a list of both the encapsulated and the nonencapsulated general sense receptors. Note where each is located and what it does.
4. Define the term *adaptation.* What advantages does it confer? Can you think of any disadvantages?
5. Describe the receptors for taste, explaining where they are located, what they look like, and how they operate.
6. Taste buds detect four basic flavors. What are they? How do you account for the thousands of different flavors that you can detect?
7. Describe the olfactory membrane and the structure of the receptor cells. How do these cells operate? In what ways are taste receptors and olfactory receptors similar? In what ways are they different?
8. Explain the following statement: taste and smell are complementary.
9. Describe the function of chemoreception in the animal kingdom. Give some examples. Are the structures found in animals other than vertebrates like those found in humans?
10. Draw a cross section of the human eye, and label its parts.
11. Define the following terms: retina, rods, cones, fovea centralis, optic disc, ganglion cells, bipolar neurons, and optic nerve.
12. Compare and contrast rods and cones.
13. When focusing on a nearby object, your eyes go through several changes. Describe those changes and what they accomplish.
14. Define the following terms: myopia, hyperopia, presbyopia, and astigmatism.
15. You walk into a dark movie theater and find that you can barely make out the aisle. After a while, your vision recovers. Explain both phenomena.
16. What is color blindness? What is the most common type? Explain what your world would look like if you were afflicted by red-green color blindness.
17. What is a cataract, and how is it treated?
18. Describe the disease known as glaucoma. What causes it, and how is it treated?
19. Describe the range of photoreceptor organs found in the animal kingdom. What trends do you see when progressing from invertebrates to vertebrates? Do the similarities suggest anything about evolution to you?
20. Describe the anatomy of the ear and the role of the outer, middle, and inner ears in hearing.
21. How do the semicircular canals operate? How do the utricle and saccule operate?
22. Describe the different types of deafness and how they are treated.

CHAPTER

19

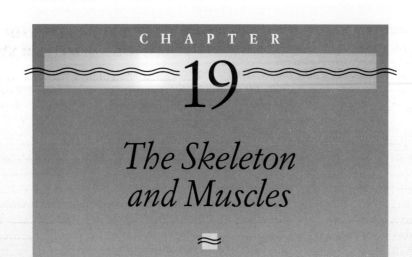

The Skeleton and Muscles

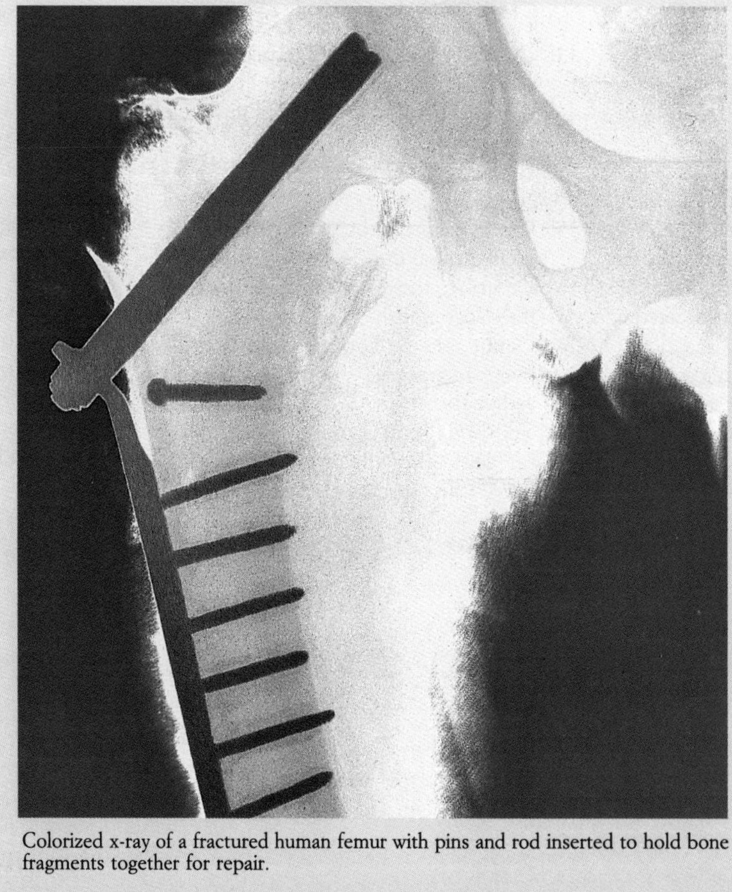

Colorized x-ray of a fractured human femur with pins and rod inserted to hold bone fragments together for repair.

A hiker on an afternoon stroll in North Carolina happens upon a white-tailed deer, munching on grass along the trail. Startled, the deer bolts away to the safety of the forest. Later, the hiker flushes a duck paddling in a quiet pond. Noisily, the duck flaps its wings, beating the water as it "runs" across the surface of the pond. When it has reached proper speed, it takes off, disappearing from sight.

These movements all depend on the skeleton and attached muscles. During evolution, these basic components have been molded by environmental forces to provide a variety of locomotive systems—the fins of fish, the wings of birds, and the legs of humans. Nonetheless, many species have a common architecture, illustrating once again that the evolutionary process is conservative; that is, it preserves what works. ▻ Figure 19–1, for example, shows the remarkable similarity of the muscles in frogs and humans.

This chapter discusses motor systems in the animal kingdom, starting with an overview of invertebrates and vertebrates. It then focuses on the human skeleton and the muscles that attach to it, the skeletal muscles. Together, the bones and muscles of vertebrates constitute the musculoskeletal system. In humans, they make up 50% to 60% of the body weight of an adult.

AN OVERVIEW OF THE EVOLUTION OF MOTOR SYSTEMS

The animal kingdom is defined in part by movement—the ability to move from one place to another and the ability to move body parts to perform tasks. Movement is fundamental to animal success. It is vital to reproduction, feeding, protection, and many other daily tasks.

As in many other systems discussed in this book, animal locomotion varies considerably, from the slithering motion of planaria to the complex movements of organisms like you and me. Despite the vast differences, two common features are present in all animals. First, all animals have muscles. Second, all animals convert chemical energy in their muscles into mechanical energy (movement). The expression of these common characteristics is determined by the nature of the environment in which an animal lives. That is to say, natural selection helped shape the physical adaptations involved in movement. Animals that fly through the air, for instance, developed wings with a large surface area to push against low-density air. Paddles, fins, and flippers, much smaller than wings, developed in water-dwelling organisms in response to a denser medium that required less push.

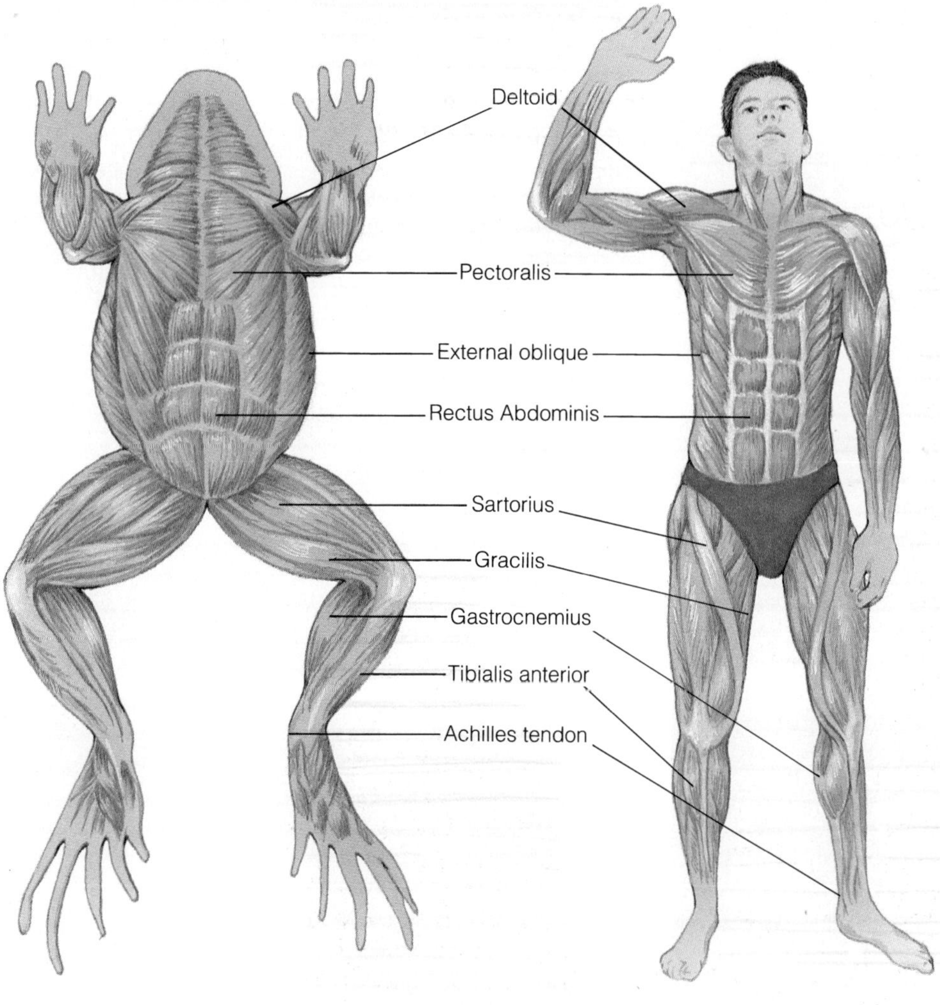

▻ FIGURE 19–1 Comparison of the Muscles of the Frog and Human A glance at the muscular system and bones (not shown here) of two very different species suggests a common evolutionary ancestry.

Invertebrates Contain Opposing Muscle Groups that Permit Primitive Movements

One of the simplest animal motor systems occurs in the sea anemone, which lives in and around coral reefs, where it attaches to the bottom (▻ Figure 19–2a). This system contains two sets of muscle fibers, circular and longitudinal. The circular muscle fibers are found in bundles that form hoops around the body axis. When they contract, the organism elongates (Figure 19–2b).The longitudinal muscle fibers occur in bundles that run the length of the animal's body (Figure 19–2c). When they contract, the organism shortens. This simple muscle system helps the anemone escape danger, as any scuba diver who has brushed past an anemone will attest.

Movement in earthworms and other annelids is also dependent on antagonistic circular and longitudinal muscles located in the body wall. When the circular muscles contract, an earthworm elongates. When the longitudinal muscles contract, the body shortens. Forward movement is aided by small bristles that protrude from the body. As the earthworm elongates, these bristles actually grip the ground, so when the body contracts, force is applied against the bristles, and the earthworm's body is pulled forward.

Interestingly, longitudinal and circular muscle fibers are found in many higher organisms, such as humans, but only in the walls of certain organs, like the small intestine, where they help propel the contents forward.

The longitudinal and circular muscles of the sea anemone and earthworm are said to be antagonistic, because they exert opposite effects. In the animal kingdom, many muscles are arranged in antagonistic groups. The muscles of your upper arm, for instance, consist of two opposing groups. The muscles in front cause the arm to flex and those in back cause it to extend. Antagonistic muscle groups, therefore, are a common characteristic of all animals.

The Presence of Exoskeletons in Invertebrates Provides a Wider Range of Movements

Anemones and annelids have no skeleton, and their movements are rather limited. Some invertebrates, like insects, crabs, and other arthropods, have hard external skeletons, or **exoskeletons,** made of chitin, proteins, and (sometimes) lipids (▻ Figure 19–3). Exoskeletons protect arthropods against prey and also reduce water loss. As you can see in Figure 19–3, the exoskeleton is not one large inflexible case but, rather, consists of several segments that connect to one another at pliable regions that act like hinges. Antagonistic muscles located internally attach to the exoskeleton and span the hinges, thus permitting movement. Exoskeletons are found in other animals as well, such as lobsters, clams, and snails.

(a)

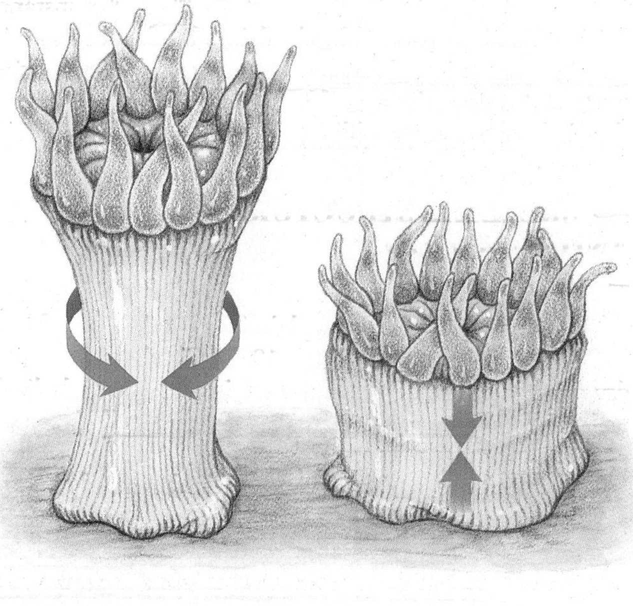
(b) (c)

▻ **FIGURE 19–2 Muscular Contraction in the Sea Anemone** *(a)* This sedentary animal attaches to coral reefs. *(b)* When the circular muscles contract, the organism elongates. *(c)* When the longitudinal muscles contract, it shortens.

⊳ **FIGURE 19–3 Arthropod Exoskeleton** The hard, nonliving exoskeleton of the scarab beetle protects it from predators and reduces water loss. The exoskeleton is hinged by pliable regions. Muscles lying internally attach to the segments and are responsible for movement of the parts of the exoskeleton.

In Vertebrates, Opposing Muscle Groups Acting on Internal Skeletons Provide a Wider Range of Movements

Internal skeletons are found in a wide array of vertebrates, including birds, fish, reptiles, amphibians, and mammals. In animals such as rays and sharks, skeletons are composed entirely of cartilage. In most vertebrates, internal skeletons are made of bones, with small amounts of cartilage, as noted in Chapter 10 (⊳ Figure 19–4).

The bones of the skeleton are attached by one of several types of joints, which permit a wide range of motions. Together, the muscles and bones form the musculoskeletal system. Besides aiding in movement, the musculoskeletal system is also involved in homeostasis. Bone, for instance, helps maintain constant blood calcium levels, necessary for muscle contraction. Muscles contract rhythmically when birds and mammals are cold, causing them to shiver and thus helping generate additional body heat.

SKELETAL STRUCTURE AND FUNCTION

The word **skeleton** is derived from a Greek word that means "dried-up body." Many people's first impression of bone isn't much different (⊳ Figure 19–5). To them, bones are dry, dead structures. **Bone,** however, is a living, metabolically active tissue. Bone tissue contains numerous cells, known as **osteocytes.** These cells are embedded in a calcified extracellular material, or matrix, which gives the bones of the body their characteristic hardness, strength, and flexibility. The extracellular material of bone consists of (1) an organic component, collagen, a protein that imparts flexibility; and (2) an inorganic component, chiefly calcium phosphate crystals that are deposited on the collagen fibers, which imparts strength. (For more on bone, you might want to review the information presented in Chapter 10.)

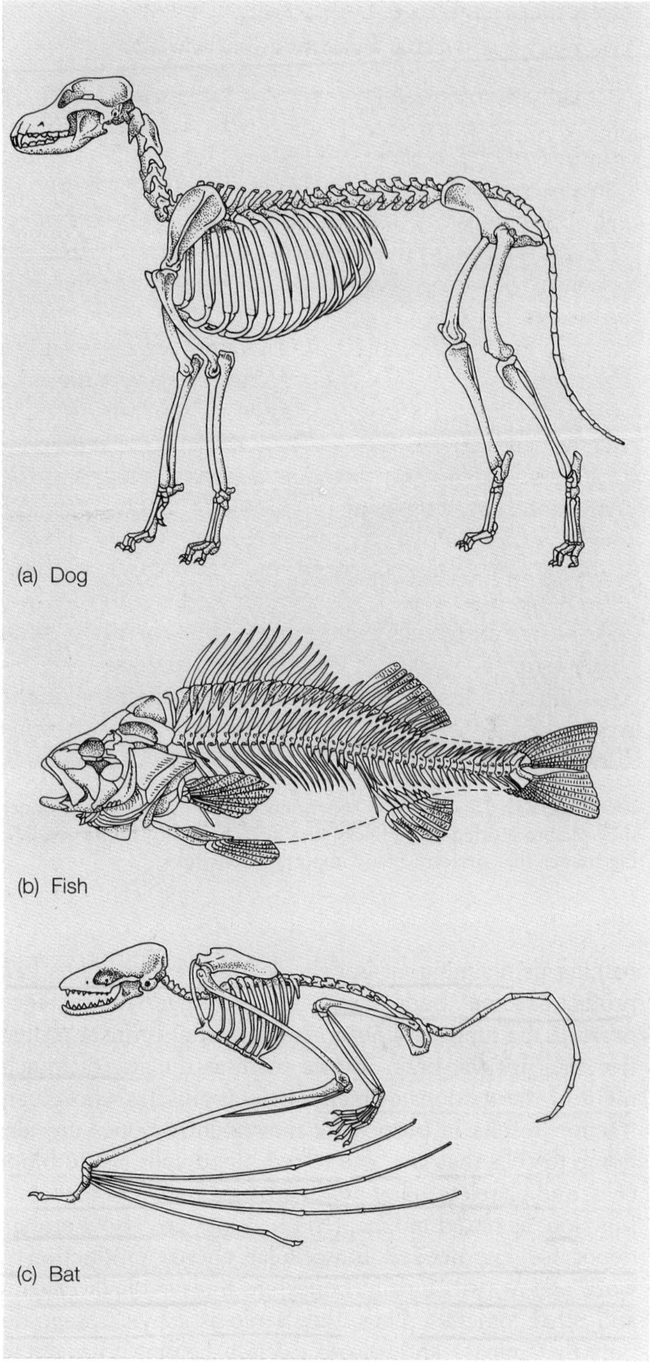

⊳ **FIGURE 19–4 Skeletal Similarities** The skeletons of the dog, fish, and bat exhibit many anatomical similarities. Each skeleton consists of a spinal column, ribs, skull, and appendages.

Bones Serve Many Functions and Play an Important Role in Homeostasis

The human skeleton consists of 206 bones, discrete structures made of bone tissue (⊳ Figure 19–6). Bones provide internal structural support, giving shape to our bodies and

▻ **FIGURE 19–5 Bone** When most people think about bones, they picture a dried-out, lifeless structure. In reality, bones are living tissues that perform many important functions.

helping us maintain an upright posture. Some bones help protect internal body parts. The rib cage, for instance, protects the lungs and heart, and the skull forms a protective shell for the brain. Bones serve as the site of attachment of the tendons of many skeletal muscles whose contraction results in purposeful movements. Bones are also home to cells that give rise to red blood cells, white blood cells, and platelets, all of which are essential to homeostatic function, as noted in Chapter 13. Bones are also a storage depot for fat, needed for cellular energy production at work and at rest. Finally, bones are a reservoir of calcium and other minerals. Thus, they serve as active storage depots that release and absorb calcium helping to maintain normal blood levels. Calcium is essential to muscle contraction, and disturbances in blood calcium levels can impair muscle contraction.

The Human Skeleton Consists of Two Parts

The human skeleton consists of two parts, the axial skeleton and the appendicular skeleton (Figure 19–6). The **axial skeleton** forms the long axis of the body. It consists of the skull, the vertebral column, and the rib cage. The **appendicular skeleton** consists of the bones of the arms and legs and the bones of the shoulders and pelvis, by which the upper and lower extremities are attached to the axial skeleton.

▻ Figure 19–7 illustrates the anatomy of the humerus, the bone found in the upper arm. As shown, the humerus, like other long bones of the body, consists of a long, narrow shaft, the **diaphysis,** and two expanded ends, the **epiphyses.** The ends of the epiphyses are coated with a thin layer of hyaline cartilage, described in Chapter 10, which reduces friction in joints. The protrusions on the bone mark the sites of muscle attachment.

All Bones Have a Hard, Dense Outer Layer that Surrounds a Less Compact Central Region

Take a moment to study the skeleton in Figure 19–6. As you examine it, you may notice that bones come in a variety of shapes and sizes. Some are long, some are short, and some are flat and irregularly shaped. Nevertheless, all bones share some characteristics. For example, all bones consist of an outer shell of dense material called **compact bone,** surrounding a spongy interior. **Spongy bone,** so named because of its spongy appearance, is less dense than compact bone and is generally concentrated in the epiphyses. As shown in Figure 19–7c, spongy bone forms a latticework.

On the outer surface of the compact bone is a layer of connective tissue, the **periosteum** ("around the bone"). The outer layer of the periosteum is composed of dense, irregularly arranged connective tissue fibers and serves as the site of attachment for many skeletal muscles (▻ Figure 19–8). The inner layer of the periosteum contains osteogenic (bone-forming) cells that participate in the production of new bone during remodeling or repair. The periosteum is richly supplied with blood vessels, which enter the bone at numerous locations. Blood vessels travel through small canals in the compact bone and course through the inner spongy bone, providing nutrients and oxygen and carrying off cellular wastes, such as carbon dioxide. The periosteum is also richly supplied with nerve fibers. The majority of the pain felt after a person bruises or fractures a bone results from pain fibers in the periosteum.

Spongy bone contains numerous small, adjoining cavities. They connect with the large, hollow interior of the diaphysis, together forming the **marrow cavity.** The marrow cavities in most of the bones of the fetus contain **red marrow.** Red marrow is a blood-cell factory, producing RBCs, WBCs, and platelets to replace those routinely lost each day. As an individual ages, most red marrow is slowly "retired" and becomes filled with fat, becoming **yellow marrow.** Yellow marrow begins to form during adolescence and, by adulthood, is present in all but a few bones. Red blood cell formation, however, continues in the bodies of the vertebrae, the hip bones, and a few others. Yellow marrow can be reactivated to produce blood cells under certain circumstances—for example, after an injury.

The Joints Permit Varying Degrees of Mobility

A gymnast races across the mat and leaps into space, twirling effortlessly before landing on her feet. As soon as she lands, she takes off again across the mat in a series of three

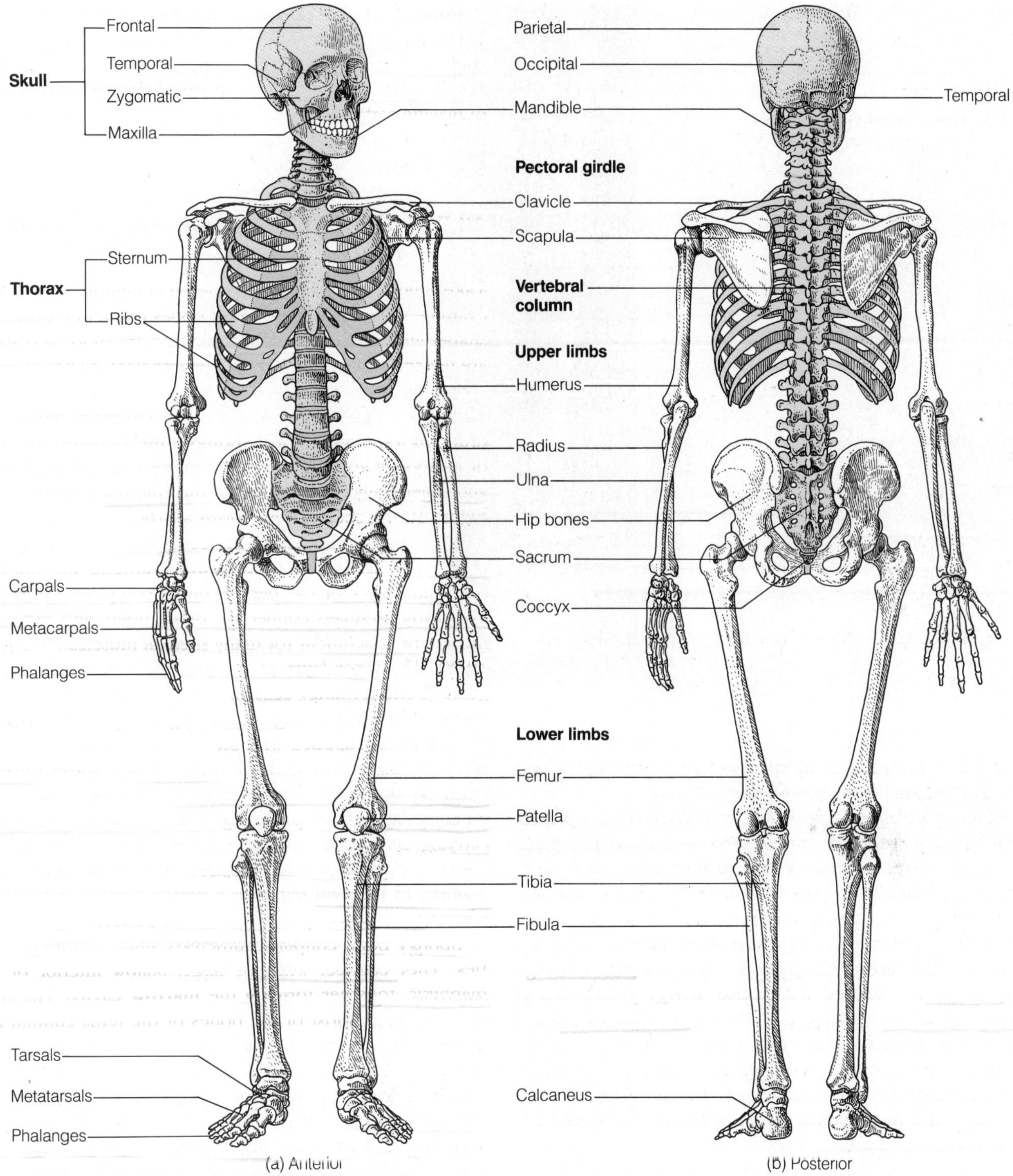

FIGURE 19–6 The Human Skeleton (*a*) Anterior view. (*b*) Posterior view. Over 200 bones of all shapes and sizes make up the skeleton. The shaded region shows the axial skeleton. The unshaded region is the appendicular skeleton.

back handsprings. These delightful movements are the result of long hours of practice and exercise and are made possible by the joints. **Joints** are the structures that connect the bones of the skeleton.

Joints can be classified by the degree of movement they permit. Those that permit no movement are called *immovable joints.* Those that permit some movement are *slightly movable joints,* and those that permit free movement are *freely movable joints.* Consider some examples.

The bones of the skull shown in ▻ Figure 19–9a are held together by immovable joints. As illustrated, opposing bones in the skull interdigitate (interlock). Fibrous connec-

▻ **FIGURE 19–7 Anatomy of Long Bones** *(a)* Photograph and *(b)* drawing of the humerus. Notice the long shaft and dilated ends. *(c)* Longitudinal section of the humerus showing the position of the compact bone, spongy bone, and marrow.

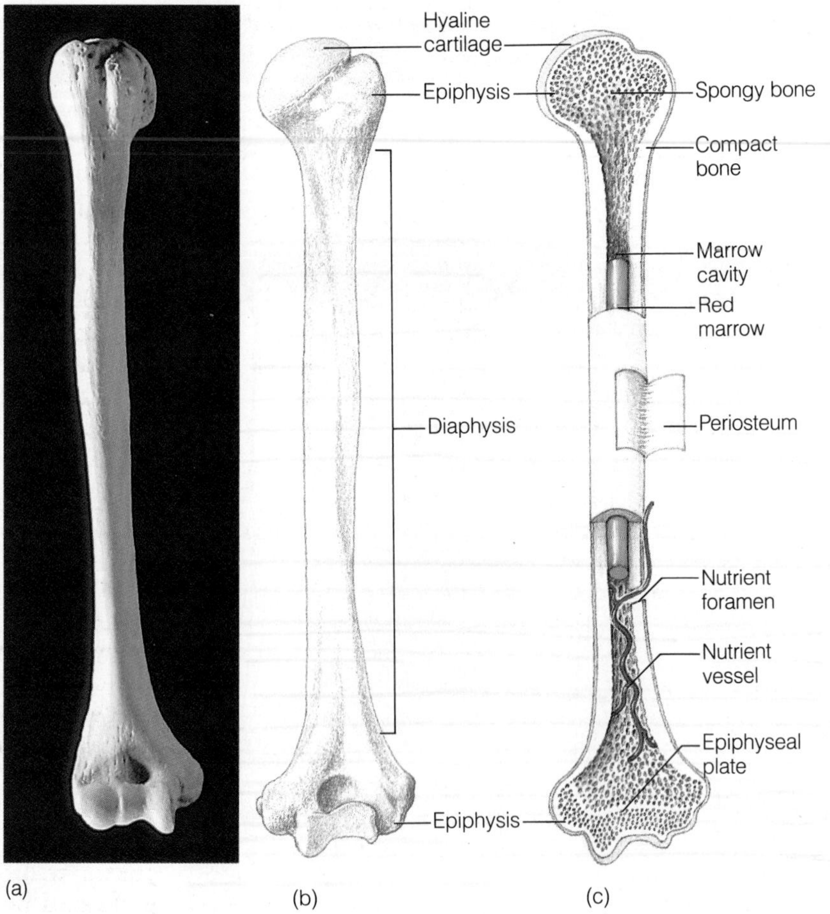

tive tissue spans the small space between the interlocking bones, holding the bones together.

The pubic symphysis is the joint formed by the two pubic bones (Figure 19–9b). These bones are held in place by fibrocartilage, and the joint is basically immovable. Near the end of pregnancy, however, hormones loosen the fibrocartilage of the pubic symphysis. This allows the pelvic outlet to widen enough to permit the baby to be delivered.

The bodies of the vertebrae are united by slightly movable joints (▻ Figure 19–10). Each vertebra is separated from its nearest neighbors by an intervertebral disc. The inner portion of the disc acts as a cushion, softening the impact of walking and running, as noted in Chapter 10. The outer, fibrous portion holds the disc in place and joins one vertebra to its nearest neighbor. The joints between the vertebrae offer some degree of movement, resulting in a fair amount of flexibility. If they did not, we would be unable to bend over to tie our shoes or unable to curl up on the couch for an afternoon snooze.

The most common type of joint is the freely movable, or **synovial, joint.** The synovial joints are more complex than other types and permit varying degrees of movement. Although synovial joints differ considerably in architecture, they share several features.

The first commonality is the hyaline cartilage located on the articular (joint) surfaces of the bones (▻ Figure 19–11a). This thin cap of hyaline cartilage reduces friction and facilitates movement. The second common feature of synovial joints is the joint capsule. The **joint capsule** is a double-layered structure that joins one bone to another in the joint (Figure 19–11a). The outer layer of the capsule consists of dense connective tissue that attaches to the periosteum of adjoining bones. Parallel bundles of dense connective tissue fibers in the outer layer of the capsule form **ligaments,** which run from bone to bone, giving additional support to the joint. As a rule, ligaments are fairly inflexible. However, some individuals have remarkably flexible ligaments and tendons. Because of this, some people can extend their thumbs well beyond the 90 degrees possible for most of us. And some can extend their fingers so much that they can touch the back of their hand. These people are said to be "double jointed."

The inner layer of the joint capsule is called the **synovial membrane,** and it consists of loose connective tissue with a generous supply of capillaries. The synovial membrane produces a fairly thick, slippery substance known as **synovial fluid.** It provides nutrients to the articular cartilage (the hyaline cartilage on the articular surfaces of the bone) and also acts as a lubricant, facilitating the movement of bones in joints. Normally, the synovial membrane produces only enough fluid to create a thin film on the articular cartilage. Injuries to a joint, however, may result in a dramatic increase in synovial fluid production, causing swelling and pain in joints.

Tendons and muscles provide additional support in some joints and are commonly associated with synovial joints. In the shoulder joint, for example, the muscles of the shoulder help hold the head of the humerus in the socket (formed by the scapula). Muscles in the hip also help hold the head of the femur in place. Because muscles strengthen joints, individuals who are in poor physical shape are much more likely to suffer a dislocation on a ski trip or during exercise than someone who is in good shape. **Dislocation** is an injury in which a bone is displaced from its proper position in a joint due to a fall or some other unusual body movement. In some cases, bones will slip out of place, then back in without assistance. In others, the bone can be put back in place only by a physician or other trained health-care worker.

Synovial joints come in many shapes and sizes and are classified on the basis of structure. Two of the most common are the hinge joint and the ball-and-socket joint. The knee joint is a **hinge joint,** as are the joints in the fingers. Hinge joints open and close like hinges on a door and therefore provide for movement in only one plane. The hip and shoulder joints are **ball-and-socket joints.** They provide a wider range of motion. (Compare the movements permitted by the shoulder and hip joints to those permitted by the knee joint.)

Arthroscopy Permits Surgeons to Repair Injured Joints with a Minimum of Trauma. The acrobatics performed by modern dancers or Olympic gymnasts illustrate the wide range of movement that joints allow. The joints, however, are a biological compromise. They must allow movement, but also provide some degree of stability,

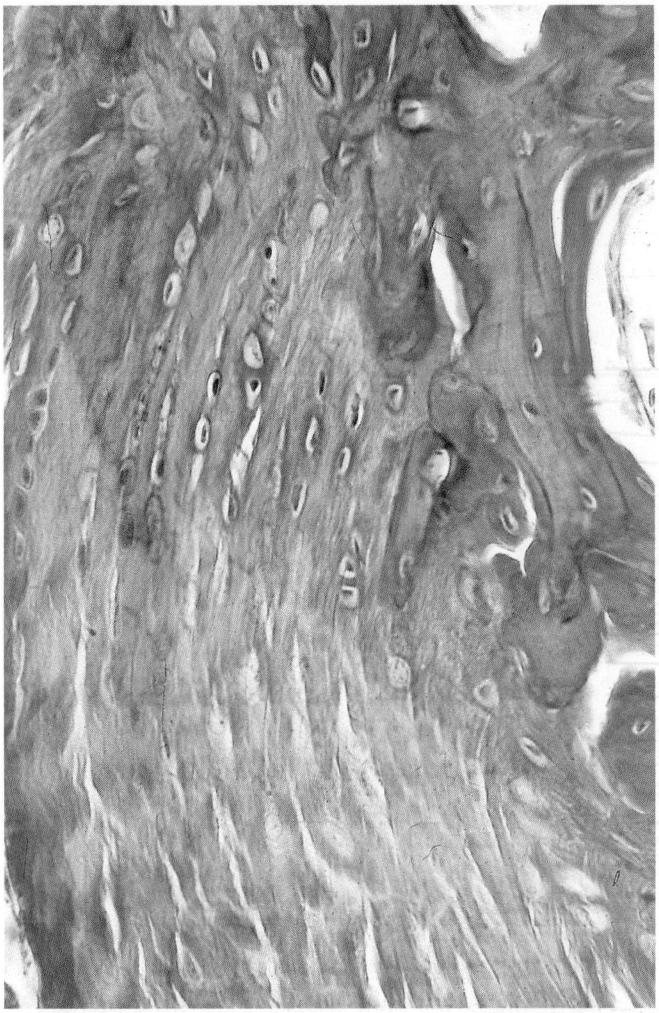

▻ **FIGURE 19–8 Tendon Attachment to Bone** The tendon attaches to the outer layer of the periosteum.

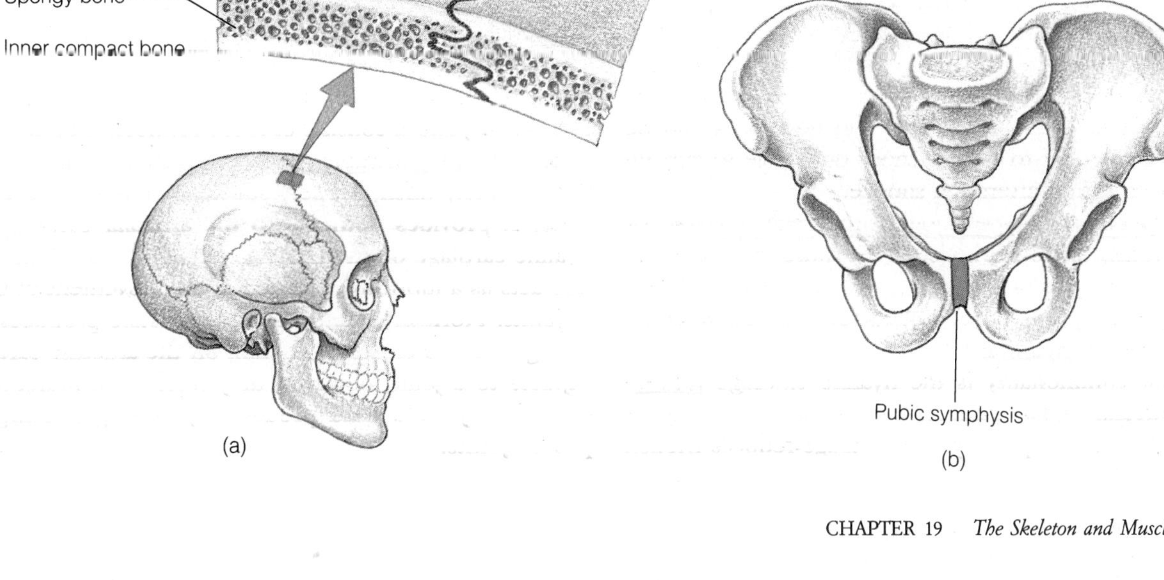

▻ **FIGURE 19–9 Two Immovable Joints** *(a)* Many of the bones of the skull are held in place by joints called sutures. The bones are linked by fibrous tissue, and the joints are immovable. *(b)* The pubic symphysis is another immovable joint. During childbirth it softens and expands to permit birth.

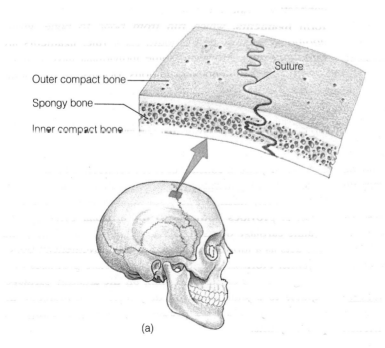

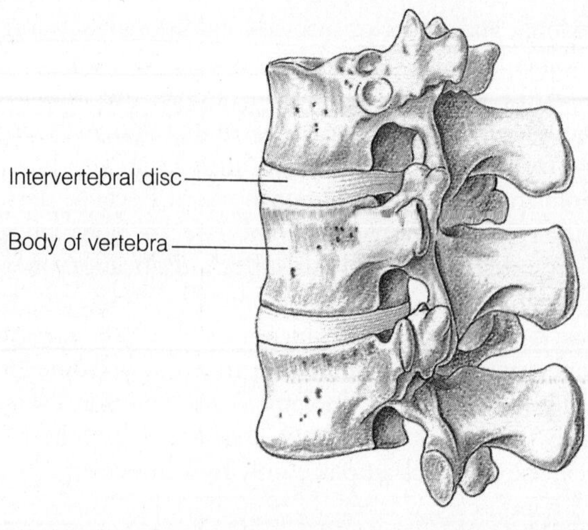

▻ **FIGURE 19–10 A Slightly Movable Joint** The intervertebral discs allow for some movement, giving the vertebral column flexibility.

helping hold bones in place. Full mobility would compromise strength, and absolute strength would compromise mobility.

Given the need for compromise, it is not surprising that joint injuries are common among physically active people, especially athletes (Table 19–1). A hard blow to the knee of a football player, for instance, can tear the ligaments, disabling the joint. A rough fall can dislocate a skier's shoulder. Improperly lifting an object can strain the ligaments that join the vertebrae of the back, resulting in considerable pain.

Torn ligaments, tendons, and cartilage in joints heal very slowly because they are not well endowed with blood vessels.[1] For years, joint repair required major surgery that was so traumatic it put patients out of commission for several months. Today, however, new surgical techniques allow physicians to repair joints with a minimum of trauma (▻ Figure 19–12). Through small incisions in the skin over the joint, surgeons insert a device called an **arthroscope.** It allows them to view the damage inside the joint cavity and also insert special instruments to snip off damaged cartilage. Consequently, a surgeon can repair damaged cartilage without opening the joint. This reduces damage caused by large incisions and allows athletes to be back on their feet and on the playing field in a matter of weeks. New surgical techniques are also used to rebuild torn ligaments, thus returning joints to nearly their original state.

[1]Because much of it is avascular, cartilage may not heal at all.

Osteoarthritis Is a Degenerative Bone Disease Caused by Wear and Tear on the Articular Cartilages. Virtually every time you move, you use one or more of your joints. Problems in joints, therefore, are often quite noticeable. One of the most common problems is called **degenerative joint disease** or **osteoarthritis.** Although its cause is not known, degenerative joint disease may simply result from wear and tear on a joint. Over time, excess wear may cause the articular cartilage on the ends of bones to flake and crack. As the cartilage degenerates, the bones come in contact, grinding against each other during movement and causing considerable pain and discomfort. Swelling usually accompanies these changes, and swelling in a joint tends to reduce mobility.

Osteoarthritis occurs most often in the weight-bearing joints—the knee, hip, and spine—which are subject to the most wear over time. Osteoarthritis may also develop in the finger joints. The amount of swelling in victims of osteoarthritis varies considerably. Some patients experience virtually no swelling; in others, the joints become enlarged and disfigured.

Osteoarthritis is extremely common. X-ray studies of

▻ **FIGURE 19–11 A Synovial Joint** *(a)* A cross section through the hip joint (a ball-and-socket joint) showing the structures of the synovial joint. *(b)* Ligaments help support the joint.

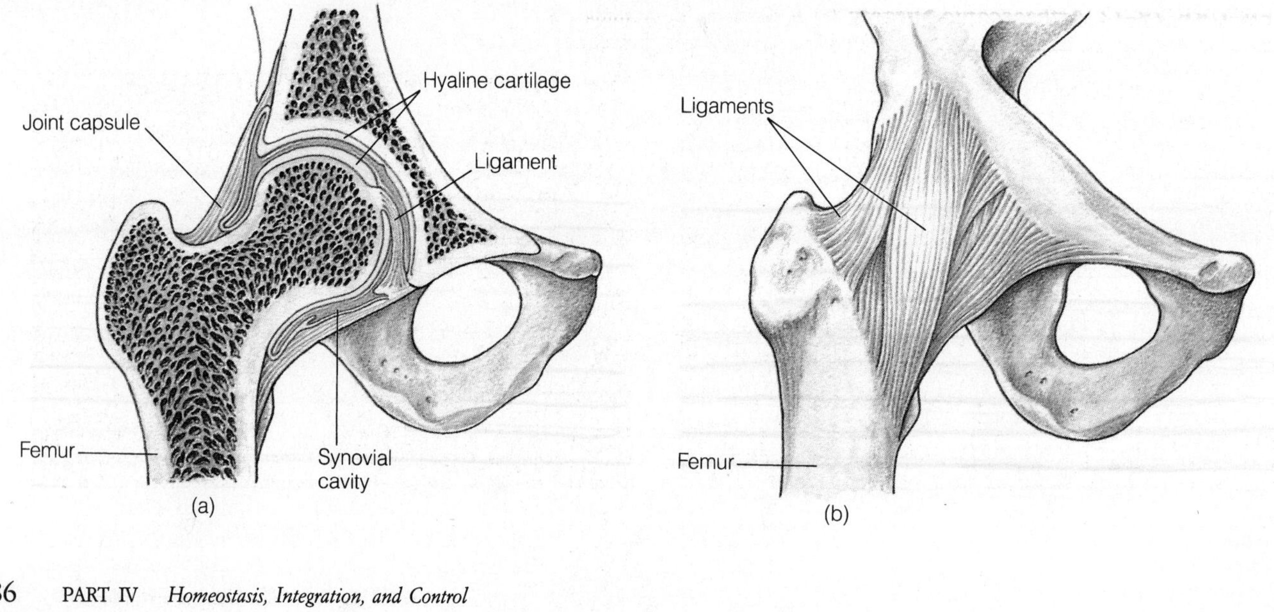

TABLE 19–1 Common Injuries of the Joints

INJURY	DESCRIPTION	COMMON SITE
Sprain	Partially or completely torn ligament; heals slowly; must be repaired surgically if the ligament is completely torn.	Ankle, knee, lower back, and finger joints
Dislocation	Occurs when bones are forced out of a joint; often accompanied by sprains, inflammation, and joint immobilization. Bones must be returned to normal positions.	Shoulder, knee, and finger joints
Cartilage tears	Cartilage may tear when joints are twisted or when pressure is applied to them. Torn cartilage does not repair well because of poor vascularization. It is generally removed surgically; this operation makes the joint less stable.	Cartilage in the knee is the most common

people over 40 years of age show that most people have some degree of degeneration in one or more joints. Fortunately, many people do not even notice the problem, and the disease rarely becomes a serious medical problem.

Wear and tear on joints is worsened by obesity. The extra pressure on the joints apparently wears the cartilage away more quickly. Thus, weight control can help reduce the rate of degeneration in people already suffering from the disease.[2] Painkillers, such as aspirin, and other anti-inflammatory drugs can be used to treat the symptoms (pain and swelling) that accompany osteoarthritis. Injections of steroids may help reduce inflammation, although repeated injections often damage the joint.

Rheumatoid Arthritis Is an Autoimmune Disease. Another common disorder of the synovial joint is rheumatoid arthritis. **Rheumatoid arthritis** is the most painful and crippling form of arthritis. It is caused by an inflammation of the synovial membrane. Inflammation often spreads to the articular cartilages, causing considerable damage. If the condition persists, rheumatoid arthritis causes degeneration of the bones. The thickening of the synovial membrane and degeneration of the bone often disfigure the joints, reduce mobility, and cause considerable pain (▻ Figure 19–13). In some cases, afflicted joints may be completely immobilized. In severe cases, the bones may become dislocated, causing the joints to collapse.

Rheumatoid arthritis generally occurs in the joints of the wrist, fingers, and feet. It can also affect the hips, knees, ankles, and neck. In many cases, inflammation also occurs in the heart, lungs, and blood vessels.

Research suggests that rheumatoid arthritis results from an **autoimmune reaction**—that is, an immune response to the cells of one's own synovial membrane (Chapter 15). Rheumatoid arthritis occurs in people of all ages but most commonly appears in individuals between the ages of 20 and 40. Rheumatoid arthritis is usually a permanent condi-

[2] Weight control is also a preventive measure that helps people avoid the problem in the first place.

▻ **FIGURE 19–12 Arthroscopic Surgery** (*a*) A physician performing arthroscopic surgery. (*b*) Inside view of the knee joint through an arthroscope.

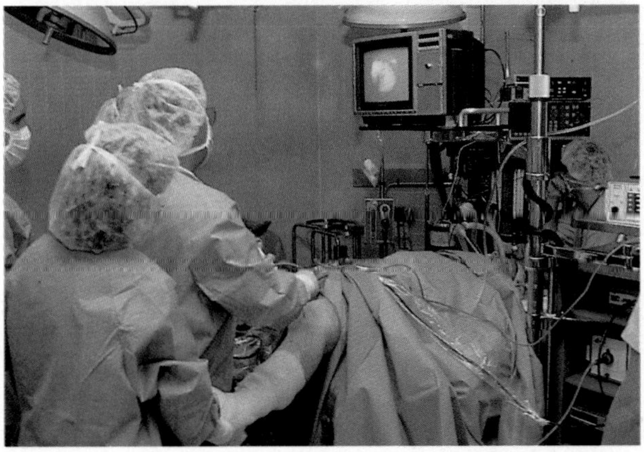

(a)

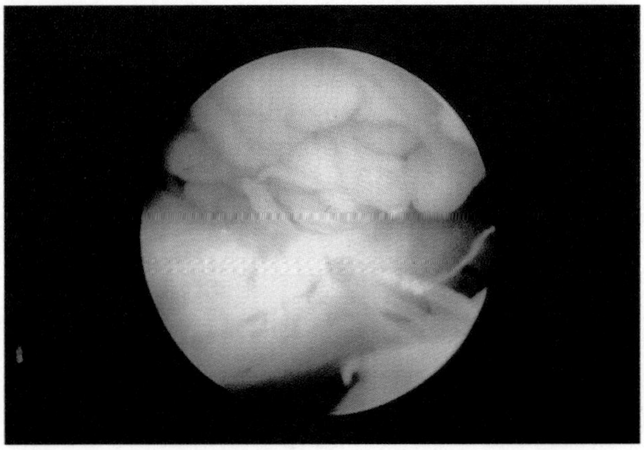

(b)

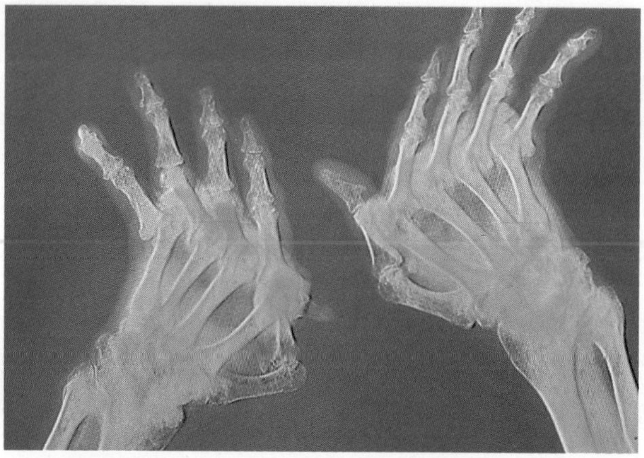

▻ **FIGURE 19–13 X-Ray of Hand Disfigured by Rheumatoid Arthritis** The image has been color-enhanced by computer.

tion, although the degree of severity varies widely. Patients suffering from it can be treated with physical therapy, painkillers, anti-inflammatory drugs, and even surgery.

Diseased joints can also be replaced by artificial ones, or **prostheses,** restoring mobility and reducing pain. Plastic joints can be used to replace the finger joints. These prostheses greatly improve the appearance of the hands of a person with rheumatoid arthritis, eliminating the gnarled, swollen joints. Moreover, patients regain the use of previously crippled fingers. Day-to-day chores (buttoning a shirt) that often required assistance become noticeably easier. Tasks that had once been impossible because of arthritis—for example, opening screw-top jars and picking up coins—once again become possible. Severely damaged knee and hip joints are replaced with special steel or Teflon substitutes, which, if fitted properly, may last 10 to 15 years (▻ Figure 19–14). To put a new hip or knee joint in place, the surgeon first cuts away the degenerating bone. Then the prosthesis is inserted into the shaft of the bone.

Embryonic Development and Bone Growth

Most of the bones of the human skeleton start out as hyaline cartilage. During embryonic development, hyaline cartilage forms in the arms, legs, head, and torso where bone will eventually be (▻ Figure 19–15b). In short order, the cartilage is converted to bone. This process is known as **endochondral ossification.**

As shown in Figure 19–15a, endochondral ossification in cartilage begins in a region known as the **primary center of ossification.** Here, cartilage cells inside the template enlarge, compressing the extracellular material between them. Calcium crystals are then deposited on the extracellular material of the cartilage, and the cartilage cells soon die.

The primary center of ossification expands like a fire spreading in all directions, and a thin layer of bone is deposited around the periphery of the cartilage mass by cells of the **perichondrium,** the connective tissue layer surrounding the cartilage. As soon as bone is deposited, the perichondrium becomes the periosteum. The thin shell of bone laid down on the periphery of the cartilage will eventually become a layer of compact bone.

Blood vessels soon invade the primary center of ossification from the periosteum, carrying with them stem cells that will eventually give rise to a group of bone-forming cells called **osteoblasts.** Other stem cells carried into the primary center of ossification with the blood vessels will give rise to RBCs, WBCs, and platelets. Osteoblasts proliferate on the spicules of calcified cartilage in the interior of the embryonic bone and soon begin to secrete collagen fibers. Collagen fibers are deposited on the slightly calcified spicules, and additional calcium is then deposited on the collagen fibers. This results in the formation of a mass of spongy bone in the primary center of ossification.

Ossification centers also form in the ends (epiphyses) of the bone, but slightly later. These regions of bone formation are called the **secondary centers of ossification.** As

▻ **FIGURE 19–14 Artificial Joints** An artificial knee joint (*a*) and hip joint (*b*).

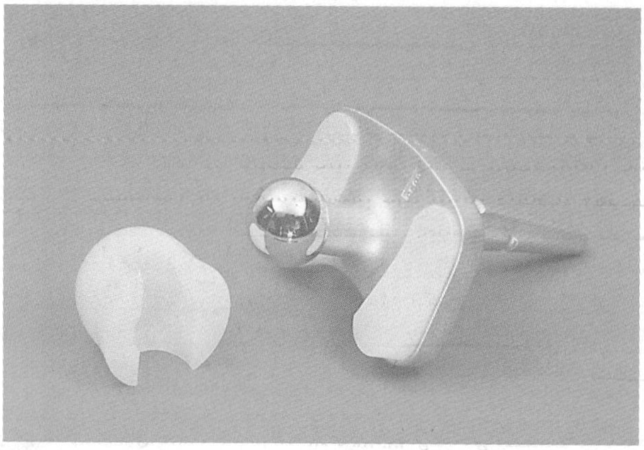

(a)

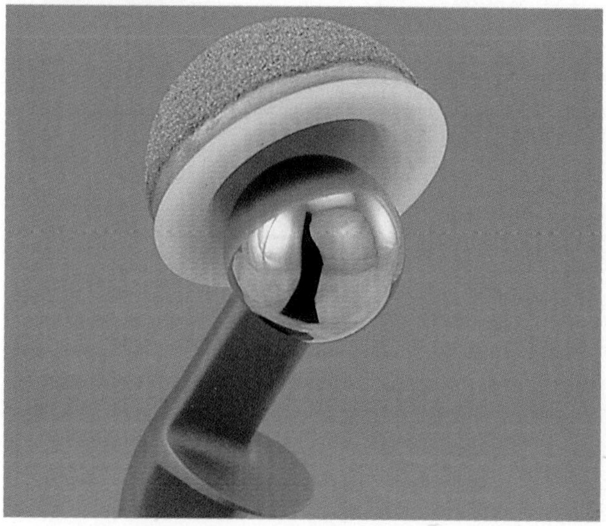

(b)

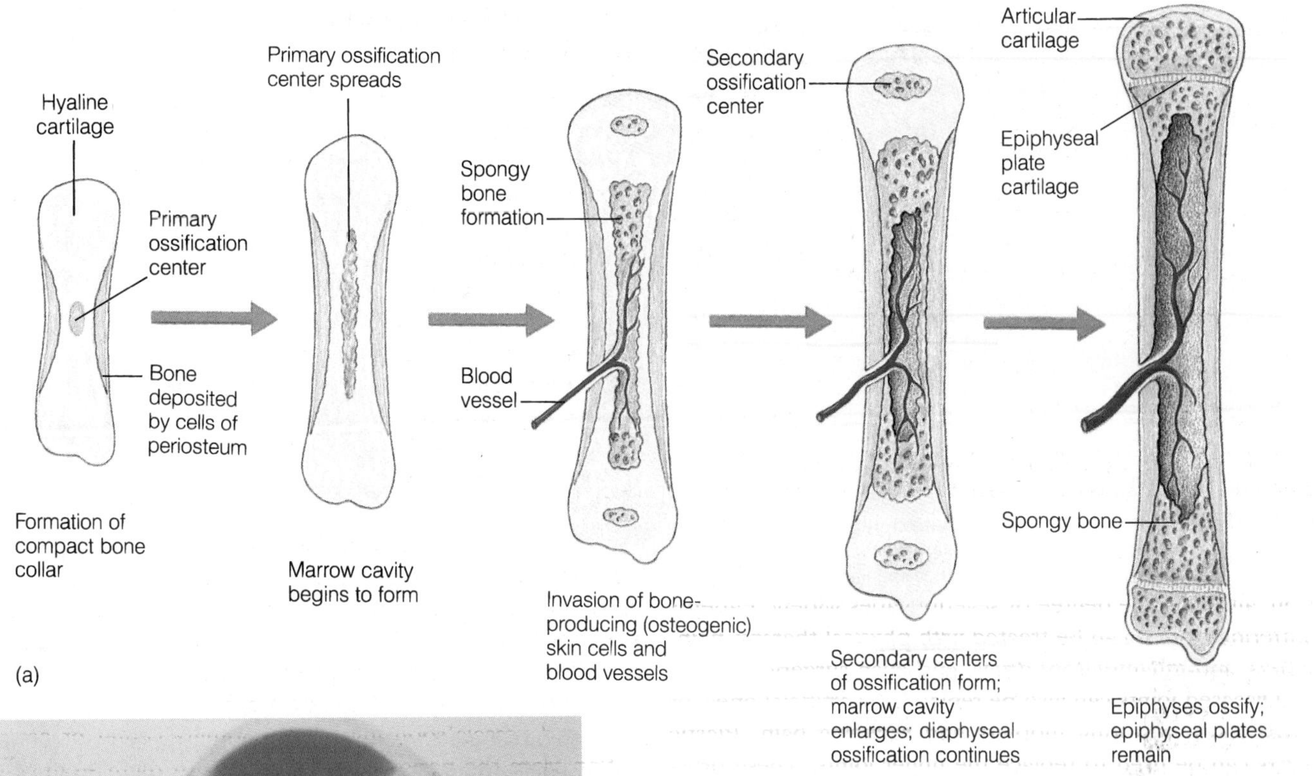

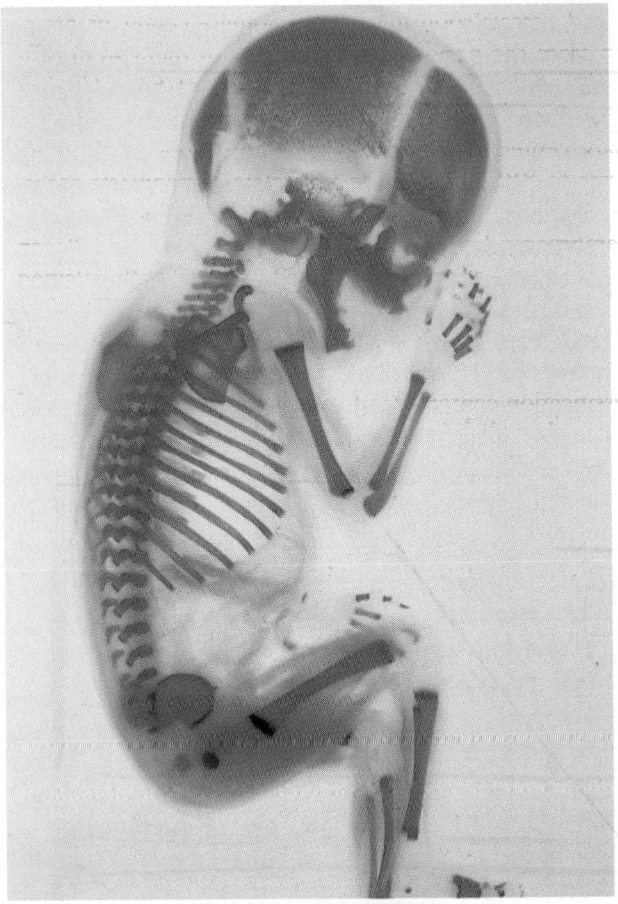

▻ **FIGURE 19–15 Endochondral Ossification**
(*a*) Stages of bone formation. (*b*) Eighteen-week-old human embryo showing bones forming by endochondral ossification.

shown in Figure 19–15a, the primary and secondary centers of ossification spread outward, and with the aid of the periosteal bone deposition, described above, the entire cartilage mass is eventually converted to bone.

As the bone develops, much of the calcified material in the shaft is removed by bone-digesting cells, or **osteoclasts,** soon after it is formed. Osteoclasts are multinucleated cells that, when activated, digest the extracellular material of bone (▻ Figure 19–16). They hollow out the center of the bone, forming the marrow cavity. Stem cells brought in with the blood vessels proliferate and fill the marrow. The stem cells remain in the marrow, dividing and differentiating to form RBCs, WBCs, and platelets.

When the bone is completely formed, all that remains of the cartilage is two narrow bands located between the shaft of the bone and its two ends (▻ Figure 19–17). Called the **epiphyseal plates,** these bands of cartilage contain actively dividing cells that permit bone to elongate. The process of bone elongation is beyond the scope of this book, but it basically results from the proliferation of cartilage on one side of the plate and the ossification on the other. The epiphyseal plates remain active in children and adolescents until their long bones stop growing.[3] Eventually, the increasing levels of sex steroids cause the plates to be converted to bone.

[3] Boys may continue growing until they are 20 or 21. Girls generally stop by age 17.

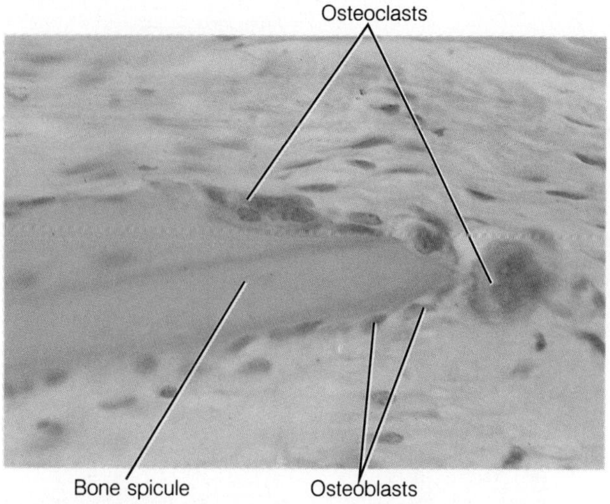

▻ **FIGURE 19-16 The Bone-Destroying Osteoclast** Osteoclasts eat away at bone, releasing calcium to restore blood levels and helping remodel bone to meet changing needs.

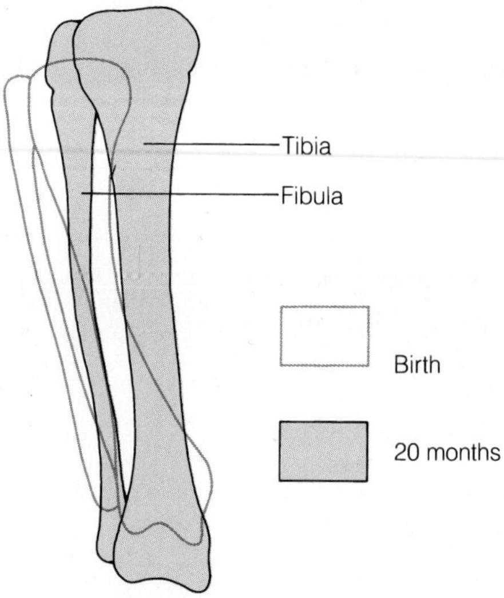

▻ **FIGURE 19–18 Remodeling of a Baby's Leg Bones** Notice how dramatically the bones change to meet the changing needs of the toddler.

Bones Are Constantly Remodeled in Adults to Meet Changing Stresses Placed on Them

Bones are dynamic structures that undergo considerable remodeling in response to changes in our lives. In a newborn baby, for example, the bones of the leg (the tibia and fibula) are quite bowed. Cramped inside the mother's uterus, the bones do not grow very straight. During the first two years of life, however, as the child begins to walk and run, the leg bones generally straighten (▻ Figure 19–18). The bones are literally remodeled to meet the markedly different stresses placed on them by upright posture.

Remodeling occurs throughout adult life as well. During sedentary periods, for example, compact bone decreases in thickness. Increasing one's level of activity, however, causes compact bone to thicken, thus helping the bone withstand stresses placed on it by walking, running, or standing. Spongy bone also undergoes considerable remodeling. Thus, even the internal architecture of a bone changes to meet new stresses.

Two cells are responsible for bone remodeling: osteoclasts and osteoblasts. Osteoclasts are akin to the wrecking crew that comes into a house to tear out walls before a remodeling job. Osteoblasts lay down new bone during the remodeling phase and are like the carpenters who rebuild new walls after they have been torn down. As noted in Chapter 10, when an osteoblast becomes surrounded by calcified extracellular material, it is referred to as an **osteocyte.**

▻ **FIGURE 19–17 Epiphyseal Plate** The epiphyseal plate allows for bone elongation. Cartilage is added at the epiphyseal end while new bone is forming at the diaphyseal end, thus elongating the bone.

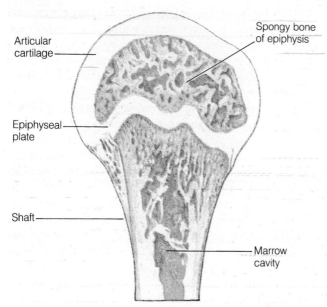

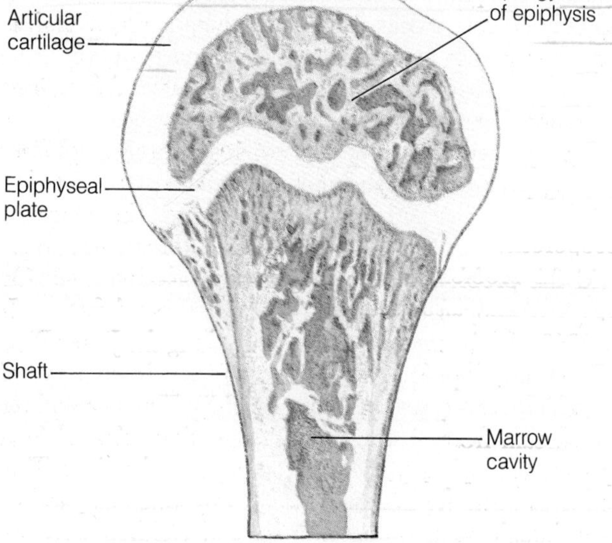

Bone Is a Homeostatic Organ that Helps Maintain Proper Levels of Calcium in the Blood, Tissue Fluids, and Cells

In addition to their role in remodeling bone, osteoblasts and osteoclasts help control blood calcium levels. These cells are therefore part of a homeostatic mechanism and they are controlled in large part by two hormones: parathormone and calcitonin. When blood calcium levels fall, for example, the parathyroid glands (small glands embedded in the thyroid glands in the neck) release **parathormone (PTH).** PTH travels throughout the body in the bloodstream. When it reaches the bone, it stimulates the osteoclasts, causing them to digest bone in their vicinity. The calcium released by the activity of the osteoclasts replenishes blood calcium levels.

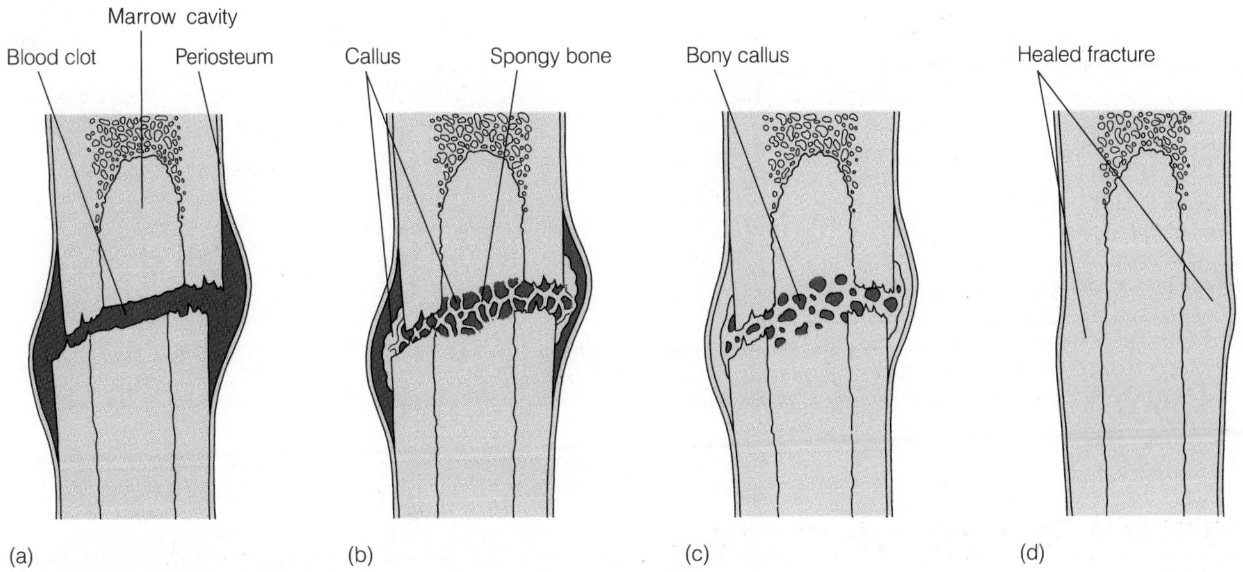

▻ **FIGURE 19–19 Fracture Repair** (*a*) A blood clot forms. (*b*) The blood clot is invaded by fibroblasts and other cells, forming the callus. (*c*) Calcium is deposited in the callus, knitting the ends together. (*d*) The fracture is repaired.

When calcium levels rise—for example, after a meal—the thyroid releases a hormone that exerts an opposite effect. This hormone, **calcitonin,** inhibits osteoclasts, stopping bone destruction. It also stimulates osteoblasts, causing them to deposit new bone. The inhibition of osteoclasts and the formation of new bone help lower blood calcium levels, returning them to normal.

Bone Fractures Are Repaired by Fibroblasts and Osteoblasts

Bone fractures can vary considerably in their severity. Some may involve hairline cracks, which mend fairly quickly. Others involve considerably more damage and take longer to repair. In order for repair to occur, the broken ends must often be aligned and immobilized by a cast. Severe fractures may require surgeons to insert steel pins to hold the bones together.

Like other connective tissues, bones are capable of self-repair, an evolutionary adaptation essential to all animals with internal skeletons. As ▻Figure 19–19 illustrates, blood from broken blood vessels in the periosteum and marrow cavity pours into the fracture, forming a clot. Within a few days, the blood clot is invaded by fibroblasts, connective tissue cells from the periosteum. Fibroblasts produce and secrete collagen fibers, thus forming a mass of cells and fibers, the **callus,** that bridges the broken ends. The callus protrudes at the surface of the bone.

The callus is next invaded by osteoblasts from the periosteum. Osteoblasts convert the callus to bone, knitting the two ends together. As Figure 19–19 shows, the callus is initially much larger than the bone itself. Excess bone, however, is gradually removed by osteoclasts, so much of the bony callus disappears.

Osteoporosis Involves a Loss of Calcium, which Results in Brittle, Easy-To-Break Bones

Helen Brockman, who is 70, gets out of bed one morning, steps on the hardwood floor, and feels a sharp pain in her back. Unknown to her, she has just fractured one of her vertebrae, the small bones that compose her spine. Like 20 million other Americans, Brockman has **osteoporosis** ("porous bone"), a condition characterized by a progressive loss of calcium from the bones of the skeleton (▻ Figure 19–20). In some individuals, calcium loss may be so severe that bones become porous and brittle—so much so that even normal activities such as getting out of bed in the morning or doing housework may cause fractures.

Osteoporosis occurs most often in postmenopausal women. Menopause occurs between 45 and 55 years of age and results from a shutdown of ovarian estrogen production. Estrogen is a reproductive hormone, but like many other hormones in the body, it performs several functions. In women, it reduces bone loss.

Osteoporosis also occurs in people who are immobilized for long periods. Hospital patients who are restricted to bed for two or three months, for example, show signs of osteoporosis. For a discussion of the way hibernating bears avoid this problem, see Spotlight on Evolution 19-1. Osteoporosis may also result from environmental factors. In the 1960s, for instance, women living along the Jintsu River in Japan developed a painful bone disease known as *itai-itai* (which literally means "ouch, ouch"). The women lived downstream from zinc and lead mines that released large amounts of the heavy metal cadmium into the river. They used river water for drinking and for irrigating rice paddies. Even though men, young women, and children were ex-

▻ **FIGURE 19–20 Osteoporosis** The loss of estrogen or prolonged immobilization weakens the bone. Bone is dissolved and becomes brittle and easily breakable. (*a*) A section of the body of a lumbar vertebra from a 29-year-old woman. (*b*) Some thinning is evident in a vertebra of a 40-year-old woman. (*c*) Bone loss is severe in an 84-year-old woman. (*d*) Bone loss is most severe in a 92-year-old woman. Osteoporosis is not inevitable and can be prevented by exercise, calcium supplements, estrogen supplements, and other measures.

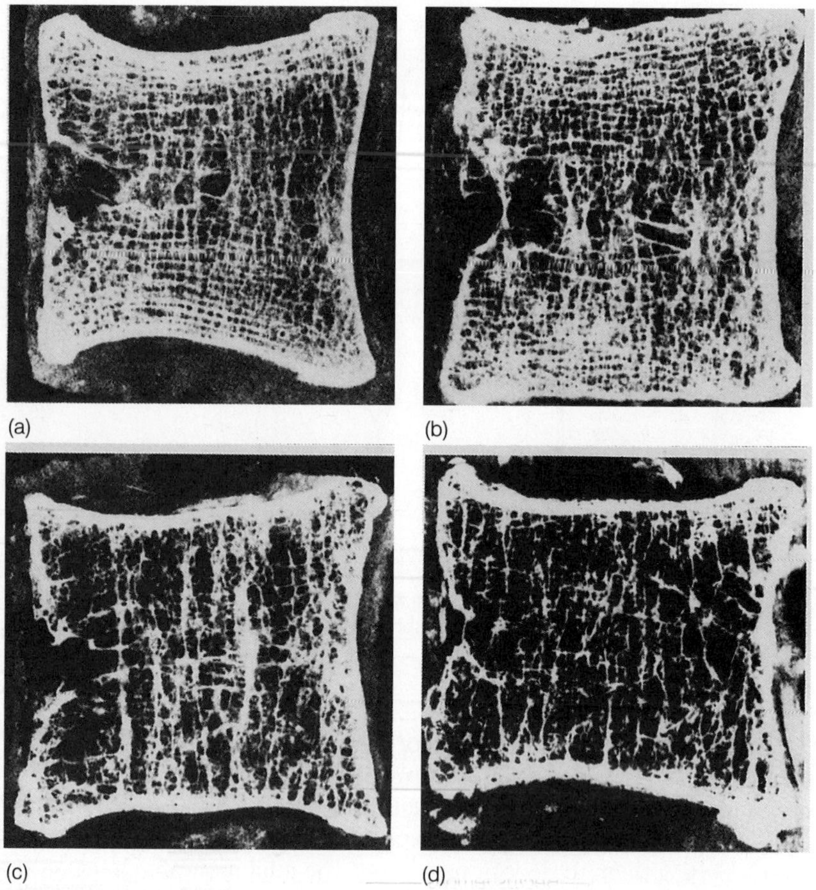

posed to cadmium, 95% of the cases occurred in postmenopausal women.

Researchers at the Argonne National Laboratory believe that cadmium may have accelerated bone loss in the postmenopausal Japanese women. To study the connection, they fed mice diets containing various levels of cadmium chloride. The group receiving the highest dose showed significant reductions in bone calcium when their ovaries had been removed.

These findings may also help explain why older women who smoke are more likely to suffer from osteoporosis. Cadmium is one of several harmful substances present in cigarette smoke. To test the connection between smoking and osteoporosis, researchers cultured fetal bone tissue in a medium containing cadmium levels similar to those found in a smoker's blood and compared the results with a control group of cells grown in a medium containing normal cadmium levels. The bone tissue exposed to higher levels of cadmium exhibited a 70% reduction in calcium content, compared with a 25% loss in the control samples, thus supporting the hypothesis that cadmium in cigarette smoking increases the incidence of osteoporosis.

The new findings provide a plausible explanation for the fact that female smokers experience more bone fractures and tooth loss than nonsmokers. Smoking, however, may also act by decreasing estrogen levels, making it doubly dangerous to a woman's health. For a discussion of ways to prevent osteoporosis, see Health Note 19–1.

THE SKELETAL MUSCLES

Purposeful movement is one of the most distinctive features of animal life. As noted in the introduction to this chapter, in most vertebrates, skeletal muscles acting on bones permit movement. Most skeletal muscles are under the control of the nervous system.[4]

▻ Figure 19–21 shows some of the skeletal muscles of the body. A glance at this figure reveals that most skeletal muscles cross one or more joints. Therefore, when they contract, they produce movement. Muscles generally work in groups, rather than alone, to bring about various body movements. Groups of muscles are often arranged in such a way that one set causes one movement and another set on the opposite side of the joint causes an opposing movement. For example, the biceps is a muscle in the upper arm. When it and other members of its group contract, they cause the arm to bend, a movement called flexion. On the back side of the upper arm is another muscle, the triceps. When it contracts, the triceps causes the arm to straighten, or extend. Opposing muscles are called **antagonists.** Generally, when one muscle contracts to produce a movement, the antagonistic muscles relax. Antagonistic sets of muscles are under the control of the cerebellum.

Not all skeletal muscles are arranged so that they can

[4]Chapter 10 discussed all three types of muscle: skeletal, cardiac, and smooth.

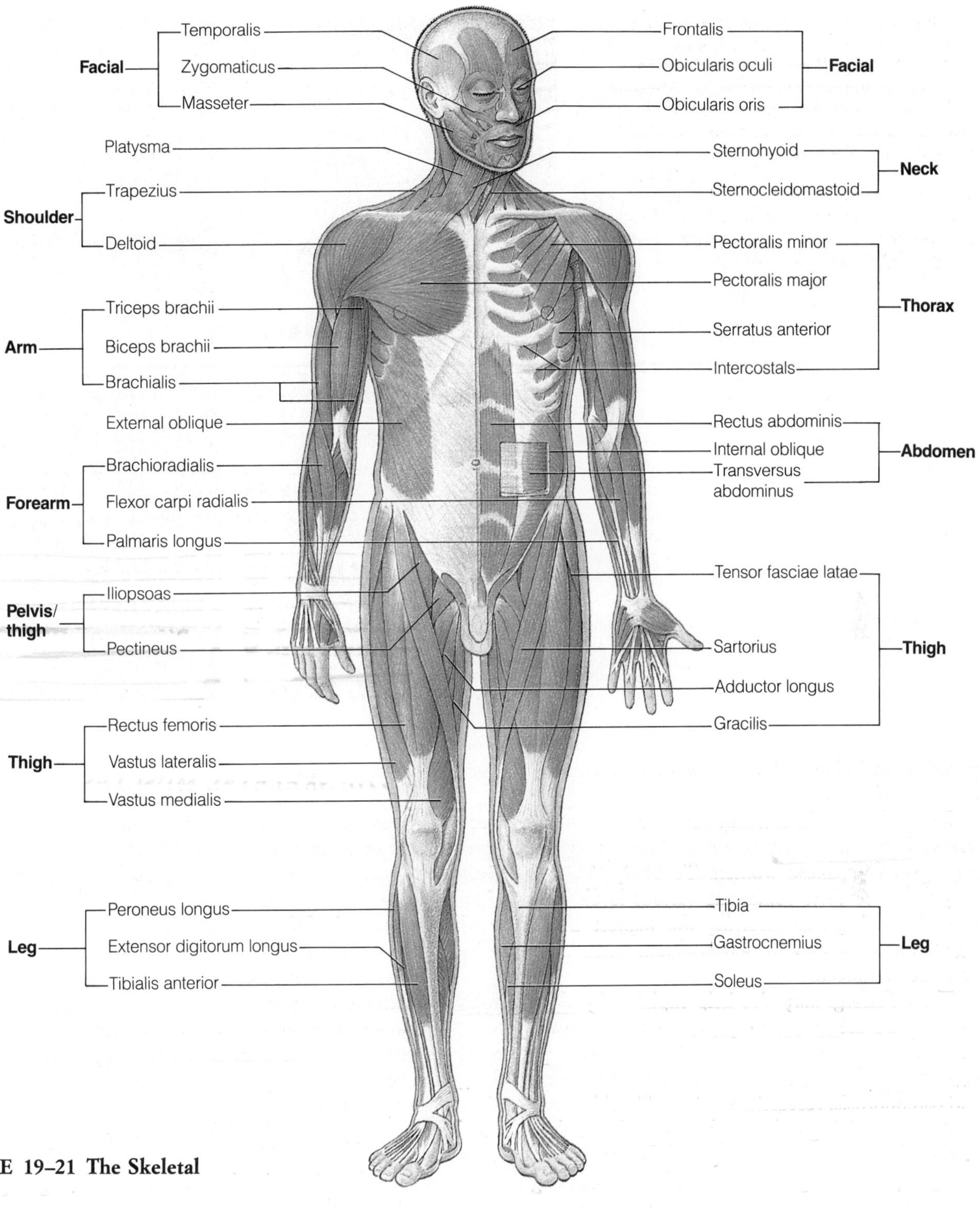

▻ **FIGURE 19–21 The Skeletal Muscles**

make bones move. Some muscles simply steady joints, allowing other muscles to act. These stabilizing muscles are known as **synergists.** Muscles in the face are another example of those that do not make bones move. Anchored to the bones of the skull and to the skin of the face, these muscles allow us to wrinkle our skin, open and close our eyes, and move our lips.

Muscles help us move about in the environment. Although few of us are aware of it, our muscles also constantly work to maintain our posture, helping us stand or sit upright despite the never-tiring pull of gravity.

The muscles of the body also produce enormous amounts of heat as a by-product of metabolism. When working, the muscles produce additional heat—so much that you can cross-country ski in freezing weather wearing only a light sweater.

SPOTLIGHT ON EVOLUTION 19–1

HIBERNATING BEARS REVEAL SECRETS OF SKELETAL STABILITY

It is late fall in Minnesota. Black bears have begun to hole up in their dens or settle down on the forest floor, creating a nest of twigs and leaves where they will hibernate (Figure 1). During hibernation, the bears remain immobile. They do not eat or drink, urinate or defecate for up to five full months. Even more remarkably, two months into their winter hibernation, females give birth and nurse their young for three months before waking from their slumber.

Preparations for winter hibernation actually begin in the late summer, when the bears begin to gorge themselves on grubs, carrion (dead animals), berries, nuts, and insects—virtually anything they can get their paws on. Adults consume approximately 20,000 Calories a day, five times more than is required to stay alive. A layer of fat grows, reaching 5 inches in thickness.

During hibernation, the bear's heart rate and breathing drop, although its metabolism remains quite high—about 4000 Calories per day. Moreover, although the bears are inactive during hibernation, they are not asleep. They remain alert.

During hibernation, energy is supplied by fat. Very little protein is broken down. Potentially toxic nitrogenous waste in the blood is converted into protein. Therefore, a hibernating bear actually increases its lean body mass during this long fast. This adaptation prevents uremia, the potentially lethal buildup of toxic wastes in the animal's bloodstream.

Successful hibernation in the bear also depends on another as-yet poorly understood physiological adaptation—a mechanism by which the bear prevents osteoporosis during this long period of inactivity. Except for bears, all mammals kept immobile for long periods suffer from osteoporosis. Human astronauts, bedridden patients, and inactive elderly people are good examples. The bear is different. Studies show that bone mass may shift in a hibernating bear to the main pressure points where bones support the resting bear's weight, but there is no net loss of bone calcium. All in all, bone metabolism remains fairly normal.

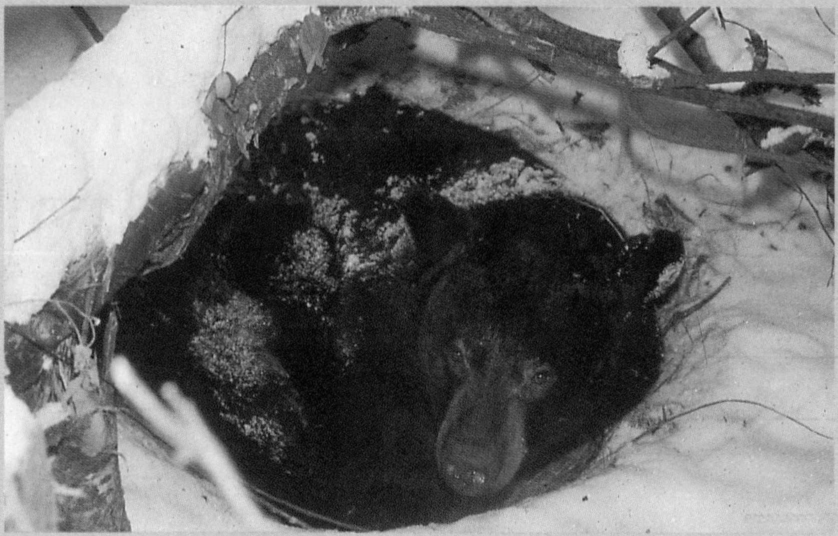

▻ **FIGURE 1 Hibernating Bear**

The conservation of bone matter is probably unique to bears. Hibernating ground squirrels, for instance, lose bone mass and get rid of the excess calcium in their urine. Scientists believe that the bear's unique ability to conserve bone is probably the result of a hormonal substance, which they have set out to isolate and purify. This substance may have important implications in the prevention and treatment of osteoporosis in humans.

Scientists also believe that the bear may produce a chemical substance, dubbed hibernation induction trigger, that sets into motion a wide array of complex physiological changes associated with hibernation. This substance, already isolated and purified, is believed to be chemically related to the opiates such as morphine, which depress some nervous system functions. Injected into nonhibernating animals, it results in a dramatic decline in heart rate and many metabolic processes.

Hibernation induction trigger may prove quite useful in treating human ailments. The army, for instance, is seeking a drug that slows metabolism and reduces head swelling in trauma victims. Surgeons might be able to use it to slow metabolism and cool the body during surgery. Preliminary studies show that this hormone prevents fibrillation and thus could become an important tool in stopping heart attacks.

While scientists probe the mysteries of the bear, another winter sets in, and this remarkably adapted animal settles down for its unique and fascinating retreat, an adaptation to the rigors of winter where food is scarce and survival would be questionable.

Skeletal Muscle Cells Are Known as Muscle Fibers and Are Both Excitable and Contractile

Skeletal muscles consist of long, unbranched cells called **muscle fibers** (▻ Figure 19–22). Muscle fibers are multinucleated structures formed during embryonic development by the fusion of many smaller cells. Viewed with the light microscope, skeletal muscle fibers appear striated.

Like nerve cells, the muscle fiber is an excitable cell. A small potential difference (about -60 mv) exists across the plasma membrane of skeletal muscle fibers, as in neurons.

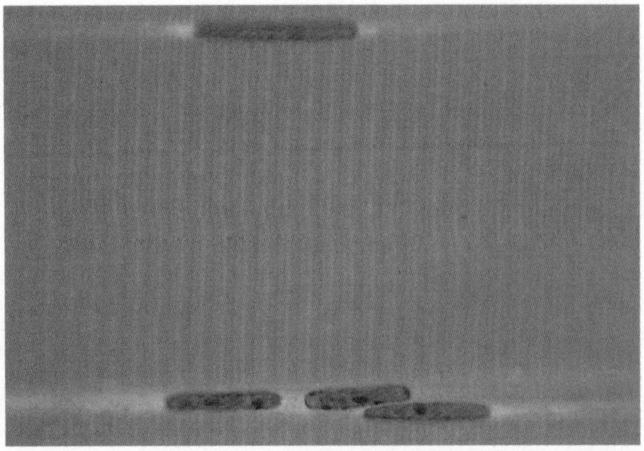

(a) (b)

▻ **FIGURE 19–22 Light Micrographs of Skeletal Muscle** (*a*) Notice the banding pattern on these muscle fibers. (*b*) Higher magnification showing nuclei and banding pattern.

When the membrane is stimulated by a neurotransmitter from the terminal bouton of a motor neuron, a bioelectric impulse is generated. The impulse travels along the membrane of the muscle fiber in the same way a nerve impulse travels along an unmyelinated axon or dendrite.

Muscle fibers are also contractile. When stimulated, the contractile proteins inside the fibers cause the cells to shorten (explained in more detail shortly). Muscle fibers are also elastic, capable of returning to normal length after a contraction has ended.

Each muscle fiber in a skeletal muscle is surrounded by a delicate layer of connective tissue, the **endomysium** (▻ Figure 19–23). Individual fibers are joined in groups known as **fascicles.** Fascicles are also held together by a connective tissue, the **perimysium.** Numerous fascicles are bound by a connective tissue sheath that surrounds the entire muscle, the **epimysium.** This arrangement provides support and protection for muscle cells. In many muscles, the epimysium (the outermost layer) fuses at the ends of the muscle to form tendons. Tendons often attach to the periosteum of bones. Because the tendon is continuous with the epimysium and because the perimysium and endomysium are also connected to the epimysium, muscle contraction can exert a powerful force on the point of attachment.

Muscle Fibers Contain Many Small Bundles of Contractile Filaments Known as Myofibrils

Muscle cells are uniquely adapted to perform their function. Understanding how a muscle contracts, however, first requires a careful look at the muscle fiber. To begin, imagine that you could tease a single skeletal muscle fiber (cell) free from its fascicle. Under a microscope, you would find that each muscle fiber is a long cylinder wrapped in plasma membrane and containing many nuclei. Each muscle fiber is characterized by a series of dark and light bands (▻ Figure 19–24b). Inside each muscle fiber are numerous bundles of threadlike filaments, mostly actin and myosin, the contractile proteins. Each bundle of filaments in the muscle fiber is known as a **myofibril** (Figure 19–24b); each muscle cell contains numerous myofibrils.

Myofibrils contain contractile filaments and, as illustrated in Figure 19–24c, are striated. The wide, dark bands

▻ **FIGURE 19–23 Connective Tissue Layers Investing Skeletal Muscle** Individual muscle fibers are surrounded by the endomysium. Muscle fibers are bundled together by the perimysium to form a fascicle. Fascicles are held together by the epimysium, which also forms the tendons.

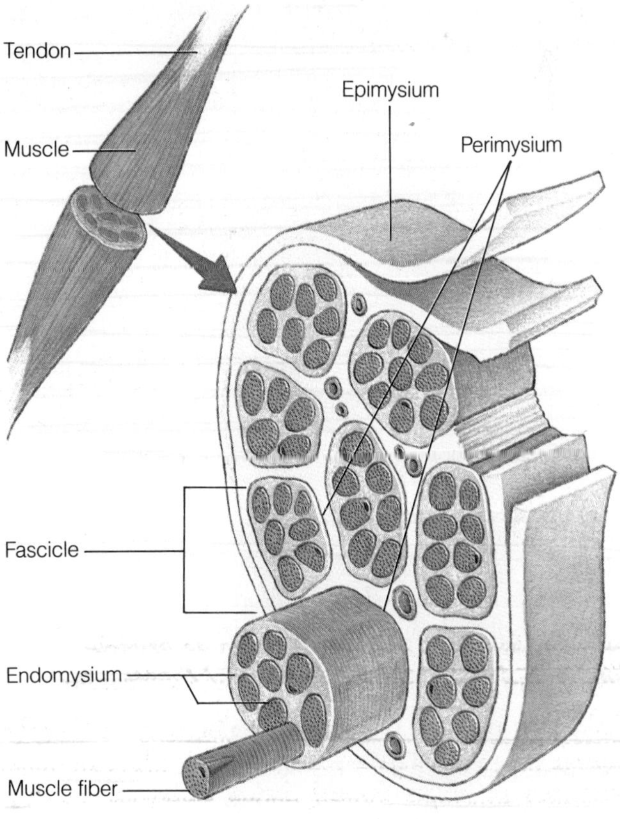

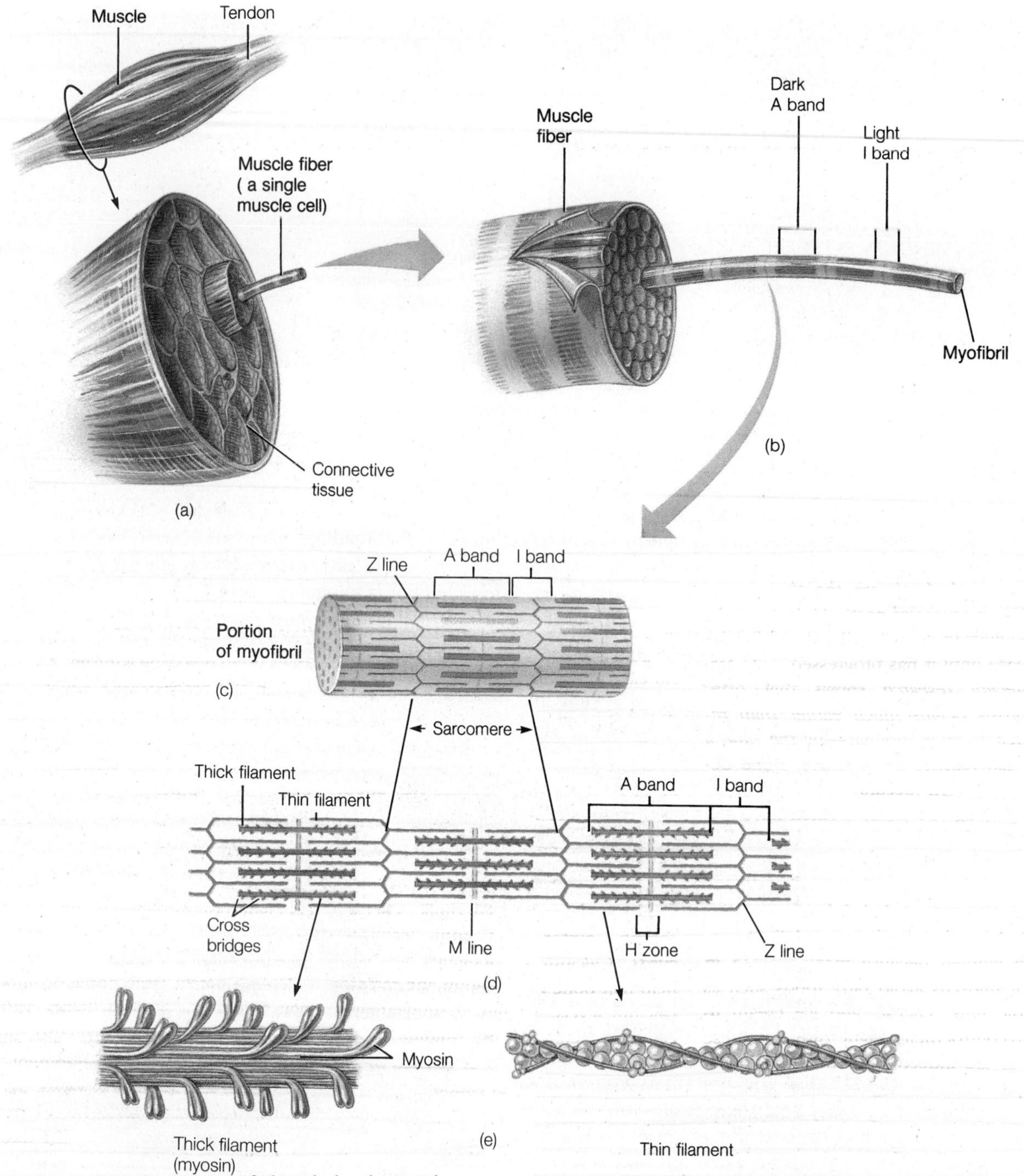

▻ FIGURE 19–24 Structure of the Skeletal Muscle Fiber, Myofibril, and Sarcomere (*a*) A single muscle fiber teased out of the muscle. (*b*) Each muscle fiber consists of many myofibrils. (*c*) Note the banded pattern of the myofibril. (*d*) Sarcomeres consist of thick and thin filaments, as shown here. (*e*) Molecular structure of the thick (myosin) and thin (actin) filaments.

of the myofibril are called **A bands;** the narrower, light bands are called **I bands.**[5] In the myofibril, the light and dark bands are arranged in a uniform pattern, which gives the myofibril and muscle fiber a striated appearance. Also shown in Figure 19–24c is a fine line that runs down the center of each I band. This jagged line looks like many letter Zs stacked on one another and is called the **Z line.**

[5]The words *dark* and *light* may help you remember which is which. Dark contains an *a,* and light contains an *i.*

The region of the myofibril between two adjacent Z lines is known as a **sarcomere** and is considered the functional unit of the muscle cell. As shown in Figure 19–24d, the sarcomere contains thick and thin filaments. The thick filaments are made of the protein myosin and lie in the middle of the sarcomere. The thin filaments are composed primarily of the protein actin and extend from the Z line toward the center of the sarcomere but do not join in the middle.

HEALTH NOTE 19–1

PREVENTING OSTEOPOROSIS: A PRESCRIPTION FOR HEALTHY BONES

Twenty million Americans, mostly women, suffer from a painful, debilitating disease called osteoporosis (▻ Figure 1). Caused by a gradual deterioration of the bone, osteoporosis results in nagging pain and discomfort. Bones fracture easily.

If current trends continue, one of every two American women will develop postmenopausal osteoporosis. Each year, nearly 60,000 Americans—mostly women—will die from complications resulting from the disease. Hemorrhage; fat embolisms (globules), sometimes released from the yellow marrow of broken bones; and shock are the three most common causes of death. Unfortunately, most women do not realize they have the disease until it has progressed quite far.

Recent research shows that osteoporosis begins much earlier than researchers once thought—by the time a woman reaches her mid-20s. Bone demineralization occurs very rapidly. In fact, by age 30 many women have lost one-third of their bone calcium! Between the ages of 30 and 50, many women's bones continue to deteriorate, becoming extremely brittle.

Calcium loss begins so early in American women because many women in their mid-20s avoid fatty foods such as whole milk, cheese, and ice cream to help control their weight. Although these foods are indeed fatty, they are also a major source of calcium. Milk products are also shunned by many adults who develop an intolerance to lactose (a sugar) in milk and other dairy products. Lactose intolerance results from a sharp reduction in the production and secretion of the intestinal enzyme lactase as one ages, making it difficult to digest lactose. Because of these and other factors, adult women often consume only about one-half of the 1000 to 1500 milligrams of calcium they need every day.

▻ FIGURE 1 Bone Deterioration
An elderly woman suffering from osteoporosis. Notice the hunched back due to the collapse of vertebrae.

Osteoporosis may be prevented and even reversed by eating calcium-rich foods, such as spinach, milk, cheese, shrimp, oysters, and tofu (soybean curd). Calcium supplements can also help halt bone deterioration and restore calcium levels. Vitamin D supplements can also help, because vitamin D increases the absorption of calcium in the intestines. (A word of caution, however: excessive vitamin D—five times the RDA—can be toxic.)

Studies also suggest that osteoporosis can be prevented by exercise. Aerobics, jogging, walking, and tennis in conjunction with the dietary changes noted above all help prevent the disease.

Research shows that osteoporosis can be reversed by exercise even after the disease has reached the dangerous stage. Forty-five minutes of moderate exercise (walking) three days a week, for example, greatly decreases the rate of calcium loss in older individuals. In addition, this exercise regime stimulates the rebuilding process, replacing calcium lost in previous years. Continued exercise increases bone calcium levels and decreases the rate of bone fractures and fatal complications noted earlier.

Another effective treatment for postmenopausal women is estrogen. Low doses of estrogen halt bone demineralization and promote bone formation. Because women who are given estrogen suffer an increased risk of endometrial cancer (cancer of the uterine lining), physicians often prescribe a mixed dose of estrogen and progesterone, which reduces the likelihood of this type of cancer.

Studies have also shown that high doses of fluoride and calcium stimulate bone development. Calcium fluoride treatment increases bone mass approximately 3% to 6% per year and decreases bone fractures. The average patient in one study experienced one fracture every eight months before treatment. After treatment, that figure dropped to one fracture every 4.5 years.

Unfortunately, large doses of fluoride may erode the stomach lining, causing internal bleeding. They may also stimulate abnormal bone development and cause pain and swelling in joints. To offset these problems, researchers have developed a pill that releases the fluoride gradually.

For millions of young women, early detection and sound preventive measures, including exercise, vitamin D, dietary improvements, and fluoride treatments, can prevent osteoporosis.

During Muscle Contraction, the Actin Filaments Slide Inward, Causing the Sarcomeres to Shorten

Actin and myosin filaments are surprisingly delicate yet are responsible for all muscle contraction. When a muscle contracts, each sarcomere shortens. The actin filaments slide toward the center of the sarcomere, sometimes even touching in the middle. The actin filaments are pulled inward by the myosin molecules. As ▻ Figure 19–25 shows, myosin filaments consist of numerous golf-club-shaped myosin molecules, which are arranged with their "club ends," or heads, projecting toward the actin filaments. During muscle contraction, the heads of the myosin molecules attach

▻ **FIGURE 19–25 Structure of Myosin and Actin Filaments**
(*a*) Myosin molecules join to form a myosin filament. Note the presence and orientation of the heads of the myosin molecules. (*b*) Globular actin molecules form intertwining strands. Note the presence of the myosin binding sites. (*c*) Tropomyosin attaches to the binding sites, covering them. Tropomyosin molecules are held in place by the troponin.

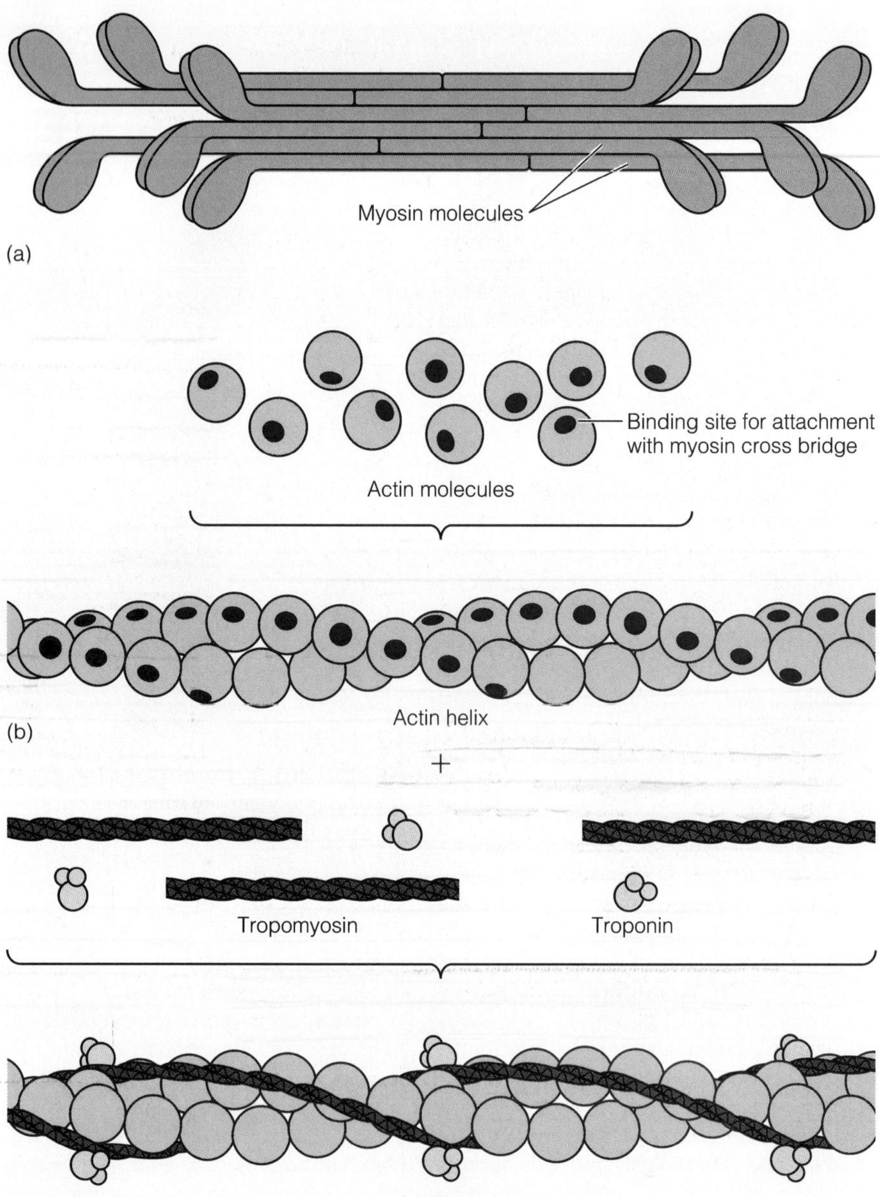

to actin filaments, forming **cross bridges,** which tug the filaments inward, causing the sarcomere to shorten.

How do the myosin cross bridges function? To answer this question, we must first take another look at Figure 19–25. This illustration shows the molecular makeup of the actin and myosin filaments. As illustrated in Figure 19–25b, each actin filament consists of two strands of globular actin molecules joined in a double helix like two bead necklaces. Each actin molecule contains a binding site to which the club ends of the myosin molecules attach. The actin filament therefore contains many binding sites. When the heads of the myosin molecules attach to the binding sites, they undergo a change in shape that causes them to pull the actin filament toward the center of the sarcomere.

In the resting state, the binding sites on the actin molecules of the actin filament are blocked by long, stringlike protein molecules known as **tropomyosin,** shown in Figure 19–25b. Tropomyosin prevents the heads of the myosin molecules from binding to the actin filaments when a muscle is at rest. As shown in Figures 19–25b and 19–25c, the tropomyosin molecules are held in place by another protein, **troponin.** The troponin molecules act like thumbtacks that secure the tropomyosin molecules.

Muscle contraction is stimulated when the tropomyosin is removed from the actin filaments, freeing up the binding sites on the actin filaments. In order to understand how the tropomyosin "guard" is removed, let's examine the sequence of events that occurs when a nerve impulse arrives at the muscle cell. ▻ Figure 19–26 illustrates the **neuromuscular junction,** the synapse between the terminal bouton of a motor neuron and a muscle fiber. When the nerve impulse arrives at the terminal bouton, it triggers the re-

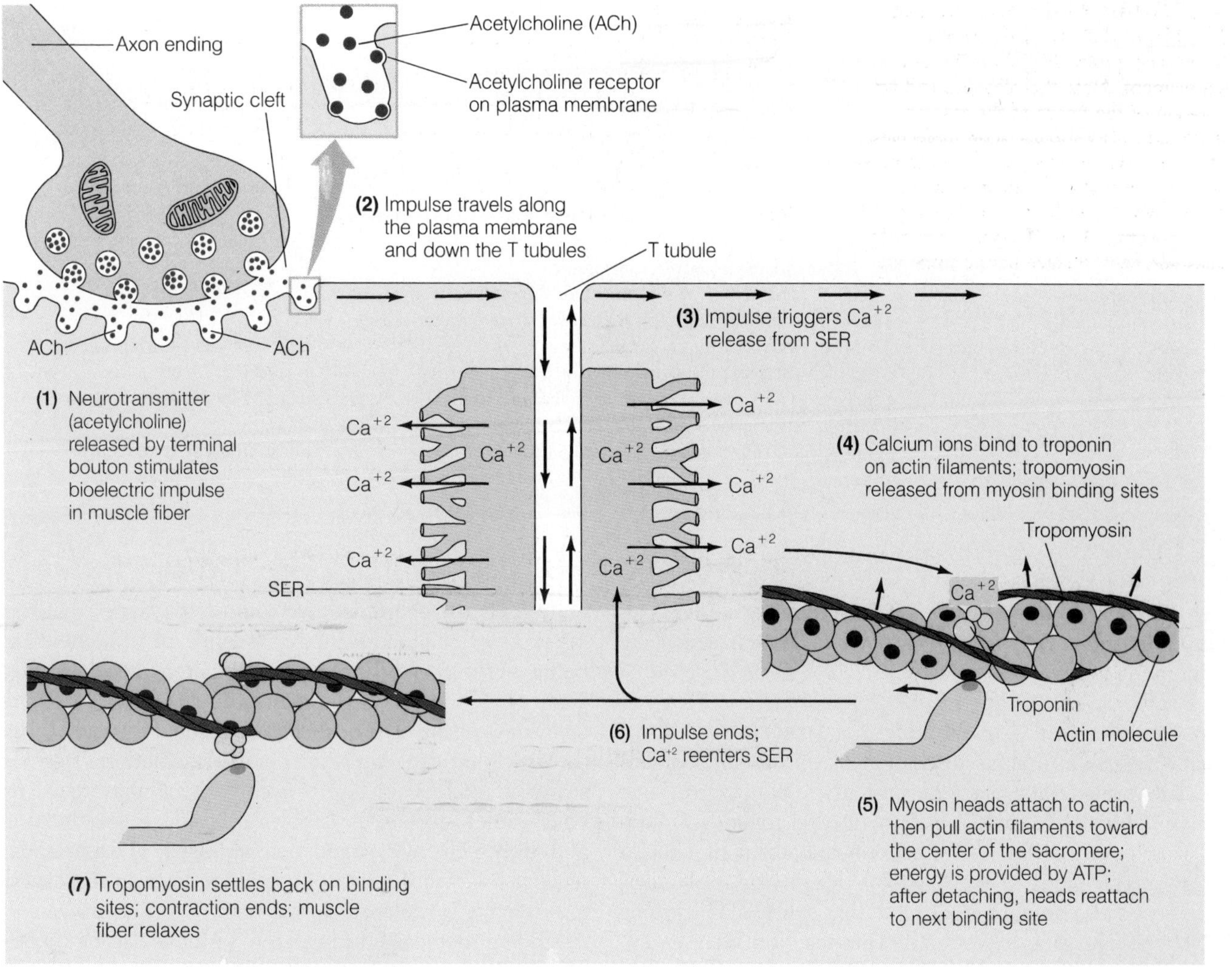
Axon ending
Synaptic cleft
Acetylcholine (ACh)
Acetylcholine receptor on plasma membrane
(2) Impulse travels along the plasma membrane and down the T tubules
T tubule
ACh
ACh
(3) Impulse triggers Ca+2 release from SER
(1) Neurotransmitter (acetylcholine) released by terminal bouton stimulates bioelectric impulse in muscle fiber
Ca+2
Ca+2
Ca+2
Ca+2
Ca+2
Ca+2
Ca+2
Ca+2
Ca+2
(4) Calcium ions bind to troponin on actin filaments; tropomyosin released from myosin binding sites
Tropomyosin
SER
Ca+2
Troponin
Actin molecule
(6) Impulse ends; Ca+2 reenters SER
(5) Myosin heads attach to actin, then pull actin filaments toward the center of the sacromere; energy is provided by ATP; after detaching, heads reattach to next binding site
(7) Tropomyosin settles back on binding sites; contraction ends; muscle fiber relaxes

▻ **FIGURE 19–26 The Neuromuscular Junction and Muscle Contraction** Axons from motor neurons terminate on the surface of the muscle fiber, forming the neuromuscular junction.

lease of the neurotransmitter acetylcholine. Acetylcholine diffuses into the synaptic cleft, the space between the terminal bouton and the plasma membrane of the muscle fiber. It then binds to receptors in the plasma membrane.

The binding of acetylcholine to the receptors stimulates changes in the membrane permeability of the muscle fiber, resulting in membrane depolarization caused by the rapid influx of sodium ions. A wave of depolarization travels along the plasma membrane of the muscle fiber. As illustrated in Figure 19–26, the plasma membrane periodically "dips" into the muscle fiber. These deep invaginations, called **T tubules (transverse tubules),** conduct the impulse to the interior of the cell, facilitating contraction of the muscle. As it travels inward, the impulse stimulates the release of calcium ions stored inside the smooth endoplasmic reticulum (SER) of the muscle cell. In muscle fibers, the SER is called the **sarcoplasmic reticulum.** The sarcoplasmic reticulum lies close to the T tubules. Calcium ions released from the sarcoplasmic reticulum diffuse outward into the myofibril and attach to the troponin molecules, which hold the tropomyosin molecules in place over the binding sites on the actin filaments. When calcium binds to the troponin, the troponin molecules are released, and the tropomyosin molecules slide off the binding sites on the actin filaments. This allows the heads of the myosin filaments to attach to the actin. The cross bridges thus formed contract and pull the actin filaments inward, causing the myofibrils of the sarcomere and the muscle cells to contract. Myosin cross bridges give a brief tug on the actin filaments, then detach, becoming available to bind again. This cycle repeats 50 to 100 times during each muscle contraction.

The actin filaments are pulled inward in much the same way that you would pull a boat tied to a rope toward shore. Contraction ends when the calcium ions are actively transported back into the sarcoplasmic reticulum. When calcium levels fall, tropomyosin slips back into place over the binding sites on the actin filament.

This description of the contraction of myofibrils by the inward movement of the actin filaments is called the **slid-**

TABLE 19–2 Components of Muscle Contraction

MUSCLE FIBER COMPONENT	FUNCTIONAL ROLE
Plasma membrane	Conducts impulse from terminal bouton of motor neuron
T tubule	Conducts impulse into the interior of the muscle fiber
Sarcoplasmic reticulum	Releases stored calcium, which stimulates contraction; absorbs calcium to end contraction
Tropomyosin	Blocks binding sites on actin filament, preventing contraction
Troponin	Holds tropomyosin in place on actin filament, blocking contraction; binds to calcium, releasing tropomyosin to permit contraction
Actin filaments	Slide toward center of sarcomere during contraction
Myosin filaments	Pull the actin filaments toward center of sarcomere during contraction
Heads of myosin molecules (cross bridges)	Bind to actin and pull actin filaments; contain binding site for ATP; contain myosin ATPase, which catalyzes the breakdown of ATP
Calcium ions	Released from the sarcoplasmic reticulum; bind to troponin, causing it to release tropomyosin from binding sites on the actin filaments
ATP	Binds to cross bridges of myosin filaments; broken down by ATPase in cross bridges, providing energy for muscle contraction

ing filament theory. Table 19–2 summarizes the roles played by the various components of the muscle cell. Take a moment to review them now.

Energy Needed for Muscle Contraction Is Provided by ATP. In this discussion of the fascinating molecular events taking place during muscle contraction, one factor is missing—energy. Energy needed for muscle contraction comes from ATP, the principal form of cellular energy. ATP binds to the heads of the myosin molecules. The heads also contain an enzyme that splits ATP, forming ADP and inorganic phosphate. This reaction releases energy. When ATP is converted to ADP, the energy released is captured and stored in the myosin molecules momentarily like energy stored in a compressed spring. The energy is released when the head of the myosin molecules bind to actin, which causes the heads of the myosin molecules to change shape, drawing the actin filaments toward the center of the sarcomere.

ATP Is Regenerated by One of Several Mechanisms. ATP is quickly regenerated by a high-energy molecule stored in muscle. Known as **creatine phosphate,** this molecule contains a high-energy bond, indicated by the squiggly line (below). Stored in muscle in high concentrations, creatine phosphate reacts with ADP as follows:

$$\text{creatine} \sim \text{P} + \text{ADP} \rightarrow \text{ATP} + \text{creatine}$$

Creatine phosphate replenishes ATP used during muscle contraction, but supplies last only 30 seconds or so. ATP is also generated by glycolysis, the citric acid cycle, and the electron transport system, as discussed in Chapter 6.

Because the electron transport process requires oxygen, vigorous muscle contraction causes oxygen supplies inside muscle cells to fall. If the circulatory system cannot replace oxygen as quickly as it is being used, the citric acid cycle and the electron transport system are shut down. To generate ATP, the cell must turn to fermentation—that is, the breakdown of glucose in the absence of oxygen. Besides being inefficient, this process results in the buildup of lactic acid. The shortage of ATP and the buildup of lactic acid occurring during vigorous exercise result in **muscle fatigue.**

Muscle fatigue also results from a depletion of glycogen stores in skeletal muscle. Glycogen, you may recall from your earlier studies, is a polysaccharide found in muscle and liver cells. Glycogen is composed of thousands of glucose molecules. Glycogen broken down in muscles during exercise is replaced during rest.

Oxygen depleted during exercise must also be replaced. This replacement occurs rather quickly. The muscle oxygen deficiency, called the **oxygen debt,** is often largely replaced right after you exercise, explaining why you keep breathing hard for a while after you stop exercising.

Individual Skeletal Muscle Fibers Contract Fully when Stimulated

Individual skeletal muscle fibers obey the **all-or-none law;** that is, when activated by an action potential, a muscle fiber contracts fully. A single contraction, followed by relaxation, is called a **twitch.** The force of that contraction is shown in ▻ Figure 19–27. You will notice that there is a brief lag after the impulse is generated in the membrane of the muscle fiber and contraction begins. Known as the **latent period,** it results from at least three factors: (1) the time required for the action potential to travel into the T tubules, (2) the time required for calcium to diffuse out of the sarcoplasmic reticulum and to bind to troponin, and (3) the time required for the filaments to begin sliding.

Peak tension in a single muscle cell occurs sometime after the impulse arrives. Relaxation requires an equal amount of time as calcium ions are pumped back into the sarcoplasmic reticulum.

Even though individual muscle fibers contract fully

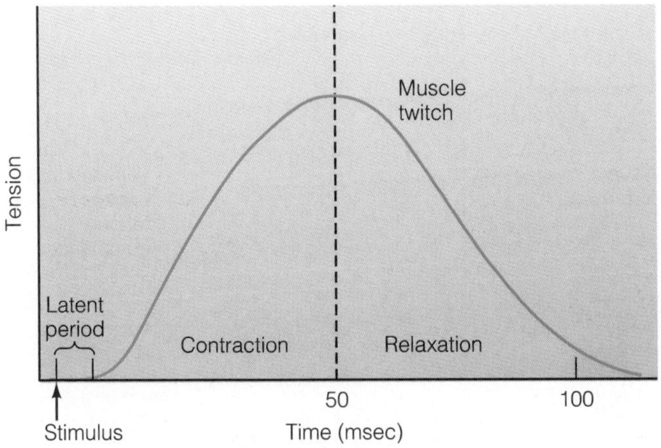

▻ **FIGURE 19–27 The Muscle Twitch** A graph of muscle fiber contraction, showing the tension generated and the time required to reach maximum tension and relaxation.

when stimulated, whole skeletal muscles are capable of producing contractions of varying strength, referred to as **graded contractions.** Graded contractions result from two different processes: recruitment and wave summation.

The Strength of Muscle Contraction Can Be Increased by Stimulating (Recruiting) Additional Muscle Fibers to Contract.

To generate the force needed to move arms and legs or even eyelids requires the action of more than one muscle cell. The engagement of additional muscle fibers during muscle contraction is called **recruitment.** It is akin to what the U.S. Army might do before it goes to war.

To understand how recruitment occurs in muscle, we must take a look at the innervation of muscles by motor neurons. As shown in ▻ Figure 19–28a, the axons of motor neurons, on reaching the muscles they supply, form many branches. A single motor neuron, in fact, may innervate dozens of individual muscle fibers in a skeletal muscle. A motor neuron and the muscle fibers it supplies constitute a **motor unit.** The fewer muscle fibers in a motor unit, the finer the control. The less control, the more muscle fibers in a motor unit. Thus, the neurons that supply the strong muscles of the leg, which produce crude but strong propulsive force, may end on as many as 2000 muscle fibers. Each time an impulse is delivered to the motor unit, it stimulates all 2000 muscle fibers. In contrast, muscles involved in fine motor movements, such as the extrinsic muscles of the eye or the muscles of the hand, contain "smaller" motor units with a few dozen muscle fibers per axon, providing a greater degree of control.

How does this process relate to recruitment? To increase the force of contraction in a skeletal muscle, the

▻ **FIGURE 19–28 The Motor Unit** (*a*) Light micrograph of axon branching to terminate on many muscle fibers. (*b*) Each axon branches at its termination, supplying a few dozen to many thousand muscle fibers. A single axon and its muscle fibers constitute a motor unit.

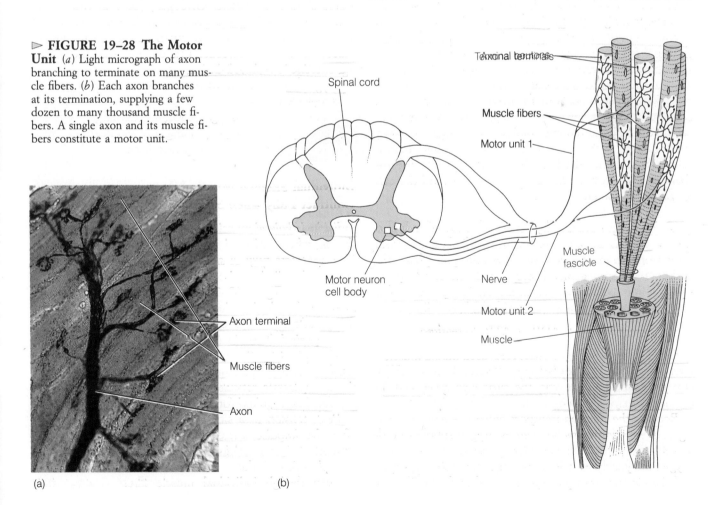

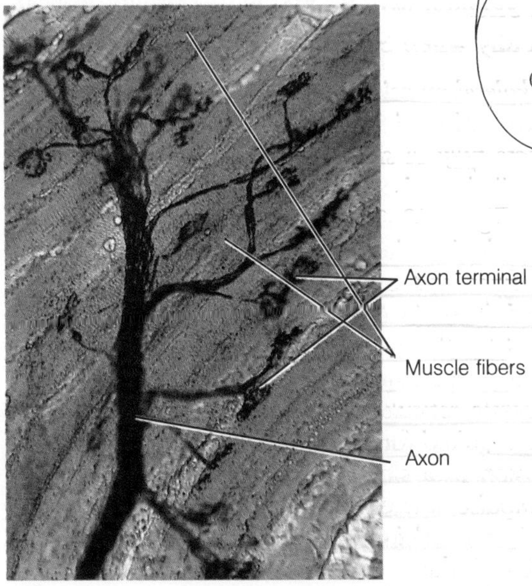

(a)

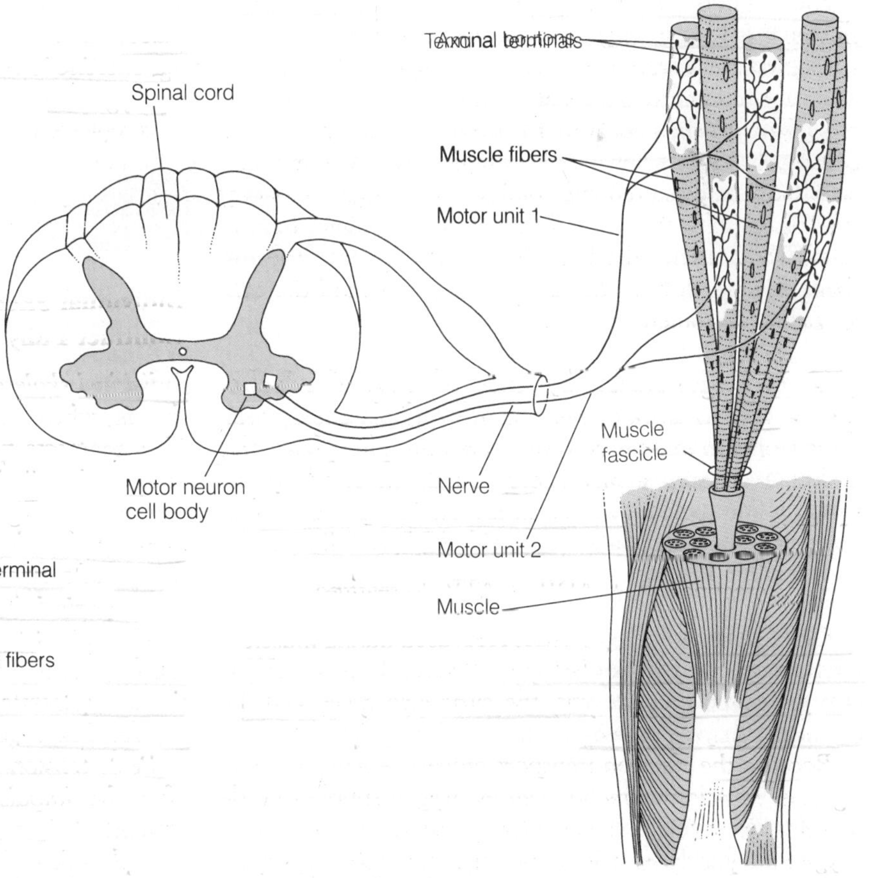

(b)

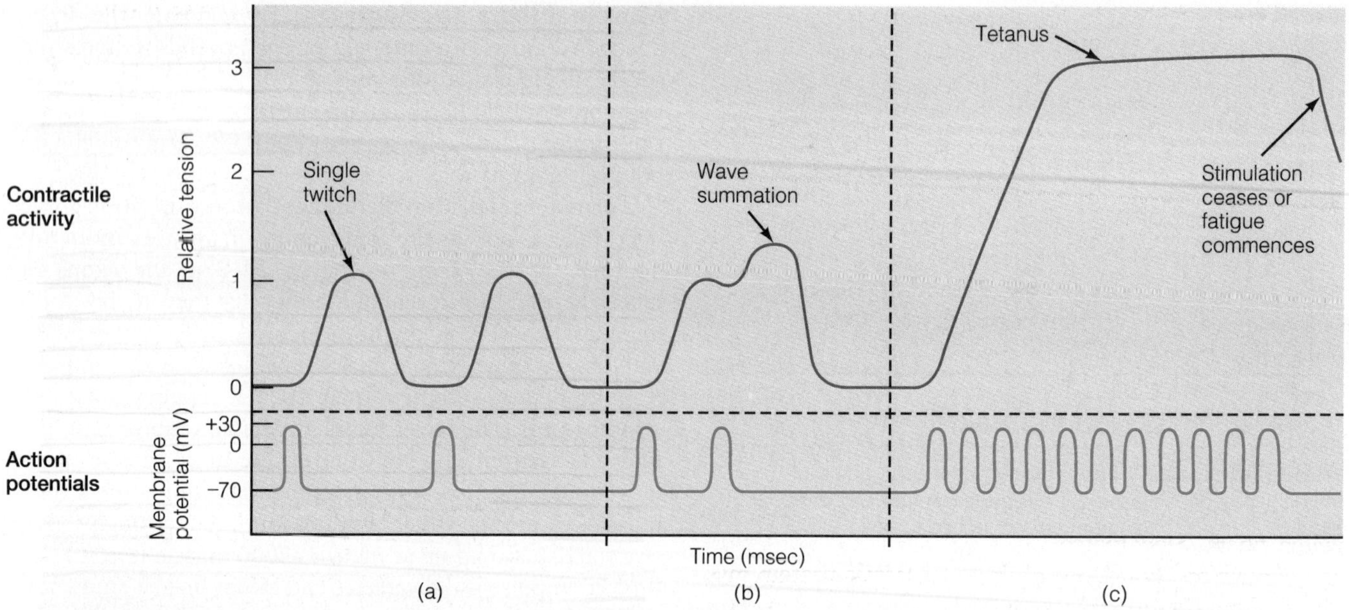

The duration of the action potentials is not drawn to scale, but is exaggerated.

▻ **FIGURE 19–29 Graph of Muscle Contraction** (*a*) The force generated by two separate stimuli (action potentials from motor neuron). (*b*) The force generated by two closely timed stimuli. (*c*) The force generated by many closely spaced stimuli.

central nervous system recruits additional motor neurons. The more motor neurons that are stimulated, the stronger the force of contraction. Thus, if you were going to lift a small piece of styrofoam, your nervous system would recruit only a tiny fraction of the muscle fibers in your arms, but if you were going to lift a 40-pound bag of dog food, your brain would recruit many more motor neurons and muscle fibers.

Additional Tension May Be Created in a Muscle Fiber by Nerve Impulses Arriving While the Muscle Fiber Is Still Contracted. Graded contractions also result from increasing the contractile force that each fiber generates. Each time a muscle fiber is stimulated, it contracts fully. ▻ Figure 19–29a shows the strength of contractions resulting from nerve impulses arriving after muscle fibers contract and relax. If, however, nerve impulses reach the muscle fiber before it has relaxed—that is, while it is still somewhat contracted—additional tension is created (Figure 19–29b). In other words, the muscle fibers contract more forcefully. In a sense, then, the second contraction "piggybacks" on the first; this process is technically referred to as **wave summation.**

If the nerve impulses arrive frequently at enough skeletal muscle fibers, a smooth, sustained contraction will occur in the muscle (Figure 19–29c). When you carry a bag of groceries in your arms, for example, your arm muscles contract to support the weight and remain contracted throughout the activity. A sustained contraction at maximal strength is called **tetanus** (not to be confused with the serious, often fatal bacterial infection of the same name).[6] Tetanic contractions eventually cause muscle fatigue. The muscle stops contracting, even though the neural stimuli may continue.

Muscle Tone Results from the Contraction of a Small Number of Muscle Fibers that Keep Muscles Slightly Tense

Touch one of your muscles. Even if you are not in peak physical condition, you will notice that the muscle is firm. This firmness is called **muscle tone.** Muscle tone is essential for maintaining posture. Without it, you would literally fall into a heap on the floor when you stood up. It also helps generate heat in warm-blooded animals.

Muscle tone results from the contraction of muscle fibers during periods of inactivity. But not all fibers contract, just enough of them to keep the muscles slightly tense.

Muscle tone is maintained, in part, by the muscle spindle, a receptor that monitors muscle stretching (Chapter 18). The spindle "alerts" the brain and spinal cord to the degree of stretching. When muscles relax, signals travel to the spinal cord, then back out motor axons, which stimulate a low level of muscle contraction, maintaining muscle tone.

[6]It is called tetanus because the toxins from the bacterial infection cause the muscle to lock.

Two Types of Muscle Fibers Are Found in Skeletal Muscle, Slow- and Fast-Twitch

Physiological studies have revealed the presence of two types of skeletal muscle fibers: fast-twitch and slow-twitch fibers.[7] **Slow-twitch muscle fibers** contract relatively slowly but have incredible endurance. Endurance athletes (for example, long-distance runners) perform for long periods without tiring because their leg muscles have a higher proportion of slow-twitch fibers than those of us who peter out quickly (▻ Figure 19–30). The flight muscles of birds, which can travel long distances without stopping, are similar.

Fast-twitch muscle fibers contract swiftly. The muscles of sprinters and other athletes whose performance depends on quick bursts of activity contain a higher proportion of fast-twitch fibers.

Slow-twitch fibers are anatomically distinct from fast-twitch fibers. Slow-twitch fibers, for example, are smaller and contain an abundance of the cytoplasmic protein myoglobin. **Myoglobin** binds to oxygen as does its counterpart in RBCs, hemoglobin, and releases oxygen when it is needed during exercise. Slow-twitch fibers also contain a slow-acting **myosin ATPase.** This enzyme forms the heads of the myosin molecules and splits ATP during muscle contraction; it is largely responsible for the slow-twitch fiber's physiological characteristics.

Fast-twitch fibers are larger than slow-twitch fibers.[8] They contain a fast-acting myosin ATPase, which permits rapid contraction.

Skeletal muscles generally contain a mixture of slow- and fast-twitch fibers, giving each muscle a wide range of performance abilities. However, a muscle that performs one type of function more often than another may have a disproportionately higher number of fibers corresponding to the type of activity it performs. The muscles of the back, for example, contain a larger number of slow-twitch fibers. These muscles operate throughout the waking hours to help maintain posture. They do not need to contract quickly but must be resistant to fatigue. In contrast, the muscles of the arm are used for many quick actions—waving, playing tennis, grasping falling objects. Fast-twitch fibers are more common in the muscles of the arm.

Genetic Differences May Account for Differences in Athletic Performance. New research suggests that one of the reasons some people excel in certain sports whereas others do not may lie in the relative proportion of fast- and slow-twitch fibers in their skeletal muscles. As mentioned earlier, a study of the skeletal muscles of world-class long-distance runners suggests that their physical endurance is primarily due to a high proportion of slow-twitch fibers, a trait that may be genetically determined. Biochemical studies also show that the muscle cells in endurance athletes have a higher level of ATP, both at rest and during exercise, thus providing more energy for muscle contraction. Endurance athletes start out with a larger storehouse of energy and maintain a larger supply throughout exercise.

▻ **FIGURE 19–30 Built to Last** An abundance of slow-twitch fibers gives the Olympic long-distance runner endurance.

Exercise Builds Muscles and Increases Endurance

Milo of Croton was a champion wrestler in ancient Greece who stumbled across a revolutionary way to build muscle. His method was simple. Milo secured a newborn calf, then proceeded to carry the animal around every day. Day after day, he faithfully followed this routine, so by the time the calf had become a bull, so had he. Milo's program worked, for he went on to win six Olympic championships in wrestling.

The Greek wrestler discovered a principle of muscle physiology familiar to weight lifters today: when muscles are made to work hard, they respond by becoming larger and stronger. The increase in size and strength results from an increase in the amount of contractile protein inside muscle cells.

[7] In truth, there are three types—one kind of slow-twitch and two types of fast-twitch. For ease of discussion, only two types will be discussed in this chapter.

[8] Studies show that one type of fast-twitch fiber fatigues rather easily whereas the other type is fatigue resistant.

Muscle protein is quickly made and quickly destroyed. In fact, half of the muscle you gain in a weight-lifting program is broken down two weeks after you stop exercising. In order to build muscle, then, the rate of formation must exceed the rate of destruction. If you don't keep up with your exercise program, your newly developed biceps will disappear quickly.

High intensity exercise, such as weight lifting, builds muscle. It takes surprisingly little exercise to have an effect. Working out every other day for only a few minutes will result in noticeable changes in muscle mass. According to some sources, 18 contractions of a muscle (in three sets of six contractions each) are enough to increase muscle mass, if the contractions force your muscles to exert over 75% of their maximum capacity.

Low-intensity exercise, such as aerobics and swimming, tends to burn calories but not build muscle bulk. Aerobic exercise therefore helps increase endurance; that is, it increases one's ability to sustain muscular effort. Stamina or endurance results from numerous physiological changes. One of the most important changes occurs in the heart.

The heart responds to exercise like any other muscle—it grows stronger and even enlarges. A well-exercised heart beats more slowly but pumps more blood with each beat, as explained in Chapter 12. The net result, then, is that the heart works more efficiently, delivering more oxygen to skeletal muscles.

Increased endurance also results from improvements in the function of the respiratory system. For example, exercise increases the strength of the muscles involved in breathing. These muscles become stronger and can operate longer without tiring. Breathing during exercise becomes more efficient.

Increased endurance may also be attributed to an increase in the amount of blood in the body. An increase in blood volume, in turn, results in an increase in the number of RBCs in the body, thus increasing the amount of oxygen available to body cells. This improvement, combined with others, allows an individual to work out longer without growing tired.

When you set out on an exercise program, it is important to establish your goals first. If you are interested in increasing your endurance, you should pursue an exercise regime that works the heart and muscles at a lower intensity over longer periods. If you are after bulk, the answer lies in high-intensity exercises that result in increased muscle mass.

Many health clubs and university gyms offer programs or advice on ways to achieve your goals. And many offer a variety of machines to help you build muscles or simply tone them up. Most of the exercise machines work one particular muscle group—for example, the muscles of the upper arm or the muscles of the chest. In exercise clubs, a half-dozen or more machines, each designed for a specific muscle group, are usually placed in a line. You simply go down the line, working one set of muscles, then another, until you have exercised your entire body.

Exercise machines are popular for several reasons. First of all, they are safer than free weights—barbells and dumbbells. Progressive resistance machines eliminate the chances of your dropping a weight on your toes—or someone else's, for that matter. Moreover, it is almost impossible to strain your back if you make a mistake using one. In contrast, lifting free weights requires care and training as well as brawn.

These machines also reduce the amount of time a person needs to exercise by about half. Why? They require work when flexing and extending a joint. The biceps machine, for example, requires you to pull the weights up, then let them return slowly. In both directions, your muscles are being forced to work. The machines also allow you to lift heavier weights. The reason for this is simple: With free weights, you can only lift a weight that can safely be moved through the part of the exercise where your muscle is the weakest. Any more, and you can tear a muscle or drop a weight. One popular brand of progressive-resistance machine, the Nautilus, alters the resistance automatically as you perform an exercise. In the weakest phase of the exercise, the machine reduces the resistance to prevent damage. Throughout the rest of the exercise, however, the resistance is full.

ENVIRONMENT AND HEALTH: ATHLETES AND STEROIDS

In recent years, many Americans have become increasingly troubled by the use of illegal drugs. Even athletes have come under scrutiny—some for using cocaine and other addictive drugs and others for using anabolic steroids.

Anabolic steroids are synthetic hormones that resemble the male sex hormone, testosterone. When taken in large doses, anabolic steroids stimulate muscle formation (anabolism). They increase muscle size and strength by stimulating protein synthesis in muscle cells. High doses may also reduce inflammation that frequently results from heavy exercise, allowing athletes to work out harder and longer.

When it comes to building muscle, steroids and exercise are an unbeatable combination. Some users think that steroids may even increase aggression, which may be helpful to football players and other competitive athletes on the playing field. Despite steroids' apparent benefits, physicians are concerned about their adverse health effects.

Steroids, for example, may result in psychiatric (mental) and behavioral problems. In interviews with 41 athletes who used steroids, researchers found that one-third of them developed severe psychiatric complications. These athletes routinely took steroids in doses 10 to 100 times greater than those used in medical studies of the drugs. The athletes also reported using as many as five or six steroids simultaneously in cycles lasting 4 to 12 weeks. This practice, known as "stacking," is quite common and may be responsible for the psychiatric effects.

Athletes in the study reported episodes of severe depression during and especially after steroid use. Some reported feelings of invincibility. One man, in fact, deliberately drove into a tree at 40 miles per hour while a friend videotaped him. Some subjects reported psychotic symptoms in association with steroid use, including auditory hallucinations (hearing voices). Withdrawal from steroids results not only in depression but also in suicidal tendencies.

Making matters worse, steroids may also damage the heart and kidneys and frequently reduce testicular size in men. In women, steroids deepen the voice and may cause enlargement of the clitoris. Steroids also cause severe acne and may cause liver cancer. Unfortunately, there are no scientific studies on the long-term health effects of steroids.

Despite the fact that anabolic steroids are banned by the National Football League, the International Olympic Committee, and college athletic programs, athletes continue to use them. Most steroids used in the United States are imported illegally from Mexico and Europe. Because of a federal crackdown on the importation of steroids, some experts believe that the inflow may be slowing. Others are not so optimistic, contending that the $100-million-a-year black market will not be easily thwarted.

A recent survey of 46 public and private high schools across the United States involving over 3000 teenagers suggests that steroid use is especially prevalent in high school seniors. About 1 of every 15 senior boys reported taking anabolic steroids.

The study also suggests that the use of anabolic steroids begins early in junior high school. Two-thirds of the students surveyed said that they had started using them by age 16. Nearly half the steroid users said they took the drugs to boost athletic performance. Twenty-seven percent said their primary motive was to improve their appearance. Researchers say that adolescents who use steroids may be putting themselves at risk of stunted growth, infertility, and psychological problems.

Steroid use in the United States illustrates our dependence on quick fixes. It also illustrates the almost obsessive focus on performance and achievement in our highly competitive society, a social environment that may be endangering the health of our children and our athletes.

SUMMARY

AN OVERVIEW OF THE EVOLUTION OF MOTOR SYSTEMS

1. Animals are partly characterized by their ability to move from one place to another and to move body parts, which allows them to perform many tasks.
2. Animal locomotion varies considerably. Despite the vast differences, all animals have two common features: they possess muscles and convert chemical energy in their muscles into movement.
3. The simplest animal motor systems are found in the sea anemones and annelids, which contain two sets of antagonistic muscle fibers, circular and longitudinal.
4. The presence of exoskeletons in invertebrates, such as arthropods, crabs, snails, and clams, provides a wider range of movements.
5. In many vertebrates, opposing muscle groups acting on internal skeletons permit an even wider range of movements. Internal skeletons are found in vertebrates, including birds, fish, reptiles, amphibians, and mammals.

SKELETAL STRUCTURE AND FUNCTION

6. In vertebrates, bones provide internal support, allow for movement, help protect internal body parts, produce blood cells and platelets, store fat, and help regulate blood calcium levels.
7. Most vertebrate bones have an outer layer of compact bone and an inner layer of spongy bone. Inside the bone is the marrow cavity, filled with either fat cells (yellow marrow) or blood cells and blood-producing cells (red marrow) or with combinations of the two.
8. The joints unite bones. Three types of joints are found: immovable, slightly movable, and freely movable. The movable joints allow for flexion, extension, and other important movements.
9. Joints are often subject to injury and disease. Torn ligaments and ripped cartilage, for example, are common injuries among athletes. Wear and tear on some joints may cause the articular cartilage to crack and flake off, resulting in degenerative joint disease, or osteoarthritis. Rheumatoid arthritis results from an autoimmune reaction that produces inflammation and thickening of the synovial membrane, disfiguring and stiffening joints and causing pain.

EMBRYONIC DEVELOPMENT AND BONE GROWTH

10. Most of the bones form from hyaline cartilage.
11. Bone is constantly remodeled after birth to accommodate changing stresses. During bone remodeling, osteoclasts destroy bone. Osteoclasts are stimulated by parathormone, produced by the parathyroid glands.
12. Osteoclasts also participate in the homeostatic control of blood calcium levels. When activated, these cells free calcium from the bone, raising blood calcium levels.
13. Osteoblasts are bone-forming cells stimulated by calcitonin from the thyroid gland. Calcitonin secretion helps decrease blood calcium levels and increases the amount of calcium in bones.
14. Osteoporosis is a disease of the bone caused by progressive decalcification. The bones become brittle and easily broken. Osteoporosis is most common in postmenopausal women and results from the loss of the ovarian hormone estrogen. Osteoporosis also occurs in people who are immobilized for long periods.
15. Exercise, calcium and fluoride supplements, calcium-rich foods, vitamin D, and estrogen supplements can all help prevent or reverse this potentially fatal disease.

THE SKELETAL MUSCLES

16. Skeletal muscles are involved in body movements, help maintain our posture, and produce body heat both at rest and while we are working or exercising.
17. Skeletal muscles consist of long, unbranched, multinucleated cells known as muscle fibers, which are excitable and contractile.
18. Inside each muscle fiber are numerous myofibrils, bundles of the contractile filaments actin and myosin.
19. Contraction occurs when the heads of the myosin filaments attach to binding sites on the actin filaments, then pull the actin filaments inward.
20. Muscle contraction is stimulated by nerve impulses from motor neurons that cause the release of acetylcholine.
21. The energy for muscle contraction comes from ATP. ATP is replenished by creatine phosphate, glycolysis, and cellular respiration.
22. Individual muscle fibers obey the all-or-none law. When activated by an action potential, they contract fully. Contractions of varying strength (graded contractions) can be generated by recruitment and by wave summation.
23. Recruitment results from the engagement of many motor units. A motor unit is a motor axon and all of the muscle cells it innervates.
24. Wave summation is a piggybacking of muscle fiber contractions occurring when stimuli arrive before the fiber relaxes.
25. Muscle tone is the rigidity of resting muscle caused by low-level contraction of some muscle fibers.
26. The body contains two types of skeletal muscle fibers: slow-twitch and fast-twitch fibers. Skeletal muscles generally contain a mixture of the two types, but fast-twitch fibers are found in greatest number in muscles that perform rapid movement. Slow-twitch fibers are found in muscles such as those of the back that perform slower motions or are involved in maintaining posture.
27. Muscle mass can be increased by exercise. An increase in muscle mass results from an increase in the amount of contractile protein (actin and myosin) in muscle fibers.
28. To build mass, one must generally use more weight and do fewer repetitions.
29. Building endurance requires less weight and more repetitions. Endurance is a function of at least three factors: the condition of the heart, the condition of the muscles of the respiratory system, and the blood volume. Improvement in all three factors increases the efficiency of oxygen supply to muscles.

ENVIRONMENT AND HEALTH: ATHLETES AND STEROIDS

30. Many athletes are using synthetic anabolic steroids to improve performance and build muscle.
31. Unfortunately, massive doses of steroids have many harmful effects. They can increase aggression, cause psychiatric imbalance, such as severe depression, and result in damage to the heart and kidneys.

EXERCISING YOUR CRITICAL THINKING SKILLS

1. In a study of osteoporosis in Australian women, researchers concluded that the best prevention against bone fractures was estrogen-replacement therapy. In their study, the researchers recruited 120 nonsmokers ranging in age from 50 to 60. During the two-year study, the researchers studied the density of bone in the forearms of the women at three-month intervals. Early measurements showed very low levels of calcium—levels that indicated a high risk of fractures.

   The subjects were divided into three groups. Women in the first group received a placebo. Women in the second group received 1 gram of calcium per day, and women in the third group received daily doses of estrogen and progesterone (sex steroid hormones). In this study all women were encouraged to take two brisk 30-minute walks every day and engage in one low-impact aerobics class each week.

   The study showed that bone loss continued in the control group (exercise only) at a rate of 2.5% per year. Bone loss also continued in the group that received a daily dose of calcium, but at a slower rate (0.5% to 1.3% per year). In the group that exercised and received hormone therapy, bone density increased from 0.8% to 2.7% per year.

   Given the potential side effects of hormone therapy, the researchers recommended that only women with lowest bone densities should receive hormone therapy and that others should exercise and take calcium to reduce their risk of osteoporosis.

   Going back over the experimental design, can you pinpoint any potential flaws? If you noted that no mention was made of dietary calcium intake by the various groups, you're correct. Unfortunately, the diets of the women in this study were not monitored to determine how much calcium was being ingested through food and beverages. Second, this study focused on the bones of the forearm, not the load-bearing bones that are at greatest risk of fractures. Walking would probably influence the load-bearing bones of the legs, hips, and back, and to accurately assess the effects of the exercise regime prescribed in this study, the researchers probably should have looked at bone density in all three sites, not the arms. Given these inadequacies, do you think this study's conclusions are reliable?
2. A friend is selling an all-natural supplement guaranteed to give you more energy and help you sleep better at night. She cites a number of examples of users who felt energized and could sleep better. She also notes that the developer of the product has a Ph.D. in nutrition. She wants you to try the product, which costs $40 a bottle. Using the critical thinking rules you learned in Chapter 1, describe what you might do in this situation. How would you evaluate the evidence your friend has given supporting the use of this supplement? What additional information would you want?

TEST OF TERMS

1. The skull, ribs, and ______________ form the ______________ skeleton.
2. The ______________ skeleton consists of the bones of the legs and arms and the bones of the shoulders and pelvis to which they attach.
3. The humerus and femur are examples of ______________ bones. The shaft, or ______________ , of this type of bone consists of a layer of __________ bone surrounding the ______________ cavity.
4. Inside the epiphyses of the humerus is a network of bony spicules forming ______________ bone.
5. The knee and shoulder joints are examples of ______________ joints. They contain ______________ fluid, which allows for easy movement of the bones. These joints are supported by four structures: the ___ ______________, ligaments, muscles, and ______________ .
6. Closing a joint is called ______________ ; opening it is called ______________ .
7. Knee surgery can be performed using a(n) ______________ , a device that reduces the trauma and allows surgeons to see what they are doing.
8. Degenerative joint disease, or ____________ , results from wear and tear on a joint. It occurs most often in the weight-bearing joints.
9. ______________ arthritis is believed to be an autoimmune disease. It results in a thickening of the ________ membrane and degeneration of the joint.
10. An artificial joint or body part is called a(n) ______________ .
11. Most bone in the body is formed during embryonic development from ____________ ____________ . The first location of bone formation in the diaphysis is called the ______________ ____ of ossification.
12. Bone is broken down by large, multinucleated cells called ________ ; it is laid down by ______________ .
13. The zone of cartilage lying between the epiphysis and diaphysis of a growing bone is called the ____________ ____________ .
14. Bone deposition is stimulated by the hormone ____________ secreted by cells in the ______________ gland. Bone dissolution is stimulated by ____________ .
15. Prolonged bone degeneration, which results in brittle bones and occurs chiefly in postmenopausal women, is called ______________ .
16. A skeletal muscle fiber is formed during embryonic development by the fusion of numerous embryonic muscle cells. Skeletal muscle fibers are long, unbranched cells containing many ______________ . Viewed with the light microscope, they appear banded, or ______________ .
17. Each muscle fiber is surrounded by a thin layer of connective tissue, the ______________, which holds it in place.
18. The entire muscle is surrounded by a layer of connective tissue called ______________, which forms the tendons that attach the muscle to bone.
19. Inside the muscle fiber are numerous bundles of contractile filaments. Each bundle is called a(n) ______________ .
20. The ______________ is the functional unit of the muscle fiber. It extends from one Z line to the next and contains a dark central band, the ______________ band, and two lighter bands on either end, the ______________ bands.
21. Contraction results when ____________ ions are released from the ____________ reticulum. These ions bind to ______________ molecules, thus releasing tropomyosin from the binding sites on the ____________ filament.
22. The heads of the ______________ filaments pull the actin filaments inward, causing contraction.
23. The neurotransmitter ______________ is released at the neuromuscular junction. It causes an action potential to be set up in the muscle membrane. The action potential penetrates the interior of the muscle fiber via the ______________ ______________ .
24. ______________ ______________ occurs when muscle depletes its supplies of glycogen and ATP and when ______________ ______________ levels increase.
25. The contraction of a single muscle cell is called a muscle __________. It is an ______________-______________-______________ response to stimulation, which results in a complete contraction.
26. The strength of contraction can be increased by ______________ summation, the piggybacking of contractions, and by recruiting more ______________ units.
27. A world class long-distance runner would probably have an abundance of ____________-____________ muscle fibers in his or her leg muscles. These fibers have a large amount of ____________, which holds oxygen and releases it during activity.
28. A sprinter might have an abundance of ______________-______________ fibers. They contain a fast-acting ______________ ______________, which splits ATP, providing energy for muscle contraction.
29. Synthetic hormones called ____________ ______________ can help an athlete increase muscle mass but have many deleterious side effects.

Answers to the Test of Terms are found in Appendix B.

TEST OF CONCEPTS

1. Describe the functions of bone. In what ways does bone participate in homeostasis?
2. The synovial joints move relatively freely. What structures support the joint, helping keep the bones in place?
3. A young patient comes to your office with swollen joints and complains about pain and stiffness in the joints. Friends have suggested that the boy has arthritis, but his parents argue that he is too young. Only old people

get arthritis, they say. How would you answer them?

4. Describe the process of bone formation. Be sure to define the following terms: primary and secondary centers of ossification, osteoclasts, perichondrium, osteoblasts, and osteocytes.
5. Bone is constantly remodeled, from infancy through adulthood. Explain when bone remodeling occurs and what cells and hormones participate in the process.
6. Using what you know about bone, explain why an office worker who exercises very little is more likely to break a bone on a skiing trip than a counterpart who works out every night after work.
7. Describe how a bone heals after being fractured.
8. A 30-year-old friend of yours who smokes, exercises very little, and avoids milk products because of her diet says: "Why should I worry about osteoporosis? That's a disease of old women." Based on what you know about bone, how would you respond to her?
9. Describe the major functions of skeletal muscle.
10. Describe the detailed structure of a skeletal muscle fiber.
11. Describe the molecular events involved in muscle contraction.
12. Define the term *graded contraction* and describe how they are achieved.
13. Describe the two types of skeletal muscle fibers and explain how they differ.
14. A friend comes to you complaining that he can't seem to lose weight. He works out on barbells three times a week for an hour or so each time. His diet hasn't changed much, but with the increase in exercise, he thinks he should be losing weight, rather than staying even. Why do you think he isn't losing weight?
15. A friend is thinking about using anabolic steroids to improve his performance in gymnastics. He argues that he is young and will be taking the drug only for a year or so. He points out that other young men are using steroids and they really help build muscle and endurance. What advice would you give him?

CHAPTER

20

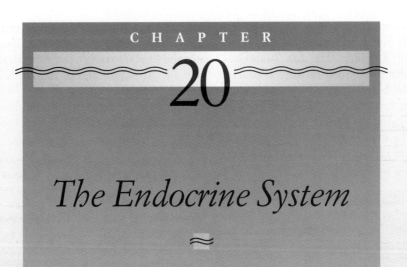

The Endocrine System

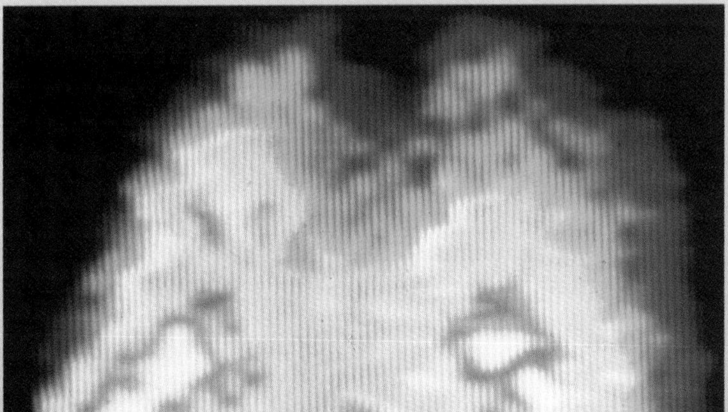

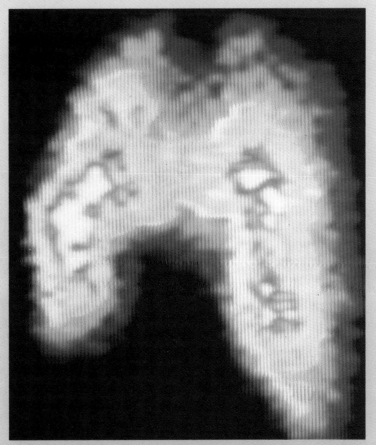

NMR scan of a normal (non-diseased) human thyroid gland. The thyroid gland is located in the neck.

One day while on his morning jog, President George Bush began to feel weak. His heartbeat became irregular. He felt breathless. Fearing a heart attack, his security entourage rushed the president to the hospital. After thorough testing showed no sign of a heart attack, a blood test revealed elevated levels of a hormone in the president's blood called thyroxine. Produced by the thyroid gland, thyroxine causes a general acceleration of all chemical reactions in the body, affecting a number of mental as well as physical processes. Hyperthyroidism, or Grave's disease (after its discoverer, not the final outcome), is a rare condition that can occur at any age. It is just one of many diseases caused by a hormonal imbalance. This chapter discusses the endocrine system of invertebrates and vertebrates and presents some interesting examples of homeostatic imbalance caused by endocrine disorders in vertebrates.

≈ PRINCIPLES OF ENDOCRINOLOGY

The vertebrate **endocrine system** consists of numerous small glands scattered throughout the body (▻ Figures 20–1a and 20–1b). These highly vascularized glands produce and secrete chemical substances known as hormones, a word that comes from the Greek *hormon,* which means "to stimulate" or "excite." A **hormone** is a chemical produced and released by cells or groups of cells that consti-

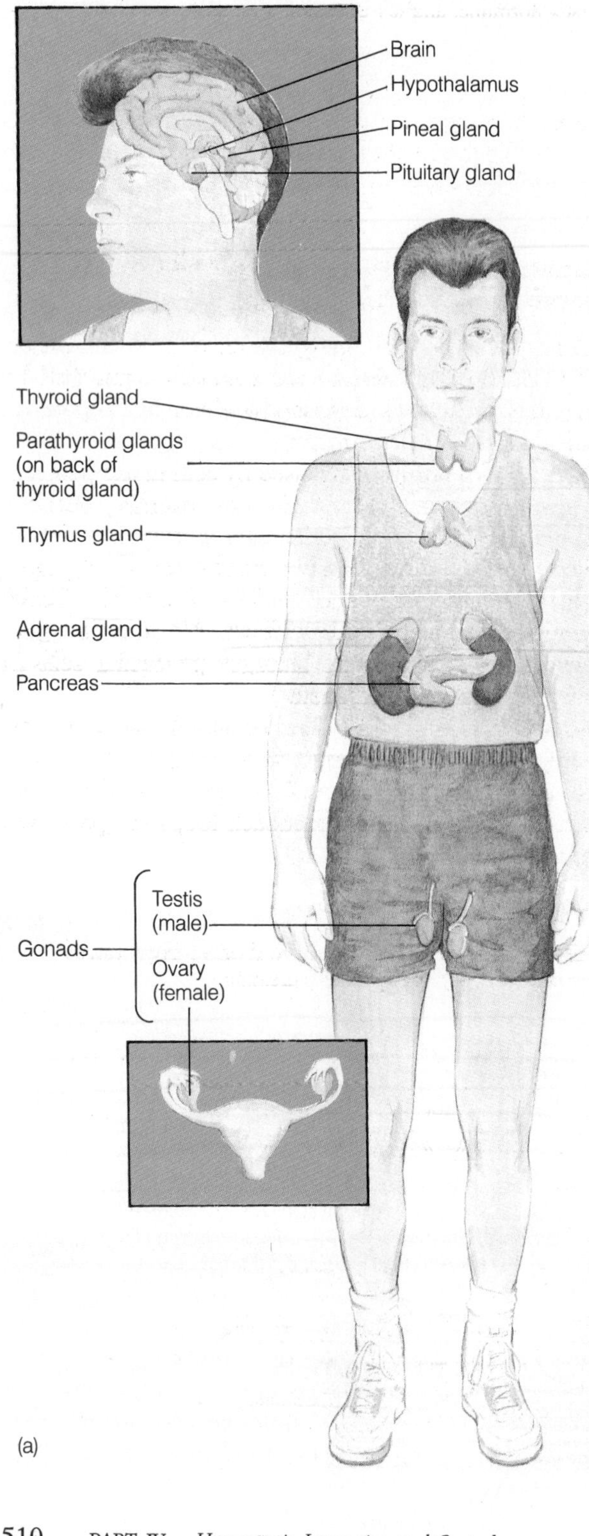

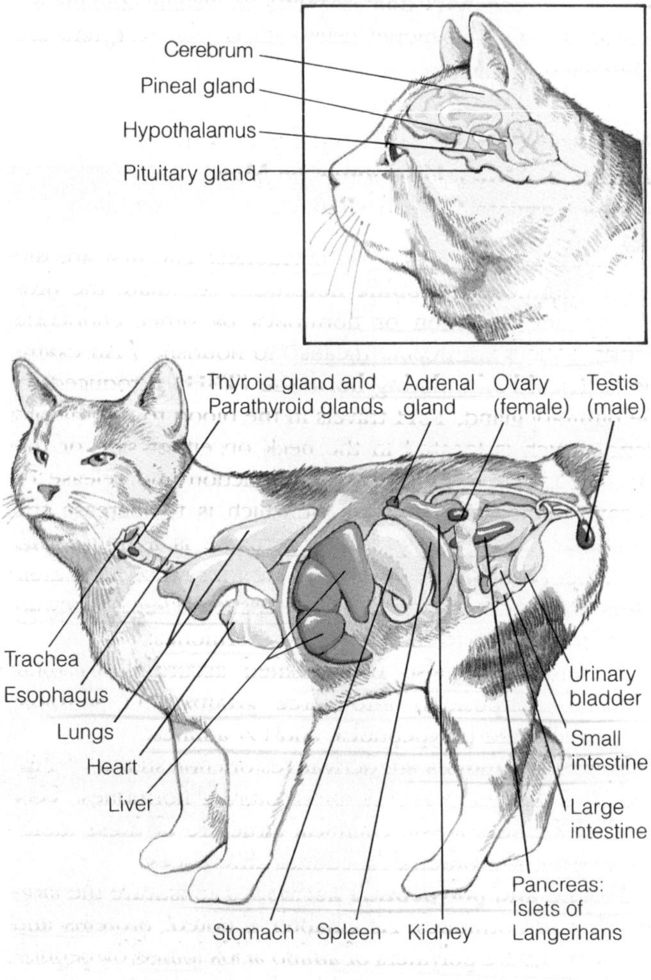

▻ **FIGURE 20–1 The Vertebrate Endocrine System**
(*a*) The endocrine system consists of a scattered group of glands that produce hormones, which help regulate growth and development, homeostasis, reproduction, energy metabolism, and behavior. (*b*) The endocrine system of the cat.

tute the **endocrine (ductless) glands.** Hormones travel in the blood to distant sites, where they elicit their response(s). The hormone insulin, for example, is produced by the pancreas and is transported in the blood to skeletal muscle and other body cells, where it stimulates glucose uptake and glycogen synthesis. The cells a hormone affects are called its **target cells.** For a discussion of some of the early research that contributed to our understanding of the endocrine system, see Scientific Discoveries 20-1.

Hormones function in five principal areas: (1) homeostasis, (2) growth and development, (3) reproduction, (4) energy production, storage, and use, and (5) behavior. At any one moment, the blood carries dozens of hormones. The cells of the body are therefore exposed to many different chemical stimuli. How does a cell keep from responding to all the signals?

Target Cells Contain Receptors for Specific Hormones

Target cells respond only to specific hormones. The "selection process" depends on protein **receptors** in the plasma membrane and in the cytoplasm of target cells. Receptors bind specifically to one type of hormone, in the same way that enzymes bind to only one substrate (Chapter 4). Thus, the cells in the body that respond to insulin contain receptors for this hormone; cells without the receptors are unaffected.

Hormones Stimulate the Synthesis and Release of Other Hormones or May Simply Activate Cellular Processes

Hormones fit into two broad categories. The first are the trophic hormones. **Trophic hormones** stimulate the production and secretion of hormones by other endocrine glands. (The word *trophic* means "to nourish.") An example is **thyroid-stimulating hormone (TSH).** Produced by the pituitary gland, TSH travels in the blood to the thyroid gland, which is located in the neck on either side of the larynx. Here, it stimulates the production and release of thyroxine, the chief function of which is to increase the metabolic rate of body cells. Thyroxine is a nontrophic hormone. **Nontrophic hormones** are those that exert their effect principally on nonendocrine tissues. That is, they *do not* stimulate the synthesis of other hormones.

Hormones can also be classified according to their chemical composition into three groups: (1) steroids, (2) proteins and polypeptides, and (3) amines.

Steroid hormones are derivatives of cholesterol. ▻ Figure 20–2 illustrates two common steroid hormones. Very small differences in the chemical structure of these molecules result in profound functional differences.

Protein and polypeptide hormones constitute the largest class of hormones. As Chapter 2 noted, proteins and polypeptides are polymers of amino acids joined by peptide bonds. Growth hormone and insulin are two examples.

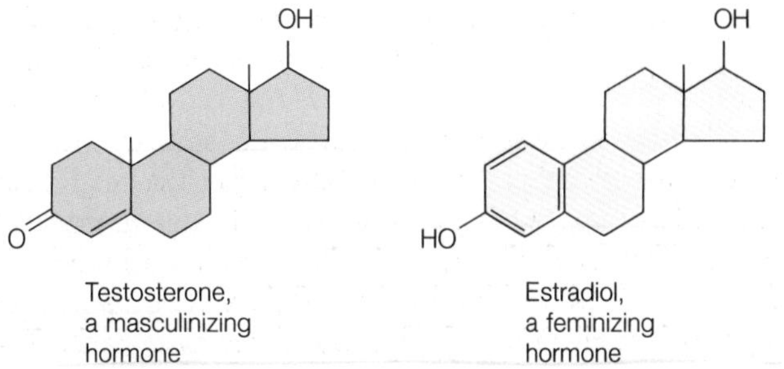

▻ **FIGURE 20–2 Two Common Steroids** (*a*) Testosterone, a male sex hormone, and (*b*) estradiol, a female sex steroid.

Amine hormones are derivatives of the amino acid **tyrosine.** ▻ Figure 20–3 shows the structure of two of the four amine hormones produced in the body.

Hormone Secretion Is Often Controlled by Negative Feedback Mechanisms

Hormones help control many homeostatic mechanisms. Their production and release are generally controlled by negative feedback loops, described in detail in Chapter 10. Consider the hormone glucagon.

Glucagon is a hormone released by cells in the pancreas when glucose concentrations fall—for example, between meals. Released into the bloodstream, glucagon travels to the liver, where it stimulates the breakdown of glycogen, causing the release of glucose molecules into the blood. This helps restore blood glucose levels. When normal glucose levels are achieved, the glucagon-producing cells in the pancreas end their secretion.

Not all feedback loops are as simple as this one, however; some involve intermediary compounds. Nevertheless, all operate on the same basic principle. (Examples of other endocrine negative feedback loops are presented throughout this chapter.)

▻ **FIGURE 20–3 Representative Amine Hormones** (*a*) Thyroxine and (*b*) epinephrine (adrenalin).

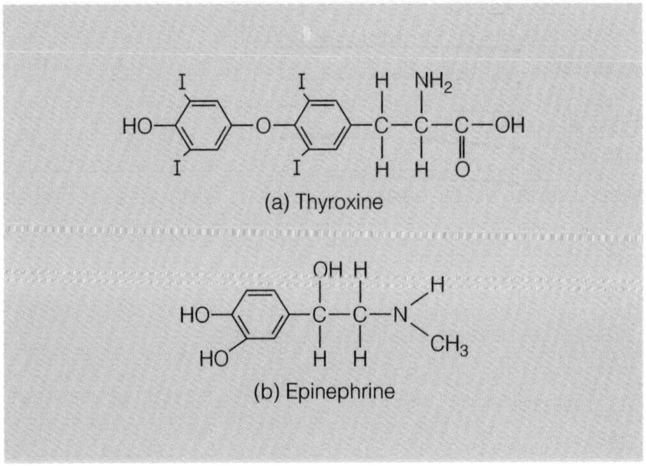

SCIENTIFIC DISCOVERIES THAT CHANGED THE WORLD 20-1

PANCREATIC FUNCTION: IS IT CONTROLLED BY NERVES OR HORMONES?

Featuring the work of Bayliss and Starling

The concept of chemical control of body functions, like many other concepts in biology, is rooted in a large number of experiments. Animal experiments in the mid-1800s, for example, showed that one could remove the testes and transplant them to another body location without the loss of secondary sex characteristics. This finding suggested that the testes produced a blood-borne substance that affected other tissues. This study and dozens of others contributed bits of information that, within 100 years, gave rise to many generalizations about hormone production, structure, and function.

Of the many experiments crucial to our understanding of endocrinology, one by two British physiologists, W. M. Bayliss and E. H. Starling, stands out. Published in 1902, this experiment set to rest a debate about the mechanism by which food entering the small intestine from the stomach stimulated the production and release of pancreatic secretions.

Several experiments before the publication of this work suggested that hydrochloric acid in chyme stimulated nerve endings in the small intestine, triggering a reflex that resulted in the release of pancreatic juices, which contain sodium bicarbonate. In these studies, researchers injected substances into the duodenum of anesthetized dogs, then studied the release of pancreatic secretions. Even though severing the nerves that supplied the small intestine did not affect the results, researchers still argued that the nervous system was involved in the control. They hypothesized that clumps of nerve cells (ganglia) and their network of interconnected fibers in the wall of the intestine participated in local reflex arcs—illustrating the influence of bias in the interpretation of scientific research.

Bayliss and Starling repeated these experiments but were careful to sever all nerve connections to the small intestine. They then tied a piece of the small intestine off and injected small amounts of hydrochloric acid into it. This treatment resulted in the production of pancreatic secretion at a rate of one drop every 20 seconds over a period of about 6 minutes. Because it was previously known that acid introduced into the bloodstream had no effect on the release of pancreatic secretions, the researchers concluded that "the effect was produced by some chemical substance finding its way into the veins of the loop of jejunum in question and being carried in the bloodstream to the pancreatic cells."

To verify their hypothesis, they tried another experiment. In this one, they scraped off the cells lining the section of the small intestine, exposed them to acid, ground them up, and then extracted the fluid, which they injected into the bloodstream. After a brief latent period, the pancreas began secreting. This experiment led to the eventual abandonment of the "nervous control hypothesis."

Bayliss and Starling referred to the mystery substance as "secretin," because it stimulated pancreatic secretion. This name remains in use today. Thankfully, most other hormones discovered since that time have far more specific names.

Besides settling the debate over the control of pancreatic function, Bayliss and Starling helped clarify our understanding of endocrinology, furnishing definitions for some of the most important terms in the field. One good example is the term *endocrine gland,* which they defined as an organ that secretes into the blood specific substances that affect some other organ or process located at a distance from the gland.

Positive feedback loops are also encountered in the endocrine system, but rarely. A positive feedback loop results when the product of a cell or organ stimulates additional production. Positive feedback loops perform some specialized functions. Ovulation, the release of the ovum from the ovary, for example, is stimulated by a positive feedback loop that is discussed in the next chapter. Fortunately, each positive feedback loop has a built-in mechanism that ends the escalating cycle, preventing the response from getting out of hand.

Most Hormones Undergo Periodic Fluctuations in Their Release

The fact that hormones are controlled by negative feedback loops does not mean that hormone concentrations in the blood are constant 24 hours a day, 365 days a year. In fact, virtually all hormones undergo periodic fluctuations in their release, causing many body functions also to fluctuate. These natural fluctuations in body function are called **biological cycles,** or **biorhythms,** and were described in Chapter 10.

Biological cycles vary in length. Some hormones, for example, are released in hourly pulses, or cycles. Others are released in daily cycles. For example, cortisol is a steroid hormone produced by the adrenal cortex in a 24-hour, or **circadian rhythm,** as shown in ▻ Figure 20–4. Cortisol increases glucose, among other functions. Its secretion increases during the night, reaching a peak just before one wakes up, then falls sharply during the day.

Other hormones are released in monthly cycles. The 28-day menstrual cycle in women, controlled by hormones,

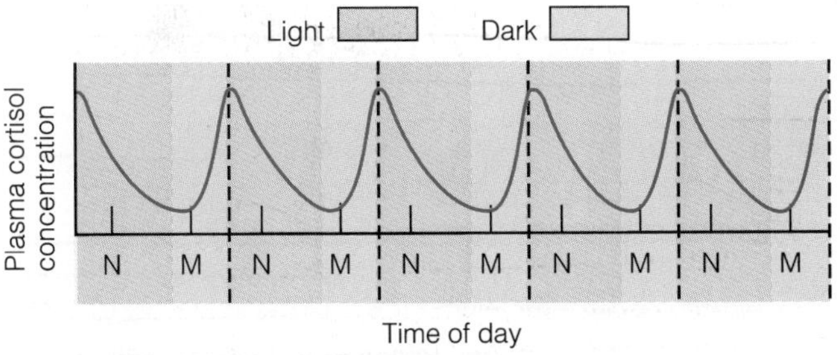

N is noon; M is midnight.

▻ **FIGURE 20–4 Cortisol Secretion** Cortisol secretion follows a diurnal (daily) rhythm with highest levels occurring in the night just before waking. Adapted with permission from George A. Hedge, Howard D. Colby, and Robert L. Goodman, *Clinical Endocrine Physiology* (Philadelphia: W. B. Saunders Company, 1987), Figure 4–4, p. 80.

is a good example. During the cycle, levels of the reproductive hormones vary considerably.

Still other hormones are released in seasonal cycles. Thyroid hormone in people, for instance, is released in greater amounts during the winter months than during the summer. Thyroid hormone stimulates metabolism, and elevated levels raise body temperature. This cycle probably explains why you feel cold on the first 50-degree day in the fall but feel warm when the first 50-degree day arrives in spring.

Most hormonal cycles are controlled by the biological clocks, regions of the brain that regulate biological cycles (Chapter 10). Seasonal cycles, however, are usually controlled by environmental conditions—for example, temperature or photoperiod (day length).

The Chemical Nature of a Hormone Determines How It Is Transported in the Blood and How It Acts on Cells

Protein and polypeptide hormones are water-soluble (hydrophilic). Consequently, they dissolve in the plasma of the blood, which transports them to their target cells. In contrast, steroid hormones are lipids and do not dissolve in water. To be transported, they must be bound to much larger plasma proteins, like albumin.

Protein and Polypeptide Hormones Stimulate Changes in Target Cells by Activating a Second Messenger Inside the Target Cell. Protein molecules travel freely in the blood, but they cannot penetrate the plasma membrane of their target cells. Protein hormones must trigger internal cellular changes from outside their target cells.

A hypothesis explaining how protein and polypeptide hormones trigger intracellular changes was first presented by the late Earl Sutherland, a medical researcher at Vanderbilt University. His proposal, called the **second messenger theory,** won him a Nobel Prize in 1971. The theory states that protein and polypeptide hormones (first messengers) stimulate intracellular changes through a second messenger, an intermediary synthesized inside the cell when the hormone binds to the plasma membrane of the target cell. The second messenger, in turn, triggers internal anatomical and physiological changes.

Since his early work, researchers have found that some amine hormones—adrenalin (epinephrine) and noradrenalin (norepinephrine) from the adrenal medulla—also bring about their effects by the second messenger mechanism.

To understand how the second messenger operates, we will study the process by steps.

1. Polypeptide hormones (and certain amines) first bind to protein receptors in the plasma membrane of the target cell (▻ Figure 20–5).
2. These receptors are linked to the enzyme **adenylate cyclase** on the inner surface of the plasma membrane.
3. Binding of the polypeptide hormone to the receptor activates adenylate cyclase.
4. The activated enzyme catalyzes the conversion of intracellular ATP to **cyclic AMP (cAMP),** a special nucleotide shown in ▻ Figure 20–6.
5. Cyclic AMP is the **second messenger.** This water-soluble compound is free to move about in the cytoplasm, where it activates another enzyme, **protein kinase.**
6. Protein kinase catalyzes the addition of phosphate groups to other enzymes. This process activates some enzymes and deactivates others. Because enzymes control chemical reactions in the cell, cyclic AMP can have profound effects on cellular function.

Cyclic AMP is not the only second messenger in body cells. In some cells, calcium ions serve as an intracellular second messenger. In other cells, cyclic GMP (guanosine monophosphate) is a second messenger. Both function independently of cAMP.

Steroid Hormones Enter the Cytoplasm of Their Target Cells and Bind to Receptors that Diffuse into the Nucleus, Activating Genes Directly. Steroid hormones stimulate intracellular change by acting directly on the genes, rather than via enzymes. The process is known as the **two-step mechanism.** Here's how it works:

1. Because of their lipid solubility, steroid hormones readily penetrate the plasma membrane of the target cell (▻ Figure 20–7).
2. Inside the cell, steroid hormones bind to protein receptors, typically found in the cytoplasm but sometimes found in the nucleus. Like their counterparts in the plasma membrane, these receptors recognize and bind to one type of steroid hormone, thus conferring specificity.
3. When a steroid binds to a cytoplasmic receptor, it causes the protein to change its shape slightly. This change activates the receptor-hormone complex. The activated receptor-hormone complex enters the nucleus.
4. The activated complex binds to the chromatin at a

▻ **FIGURE 20–5 The Second Messenger Theory** Protein and polypeptide hormones bind to a membrane-bound receptor, thus stimulating the production of intracellular cyclic AMP by adenylate cyclase. Cyclic AMP activates protein kinase, which adds phosphates to intracellular enzymes, activating some and deactivating others. Phosphodiesterase (PDE) converts cyclic AMP into AMP. Adapted with permission from George A. Hedge, Howard D. Colby, and Robert L. Goodman, *Clinical Endocrine Physiology* (Philadelphia: W. B. Saunders Company, 1987), Figure 1–8, p. 18.

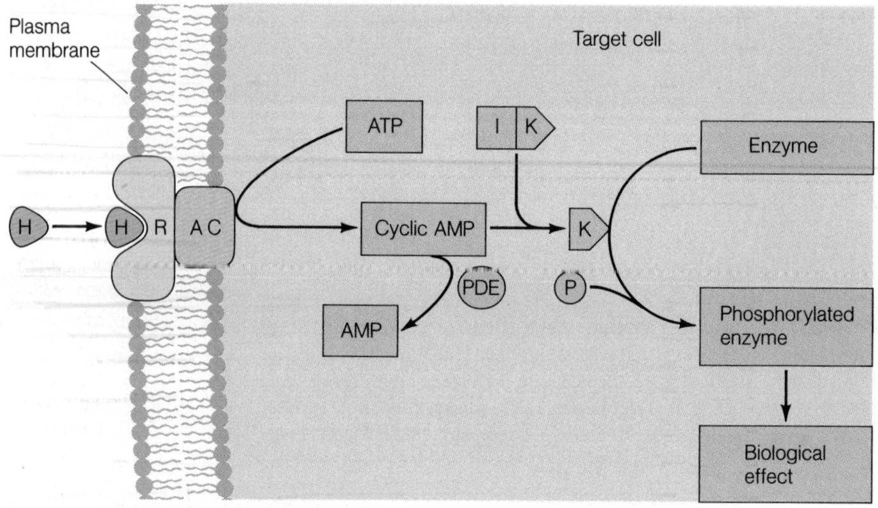

H = Free hydrophilic hormone
R = Surface receptor
AC = Adenylate cyclase
ATP = Adenosine triphosphate
AMP = Adenosine monophosphate
PDE = Phosphodiesterase
IK = Inactive protein kinase
K = Active protein kinase
P = Phosphate

specific site of attachment on the DNA known as an **acceptor site.**[1]

5. The binding of the receptor-steroid complex to the chromatin turns on certain genes, resulting in the transcription of DNA—that is, the production of mRNA (Chapter 9).
6. Messenger RNA, in turn, provides a template for the production of structural proteins and/or enzymes. Thus, steroid hormones affect the structure and function of target cells.

The second-messenger and two-step mechanisms will require a bit of study to master, but as you work through the details, take a moment to marvel at these remarkable systems of control. You may also want to pause to recall that they developed during evolution, a process of trial and error that has produced so many fascinating structures and functions.

[1] Thyroid hormones also effect intracellular change through the two-step mechanism, but they bind directly to DNA.

The Endocrine System and Nervous System Are Both Control Systems, but They Exhibit Marked Differences

Now is also a good time to reflect on the similarities and differences between the body's two major control systems, the endocrine and nervous systems. Although similar in some respects—they both send signals to cells that help regulate their function and help coordinate body function—these systems differ in several key respects. First, the nervous system often elicits very rapid, short-lived responses (for example, a muscle contraction). Although there are exceptions to this rule, the endocrine system generally brings about much slower, longer lasting responses (a change in body temperature). The differences can be attributed to the

▻ **FIGURE 20–6 Cyclic AMP, the Second Messenger**
(*a*) Cyclic AMP compared with (*b*) ATP.

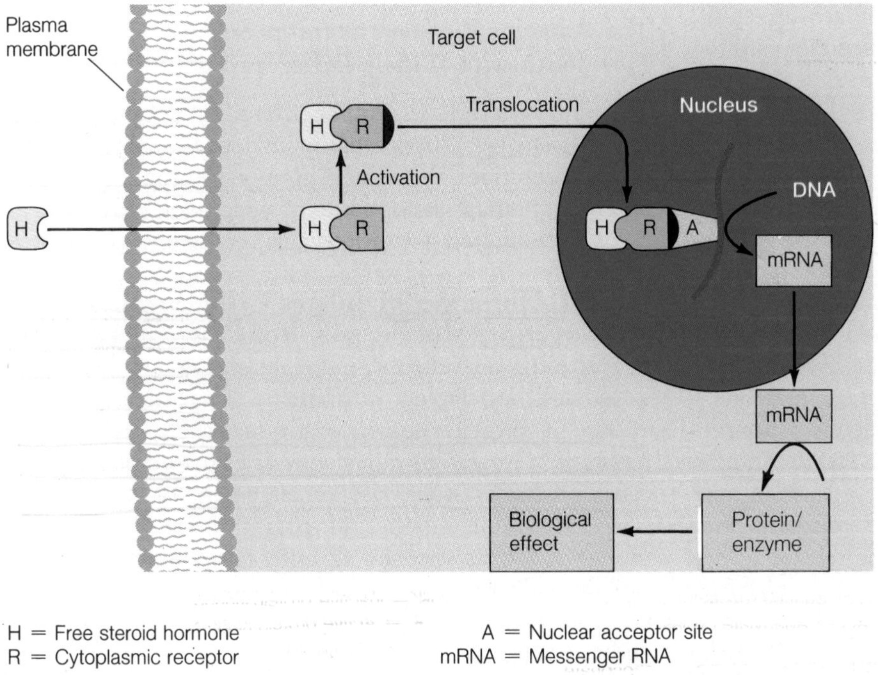

▻ **FIGURE 20–7 The Two-Step Mechanism** Steroid hormones bind to specific receptors in the cell, usually in the cytoplasm. The receptor-steroid complex then binds to the acceptor sites on the nuclear DNA, activating the genes. Adapted with permission from George A. Hedge, Howard D. Colby, and Robert L. Goodman, *Clinical Endocrine Physiology* (Philadelphia: W. B. Saunders Company, 1987), Figure 1–9, p. 20.

type of signal found in each system. In the nervous system, messages are conveyed by bioelectric impulses that travel along the nerves of the body. In the endocrine system, messages are chemical in nature and are transmitted through the bloodstream.

Despite these differences, the endocrine and nervous systems work in conjunction to ensure homeostasis. And, as you will soon see, these systems sometimes work in concert.

THE PITUITARY AND HYPOTHALAMUS

Attached to the underside of the brain by a thin stalk is the **pituitary gland** (▻ Figure 20–8). About the size of a pea, the pituitary lies in a depression in the base of the skull, the **sella turcica** (literally, "Turkish saddle").

The pituitary gland is one of the most complex of all the endocrine organs. It is divided into two major parts: the anterior pituitary and the posterior pituitary. Together, they

▻ **FIGURE 20–8 The Pituitary Gland**
(*a*) A cross section of the brain showing the location of the pituitary and hypothalamus. (*b*) The structure of the pituitary gland.

(*c*) Releasing and inhibiting hormones travel via the portal system from the hypothalamus to the anterior pituitary, where they affect hormone secretion.

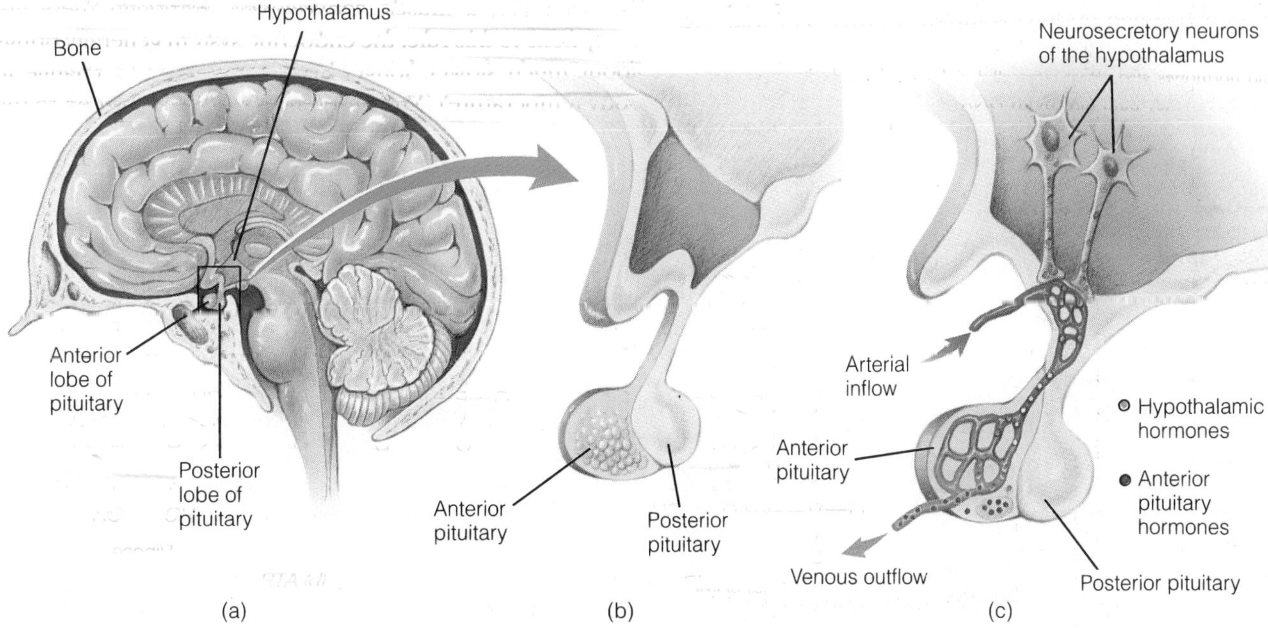

secrete a large number and variety of hormones, affecting a great many of the body's functions. The anterior pituitary, for example, produces seven protein and polypeptide hormones, six of which are discussed in this chapter (Table 20–1).

The release of the hormones from the anterior pituitary is controlled by a region of the brain, the **hypothalamus,** lying just above the pituitary gland. The hypothalamus contains receptors for a variety of blood-borne chemical substances. These receptors monitor blood levels of hormones, nutrients, and ions. When activated, the receptors stimulate specialized nerve cells within the hypothalamus. These nerve cells are called **neurosecretory neurons**—so named because they synthesize and secrete hormones, which act on the anterior pituitary (Figure 20–8). Some of the hypothalamic hormones stimulate the release of anterior pituitary hormones and are called **releasing hormones,** designated RH. Others inhibit the release of hormones from the anterior pituitary and are called **inhibiting hormones,** designated IH.

The releasing and inhibiting hormones travel down the axons of the neurosecretory cells, which terminate in the lower part of the hypothalamus, just above the pituitary gland. These hormones are then stored in the axon terminals of the neurosecretory cells. When released from the axon terminals, they diffuse into nearby capillaries. As shown in Figure 20–8, these capillaries drain into a series of veins in the stalk of the pituitary. The veins, in turn, empty into a capillary network in the anterior pituitary. Thus, hypothalamic hormones are transported directly to their target cells. This unusual arrangement of blood vessels in which a capillary bed drains to a vein, which drains into another capillary bed, is called a **portal system.**[2]

[2] Another portal system is associated with the liver and digestive tract.

The Anterior Pituitary Secretes Seven Hormones with Widely Different Functions

The pituitary produces a number of other hormones that in sickness and health profoundly influence our bodies. This section describes the major hormones produced by the anterior pituitary gland, some of which you may have already encountered in previous chapters or in class lectures.

Growth Hormone Stimulates Cell Growth, Primarily Targeting Muscle and Bone. We all know that people differ considerably in height and body build. These differences are largely attributable to **growth hormone (GH),** a protein hormone produced by the anterior pituitary.[3] Growth hormone stimulates growth in the body, causing cellular *hypertrophy* (enlargement) and *hyperplasia* (increase in the number of cells through division). Although it affects virtually all body cells, growth hormone acts primarily on bone and muscle. In muscle, it stimulates the uptake of amino acids and protein synthesis. In bone and cartilage, it acts through an intermediary. That is, stimulates the production of several small proteins by the liver. Known as **somatomedins,** these proteins stimulate cartilage and bone to grow. Thus, the more growth hormone produced during the growth phase of an individual, the taller and heftier he or she will be. In men, body growth is also stimulated by testosterone, an anabolic steroid produced by the testes. Testosterone stimulates bone and muscle growth, thus explaining why men are generally taller and more massive than women.

Growth hormone secretion undergoes a diurnal (daily) cycle. Like cortisol, the highest blood levels are present during sleep (▻ Figure 20–9). During the day, the level in

[3] Growth hormone levels are determined by the genes.

TABLE 20–1 Hormones Secreted by the Pituitary Gland

HORMONE	FUNCTION
Anterior pituitary	
Growth hormone (GH)	Stimulates cell growth. Primary targets are muscle and bone, where GH stimulates amino acid uptake and protein synthesis. It also stimulates fat breakdown in the body.
Thyroid-stimulating hormone (TSH)	Stimulates release of thyroxine and triiodothyronine
Adrenocorticotropic hormone (ACTH)	Stimulates secretion of hormones by the adrenal cortex, especially glucocorticoids
Gonadotropins (FSH and LH)	Stimulate gamete production and hormone production by the gonads
Prolactin	Stimulates milk production by the breast
Melanocyte-stimulating hormone (MSH)	Function in humans is unknown.
Posterior pituitary	
Antidiuretic hormone (ADH)	Stimulates water reabsorption by nephrons of the kidney
Oxytocin	Stimulates ejection of milk from breasts and uterine contractions during birth

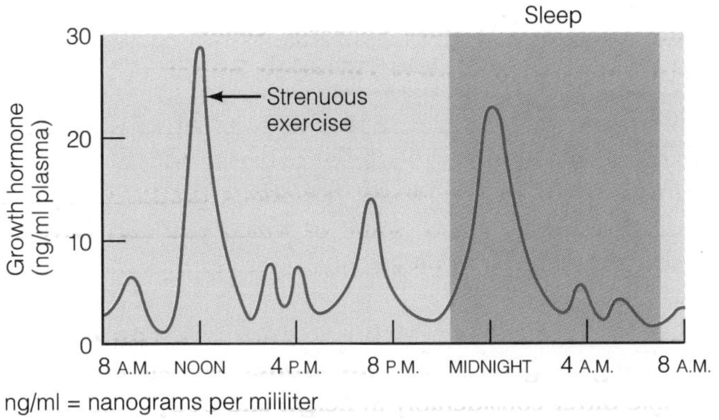

▻ **FIGURE 20–9 Growth Hormone Secretion in an Adult** Growth hormone is released during exercise and helps promote muscle growth. It is also released at night in a circadian rhythm.

the blood declines. It is no wonder that sleep is so important to a growing child. Growth hormone secretion declines gradually as we age.

Like the secretion of many other hormones of the anterior pituitary, that of growth hormone is controlled by a releasing hormone (GH-RH) *and* an inhibiting hormone (GH-IH), both produced by the hypothalamus. Levels of growth hormone in the blood participate in a negative feedback loop. Growth hormone release can also be stimulated directly through the nervous system (hypothalamus). Stress and moderate exercise, for example, stimulate the hypothalamus to release GH-RH (Figure 20–9).

Deficiencies in growth hormone can result in dramatic changes in body shape and size, depending on when the deficiency occurs. Undersecretion, or **hyposecretion,** occurring during the growth phase of a child is one cause of stunted growth (dwarfism) (▻ Figure 20–10a). Oversecretion, or **hypersecretion,** results in **giantism** if the excess occurs during the growth phase (Figure 20–10b). If the pituitary begins producing excess growth hormone after growth is complete, the result is a relatively rare disease called **acromegaly.** Facial features become coarse, and hands and feet continue to grow throughout adulthood (▻ Figure 20–11). Growth of the vertebrae results in a hunched back. Organs also grow in response to continued secretion of growth hormone. Thus, the tongue, kidneys, and liver often become quite enlarged in patients with acromegaly.

Thyroid-Stimulating Hormone Stimulates the Thyroid Gland to Produce Thyroxine. Thyroid-stimulating hormone (TSH) is a protein hormone produced by the anterior pituitary; its release is controlled by **TSH-RH (TSH-releasing hormone)** from the hypothalamus. TSH-RH secretion is stimulated by cold and stress. Its secretion is also regulated by the level of thyroxine in the blood in a classical negative feedback loop. Receptors in the hypothalamus detect the level of circulating thyroxine. When circulating levels of the hormone are low, the hypothalamus releases TSH-RH. When the level of thyroxine increases, TSH-RH secretion declines (▻ Figure 20–12).

TSH-RH stimulates the production and release of TSH in the anterior pituitary. Thyroid-stimulating hormone, in turn, then travels in the blood to the thyroid gland, located in the neck (see Figure 20–1), where it stimulates the production and release of thyroxine and triiodothyronine.

These hormones enter the bloodstream and affect a great many cells. One of their chief functions is to stimulate the catabolism (breakdown) of glucose by body cells. Because glucose catabolism produces energy and heat, in-

▻ **FIGURE 20–10 Disorders of Growth Hormone Secretion** (*a*) Pituitary dwarves. (*b*) Pituitary giant.

(a)

(b)

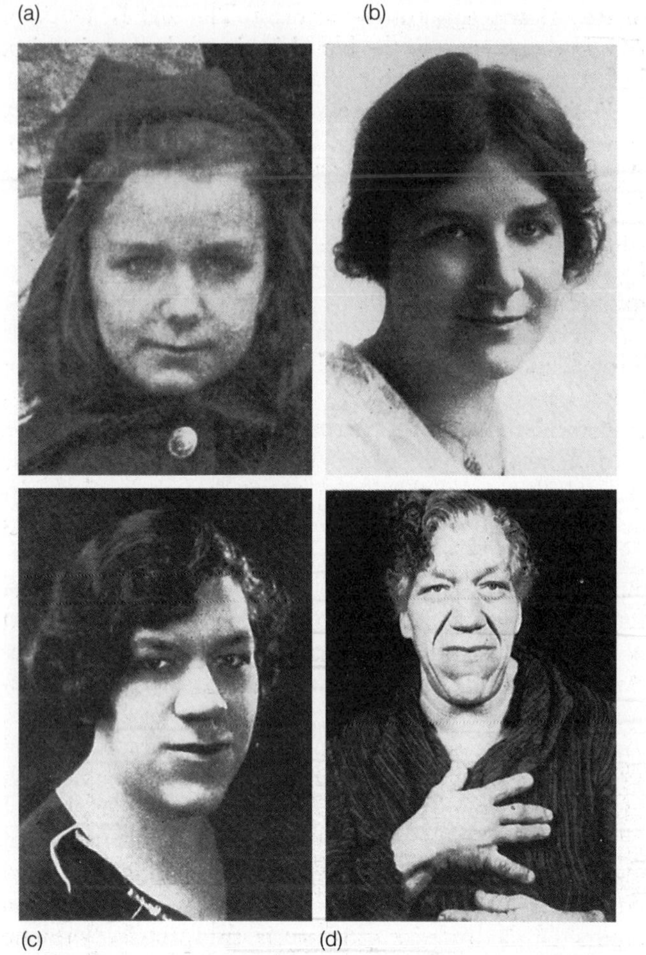

▻ **FIGURE 20–11 Acromegaly** Hypersecretion of growth hormone in adults results in a gradual thickening of the bone, which is especially noticeable in the face, hands, and feet. There is no sign of the disorder at age 9 (*a*) or age 16 (*b*). Symptoms are evident at age 33 (*c*) and age 52 (*d*).

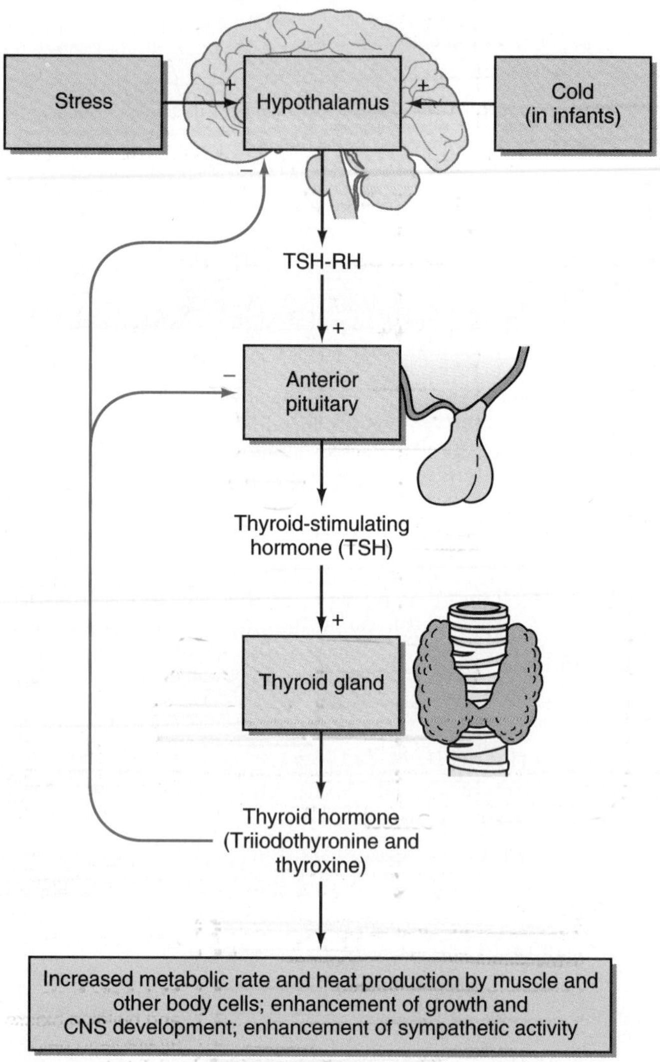

▻ **FIGURE 20–12 Negative Feedback Control of TSH Secretion** Triiodothyronine (T_3) and thyroxine regulate hypothalamic and pituitary activity. Other factors, such as stress and cold, influence the release of TSH via the hypothalamus. (A + denotes stimulation; a − denotes inhibition.)

creased levels of these hormones also raise body temperature. Interestingly, in Grave's disease, from which Bush suffers, TSH levels are normal. For reasons not well understood, the thyroid gland produces excess thyroxine without the urging of the pituitary.

Adrenocorticotropic Hormone Stimulates the Release of Hormones from the Adrenal Cortex. A relative of yours gradually loses weight and complains of chronic fatigue and weakness. Her skin grows progressively darker, and she suffers occasional bouts of diarrhea, constipation, or mild indigestion. A blood test reveals that she is suffering from Addison's disease, which is believed to be caused by an autoimmune reaction against the **adrenal cortex,** the outer layer of the adrenal gland (see Figure 20–1). A few months after she receives steroid tablets, which replace the lost hormones, her symptoms vanish.

Like many other endocrine glands, the adrenal cortex is under control of the anterior pituitary. More specifically, it is under the control of **adrenocorticotropic hormone (ACTH),** a polypeptide hormone released by the anterior pituitary. In response to ACTH, the adrenal cortex produces and secretes a group of steroid hormones known as the **glucocorticoids.** Glucocorticoids (like the hormone cortisol) increase blood glucose levels, thus helping maintain homeostasis (more on this later).

ACTH release is under the control of the hypothalamic releasing hormone, ACTH-RH. ACTH-RH secretion is controlled by at least three factors. The first is the level of circulating glucocorticoids, which participate in a negative feedback loop (▻ Figure 20–13). As the level of glucocorticoid decreases, ACTH-RH secretion climbs. This rise causes an increase in the release of ACTH by the pituitary.

ACTH-RH secretion is also controlled by stress, acting through the nervous system. A stress-stimulated increase in

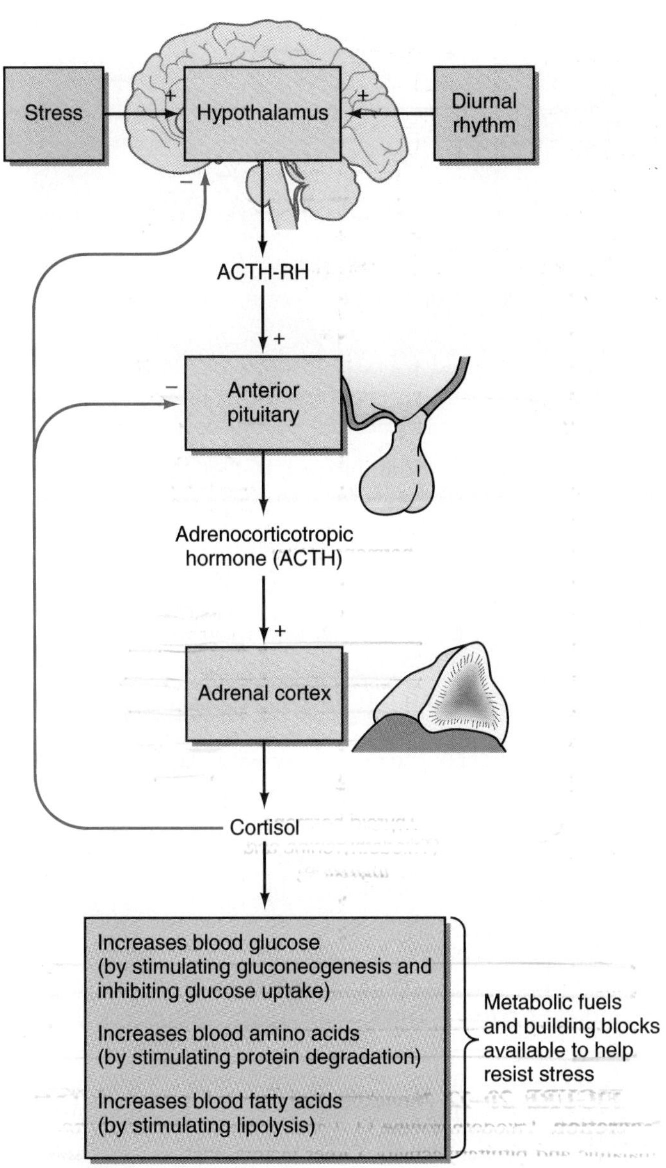

▻ FIGURE 20–13 Feedback Control of ACTH Cortisol regulates hypothalamic and pituitary activity, but stress and the biological clock also influence the release of ACTH-RH.

ACTH-RH results in a rise in ACTH release, which, in turn, stimulates an increase in the release of glucocorticoids by the adrenal cortex. Glucocorticoids increase blood glucose levels, and this mechanism is an adaptation that ensures additional energy for cells (especially muscle) when the body is under stress.

ACTH secretion is also affected by the light-dark cycle through a biological clock in a region of the brain that monitors day length and controls ACTH-RH secretion. At night, the clock overrides the negative feedback mechanism involving ACTH-RH and glucocorticoids, which serves to keep levels fairly constant. The biological clock increases ACTH-RH secretion while you sleep. The increased production of ACTH-RH results in elevated levels of ACTH and also adrenal steroids. The highest levels of ACTH are present in the blood in the early morning, just before you awaken.

Besides producing glucocorticoids, the adrenal cortex synthesizes and releases sex steroids and aldosterone. Aldosterone was described in Chapter 16 because it plays an important role in regulating kidney function (sodium reabsorption) and blood volume. The sex steroids produced by the adrenals are released in seemingly insignificant quantities.

The Gonadotropins Are Hormones that Stimulate the Ovaries and Testes. Reproduction in both males and females is under control of the anterior pituitary. The anterior pituitary produces two hormones that affect the gonads. Known as **gonadotropins,** these hormones are discussed in detail in the next chapter. The two gonadotropins are follicle-stimulating hormone and luteinizing hormone. **Follicle-stimulating hormone (FSH)** promotes gamete formation in men and women. **Luteinizing hormone (LH)** is a trophic hormone that stimulates gonadal hormone production. In men, LH stimulates the production of testosterone, the male sex steroid, by the testes. In women, LH stimulates progesterone secretion by the ovaries. Both FSH and LH are under the control of **gonadotropin releasing hormone,** produced and secreted by the hypothalamus.

Prolactin Stimulates Milk Production in the Breasts of Women. One of the most intriguing hormones of the anterior pituitary is the protein hormone **prolactin.** Prolactin performs many functions in vertebrates. In birds, for example, it stimulates migratory behavior. In fish, it helps regulate electrolyte balance. In women, prolactin is secreted by the anterior pituitary at the end of pregnancy and stimulates milk production by the mammary glands. Suckling prolongs the release of prolactin for at least 3 to 12 months, sometimes much longer, thus maintaining milk production necessary for breast feeding.

Prolactin secretion is controlled by a **neuroendocrine reflex,** a reflex involving both the nervous and endocrine systems. As ▻ Figure 20–14 shows, suckling stimulates sensory fibers in the breast. Nerve impulses travel to the hypothalamus via sensory neurons. In the hypothalamus, these impulses stimulate the release of prolactin releasing hormone (PRH). This hormone, in turn, travels to the anterior pituitary in the bloodstream, where it stimulates the secretion of prolactin. Prolactin, in turn, stimulates milk production by the breasts.

The neuroendocrine reflex is the basis of commercial milk production. When a dairy cow gives birth, it produces milk to feed its calf. On dairy farms, however, calves are often weaned (separated from their mothers) fairly early. Left alone, the mothers would stop producing milk. Farmers prolong milk production by milking their cows, either manually or by machine (▻ Figure 20–15). Milking machines and hand milking stimulate the neuroendocrine reflex, continuing the production of milk.

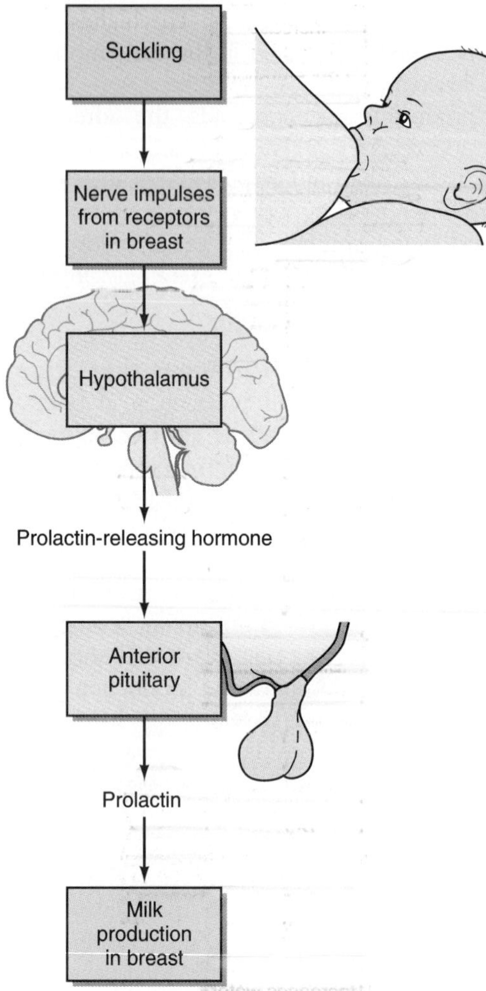

▻ **FIGURE 20–14 Neuroendocrine Reflex and Prolactin Secretion** Suckling stimulates prolactin release by the anterior pituitary.

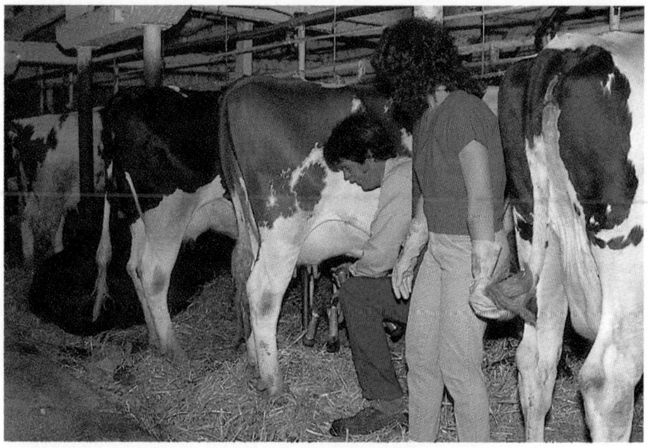

▻ **FIGURE 20–15 Stimulating a Reflex** Milking by hand or machine prolongs milk production long after a calf has been weaned from its mother.

The Posterior Pituitary Secretes Two Hormones

Have you ever wondered why alcoholic beverages make a person urinate so much or what causes milk to flow from a woman's breast? Have you ever wondered why a woman's uterus contracts when she delivers a baby? The answer to these puzzles is rather simple: hormones. More specifically, these phenomena can be attributed to two hormones from the posterior pituitary. Like the hypothalamus, the **posterior pituitary** is a neuroendocrine gland—that is, a gland made of neural tissue that produces hormones. Derived from brain tissue during embryonic development, the posterior pituitary remains connected to the brain throughout life.

The posterior pituitary consists of the axons and terminal ends of neurosecretory cells whose cell bodies are located in the hypothalamus (▻ Figure 20–16). The neurosecretory cells produce two hormones, antidiuretic hormone and oxytocin, each of which consists of nine amino acids. Oxytocin and antidiuretic hormone are synthesized in the cell bodies of the neurosecretory cells and travel down the axons of these cells into the posterior pituitary. The hormones are then stored in the axon terminals of the cells and released into the surrounding capillaries.

Antidiuretic Hormone Increases Water Absorption in the Kidney. Antidiuretic hormone(ADH) regulates water balance in humans and, as you will soon see, is noticeably affected by alcohol (see also Chapter 16). ADH travels in the bloodstream to the kidney and there increases water absorption by the distal convoluted tubules and collecting tubules (▻ Figure 20–17). As a result, water reenters the bloodstream, increasing blood volume and maintaining the normal osmotic concentration of the blood.

ADH secretion is controlled by **osmoreceptors** in the hypothalamus. These receptors monitor the concentration of dissolved substances (especially sodium ions) in the blood. When the concentration exceeds the normal level—for example, during dehydration—the osmoreceptors activate the ADH-producing cells, causing them to release ADH into the bloodstream. ADH, in turn, causes water to be reabsorbed by the kidney, diluting the blood. When the osmotic concentration of the blood approaches homeostatic levels, ADH secretion ceases (Figure 20–17).

As you might have guessed, the secretion of ADH is inhibited by ethanol in alcoholic beverages. The reduction in ADH decreases water reabsorption in the kidneys, thus increasing urine production. The increase in urinary output, in turn, can lead to dehydration—the dry mouth and intense thirst a person may experience as part of a hangover.

ADH secretion may also decline as a result of trauma. A blow to the head, for example, can damage the hypothalamus and cause a sharp decline in ADH output. In the absence of ADH, the kidneys produce several gallons of urine a day. To keep up with the water loss and to prevent dehydration, an individual must drink enormous quantities of water. This condition is known as **diabetes insipidus** (Chapter 16).

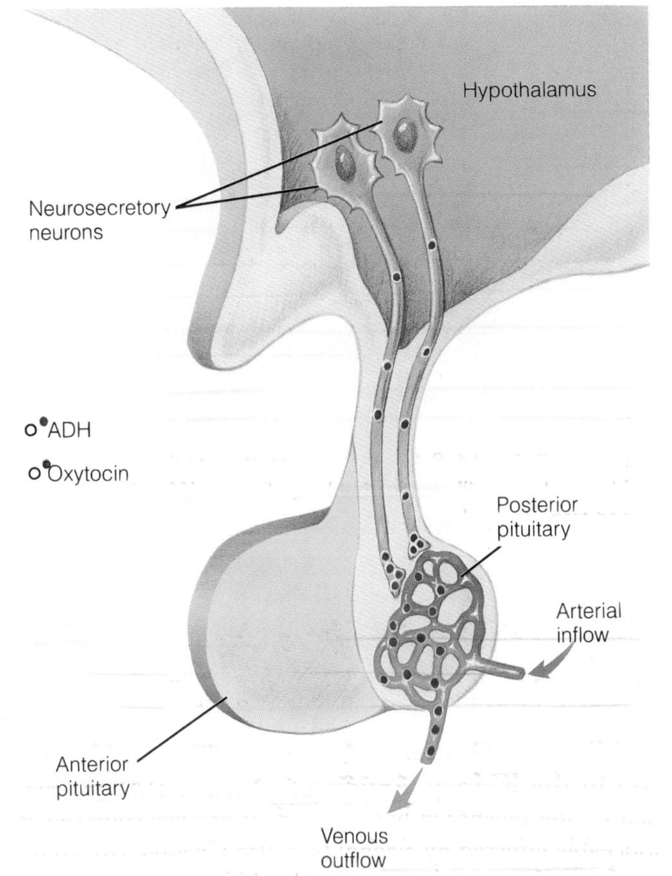

▻ **FIGURE 20–16 The Posterior Pituitary** Neurosecretory neurons that produce oxytocin and ADH originate in the hypothalamus and terminate in the posterior pituitary. Hormones are produced in the cell bodies of the neurons and are stored and released into the bloodstream in the posterior pituitary.

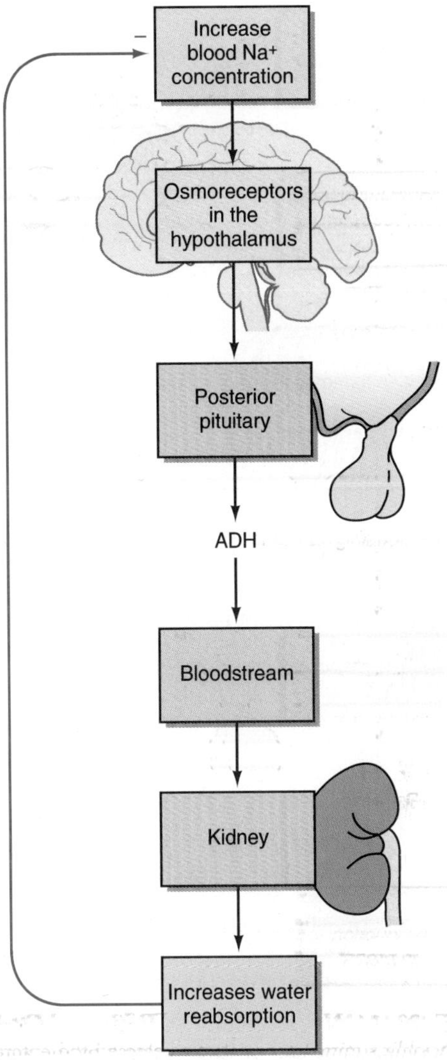

▻ **FIGURE 20–17 Role of ADH in Regulating Fluid Levels** ADH secretion is stimulated by an increase in the blood concentration of sodium caused by dehydration. ADH increases water reabsorption in the kidney, thus eliminating the stimulus for ADH secretion.

Because of its ability to increase arterial blood pressure, ADH is sometimes referred to as **vasopressin.** Under most conditions, however, it is doubtful that enough ADH is secreted to increase blood pressure. ADH secretion in levels sufficient to exert a vasopressive effect occurs when an individual suffers an extreme loss of blood—for example, in an automobile accident. When released in massive quantities, ADH constricts the walls of severed arteries, reducing blood loss. Clotting factors in the blood also help seal off damaged vessel walls. ADH also causes vasoconstriction throughout the cardiovascular system, which helps maintain blood pressure. This process is important because a large decrease in blood pressure can reduce the flow of blood to body cells and cause death.

Oxytocin Facilitates Birth and Stimulates Milk Let-down. Oxytocin from the posterior pituitary is a dual-purpose hormone. It stimulates the contraction of the smooth muscle of the uterus and stimulates contraction of the smooth-muscle-like cells around the glands in the breast.

Oxytocin release, like that of prolactin, is controlled by a neuroendocrine reflex. During childbirth, for example, the walls of the uterus and cervix stretch. Impulses from the stretch receptors in the uterus travel to a region of the hypothalamus containing the neurosecretory cells that produce oxytocin. Impulses from the stretch receptors cause the neurosecretory cells to release oxytocin into the blood of the posterior pituitary (where the neurosecretory cells terminate). Oxytocin travels in the blood to the uterus, where it stimulates smooth muscle contraction, aiding in the expulsion of the baby.

Oxytocin release is stimulated by another neuroendocrine reflex, which is activated by suckling (▻ Figure 20–18). Sensory fibers in the breast conduct bioelectric impulses to the hypothalamus, triggering the release of oxytocin. Oxytocin travels in the blood to the breast, where it stimulates **milk let-down**—the ejection of milk from the glands soon after suckling begins. Milking machines and hand milking stimulate the reflex in dairy cattle. As you may recall, prolactin from the anterior pituitary is also

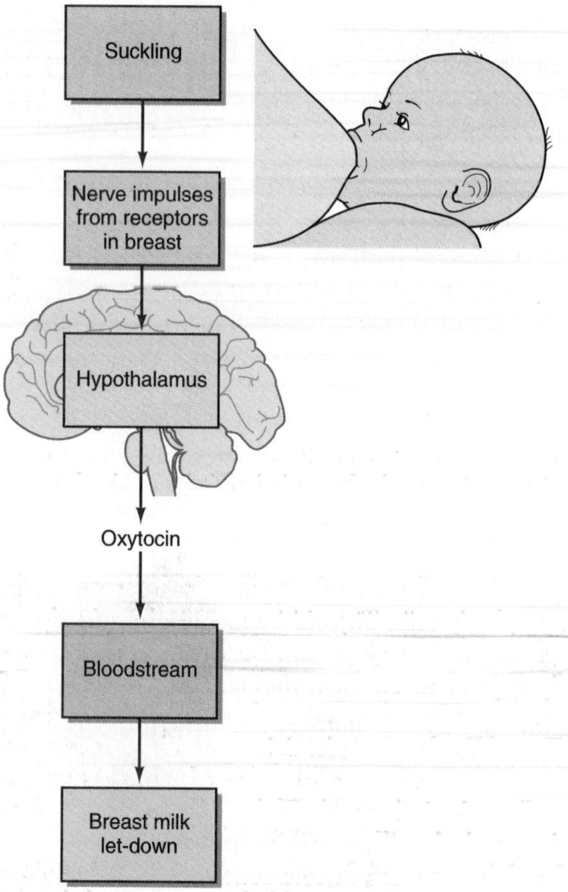

▻ **FIGURE 20–18 Milk Let-Down Reflex** Suckling stimulates oxytocin release by the posterior pituitary, causing the expulsion of milk from the glands of the breast.

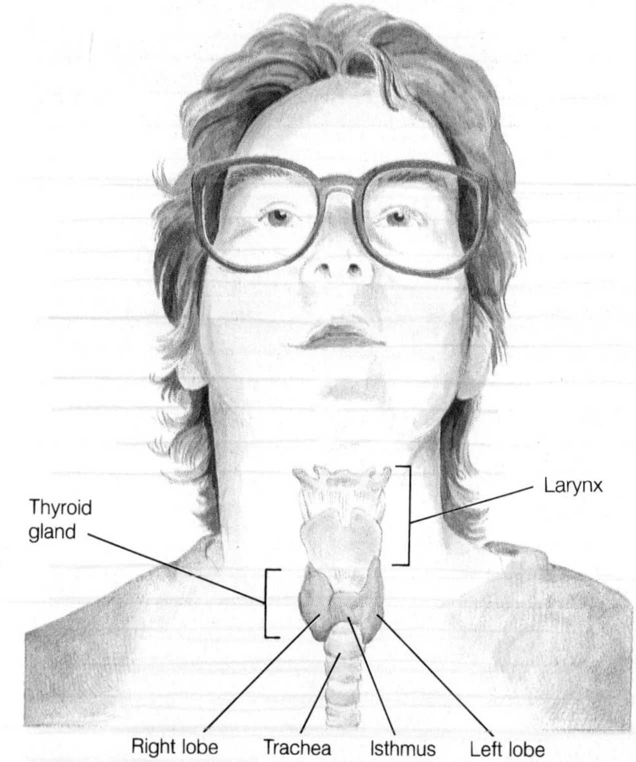

▻ **FIGURE 20–19 Location of the Thyroid Gland**

released during suckling. This hormone stimulates milk production.

THE THYROID GLAND

On a 50°F day in the fall, you bundle up in a sweater but still feel cold. However, a day with the same temperature in the spring feels warm, even if you don't wear a sweater. Why the difference?

The answer lies in the thyroid gland and two of its hormones. The **thyroid gland** is a U- or H-shaped gland (it varies from one person to the next) located in the neck (▻ Figure 20–19). The thyroid gland produces three hormones: (1) thyroxine (tetraiodothyronine, or T_4), (2) a chemically similar compound, triiodothyronine (T_3), and (3) calcitonin. The first two are involved in controlling metabolism and heat production; the last helps regulate blood levels of calcium.

As ▻ Figure 20–20a shows, the thyroid gland consists of large, spherical structures known as **follicles.** Each follicle consists of a central region containing a gel-like material called **thyroglobulin,** which is surrounded by a single layer of cuboidal **follicle cells.** Thyroglobulin is a glycoprotein and is produced by the follicle cells. Thyroxine and triiodothyronine are derived from thyroglobulin. How?

Thyroglobulin molecules contain certain amino acids to which iodine ions are attached. Thyroxine and triiodothyronine are produced when two of these iodinated amino acids are covalently linked. To make T_3 and T_4, the thyroid requires large quantities of iodide, I^-. This ion must be actively transported into the cell, where it is added to certain amino acids in the thyroglobulin molecules.

Thyroglobulin is the storage form of T_3 and T_4. When the blood levels of these hormones fall, TSH is secreted by the anterior pituitary. TSH, in turn, stimulates the follicle cells to engulf a portion of the thyroglobulin reserve. Lysosomes in the cytoplasm of the follicle cells break down the ingested material, releasing T_3 and T_4. These hormones then diffuse out of the cell into the bloodstream.

Because the thyroid requires iodide, persistent dietary deficiencies of this nutrient can alter the thyroid's function, resulting in a condition known as **goiter.** Goiter is an enlargement of the thyroid gland (▻ Figure 20–21). A shortage of iodide in the diet results in a decline in the levels of thyroxine and triiodothyronine in the blood. The hypothalamus responds by producing more TSH. TSH causes the thyroid to increase thyroglobulin production. Without sufficient iodide, however, the thyroid gland cannot produce thyroxine and triiodothyronine. As a result, the thyroid gland continues to produce thyroglobulin. This causes the gland to hypertrophy (hy-per-truh-phee), or enlarge, sometimes forming softball-sized enlargements on the neck.

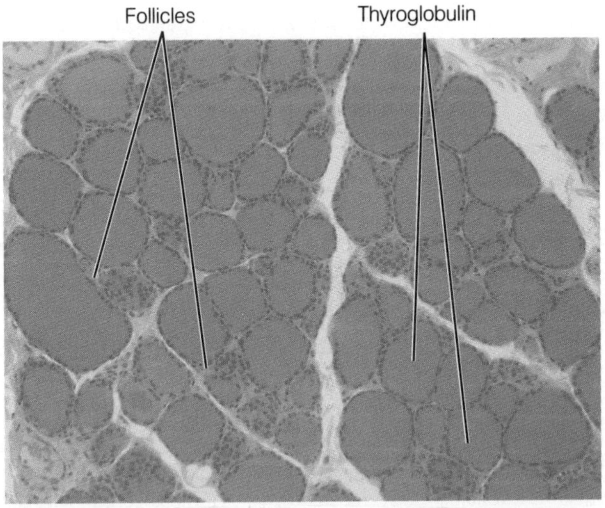

(a)

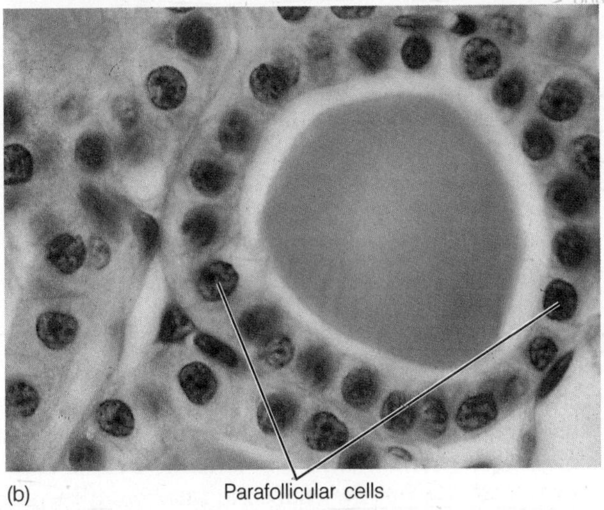

(b)

▻ **FIGURE 20–20 Histologic Structure of the Thyroid Gland** (*a*) The thyroid follicles produce thyroxine and triiodothyronine. They are stored in the colloidal material called thyroglobulin. (*b*) Enlargement showing calcitonin-producing cells.

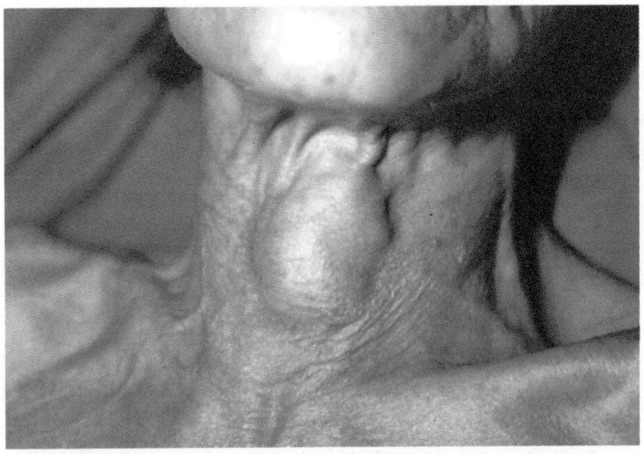

▻ **FIGURE 20–21 Goiter** An enlargement of the thyroid gland most often results from a lack of iodine in the diet.

Goiter was once common in areas of Europe and the United States where iodine had been leached out of the soil by rain—for example, near the Great Lakes. Crops grown in iodine-poor soils failed to provide sufficient quantities. Iodine was first added to table salt, in fact, to help prevent the problem. Few people in developed countries develop goiter now because they eat iodized salt or receive sufficient quantities of iodine in their food or through mineral supplements.

Thyroxine and Triiodothyronine Accelerate the Breakdown of Glucose and Stimulate Growth and Development

As ▻ Figure 20–22 illustrates, thyroxine and triiodothyronine are virtually identical. As a result, the two hormones have nearly identical functions. Both hormones, for example, accelerate the rate of mitochondrial glucose catabolism in most body cells. Thyroid hormones also stimulate cellular growth and development. Bones and muscles are especially dependent on them during the growth phase. Even normal reproduction requires these hormones. A deficiency of T_3 and T_4, for example, delays sexual maturation in both sexes. In children, depressed thyroid output stunts mental as well as physical growth. If the deficiency is not detected and treated, the effects will be irreversible.

In adults, reduced thyroid activity, or **hypothyroidism,** is less severe and is fully reversible because growth has been completed. Hypothyroidism, however, decreases the metabolic rate, making a person feel cold much of the time. People suffering from hypothyroidism also feel tired and worn out. Even simple mental tasks become difficult. Their heart rate may slow to 50 beats per minute. Hypothyroidism is treated by pills containing artificially produced thyroid hormone.

Excess thyroid activity, **hyperthyroidism,** in adults results in elevated metabolism, excessive sweating (due to overheating), and weight loss, despite increased food intake. The increase in thyroid hormone levels results in increased mental activity, resulting in nervousness and anxiety. People suffering from hyperthyroidism often find it difficult to sleep. Their heart rate may accelerate, as in the case of President Bush, and individuals may lose their sensitivity to cold. Some people suffering from hyperthyroidism exhibit a condition called **exophthalmos,** or bulging eyes. The eyes may protrude so much that the eyelids cannot close completely. Exophthalmos may cause double vision or blurred vision.

Hyperthyroidism is treated in a number of ways. Patients may be given antithyroid medications, drugs that antagonize the effects of thyroid hormones. Surgery may be required to remove part or all of the gland if it has become cancerous. The most common treatment for hyperthyroidism, however, is radioactive iodine. As noted earlier, iodine is concentrated in the thyroid gland. Radioactive iodine accumulated in the cells of the thyroid follicle irra-

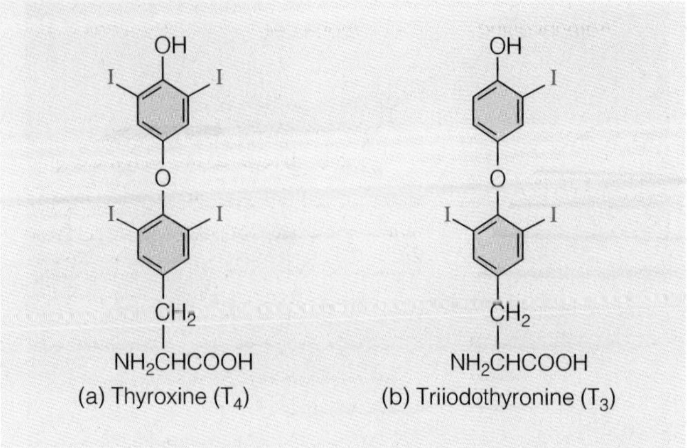

▻ **FIGURE 20–22 Structure of the Thyroid Hormones**
(*a*) Thyroxine; (*b*) triiodothyronine.

diate overactive follicle cells, damaging them and thus reducing their output of thyroid hormones. This procedure, while effective, may lead to other problems (notably cancer) later in life.

Calcitonin Decreases Blood Levels of Calcium

Large, round cells found in the perimeter of the thyroid follicles produce **calcitonin,** or **thyrocalcitonin,** a polypeptide hormone that lowers the blood calcium level (see Figure 20–20b). Calcitonin has three effects: (1) It inhibits osteoclasts, bone-resorbing cells, thus reducing the release of calcium from bone. (2) It stimulates bone-forming cells, osteoblasts, causing calcium to be deposited in bone and thus reducing blood levels. (3) It increases the excretion of calcium (and phosphate) ions by the kidneys. All three effects help lower blood calcium.

Calcitonin is involved in a simple negative feedback loop with calcium ions in the blood. When the calcium ion concentration increases, calcitonin secretion increases. As calcium concentrations fall, calcitonin secretion falls.

THE PARATHYROID GLANDS

The **parathyroid glands** are four small nodules of tissue embedded in the back of the thyroid gland. These glands, once mistaken by anatomists for undeveloped thyroid tissue, are independent endocrine glands. They produce a polypeptide hormone known as **parathyroid hormone,** or **parathormone (PTH).**

As you may recall from Chapter 19, parathyroid hormone secretion is stimulated when calcium levels in the blood drop. PTH quickly reverses the decline and restores blood calcium levels by (1) increasing intestinal absorption of calcium, (2) stimulating bone destruction by osteoclasts, and (3) increasing calcium reabsorption in the kidney.

As illustrated in previous examples, hormones often act on two or more targets to bring about a desired effect. This redundancy no doubt evolved because it increased the level of homeostatic control—in the same way that redundant safety systems in a nuclear power plant provide a greater degree of control in the case of malfunction.

Calcium homeostasis is also influenced by vitamin D, which greatly increases calcium absorption in the intestine when blood levels fall. Interestingly, vitamin D also increases the responsiveness of bone to PTH. In calcium homeostasis, PTH plays a major role, supported by vitamin D.

As in other glands, the parathyroid may malfunction. **Hyperparathyroidism,** excess secretion of parathyroid hormone, is the most common condition. Hyperparathyroidism may result from a tumor of the parathyroid glands, which causes the secretion of excess PTH. Excess PTH results in a loss of calcium from the bones and teeth. It also upsets several metabolic processes, resulting in indigestion and depression. Because bones contain enormous amounts of calcium, most symptoms of hyperparathyroidism (except high blood calcium) do not appear until two to three years after the onset of the disease. Therefore, by the time the disease is discovered, kidney stones may already have formed from calcium, cholesterol, and other substances, and bones may have become more fragile and susceptible to breakage. To prevent further complications, parathyroid tumors must be removed.

THE PANCREAS

A teenage girl comes to her doctor's office. She complains of frequent urination, bladder infections, fatigue, and weakness. Tests show that she is suffering from **diabetes mellitus,** a disorder of the endocrine function of the pancreas. Diabetes mellitus has several causes, but in young people it is generally caused by a lack of insulin output.

Insulin is the glucose-storage hormone; that is, it stimulates the uptake of glucose by body cells. It also stimulates the synthesis of glycogen in liver and muscle cells. Glycogen is the storage form of glucose. Supplies of glycogen in the liver increase immediately after meals, then decline in the period between feedings as the body uses these stores to supply body cells with glucose. Muscles use glycogen supplies principally during exercise.

Insulin is produced by the **pancreas,** a dual-purpose organ located in the abdominal cavity and described in Chapter 11 (▻ Figure 20–23). The head of the pancreas lies in the curve of the duodenum and its tail stretches to the left kidney. Most of the pancreas consists of tiny clumps of cells (acini) that produce digestive enzymes and sodium bicarbonate. Scattered throughout these enzyme-producing cells are small islands of endocrine cells, the **islets of Langerhans** (Figures 20–23 and ▻ 20–24). About 2 million islets are found in the human pancreas, and each islet contains about 200 cells. The islets contain four cell types, two of which will be discussed in this chapter, the alpha cell and the beta cell.

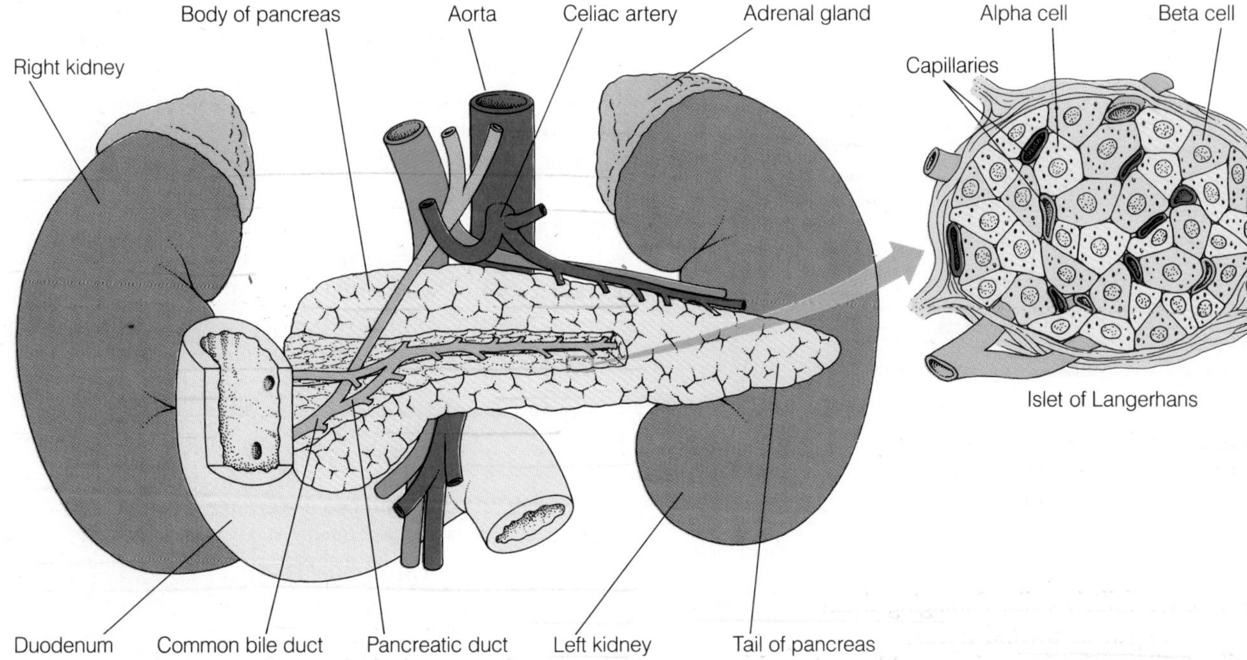

▻ **FIGURE 20–23 The Pancreas** This dual-purpose organ is located in the abdominal cavity. It produces digestive enyzmes, which it releases into the small intestine via the pancreatic duct, and two hormones, which it releases into the bloodstream.

Insulin Is a Glucose-Storage Hormone and Is Produced by the Beta Cells

The **beta cells** produce insulin. This protein hormone is released within minutes after glucose levels in the blood begin to rise. Like many other hormones, insulin plays a major role in homeostasis. It affects a number of cellular processes and a number of different cells. Its principal targets, however, are skeletal muscle cells, liver cells, and fat cells. We will examine a few of its major functions, beginning with those affecting skeletal muscles.

Skeletal muscle cells are virtually impermeable to glucose in the absence of insulin. When insulin is present, however, the transport of glucose into muscle cells increases dramatically. Glucose uptake rises because insulin stimulates facilitated transport. It also increases glycogen synthesis in skeletal muscle cells, which not only helps store glucose for later use but also lowers intracellular concentrations, accelerating diffusion. Insulin also increases the uptake of amino acids by muscle cells and stimulates protein synthesis in them, thus promoting muscle formation.

In contrast, the plasma membrane of liver cells is quite permeable to glucose, so glucose enters with great ease whether or not insulin is present. Nevertheless, insulin still increases the uptake of glucose by liver cells. It does so by stimulating the addition of phosphate groups to glucose molecules that have entered the cytoplasm. This traps glucose in the liver cell, because phosphorylated glucose cannot diffuse through the plasma membrane. Insulin also increases glycogen formation, helping store glucose for times of need (▻ Figure 20–25). Phosphorylation and glycogen synthesis decrease the cytoplasmic concentrations of glucose, thus helping maintain the concentration gradient between the blood and the cytoplasm. This helps ensure a steady influx of glucose into the liver cell.

In fat cells, insulin increases glucose uptake and also stimulates lipid synthesis, thus helping store foodstuffs for times of need.

▻ **FIGURE 20–24 The Islets of Langerhans** Scattered among the acini of the pancreas (which produce digestive enzymes) are small islands of cells, the Islets of Langerhans. They produce two hormones: glucagon, which increases blood glucose, and insulin, which lowers blood glucose.

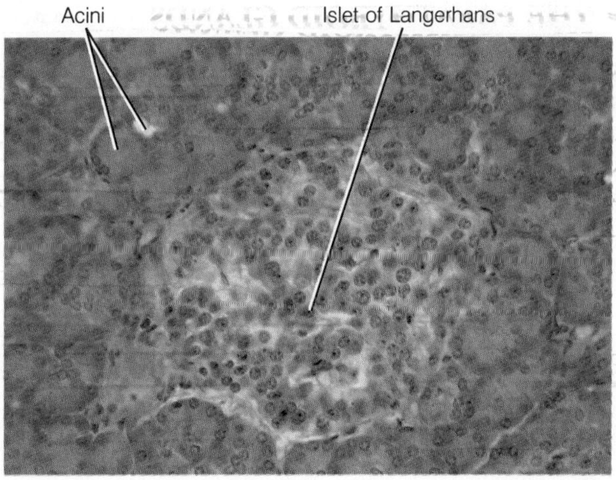

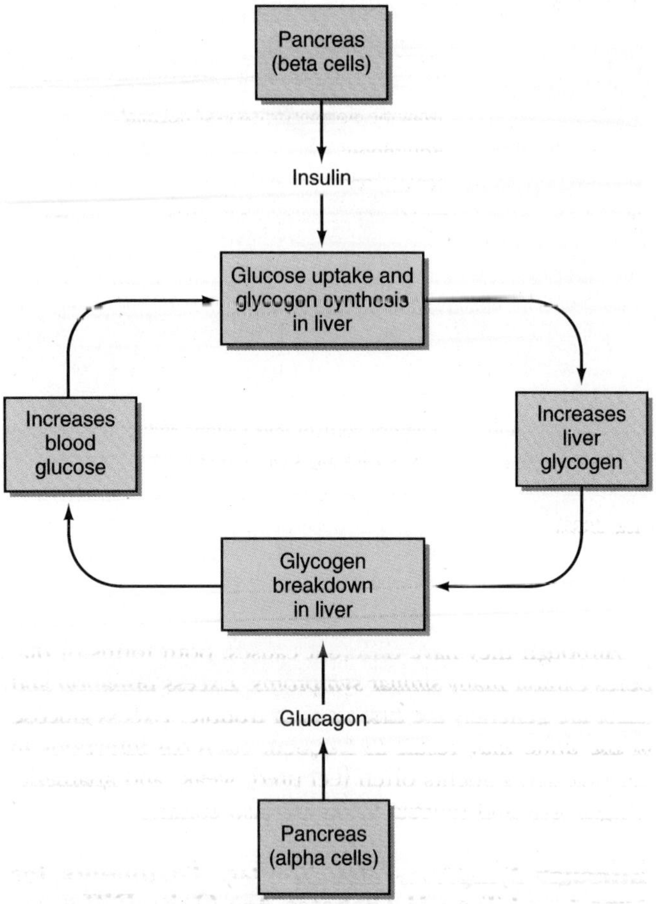

▻ **FIGURE 20–25 Role of Liver and Pancreas in Controlling Blood Glucose Levels** Glucagon and insulin are antagonistic hormones that regulate blood glucose levels through different mechanisms.

Glucagon Increases Blood Levels of Glucose, Thus Opposing the Actions of Insulin

A successful homeostatic mechanism often requires antagonistic—or opposing—controls. (In a car we have an accelerator and brakes to control speed.) In glucose homeostasis, the antagonistic control is provided by **glucagon.** Produced by the **alpha cells** of the Islets of Langerhans, glucagon is a polypeptide hormone that increases blood levels of glucose by stimulating the breakdown of glycogen in liver cells (Figure 20–25). This process, called **glycogenolysis,** (gly-co-je-nol-i-sis) helps maintain proper glucose levels in the blood between meals. One molecule of glucagon causes 100 million molecules of glucose to be released.

Glucagon also elevates serum glucose levels by stimulating a process known as **gluconeogenesis,** the synthesis of glucose from amino acids and glycerol molecules derived from the breakdown of triglycerides. Gluconeogenesis occurs in the liver.

Glucagon secretion, like that of insulin, is controlled principally by glucose concentrations in the blood. When

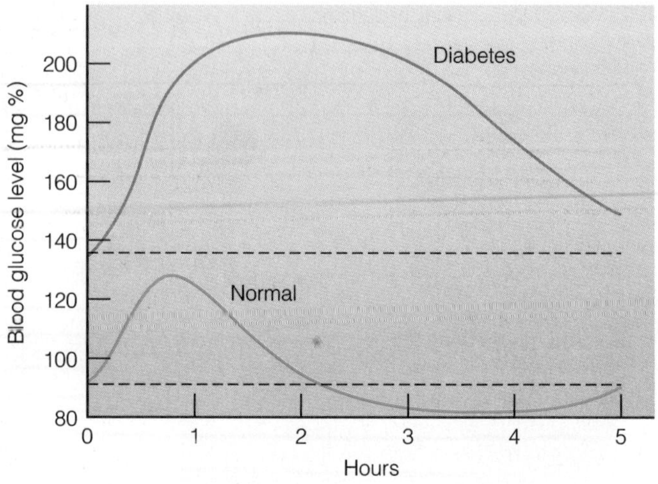

▻ **FIGURE 20–26 Glucose Tolerance Test** In a diabetic, blood glucose levels increase rapidly and remain high after ingestion of glucose. In a normal patient, blood glucose levels rise but are quickly reduced by insulin.

glucose levels fall, the alpha cells release the hormone into the bloodstream. Glucagon release is also stimulated by an increase in the concentration of amino acids in the blood. Because glucagon stimulates gluconeogenesis, this process ensures that excess amino acids in high-protein meals are used to produce glucose.

Diabetes Mellitus Is a Disease Resulting from an Insulin Deficiency or a Decrease in Tissue Sensitivity to Insulin

▻ Figure 20–26 shows a graph of blood glucose levels in normal and diabetic people. The results were obtained during a **glucose tolerance test.** During this test, physicians give their patients an oral dose of glucose, then check blood levels at regular intervals.

The bottom line on the graph in Figure 20–26 represents the blood glucose levels in a normal patient with a healthy pancreas. As illustrated, glucose levels increase slightly after the administration of glucose, but within one to two hours, glucose levels have decreased to normal, thanks to insulin secretion.

The top line illustrates the response in a diabetic patient, who produces insufficient amounts of insulin or whose cells have become unresponsive to insulin. Glucose levels rise considerably in these patients after they are given an oral dose of glucose and remain elevated for four to five hours. A similar response is seen in untreated diabetics after they eat a meal. However, blood glucose levels remain higher much longer because there is little place for the glucose to go. The absence of insulin means that virtually no glucose will enter skeletal muscle cells and that liver uptake will be greatly reduced. Glycogen synthesis in liver and muscle cells is also greatly reduced.

Glucose levels eventually decline as this nutrient is used by brain cells, red blood cells, and others. The kidneys also

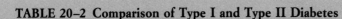

TABLE 20–2 Comparison of Type I and Type II Diabetes

CHARACTERISTIC	TYPE I DIABETES	TYPE II DIABETES
Level of insulin secretion	None or almost none	May be normal or exceed normal
Typical age of onset	Childhood	Adulthood
Percent of diabetics	10% to 20%	80% to 90%
Basic defect	Destruction of beta cells	Reduced sensitivity of insulin's target cells
Associated with obesity?	No	Usually
Genetic and environmental factors important in precipitating overt disease?	Yes	Yes
Speed of development of symptoms	Rapid	Slow
Development of ketosis	Common if untreated	Rare
Treatment	Insulin injections; dietary management	Dietary control and weight reduction; occasionally oral hypoglycemic drugs

From L. Sherwood, *Human Physiology: From Cells to Systems* (St. Paul, Minn.: West, 1989).

excrete excess glucose, helping lower blood levels. Unfortunately, this process wastes a valuable nutrient and causes the kidneys to lose large quantities of water. Excess urination and constant thirst are the chief symptoms of untreated diabetes. Diabetics may have to urinate every hour, day and night.

Diabetes Is Really Two Diseases with Similar Symptoms but Different Causes. Physicians recognize two types of diabetes (Table 20-2). **Type I diabetes** occurs mainly in young people and results from a deficiency in insulin production. Type I diabetes is also called **juvenile diabetes** or **early-onset diabetes.** Type I diabetes is believed to be caused by damage to the insulin-producing cells of the pancreas. One leading theory suggests that it is caused by an autoimmune reaction. Viral infection of the pancreas may also cause Type I diabetes.

In patients with Type I diabetes, insulin production varies. In some, it is only slightly depressed; in others, it is completely suppressed. In the absence of insulin, body cells are starved for glucose. Energy must be provided by the breakdown of fats.

Type II diabetes usually occurs in people over 40 and is sometimes called **late-onset diabetes** (Table 20–2). In this disease, the beta cells produce normal or above-normal levels of insulin. However, problems arise because of a decline in the number of insulin receptors in target cells, which makes them unresponsive to insulin.

Type II diabetes is commonly associated with obesity. Heredity may also be an important contributor, as it is in Type I diabetes. Studies show that about one-third of the patients with Type II diabetes have a family history of the disease. Critical thinking rules in Chapter 1 warn us to look for other explanations. When we do, we find that it may be obesity that is inherited, rather than the tendency to develop diabetes.

Although they have different causes, both forms of diabetes exhibit many similar symptoms. Excess urination and thirst are generally the first signs of trouble. Excess glucose in the urine may result in frequent bacterial infections in the bladder. Patients often feel tired, weak, and apathetic. Weight loss and blurred vision are also common.

Although Symptoms Are Similar, Treatments for Type I and Type II Diabetes Are Quite Different. Early-onset diabetes (Type I) is treated with insulin injections and therefore is often called **insulin-dependent diabetes.** Patients receive regular injections of insulin—usually two to three times per day. Patients are also required to eat meals and snacks at regular intervals to maintain constant glucose levels in the blood and to ensure that regular insulin injections always act on approximately the same amount of blood glucose.

Insulin injections are tailored to an individual's lifestyle and body demands. To help mimic the body's natural release, medical researchers have developed a device called an **insulin pump,** which delivers predetermined amounts of insulin when needed. Because the device cannot measure blood glucose levels, it may deliver an inappropriate amount under certain circumstances. Health Note 20–1 describes the device in more detail. Medical researchers are also experimenting with ways to transplant healthy beta cells in the pancreas of a diabetic. (Chapter 3 and the Point/Counterpoint in that chapter discuss fetal cell transplantation.)

In a recent study, researchers from the University of Florida reported findings that may help prevent Type I diabetes. The researchers withdrew blood from 5000 young children who were asymptomatic—that is, showed no symptoms of early-onset diabetes. They then analyzed the blood for antibodies produced by an autoimmune reaction and followed the children's health status over time to de-

HEALTH NOTE 20–1

ON THE ROAD TO AN ARTIFICIAL PANCREAS

Diabetics take insulin injections two or three times a day. If they do not, they could easily go into a diabetic coma. These injections allow most victims of juvenile diabetes to live a fairly normal life. Occasionally things go awry, however. Too much insulin or too much exercise causes blood glucose levels to fall, making a diabetic feel dizzy and weak. Sometimes the balance is tipped the other way. An excess of blood sugar brings on hyperglycemia (high blood sugar) and with it depression, fatigue, irritability, and weakness.

Besides worrying over insulin doses and playing a continual balancing act with their blood sugar, most diabetics develop complications 20 to 30 years after the onset of the disease. Blindness and lethal or disabling diseases of the kidney, nervous system, and cardiovascular system are common. These complications probably result from periodic elevations in blood glucose levels, which occur despite good insulin management. High glucose levels damage blood vessels and nerves, injuring critical organs.

To prevent complications, physicians try to mimic normal insulin secretion patterns through injections three times a day just before meals. But mimicking the body's homeostatic system with regular insulin injections is crude. The pancreas normally monitors blood glucose levels minute by minute. A conscientious diabetic can measure blood glucose only three or four times a day. He or she may make adjustments for large meals or additional exercise, but such accommodations are primitive in comparison with the body's elaborate system of glucose homeostasis, a product of evolution vital to the survival and successful propagation of many multicellular animals.

Consequently, medical researchers are looking for ways to replace the daily injections. One hopeful possibility is a device called an insulin infusion pump, which delivers tiny amounts of fast-acting insulin to the body day and night via a long plastic tube and needle inserted into the skin of the thigh or abdomen (▻ Figure 1). The needle and tubing are usually replaced every three to four days. The insulin pump provides baseline insulin levels needed to maintain proper blood glucose concentrations. Worn outside the body, the pump also delivers a surge of insulin at mealtimes to offset the rise in blood sugar that accompanies a meal. To do this, one simply presses a button on the pump 30 minutes before eating. The pump delivers a preprogrammed amount of insulin. If the meal is going to be larger than anticipated, a small adjustment can be made to protect against hyperglycemia.

▻ **FIGURE 1 Insulin Pump** This device delivers preprogrammed doses of insulin to diabetics, mimicking the pancreas.

The newer insulin pumps capitalize on computer technology to regulate insulin flow day and night. They offer great flexibility, allowing an individual to accommodate differing levels of exercise and meals of varying size. They come equipped with a memory to store information on the exact doses given over a certain period. Using this information, physicians can fine-tune the program to an individual's needs.

The insulin pump is not a popular item among diabetics when it comes to aesthetics and comfort. When it comes to controlling blood sugar, however, the device receives high praise. For many, the freedom from daily injections outweighs the discomfort. For pregnant diabetic women, the pump may mean the difference between a normal child and no child at all, for even mild hyperglycemia can cause fetal death.

While refinements are being made to the insulin pump, researchers are also testing a biodegradable wafer that can be impregnated with insulin and inserted under the skin of a diabetic. The wafer also contains a sugar-sensitive enzyme. As blood sugar levels rise, the enzyme is activated. This action, in turn, causes a slight increase in acidity, which increases the solubility of the insulin in the wafer, allowing it to be released into the bloodstream. As a result, insulin levels respond very closely to blood glucose levels. Insulin is released only as needed in response to rising blood sugar levels.

This innovative approach could prove even more effective than the insulin pump, because it operates on a negative feedback principle as do the beta cells in the pancreas.

termine which ones developed diabetes. During the study, 12 of the subjects developed early-onset diabetes. Interestingly, blood samples from these individuals all contained an antibody that attacks a protein on the surface of the beta cells of the pancreas. The antibody was found in blood samples taken as many as seven years before the onset of diabetes.

The antibody to the surface protein of the beta cell may provide a biochemical marker, an advance notice of the disease. Eventually, researchers hope to find ways to attach chemicals to the surface proteins of the beta cells that would neutralize or destroy the antibodies that bind to them, thus preventing the disease from developing. If this research proves fruitful, regular blood tests could be used to screen children. When the antibodies are found, special treatment could be used to destroy them before they have time to destroy the islet cells. This treatment could free millions of people from a lifetime of insulin injections and the risk of blindness and other serious medical problems later in life.

Type II diabetes is often called **insulin-independent diabetes,** because insulin injections are useless given that the disease may stem from a lack of insulin receptors in target cells. In many patients, Type II diabetes can be eliminated by weight loss. In others, the disease can be controlled by diet and exercise. Physicians restrict carbohydrate intake of their patients and ask them to eat small meals at regular intervals during the day. Candy, sugar, cakes, and pies are off limits. Glucose must be supplied by complex carbohydrates, such as starches.

The treatments described above have dramatically changed the prognosis for diabetics. At one time the disease was fatal. Today, patients can live healthy, fairly normal lives. Risks are still present. Type I diabetics, for example, still suffer from diabetic comas, or unconsciousness, caused when they receive insufficient amounts of insulin—for example, if they forget their insulin injection or inject too little. Without insulin, the body cells become starved for glucose (even though blood levels are high) and begin breaking down fat. Excessive fat catabolism releases toxic chemicals (called ketones) that cause the patient to lose consciousness. Early- and late-onset diabetics may also suffer from loss of vision, nerve damage, and kidney failure 20 to 30 years after the onset of the disease, even if they are being treated. Damage to the circulatory system may cause gangrene, requiring amputation of limbs, especially the lower extremities. These serious complications result from the inevitable periodic elevations in blood glucose levels that occur over the years.

THE ADRENAL GLANDS

You are standing on the banks of a raging river. Your raft is tied to a tree and floating in the calm water of an eddy above the white water. As you watch kayakers and rafters head into the rapids, your heart starts to race, and your intestines churn in excitement (and fear). Your turn is coming up.

This natural response results from the secretion of two of the many hormones produced by the **adrenal glands.** As ▻ Figure 20–27 shows, the adrenal glands perch atop the kidneys, and each consists of two zones. The central region, or **adrenal medulla,** produces the hormones that increase the heart rate and accelerate breathing when a person is excited or frightened. The outer zone, the **adrenal cortex,** produces a number of steroid hormones mentioned earlier and discussed in more detail shortly.

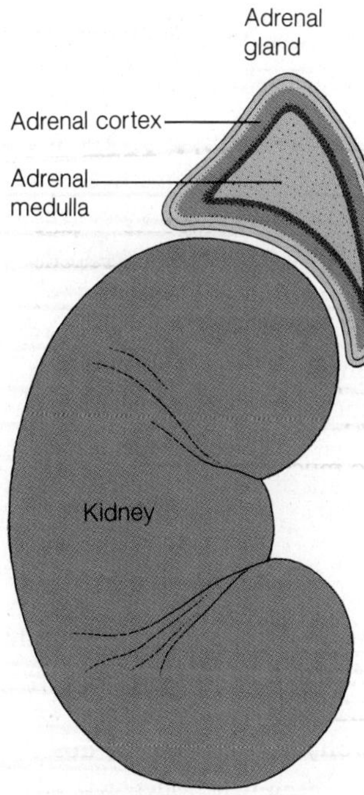

▻ **FIGURE 20–27 Adrenal Gland** The adrenal glands sit atop the kidney and consist of an outer zone of cells, the adrenal cortex, which produces a variety of steroid hormones, and an inner zone, the adrenal medulla. The adrenal medulla produces adrenalin and noradrenalin, the secretion of which is controlled by the autonomic nervous system.

The Adrenal Medulla Produces Stress Hormones

The adrenal medulla produces two hormones: **adrenalin** (epinephrine) and **noradrenalin** (norepinephrine). In humans, about 80% of the adrenal medulla's output is adrenalin. Helping animals meet the stresses of life, adrenalin and noradrenalin are instrumental in the fight-or-flight response—that is, the physiological reactions that take place when an animal is threatened and facilitate its ability to fight or flee the scene (Chapter 17).

Adrenalin and noradrenalin are secreted under all kinds of stress—for example, when an angry dog leaps out at you, when a careless driver cuts in front of you in heavy traffic, or even as you wait outside a lecture hall, anticipating a final exam. Nerve impulses traveling from the brain to the adrenal medulla trigger the release of adrenalin and noradrenalin. The response is part of the autonomic nervous system (Chapter 17).

The physiological changes these hormones cause are many. For example, adrenalin and noradrenalin elevate blood glucose levels, making more energy available to body cells, particularly skeletal muscle cells. They also increase one's breathing rate, which provides additional oxygen to skeletal muscles and brain cells. These hormones cause the heart rate to accelerate as well. This action increases circulation and ensures adequate glucose and oxygen for body cells that might be called into action. Adrenalin and noradrenalin also cause the bronchioles in your lungs to dilate, permitting greater movement of air in and out of the lungs. Under the influence of these hormones, blood vessels in the intestinal tract constrict, putting digestion on temporary hold, while the blood vessels in the skeletal muscles dilate, increasing flow through them. Mental alertness increases as a result of increased blood flow and hormonal stimulation. You are ready to fight or flee.

The Adrenal Cortex Produces Three Types of Hormones with Markedly Different Roles

Surrounding the medulla is the adrenal cortex (Figure 20–27). The adrenal cortex produces three types of steroid hormones, each of which has a markedly different function. The first group, the **glucocorticoids,** affect glucose metabolism and help maintain blood glucose levels—hence the name. The second group, the **mineralocorticoids,** help regulate the ionic concentration of the blood and tissue fluids. The final group, the **sex steroids,** are identical to the hormones produced by the ovaries and testes. In healthy adults, the amount of adrenal sex steroids synthesized and released is insignificant compared to the amounts produced by the gonads.

Glucocorticoids. The prefix *gluco* reflects the fact that these steroid hormones affect carbohydrate metabolism. (Glucose, of course, is one of the many carbohydrates found in living things.) The secretion of glucocorticoid is governed by ACTH, a hormone of the anterior pituitary, as discussed earlier in the chapter. Several chemically distinct glucocorticoids are secreted, but the most important is **cortisol.** Cortisol increases blood glucose by stimulating gluconeogenesis (the synthesis of glucose from amino acids and glycerol). This action makes more glucose available for energy production in times of stress. Cortisol also stimulates the breakdown of proteins in muscle and bone, freeing amino acids that can then be chemically converted to glucose molecules in the liver.

In **pharmacological doses,** levels beyond those seen in the body, cortisol inhibits inflammation, the body's response to tissue damage. Pharmacological doses of glucocorticoids also depress the allergic reaction. Cortisol brings about its effects by inhibiting the movement of white blood cells across capillary walls, thus impeding their migration into damaged tissue. Cortisol also reduces the number of circulating lymphocytes by destroying them at their site of formation.

Because they reduce inflammation, cortisol and other glucocorticoids (particularly cortisone, a synthetic glucocorticoid-like hormone) can be used to treat inflammation resulting from diseases such as rheumatoid arthritis or physical injuries—for example, damage to a knee during a basketball game. However, the benefits must be weighed carefully against the damage that can be caused by upsetting the body's homeostatic balance (discussed below).

Mineralocorticoids. As their name implies, the mineralocorticoids are involved in electrolyte or mineral salt balance. The most important mineralocorticoid is **aldosterone** (described in Chapter 16). It is the most potent mineralocorticoid and constitutes 95% of the adrenal cortex's hormonal output.

Although mineralocorticoids regulate the level of several ions, their main function is to control sodium and potassium concentrations, and their chief target is the kidney. As noted in Chapter 16, aldosterone increases the movement of sodium ions out of the nephron and into the blood. This process is called tubular reabsorption. Aldosterone also stimulates sodium ion reabsorption in sweat glands and saliva and potassium excretion by the kidney. When sodium is shunted back into the blood in the kidney and the skin, water follows. Aldosterone therefore helps conserve body water.

As also noted in Chapter 16, aldosterone secretion is controlled by the sodium ion concentration in a negative feedback loop. As sodium levels fall, aldosterone secretion increases. As sodium levels are restored, aldosterone secretion declines. As you might expect, aldosterone secretion is also stimulated when potassium levels climb and when blood volume and blood pressure decline.

Diseases of the Adrenal Cortex. ▻ Figure 20–28 shows a patient with **Cushing's syndrome.** Cushing's syndrome generally results from pharmacological doses of cortisone, a synthetic glucocorticoid used to treat rheumatoid arthritis or asthma. In a few instances, the disease may be caused by a pituitary tumor that produces excess ACTH or a tumor of the adrenal cortex that secretes excess glucocorticoid.

Patients with Cushing's disease often suffer persistent **hyperglycemia**—high blood sugar levels—because of the presence of high levels of glucocorticoid. Bone and muscle protein may also decline sharply, because glucocorticoids stimulate the breakdown of protein. Individuals complain

▻ **FIGURE 20–28 Cushing's Syndrome** This disease results from an excess of glucocorticoid hormone, either cortisol or cortisone. It is most often caused by cortisone treatment for allergies or inflammation. The most common symptoms are a round face due to edema and excess fat deposition.

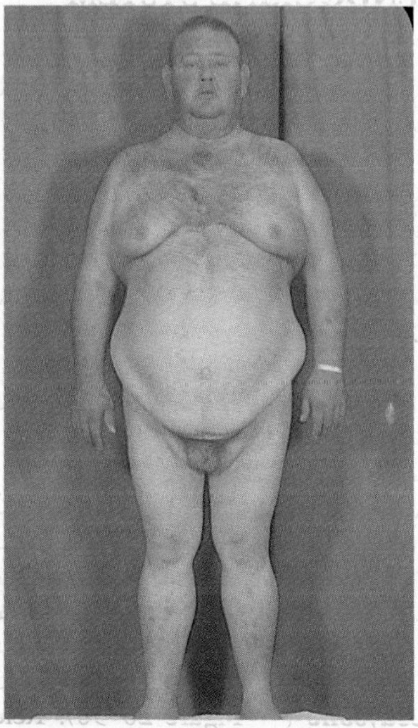

of weakness and fatigue. Loss of bone protein increases the ease with which bones fracture.

Water and salt retention are also common in Cushing's patients, resulting in tissue edema (swelling). Cushing's patients have "moon face," a rounded face resulting from edema. These symptoms occur because glucocorticoids, in high concentrations, have mineralocorticoid effects.

Because most cases of Cushing's syndrome result from steroids taken for health reasons, treatment is a simple matter. By gradually reducing the glucocorticoid dose, a physician can eliminate the symptoms. Tumors in the pituitary and the adrenal cortex can be treated with radiation or surgery.

Another disease of the adrenal cortex mentioned earlier in this chapter but worth studying more closely is **Addison's disease.** Most cases of Addison's disease are thought to be autoimmune reactions in which the cells of the adrenal cortex are recognized as foreign and destroyed by the body's own immune system.

Addison's disease is characterized by a variety of symptoms resulting from the loss of hormones from the adrenal cortex. These symptoms include loss of appetite, weight loss, fatigue, and weakness. Although insulin and glucagon are still present, the absence of cortisol upsets the body's homeostatic mechanism for controlling glucose. The body's reaction to stress is also impaired. The lack of aldosterone results in electrolyte imbalance. Because aldosterone helps maintain sodium levels and blood pressure, patients with Addison's disease have low blood sodium levels and low blood pressure. Addison's disease can be treated with steroid tablets that replace the hormones the adrenal cortex is no longer producing. Treatment allows patients to lead a fairly normal, healthy life.

THE ENDOCRINE SYSTEM OF INVERTEBRATES

Biologists have discovered hormones in arthropods and a variety of other invertebrates. Even protozoans, sponges, plants, and possibly some prokaryotes contain chemical messengers. These hormones control a wide variety of functions. Studies of the endocrinology of invertebrates suggest that vertebrate hormones evolved from those found in invertebrates but often perform quite different tasks today.

The best understood invertebrate hormones are those in insects that control metabolism, molting (periodic shedding of the exoskeleton), and reproduction. Hormones that control both fat and sugar levels have been identified in insects and, in one case, chemically characterized.

Molting occurs in many insects and crustaceans and is a process in which they shed their exoskeleton and reconstruct a new one at regular intervals to accommodate growth (▻ Figure 20–29). In insects and crustaceans, the immediate stimulus for molting is a steroid hormone known as **ecdysone** (▻ Figure 20–30). Released from a pair of glands in the bodies of insects and crustaceans, ecdysone stimulates the cells of the epidermis, causing them to secrete a new exoskeleton. The old exoskeleton is partially digested by enzymes and eventually splits apart as the new one develops underneath it.

▻ **FIGURE 20–29 Molting** Insects and other invertebrates periodically shed their exoskeletons to permit growth.

In insects, ecdysone is controlled by a hormone from the brain appropriately named **brain hormone.** This hormone stimulates the prothoracic gland of the insect to release ecdysone. In crustaceans, ecdysone is controlled by a slightly different mechanism, an inhibitory hormone appropriately known as **molt-inhibiting hormone (MIH).** When MIH secretion stops, ecdysone is produced, and a molt occurs.

The secretion of molt-inhibiting hormone and brain hormone is controlled by a variety of internal and external factors, depending on the species. Day length, temperature, and nutritional status are the most important controls. These factors help synchronize molting with the animal's growth and development or with seasonal environmental factors such as food availability.

Hormones also control the complex life cycles of some insects and other animals. Many animals pass through two or more morphologically distinct stages in their postembryonic life (▻ Figure 20–31). Frogs, for example, develop from tadpoles. The transformation of an animal from one stage to another is called **metamorphosis.** This strategy allows an organism to exploit different habitats. In differ-

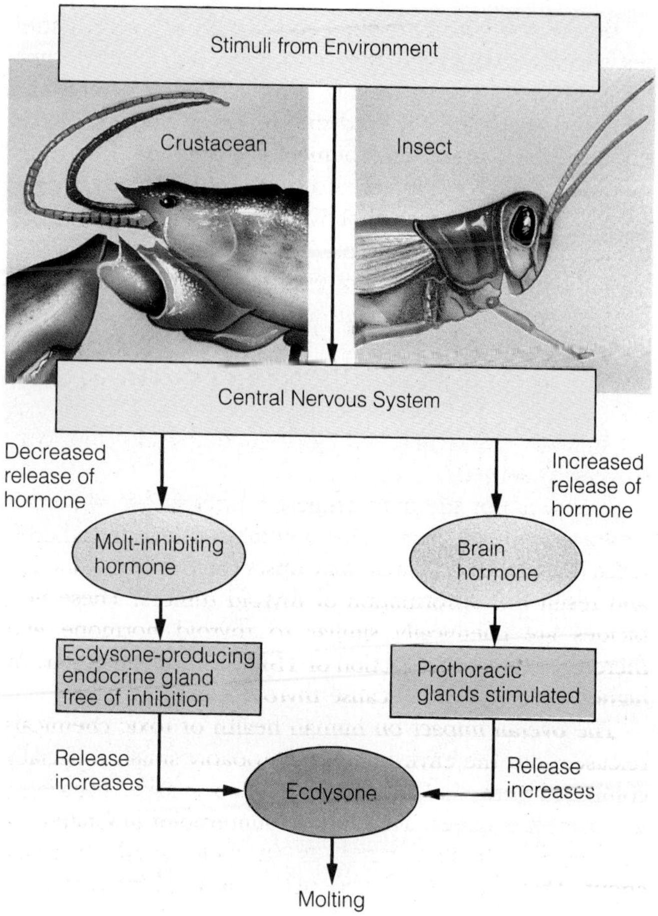

▻ **FIGURE 20–30 Hormonal Control of Molting** Molting in insects and crustaceans is controlled by the hormone ecdysone. However, ecdysone is controlled by brain hormone in insects and by molt inhibiting hormone in crustaceans.

ent stages, organisms use very different modes of locomotion and feeding.

Metamorphosis is regulated by hormones. As shown in Figure 20–31, moths and butterflies develop from caterpillars (larvae). The caterpillars pass through a series of stages (larval instars), then form cocoons, or **pupae.** Inside the cocoon, radical changes occur that produce a moth or butterfly. Ecdysone and brain hormone are involved in the entire process. Another hormone, known as **juvenile hormone (JH),** is also involved. As illustrated in Figure 20–31, JH is secreted in large amounts early on. As long as JH levels are high, larva-to-larva molts occur. When metabolic changes cause JH levels to fall, though, the larva forms a cocoon in the presence of ecdysone. The ratio of juvenile hormone to ecdysone, therefore, determines the pattern of development. Interestingly, some plants have evolved chemical analogues of juvenile hormone, which when ingested by hungry larvae prevent them from pupating. This mechanism prevents the larvae from maturing and going on to produce eggs and additional generations of larvae.

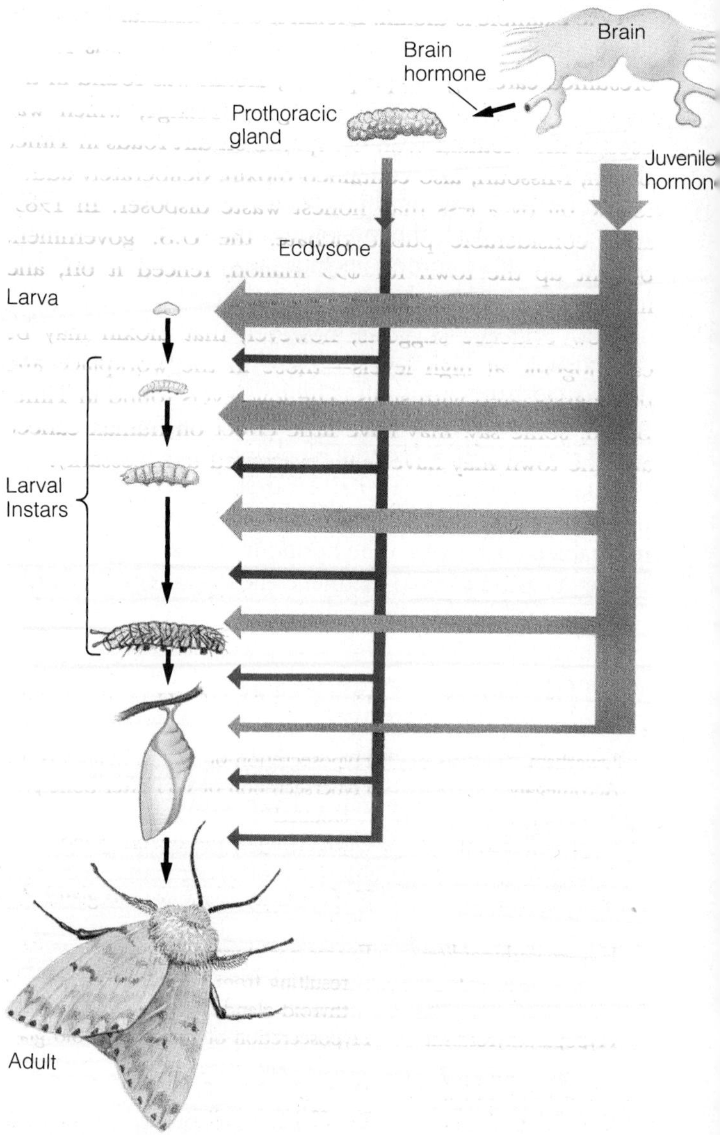

▻ **FIGURE 20–31 Life Cycle of the Moth** Note that the moth larvae, which form from eggs, pass through several instar (larval) stages. In order to grow, the larvae must molt. Molting is controlled by the hormone ecdysone when juvenile hormone levels are high. When juvenile hormone levels decline, ecdysone which is present throughout the life cycle stimulates pupation and differentiation into an adult.

ENVIRONMENT AND HEALTH: HERBICIDES AND HORMONES

Hormones orchestrate an incredible number of body functions, creating a dynamic balance that is necessary for good health. When this balance is altered, our health suffers (Table 20–3). A major participant in homeostasis, the endocrine system, like other systems, is sensitive to outside, or environmental, factors. Stress, for example, can lead to an imbalance in adrenal hormones, resulting in high blood pressure and other complications. High blood pressure, in turn, puts strain on the heart. Hormonal balance can also be altered by toxic pollutants.

One example is dioxin. Dioxin is a contaminant found in some herbicides and in paper products. Infamous for its presumed carcinogenic properties, dioxin was found in the chemical defoliant known as Agent Orange, which was used in the Vietnam War. Oil spread on dirt roads in Times Beach, Missouri, also contained dioxin, deliberately added to the oil by a less than honest waste disposer. In 1983, after considerable public debate, the U.S. government bought up the town for $35 million, fenced it off, and moved the residents out.

New evidence suggests, however, that dioxin may be carcinogenic at high levels—those in the workplace and those associated with spills. The low levels found in Times Beach, some say, may have little effect on human cancer, and the town may have been evacuated unnecessarily.

Before you get too excited about this view, recent studies suggest that dioxin may also suppress immune functions. It may bind to plasma membrane and cytoplasmic receptors that normally bind to hormones. Some researchers are calling it an "environmental hormone." One form of dioxin, TCDD, suppresses the immune system in mice at least 100 times more effectively than corticosterone, one of the body's glucocorticoids.

TCDD causes a variety of biological responses. In some cells, it causes rapid cell growth. In others, it may alter cellular differentiation. Nonetheless, its effect on the immune system may be far more important than its impact on cancer. As for the former residents of Times Beach, they might just want to stay away a little longer until the controversy is settled.

Dioxin is not the only dangerous chemical in the environment. Studies suggest that a number of common herbicides (the thiocarbamates) may upset the thyroid's function and result in the formation of thyroid tumors. These herbicides are chemically similar to thyroid hormone and therefore block the secretion of TSH, resulting in goiter. At higher levels, they may cause thyroid cancer.

The overall impact on human health of toxic chemicals released into the environment is probably small, especially compared with the impact of tobacco smoke. However, not everyone agrees. The Point/Counterpoint in Chapter 9 debates this subject, which is an issue worth thinking about. This important debate will be around for years.

TABLE 20–3 Summary of Some Endocrine Disorders

DISEASE	CAUSE	SYMPTOMS
Gigantism	Hypersecretion of GH starting in infancy or early life	Excessive growth of long bones
Dwarfism	Hyposecretion of GH in infancy or early life	Failure to grow
Acromegaly	Hypersecretion of GH after bone growth has stopped	Facial features become coarse; hands and feet enlarge; skin and tongue thicken.
Hyperthyroidism	Overactivity of the thyroid gland	Nervousness; inability to relax; weight loss; excess body heat and sweating; palpitations of the heart
Hypothyroidism	Underactivity of the thyroid gland	Fatigue; reduced heart rate; constipation; weight gain; feel cold; dry skin
Hyperparathyroidism	Excess parathyroid hormone secretion, usually resulting from a benign tumor in the parathyroid gland	Kidney stones; indigestion; depression; loss of calcium from bones
Hypoparathyroidism	Hyposecretion of the parathyroid glands	Spasms in muscles; numbness in hands and feet; dry skin
Diabetes insipidus	Hyposecretion of ADH	Excessive drinking and urination; constipation
Diabetes mellitus	Insufficient insulin production or inability of target cells to respond to insulin	Excessive urination and thirst; poor wound healing; urinary tract infections; excess glucose in urine; fatigue and apathy
Cushing's syndrome	Hypersecretion of hormones from adrenal cortex or, more commonly, from cortisone treatment	Face and body become fatter; loss of muscle mass; weakness; fatigue; osteoporosis
Addison's disease	Gradual decrease in production of hormones from adrenal gland; most common cause is autoimmune reaction	Loss of appetite and weight; fatigue and weakness; complete adrenal failure

SUMMARY

PRINCIPLES OF ENDOCRINOLOGY

1. The endocrine system consists of a widely dispersed set of highly vascularized ductless glands that produce hormones.
2. Hormones affect five vital aspects of an animal's life: (a) homeostasis, (b) growth and development, (c) reproduction, (d) energy production, storage, and catabolism, and (e) behavior.
3. Hormones act on specific cells. Specificity results from the presence of hormone receptors on target cells. Steroid hormone receptors are generally located in the cytoplasm, and protein and polypeptide hormone receptors are located in the plasma membrane.
4. Three types of hormones are produced in the body: (a) steroids, (b) proteins and polypeptides, and (c) amines.
5. Hormone secretion is controlled principally by negative feedback loops, many of which involve the hypothalamus.
6. Water-soluble hormones (proteins and polypeptides) act on cells via a second messenger.
7. Steroid (lipid-soluble) hormones act via the two-step mechanism.
8. The endocrine and nervous systems are similar in several respects. They both send signals that regulate cell structure and function. Both also help coordinate body functions.
9. The endocrine and nervous systems differ in several key respects. The nervous system elicits rapid, generally short-lived responses. The endocrine system elicits slower, longer lasting responses.

THE PITUITARY AND HYPOTHALAMUS

10. The pituitary is a pea-sized gland suspended from the hypothalamus by a thin stalk. It consists of two parts: the anterior pituitary and the posterior pituitary.
11. The anterior pituitary produces seven protein and polypeptide hormones. Their release is controlled by releasing and inhibiting hormones produced by the hypothalamus, which are transported to the anterior pituitary via a portal system.
12. The hypothalamic hormones are produced by neurosecretory neurons. Their release is controlled by chemical stimuli and nerve impulses.
13. The hormones of the anterior pituitary and their functions are summarized in Table 20–1.
14. The posterior pituitary is derived from brain tissue during embryonic development and consists of axons and terminal ends of neurosecretory cells whose cell bodies are in the hypothalamus. Hormones are produced in the cell bodies and are transported down the axons to the posterior pituitary, where they are stored in the axon terminals until released.
15. The posterior pituitary produces two hormones, antidiuretic hormone (ADH) and oxytocin, whose functions are also summarized in Table 20–1.

THE THYROID GLAND

16. The thyroid gland is located in the neck, on either side of the trachea near its junction with the larynx. The thyroid produces three hormones: thyroxine (T_4), triiodothyronine (T_3), and calcitonin.
17. Thyroxine and triiodothyronine accelerate the rate of glucose breakdown in most cells, increasing body heat. These hormones also stimulate cellular growth and development.
18. Thyroxine and triiodothyronine both require iodine for their synthesis. A deficiency of this element in the diet results in goiter, an enlargement of the thyroid gland.
19. Calcitonin lowers blood calcium levels by inhibiting osteoclasts, thus reducing bone destruction and the release of calcium from bone. Calcitonin also increases formation of bone by osteoblasts and increases the excretion of calcium in the kidneys.

THE PARATHYROID GLANDS

20. The parathyroid glands are located on the back of the thyroid gland and produce a polypeptide hormone called parathyroid hormone (PTH), or parathormone.
21. PTH is released when calcium levels in the blood drop. This hormone increases blood calcium levels by stimulating bone reabsorption by osteoclasts, increasing intestinal absorption, and increasing renal reabsorption of calcium.

THE PANCREAS

22. The pancreas produces two hormones, insulin and glucagon, from the islets of Langerhans.
23. Insulin is the glucose-storage hormone. It stimulates the uptake of glucose by many body cells and stimulates the synthesis of glycogen in muscle and liver cells. Insulin also increases the uptake of amino acids and stimulates protein synthesis in muscle cells, thus promoting muscle formation.
24. Glucagon is an antagonist to insulin. It raises glucose levels in the blood in the period between meals by stimulating glycogen breakdown and the synthesis of glucose from amino acids and fats (gluconeogenesis).
25. Diabetes mellitus is a disease involving insulin and blood glucose. It has two principal forms: Type I and Type II. Type I, also called early-onset diabetes, occurs early in life and may be caused by an autoimmune reaction that destroys the beta cells of the pancreas. It can be treated by insulin injections.
26. Type II, or late-onset, diabetes, results from a reduction in the number of insulin receptors on target cells. It may be caused by obesity and genetic factors and can often be treated successfully by dietary management.

THE ADRENAL MEDULLA AND ADRENAL CORTEX

27. The adrenal glands lie atop the kidneys and consist of two separate portions: the adrenal medulla, at the center, and the adrenal cortex, a surrounding band of tissue.
28. The adrenal medulla produces two hormones under stress: adrenalin and noradrenalin. These hormones stimulate heart rate and breathing, elevate blood glucose levels, constrict blood vessels in the intestine, and dilate blood vessels in the muscles.
29. The adrenal cortex produces three classes of hormones: glucocorticoids, mineralocorticoids, and sex steroids.
30. The glucocorticoids affect carbohydrate metabolism and tend to raise blood glucose levels. The principal glucocorticoid is cortisol.
31. In pharmacological doses, cortisol inhibits the immune system and allergic reactions and is used to treat allergies or inflammation caused by injury and infections. High doses, however, have many adverse impacts on the body.

32. The chief mineralocorticoid is aldosterone. It acts on the kidneys, sweat glands, and salivary glands, causing sodium and water retention and potassium excretion.

THE ENDOCRINE SYSTEM OF INVERTEBRATES

33. Hormones exist in a large number of invertebrates, where they control a wide variety of functions.
34. The best understood invertebrate hormones are those in insects that control metabolism, molting, and reproduction.
35. Many insects and crustaceans molt, shedding their exoskeleton and reconstructing a new one at regular intervals to accommodate growth.
36. In insects and crustaceans, the immediate stimulus for molting is the steroid hormone ecdysone, which stimulates the cells of the epidermis, causing them to secrete a new exoskeleton.
37. In insects, ecdysone is controlled by brain hormone; in crustaceans it is controlled by molt-inhibiting hormone. The secretion of both hormones is controlled by a variety of internal and external factors, among them day length, temperature, and nutritional status.
38. Hormones also control metamorphosis, the transformation of an animal from one stage to another. Ecdysone is vital to the transformation of larvae to adult stages, but its effects are modulated by juvenile hormone. Thus, the ratio of juvenile hormone to ecdysone determines the pattern of development.

ENVIRONMENT AND HEALTH: HERBICIDES AND HORMONES

39. Hormones orchestrate an incredible number of body functions, creating a dynamic balance necessary for good health. When this balance (homeostasis) is altered, our health suffers.
40. The endocrine system, like others, is sensitive to upset from environmental factors, including pollutants.
41. Dioxin, a contaminant in some herbicides and paper products, may exert most of its effects through the endocrine system.
42. A number of common herbicides (the thiocarbamates) upset the thyroid's function and may even cause thyroid tumors.

EXERCISING YOUR CRITICAL THINKING SKILLS

Find a copy of the *New England Journal of Medicine* in the library with an article on a discovery in endocrinology. How is the article organized? Analyze the article, using your critical thinking skills. Was the study performed on humans or laboratory animals? What were the major conclusions?

In your view, was the study performed correctly? Did the researchers include a control group? Was the control group similar to the experimental group? How many subjects were used? Should more have been used? Why? Can you tell whether the conclusions supported the data? Can you think of any alternative explanations for the data?

TEST OF TERMS

1. A ____________ is a chemical produced in an endocrine gland that travels in the bloodstream to its ____________ cells, where it elicits a specific response.
2. In the endocrine system, specificity is conferred by the ____________ found in the cytoplasm and in the plasma membrane.
3. As a general rule, ____________ hormones are those that stimulate the production and secretion of other hormones.
4. The body produces three types of hormones: ____________, proteins and polypeptides, and ____________.
5. Protein and polypeptide hormones activate cells through a ____________ ____________, usually cyclic AMP. It is produced inside cells from ____________ in a reaction catalyzed by a membrane-bound enzyme called ____________ ____________. Cyclic AMP activates another enzyme, protein kinase. It in turn adds ____________ to other enzymes inside the cell, turning some on and turning others off.
6. Hormones like testosterone activate cells by the ____________ - ____________ mechanism. Steroid hormones activate the ____________.
7. The ____________ gland, located beneath the ____________, produces more hormones than any other endocrine gland.
8. The ____________ ____________ gland is controlled by ____________ and inhibiting hormones produced by the ____________ cells of the ____________.
9. ____________ hormone is a protein that stimulates cellular ____________ and hyperplasia. In muscle, this hormone stimulates the uptake of ____________ and the synthesis of protein.
10. ____________ is a term that describes a condition in which an endocrine gland produces less hormone than needed.
11. ____________ is a hormone that stimulates the adrenal cortex to release its hormones.
12. The ____________ are hormones that stimulate gamete formation and endocrine production in the gonads in both males and females.
13. Milk production in humans is stimulated by a protein hormone called ____________. Milk production begins at the end of pregnancy and can be continued by ____________.
14. Hormone secretion that is stimulated by neural impulses is called a ____________ reflex.

15. The posterior pituitary is a ____________ gland. It consists of nervous tissue and releases two hormones: ____________ and ____________.
16. The thyroid gland produces three hormones: ____________, triiodothyronine, and ____________.
17. The thyroid follicles contain a colloidal material called ____________.
18. ____________ is a condition that results from a dietary deficiency of iodine.
19. Two hormones control the level of calcium in the blood. The hormone that raises serum calcium levels is ____________, and it is produced by the ____________ glands. The hormone that reduces serum calcium is ____________ and is produced by the ____________ gland.
20. ____________ is produced by the pancreas. It stimulates the uptake of ____________ by muscle cells and stimulates the synthesis of ____________.
21. The synthesis of glucose from amino acids and fatty acids is called ____________.
22. Early-onset diabetes may result from a(n) ____________ reaction.
23. Type II diabetes, or ____________ - ____________ diabetes, occurs in older people and is often associated with ____________. It is controlled by managing ____________.
24. The adrenal medulla produces two hormones, ____________ and ____________, which stimulate heart rate and breathing.
25. Glucocorticoids are produced by the ____________ ____________; they increase blood glucose by stimulating ____________. In high concentrations, they repress the ____________ system function.
26. ____________ is the principal mineralocorticoid produced by the ____________ ____________. It affects sodium and potassium ion concentrations.

Answers to the Test of Terms are found in Appendix B.

TEST OF CONCEPTS

1. Define the following terms: endocrine system, hormone, and target cell.
2. Hormones function in five principal areas. What are they? Give some examples of each.
3. Describe the concept of specificity. How is it created in the nervous system? How is it created in the endocrine system?
4. Compare and contrast the functions of the nervous and endocrine systems.
5. The endocrine system elicits slower, longer-lasting responses than the nervous system. Do you agree or disagree? Explain your reasons.
6. Define the terms *trophic* and *nontrophic hormones,* and give several examples of each.
7. Give two examples of negative feedback loops in the endocrine system, a simple feedback mechanism and a more complex one that operates through the nervous system.
8. Define the term *neuroendocrine reflex,* and give some examples.
9. Describe the second messenger theory.
10. Describe the two-step mechanism of hormone action.
11. How are the second messenger theory and the two-step mechanism similar, and how are they different?
12. Give several biological reasons for the following observations: (a) the endocrine response tends to be delayed; (b) the endocrine response tends to be prolonged; (c) some hormones perform several different functions.
13. List the hormone(s) involved in each of the following functions: blood glucose levels, growth, milk production, milk let-down, calcium levels, and metabolic rate.
14. Describe the role of the hypothalamus in controlling anterior pituitary hormone secretion.
15. Describe the ways in which the posterior pituitary differs from the anterior pituitary.
16. Explain the hormonal reasons for each of the following conditions: acromegaly, dwarfism, and giantism. Acromegaly and giantism are both caused by the same problem. Why are these conditions so different?
17. ACTH is controlled by levels of glucocorticoid and by a biological clock. How are the controls different?
18. Describe the neuroendocrine reflex involved in prolactin secretion.
19. Offer some possible explanations for the following experimental observation: Milk production occurs late in pregnancy and is thought to be stimulated by the hormone prolactin. A nonpregnant rat is injected with prolactin but does not produce milk.
20. Where is ADH produced? Where is it released? Describe how ADH secretion is controlled. What effects does this hormone have?
21. Where is oxytocin produced? Where is it released? Describe how oxytocin secretion is controlled. What effects does this hormone have?
22. A patient comes into your office. She is thin and wasted and complains of excessive sweating and nervousness. What tests would you run?
23. A patient comes into your office. He is suffering from indigestion, depression, and bone pain. An X-ray of the bone shows some signs of osteoporosis. You think that the disorder might be the result of an endocrine problem. What test would you order?
24. How are the two basic types of diabetes mellitus different? How are they similar? How are they treated, and why are these treatments chosen?
25. Describe the physiological changes that occur under stress. What hormones are responsible for them?
26. What is gluconeogenesis? What hormones stimulate the process?
27. Cortisone depresses the allergic response. A patient comes to your office and asks that you treat her allergies with cortisone. What would you tell her?
28. Aldosterone is a mineralocorticoid. Describe its chief functions. How does it help retain body fluid? Under what conditions is aldosterone secreted?

PART

V

The Continuation of Life

Reproduction and Development

CHAPTER

21

Reproduction in Animals

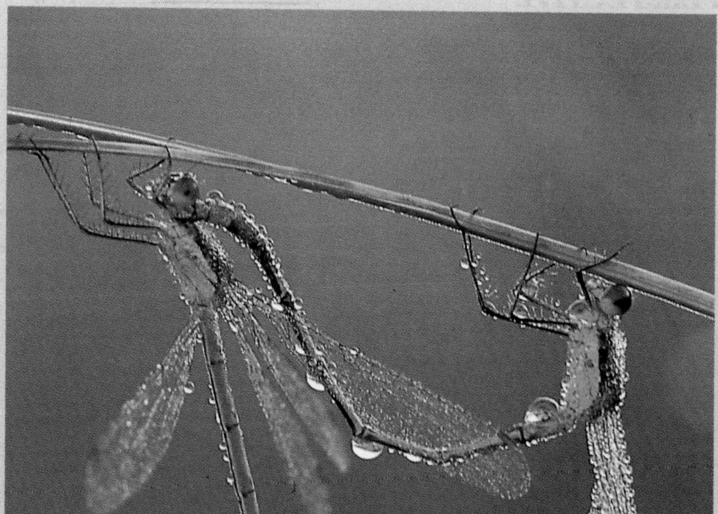

Damselflies mating. Male injects sperm into female, which fertilize the eggs.

Sandra Collins woke one day with a terrible pain in her abdomen, which persisted throughout the morning. Instead of calling her doctor, though, she shrugged off the pain and went shopping. While the young woman was browsing through a bookstore, the pain grew worse, she blacked out, and she was rushed to the emergency room by her husband. There, it was found that Sandra was suffering from severe internal bleeding caused by an ectopic pregnancy—that is, a fertilized ovum that had developed in the upper part of her reproductive tract, outside the uterus.

Fortunately for Sandra, physicians were able to counteract her severe blood loss with a transfusion. They then whisked her into the operating room, where they surgically removed the fetus, placenta, and surrounding tissue and repaired torn blood vessels.

Reproduction is one of the most basic body functions. And as this account shows, it doesn't always operate smoothly. Like many other body functions, reproduction is controlled by the endocrine and nervous systems. This chapter describes reproduction in the animal kingdom, with special emphasis on vertebrate reproduction as exemplified by humans. It will provide a wealth of information and insights that will help you understand Chapter 22 which deals with fertilization and development. We begin our study with a look at reproductive strategies in the animal kingdom.

REPRODUCTIVE STRATEGIES: AN EVOLUTIONARY PERSPECTIVE

A friend of mine was examining a copy of a book on human biology I had written and remarked that there seemed to be an awful lot of information on reproduction. (Two chapters out of 22 were on reproduction and development.) "Isn't your coverage excessive?" he wondered out loud. Before I could answer, he retracted his question. Of course there's a lot on reproduction. It's central to the lives of all organisms!

The reproductive system of animals is not involved in the day-to-day survival of organisms, as are the digestive, nervous, and endocrine systems. It is involved with the survival of entire species. Successful reproduction, in fact, is the measure of the success of any adaptation, whether it occurs in plants, animals, or bacteria.

As you learned in the first chapter of this book, organisms reproduce in one of two ways: asexually or sexually. This section describes each of these strategies in animals and describes their importance.

Asexual Reproduction Involves only One Parent and Tends to Produce Offspring that Are Genotypically and Phenotypically Identical to the Parent

Asexual reproduction occurs when an organism produces an offspring by itself—for example, by budding or fragmenting. Commonly encountered in invertebrates, such as sponges and flatworms, asexual reproduction produces offspring that are genetically identical to their parents and their siblings.

Budding occurs when small outgrowths, or buds, form on the body of the parent (▻ Figure 21–1). These buds often separate from the parent's body and develop into free-living forms. In some cases, however, buds remain attached and produce colonies.

Fragmentation occurs when a body part is lost due to injury. Sea stars, for example, can shed body parts when captured by predators, which permits them to escape death. The lost arm, if still attached to a section of the central disc, may form a new individual. A new arm will form to replace the one lost in the original animal's escape.

Certain free-living flatworms and annelids fragment into two or more pieces, each of which can develop into a new individual (▻ Figure 21–2). Sponges and many sea stars also fragment into numerous pieces, many of which develop into adult forms.

Like many other aspects of life, asexual reproduction has advantages and disadvantages. For example, asexual reproduction eliminates the need for males and females to find each other when mating time approaches. Energy that might otherwise have been spent finding and selecting a mate and mating is devoted to the production of offspring. Consequently, organisms that reproduce asexually often form large populations that rapidly exploit locally available resources.

The most notable disadvantage of asexual reproduction is that it produces genetically similar offspring. As men-

▻ **FIGURE 21–1 Asexual Reproduction by Budding**

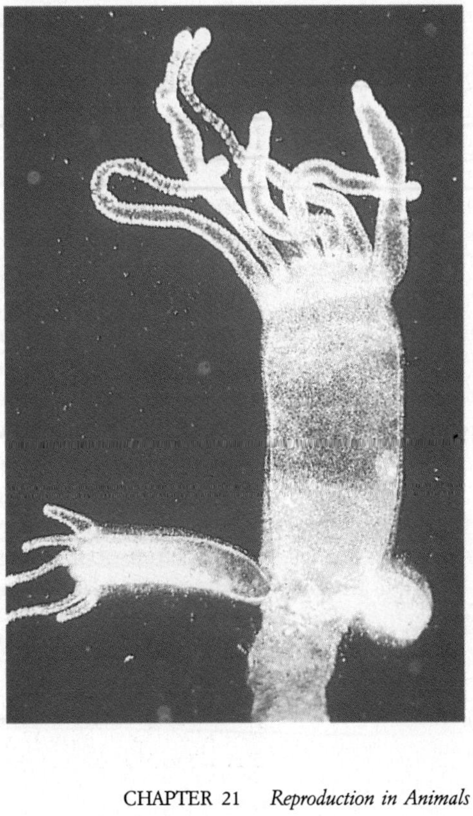

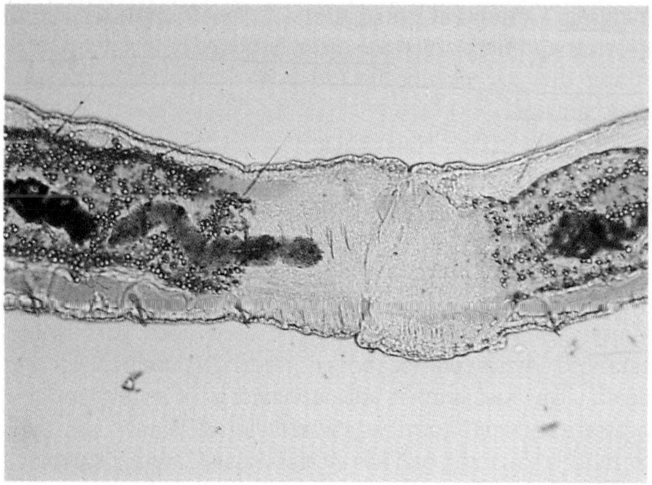

▻ **FIGURE 21–2 Asexual Reproduction by Fragmenting** This marine worm is beginning to fragment, as evidenced by the constriction shown in the center of the photo.

▻ **FIGURE 21–3 The Peccary** This animal which lives in South America, reproduces throughout the year but produces the largest number of offspring during the rainy season when there's ample water for milk production.

tioned, mutations are the only source of genetic variation. Although genetic similarity creates few problems in stable environments, it is a disadvantage when conditions change. If there are no organisms adapted to the change, an entire population can be eliminated.

Sexual Reproduction Involves Gametes that Combine to Form a New Individual

Sexual reproduction involves sex cells, or gametes, each containing half the chromosomes of the adult in which they are produced. The gametes of sexually reproducing organisms are referred to as the egg and sperm. Sperm are typically small, motile cells, and eggs are usually large, nonmotile cells. In most cases, the egg and sperm are produced by separate individuals.

Gametes combine to form a **zygote,** a cell containing a full set of chromosomes, half from the egg-producing female and half from the sperm-producing male. As a result, the offspring are genetically unlike the parents. In other words, the offspring have a different combination of genes. Thus, sexual reproduction produces considerable genetic variation in the offspring, which, in turn, provides an advantage in evolution. Unlike asexually reproducing organisms, sexually reproducing organisms are more likely to survive if environmental conditions change. That is because some members of the population will probably have the right genetic combination and subsequent adaptations that permit them to endure the change.

If you ever worked on a farm or at a zoo, you probably noticed that most sexually reproducing animals breed only once a year. Breeding is timed so that offspring are born into the world at a time when their survival is enhanced. In temperate regions, like the United States, most animals breed in the fall and give birth in the spring.

The timing of breeding in vertebrates and many invertebrates is often regulated by day length, especially in temperate environments where climate varies greatly from one season to the next. Elk and deer in North America, for example, breed in the fall as days shorten. Day length affects the level of a hormone called **melatonin,** which is produced by a small endocrine gland in the brain, the **pineal gland.** Melatonin levels affect reproductive hormones. In environments with more stable conditions such as the tropics, many animals are reproductively active all year.

Other environmental factors may also influence breeding and reproductive success, notably temperature and the availability of food and water. Peccaries, for instance, are piglike animals that live in the arid regions and forests of South America (▻ Figure 21–3). Although they breed year round, peccaries produce the largest number of young in the rainy seasons, when there is ample water for milk production. The breeding cycles of aquatic animals, such as fish and frogs, are influenced by water temperature. Amphibians, for instance, often breed when water temperatures increase.

Sexual reproduction may involve internal or external fertilization. In birds, mammals, and reptiles, for example, sperm are deposited inside the female reproductive system, where fertilization occurs. The fertilized ovum may develop inside the female, as in mammals, which give birth to live young, or it may be deposited in an egg, as in the case of birds and many reptiles (▻ Figure 21–4).

In other vertebrates, such as fish and amphibians, fertilization typically occurs after the eggs are laid, although there are some exceptions (▻ Figure 21–5). In external fertilization, organisms usually shed large numbers of gametes into the water at the same time. The fertilized eggs are often abandoned, so large numbers are necessary to ensure survival. For a unique form of external fertilization see Spotlight on Evolution 21–1.

(a)

(b)

▻ **FIGURE 21–4 Live Birth or Hatching** (*a*) In many animals, eggs are fertilized internally, and the offspring develop inside the female's reproductive system. Babies are born live. (*b*) In others, such as birds, fertilization occurs internally, but eggs are encased in a hard shell and deposited externally. Offspring hatch from the egg after a period of incubation.

▻ **FIGURE 21–5 External Fertilization** Some vertebrates, like these frogs, lay their eggs externally, where they are fertilized.

HUMAN REPRODUCTION

The Male Reproductive System Produces Sperm

In mammals, the male reproductive system consists of seven basic components: (1) the two **testes,** which produce sex steroid hormones and sperm; (2) the two **epididymises,** which store sperm produced in the testes; (3) a pair of ducts, each known as a **vas deferens,** that conducts sperm from the epididymis of each testis to the urethra; (4) **sex accessory glands,** which produce secretions that make up the bulk of the ejaculate; (5) the **urethra,** which conducts sperm to the outside; (6) the **penis,** the organ of copulation, and (7) the scrotum, a sac that houses the testes (▻ Figure 21–6; Table 21–1).

The Testes Are Formed Inside the Body Cavity During Embryonic Development, then Migrate into the Scrotum. The testes are suspended in a pouch known as the **scrotum.** As Figure 21–6 shows, the scrotum is attached to the body below the attachment of the penis. Although the testes reside in the scrotum throughout a man's life, they do not originate there. During embryological development, the testes form inside the abdominal cavity near the kidneys. Toward the end of development, the testes descend into the scrotum through a small tunnel, known as the **inguinal canal,** which links the body cavity with the scrotum (▻ Figure 21–7). The testes are guided into the scrotum with the aid of a ligament that disappears soon afterward.

By the end of the eighth month of development, the testes generally complete their migration into the scrotum. In some males, however, the testes fail to descend, resulting in a condition known as **cryptorchidism.** In many of these boys, the testes descend during the first two years of life. If the testes have not descended on their own by the time a boy is 5 years old, they must be moved into the scrotum surgically to prevent permanent sterility.

▻ **FIGURE 21–6 Anatomy of the Male Reproductive System**

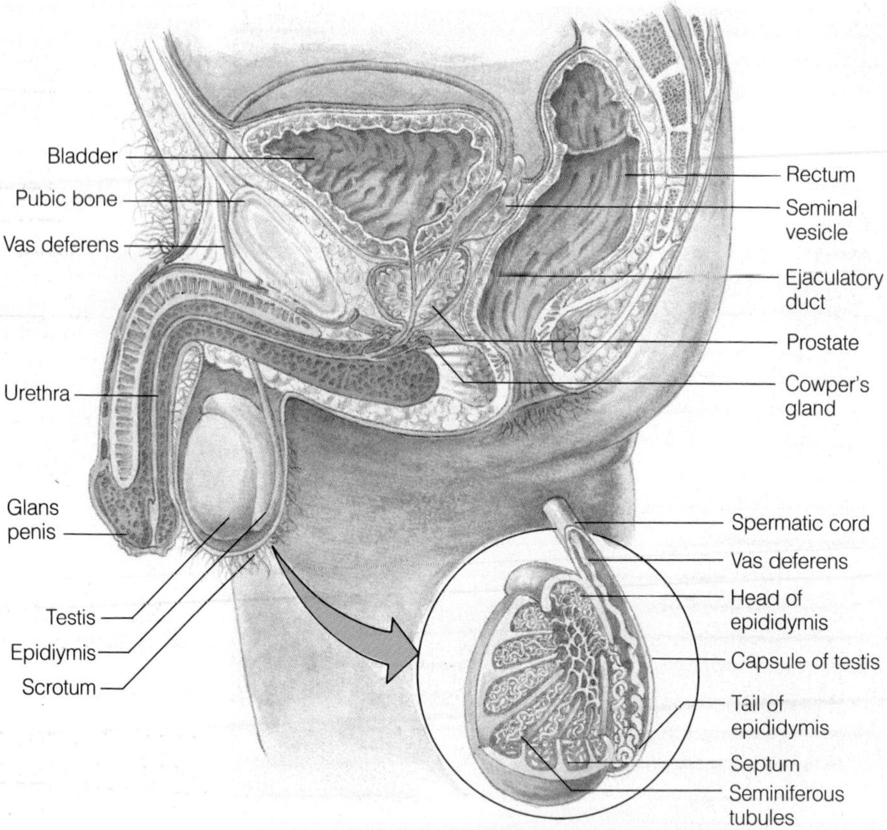

TABLE 21–1 The Male Reproductive System

COMPONENT	FUNCTION
Testes	Produce sperm and male sex steroids
Epididymes	Store sperm
Vasa deferentia	Conduct sperm to urethra
Sex accessory glands	Produce seminal fluid that nourishes sperm
Urethra	Conducts sperm to outside
Penis	Organ of copulation
Scrotum	Provides proper temperature for testes

The inguinal canal through which the testes descend is a biological compromise: it is necessary for the movement of the testes into the scrotum, but remains a weak point in the lower abdomen of men. If the canal does not close properly, organs in the abdomen push aside the muscles that normally keep the canal closed. Loops of intestine may then descend into the inguinal canal, causing pain and discomfort (▻ Figure 21–8). This condition, known as an **inguinal hernia,** can be corrected by surgery in which the weakened muscles are sewn together, thus blocking the canal and preventing intestines from reentering.

The Scrotum Helps Keep the Testes Cool. The scrotum provides an environment whose temperature is

▻ **FIGURE 21–7 The Inguinal Canal** The testis descends through the inguinal canal, an opening through the musculature in the lower abdominal wall. In adults, the inguinal canal provides a route for the vas deferens, blood vessels, and nerves that supply each testis.

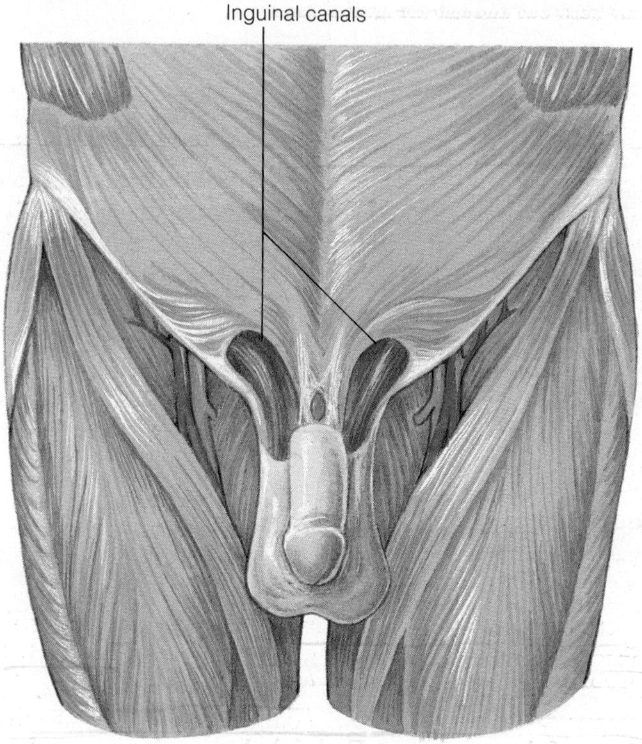

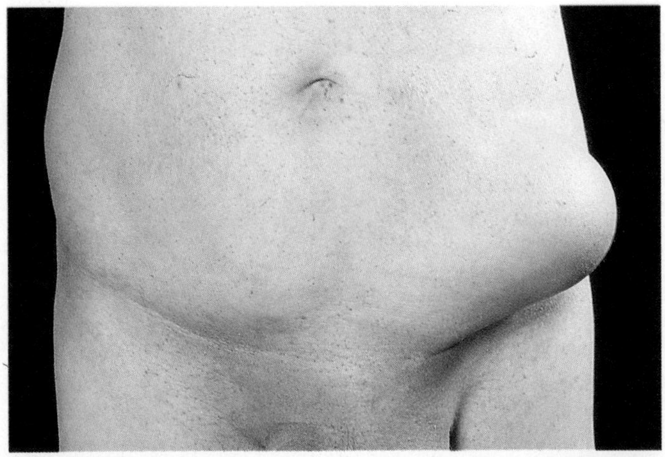

(a)

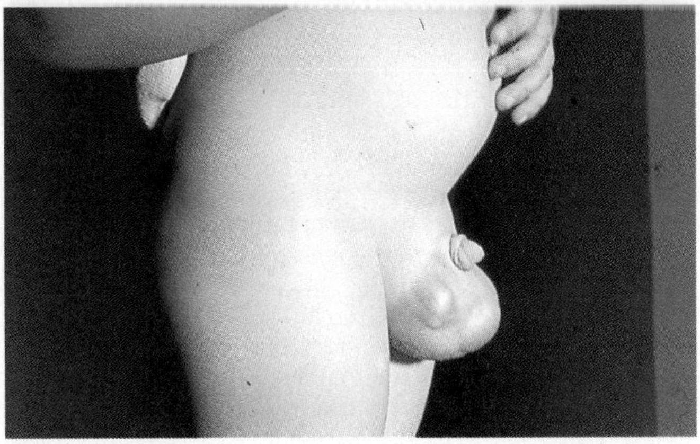

(b)

▻ **FIGURE 21–8 Inguinal Hernia** (*a*) Loops of intestine may push through the weakened musculature surrounding the inguinal canal. (*b*) In some instances, large sections of the intestine may push out through the iguinal canal.

suitable for sperm development. Inside the body cavity, the temperature is too high for sperm development, explaining why men whose testes fail to descend are usually sterile.

The influence of body heat on sperm development is illustrated by the plight of male long-distance runners, who run hundreds of miles a month. This level of exercise elevates their body temperature, and even though the testes are suspended in the scrotum, scrotal temperature may be so high that sperm formation declines. Long-distance runners may become temporarily sterile. Tight-fitting pants or shorts may have the same effect.

Sperm Are Produced in the Seminiferous Tubules and Stored in the Epididymis. As ▻ Figure 21–9 shows, each testis is surrounded by a dense layer of connective tissue. This layer is invested with numerous pain fibers, a fact to which most men will attest. The testis is divided into 200 to 400 compartments, or **lobules,** by fibrous tissue partitions that connect with the connective tissue of the outer coat. Each lobule contains one to four

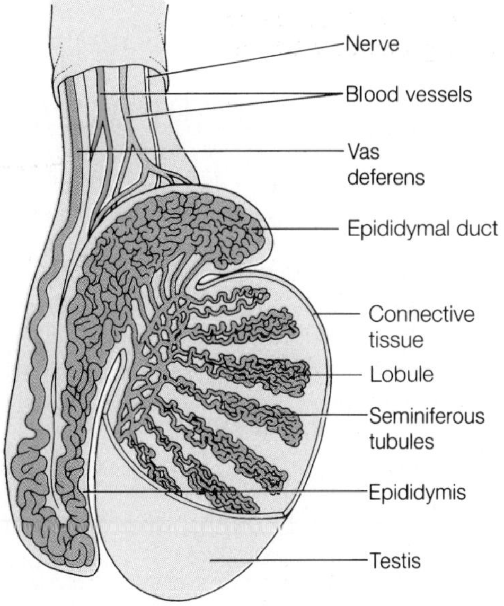

▻ **FIGURE 21–9 Interior View of the Testis** The lobules and seminiferous tubules are shown.

highly convoluted tubes, the **seminiferous tubules,** in which sperm are formed. Stretched end to end, the seminiferous tubules of each testis would extend for about half a mile.

Sperm produced in the seminiferous tubules empty into a network of connecting tubules in the "back" of the testes. These tubules, in turn, empty into the **epididymal duct,** where sperm are stored until released during **ejaculation,** the ejection of sperm. As Figure 21–9 shows, the epididymal duct, which forms the epididymis, empties into the **vas deferens** (plural, **vasa deferentia**). These muscular ducts course upward, passing from the scrotum into the body cavity through the inguinal canals. Inside the pelvic cavity, the vasa deferentia join with the urethra. During ejaculation, the smooth muscles in the walls of the epididymal duct and vasa deferentia contract, propelling sperm to the urethra.

Semen Consists Mostly of Fluids Produced by the Sex Accessory Glands. During ejaculation, sperm are joined by fluids produced by the sex accessory glands. The **sex accessory glands** are located near the neck of the urinary bladder and include the two seminal vesicles, the prostate gland, and the two Cowper's glands, as shown in Figure 21–6. The secretions produced by the sex accessory glands empty into the urethra and the vasa deferentia and make up 99% of the volume of the **ejaculate,** or **semen.** Semen is a fluid that contains sperm and sex accessory gland secretions. The semen contains fructose, a monosaccharide that is used by sperm mitochondria to generate energy needed to help propel the sperm through the female reproductive tract. Semen also contains a buffer that helps neutralize the acidic secretions of the female reproductive tract. Prostaglandin, a chemical substance that

SPOTLIGHT ON EVOLUTION 21–1

THE MOUTH-BREEDING FISH OF LAKE MALAWI

Angel fish are members of one of the largest families of fish in the world, known as cichlids (sick-lids). Members of this large and diverse family can be found in Central America, South America, and Africa. Although they are all interesting, one of the most intriguing is a group from eastern Africa's Lake Malawi (▷ Figure 1).

Located in the Great Rift Valley near the equator, Lake Malawi is the sixth largest in the world and reportedly contains the largest number of fish species, most of them belonging to the cichlid family. The cichlids of Lake Malawi are some of the most colorful freshwater fish known to science. Brilliant blues and yellows decorate some of these remarkable creatures. Two favorites are the Cobalt Blue and the Peacock, shown in the figure.

The diversity of Lake Malawi is often attributed to the lake's size and to the variety of geographically isolated habitats in this massive body of water. At one time, the lake was much lower, and as water levels rose, the fish spread into new habitats. In new habitats, they evolved unique coloration and unique modes of feeding, reflected largely in differences in jaw structure.

One feature that sets the African cichlid apart from most other fish is its breeding behavior. Males and females engage in a fascinating courtship unlike virtually any other fish species. Males actively pursue females, then attract their attention by holding their fins erect and vibrating wildly. If the female is ready, she approaches, aligning herself toward the male's tail. She often pecks at prominent eggs spots on the male's anal fin. Aligned head to tail, the male and female form a circle with their bodies. Then, the female drops an egg. The male and female quickly change positions, with the male situated over the egg. Hovering above it, he releases sperm. The female then picks up the fertilized egg in her mouth and deposits another one. The mating dance goes on until all of the eggs have been laid and fertilized.

Most fish produce large quantities of eggs, which are fertilized en masse by the males. After fertilization, both parents generally abandon the eggs. The lack of parental care is offset by the large number of eggs. In sharp contrast, African cichlids produce only about 5 to 30 eggs. (Larger females may lay more.) Mouth-breeding may have evolved as a means of compensation for the small number of eggs, which are held in the female's mouth throughout development. Seven to 10 days after fertilization, the eggs hatch inside the mother's mouth. The tiny fry remain in her mouth, where they live off their yolk sac and feed off algae and bacteria in the water. Ten days after hatching, the young are fully formed but still remain in the protective custody of the mother's mouth another two weeks. At this time, the young are "spit out", but they are often free to seek shelter inside their mother's mouth should danger arise. An aquarist approaching the tank will often send the swarm of fry back into the mother's mouth in a flash as if they were being vacuumed up.

During this entire development period, females of a few species nibble on food without swallowing their young, but most African cichlid females refuse to eat. During their long fast, the female's abdomen becomes concave, but she refuses food—this despite the fact she is carrying a mouthful of eggs and fry, a delicacy among fish.

The care given to the young of African cichlids may be an evolutionary adaptation to the small number of eggs and to the conditions of the lake—perhaps a reflection of heavy predation when water levels were lower. Whatever the reason, the African cichlids have evolved one of the most amazing forms of maternal behavior in the animal kingdom, a source of fascination to countless fish fanciers the world over.

▷ **FIGURE 1 African Cichlids** (*a*) Cobalt Blue and (*b*) Peacock.

causes the muscle of the uterus to contract, is also found in the ejaculate. Muscular contraction is believed to be largely responsible for the movement of sperm up the female tract.

Interestingly, lymphocytes "patrol" the testes and seminal vesicles. Some lymphocytes are expelled with the ejaculate, explaining why AIDS is transmitted in the semen of men.

The paired **seminal vesicles,** which empty into the vasa deferentia, produce the largest portion of the ejaculate.

The **prostate gland,** whose contents empty directly into the urethra, surrounds the neck of the bladder. Routine medical examinations of men over the age of 45 show that nearly all of them have enlarged prostates. Prostatic enlargement results from the formation of small nodules inside the gland. These nodules form by the condensation of prostatic secretions inside the gland (▻ Figure 21–10). Although they usually cause no trouble, in some men the nodules grow quite large, blocking the flow of urine and making urination painful. In such cases, the prostate may be reamed out by a device inserted through the penis or may be removed through surgery. The prostate is also a common site for cancer in men and therefore should be checked regularly by a physician. The **Cowper's glands,** a pair of small glands located below the prostate on either side of the urethra, are the smallest of the sex accessory glands.

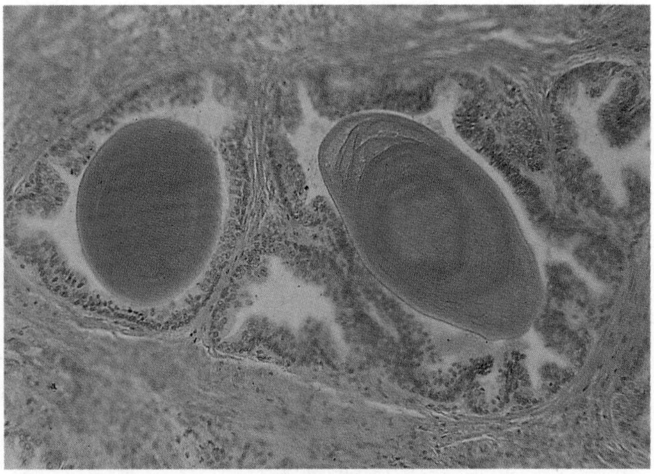

▻ **FIGURE 21–10 Prostatic Accretion**

Sperm Are Formed from Stem Cells Known as Spermatogonia. Sperm are produced in the seminiferous tubules. ▻ Figure 21–11a shows a cross section through two seminiferous tubules. The lining of the wall of the tubule is known as the **germinal epithelium,** because its cells give rise to male germ cells, sperm. The formation

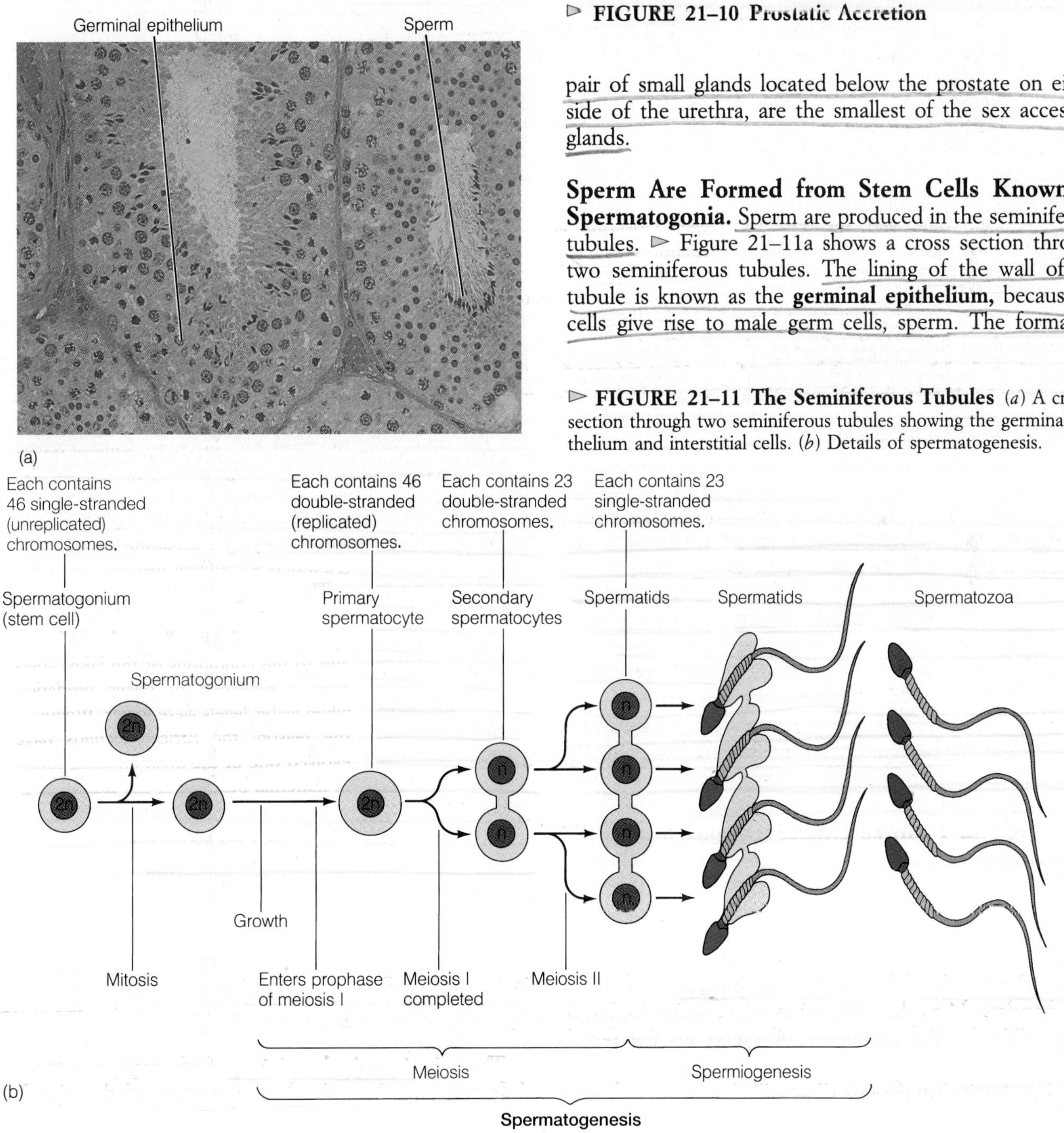

▻ **FIGURE 21–11 The Seminiferous Tubules** (*a*) A cross section through two seminiferous tubules showing the germinal epithelium and interstitial cells. (*b*) Details of spermatogenesis.

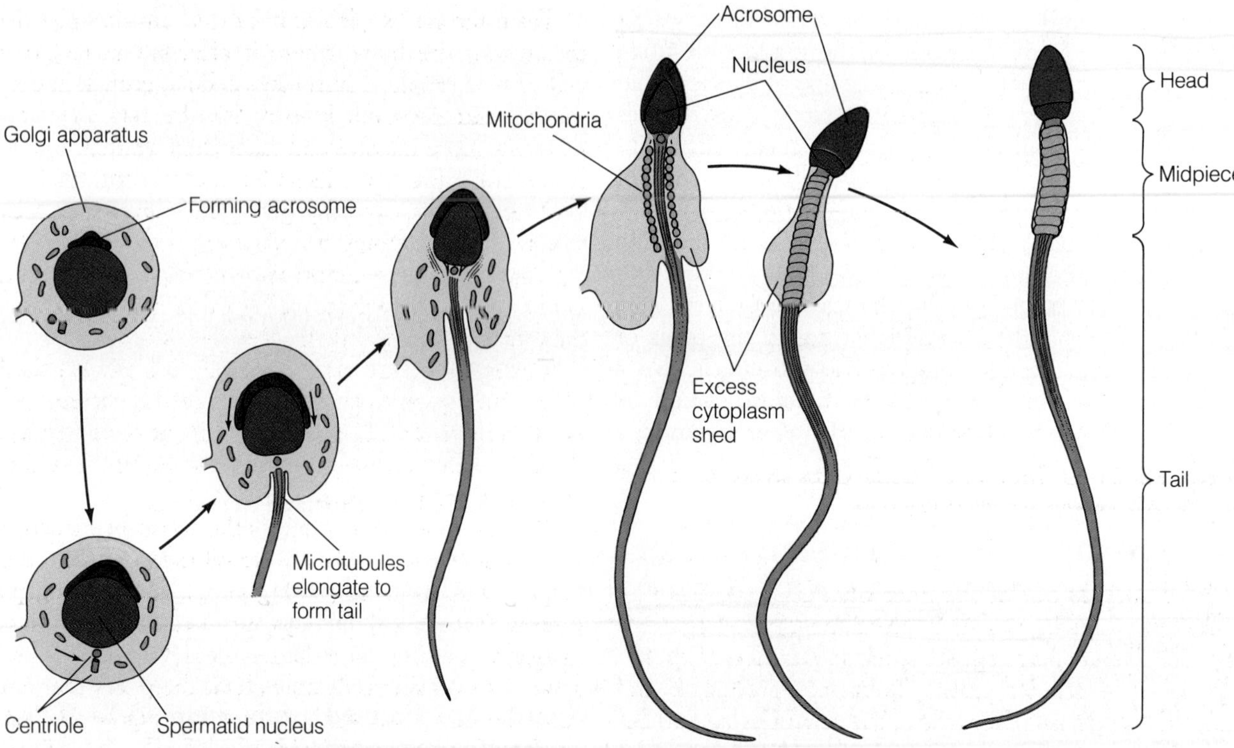

▻ **FIGURE 21–12 Sperm Formation** Sperm form from spermatids in the germinal epithelium. Note the following changes: nuclear condensation, loss of cytoplasm, tail formation, alignment of the mitochondria, and acrosome formation.

of sperm is known as **spermatogenesis.** This phenomenon involves two processes: (1) a special type of cell division known as **meiosis** (described in Chapter 8) and (2) **spermiogenesis,** a process of cellular differentiation (Figure 21–11b).

Spermatogenesis begins with spermatogonia. Located in the periphery of the seminiferous tubule in the germinal epithelium, **spermatogonia** divide mitotically, ensuring a constant supply of sperm-producing cells. Some of the spermatogonia formed during cellular division, however, enlarge and become **primary spermatocytes.** Two meiotic divisions follow in the formation of sperm.

As you may recall from Chapter 8, the first division in meiosis is called meiosis I. During meiosis I, primary spermatocytes divide to form two **secondary spermatocytes** (Figure 21–11b). Each secondary spermatocyte contains 23 double-stranded chromosomes. During **meiosis II,** secondary spermatocytes divide, forming four spermatids. Each spermatid contains 23 single-stranded (unreplicated) chromosomes.

Spermatids soon develop into sperm. During this process, shown in ▻ Figure 21–12, the nuclear material of the spermatid condenses and most of the cytoplasm is lost, thus streamlining the cell. The sperm tail forms from the centriole, providing a means for locomotion. The mitochondria of the spermatid congregate around the first part of the tail, where they provide energy for propulsion. The Golgi apparatus enlarges and forms an enzyme-filled cap over the condensed nucleus, the head of the sperm. Called the **acrosome,** this cap will help the sperm digest its way through the coatings surrounding the ovum during fertilization (Chapter 22). The **spermatozoan,** or mature sperm, is a marvel of biological architecture. It is rid of excess cytoplasmic baggage and is streamlined for relatively swift movement.

On average, men produce 200 million to 300 million sperm every day and the average 3-milliliter ejaculate contains 240 million or more—about as many people as there are in the U.S. population. Such large numbers evolved because many sperm are eliminated as they travel through the female reproductive tract and because many sperm are required to dissolve away the ovum's outer coatings, so that one sperm cell can reach the ovum and fertilize it.

In humans, each sperm formed during meiosis contains 23 single-stranded (unreplicated) chromosomes—half the number in a normal somatic cell. Thus, when the sperm unites with an ovum (also containing 23 unreplicated chromosomes) during fertilization, they produce a zygote that contains 46 single-stranded chromosomes. One half of its chromosomes come from each parent.

Cells Lying Between the Seminiferous Tubules of the Testis, Called Interstitial Cells, Produce the Male Sex Steroid Testosterone. The spaces between the seminiferous tubules contain clumps of large cells known as **interstitial cells** (▻ Figure 21–13). These cells produce **androgens,** steroids that exert a masculinizing effect. The most important androgen is **testosterone.** Tes-

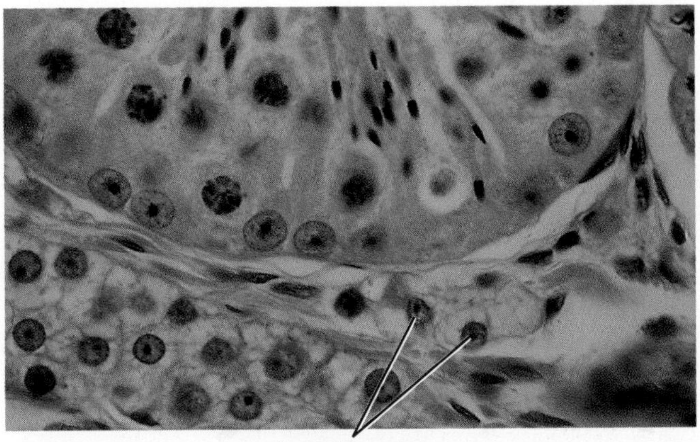

▻ **FIGURE 21–13 Interstitial Cells** Cross section of seminiferous tubules showing the interstitial cells.

tosterone diffuses out of the interstitial cells and into the seminiferous tubules, where it stimulates spermatogenesis (prophase I) and spermiogenesis, the maturation of sperm. In the absence of testosterone, sperm cell production declines, then stops, and the walls of the seminiferous tubules shrink.

▻ **FIGURE 21–14 Pattern Baldness** Some men are genetically predisposed to develop pattern baldness.

Testosterone is also transported in the bloodstream throughout the body, where it affects a variety of target cells. For example, it stimulates cellular growth in bone and muscle and accounts in part for the fact that men are generally more massive and taller than women. In addition, testosterone promotes facial hair growth and thickening of the vocal cords, typically giving men deeper voices than women. Testosterone stimulates growth of the laryngeal cartilage, producing the prominent bulge called the Adam's apple. It also stimulates cell growth in the skin, making most men's skin slightly thicker than women's.

Testosterone also affects the hair follicles on the heads of genetically predisposed men, causing pattern baldness (▻ Figure 21–14). It is not the absence of testosterone, as some believe, but the presence of testosterone and certain genes that lead to this condition.

Testosterone also stimulates the sebaceous glands of the skin. **Sebaceous glands** secrete oil (**sebum**) onto the skin, helping moisturize it (▻ Figure 21–15). During puberty (sexual maturation) in boys, testosterone levels rise dramatically, causing a marked increase in sebaceous gland activity. Dead skin cells may block the pores that normally carry the oil to the skin's surface (▻ Figure 21–16). As a result, sebum collects inside the glands. Bacteria on the skin often invade and proliferate in the small pools of oil, resulting in inflammation, pus formation, and swelling. The skin protrudes, forming an **acne pimple.**

Mild acne can be treated by washing the skin twice a day with unscented soap. Women should avoid makeup that has an oily base or use a nonoily type of foundation and wash their faces thoroughly each night. Sunlight also

▻ **FIGURE 21–15 Sebaceous Gland** The sebaceous glands are associated with hair follicles. They produce oil, which seeps onto the skin surface from the follicles.

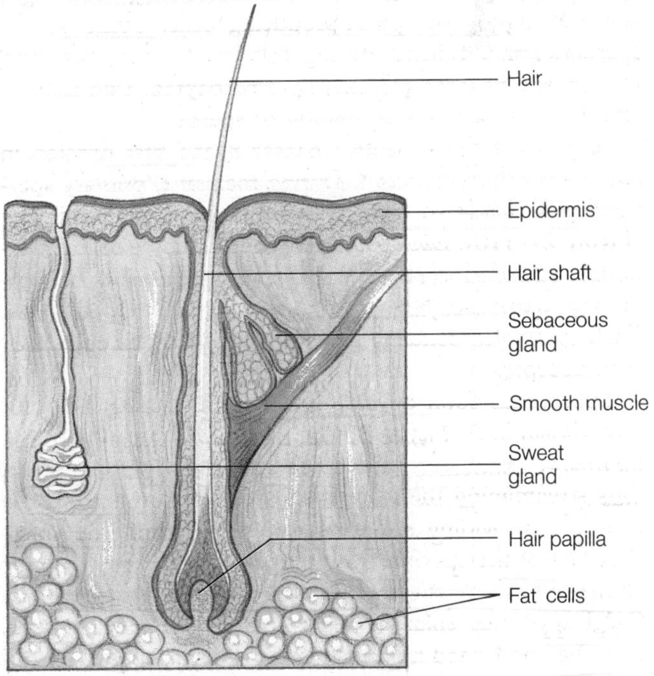

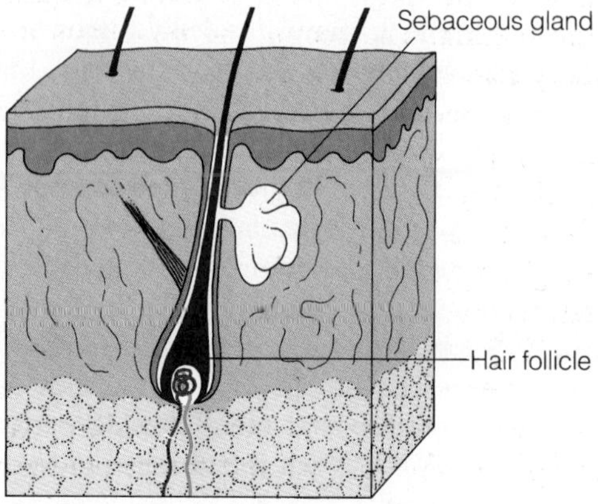

(a) Sebaceous glands associated with hair follicles secrete sebum, an oily substance that lubricates the skin and hair.

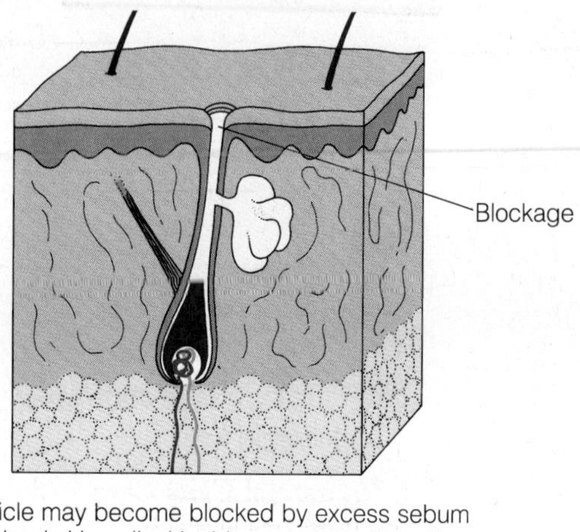

(b) A follicle may become blocked by excess sebum and dead skin cells. Unable to escape, the sebum builds up in the hair follicle.

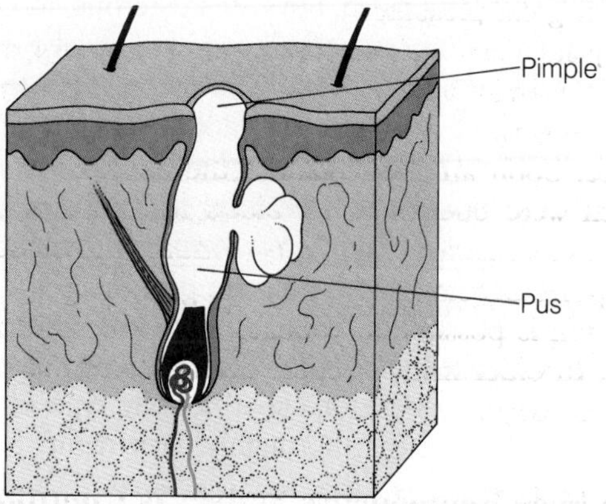

(c) Bacteria present on the skin may infect the sebum, causing inflammation; pus and swelling form an acne pimple.

▻ **FIGURE 21–16 Formation of an Acne Pimple** (*a*) Testosterone stimulates oil production in the sebaceous glands. (*b*) If the outlet is blocked, sebum builds up in the gland and (*c*) may become infected.

helps clear up acne, because it dries the oil on the skin and kills skin bacteria. Severe acne can be treated by special ointments and antibiotics.

The Penis Contains Erectile Tissue that Fills with Blood During Sexual Arousal. In order for fertilization to occur, many millions of sperm must be deposited in the female reproductive tract. The **penis** serves as the copulatory organ, which in humans deposits sperm inside the woman's vagina. As illustrated in ▻ Figure 21–17, the penis consists of a shaft of varying length and an enlarged tip, known as the **glans penis.** The glans is covered by a sheath of skin at birth, the **foreskin.** The foreskin gradually becomes separated from the glans in the first two years of life. At puberty, the inner lining of the foreskin begins to produce an oily secretion called smegma. Bacteria can grow in the protected, nutrient-rich environment created by the foreskin, so special precautions must be made to keep the area clean.[1]

Because of potential health problems or religious reasons, parents may opt to have the foreskin removed in the first few days of their son's life. The operation, called **circumcision** (literally, "to cut around"), may help reduce penile cancer in men and may also reduce cervical cancer in the wives or sexual partners of circumcised men, as explained in the Point/Counterpoint in this chapter.

The penis becomes rigid, or erect, during sexual arousal. Erection is mediated by neurons belonging to the autonomic nervous system. The autonomic nervous system, discussed in Chapter 17, consists of two functionally and anatomically different divisions. The parasympathetic division is responsible for erection, among other functions. The sympathetic division is responsible for ejaculation.

During sexual arousal, nerve impulses in the parasympathetic division of the autonomic nervous system cause arterioles in the penis to dilate. Blood flows into a spongy **erectile tissue** in the shaft of the penis, making it harden. The growing turgidity compresses a large vein on the dorsal surface of the penis, blocking the outflow of blood and further stiffening the organ.

Coursing through the penis is the urethra, a duct that carries urine from the bladder to the outside of the body during urination. The urethra also transports semen, sperm, and secretions of the sex accessory glands during ejaculation.

Some men lose their ability to become erect or to sus-

[1] Parents must routinely clean the area in children once the foreskin becomes separated from the glans.

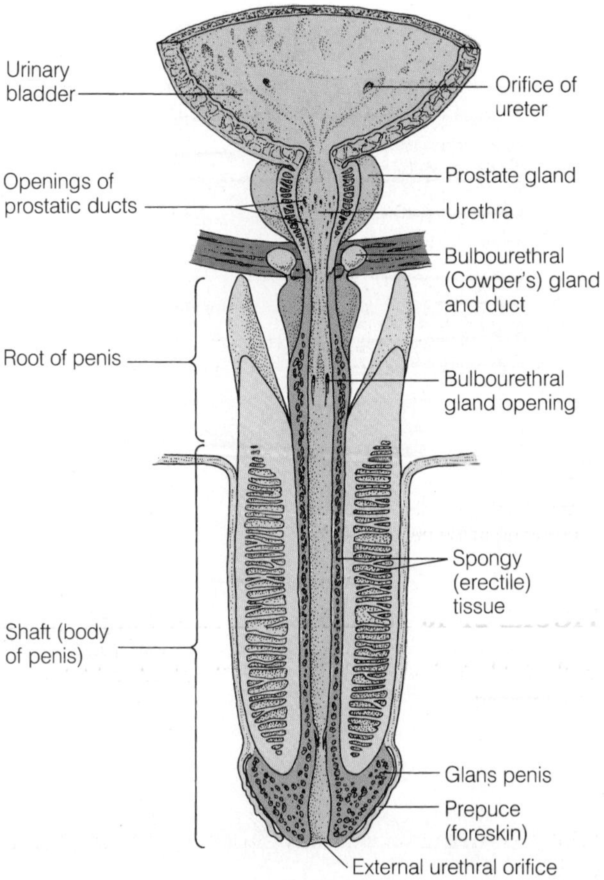

▻ **FIGURE 21–17 Anatomy of the Penis** The penis consists principally of spongy tissue that fills with blood during sexual arousal. The urethra passes through the penis, carrying urine or semen.

tain an erection. This condition, known as **impotence,** may be caused by psychological, physical, or physiological problems. In most cases, pinpointing the exact cause is difficult. For example, marital conflict, stress, fatigue, and anxiety can all lead to impotence. If the problem is psychological, therapy is often advised. Patients with nerve damage, however, are not so lucky; permanent impotence is likely. Nerve damage may result from diabetes mellitus or from traumatic accidents. For patients with irreversible impotence, urologists can surgically insert an inflatable plastic implant in the penis. The penile implant is attached to a small, fluid-filled reservoir in the scrotum. The fluid is manually pumped into the implant upon demand, making the penis erect and permitting sexual intercourse.

Ejaculation Is a Reflex Mechanism. When sexual stimulation becomes intense, sensory nerve impulses traveling to the spinal cord activate motor neurons there. These neurons send impulses to the smooth muscle in the walls of the epididymes and vasa deferentia, causing them to contract. This action, in turn, propels sperm into the urethra. Nerve impulses emanating in the motor neurons of the spinal cord also stimulate the smooth muscle in the walls of the sex accessory glands to contract, causing these glands to empty their secretions into the vasa deferentia and the urethra. The sperm and secretions from the sex accessory glands combine and form semen.

Semen is then propelled onward by smooth muscle contractions in the walls of the urethra, which cause the sperm to be released in spurts. The contractions in the urethra, like those in the epididymes, vasa deferentia, and sex accessory glands, are caused by nerve impulses from motor neurons involved in this spinal cord reflex.

According to the world-renowned sex researchers William Masters and Virginia Johnson, the male sexual response consists of four parts: the excitement phase, the plateau phase, the orgasm, and the resolution phase (explained in Table 21–2). Sexual arousal in the first two phases is accompanied by a tensing of the body muscles, rapid breathing, and increased blood pressure. Ejaculation occurs during the orgasm, or orgasmic phase. The "frenzy" of muscle contraction that occurs during ejaculation brings with it great pleasure.

Ejaculation is quickly followed by muscular and psychological relaxation (which explains why many men fall asleep after orgasm). This relaxation is part of the resolution phase. Soon after ejaculation, the arterioles in the penis, which were opened to let blood flow in during erection, begin to constrict. This action reduces the blood flow, and the penis becomes flaccid once again. In general, another erection is possible in younger men within 10 to 15 minutes. In older men, a repeat performance may take hours or even days.

The Male Reproductive System Is Controlled by Three Hormones, Testosterone, Luteinizing Hormone, and Follicle-Stimulating Hormone

As noted earlier, the testes produce sex steroid hormones—notably, testosterone. Besides influencing spermatogenesis, the male sex steroid hormones are responsible for **secondary sex characteristics**—that is, male physical features such as facial hair growth, greater muscle and bone development, and deeper voice.

Testosterone secretion by the interstitial cells is controlled by luteinizing hormone (LH). LH in males is also known as **interstitial cell–stimulating hormone (ICSH),** for obvious reasons. ICSH secretion, in turn, is controlled by a releasing hormone produced by the hypothalamus, known as gonadotropin-releasing hormone (GnRH).

As ▻ Figure 21–18 shows, the secretion of GnRH and ICSH is controlled by testosterone levels in the blood in a classic negative feedback loop. A decline in testosterone levels in the blood, for example, signals an increase in GnRH secretion, resulting in an increase in ICSH secretion. But when testosterone levels return to normal, GnRH and LH release subside. This feedback loop explains why athletes who use synthetic anabolic steroids (androgens) experience a decline in testicular size (Chapter 19).

The pituitary also produces the gonadotropin follicle-

TABLE 21–2 Male Responses During the Sexual Response Cycle

Excitement Phase
1. Vasocongestion (accumulation of blood) erects the penis.
2. Scrotal skin tightens.
3. Testes start to increase in size.
4. Testes and scrotum become elevated.
5. Nipples may become erect.
6. Some increase occurs in muscle tension, heart rate, and blood pressure.

Plateau Phase
1. There is a slight increase in the area of the glans penis.
2. Vasocongestion purples the glans.
3. Testes continue to elevate up into the scrotal sac until they are positioned close against the body.
4. Testes increase in size as much as 50%.
5. Cowper's glands secrete a few drops of fluid.
6. Possible sex flush and muscle tension are present, breathing is rapid, and heart rate increases (100 to 160 beats per minute).

Orgasm
1. Contractions of the vas deferens, seminal vesicles, ejaculatory duct, and prostate gland cause semen to collect in the base of the urethra.
2. Collection of semen in the base of the urethra produces feelings of ejaculatory inevitability.
3. The internal sphincter in the prostate contracts, preventing passage of urine.
4. The external sphincter in the prostate relaxes, allowing passage of semen.
5. Contractions in the urethra propel the semen out of the penis. The contractions occur four or five times at intervals of eight-tenths of a second.
6. Muscles go into spasm throughout the body, respiration increases, and blood pressure and heart rate reach a peak (about 180 beats per minute).

Resolution Phase
1. Body returns gradually to its prearoused state.
2. Male gradually loses his erection.
3. Testes and scrotum return to normal size.
4. Scrotum regains its wrinkled appearance.
5. Male enters a refractory period (unresponsive to further sexual stimulation).
6. Blood pressure, heart rate, and respiration become normal.
7. About one-third of males find their palms and soles or entire body covered with perspiration.

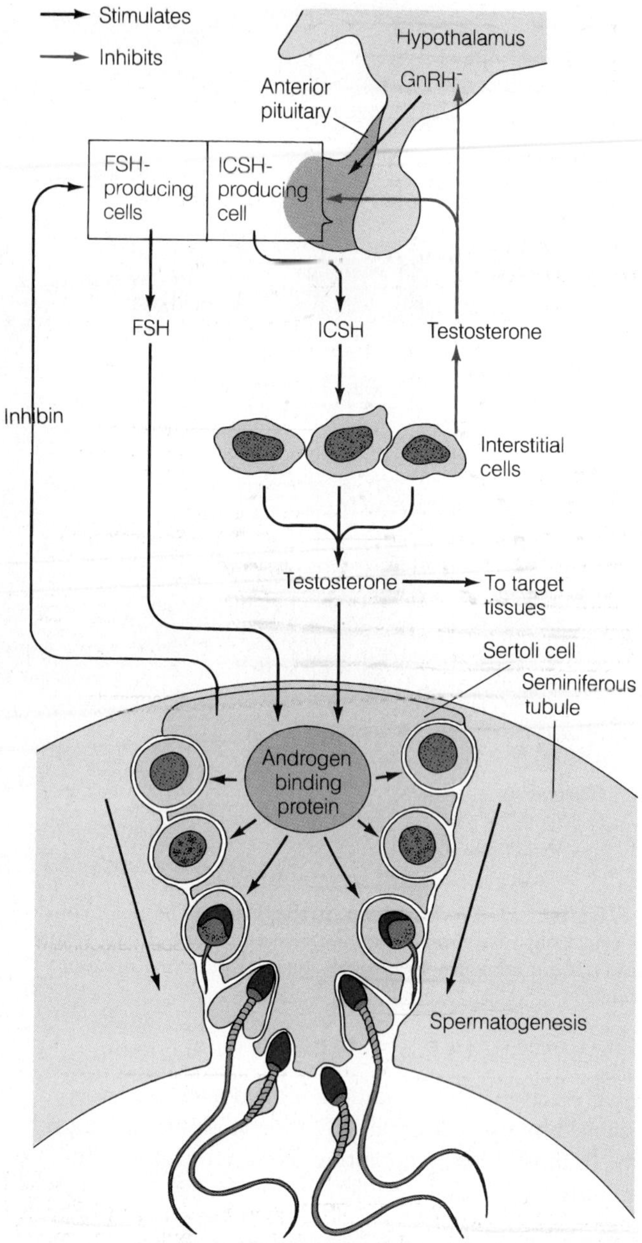

▻ **FIGURE 21–18 Hormonal Control of Testicular Function** Testosterone, FSH, and ICSH participate in a negative feedback loop. The testes also produce a substance called inhibin, which controls GnRH secretion.

stimulating hormone (FSH). As shown in Figure 21–18, FSH stimulates spermatogenesis in conjunction with testosterone. FSH does not act directly on the spermatogenic cells, however. Instead, it exerts its effects through another cell in the germinal epithelium of the seminiferous tubule, the Sertoli cell. **Sertoli cells,** shown in ▻ Figure 21–19, are large "nurse cells." The spermatogenic cells (spermatogonia, spermatocytes, and spermatids) divide and differentiate within folds in the plasma membrane of the Sertoli cell, moving slowly to the surface of the germinal epithelium. The spermatids produced during spermatogenesis remain attached to the Sertoli cells and there differentiate into sperm.

FSH stimulates the Sertoli cells to produce a cytoplasmic receptor protein that binds to androgens, male sex steroid hormones. Called **androgen-binding protein,** this cytoplasmic receptor concentrates testosterone within the Sertoli cell. Testosterone, in turn, stimulates spermatogenesis.

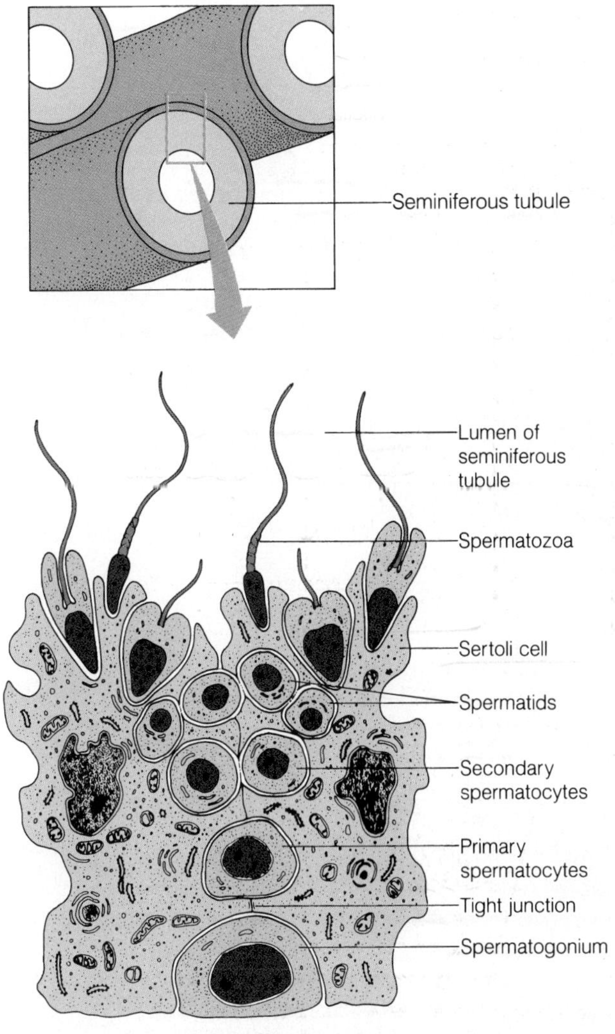

▻ **FIGURE 21–19 Sertoli Cells** These cells encompass the spermatogenic cells as they develop in the germinal epithelium.

Like the release of LH, FSH secretion is controlled by GnRH produced by the hypothalamus. When GnRH is released from the hypothalamus, it travels to the anterior pituitary and there stimulates the release of FSH. Interestingly, the FSH-producing cells are also controlled by a peptide hormone produced by Sertoli cells (Figure 21–18). This hormone, called **inhibin,** apparently inhibits the activity of FSH-secreting cells in the anterior pituitary, thus blocking the action of GnRH. When inhibin levels are high, FSH secretion is low.

The Female Reproductive System Consists of Two Parts: the External Genitalia and the Genital, or Reproductive, Tract

▻ Figure 21–20 illustrates the female reproductive system, which consists of two parts: the external genitalia and the reproductive tract. The **female reproductive tract** consists of four structures: (1) the ovaries, (2) the uterine tubes, (3) the uterus, and (4) the vagina. Table 21–3 offers a quick summary of the role of each structure.

The **uterus** is a pear-shaped organ about 7 centimeters (3 inches) long and about 2 centimeters (less than 1 inch) wide at its broadest point in nonpregnant women.[2] The wall of the uterus contains a thick layer of smooth muscle cells, the **myometrium.** The uterus houses and nourishes the developing embryo and fetus.

Attached to the uterus are two tubes, the **uterine tubes,** or **oviducts.** In humans, the uterine tubes are usually referred to as the **Fallopian tubes.** Ova are produced by the **ovaries,** paired, almond-shaped organs that are attached to the uterus by the ovarian ligaments. As Figure 21–20a shows, the ends of the uterine tubes are widened like a catcher's mitt and fit loosely over the ovaries. Currents created by cilia in the lining of the uterine tubes draw the ova inside and down the tubes to the uterus.

Fertilization occurs in the upper third of the uterine tubes. The fertilized ovum is then transported down the uterine tubes to the uterus. In the uterus, the embryo attaches to the lining, the **endometrium,** and embeds itself there, remaining for the duration of pregnancy.

At birth, the fetus is expelled from the uterus through the **cervix,** the lowermost portion of the uterus. As Figure 21–20 shows, the cervix protrudes into the **vagina,** a distensible, 3-inch, tubular organ that leads to the outside of the body.[3] At birth, the cervix stretches to allow the passage of the baby into the vagina. The vagina also serves as the receptacle for sperm during sexual intercourse. To reach the ovum, sperm must travel through a tiny opening and narrow canal of the cervix that leads into the lumen (cavity) of the uterus. From here, sperm move up both uterine tubes.

The **external genitalia** consist of two flaps of skin on either side of the vaginal opening (▻ Figure 21–21). The outer folds are the **labia majora.** These large folds of skin are covered with hair on the outer surface and contain numerous sebaceous glands on the inside. The inner flaps are the **labia minora.** Anteriorly, they meet to form a hood over a small knot of tissue called the **clitoris.** The clitoris

[2] The uterus is slightly larger in women who have had children and grows considerably during pregnancy to accommodate the growing embryo and fetus.

[3] The vagina is often called the birth canal.

TABLE 21–3 The Female Reproductive System

COMPONENT	FUNCTION
Ovaries	Produce ova and female sex steroids
Uterine tubes	Transport sperm to ova; transport fertilized ova to uterus
Uterus	Nourishes and protects embryo and fetus
Vagina	Site of sperm deposition, birth canal

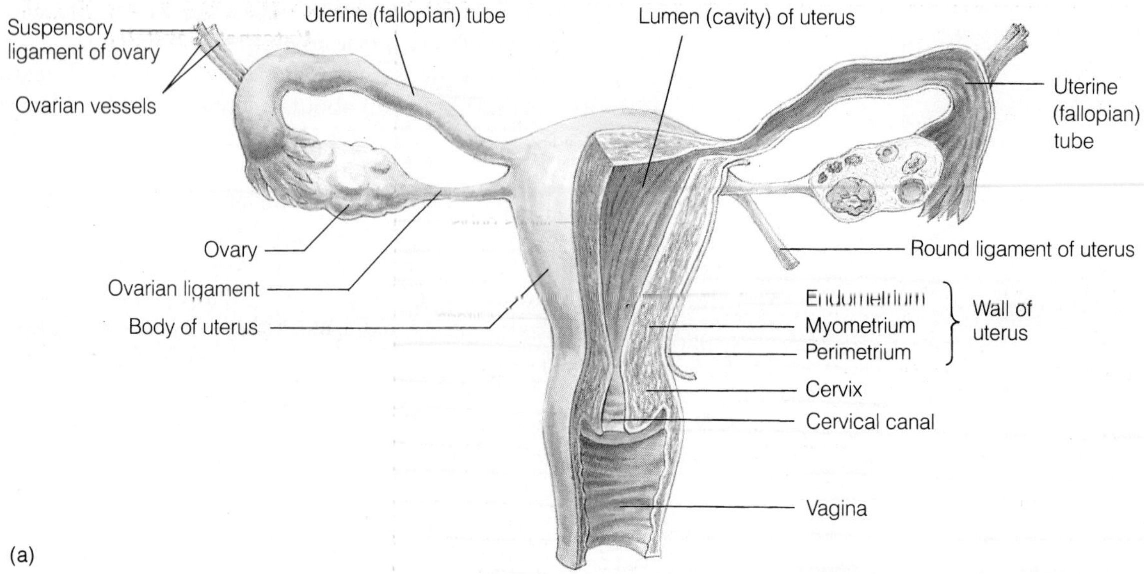

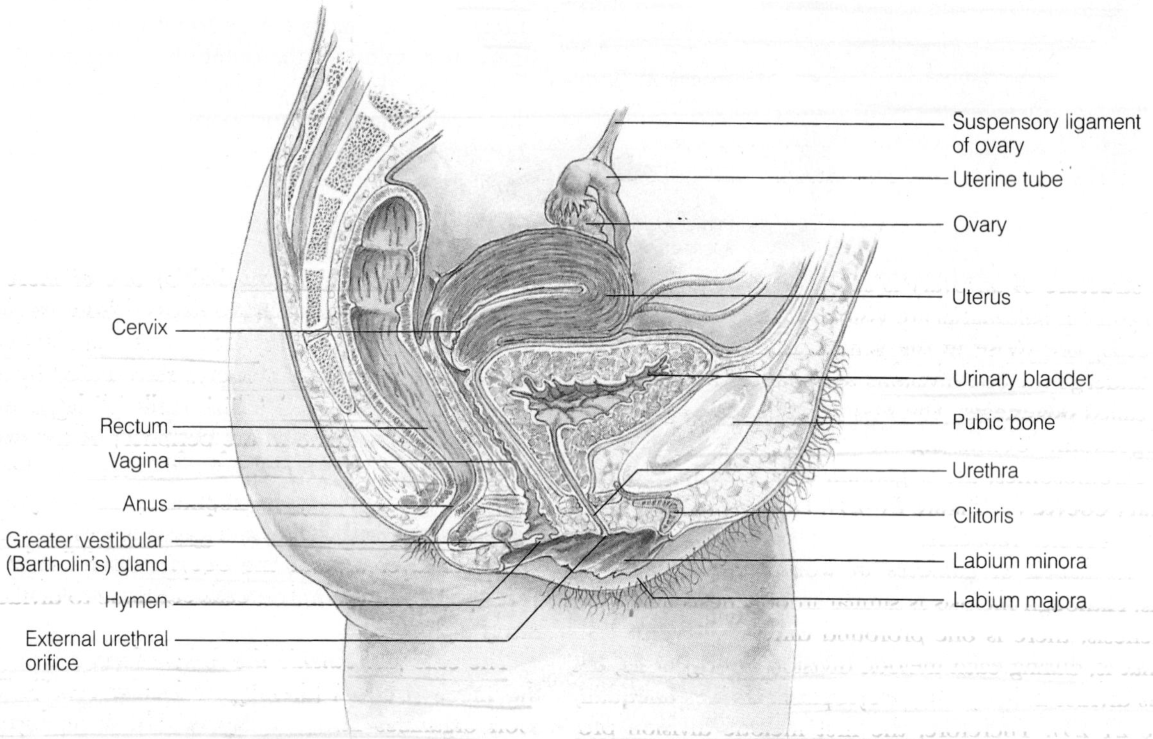

▻ **FIGURE 21–20 Anatomy of the Female Reproductive Tract** (*a*) Frontal view. (*b*) Midsagittal view.

consists of erectile tissue and is a highly sensitive organ involved in female sexual arousal. It is formed from the same embryonic tissue as the penis. In fact, a woman occasionally will be born with a greatly elongated clitoris.

The Ovaries Produce Ova and Release Them During Ovulation. During each menstrual cycle, one of the ovaries releases an ovum, the female gamete. The ovum oozes from the ovary and is drawn into the uterine tube. The release of an ovum is called **ovulation.**[4] Ovulation occurs approximately once a month in women during their reproductive years—that is, from puberty (age 11–15) to menopause (age 45–55). Ovulation is temporarily halted when a woman is pregnant and may even be suppressed by emotional and physical stress.

[4]The release of the ovum is probably not an explosive event, although many women feel a sharp pain when it occurs.

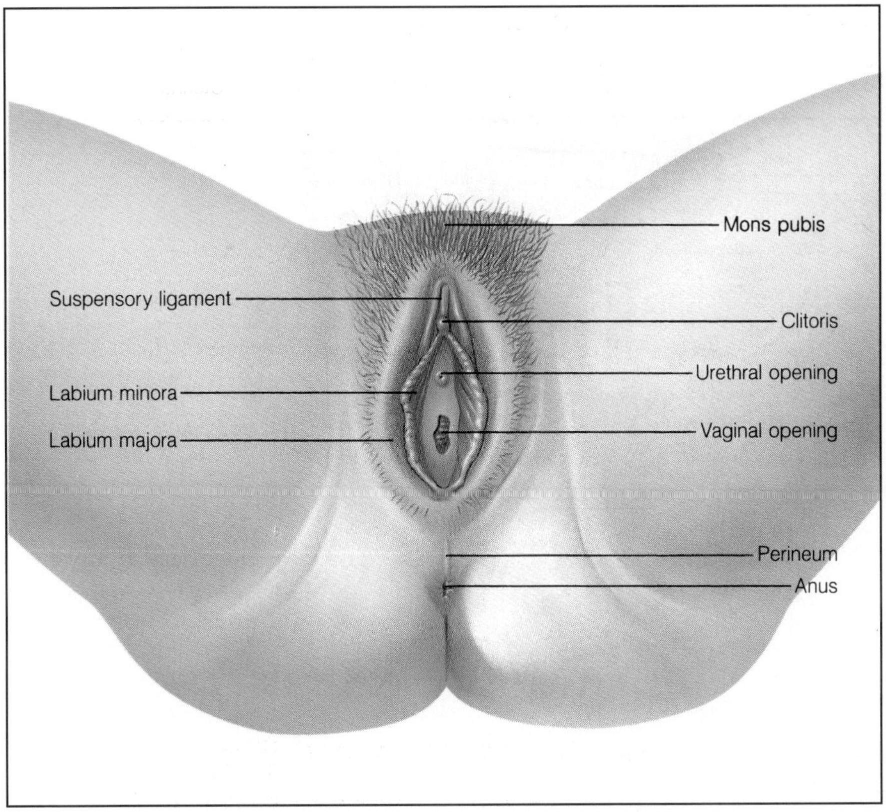

▻ **FIGURE 21–21 Female External Genitalia**

The structure of an ovary is shown in ▻ Figure 21–22. Several ovarian landmarks are visible. One is the germ cell. Germ cells, like those in the seminiferous tubules of the testes, undergo meiotic divisions to produce ova. This process is called **oogenesis.** The first germ cell in oogenesis is the **oogonium.** Containing 46 single-stranded (unreplicated) chromosomes, the oogonium enlarges and becomes a **primary oocyte** (▻ Figure 21–23). During this transition, the chromosomes replicate.

The formation of gametes in women also occurs by meiosis. Although meiosis is similar in oogenesis and spermatogenesis, there is one profound difference worth noting. That is, during each meiotic division in oogenesis, the nucleus divides in half, but the cytoplasm divides unequally (Figure 21–23). Therefore, the first meiotic division produces only one cell, the **secondary oocyte,** and a small package of discarded nuclear material containing 23 double-stranded chromosomes, called the **first polar body,** and a tiny amount of cytoplasm.

During the second meiotic division, the cytoplasm of the secondary oocyte also divides unequally. This "unequal division" results in the formation of an **ovum,** containing 23 single-stranded chromosomes and another "nuclear discard," the **second polar body.** The second meiotic division occurs only after a sperm penetrates the secondary oocyte. In humans and virtually all other animals, the first polar body usually does not divide.

Germ cells are housed in the ovary in special structures called follicles, shown in Figure 21–22a. A **follicle** consists of a primary oocyte surrounded by one or more layers of **follicle cells.** Follicle cells are derived from the loose connective tissue of the ovary. The most abundant of all follicles consists of a primary oocyte surrounded by an incomplete layer of flattened follicle cells. A large number of these follicles is found in the periphery of the ovary.

Each month, a dozen or so of these follicles begin to develop. During early development, the oocyte enlarges. The follicle cells divide and grow, eventually forming a complete layer around the oocyte. As follicular development proceeds, the follicle cells continue to divide, forming many layers.

The cells just outside the follicle form a layer known as the **theca folliculi** (literally, "follicular coat"). The theca soon organizes into two layers. The inner layer of cells contains many capillaries and produces a small amount of androgen (male hormone) that diffuses into the follicles, where it is converted into estrogen by the follicle cells. The outer layer is composed of tightly packed connective tissue cells.

In the largest follicles, a clear liquid begins to accumulate between the follicle cells. The fluid creates small spaces among the follicle cells, which enlarge as additional fluid is generated. Eventually, the cavities coalesce, forming one central cavity. At this point, the follicle is called an **antral follicle.**

As noted, a dozen or so follicles begin developing during each cycle. However, all but one usually degenerate. The follicle or follicles that escape degeneration, however, con-

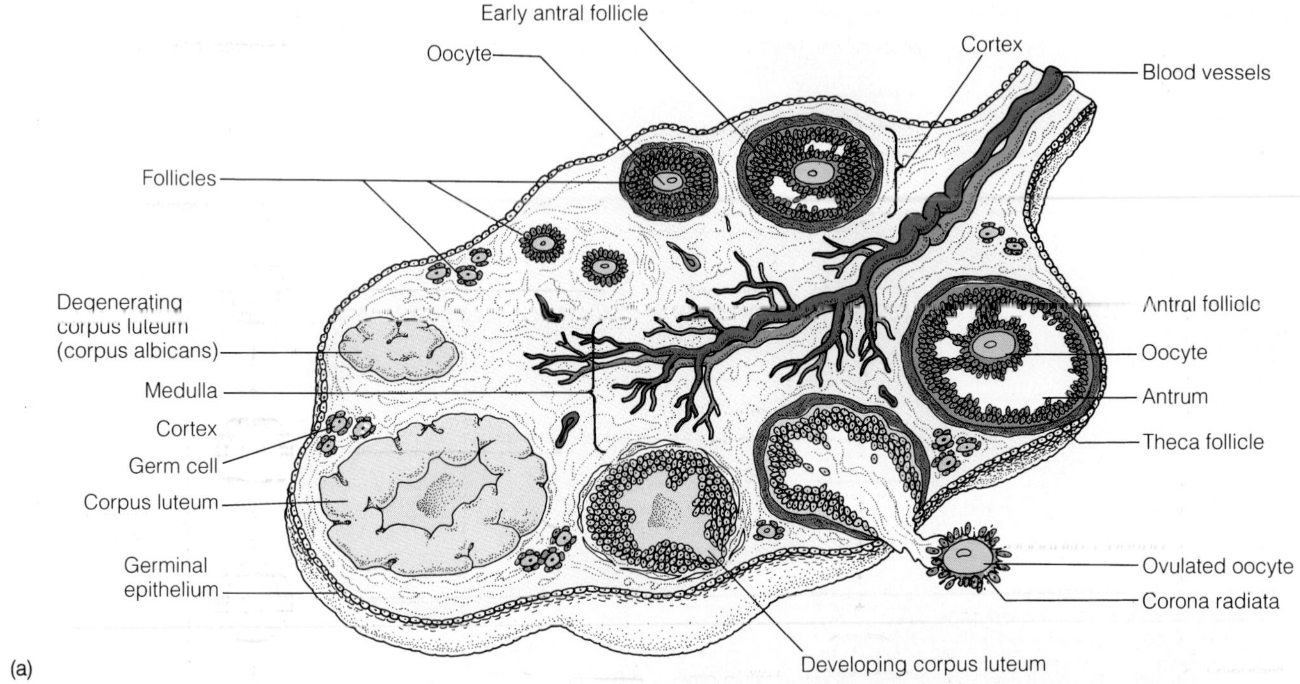

▻ **FIGURE 21–22 Structure of the Ovary** (*a*) This drawing illustrates the phases of follicular development and also shows the formation and destruction of the corpus luteum (CL). Antral follicles give rise to the CL. A fully formed CL and antral follicle would not be found in the ovary at the same time. (*b*) Antral follicle.

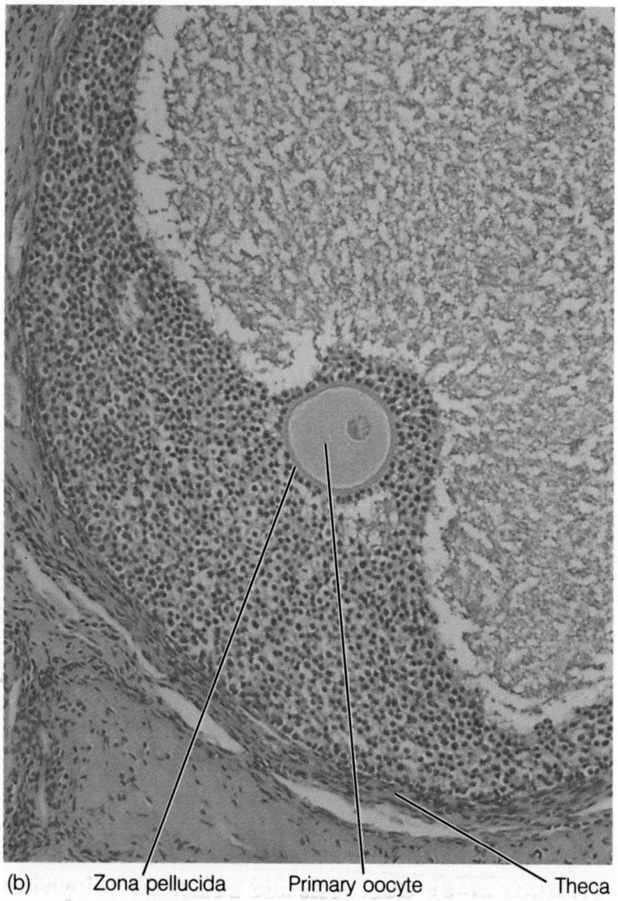

tinue to enlarge. As fluid accumulates, the follicle bulges from the surface of the ovary like a pimple. The pressure exerted on the outside of the ovary causes the ovary's surface to stretch. Blood vessels supplying the tissue may be compressed, resulting in a region of cellular necrosis (death). This weakens the wall. Enzymes released from ovarian cells in the region are then thought to begin to digest the tissue at the weak point. Eventually, the wall of the follicle breaks down, and the oocyte is released.

Around the time of ovulation, the primary oocyte in the antral follicle completes the first meiotic division (Figure 21–23). It is then called a secondary oocyte. As illustrated in ▻ Figure 21–24, the secondary oocyte released from the ovary is surrounded by a layer of follicle cells. Immediately surrounding the oocyte is a fairly thick layer of gel-like material, called the **zona pellucida** ("clear zone"), shown in Figure 21–24. Incoming sperm must penetrate the follicle cell layer and the zona pellucida in order for fertilization to occur. Enzymes released from the acrosome of the many sperm that arrive at the site of fertilization "digest" the molecules that hold the follicle cells together, permitting sperm access to the ovum. Consequently, fertilization is a cooperative venture at first. Large numbers of sperm are needed to get through the follicle-cell barrier. Acrosomal enzymes of each sperm also digest the zona pellucida, helping each sperm penetrate this barrier. Thus, once sperm reach the zona pellucida, fertilization becomes a race. The first sperm to reach the plasma membrane of the ovum wins!

After ovulation, the ovulated follicle collapses. The remaining follicle cells enlarge and multiply. The thecal cells invade the interior of the collapsed follicle, where they proliferate and grow. The follicle and thecal cells transform

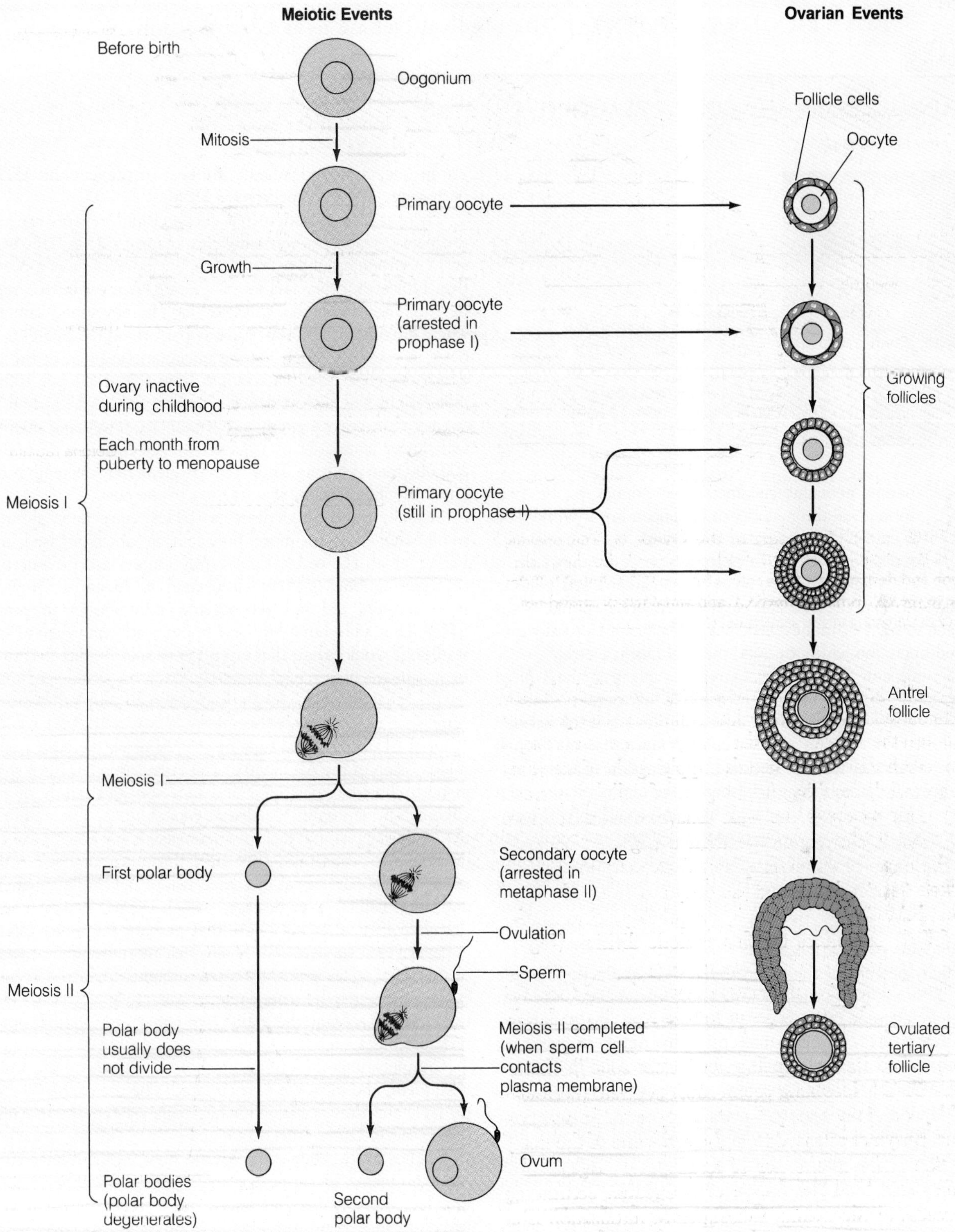

▻ **FIGURE 21–23 Oogenesis and Follicle Development**

the collapsed follicle into the **corpus luteum (CL)** ("yellow body")—so named because of the pigment it contains in cows and pigs, but not humans.

The CL produces two sex hormones, estrogen and progesterone, but the fate of the CL ultimately depends on the fate of the oocyte. If it is fertilized, the CL will remain active for several months, producing estrogen and progesterone needed for a successful pregnancy. If fertilization

AN UNNECESSARY AND COSTLY PRACTICE

Dr. Thomas Metcalf

Dr. Thomas J. Metcalf is a clinical associate professor of pediatrics at the University of Utah and practices at Willow Creek Pediatrics in Salt Lake City.

Routine neonatal circumcision continues to be performed on the majority of American boys. While data is currently emerging on the benefits of circumcision, I remain convinced that routine circumcision of the newborn is not needed if adequate hygiene of the uncircumcised penis is maintained, that circumcision costs the American public a significant amount of money, and that it remains a culturally motivated operation without a valid medical raison d'être.

Circumcision for nonreligious reasons began in the United States in the 1870s. A theory emerged that masturbation was the cause of many illnesses. Genital surgery of both sexes became established as a preventive and/or cure for masturbation. Though genital surgery of females declined, circumcision of males persisted, and became firmly established in the United States, even as it fell into disfavor in Europe. While most medical texts eventually stopped advocating the procedure, it was not until 1949 that various authors began to speak out against newborn circumcision. Amid continued controversy over the benefits and risks of circumcision, the medical community gradually came to the position that "there is no absolute medical indication for routine circumcision of the newborn."

Eighty percent of the world remains uncircumcised. In the U.S., 60% to 90% of male newborns are circumcised. The cost for a circumcision in Utah is $85; in New York the physician's fee alone is $175. Thus the total cost to the American public for newborn circumcisions is somewhere between $153 million and $360 million. Are there compelling reasons to circumcise every newborn boy at such a cost?

Wiswell and other authors have documented a greater incidence of urinary tract infections (UTIs) in uncircumcised male infants, roughly 1% to 2% versus 0.1% in circumcised infants. This may be due to more frequent and heavier colonization of the periurethral area by pathogens in uncircumcised males. However, circumcision cannot be thought of as protective against UTI—if a male infant presents with an illness and fever for which no obvious source of infection is found, the pediatrician should obtain a urine analysis to rule out the presence of a UTI, whether the child is circumcised or not. Circumcision does not do away with the need for this test. The cost to evaluate a possible UTI is currently $86. Given a UTI incidence of 1% to 2% in uncircumcised infants, the cost to prevent this UTI by performing 100 circumcisions is $8500.

There is little question that circumcision prevents cancer of the penis, which has mortality rate of up to 25%. This would seem a strong argument for neonatal circumcision as a preventive. However, other factors play a role. According to a report by the Task Force on Circumcision, in developed countries where circumcision is not routinely performed (and parents and boys are used to caring for the uncircumcised penis), the incidence of penile carcinoma ranges from 0.3 to 1.1 per 100,000 men, about half the incidence in uncircumcised U.S. men, but greater than that in circumcised men. In developing countries with lower standards of hygiene, the incidence is 3 to 6 per 100,000 men per year. Thus, good hygiene may make up for the effect of circumcision, at a fraction of the cost.

Sexually transmitted diseases (STDs) have been shown in some studies to be more prevalent in uncircumcised state. Parker et al. showed a significantly higher risk of four types of STDs among uncircumcised men in Australia. However, these authors did not recommend circumcision to prevent STDs. They stated that "if these findings are confirmed in other studies, it would seem that attention should be directed to the improvement of personal hygiene among uncircumcised men." In the final analysis, pediatricians should not advocate circumcision to prevent STDs.

Lack of circumcision may be related to an increased risk of transmission of the AIDS virus. This correlation has been shown in both clinic-based studies and in a statistical study of male circumcision status in 37 African capital cities. The authors of the study suggest that lack of circumcision is a cofactor in HIV infection, not causative. They state that uncircumcised African males "are apparently at increased risk of developing chancroid and other genital-ulcer disease," which in turn facilitates infection with HIV; they also write that perhaps the intact foreskin enhances viral survival and, finally, that more frequent infection of the glans of the penis increases susceptibility to HIV. Again, while all this may be true, circumcision would not be an effective way of solving the AIDS epidemic and should not be promoted as such.

Complications of routine newborn circumcision are indeed infrequent, and most are easily treated. The use of the xylocaine for local anesthesia, while inflicting pain, renders the remainder of the circumcision procedure painless.

Studies in the United States and New Zealand have shown more problems in caring for the uncircumcised penis in the first few years of life, but data from Europe suggest that the uncircumcised penis presents few problems for parents and boys used to dealing with it. In any event, problems that arise during care of the circumcised *or* uncircumcised penis are generally minor, requiring only one medical visit for correction.

In sum, routine newborn circumcision presents a very significant financial cost to society. Adequate hygiene of the uncircumcised penis appears to do away with the need for circumcision.

A SAFE AND BENEFICIAL PROCEDURE

Dr. Thomas Wiswell

Dr. Thomas E. Wiswell is chief of the Newborn Medicine Service at Walter Reed Army Medical Center in Washington, D.C.

Sixty percent to 90% of newborn boys (1.2 to 1.8 million) are circumcised annually in the United States. Several issues have convinced me of the benefits of the procedure: (1) the prevention of urinary tract infections (UTIs) and complications from them; (2) the prevention of penile cancer; (3) the lower incidence of sexually transmissible diseases in uncircumcised males; (4) the low risk for complications from the operation; (5) the greater incidence of penile problems among "intact" boys; and (6) recent evidence showing that circumcision protects against AIDS.

Circumcised boys are 10 to 39 times *less* likely to have UTIs during infancy. In a population of more than 400,000 children during a 10-year period, we found the higher incidence of UTIs in males compared with females to be primarily due to a lack of circumcision. From 1975 to 1984 the circumcision frequency rate decreased substantially from more than 86% to less than 71%. However, there was a concomitant significant rise in the number of male infants with UTIs, a rise solely attributable to the presence of more uncircumcised boys. We have also found uncircumcised boys aged 1 to 15 years to be 2.5 times as likely to develop UTIs compared with their circumcised counterparts. Urinary tract infections are not benign. More than 36% of 88 boys below 1 month of age with a UTI had concurrent infection in their blood stream. Furthermore, 3 of these infants had concomitant meningitis, 2 had renal failure, and 2 died. Littlewood has reported that 11% of children with a UTI during the first month of life may die. There are longer term effects of UTIs in children. Ten percent to 15% of infected infants will subsequently demonstrate kidney scarring. Of these infants, approximately 10% will develop high blood pressure, and 2% to 3% will ultimately require dialysis or kidney transplantation.

Penile cancer is the only malignancy that can be prevented categorically by a prophylactic procedure, neonatal circumcision. Of the more than 60,000 cases of penile cancer that have occurred in the United States since 1930, fewer than *10* have been in circumcised men. More than 1000 men develop penile cancer each year, and 225 to 317 annually die from it. The basic therapy for this malignancy is amputation of the penis.

Virtually all sexually transmissible diseases (STDs) have been found to occur more frequently among uncircumcised men. Fink has enumerated more than 60 references that have found STDs to occur more often among "intact" individuals. I am struck by the paucity of contrary reports. There is only one report of a venereal disease (nongonococcal urethretitis) being more common in circumcised men. However, in this population more than 60% of the cases of another STD (gonorrhea) occurred in uncircumcised men.

Serious complications from routine foreskin removal are infrequent and relatively minor. We have recently examined a population of more than 100,000 circumcised boys and found complications in fewer than 2 per 1000. Two other large investigations reported complications from circumcision in 0.06% and 0.20% of circumcised boys, respectively. The majority of the complications are easily treated bleeding and minor infections. Atypical complications of the procedure (glans loss, staphylococcal scalded skin syndrome, etc.) occur infrequently and receive note due to their uniqueness. Death rarely occurs as a complication of circumcision. To date there have been a total of *3* reported deaths in the United States since 1954 that can be ascribed to neonatal circumcision. This contrasts sharply with the potentially preventable 7,500 to 11,500 deaths from penile cancer that have occurred during the same period.

Herzog and Alvarez found uncircumcised boys aged 4 months to 12 years to be more likely to have "penile problems" than were circumcised boys. Fergusson et al. similarly described uncircumcised boys as having more problems than their circumcised counterparts during the first 8 years of life. In both investigations, the "problems" largely consisted of balanitis (infection in and around the head of the penis) and phimosis (an abnormal constriction of the foreskin that prevents urine from being excreted). Finally, we found the risks from circumcision during the first month of life to be fewer than the risks from the uncircumcised state.

Reports have recently appeared suggesting that circumcision protects against AIDS. Simonsen et al. found uncircumcised men to be 9½ times as likely to become infected following exposure to the human immunodeficiency virus (HIV). A recent review in *Science* reported three corroborative studies (from Africa and the United States) that suggest a link between the lack of circumcision and AIDS virus infection.

As a pediatrician, I am a child advocate. I have pondered this issue for many years. I understand that we would have to circumcise "the many" to protect "the few." However, we have no way of identifying "the few." Neonatal circumcision is a rapid and generally safe procedure that must be performed by experienced caretakers. With the low complication rate and the many benefits of the procedure, I personally believe we should routinely circumcise newborn boys.

SHARPENING YOUR CRITICAL THINKING SKILLS

1. Summarize the key points of each author, then list the data they use to support their main points.

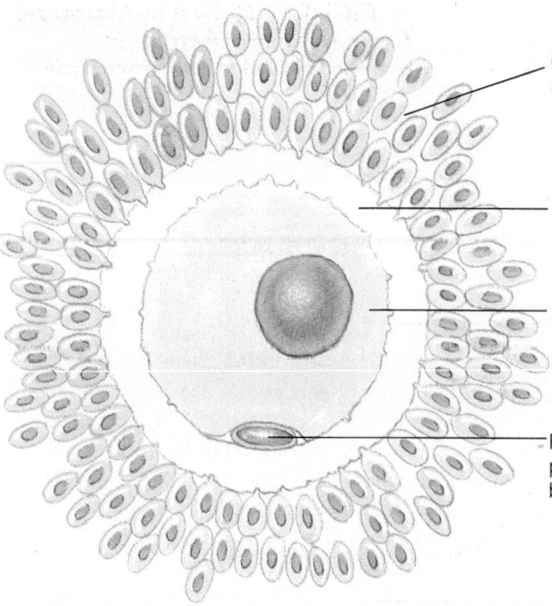

▻ **FIGURE 21–24 Ovulated Oocyte** Light micrograph of a recently ovulated follicle showing the corona radiata and zona pellucida.

does not occur, the CL soon disappears, and the process begins all over. (More on this process in Chapter 22.)

Hormonal Changes Taking Place in Women Produce Cyclic Variations in the Ovary and Uterus

Women of reproductive age undergo a series of interdependent hormonal, ovarian, and uterine changes each month known as the **menstrual cycle.** The length of the menstrual cycle varies from one woman to the next. In some it lasts 25 days, and in others it may last up to 35 days.[5] On average, however, the cycle repeats itself every 28 days. Ovulation usually occurs approximately at the midpoint of the 28-day cycle, or about 14 days before the onset of menstruation.

As noted above, the menstrual cycle involves three interdependent cycles (▻ Figure 21–25). The first is a hormonal cycle. The hormonal cycle, in turn, produces cyclic changes in the ovary (the ovarian cycle) and the uterus (the uterine cycle). Understanding the menstrual cycle is easiest if we begin with the hormonal and ovarian cycles.

The Hormonal and Ovarian Cycles. The first half of the menstrual cycle is known as the **follicular phase,** for during this time the follicles grow toward ovulation. The second half is called the **luteal phase,** so named because the corpus luteum forms during this time. To understand what drives these changes, we begin with the pituitary.

As shown in Figure 21–25a, FSH and LH are released from the anterior pituitary during the first half of the cycle, then peak in the middle, just before ovulation. FSH stimulates follicular development during the follicular phase by promoting mitosis of the follicle cells. LH stimulates estrogen production. Estrogen released from the follicle during the follicular phase also stimulates mitotic division of follicle cells, helping them grow.

As illustrated in ▻ Figure 21–26, estrogen is produced by the ovary from cholesterol. Its production begins in the cells of the theca. As illustrated, these cells convert cholesterol to androgen. This conversion requires small amounts of the hormone LH. Androgen diffuses out of the theca cells and into the follicle cells, where it is converted to estrogen. FSH is necessary for this conversion.

Estrogen secreted during the follicular phase of the menstrual cycle controls the release of both FSH and LH by inhibiting the release of GnRH from the hypothalamus in a classical negative feedback mechanism. As a result, throughout most of the follicular phase, LH and FSH levels are low and fairly constant. Just before ovulation, however, both LH and FSH secretion by the pituitary increases dramatically. These surges in LH and FSH secretion are the result of one of the body's rarest events, a positive feedback loop. Here's how it is triggered. As illustrated in Figure 21–25a, during the follicular phase of the menstrual cycle the amount of estrogen in a woman's blood creeps up fairly slowly. When estrogen reaches a certain critical level, however, both the hypothalamus and the anterior pituitary respond with a sudden outpouring of LH and FSH.

The LH surge has at least four effects: (1) it causes the primary oocyte to complete its first meiotic division, forming a secondary oocyte; (2) it stimulates the release of the enzymes that break down the ovarian wall, resulting in ovulation; (3) it stimulates estrogen production and release; and (4) it converts the collapsed follicle into a corpus luteum. The role of the preovulatory surge of FSH, if any, is not known.

During the second half of the menstrual cycle, the luteal phase, LH secretion gradually declines (Figure 21–25a). What LH is present during the luteal phase, however, stimulates the corpus luteum to produce estrogen and progesterone. As LH levels decline, though, estrogen and progesterone secretion from the CL also decline. If pregnancy does not occur, the CL stops producing hormones and degenerates (Figure 21–25b). Only a hormonal signal (discussed below) from the newly formed embryo can save it. Otherwise, the decline in levels of estrogen and progesterone permits a new cycle to begin.

The Uterine Cycle. The uterine lining, or endometrium, also undergoes cyclic changes during the menstrual cycle (Figure 21–25c). These changes result from cyclic changes in ovarian hormones, which, in turn, are controlled by changes in pituitary hormone secretion. As Figure 21–25c shows, the endometrium thickens throughout much of the cycle in preparation for pregnancy. In the absence of fertil-

[5]The length of the menstrual cycle may also vary from month to month in the same woman.

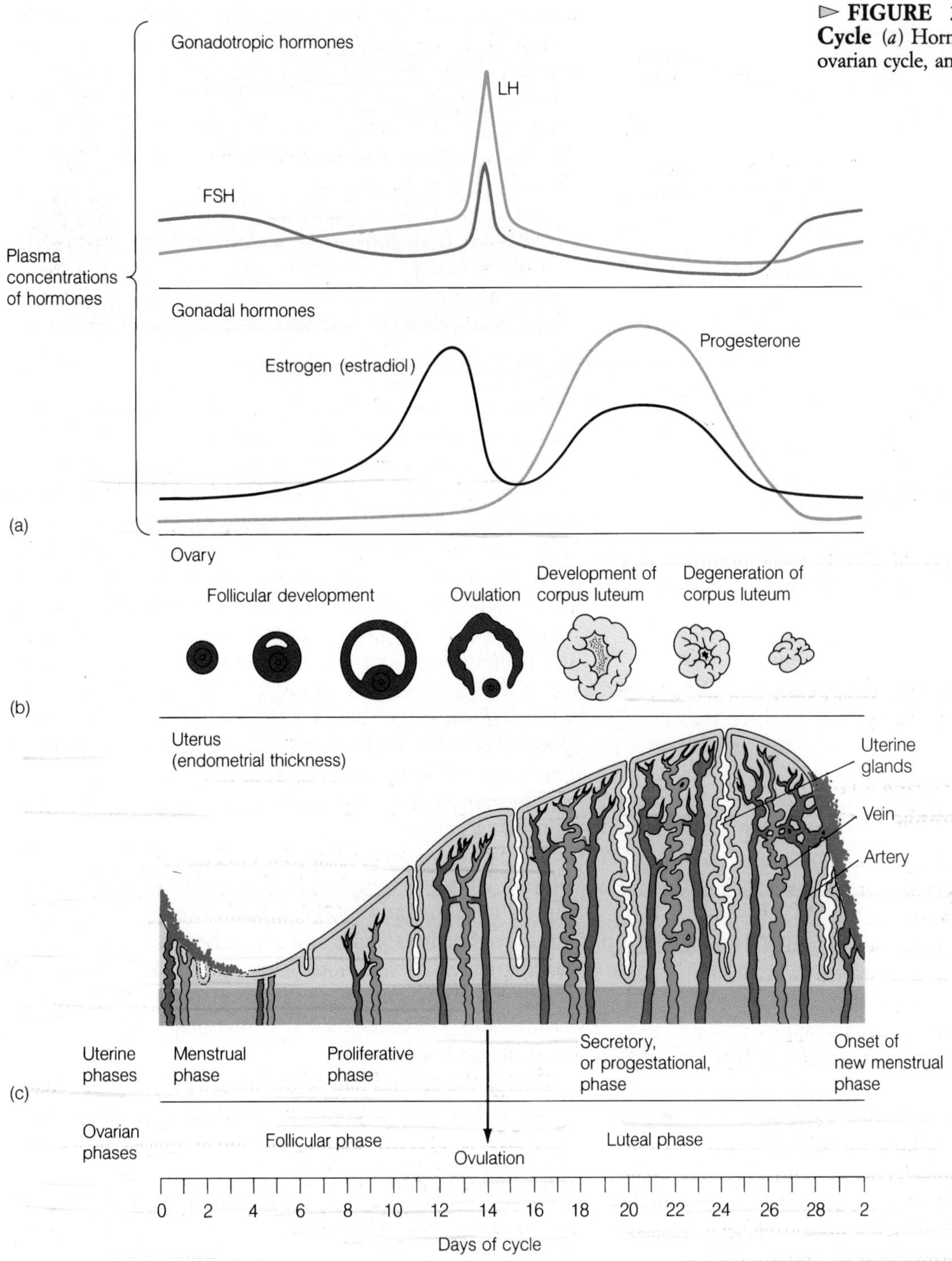

▻ **FIGURE 21–25 The Menstrual Cycle** (*a*) Hormonal cycles, (*b*) the ovarian cycle, and (*c*) the uterine cycle.

ization, however, most of the thickened endometrium is shed, a process called **menstruation.**

To see how the endometrium responds to hormonal changes, we will begin on day 1 of the average 28-day menstrual cycle. Day 1 of the cycle is the first day of menstruation. During the first four or five days of the cycle, the uterine lining is shed. Tissue that formed in the previous menstrual cycle sloughs off from the lining and passes out of the uterus into the vagina along with a considerable amount of blood—on average, about 50 to 150 milliliters. The loss of blood during menstruation is the main reason that women are more prone to develop anemia than men and why women should eat iron-rich foods or take iron supplements (Chapter 11).

As soon as the endometrium has been shed, the lining of the uterus begins to rebuild. Initial regrowth in the follicular phase is stimulated by ovarian estrogen. Estrogen stimulates the growth of glands **(uterine glands)** in the endometrium and also promotes cell division in the basal layer (deepest layer) of the endometrium—all that is left after menstruation. During the regrowth phase, also known as the **proliferative phase,** the uterine glands begin to fill

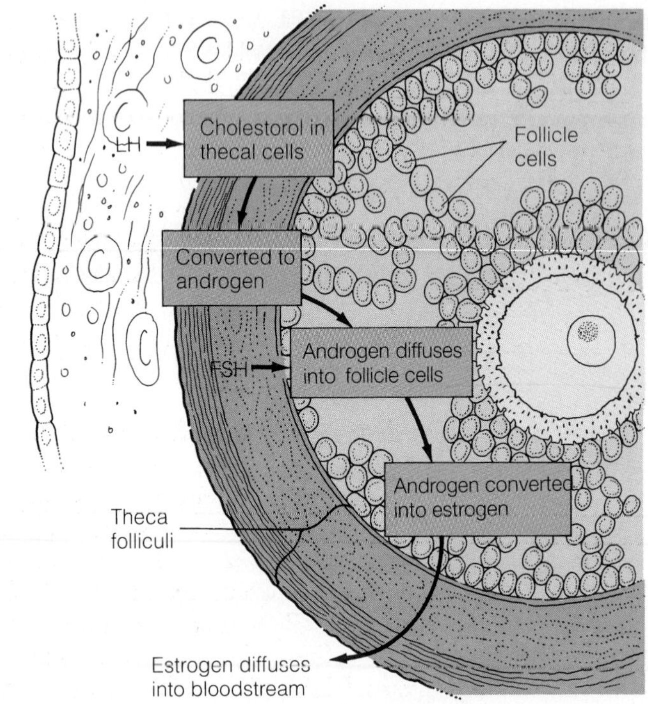

▻ **FIGURE 21–26 Estrogen Production in the Ovary**

▻ **FIGURE 21–27 Home Pregnancy Tests**

with a nutritive secretion, which will help nourish an embryo should fertilization occur.

After ovulation the endometrium continues to thicken under the influence of both estrogen and progesterone. The uterine glands become distended with a glycogen-rich secretion. The last half of the uterine cycle is therefore called the **secretory phase.**

If fertilization does not occur, the uterine lining starts to shrink approximately four days before the end of the cycle, then begins to slough (pronounced "sluff") off, starting menstruation.

The shedding of the uterine lining (menstruation) is triggered by a decline in estrogen and progesterone concentrations in the blood. Progesterone acts as a uterine tranquilizer, inhibiting smooth muscle contraction in the myometrium. When progesterone levels fall, the uterus begins to undergo periodic contractions. These contractions propel the sloughed endometrium out of the uterus and are responsible for the cramps that many women experience during menstruation.

If fertilization occurs, the newly formed embryo produces a hormone called **human chorionic gonadotropin (HCG).** HCG is an LH-like hormone that stimulates the corpus luteum, maintaining its structure and function. When HCG is present, ovarian estrogen and progesterone continue to be secreted, and the uterine lining remains intact and suitable for embryonic attachment. A newly formed embryo arrives in the uterus and attaches to the lining approximately four days after fertilization. It then embeds in the thickened endometrium, from which it derives its nutrients.

HCG maintains the corpus luteum for approximately six months and shows up in detectable levels in a woman's blood and urine about 10 days after fertilization. Pregnancy tests available through a doctor's office or drugstore detect HCG in a woman's urine. The tests use a commercially prepared antibody to HCG, which binds to the hormone (▻ Figure 21–27). The home pregnancy tests are relatively inexpensive, fairly reliable, and fast.

The Sexual Responses of Women and Men Are Similar in Many Respects. Like that of men, the sexual response in women involves four stages (Table 21–4). During the excitement phase, cervical glands produce a secretion that lubricates the vagina. Erectile tissue beneath the labia fills with blood, and the external genitalia expand. Arousal causes the nipples to become erect. Heart rate, muscle tension, and blood pressure increase.

During the next phase, breathing, heart rate, and muscle tension increase even more. The outer third of the vagina constricts due to the engorgement of blood vessels in the vaginal wall. The clitoris also becomes engorged. The clitoris is invested with numerous sensory nerve fibers, and it yields considerable pleasure when stimulated. The nipples become more erect, and the glands near the opening of the vagina are activated, producing a lubricant that facilitates the insertion of the penis.

Sexual arousal can lead to orgasm, rhythmic contractions of the uterus and vagina. These contractions are accompanied by intense physical pleasure. Women often report feelings of warmth throughout their bodies after orgasm.

After orgasm, the body relaxes. Blood drains from the clitoris and labia. Blood pressure, heart rate, and respiration return to normal. Unlike men, women generally do not experience a refractory period (a time when orgasm is not possible) after orgasm. Thus, a woman can sometimes experience multiple orgasms during sexual intercourse.

Estrogen and Progesterone Help to Regulate the Menstrual Cycle, but also Exert Numerous other Effects. Like testosterone in boys, estrogen secretion in

TABLE 21–4 Female Responses During the Sexual Response Cycle

Excitement Phase

1. Vaginal lubrication begins.
2. Vasocongestion swells the external genitalia.
3. Labia majora flatten and retract from the vaginal opening.
4. Inner two-thirds of the vagina expands.
5. Vaginal walls thicken.
6. Breasts enlarge, and blood vessels near the surface become more prominent.
7. Nipples become erect.
8. Muscle tension, heart rate, and blood pressure increase somewhat.

Plateau Phase

1. Vasocongestion produces a narrow vaginal pathway in the outer third of the vagina.
2. Inner part of the vagina expands fully.
3. Uterus becomes elevated.
4. Clitoris withdraws beneath the clitoral hood.
5. Rosy appearance may occur on the stomach, thighs, and back.
6. Nipples become more erect.
7. Muscles tense, breathing is rapid, and heart rate increases (100 to 160 beats per minute).

Orgasm

1. Swelling of the tissues of the outer part of the vagina constricts the vaginal opening.
2. Contractions begin in the outer third of the vagina. First contractions may last 2 to 4 seconds, and later ones may last 3 to 15 seconds. They occur at intervals of 0.8 of a second.
3. Inner two-thirds of the vagina expands slightly.
4. Uterus contracts.
5. Muscles go into spasm throughout the body, respiration increases, and blood pressure and heart rate reach a peak (about 180 beats per minute).

Resolution Phase

1. Blood is released from engorged areas.
2. Rosy appearance disappears.
3. Clitoris descends to normal position.
4. Vagina, uterus, and labia gradually shrink to normal size.
5. Blood pressure, heart rate, and respiration become normal.
6. About one-third of females find their palms and soles or entire body covered with perspiration.

girls increases dramatically at puberty. As the levels of estrogen in the blood increase, the hormone begins to stimulate follicle development in the ovaries. Estrogen is also an anabolic hormone—a hormone that stimulates anabolic (synthesis) reactions. Estrogen's anabolic effects result in the pubertal growth of the external genitalia (for example, the breasts). Estrogen also stimulates growth of the internal reproductive structures: the uterus, uterine tubes, and vagina.

Estrogen promotes rapid bone growth in the early teens. Because estrogen secretion in girls usually occurs earlier than testosterone secretion in boys, girls typically go through their growth spurt earlier. However, estrogen also stimulates the closure of the epiphyseal plates, described in Chapter 19, ending the growth spurt fairly early. Most girls, in fact, reach their full adult height by the age of 15 to 17. Boys experience their most rapid growth later in adolescence and continue growing until the age of 19 to 21.

Estrogen also stimulates the deposition of fat in women's hips, buttocks, and breasts, giving the female body its characteristic shape. In addition, estrogen stimulates the growth of ducts in the mammary glands. Progesterone works with estrogen to stimulate breast development.

Premenstrual Syndrome Is a Condition Afflicting Many Women. For reasons not yet fully understood, many women suffer premenstrual irritability, depression, fatigue, and headaches. Many complain of bloating, swelling and tenderness of the breasts, tension, and joint pain. Together these symptoms constitute a condition called **PMS,** or **premenstrual syndrome.** Premenstrual syndrome is a clinically recognizable condition characterized by one or more of these symptoms. PMS strikes 4 of every 10 women of reproductive age. All told, women report more than 150 different physical and psychological symptoms that emerge before menstruation begins.

Despite the prevalence of PMS, it may be years before medical scientists can pinpoint the cause or causes of PMS. Nevertheless, dozens of "cures," ranging from massive doses of progesterone to vitamin B-6 to L-tryptophan, have been prescribed by clinics specializing in PMS.[6] Buyers should beware, however, for very little good scientific evidence is available to indicate which, if any, of the "cures" really work. Most of the evidence consists of testimonials—individual accounts. The critical thinking skills you learned in Chapter 1 suggest that anecdotal information such as this is no substitute for controlled studies.

Work is now under way to test various treatments, but the results are not expected for several years. In the meantime, physicians recommend that women suffering from PMS see their family doctor to be certain that the symptoms are not in fact caused by some other problem.

Menopause Is the Cessation of Menstruation. The menstrual cycle continues throughout the reproductive years, but after a woman reaches 20, the ovaries gradually start to become less responsive to gonadotropins (▻ Figure 21–28). Responsiveness declines slowly. But as it declines, estrogen levels gradually fall off. Ovulation and menstruation become increasingly more erratic and eventually stop. This cessation is called **menopause.**

Menopause is attributed to a reduction in the number of ovarian follicles. At about age 45, most of the follicles

[6] As noted earlier in the book, L-tryptophan has been implicated in the paralysis of body muscles and death of over a dozen people. At this writing, preliminary studies suggest that the amino acid may not be at fault but, rather, that the pills taken by victims contained a contaminant that caused the ill effects.

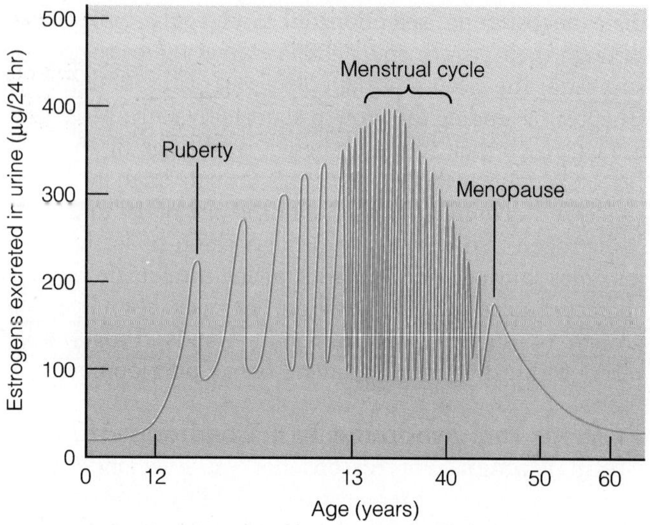

▻ **FIGURE 21–28 Ovarian Hormone Secretion** Notice that at about age 20, ovarian estrogen secretion begins a gradual decline. Adapted with permission from A. C. Guyton, *Textbook of Medical Physiology*, 7th ed. (Philadelphia: W. B. Saunders Company, 1986), Figure 81–8, p. 979.

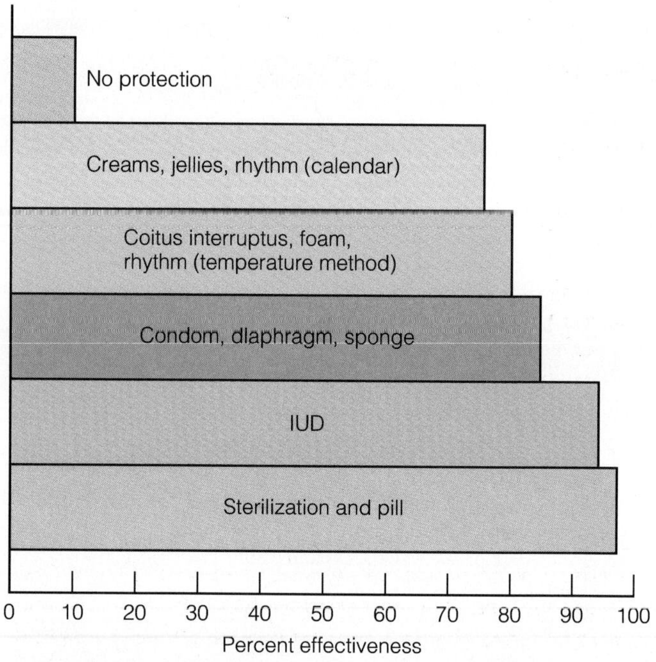

▻ **FIGURE 21–29 Effectiveness of Contraceptive Measures** Percent effectiveness is a measure of the number of women in a group of 100 who will not become pregnant in a year.

that were in the ovary at puberty have been stimulated to grow and have either degenerated or ovulated. Consequently, FSH and LH from the pituitary have no follicles to stimulate, and the production of ovarian estrogen plummets.

Ovulation and menstruation generally cease between the ages of 45 and 55. The dramatic alteration in the hormonal climate results in several important physiological changes. For example, the decline in estrogen secretion causes the breasts and reproductive organs, such as the uterus, to begin to atrophy. Vaginal secretions often decline, and in some women sexual intercourse may become painful.

The decline in estrogen levels may also result in behavioral disturbances. Many women become more irritable and suffer bouts of depression. Three quarters of all women suffer "hot flashes" and "night sweats" induced by massive vasodilation of vessels in the skin. Fortunately, these symptoms usually pass.

As noted in Chapter 19, declining estrogen levels also accelerate osteoporosis. To counter osteoporosis and other impacts of the decline in ovarian function, physicians sometimes prescribe pills containing small amounts of estrogen and progesterone, as well as a program of exercise and a diet rich in calcium and vitamin D (see Health Note 19–1).

BIRTH CONTROL

Few topics in modern society create as much controversy as birth control. **Birth control** is any method or device that prevents births. Birth control measures fall into two broad categories: (1) **contraception,** ways of preventing pregnancy, and (2) **induced abortion,** the deliberate expulsion of a fetus.

Contraceptive Measures Help Prevent Pregnancy. ▻ Figure 21–29 summarizes the effectiveness of the most common means of contraception. Effectiveness is expressed as a percentage. A 95% effectiveness rating means that 95 women out of 100 using a certain method in a year will not become pregnant.

Abstinence. Not listed in the figure is a form of birth control that many of us forget to talk about, **abstinence,** refraining from sexual intercourse. This form of birth control is appropriate for many people and should not be overlooked as a strategy of reducing unwanted pregnancy and preventing the transmission of AIDS and other diseases (discussed later).

Sterilization. Except for complete abstinence, sterilization and the pill (discussed shortly) are the most effective birth control measures (Figure 21–29). In 1982, sterilization became the leading method of contraception practiced by married couples in the United States.

In women, sterilization is performed by cutting the uterine tubes. This process is called **tubal ligation** and rarely involves an overnight stay in the hospital (▻ Figure 21–30a). Surgeons usually make two small incisions in the abdomen just beneath the navel. An instrument called a **laparoscope** is inserted through each incision. The surgeon locates the uterine tubes through a special lighted viewing lens. An attachment to the laparoscope is then used to cut the uterine tubes. Surgeons then either tie off the cut ends, or cauterize them—that is, burn them using an electrical current. In some cases, surgeons clamp the uterine tubes shut with plastic or metal rings.

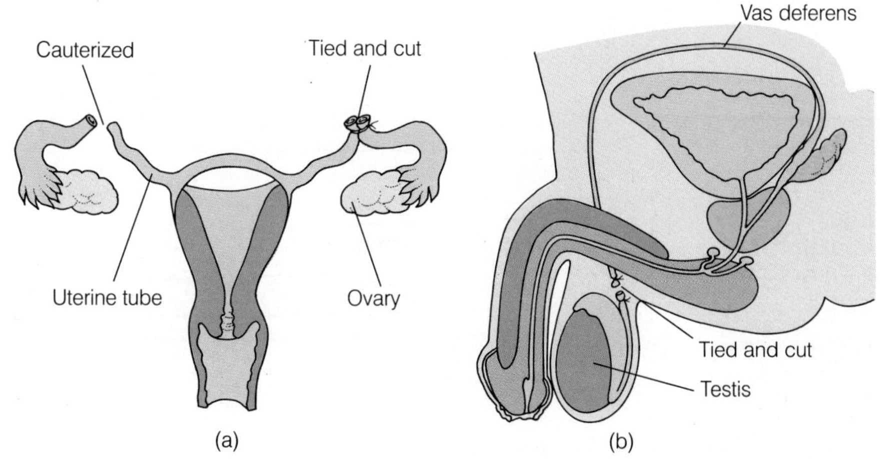

▻ **FIGURE 21–30 Sterilization Methods** (*a*) In a tubal ligation, the uterine tubes are cut, then tied off or cauterized. (*b*) In a vasectomy, the vasa deferentia are cut, then tied off.

Male sterilization requires a far less traumatic surgical procedure called a **vasectomy,** which can be carried out in a physician's office under local anesthesia (Figure 21–30b). To perform a vasectomy, a physician makes a small incision in the scrotum. Each vas deferens is exposed, then cut, and the free ends are tied off or cauterized.

A quick review of male anatomy will show that vasectomies only prevent the sperm from passing into the urethra during ejaculation. They do not impair sex drive and have virtually no effect on ejaculation, because 99% of the volume of the ejaculate is produced by the sex accessory glands.

Vasectomy and tubal ligation are essentially irreversible. However, special surgical methods, called **microsurgery,** can be used to reconnect the uterine tubes and the vasa deferentia. This procedure is costly and not always successful.

The Pill. The **birth control pill** is the most effective *temporary* means of birth control available (▻ Figure 21–31). Birth control pills come in several varieties, but the most common is the combined pill. It contains synthetic estrogen and progesterone, which inhibit the release of LH and FSH through the pituitary and hypothalamus. As a result, follicle development and ovulation are inhibited. A minipill containing progesterone alone is also available. Even though it is less effective than the combined pill, the minipill is more suitable for some women because it results in fewer side effects.

Birth control pills must be taken throughout the menstrual cycle. Skipping a few days may release the pituitary and hypothalamus from the inhibitory influences of estrogen and progesterone, resulting in ovulation and possible pregnancy.

Although effective, birth control pills have some adverse health effects worth noting. Even though the incidence of these adverse side effects is small, a woman considering different birth control options should carefully study them before making a decision.

One "adverse" effect is death. Table 21–5 compares the risk of death from taking birth control pills (and using other contraceptives) with a number of common risk factors. As shown, the risk of a nonsmoker dying from taking birth control pills is 1 in 63,000 in any given year, whereas the risk of dying in an auto accident is 1 in 6,000.

Death from birth control pills may result from a heart attack, stroke, or blood clot. The incidence of these life-threatening side effects is lowest in nonsmoking women under the age of 30. To reduce the risk even more, pharmaceutical companies have dramatically lowered the estrogen content of the combined pill, because estrogen is responsible for most of the adverse side effects.

Early studies showed a positive correlation between the use of birth control pills and cancers of the breast and cervix. More recent studies suggest that the new generation

▻ **FIGURE 21–31 The Birth Control Pill** One of the most effective means of birth control, the pill consists of a mixture of estrogen and progesterone, which is taken throughout the menstrual cycle to block ovulation. Birth control pills are packaged in numbered containers to help women keep track of them.

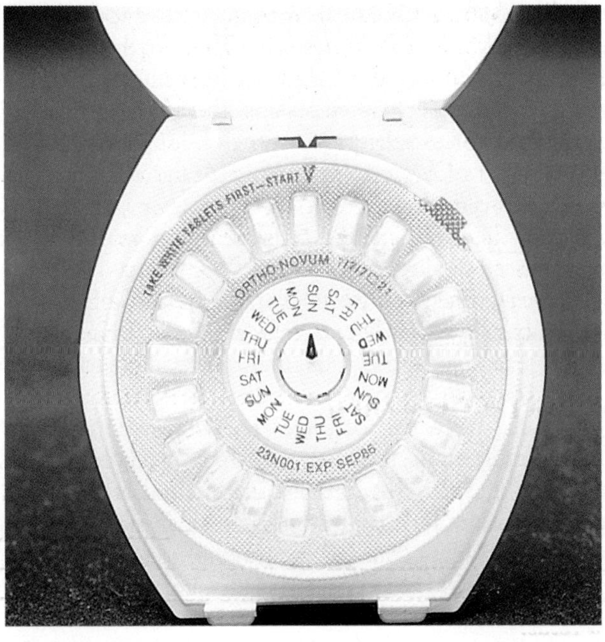

TABLE 21–5 Comparison of Risk of Some Voluntary Activities

RISK	CHANCE OF DEATH IN A YEAR (UNITED STATES)
Smoking	1 in 200
Motorcycling	1 in 1,000
Automobile driving	1 in 6,000
Power boating	1 in 6,000
Rock climbing	1 in 7,500
Playing football	1 in 25,000
Canoeing	1 in 100,000
Using tampons (toxic shock syndrome)	1 in 350,000
Contracting reproductive tract infections through sexual intercourse	1 in 50,000
Preventing pregnancy:	
Oral contraception—nonsmoker	1 in 63,000
Oral contraception—smoker	1 in 16,000
Using intrauterine devices (IUDs)	1 in 100,000
Using barrier methods	None
Using natural methods	None
Undergoing sterilization:	
Laparoscopic tubal ligation	1 in 20,000
Hysterectomy	1 in 1,600
Vasectomy	None
Pregnancy:	1 in 10,000
Nonlegal abortion	1 in 3,000
Legal abortion:	
Before 9 weeks	1 in 400,000
Between 9 and 12 weeks	1 in 100,000
Between 13 and 16 weeks	1 in 25,000
After 16 weeks	1 in 10,000

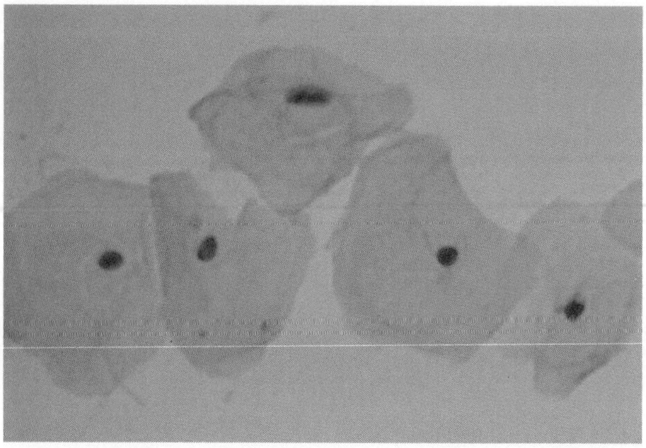

(a)

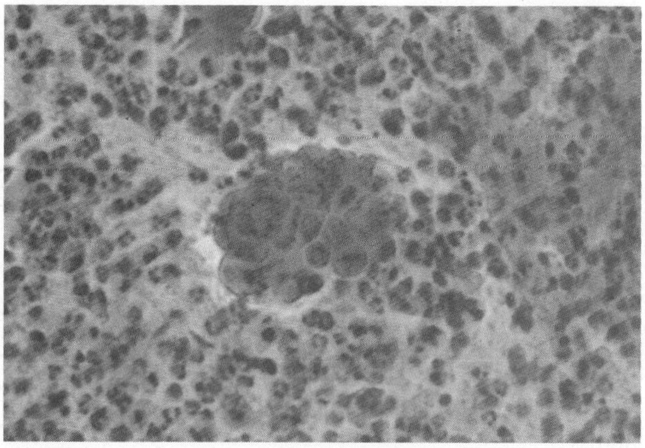

(b)

▻ **FIGURE 21–32 The Pap Smear** (*a*) A photomicrograph of a normal Pap smear showing large, flattened cells. (*b*) A photomicrograph of a cancerous smear, showing many small cancer cells.

of low-estrogen pills is less likely to cause cancer of the breast or cervix. Even with reduced estrogen levels, however, women who take birth control pills are more likely to develop cervical cancer than women who do not. Physicians, therefore, recommend annual Pap smears for women who are on the pill. During a **Pap smear,** the cervical lining is swabbed. The swab picks up cells sloughed off by the epithelium, which are later examined under a microscope for signs of cancer (▻ Figure 21–32). This procedure helps physicians diagnose cervical cancer early, increasing a woman's chances of survival.

Smoking increases the likelihood of side effects from birth control pills. If a woman is a smoker and takes the pill, for example, she is four times more likely to die from a heart attack or stroke than a nonsmoker. The risk of side effects also increases with age. To reduce the chances of developing serious side effects, women over the age of 35 who smoke should either use an alternative method of birth control or should give up smoking. Birth control pills are also not advised for women with a medical history of blood clots, high blood pressure, diabetes, uterine cancer, and cancer of the breast.

Birth control pills do have beneficial effects. First of all, they prevent pregnancy. National statistics show that one of every 10,000 women who becomes pregnant and delivers will die from complications, usually during delivery. Thus, even with the risks associated with the pill, using this mode of contraception is six times safer than pregnancy.

Birth control pills also reduce the incidence of ovarian cysts, breast lumps, anemia, rheumatoid arthritis, osteoporosis, and pelvic infection. Although birth control pills may increase the risk of cervical and breast cancer, they apparently protect a woman from cancer of the ovary and of the uterus, perhaps for life.

Intrauterine Device. The next most effective means of birth control is the **intrauterine device (IUD)** (▻ Figure 21–33). The IUD is a small plastic or metal object with a string attached to it. IUDs are inserted into the uterus by a physician, usually during menstruation, because the cervical canal is widest then and because menstrual bleeding indicates that the woman is not pregnant.

No one knows exactly how the IUD works, but there are at least two major hypotheses. Some researchers think

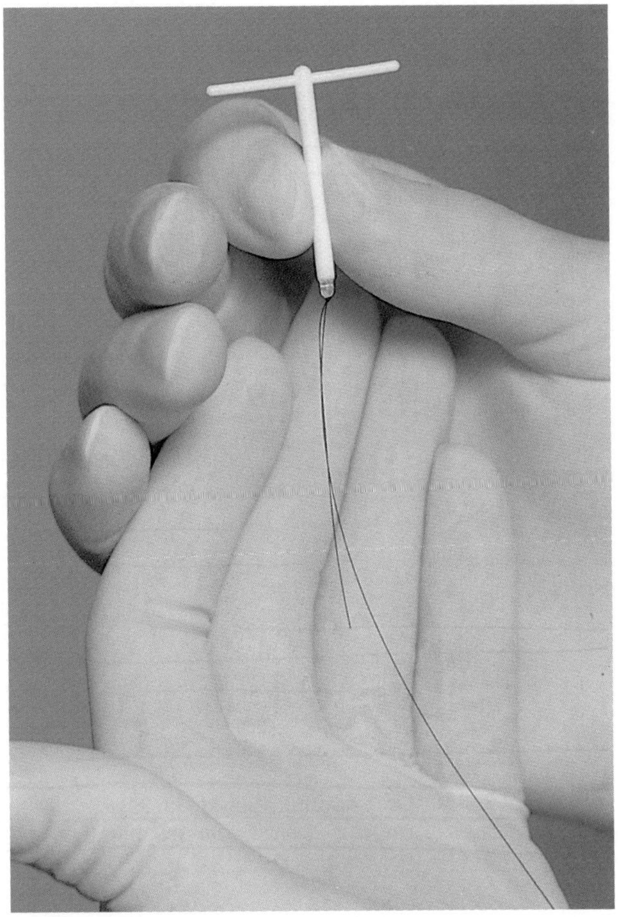

▻ **FIGURE 21–33 The IUD** IUDs come in a variety of shapes and sizes and are inserted into the uterus, where they prevent implantation. Only one type is currently legal in the United States.

that the IUD increases uterine contractions, making it difficult for the early embryo to attach and implant in the wall of the uterus. Others think that the IUD creates a local inflammatory reaction in the uterine lining, resulting in an environment inhospitable to a newly formed embryo. As a result, implantation is blocked, and the embryo dies. It is possible that both mechanisms are operating.

IUDs also have adverse effects. In some cases, the uterus expels the device, leaving a woman unprotected. Expulsion usually occurs within a month or two of insertion, so couples should check regularly during this period to be certain that the device is in place. The IUD may also cause slight pain and increase menstrual bleeding. These effects, however, are minor compared with two much rarer complications: uterine infections and perforation (a penetration of the uterine wall by the IUD). Women wearing an IUD are more likely to develop uterine infections than women practicing other forms of birth control. If not treated quickly, infections can spread to the uterine tubes, where scar tissue develops and blocks the transport of sperm and ova, causing sterility. Perforation of the uterus is a life-threatening condition requiring surgery to correct.

The Diaphragm, Condom, and Sponge. The next most effective means of birth control are the barrier methods—the diaphragm, condom, and vaginal sponge—all of which prevent the sperm from entering the uterus. The **diaphragm** is a rubber cup that fits over the end of the cervix ▻ Figure 21–34). To increase its effectiveness, a spermicidal (sperm-killing) jelly, foam, or cream should be applied to the rim and inside surface of the cup.

Diaphragms must be custom fitted by physicians. To be effective, the diaphragm must be inserted no more than two hours before sexual intercourse and must be worn for at least six hours afterward. If sexual intercourse is repeated, additional spermicidal cream or jelly should be injected into the vagina onto the diaphragm via a special applicator. If intercourse is desired after six hours, the diaphragm should be removed, washed, recoated, and then reinserted.

Smaller versions of the diaphragm, called **cervical caps,** are also available. Cervical caps fit over the end of the cervix. The cervical cap most often used is held in place by suction. When used with spermicidal jelly or cream, the caps are as effective as diaphragms.

Condoms are thin, latex rubber sheaths that are rolled onto the erect penis (▻ Figure 21–35). Sperm released during ejaculation are trapped inside (often in a small reservoir at the tip of the condom) and are therefore prevented from entering the vagina. Some condoms are prelubricated with a spermicidal chemical. Besides preventing fertilization, condoms also protect against sexually transmitted diseases, a benefit not offered by any other birth control measure except abstinence. Condoms are widely available and easy to obtain. They do not require a doctor's prescription. Because of the danger of sexually transmitted diseases, doctors in this country recommend that some women use vaginal condoms, latex devices that fit into the vagina.

Another method of birth control is the **vaginal sponge** (▻ Figure 21–36). This small absorbent sponge is impregnated with spermicidal jelly. Inserted into the vagina, the sponge is positioned over the end of the cervix. The sponge is effective immediately after placement and remains effective for 24 hours. Cervical sponges can be purchased without a doctor's prescription. Like the condom, one size fits all.

Withdrawal. One of the oldest, but least successful, means of birth control is **withdrawal,** or **coitus interruptus,** disengaging before ejaculation. This method requires tremendous willpower and frequently fails, for three reasons: because caution is often tossed to the wind in the heat of passion, because the penis is withdrawn too late, or because of preejaculatory leakage—the release of a few drops of sperm-filled semen before ejaculation. Thus, it is not an advisable method of birth control.

Spermicidal Chemicals. As mentioned earlier, spermicidal jellies, creams, and foams contain chemical agents that kill sperm but are apparently harmless to the woman

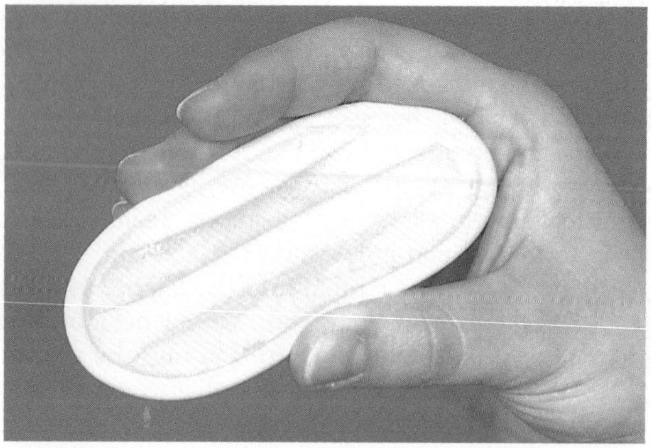

▻ **FIGURE 21–34 The Diaphragm** Worn over the cervix, the diaphragm is coated with spermicidal jelly or cream and is an effective barrier to sperm.

(▻ Figure 21–37). Spermicidal preparations are most often used in conjunction with diaphragms, condoms, and cervical caps. They can also be used alone but are only about as effective as withdrawal.

The Rhythm Method. Abstaining from sexual intercourse around the time of ovulation—the **rhythm,** or **natural, method**—can help couples reduce the likelihood of pregnancy. Because ova remain viable 12 to 24 hours after ovulation and sperm may remain alive in the female reproductive tract for up to three days, abstinence four days before and four days after the probable ovulation date should provide a margin of safety. If a couple knows the exact time of ovulation, they can time sexual intercourse to prevent pregnancy more precisely.

To practice the natural method successfully, couples must first determine when ovulation occurs (▻ Figure 21–38). Several methods are available to determine when ovulation occurs in a woman's cycle.

One approach is the **temperature method.**[7] A woman's body temperature varies throughout the menstrual cycle, as shown in ▻ Figure 21–39. In most women, body temperature rises slightly after ovulation. By taking her temperature every morning before she gets out of bed, a woman can pinpoint the day she ovulates. By keeping a temperature record over several menstrual cycles, she can determine the length of her cycle and the time of ovulation. Once the length of the cycle and the time of ovulation have been determined, days of abstinence can be determined. Erring on the safe side, some doctors recommend that couples refrain from sexual intercourse from the first day of menstruation until three days after ovulation. Translated, that means no sex for about 17 days of the 28-day cycle.

Another method used to time ovulation involves taking samples of the cervical mucus, which varies in consistency during the menstrual cycle. By testing its thickness on a

[7]The temperature method can also be used by women who want to get pregnant, because it allows them to determine the time of ovulation.

▻ **FIGURE 21–35 The Condom** Worn over the penis during sexual intercourse, it prevents sperm from entering the vagina.

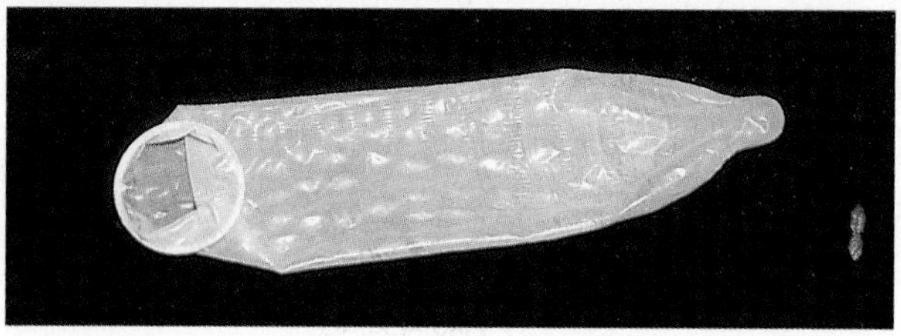

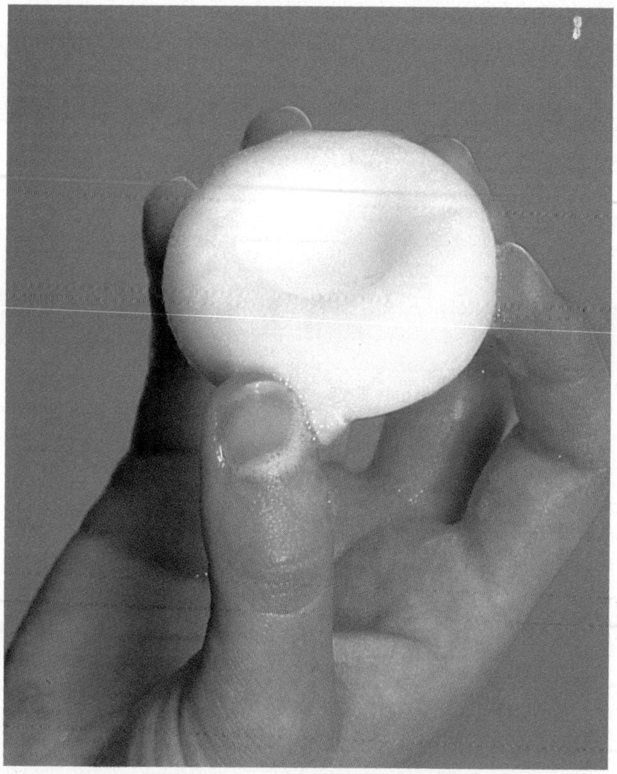

▻ **FIGURE 21–36 The Vaginal Sponge** Impregnated with spermicidal chemical, the vaginal sponge is inserted into the vagina and is effective for up to 24 hours.

▻ **FIGURE 21–37 Spermicidal Jelly** Spermicidal preparations work alone, but they are most effective when combined with another form of protection, such as a diaphragm.

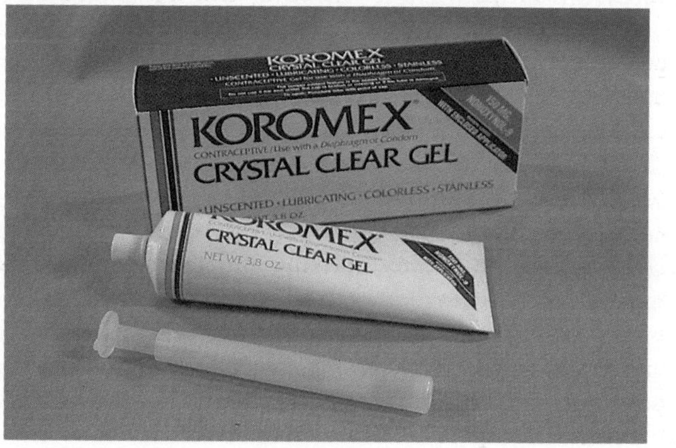

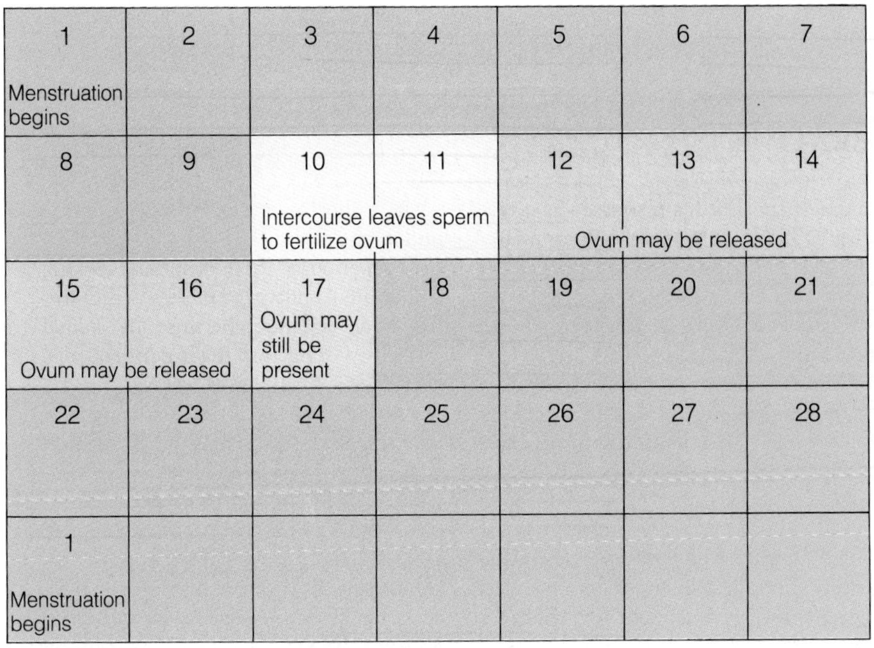
1
Menstruation begins
2
3
4
5
6
7
8
9
10
11
Intercourse leaves sperm to fertilize ovum
12
13
14
Ovum may be released
15
16
Ovum may be released
17
Ovum may still be present
18
19
20
21
22
23
24
25
26
27
28
1
Menstruation begins

▻ **FIGURE 21–38 The Natural Method** The yellow shaded areas indicate an unsafe period for sexual intercourse, assuming ovulation occurs at the midpoint of the cycle.

daily basis, a woman can fairly accurately tell when she has ovulated.

As most people practice it, the natural method of birth control requires about eight days of abstinence during each menstrual cycle—four days before ovulation and four days after ovulation. This practice minimizes the chances of a viable sperm reaching a viable ovum. Unfortunately, some women experience the greatest sexual interest around the time of ovulation. Sexual intercourse after a period of abstinence may also advance the time of ovulation. (For a discussion of new methods of birth control, see Health Note 21–1.)

Abortion Is the Surgical Termination of Pregnancy

Some couples may elect to terminate pregnancy through **abortion.** In the United States, approximately 1 million abortions are performed every year by physicians. Although many people in our society view abortion as a legitimate means of family planning, they are quick to point out that it should not be practiced as a primary means of birth control. Contraception is less costly and less traumatic.

Abortion is not suitable or morally acceptable to many people. Pro-life advocates argue that abortion should be outlawed or severely restricted—that is, allowed only in cases of rape, incest, and threat to the life of the mother. They advise unmarried women to abstain from sexual intercourse or, if they become pregnant, to give birth and either keep the baby or give it up for adoption.

Pro-choice advocates, on the other hand, think that women should have the freedom to choose whether to terminate a pregnancy or have a child. Abortion, they say, reduces unwanted pregnancies and untold suffering among unwanted infants, especially in poor families, and gives women more options than motherhood.

In the first 12 weeks of pregnancy, abortions can be performed surgically in a doctor's office via **vacuum aspiration.** The cervix is dilated by a special instrument, and the contents of the uterus are drawn out through an aspirator tube. The operation is fairly routine and relatively painless. Usually no anesthesia is given. Women bleed for a week or so after the procedure but generally experience few complications. Between 13 and 16 weeks, a larger aspirator tube is required. The lining of the uterus may have to be scraped with a metallic instrument to ensure complete removal of the fetal tissue. After 16 weeks, abortions are more difficult and more risky. Solutions of salt, urea, or prostaglandins, which stimulate uterine contractions, can be injected into the sac of fluid surrounding the embryo to induce premature labor. The hormone oxytocin

▻ **FIGURE 21–39 Body Temperature Measurements during the Menstrual Cycle**

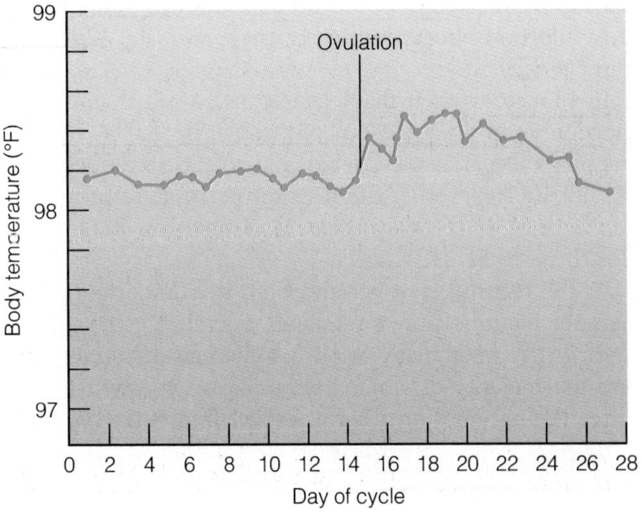

HEALTH NOTE 21–1

ADVANCES IN BIRTH CONTROL: RESPONDING TO A GLOBAL IMPERATIVE

Birth control is an issue of worldwide concern. In the wealthy, industrialized nations, many parents have chosen to limit their family size, for several reasons. First, many parents cannot afford more than one or two children. Raising a single child to college age could cost as much as $85,000. A recent study estimated that another $85,000 will be needed just to send a child born today to a public university. Considerably more is needed to send a student to a private university. Parents may also limit their family size to provide more personal care for their offspring. And some choose to have fewer children for environmental reasons. Consider some startling statistics: a child born in the United States today will require 16 pounds of coal, 3.6 gallons of oil, and 240 cubic feet of natural gas *each day* of his or her life.

This heavy dependence on resources results in enormous amounts of pollution and environmental damage. Every man, woman, and child, in fact, is responsible for over 1 ton of hazardous waste each year. Clearly, child rearing is an environmentally taxing activity.

Birth control is essential in the United States and other developed countries, many experts say, because our impact on the environment is so great. Each American, in fact, uses 25 to 40 times as much of the Earth's resources as a resident of India. In terms of environmental impact, then, the United States' 248 million people are equal to 6 to 10 billion Indians, or 1.2 to 2 times the present world population.

This is not to say that birth control is unimportant in Third World nations. Each year 90 million new residents are added to the global population, or about 3 per second. Ninety percent of the new residents are born into the Third World, where hunger and starvation abound. And these countries are faced with environmental problems of epic proportions.

Family planning is making headway in Third World nations, but the task has only just begun. To help in the process, researchers are trying to develop safer, more convenient, and even more effective methods of birth control.

For example, tests are under way on the effectiveness and safety of nasal spray contraceptives for women. Research is also proceeding quickly on transdermal contraceptive patches. The small, Band-Aid-like patches are impregnated with a blend of hormones and hormone analogs, including estrogen and progesterone. The patches are worn by a woman for a week, then replaced.

In many countries outside the United States, slow-acting, injectable contraceptives are now being used. Women are given a shot of crystalline progesterone under the skin. The crystals dissolve over a period of three months, blocking ovulation. Approval in the United States has been withheld because of animal tests suggesting that this treatment may cause some forms of cancer.

Another novel approach involves matchstick-sized capsules containing an even more potent progesterone implanted under the skin of a woman's arm; these prevent pregnancy for up to five years (▷ Figure 1). Contraceptive implants have been approved for use in 13 countries. In 1991, the Food and Drug Administration approved their use in the United States. Population experts hope that they will be widely used in Third World nations because they require virtually no effort on the part of the couple.

Researchers are also experimenting with biodegradable implants. Clinical trials are under way on a biodegradable material impregnated with progesterone that may prevent pregnancy for 18 months or more. The biodegradable material is broken down and gradually disappears.

Experimentation is continuing on a "morning-after" pill, which can be taken after sexual intercourse to prevent pregnancy. At least two morning-after pills exist. One contains a synthetic estrogen called DES (diethylstilbestrol). DES stimulates muscle contraction in the

may be administered to the woman with the same effect. Most abortions are performed by the end of the 12th week of pregnancy.

In France, women may elect to use a pill that induces abortion soon after the embryo implants. Called RU486, this pill is illegal in the United States, and some believe it will not be legalized in the near future because of opposition by the pro-life movement. (For more on RU486, see Health Note 21–1.)

In the controversy over abortion, it is important to remember that abortion is rarely an easy choice for anyone. Contrary to what many think, psychological studies show that the majority of women who choose abortion do not suffer lasting emotional harm, especially if they have had counseling. Thus, psychological counseling may be advisable before, during, and after the procedure.

SEXUALLY TRANSMITTED DISEASES

Several years ago, a group of English professors convened to discuss some pressing questions of language. One of those questions was "What is the most melodious word in the English language?" After much deliberation, they settled on a beautiful but unlikely candidate, the word "syphilis." It rolls off the tongue with ease. By most measures, however, syphilis is hardly a thing of beauty. It is a potentially crippling or deadly disease spread through sexual contact.

Infections like syphilis that are caused by bacteria and viruses and spread from one individual to another during sexual contact are known as **sexually transmitted diseases (STDs),** or, less commonly, **venereal diseases.** Bacteria and viruses transmitted by sexual contact penetrate the

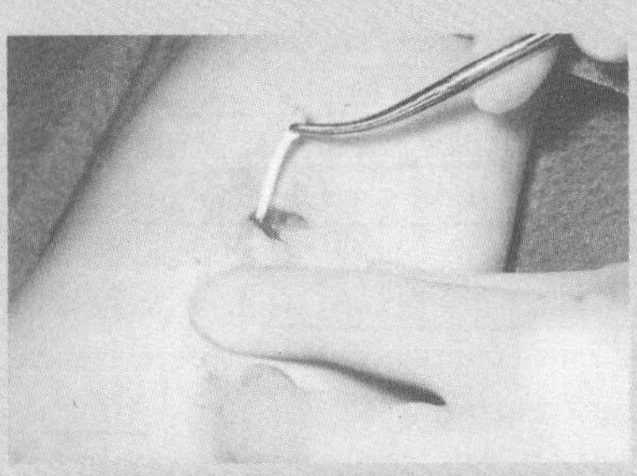

▷ **FIGURE 1 Subcutaneous Progesterone Implant** Inserted under the skin, this tiny device releases a steady stream of progesterone, blocking ovulation for months.

uterus and uterine tubes, expelling the fertilized ovum from the reproductive tract. DES is sometimes used in cases of incest or rape, but it is generally avoided because it causes vomiting and nausea.

A synthetic steroid called RU486 is also in use in France and certain other countries, but not the United States. RU486 binds to progesterone receptors in the uterus, blocking progesterone from binding. Because progesterone inhibits muscular contractions of the uterus, RU486 probably has the same effect as DES, causing an expulsion of the fertilized ovum. Vocal pro-life forces in the United States have taken a strong stand against the use of this method. At this writing, three hospitals in California are seeking permission and funding to test the drug.

You may have noticed that when it comes to birth control, there are few options for men. Why not develop a pill for men and shift some of the contraceptive burden to them?

To be effective, a pill would have to shut down spermatogenesis. Testosterone injections would do the job, because that hormone blocks the release of pituitary FSH and LH. FSH is required for spermatogenesis, and its absence would depress sperm production. Unfortunately, testosterone injections might create aggressive behavior. Complicating matters, excess androgen in males is converted to estrogen, causing feminizing side effects. Finally, androgen treatments may depress spermatogenesis, but not enough to lower sperm count to a level where a couple would feel confident.

Another route is to selectively inhibit FSH secretion. As noted in this chapter, the seminiferous tubules produce a substance called **inhibin,** which inhibits the production of FSH by the pituitary gland. If inhibin could be produced and administered to men, it might give them a better chance to participate in birth control. It could be administered in contraceptive nasal sprays.

One thing is certain, the world's population problem will not be solved through new contraceptive technologies alone. What is required is a change in the attitudes of millions of men and women throughout the world. Controlling family size must be a conscious decision followed by conscientious action.

In Africa, where population is doubling in some countries every 17 years, contraceptive use is a paltry 10% to 20%. Worldwide, only about half the women of reproductive age are using contraceptives. Education is needed to involve more people in a race to stem the swelling tide, which if unchecked will add 5 billion people to the world population in the next 40 years. Funds are needed to help pay for contraception and other family planning. When one condom costs more than the average person spends on medical care in a year, we can hardly expect widespread use. Many Third World countries, however, divert enormous amounts of money to pay for weapons and almost nothing to family planning. If the world population is to stabilize, if our children are to inherit a world worth living in, many experts agree, contraceptive use must increase.

Controlling population growth also requires improvements in education and job opportunities for men and women. Small-scale, sustainable economic development will give men and women options other than child bearing.

lining of the reproductive tracts of men and women and thrive in the moist, warm environment of the body. Most bacteria and viruses that cause STDs are spread by vaginal intercourse, but other forms of sexual contact, such as anal and oral sex, are also responsible for the spread of disease. AIDS, for example, can be transmitted by anal sex as well as vaginal and oral sex (Chapter 15). **Syphilis,** an STD caused by a bacterium, can be spread by oral, anal, and vaginal sex.

Although sexually transmitted diseases pass from one person to another during sexual contact, the symptoms are not confined to the reproductive tract. In fact, several STDs, including syphilis and AIDS, are primarily systemic diseases—that is, diseases that affect entire body systems.

One complicating factor in controlling sexually transmitted diseases is that some diseases, like gonorrhea, produce no obvious symptoms in many men and women. As a result, the disease can be transmitted without a person knowing he or she is infected. In AIDS, symptoms may not appear until several years after infection. Thus, sexually active individuals who are not monogamous can spread the AIDS virus to many people before they are even aware that they were infected.

In this section, we will examine the most common STDs, except AIDS, which was discussed in Chapter 15.

Gonorrhea Is a Bacterial Infection that May Spread to Many Organs

Gonorrhea (referred to colloquially as the "clap") is caused by a bacterium that commonly infects the urethra of men and the cervical canal of women. Painful urination and a pus-like discharge from the urethra are common complaints in men. Women may experience a cloudy vaginal discharge and lower-abdominal pain. If a woman's urethra is infected, urination may be painful. Symptoms of gonorrhea usually appear about two to eight days after sexual contact.

If left untreated, gonorrhea in men can spread to the prostate gland and the epididymis. Infections in the urethra lead to the formation of scar tissue, which narrows the urethra, making urination even more difficult. In women, the bacterial infection can spread to the uterus and uterine tubes, causing the buildup of scar tissue. In the uterine tubes, scar tissue may block the passage of sperm and ova, resulting in infertility.[8] Gonorrheal infections can also spread into the abdominal cavity through the opening of the uterine tubes. If the infection enters the bloodstream in men or women, it can travel throughout the body. Fortunately, gonorrhea can be treated by antibiotics, but early diagnosis is essential to limit the damage.

Nonspecific Urethritis Is an Extremely Common Disease Caused by Several Types of Bacteria

Nonspecific urethritis (NSU) has become the most common sexually transmitted disease known to medical science and is one of several STDs whose incidence is steadily rising in the United States. Caused by any of several different bacteria, this infection is generally less threatening than gonorrhea or syphilis, although some infections can result in sterility. Approximately one-half of the reported cases of NSU are caused by a bacterium called **chlamydia.**

Many men and women often exhibit no symptoms whatsoever and therefore can spread the disease without knowing it. In men, when symptoms occur, they resemble those of gonorrhea—painful urination and a cloudy mucous discharge from the penis. In women, symptoms resemble those of a urinary tract infection. Urination becomes painful and more frequent. NSU can be treated by antibiotics, but individuals should seek treatment quickly to avoid spread of the disease and more serious complications.

Syphilis is Caused by a Bacterium and Can Be Extremely Debilitating if Left Untreated

Despite its linguistic appeal, syphilis is a serious sexually transmitted disease caused by a bacterium that penetrates the linings of the oral cavity, vagina, and penile urethra or enters through breaks in the skin. If untreated, syphilis proceeds through three stages. In stage 1, between one and eight weeks after exposure, a small, painless red sore develops, usually in the genital area. Easily visible when on the penis, these sores often go unnoticed when they occur in the vagina or cervix of a woman. The sore heals in one to five weeks, leaving a tiny scar.

Approximately six weeks after the sore heals, individuals complain of fever, headache, and loss of appetite. Lymph nodes in the neck, groin, and armpit swell as the bacteria spread throughout the body. This is stage 2.

The symptoms disappear for several years. Then, without warning, the disease flares up. The final stage, stage 3, is an autoimmune reaction that causes paralysis, senility, or even insanity. Individuals may lose their sense of balance and may lose sensation in their legs. The bacterium can weaken the walls of the aorta, causing an aneurysm (Chapter 12).

Fortunately, syphilis can be successfully treated with antibiotics, but only if the treatment begins early. Suspicious sores in the mouth and genitals should be brought to the attention of a physician. In the late stages, antibiotics are useless. Tissue or organ damage is permanent.

Herpes Is Caused by a Virus, Is Extremely Common, and Is Essentially Incurable

Herpes is one of the most common sexually transmitted diseases. Approximately 200,000 to 300,000 people contract the disease each year. Herpes is caused by a virus, which after entering the body remains there for life. The first sign of infection is pain, tenderness, or an itchy sensation on the penis or female external genitalia, which occurs about six days after contact with someone infected by the virus. Soon afterward, painful blisters appear on the penis and female genitalia. The blisters may also form on the thighs and buttocks, in the vagina, and on the cervix.

The blisters break open and become painful ulcers that last for one to three weeks, then disappear. Because the herpes virus is a lifelong resident of the body, new outbreaks may occur from time to time, especially when an individual is under stress. Recurrent outbreaks are generally not as severe as the initial one, and in time the outbreaks generally cease altogether.

Herpes can be spread to other individuals during sexual contact, but only when the blisters are present or (as recent research suggests) just beginning to emerge. When the virus is inactive, sexual intercourse can occur without infecting a partner.

Although herpes cannot be cured, doctors may prescribe a drug called acyclovir that suppresses the virus. It reduces the incidence of outbreaks and accelerates healing of blisters.

Women who have herpes run the risk of transferring the virus to their infants at birth. Because the herpes virus can be fatal to newborns, women are often advised to deliver by cesarean section (an incision made just above the pubic bone) if the virus is active at the time of birth.

INFERTILITY

A surprisingly large percentage (about one in six) of American couples cannot conceive. The inability to conceive (to

[8] Sexually transmitted diseases are, in fact, a leading cause of infertility in women.

become pregnant) is called **infertility.** In about 50% of the couples, infertility results from problems occurring in the woman. About 30% of the cases are due to problems in the man alone, and about 20% are due to problems in both partners.

A couple who have been actively trying for a year or more to conceive should see a physician. The physician will first check obvious problems, such as infrequent or poorly timed sex, because only intercourse around the time of ovulation will be successful. If timing is not the problem, the physician will test the man's sperm count. A low sperm count is one of the most common causes of male infertility, for reasons noted earlier.

A low sperm count may result from overwork, emotional stress, and fatigue. Excess tobacco and alcohol consumption also contribute to the problem. Tight-fitting clothes and excess exercise, which raise the scrotal temperature, also reduce the sperm count. One of the most common causes of low sperm count is an enlargement of the veins draining one or both testes, a condition called a **varicocele.** The testes are also sensitive to a wide range of chemicals and drugs, and some physicians believe that the myriad of chemicals people are exposed to in everyday life may be lowering sperm production in males (see the next section).

If infertility results from a low sperm count, a couple may choose to undergo artificial insemination, using sperm from a sperm bank. These sperm are generally acquired from anonymous donors and are stored frozen. When thawed, the sperm are reactivated, then deposited in the woman's vagina or cervix around the time she ovulates.

If sperm production and ejaculation are normal, a physician will then check the woman's reproductive tract. First comes a test of ovulation. A sample of the mucus produced by the cervix and a biopsy of the uterine lining can indicate whether or not a woman is cycling. If ovulation is not occurring, **fertility drugs** may be administered. Several kinds of drugs are available. One of the more common is HCG, which, as noted earlier, is an LH-like hormone that induces ovulation. Unfortunately, fertility drugs often result in *superovulation* (the ovulation of many fertilizable ova), leaving couples with a litter, four to six babies, instead of the one child they had hoped for. In fact, most of the multiple births you hear about on the news are the result of fertility drugs.

If tests show that ovulation is occurring normally, infertility may be caused by an obstruction in the uterine tube. A previous gonorrheal or chlamydial infection may have spread into the tubes, causing scarring and obstructing the passageway. In this case, a couple may be advised to adopt a child or to try *in vitro* fertilization. During *in vitro* fertilization, ova are surgically removed from the woman, then fertilized by the partner's sperm. The fertilized ovum can be implanted in the uterus of the woman and can grow successfully to term. This procedure is expensive and time-consuming, has a low success rate, and is not widely available. It also places heavy emotional demands on the couple.

ENVIRONMENT AND HEALTH: THE SPERM CRISIS?

In September 1979, Professor Ralph Dougherty, a chemist at Florida State University, announced findings from a study of 130 healthy, male college students. In his test group, Dougherty found extremely low sperm counts of only 20 million per milliliter of semen, compared with an expected value of 60 to 100 million per milliliter. Biochemical analyses of the testes also revealed high levels of four toxic chemicals: DDT, polychlorinated biphenyls (PCBs), pentachlorophenol, and hexachlorobenzene.

An article in the Sierra Club's magazine reporting on the results proclaimed that America was facing a "sperm crisis" caused by toxic chemicals. But not everyone agrees. Health officials, in fact, have challenged these findings. Some officials contend that the low sperm counts that Dougherty recorded may have resulted from improved counting techniques. That is to say, over the years, advances in technology have allowed scientists to count sperm more accurately. As a result, estimates of the normal sperm concentration have been markedly lowered. Dougherty maintains that such improvements, while present, are not entirely responsible for the decline.

Other health officials argue that Dougherty's findings are not representative of the American public. Floridians, they say, may be exposed to high levels of pesticides used on farms. These chemicals may be contaminating the water supplies of urban and rural residents.

Additional studies in Florida and other states show that sperm counts in American men have indeed been falling since the 1950s. Prior to 1950, the average sperm count was about 110 million per milliliter. By 1980 and 1981, sperm counts had dropped to about 60 million per milli-

TABLE 21–6 Some Agents Potentially Toxic to Male and Female Reproduction

MALES	FEMALES
Natural and synthetic androgens	Natural and synthetic estrogens
Heat	Natural and synthetic progestins
Radiation	Amphetamine
Dioxin	DDT
PCBs	Parathion (insecticide)
Vinyl chloride	Carbaryl (insecticide)
Ethanol	Diethylstilbestrol
Benzene	PCBs
Diethylstilbestrol (DES)	
EDB (ethylene dibromide)	
Paraquat (herbicide)	
Carbaryl (insecticide)	
Cadmium	
Mercury	

ter. Statistical studies suggest that the decline may be related to growing pesticide use, air pollution, and other factors.

Studies in Hawaii support the belief that certain environmental chemicals may be causing a decline in sperm count. These studies show that Hawaiian men have a considerably higher sperm count than men residing in the continental United States. This observation has been attributed to a generally cleaner environment. There are, say researchers, fewer factories on the Hawaiian Islands than on the mainland. People are exposed to fewer agricultural chemicals, and frequent winds probably result in cleaner air.

Reductions in sperm count are of concern to many people because a sperm count below 20 million per milliliter is generally insufficient for fertilization. Today, low sperm counts account for a significant percentage of all infertility in U.S. couples.

Human reproduction, like other bodily processes, depends on a healthy environment. Research shows that a wide range of factors—from drugs to radiation to industrial chemicals—are toxic, or potentially toxic, to human reproduction (Table 21–6). "There has been an explosion of spermatotoxins in the environment," says Dr. Bruce Rappaport, former director of an infertility clinic in San Francisco. "The problem is environmental pollution." Today, at least 20 common industrial chemicals are known to be reproductive toxins. Ten commonly prescribed antibiotics can reduce sperm count. Even Tagamet, a drug that is used to relieve stress and treat stomach ulcers and is now the most prescribed drug in the United States, reduces sperm count by over 40%. By one estimate, at least 40 commonly used drugs depress sperm production, and thousands of other drugs and environmental pollutants have not been tested.

These facts do not necessarily mean that the United States is in a sperm crisis, but they do suggest the need for caution. Further research is needed to determine potential impacts, if any, of the many thousands of chemicals now commonly used or released into the environment. Research may prove that we need to clean up our act, or it may show that the fears are unwarranted.

SUMMARY

REPRODUCTIVE STRATEGIES

1. Organisms reproduce in one of two ways: asexually or sexually.
2. In animals, asexual reproduction occurs when an organism produces offspring by itself by budding or fragmenting. As a result, it tends to produce offspring that are genotypically and phenotypically identical to the parent.
3. Asexual reproduction permits species to proliferate quickly in the presence of adequate resources.
4. Its main disadvantage is that it results in low genetic variability in populations, a disadvantage when environmental conditions are changing.
5. Sexual reproduction involves sex cells, or gametes, each containing half the chromosomes of the adult in which they are produced.
6. Gametes combine to form a zygote whose genome is different from its parents'. Thus, sexual reproduction produces considerable genetic variation in the offspring and provides an advantage in evolution.
7. Most sexually reproducing animals breed only once a year. In temperate regions, breeding is timed so that offspring are born during a season when their survival is enhanced.
8. The timing of breeding in vertebrates and many invertebrates is regulated by day length, temperature, and the availability of food and water.
9. Sexual reproduction may involve internal or external fertilization.

HUMAN REPRODUCTION

10. The male reproductive tract consists of seven basic parts: (a) testes, (b) epididymes, (c) vasa deferentia, (d) sex accessory glands, (e) urethra, (f) penis, and (g) scrotum.
11. The testes lie in the scrotum, which provides a suitable temperature for sperm development. Each testis contains hundreds of sperm-producing seminiferous tubules. Between the seminiferous tubules are the interstitial cells, which produce testosterone.
12. Sperm produced in the seminiferous tubules are stored in the epididymis. During ejaculation, sperm pass from the epididymis to the vas deferens, then to the urethra. Secretions from the sex accessory glands are added to the sperm at this time.
13. Luteinizing hormone (LH) or interstitial cell stimulating hormone (ICSH) from the anterior pituitary stimulates the interstitial cells to produce testosterone. LH release is regulated by gonadotropin-releasing hormone from the hypothalamus.
14. Testosterone stimulates spermatogenesis, facial hair growth, thickening of the vocal cords, laryngeal cartilage growth, sebaceous gland secretion, and bone and muscle development.
15. The penis is the organ of copulation. It contains erectile tissue, which fills with blood during sexual arousal, making the penis turgid.
16. Ejaculation is under reflex control. When sexual stimulation becomes intense, sensory nerve impulses travel to the spinal cord. They stimulate motor neurons in the cord, which send impulses to the smooth muscle in the walls of the epididymes, the vasa deferentia, the sex accessory glands, and the urethra, causing ejaculation.
17. The female reproductive system consists of two basic components: the reproductive tract and the external genitalia.
18. The reproductive tract consists of (a) the uterus, (b) the two uterine tubes, (c) the two ovaries, and (d) the vagina.
19. The external genitalia consist of two flaps of skin on both sides of the vaginal opening, the labia majora and the labia minora.
20. Female germ cells are housed in follicles in the ovary. A follicle consists of a germ cell and an investing layer of follicle cells.

21. A dozen or so follicles enlarge during each menstrual cycle, but most follicles degenerate. In humans, usually only one follicle makes it to ovulation.
22. The oocyte and an investing layer of follicle cells are released during ovulation and drawn into the uterine tubes.
23. The menstrual cycle consists of a series of changes occurring in the ovaries, uterus, and endocrine system of women.
24. The first half of the menstrual cycle is called the follicular phase. It is during this period that FSH from the pituitary stimulates follicle growth and development. LH stimulates estrogen production.
25. Estrogen levels rise slowly during the follicular phase, then trigger a positive feedback mechanism that results in a preovulatory surge of FSH and LH, which triggers ovulation.
26. The ovum is expelled from the antral follicle at ovulation. The follicle then collapses and is converted into a corpus luteum (CL), which releases estrogen and progesterone.
27. In the absence of fertilization, the CL degenerates. If fertilization occurs, however, HCG from the embryo maintains the CL for approximately six months.
28. During the menstrual cycle, ovarian hormones stimulate growth of the uterine lining, which is necessary for successful implantation. If fertilization does not occur, the uterine lining is sloughed off during menstruation, which is triggered by a decline in ovarian estrogen and progesterone.
29. Like testosterone levels in boys, estrogen levels in girls increase at puberty. Estrogen promotes growth of the external genitalia, the reproductive tract, and bone. It also stimulates the deposition of fat in women's hips, buttocks, and breasts.
30. Progesterone works with estrogen to stimulate breast development. It also promotes endometrial growth and inhibits uterine contractions.
31. Many women suffer from premenstrual syndrome, characterized by irritability, depression, tension, fatigue, headaches, bloating, swelling and tenderness of the breasts, and even joint pain.
32. The menstrual cycle continues throughout the reproductive years, but after a woman reaches 20, the ovaries become progressively less responsive to gonadotropins. As a result, estrogen levels slowly decline as a woman ages. Ovulation and menstruation become erratic as a woman approaches 45.
33. Between the ages of 45 and 55, ovulation and menstruation cease. The end of reproductive function in women is known as the menopause.
34. The decline in estrogen levels results in atrophy of the reproductive organs and behavioral disturbances. Many women become irritable, suffer bouts of depression, and experience hot flashes and night sweats induced by intense vasodilation of vessels in the skin.

BIRTH CONTROL

35. Birth control refers broadly to any method or device that prevents births, and it includes two general strategies: contraception (measures that prevent pregnancy) and induced abortion (the deliberate expulsion of a fetus).
36. Figure 21–29 summarizes the effectiveness of the various birth control measures.
37. The most effective form of birth control is abstinence. Another highly effective measure is sterilization—vasectomy in men and tubal ligation in women.
38. The pill is also a highly effective means of birth control. The most common pill in use today contains a mixture of estrogen and progesterone that inhibits ovulation. In some women, however, estrogen causes adverse health effects.
39. The intrauterine device is a plastic or metal coil that is placed inside the uterus, where it prevents the fertilized ovum from implanting.
40. The diaphragm, condom, and vaginal sponge are less effective than the measures described above. The diaphragm is a rubber cap fitted over the cervix. To be fully effective, it must be coated with a spermicidal jelly, foam, or cream.
41. The condom is a thin, latex rubber sheath worn over the penis during sexual intercourse that prevents sperm from entering the vagina.
42. The vaginal sponge is a tiny, round sponge worn by the woman. It is impregnated with a spermicidal chemical.
43. One of the oldest, but least effective, methods of birth control is withdrawal, removing the penis before ejaculation. Spermicidal chemicals used alone are about as effective as withdrawal.
44. Abstaining from sexual intercourse around the time of ovulation, known as the rhythm, or natural, method, is another way to prevent pregnancy. Statistics on effectiveness show that the natural method is the least successful of all birth control measures.
45. Some couples may elect to terminate pregnancy through an abortion.

SEXUALLY TRANSMITTED DISEASES

46. Certain viruses and bacteria can be transmitted from one individual to another during sexual contact. Infections spread in this way are called sexually transmitted diseases.
47. Gonorrhea is caused by a bacterium that commonly infects the urethra in men and the cervical canal in women. Overt symptoms of the infection are frequently not present, so people can spread the disease without knowing they have it. If left untreated, gonorrhea can spread to other organs, causing considerable damage.
48. Nonspecific urethritis (NSU), the most common sexually transmitted disease, is caused by several different bacteria, but most commonly by chlamydia. It is less threatening than gonorrhea or syphilis. Many men and women show no symptoms of NSU and therefore can spread the disease without knowing it. Symptoms, when they occur, resemble those of gonorrhea.
49. Syphilis is a serious sexually transmitted disease caused by a bacterium that penetrates the linings of the oral cavity, vagina, and penile urethra. If untreated, syphilis proceeds through three stages. It can be treated with antibiotics during the first two stages, but in stage 3, when damage to the brain and blood vessels is evident, treatment is ineffective.
50. Herpes is also a very common sexually transmitted disease. It is caused by a virus. Once the virus enters the body, it remains for life. Blisters form on the genitals and sometimes on the thighs and buttocks. The blisters break open and become painful ulcers. At this stage, an individual is highly infectious. New outbreaks of the virus may occur from time to time, especially when an individual is under stress.

INFERTILITY

51. The inability to conceive is called infertility. Infertility may result from a variety of problems in men and women: poorly timed sex, low sperm count, failure to ovulate, or obstruction in the uterine tubes.

ENVIRONMENT AND HEALTH: THE SPERM CRISIS?

52. A variety of drugs and chemical pollutants affect sperm development and may be causing a decline in the sperm count of American men.

EXERCISING YOUR CRITICAL THINKING SKILLS

1. Devise an experiment or set of experiments to test the hypothesis that the United States is in the midst of a sperm crisis, a decline in male sperm production caused by chemicals in the environment. What type of evidence would support or refute your hypothesis? How can you devise your experiment to avoid bias?
2. You are a journalist for a major urban newspaper. You receive a press release from the local medical school announcing that one of the researchers has discovered that a chemical found in a common household cleaning agent reduces fertility in rats and mice. Large doses were given to both males and females before conception. The results showed a statistically significant decline in the litter size and several abnormalities. The researcher suggests that the chemical should be banned from use in homes. Using your critical thinking skills, what questions would you ask before writing your article? What other information would you seek out?

TEST OF TERMS

1. The testes lie in the ____________, a sac that provides a suitable temperature for the development of sperm. Sperm are produced inside the ____________ tubules.
2. Sperm are stored in the ____________ and delivered to the urethra during ejaculation via the ____________ ____________, two muscular ducts.
3. The ejaculate consists of secretions from three glands, known as the ____________ ____________ glands.
4. Spermatogenic cells are found in the ____________ epithelium. Spermatogenesis begins with ________, which divide mitotically. Some of these cells, however, enlarge to form the ____________ ____________, which undergo the first meiotic division, forming two ____________ ____________.
5. The second meiotic division produces four ____________, each containing ____________ single-stranded (unreplicated) chromosomes in humans.
6. The nucleus of a mature sperm is capped by an enzyme-containing structure called the ____________, which helps the sperm penetrate the barriers around the ovum.
7. Male sex steroid hormones are produced by the ____________ cells in the testes. One of the chief steroids is ____________. Its secretion is controlled by a pituitary hormone called ____________ ____________.
8. The glans penis is covered by a flap of skin at birth called the ____________. Surgical removal of this part of the penis is called ____________.
9. The penis contains ____________ tissue, which fills with blood during sexual arousal.
10. During ____________, which is under reflex control, muscle contractions in the reproductive tract cause sperm to be propelled to the outside. As the sperm pass along the tract, they are mixed with secretions from certain glands along the way. Sperm and the secretions of these glands constitute the ____________.
11. The ____________ is a pear-shaped organ that houses and nourishes the developing embryo. Ova are produced by the ovaries and picked up by the ____________ ____________.
12. Sperm are deposited in the ________ and make their way through the ____________ canal.
13. The ________ ________ are two flaps of skin covered with hair that are part of the external genitalia in women.
14. The ____________ in females is formed from the same embryological tissue that forms the penis in males.
15. Release of an ovum from the ovary is called ____________. It occurs at the midpoint of the ____________ cycle.
16. A(n) ____________ follicle consists of many layers of follicle cells and a large central cavity. The primary oocyte is surrounded by a gel-like layer called the ____________ ____________.
17. During oogenesis, the primary oocyte divides during the ____________ meiotic division, producing a secondary oocyte and the first ____________ body containing ____________ [give a number] ____________-stranded chromosomes in humans.
18. The ________ ________ is a structure in the ovary produced from a follicle that releases its ovum. It produces two steroid hormones, ______ and ____________.
19. The first half of the menstrual cycle is called the ____________ phase. LH released from the pituitary at this time stimulates ____________ production by the large follicles in the ovary, while the pituitary hormone ____________ stimulates follicle growth.
20. The lining of the uterus, called the ____________, thickens during the first half of the menstrual cycle. It continues to grow throughout the cycle, but in the absence of fertilization, it is sloughed off. The loss of blood and tissue from the lining is called ____________.
21. In women, depression, irritability, and swelling and tenderness of the breast are symptoms of ____________ ____________.
22. Ovarian function ceases in women sometime after age 45. This is called the ____________. The decline in the secretion of ____________

from the ovaries at this time results in behavioral changes and signs of atrophy in the breasts and uterus.

23. Male sterilization, a surgical procedure in which the vasa deferentia are cut, is called ____________. In women, sterilization is achieved by severing the ____________ ____________, an operation called a(n) ____________ ____________.
24. The combined birth control pill contains two synthetic femalesteroid hormones, ________ and ________.
25. Cervical cancer can be diagnosed by a ____________ ____________, a swab of the cervical lining.
26. A(n) ____________ ____________ is a plastic coil inserted in the uterus that prevents pregnancy.
27. A(n) ____________ is a rubber cup that fits over the cervix as it protrudes into the vagina. It is coated with ____________ jelly or foam.
28. A thin, latex rubber device worn over the penis during sexual intercourse to reduce disease and the chance of pregnancy is called a ____________.
29. Timing sexual intercourse to avoid the deposition of sperm in the vagina near the time of ovulation constitutes the ____________ method and is one of the least effective contraceptive techniques.
30. A woman can determine when she ovulates by taking daily ____________ measurements.
31. Gonorrhea and syphilis are ________ ____________ ____________ caused by bacteria.
32. ____________ ____________ is the most common sexually transmitted disease known to medical science and is one of several STDs whose incidence is steadily rising in the United States.
33. ____________ is a viral disease transmitted during sexual intercourse and results in small blisters in the genital region, thighs, and buttocks.

Answers to the Test of Terms are found in Appendix B.

TEST OF CONCEPTS

1. Organisms reproduce in one of two ways. What are they? How are they different? What are the advantages and disadvantages of each?
2. You are a fertility specialist. A young woman arrives in your office complaining that she has been trying to get pregnant for two years but to no avail. Describe how you would go about determining whether the problem was with her, her husband, or both of them.
3. Describe the anatomy of the male reproductive system. Where are sperm produced? Where are they stored? What structures produce the semen?
4. Describe the process of spermatogenesis, noting the cell types and the number of chromosomes in each type.
5. Why is the first meiotic division called a reduction division?
6. List the hormones that control testicular function. Where are they produced, and what effects do they have on the testes?
7. You are a family doctor. A man comes to your office complaining of impotence. What are the possible causes? How would you go about testing for the causes?
8. Trace the pathway for a sperm from the seminiferous tubule to the site of fertilization.
9. Describe the process of ovulation and its hormonal control.
10. What is the corpus luteum? How does it form? What does it produce? Why does it degenerate at the end of the menstrual cycle if fertilization does not occur?
11. What is menstruation? What triggers its onset?
12. Describe the effects of estrogen and progesterone on the reproductive tract and the body.
13. A woman comes to your office. She is 47 years old and complains of irritability and depression. She asks for the name of a reliable psychiatrist who could help her. She says that she wakes up in the middle of the night in a sweat. Would you give her the name of a psychiatrist? Why or why not? If not, what would you do?
14. Describe each of the following birth control measures, explaining what they are and how they work: the pill, IUD, diaphragm, cervical cap, condom, spermicidal jelly, and natural method.
15. Describe ways to prevent the spread of sexually transmitted diseases.

CHAPTER

22

Animal Development and Aging

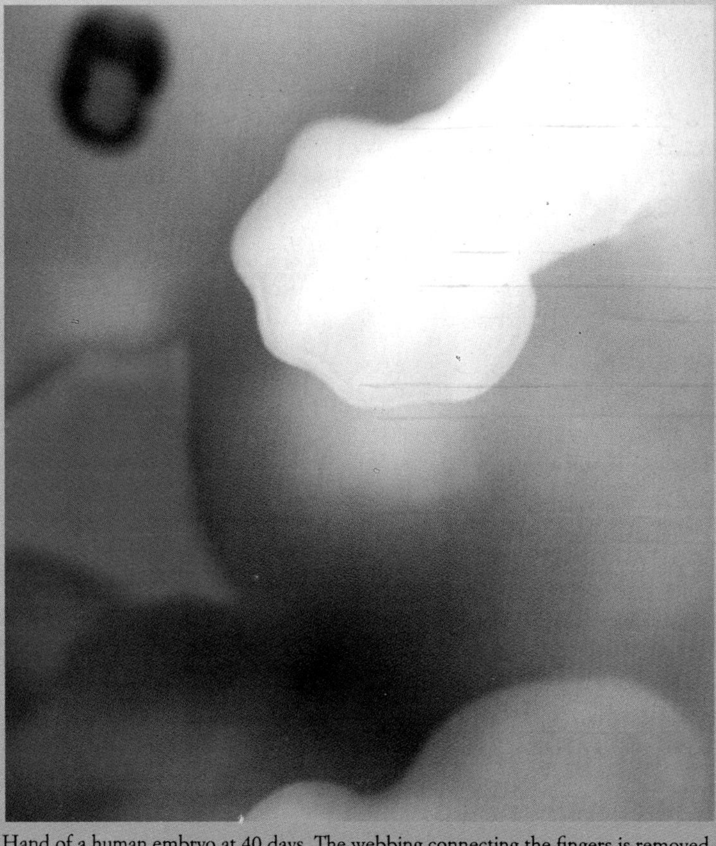

Hand of a human embryo at 40 days. The webbing connecting the fingers is removed by enzymes from lysosomes at a genetically determined stage of development.

Reproduction is one of the most varied of all biological functions. Like many other organ systems, the reproductive system has taken many twists and turns along the course of evolution, producing a fascinating array of ways to achieve the same end: procreation, or producing offspring. As you saw in the last chapter, reproductive strategies fall into one of two categories, sexual and asexual, but the variation in each major group is incredible. Among those animal species that reproduce sexually, for example, many fertilize externally. In many fishes, for instance, females lay the eggs on the bottom of a body of water or the walls of a rocky cave, and males deposit sperm on them. The parents then often abandon the fertilized eggs and move on to other activities. In other complex animal species, internal fertilization is the rule. Males deposit sperm inside the female reproductive system, where they unite with eggs. In some species, like turtles and lizards, the eggs are expelled (usually buried), then abandoned. But in others animals, the internally fertilized eggs develop internally, protecting the offspring from the harsh realities of predation, adverse weather, disease, and other factors to which their biological cousins are exposed. Internal development occurs in some species of fishes, like sharks, and in some reptiles, such as snakes. It is also found in virtually all mammals. Internal development required adaptations that permit the nourishment of offspring within the female tract for long periods.

This chapter focuses on fertilization and development, primarily on human development. It also presents a brief discussion of human growth after birth and of aging and death.

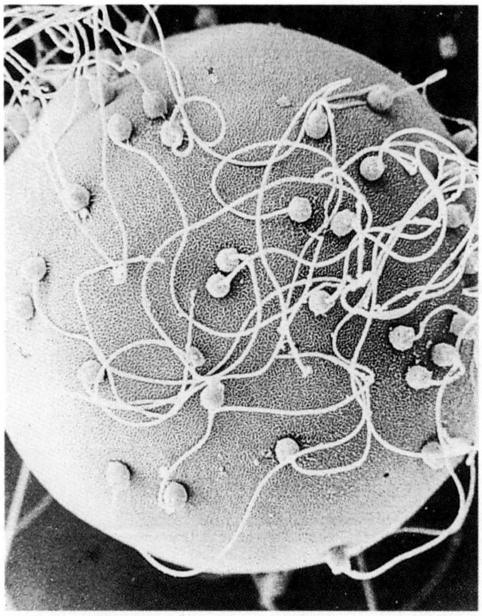

▻ **FIGURE 22–1 Fertilization** Although many sperm gather around this clam egg, only one will enter. Fertilized sperm are phagocytized by the egg.

AN OVERVIEW OF ANIMAL DEVELOPMENT

In sexually reproducing animals, new life begins when the sperm and egg unite, forming a diploid zygote. This process is referred to as **fertilization** (▻Figure 22–1). As you learned in the last chapter, the sperm and egg are produced in special reproductive organs (gonads) via meiosis.

In vertebrates and invertebrates, sperm contact the plasma membrane of eggs and are engulfed (phagocytized) by them. The nuclei of the egg and sperm cell then fuse to form a diploid zygote.

Fertilization is followed by mitotic division that converts the zygote into a two-celled structure. These cells continue to divide, producing a multicelled structure, known as a **morula** (from the Latin word *morus,* meaning "mulberry") (▻ Figure 22–2). The successive division of embryonic cells, or **cleavage,** continues for a while. In many species, such as frogs and sea urchins, a hollow cavity forms in the morula (Figure 22–2). The resulting structure is called a **blastula.**

Cleavage of the embryonic cells soon begins to slow, and the cells begin to migrate about in distinct patterns, dramatically changing the appearance of the embryo. As illustrated in Figure 22–2, some cells of the embryo invaginate to form a tube that will later become the gastrointestinal tract. This stage of development is referred to as **gastrulation,** and the embryo is called a **gastrula.** During gastrulation, three distinct layers of cells form in the gastrula: an internal layer, the **endoderm;** a middle layer, the **mesoderm;** and an outer layer, the **ectoderm.**

Following this remarkable redistribution and reorganization of cells, the three cell layers differentiate further, forming organs. The endoderm gives rise to the lining of the gut and various organs associated with the gut. The mesoderm gives rise to bone, muscle, cartilage, connective tissue, and blood. And the ectoderm gives rise to the skin and nervous system. The formation of organs is referred to as **organogenesis.**

Organogenesis is a complicated process beyond the scope of this book. What is important to remember is that once the organs form, they usually continue to grow until an individual reaches the adult stage. In some animals, such as nematodes (wormlike creatures that are often parasites), all organs grow at the same rate until adulthood is reached. In others, such as birds, reptiles, and mammals, they grow at different rates. Thus, changes in body proportions also occur in postembryonic growth and development (▻ Figure 22–3).

All animal embryos proceed pretty much the same through the morula, blastula, and gastrula stages, but there are differences in embryonic development. In chickens, for example, the egg contains so much yolk that cell division concentrates at one pole (▻ Figure 22–4). The embryonic cells therefore form a small, disklike structure atop a mass of yolk that differentiates into three layers. These layers, in turn, give rise to the organs.

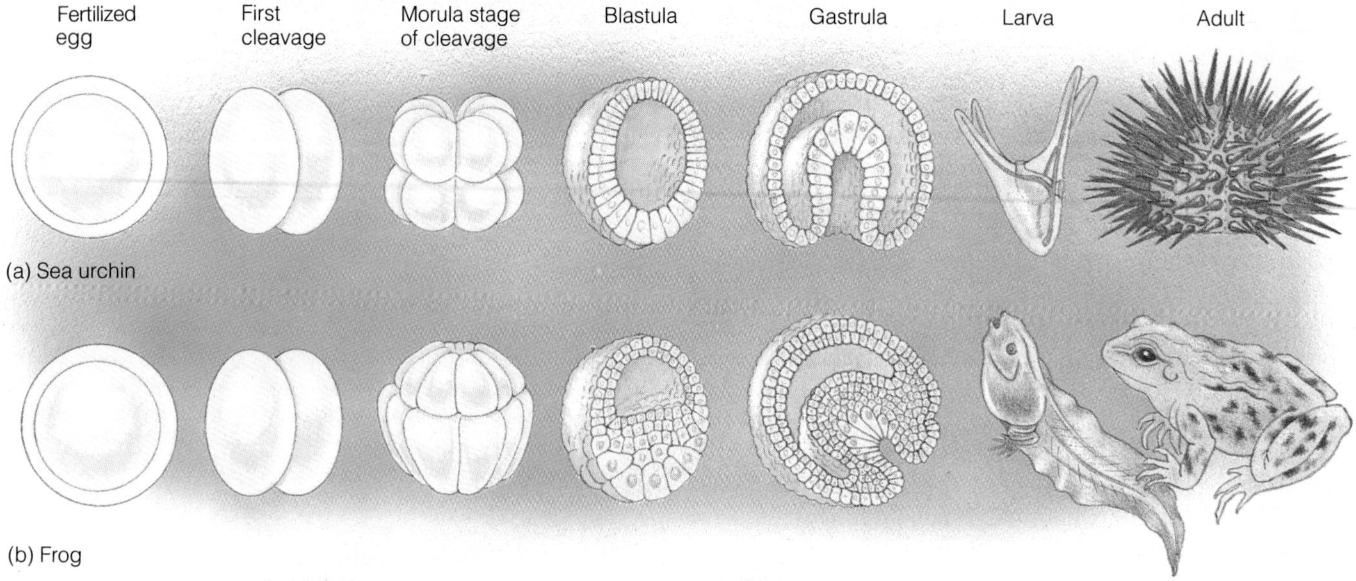

▷ **FIGURE 22–2 Early Embryonic Development** Embryonic development proceeds through several key phases as shown here in the (*a*) sea urchin and (*b*) frog.

HUMAN REPRODUCTION AND DEVELOPMENT

With this overview of animal development, we turn our attention to fertilization and development in humans.

Human Development Begins with Fertilization

In humans, the sperm and the ovum unite in the upper third of the uterine tube (▷ Figure 22–5). As you learned in the last chapter, oocytes are released from the ovary during ovulation and are drawn into the uterine tube in part by the rhythmic beating of cilia inside the tube, which create an inward-flowing current. This process is aided by fingerlike projections of the upper end of the uterine tube that, like massaging fingers, contract rhythmically and help sweep oocytes inside.

In humans, sperm are deposited in the vagina and quickly make their way up the reproductive tract (▷ Figure 22–6). Within a few minutes of ejaculation, they enter the cervical canal; 30 minutes later, they arrive at the junction of the uterus and uterine tube. Sperm then travel to the upper portion of the uterine tube, where they may encounter an ovum. Along the way, many millions of sperm die, killed by acidic secretions of the vagina and cervix. Studies based on laboratory animals suggest that only 1 in 3 sperm makes it through the cervical canal and 1 in 1000 makes it through the uterus to the uterine tubes (Figure 22–6).

Although sperm are motile, the principal driving force through the female reproductive tract is muscular contractions in the walls of the uterus and uterine tubes. Some evidence suggests that these contractions are stimulated by prostaglandins, hormonelike substances in the semen.

▷ **FIGURE 22–3 Changes in Body Proportions** In many animal species, like humans, after organ systems have formed, further growth and development involve changes in body proportions.

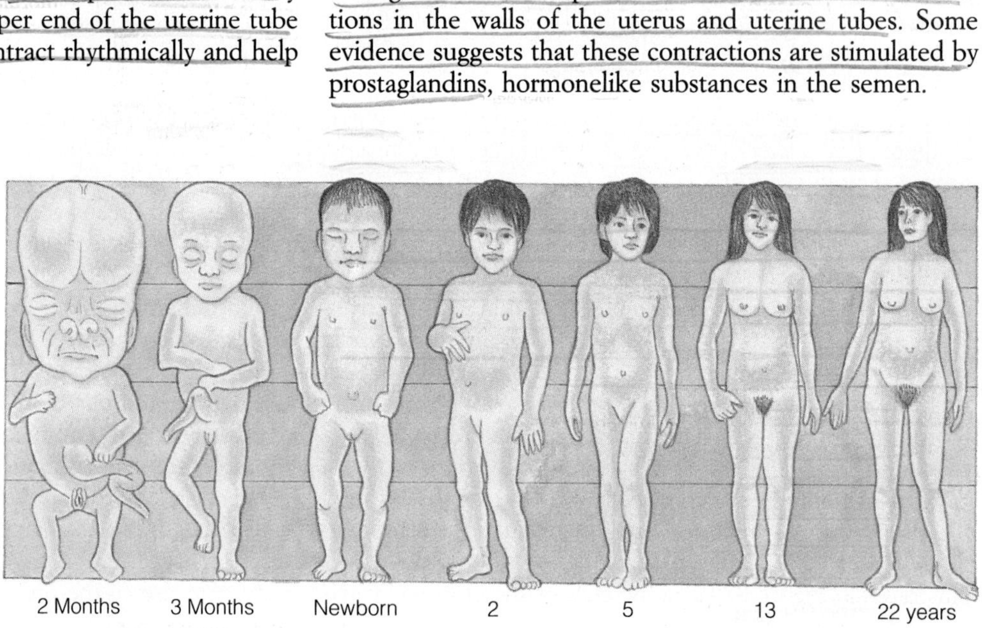

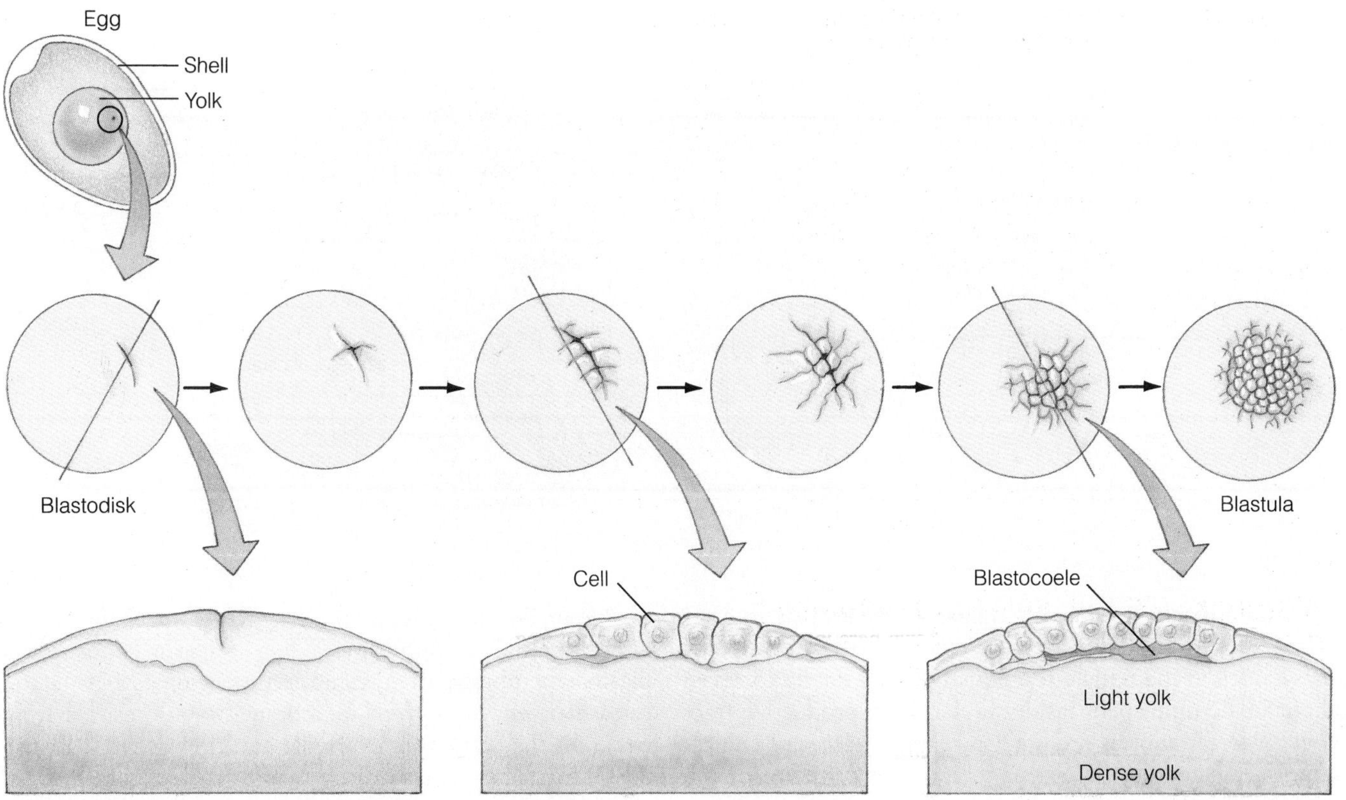

▻ **FIGURE 22–4 Early Embryonic Development in Chickens** Because of the large mass of yolk that nourishes the embryos of birds, cell division in early embryonic development is concentrated at one pole. The embryo becomes a flattened disk.

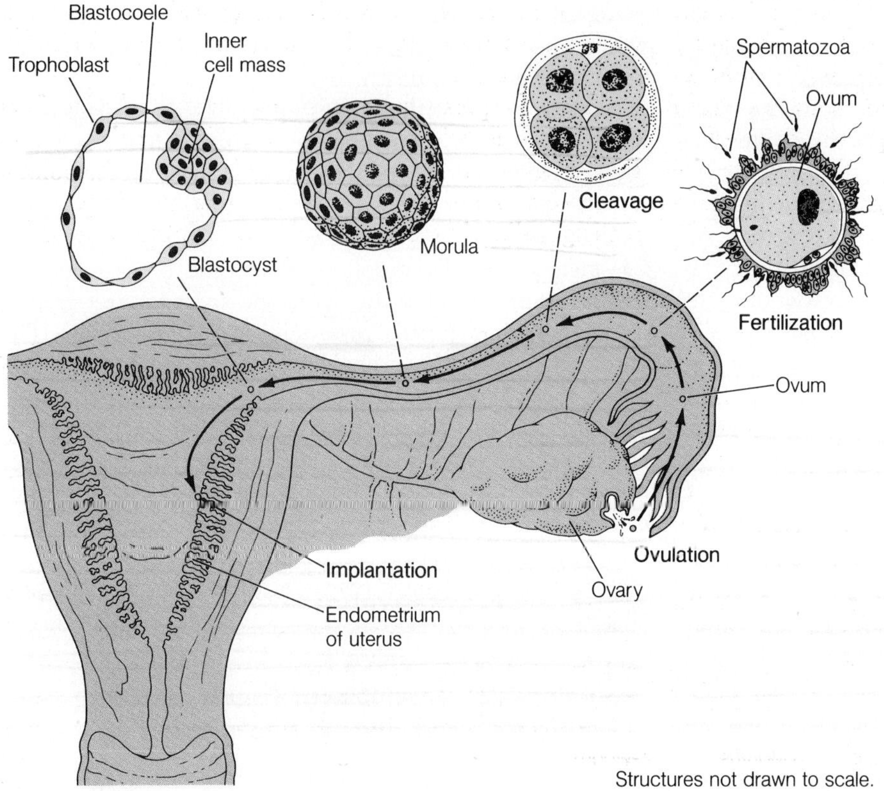

▻ **FIGURE 22–5 Fertilization and Early Embryonic Development** In humans, fertilization usually occurs in the upper third of the uterine tube. In the next three days, the zygote becomes a morula, and the morula develops into a blastocyst. The blastocyst then enters the uterus, where it will implant.

▻ **FIGURE 22–6 Sperm Transport in the Female Reproductive System** Sperm move rapidly up the female reproductive tract of humans and other vertebrates principally as a result of contractions in the muscular walls of the uterus and uterine tubes. Notice the rapid decline in sperm number along the way.

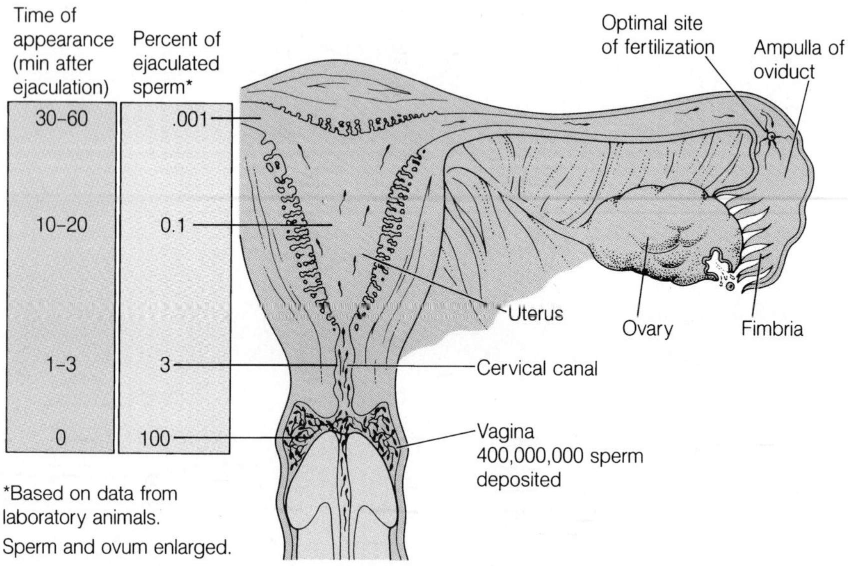

Sperm reach the site of fertilization an hour or so after ejaculation, but they cannot fertilize an ovum until they have been in the female reproductive tract for six to seven hours. The time spent inside the female reproductive tract before fertilization is required to remove a layer of cholesterol deposited on the plasma membranes of sperm by the secretions of the sex accessory glands. The cholesterol coat helps stabilize the sperm plasma membranes and protects sperm as they move through the female reproductive tract. Removal of the coat renders the sperm's membranes fragile and disruptible, a prerequisite for fertilization.

After this process, the plasma membrane over the head of the sperm fuses with the outer membrane of the acrosome, a caplike structure filled with digestive enzymes (▻ Figure 22–7). Tiny openings develop at the points of fusion, allowing acrosomal enzymes to leak out.

As Figure 22–7 shows, the follicle cells around the ovum become elongated and radiate outward like the spokes of a wheel. Acrosomal enzymes of the sperm that swarm around the oocyte like so many bees around a hive dissolve the extracellular material that binds the cells together. After passing through this layer, sperm must digest their way through the zona pellucida, a gel-like layer surrounding the oocyte, with the aid of acrosomal enzymes (Figure 22–7).

Sperm cells that traverse the zona pellucida enter the space between the plasma membrane of the oocyte and the zona. As a rule, the first sperm cell to come in contact with the plasma membrane of the oocyte will fertilize it; all other sperm are excluded.

Sperm are excluded by two known mechanisms. The first is called the **fast block to polyspermy** ("many sperm"). The fast block to polyspermy occurs when the sperm cell contacts the plasma membrane of the oocyte, triggering membrane depolarization—which, as you may recall from Chapter 16, is a change in the potential difference across the plasma membrane resulting from the influx of sodium ions. For reasons not well understood, depolarization blocks other sperm from fusing with the oocyte.

Sperm are also excluded by a slower mechanism, the **slow block to polyspermy.** When a sperm contacts the plasma membrane of a secondary oocyte, it triggers the release of enzymes from membrane-bound vesicles lying beneath the plasma membrane of the oocyte (Figure 22–7). These enzymes cause the zona pellucida to harden, which blocks other sperm from reaching the oocyte. These secretions may also cause the "extra sperm" that have attached to the plasma membrane of the oocyte to detach.

These two mechanisms are important evolutionary developments insuring that one and only one sperm penetrates an egg. Without them, fertilized eggs might quickly overload with extra nuclear material, a condition that would most likely impair subsequent cell divisions and result in embryonic death.

Sperm contact with the plasma membrane of the oocyte also triggers the second meiotic division, thus converting the secondary oocyte into an ovum. Once the sperm nucleus enters, the ovum is called a zygote.

Once inside the oocyte, the sperm cell nucleus with its 23 single-stranded (unreplicated) chromosomes swells. At this stage, the nuclei of the ovum and sperm are referred to as the male and female **pronuclei** (▻ Figure 22–8). The chromosomes in the pronuclei replicate as the pronuclei move toward the center of the ovum. A mitotic spindle assembles in the zygote. After chromosome replication is complete, the chromosomes of the male and female pronuclei condense, and the nuclear envelopes of the pronuclei disintegrate. The spindle fibers attach to the chromosomes, and the chromosomes line up on the equatorial plate. The zygote is now ready for the first mitotic division (Figure 22–8).

Pre-Embryonic Development Begins at Fertilization and Ends at Implantation

Human development is divided into three stages: pre-

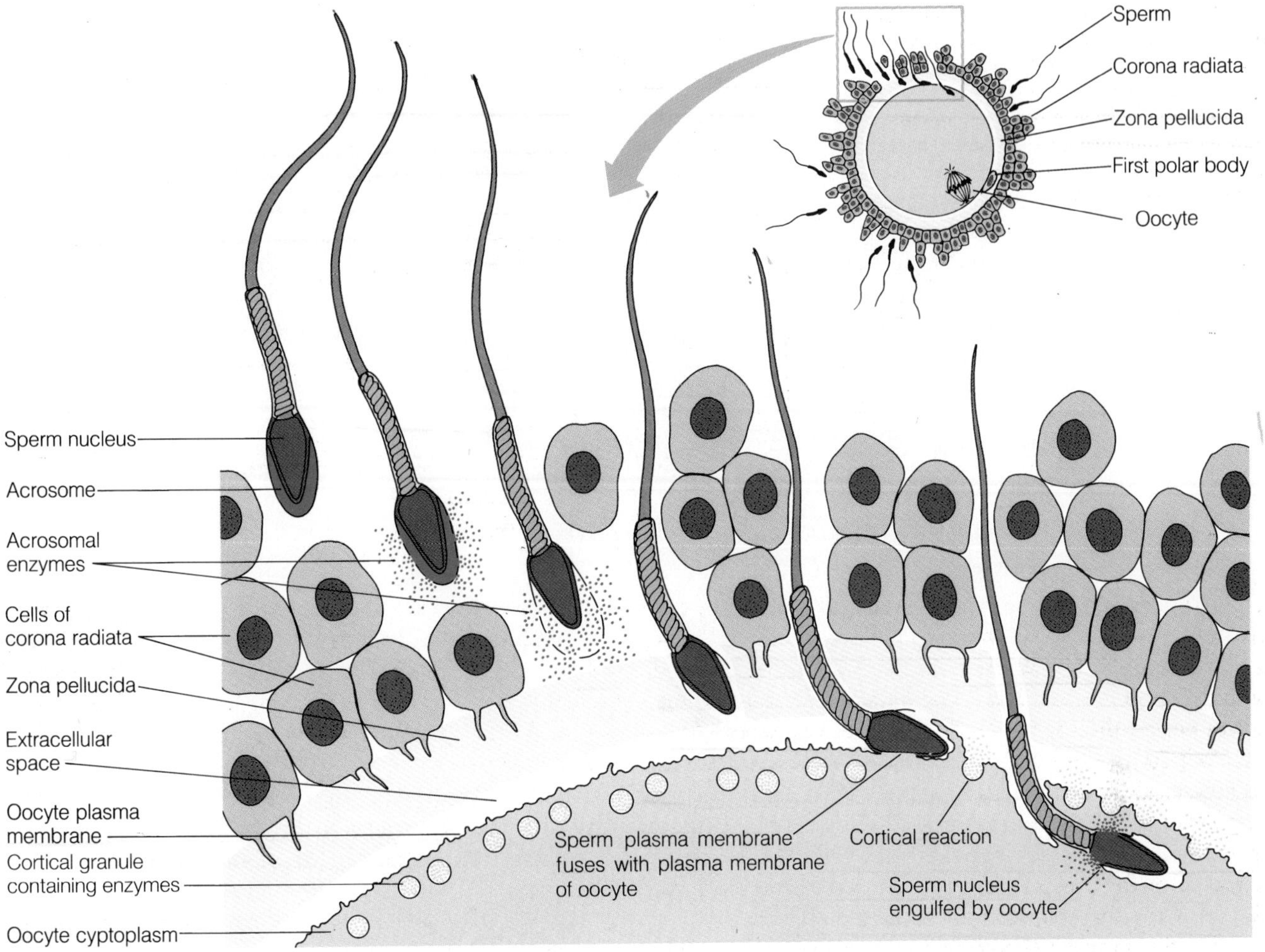

▷ **FIGURE 22–7 Fertilization and Cortical Reaction** The plasma membrane of the sperm and the outer membrane of the acrosome fuse and the membranes break down, releasing enzymes that allow the sperm to penetrate the corona radiata. Sperm digest their way through the zona pellucida via enzymes associated with the inner acrosomal membrane. Sperm are engulfed by the oocyte plasma membrane. Cortical granules are released when the sperm cell contacts the membrane. These granules cause other sperm in contact with the membrane to detach.

embryonic, embryonic, and fetal. **Pre-embryonic development** includes all the changes that occur from fertilization to the time an embryo implants in the uterine wall. During this phase, the zygote undergoes rapid cellular division and is converted into a morula, not much bigger than the fertilized ovum (▷ Figure 22–9). The morula is nourished by secretions produced by the epithelium of the uterine tubes. Approximately three to four days after ovulation, the morula enters the uterus.

Fluid soon begins to accumulate in the morula, converting it into a **blastocyst,** a hollow sphere of cells slightly larger than the morula.[1] As Figure 22–9 shows, the blastocyst consists of two parts: a clump of cells, the **inner cell mass,** which will become the embryo, and a ring of flattened cells called the **trophoblast** (meaning "to nourish the blastocyst"). The trophoblast gives rise to the embryonic portion of the **placenta,** an organ that supplies nutrients to and removes wastes from the growing embryo.

The blastocyst remains unattached in the uterine lumen for two to three days. During this period, it is nourished by secretions of uterine glands.

If the Embryo Is to Survive, It Must Embed in the Wall of the Uterus. The blastocyst attaches to the uterine lining and digests its way into the endometrium. This process, called **implantation,** begins six to seven days after fertilization. Interestingly, in some species implantation can be delayed for long periods. For a discussion of this phenomenon, see Spotlight on Evolution 22-1.

Most embryos implant high on the back wall of the uterus. The cells of the trophoblast first contact the endometrium, then adhere to it, but only if the uterine lining is healthy and properly primed by estrogen and progesterone (▷ Figure 22–10a). If the endometrium is not ready or is "unhealthy"—for example, because of the presence of an IUD or a endometrial infection—the blastocyst fails to implant. Blastocysts may also fail to implant if their cells

[1] The term blastocyst comes from the Greek *blastos* meaning germ and *kystis* meaning cyst.

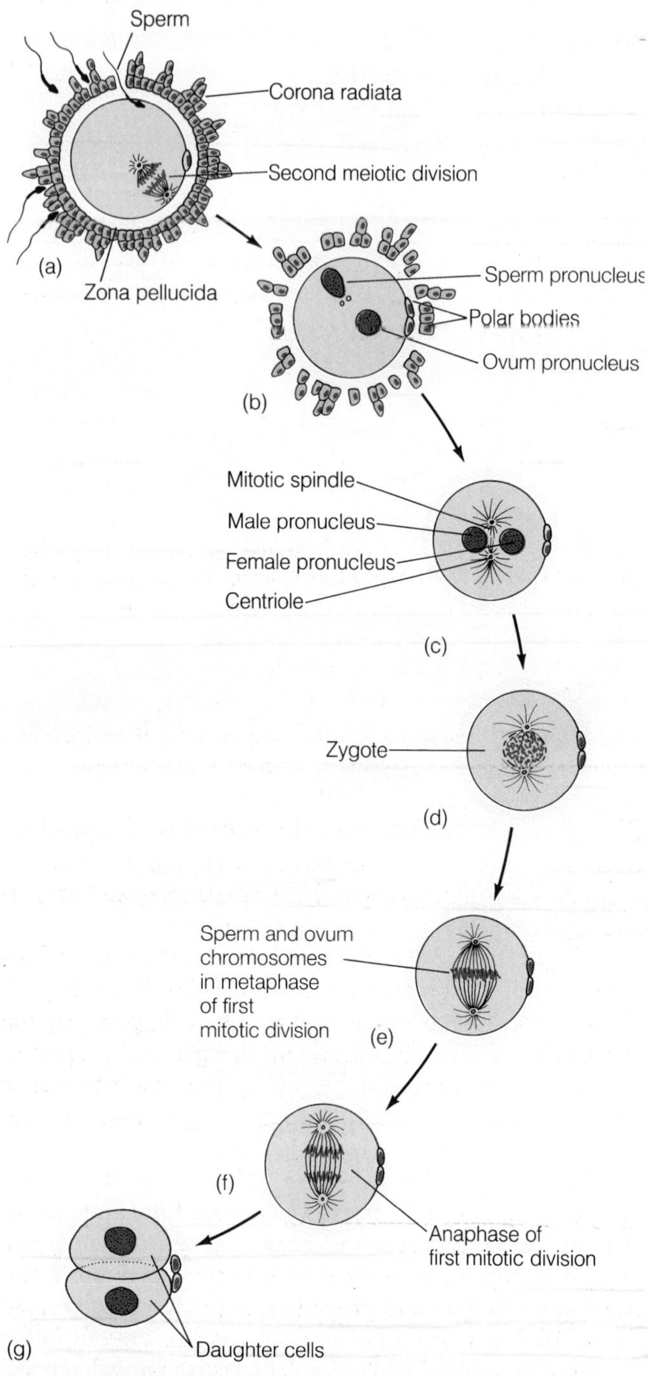

▷ FIGURE 22–8 The Zygote Prepares for Division
(*a*) The sperm contacts the plasma membrane of the oocyte. Second meiotic division takes place. (*b*) Sperm and oocyte pronuclei form. (*c*) The pronuclei migrate toward the center of the cell. Chromosomes condense, and a mitotic spindle forms. (*d*) Chromosomes condense further, and nuclear membrane breaks down. (*e*) Metaphase plate is formed. (*f*) Anaphase of the first mitotic division. (*g*) Two daughter cells form.

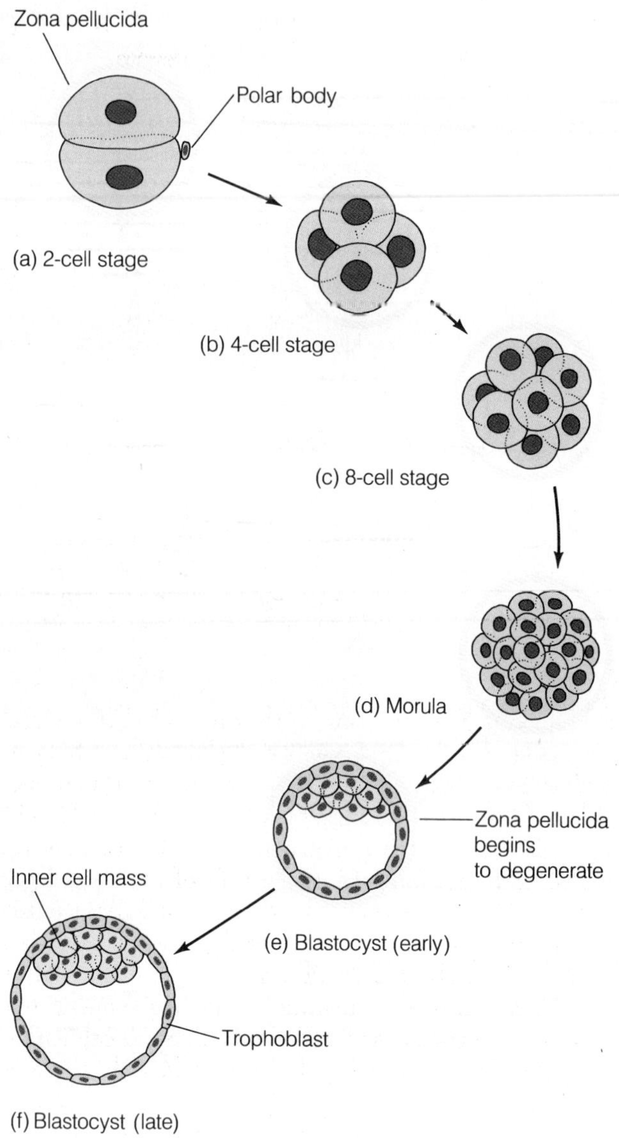

▷ FIGURE 22–9 Formation of the Morula and Blastocyst

contain certain genetic mutations. If it is unable to bind, the blastocyst perishes and is either reabsorbed (phagocytized by the cells of the endometrium) or shed during menstruation.

In cases where implantation occurs, the endometrium responds to the embryo by cellular enlargement (hypertrophy). Enzymes released by the cells of the trophoblast digest a small hole in the endometrium, and the blastocyst "bores" its way into the uterine lining (Figure 22–10b). As it "eats" its way into the layer of enlarged endometrial cells, the blastocyst feeds on nutrients released from the endometrial cells it destroys. This action helps sustain the blastocyst before the placenta forms. By day 14, the uterine endometrium grows over the blastocyst, walling it off from the uterine lumen (cavity).

Endometrial cells respond to the invasion of the blastocyst by producing prostaglandins.[2] Prostaglandins increase the development of uterine blood vessels and, therefore, help ensure an ample supply of blood and nutrients for the blastocyst.

[2]You may recall from earlier discussions that prostaglandins have many functions in the body.

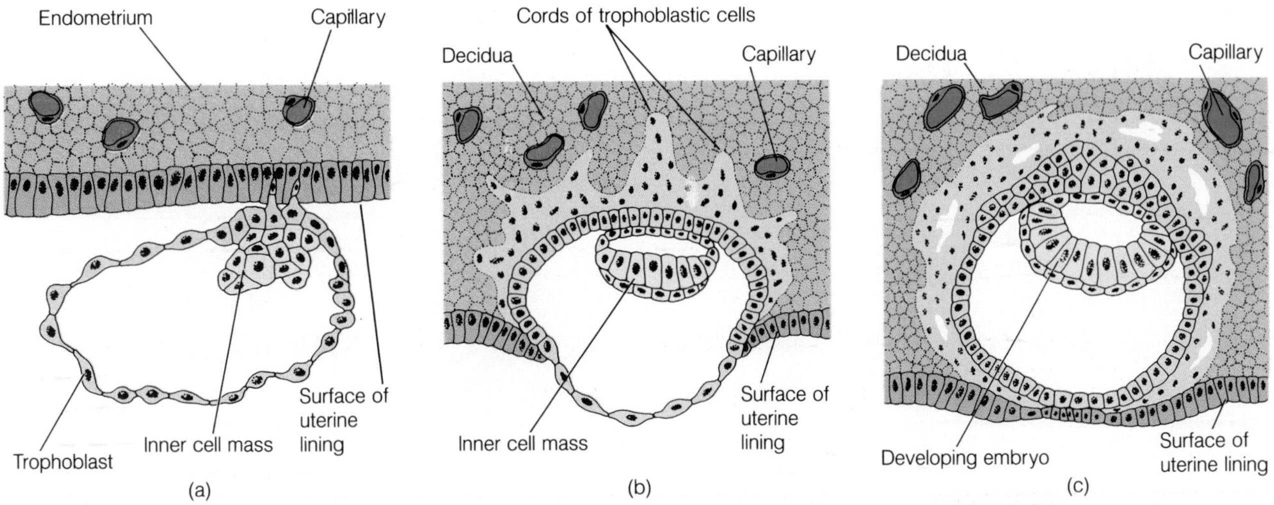

▻ **FIGURE 22–10 Implantation** (*a*) The blastocyst fuses with the endometrial lining. Endometrial cells proliferate, forming the decidua. (*b*) The blastocyst digests its way into the endometrium. Cords of trophoblastic cells invade, digesting maternal tissue and providing nutrients for the developing blastocyst. (*c*) The blastocyst soon becomes completely embedded in the endometrium.

The Placenta Forms from Embryonic and Maternal Tissue. *Placenta* is the Latin word for "cake". Shaped somewhat like a cake, the placenta forms from maternal and embryonic tissue. Figures 22–10 and ▻ 22–11 illustrate this complex process. As shown in Figure 22–11a, soon after implantation begins, the cells of the outer layer of the trophoblast begin to proliferate and invade the endometrium. Cavities form among the cells of the outer layer of the trophoblast as it digests its way into the uterine lining (Figure 22–11b). During its inward march, this layer of cells severs the walls of maternal blood capillaries. Blood pours out of the capillaries and fills the cavities (Figure 22–11c).

The inner layer of the trophoblast invades sometime later, forming fingerlike projections called placental villi (Figure 22–11c). The **placental villi** carry blood vessels from the embryo, which absorb nutrients from the pools of maternal blood in the outer layer of the trophoblast. Because the blood vessels of the placental villi connect to the developing embryo, they provide a route for nutrients to flow from the maternal blood to the embryo.

The placental villi grow and divide, increasing the total surface area for diffusion of nutrients and wastes. As they grow, they continue to invade the maternal tissue. At the same time, the blood-filled cavities enlarge, forming even bigger pools of maternal blood. Villi projecting into the blood-filled cavities absorb nutrients but also release embryonic wastes, such as carbon dioxide and urea, into the maternal blood. Because the walls of the villi are thin, wastes and nutrients can diffuse between the embryonic and maternal blood with ease. Some villi span the blood-filled cavities and anchor the embryonic portion of the placenta to the maternal tissue.

Besides providing nutrients and getting rid of embryonic wastes, the placenta produces a variety of hormones needed to maintain pregnancy. The placenta, therefore, is a respiratory, nutritive, excretory, and endocrine organ.

Placental Hormones Are Essential to Reproduction. This section discusses three placental hormones: human chorionic gonadotropin (HCG), estrogen, and progesterone (Table 22-1).

▻ Figure 22–12 shows blood levels of these three hormones during pregnancy. As illustrated, HCG levels peak in the second month of pregnancy, then drop off by the end of the third month. Levels of estrogen and progesterone, which are produced chiefly by the placenta during pregnancy, rise dramatically throughout gestation (the period between fertilization and childbirth).

HCG is an LH-like hormone produced by the embryo early in pregnancy (Chapter 21). As noted in the previous chapter, HCG prevents the corpus luteum (CL) in the ovary from degenerating when an embryo is present. It also stimulates estrogen and progesterone production from the CL, which is essential to maintaining pregnancy early in the gestational period. HCG stimulates estrogen and progesterone production by the CL for only about 10 weeks. When HCG levels decline, the CL degenerates. Estrogen and progesterone production needed to maintain pregnancy is taken over by the placenta.

High Levels of Hormones During Pregnancy May Cause Morning Sickness. Many women experience nausea (morning sickness) during the first two to three months of pregnancy. Although it often occurs in the morning hours, in some women "morning sickness" may last all day. The exact cause is not known, but some researchers believe that HCG may stimulate the brain directly, creating nausea. Other researchers blame high levels of estrogen and progesterone during pregnancy are responsible.

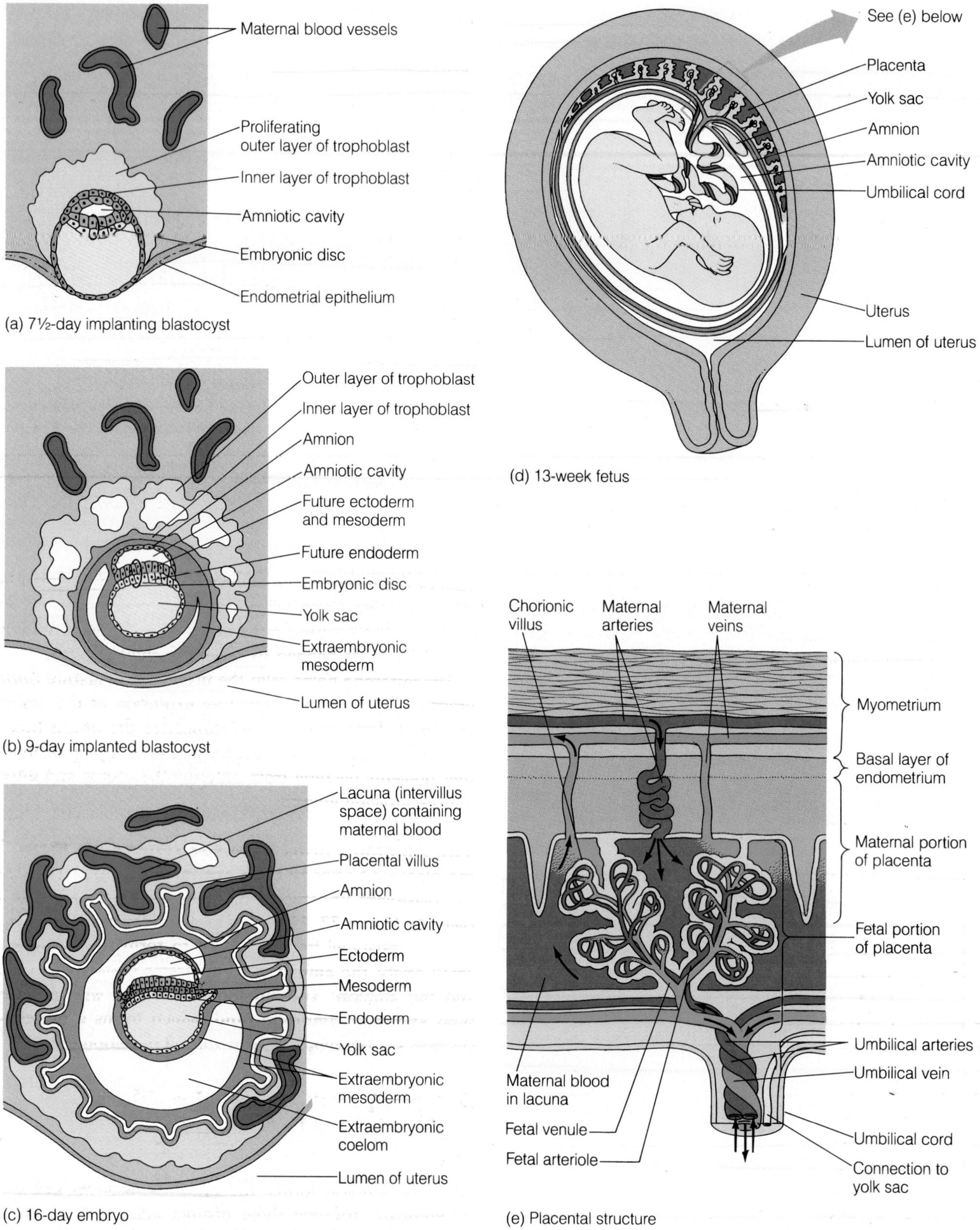

▻ **FIGURE 22–11 Placental Formation** (*a*) Invasion of the maternal tissue by the outer layer of the trophoblast. (*b*) Invasion continues. Cavities form. Note the presence of extraembryonic mesoderm from which blood vessels and blood cells will form. (*c*) The inner layer of the trophoblast and blood vessels invade, forming placental villi. (*d*) Fully formed placenta showing rich vascular supply. (*e*) Enlarged view showing the relationship of maternal and fetal blood vessels in the placenta. Note that the fetal blood vessels are in the placental villus, which is bathed in a pool of maternal blood.

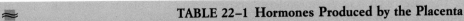

TABLE 22–1 Hormones Produced by the Placenta

HORMONE	FUNCTION
Human chorionic gonadotropin (HCG)	Maintains corpus luteum of pregnancy Stimulates secretion of testosterone by developing testes in XY embryos
Estrogen (also secreted by corpus luteum of pregnancy)	Stimulates growth of myometrium, increasing uterine strength for parturition (childbirth) Helps prepare mammary glands for lactation
Progesterone (also secreted by corpus luteum of pregnancy)	Suppresses uterine contractions to provide quiet environment for fetus Promotes formation of cervical mucus plug to prevent uterine contamination Helps prepare mammary glands for lactation
Human chorionic somatomammotropin	Helps prepare mammary glands for lactation Believed to reduce maternal utilization of glucose so that greater quantities of glucose can be shunted to the fetus
Relaxin (also secreted by corpus luteum of pregnancy)	Softens cervix in preparation for cervical dilation at parturition Loosens connective tissue between pelvic bones in preparation for parturition

SOURCE: Reprinted by permission from Lauralee Sherwood, *Human Physiology: From Cells to Systems*, Table 20-5, p. 750.

Estrogen and Progesterone, First From the Ovary and then from the Placenta, Serve a Number of Functions Essential for Pregnancy. Together, estrogen and progesterone stimulate the growth of the uterine endometrium, which is essential to successful pregnancy. By itself, estrogen stimulates growth of the smooth muscle cells of the myometrium, allowing the uterus to expand to many times its original size during pregnancy, thus accommodating the growing baby but also providing additional propulsive force needed to expel the child at birth.

Progesterone helps calm the uterine musculature during pregnancy, preventing premature expulsion of the embryo and fetus. Progesterone also stimulates the production of mucus by the cervix and the formation of a plug of mucus that prevents bacteria from entering the uterus and infecting the growing embryo.

▷ FIGURE 22–12 Blood Levels of Placental Hormones Human chorionic gonadotropin levels peak in the second month of pregnancy, then drop off by the end of the third month. Levels of estrogen and progesterone, produced chiefly by the placenta, continue to rise.

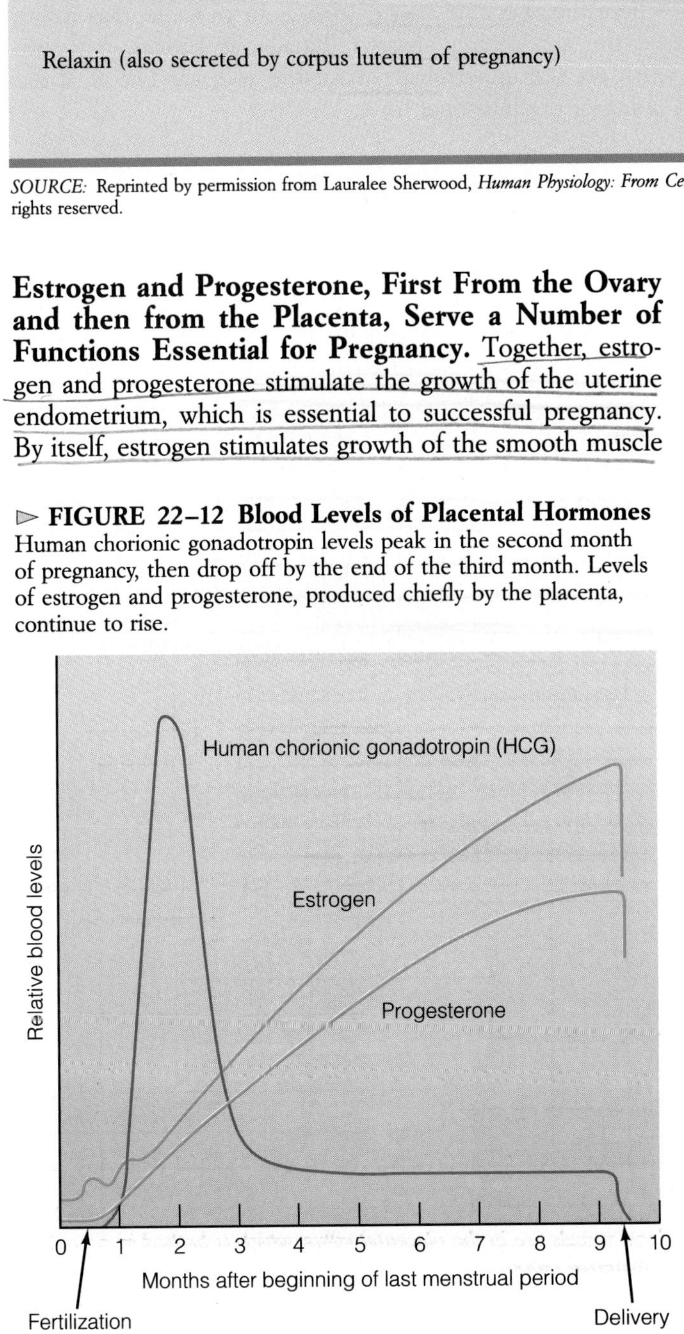

The Amnion Forms from the Inner Cell Mass. As the placenta begins to form, the inner cell mass (ICM) of the blastocyst undergoes some remarkable changes of its own. As Figure 22–11 shows, early in development a layer of cells separates from the ICM to form the **amnion.** A small cavity, the **amniotic cavity,** forms between the ICM and the amnion. The amniotic cavity fills with a watery fluid known as **amniotic fluid,** which forms a protective cushion surrounding the baby during development.[3]

Embryonic Development Begins with Implantation and Ends When the Organs Are More or Less Formed

After the amnion forms, the cells of the inner cell mass differentiate, forming three distinct germ cell layers, the ectoderm, mesoderm, and endoderm, which are known as the **primary germ layers.** The formation of the primary germ layers marks the start of **embryonic development.**

[3] A portion of the amniotic fluid is produced by the fetus. The rest apparently comes from the amniotic membranes.

The Primary Germ Layers of the Embryo Give Rise to the Organs of the Body. As noted earlier, during embryonic development the primary germ layers differentiate a begin to form organs. This process is called **organogenesis.** As shown in ▻ Figure 22–13, organogenesis begins about the third week of pregnancy.

One of the first events of organogenesis is the formation of the central nervous system (the spinal cord and brain), which is produced from ectoderm (Table 22–2). Early in embryonic development, the ectoderm invaginates (folds inward) and forms an indentation, the **neural groove,** which runs the length of the dorsal (back) surface of the embryo (▻ Figure 22–14b). The neural groove deepens and eventually closes off, forming the **neural tube** (Figure 22–14d). The walls of the neural tube thicken, eventually forming the spinal cord. Anteriorly, the neural tube expands to form the brain.

The nerves that attach to the spinal cord (spinal nerves) and brain (cranial nerves) develop from groups of ectodermal cells (the **neural crest**) that lie on either side of the neural tube throughout most of its length (Figures 22–14c and 22–14d). These cells sprout processes that extend into the body and attach to organs, muscle, bone, and skin, among others.

The middle germ layer, the mesoderm, gives rise to muscle, cartilage, bone, and other structures. Much of the mesoderm first aggregates in blocks, called the **somites,** situated alongside the neural tube (Figures 22–14d and ▻ 22–15). The somites form the vertebrae (the backbone) and the muscles of the neck and trunk. Mesoderm lateral to the somites becomes the dermis of the skin, connective tissue, and the bones and muscles of the limbs.

The endoderm, the "lowermost" germ layer of the inner cell mass, forms a large pouch under the embryo called the **yolk sac** (Figures 22–14a and ▻ 22–16). In birds, reptiles, and amphibians, the yolk sac nourishes the growing embryo, but in humans, nourishment comes from the placenta. The human yolk sac gives rise to blood cells and primitive germ cells, called **primordial germ cells.** Interestingly, these cells migrate from the wall of the yolk sac via amoeboid motion to the developing testes and ovaries, where they become spermatogonia or oogonia, depending on the sex of the embryo. Finally, as shown in Figure 22–16c, the uppermost part of the yolk sac becomes the lining of the intestinal tract.

Fetal Development Begins After the Organs Have Formed

As shown in Figure 22–13, organogenesis is well on its way by the end of the eighth week of development. All of the organ systems have begun to develop, and some such as

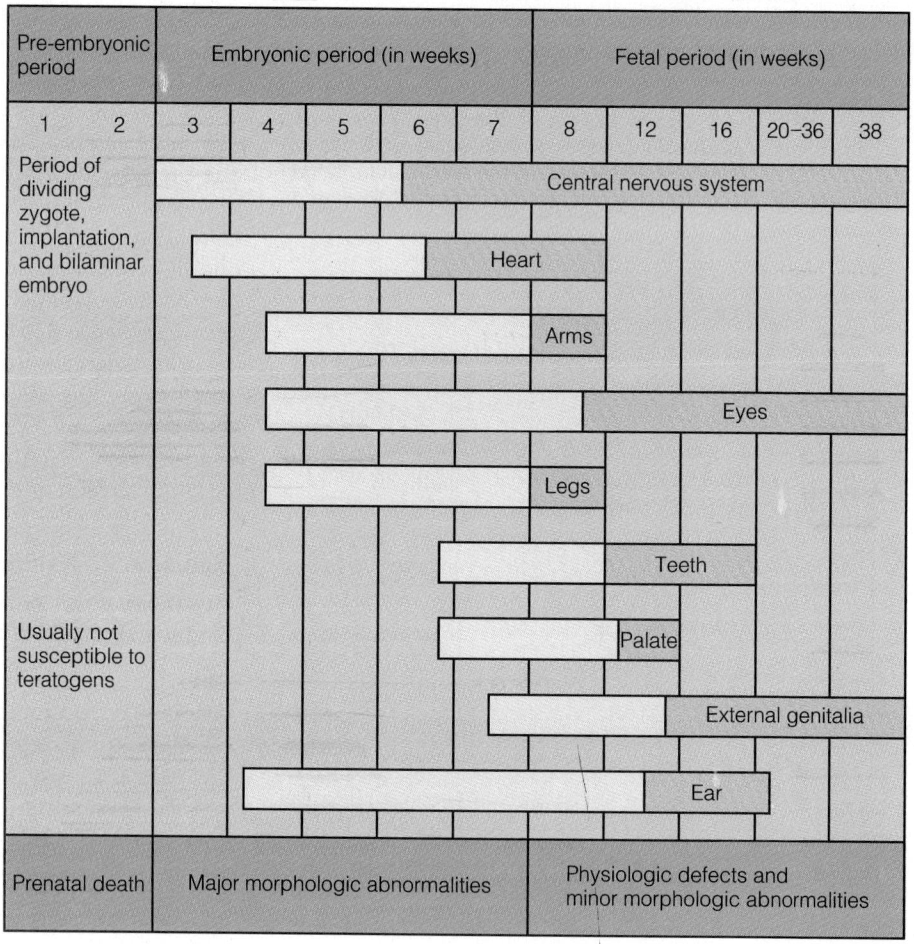

▻ **FIGURE 22–13**
Organogenesis Human development is divided into three periods, or stages: pre-embryonic, embryonic, and fetal. Organogenesis occurs during the embryonic stage. Each bar indicates when an organ system develops. The yellow shaded area indicates the periods most sensitive to teratogenic agents (agents that can cause birth defects).

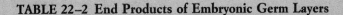

TABLE 22–2 End Products of Embryonic Germ Layers

Ectoderm	Mesoderm	Endoderm
Epidermis	Dermis	Lining of the digestive system
Hair, nails, sweat glands	All muscles of the body	Lining of the respiratory system
Brain and spinal cord	Cartilage	Urethra and urinary bladder
Cranial and spinal nerves	Bone	Gallbladder
Retina, lens, and cornea of eye	Blood	Liver and pancreas
Inner ear	All other connective tissue	Thyroid gland
Epithelium of nose, mouth, and anus	Blood vessels	Parathyroid gland
Enamel of teeth	Reproductive organs	Thymus
	Kidneys	

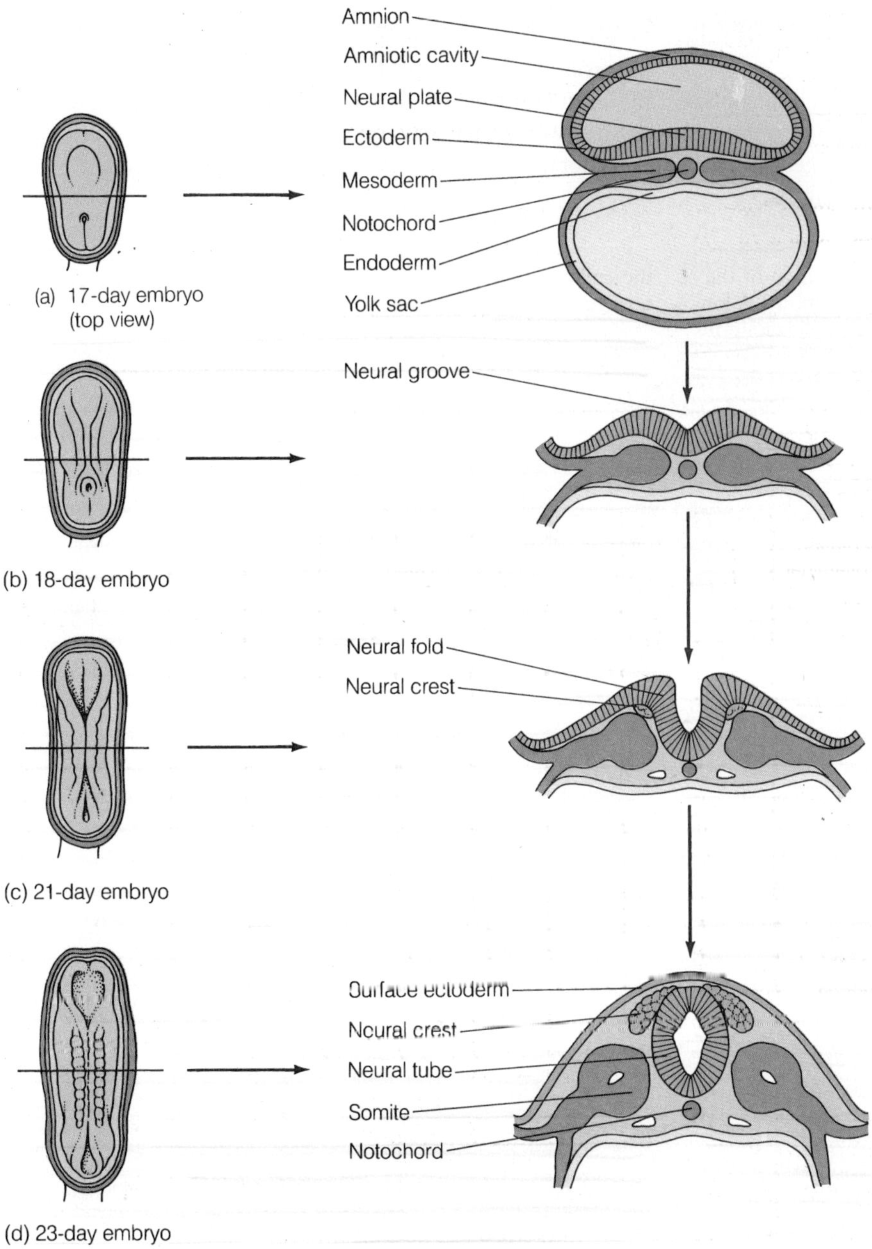

FIGURE 22–14 Formation of the Spinal Cord (*a*) A top view and cross section of a 17-day human embryo showing the relationship between the three embryonic tissues and the amnion and yolk sac. (*b*) The neural groove begins to form from ectoderm. (*c*) The neural groove deepens. (*d*) The neural tube forms. Note the presence of the neural crest, ectodermal cells that give rise to nerves.

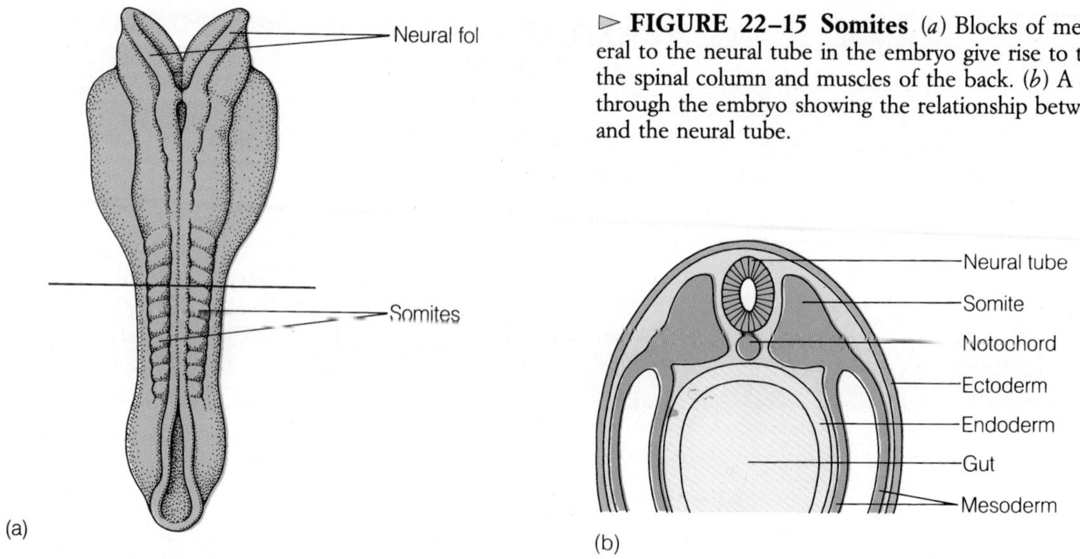

▷ **FIGURE 22–15 Somites** (*a*) Blocks of mesoderm lying lateral to the neural tube in the embryo give rise to the vertebrae of the spinal column and muscles of the back. (*b*) A cross section through the embryo showing the relationship between the somites and the neural tube.

the circulatory system (blood vessels and heart) are fully operational.

Fetal development begins in the eighth week and ends at birth. It involves two basic processes: (1) continued organ development and growth and (2) changes in body proportions—for example, elongation of the limbs.

The fetus grows rapidly during the fetal period, increasing from about 2.5 centimeters (1 inch) to 35 to 50 centimeters (14 to 21 inches) and increasing in weight from 1 gram to 3000 to 4000 grams. The fetus also undergoes considerable change in physical appearance.

▷ **FIGURE 22–16 The Yolk Sac** (*a*) A cross section of the embryo. The yolk sac forms from embryonic endoderm. (*b*) A longitudinal section showing how the upper end of the yolk sac forms the embryonic gut, (*c*) which will become the lining of the intestinal tract.

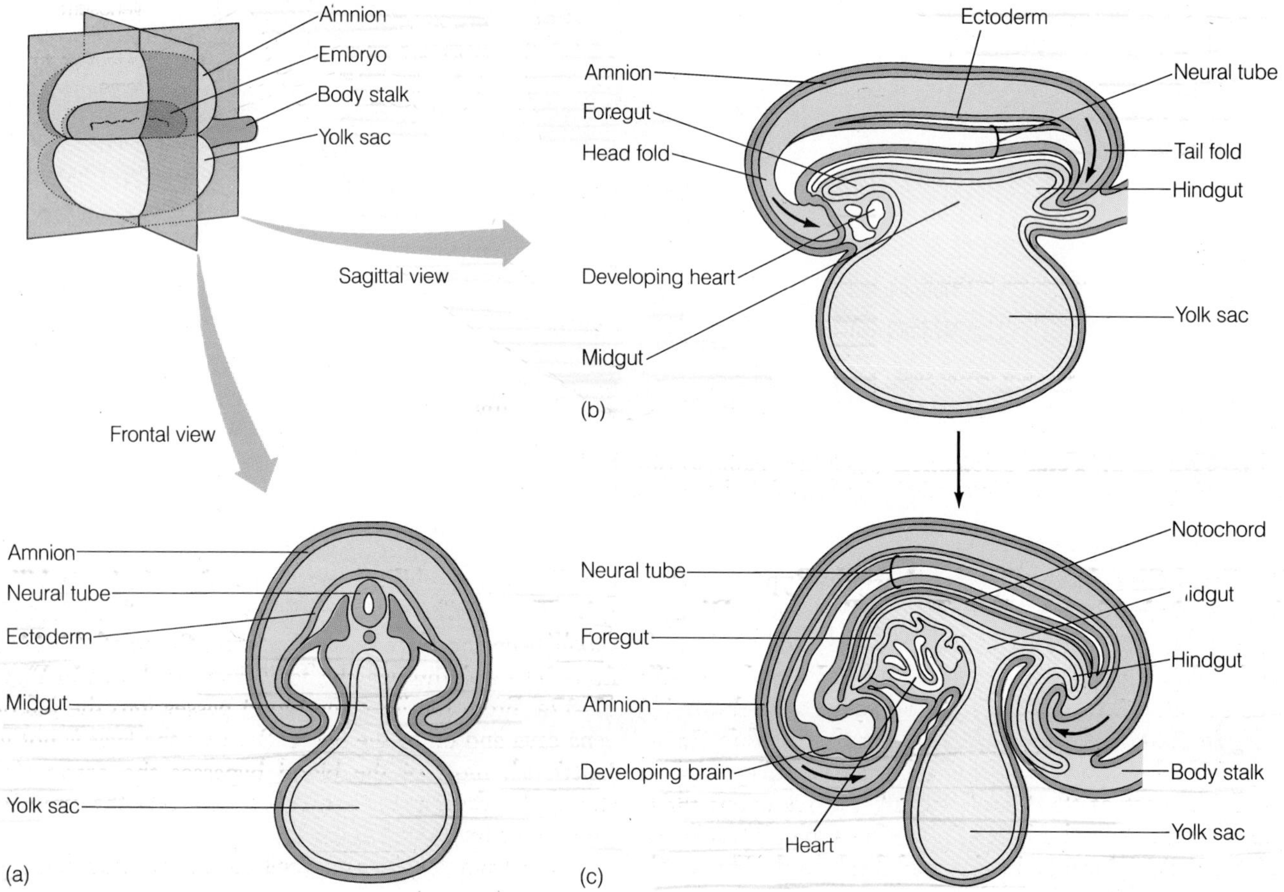

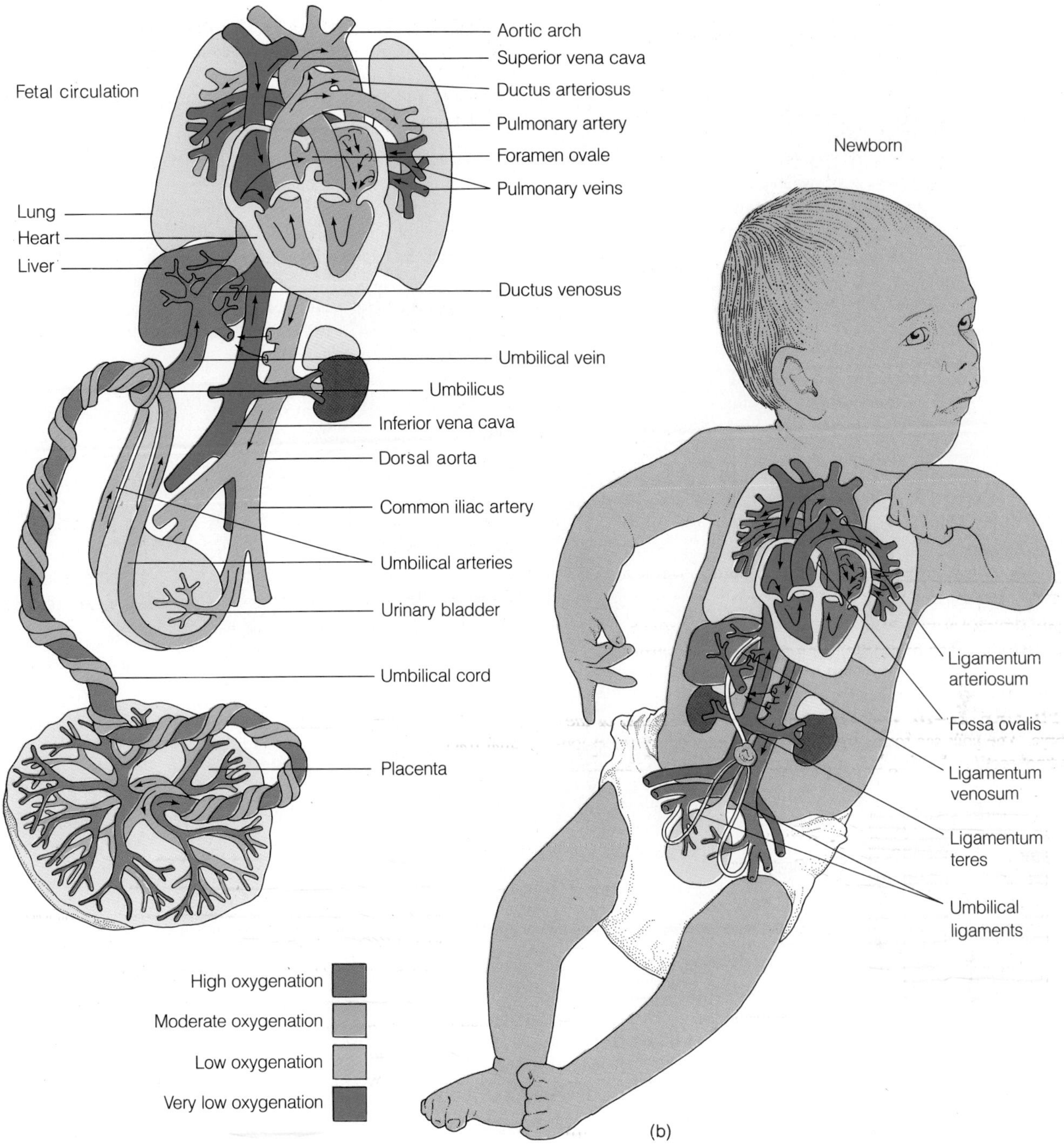

▻ **FIGURE 22–17 Fetal Circulation** (*a*) Before birth. (*b*) After birth. The umbilical arteries and umbilical vein in the fetus shrivel and disappear.

The Fetal Circulatory System Largely Bypasses the Liver and the Lungs, But These Bypasses Disappear at Birth. Space limitations prevent a discussion of the development of each organ system. Instead we will focus our attention on the circulatory system. As shown in ▻ Figure 22–17a, fetal circulation is similar to adult circulation with the exception of several bypasses (adaptations) that divert most of the blood around various organs that are not functional in the fetus. In the fetus, blood travels to and from the placenta in the umbilical cord. The cord contains two **umbilical arteries** and a single **umbilical vein.** The umbilical vein carries oxygen- and nutrient-rich blood from the placenta to the fetus. Some of this blood flows into and through the fetal liver, as shown in Figure 22–17a. From the liver, the blood passes into the inferior vena cava and on to the heart. Because the liver is not yet functional, most of the blood bypasses the organ via a shunt, the **ductus venosus,** that connects the umbilical vein to the inferior vena cava.

Blood from the inferior vena cava flows into the right

SPOTLIGHT ON EVOLUTION 22–1

DELAYED IMPLANTATION: EVOLUTION'S UNSOLVED MYSTERY

A woman and her husband who have been trying to have a baby receive a call from their doctor telling them that the latest test was positive. Joyously, they spend much of that evening talking about all the things they want to do for their child. The next week, though, the husband goes to work only to find he will be laid off in a few weeks. The company in his small town is going bankrupt and closing down all operations. With unemployment in his community running at 10%, it is unlikely that he will find a job with decent pay. His wife, who owns a restaurant that caters to workers at her husband's company, realizes that her earnings will probably drop.

Imagine, if you will, that the couple could simply put the pregnancy on hold for five or six months while they found ways to make ends meet. And, once they had gotten their lives back on track, they could restart the pregnancy. Sound like science fiction? It is. Humans can't delay pregnancy, any more than we can change the Earth's orbit around the sun.

In contrast, a number of other animals have a remarkable, but poorly understood, capacity to put pregnancy on hold. In fact, current estimates suggest that as many as 100 species experience delayed implantation, holding blastocysts in suspended animation in the uterine lumen for periods of a few days to more than a year. Among them are the black bears, western spotted skunks, weasels, otters, bats, armadillos, and kangaroos (▷ Figure 1).

Consider the fisher, a weasel-like creature that lives in the northeastern United States and much of Canada. In Maine, most female fishers give birth in mid-March. Soon after delivery, the female mates, but the blastocyst remains

▷ **FIGURE 1 Pregnancy on Hold**

atrium of the heart. In an adult, the blood flows from the right atrium into the right ventricle. In the fetus, however, blood follows two paths. Part of the blood flows from the right atrium into the right ventricle. From the right ventricle, it is pumped to the developing lungs via the pulmonary artery, supplying the growing but nonfunctional lung tissue with oxygen and nutrients and picking up waste products of cellular metabolism. The rest of the blood in the right atrium is shunted through a hole in the atrial wall, known as the **foramen ovale.** Blood entering the left atrium passes into the left ventricle, then is delivered to the head and the rest of the body through the aorta and its branches. The foramen ovale essentially diverts much of the blood from the fetal lungs, which like the liver are not yet functional.

The foramen ovale closes at birth in most infants, establishing the adult circulation pattern. In some children, however, the hole remains open and must be corrected surgically. Babies born with an open foramen ovale usually appear bluish, because much of their blood bypasses the lungs and is therefore low in oxygen. These infants are often referred to as blue babies.

Figure 22–17a also illustrates a third shunt, the ductus arteriosus. The **ductus arteriosus** lies between the pulmonary artery and the aorta. Like the foramen ovale, it helps divert blood away from the lungs. The lungs will need this blood when the baby is born and starts breathing, but during fetal development, the lungs' requirement for blood is considerably smaller.

Blood pumped to the body through the aorta and its branches returns to the heart via the inferior vena cava. Fetal blood, however, also returns to the placenta, where it can pick up nutrients and rid itself of wastes. (Remember, the placenta is serving as the fetus's lungs and digestive system.) A return route, shown in Figure 22–17a, is provided by the **umbilical arteries,** branches of the large arteries of the legs (the iliac arteries). The umbilical arteries carry blood to the placenta, where it is cleansed of much of its waste and recharged with nutrients.

in the uterine cavity for 9 to 10 months before implanting. That is, it does not implant until the following winter. This example raises an important question. Why delay implantation? Why not breed in the fall or early winter and avoid delayed implantation altogether?

Frankly, researchers are not sure. Bill Berg of the Minnesota Department of Natural Resources says that predators like the fisher have to be swift. Carrying blastocysts around instead of developing fetuses makes them lighter and better able to survive through the fall and early winter. Put in evolutionary terms, delayed implantation may have increased the survival and reproductive success of the mother during this critical period. Another wildlife biologist remarked that "delayed implantation allows animals to tailor their reproductive cycle to their yearly food and weather cycles." Again, in evolutionary terms, delayed implantation may be an adaptation to such food and climate cycles.

The answer may lie in some ecological factor or in a development that took place hundreds of thousands of years ago. Here is one purely hypothetical possibility: Perhaps at one time fishers that bred after giving birth in March did indeed give birth again in the fall. Offspring would have had to survive a harsh winter and were probably selected against. The females may also have been selected against. That is, natural selection may have weeded out those that were pregnant (and slower moving) during the summer and fall. These females may not have been able to get enough to eat or may have fallen prey to predators. Continuing with this hypothesis: Perhaps by chance, some females evolved the ability to delay implantation. Remaining sleeker through the spring, summer, and fall, these individuals had a selective advantage over their pregnant counterparts. They therefore passed their genes on more successfully, so that today all fishers are capable of delayed implantation.

Black bears also exhibit delayed implantation. Unlike fishers, bears mate in the summer, long after giving birth. Nonetheless, black bears still delay implantation until the beginning of the winter.

In black bears, delayed implantation may reduce stress on females as they prepare for winter hibernation; that is, it may help ensure that females receive enough food to get through the winter themselves. It may also increase the chances of a successful pregnancy, for if a female is not adequately nourished, she will often abort.

The advantage of delayed implantation to black bear cubs is not evident at first glance, because it puts birth in the middle of winter while females are denning. However, consider what would happen if a female black bear gave birth in the fall before she hibernated. In this case, cubs would have very little time to build up body stores of fat to make it through the long winter hibernation. Few would probably succeed and grow to reproductive age. As it is, cubs are born in the dead of winter, often in a protected den where they feed off their mother's milk while she is hibernating. When spring comes, the cubs start eating on their own. By the following fall, they are able to store up enough fat to hibernate.

As in the case of the fisher, these hypotheses, which describe the advantage of delayed implantation to males and females, fail to explain why males and females simply don't wait until the fall to breed. For now, we'll have to wait for an answer, or several answers, as biologists probe this fascinating phenomenon.

At birth, the fetal circulation pattern is replaced by the adult pattern (Figure 22–17b). The umbilical cord is tied off and cut, preventing blood flow to and from the placenta. Inside the fetus, the umbilical arteries and umbilical veins shrivel, becoming ligamentous structures. The foramen ovale closes, so that all the blood entering the right atrium now travels to the right ventricle and the lungs. The ductus venosus and the ductus arteriosus close down and become ligamentous structures, which disappear altogether as a child ages.

Most Ectopic Pregnancies Occur In the Uterine Tubes

An **ectopic pregnancy** occurs when a fertilized ovum develops outside the uterus, usually in the uterine tube.[4] Thus, most ectopic pregnancies are also called **tubal pregnancies.**

In a tubal pregnancy, the zygote implants in the lining of the uterine tube. However, placental development damages the uterine tubes, causing internal bleeding and severe abdominal pain. Because the uterine tube cannot sustain the embryo, a tubal pregnancy cannot generally proceed to term. Surgery is required to remove the embryo.

Interestingly, about 1 in every 200 pregnancies is ectopic. Usually diagnosed within the first two months, tubal pregnancies occur as a result of congenital defects in the uterine tubes or scar tissue from a previous infection, which impedes the transport of the blastocyst to the uterus.

Embryonic and Fetal Development Can Be Altered by Outside Influences, Sometimes Resulting in Birth Defects and Miscarriages

By several estimates, 31% of all fertilizations end in a

[4] Ectopic pregnancies can also occur in the abdominal cavity.

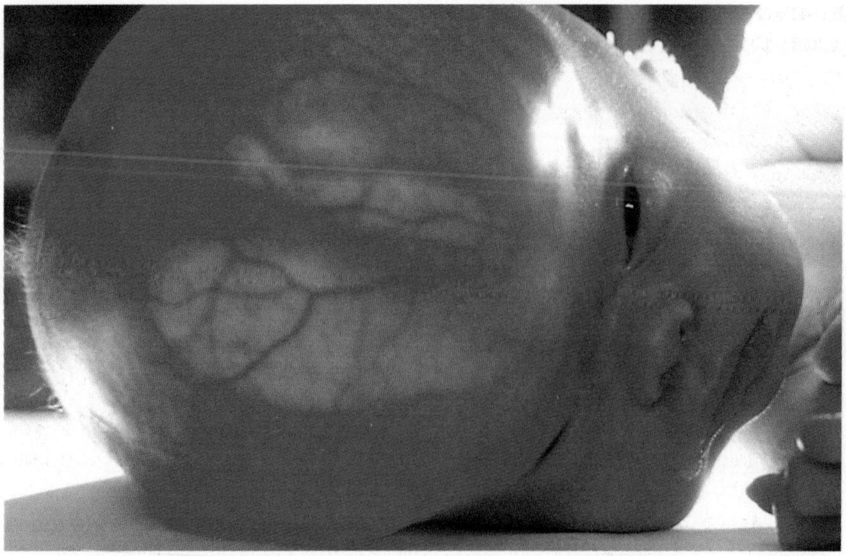

(a)

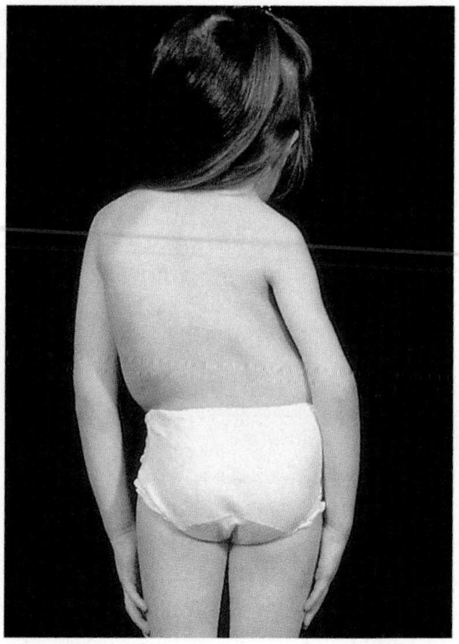

(b)

▻ **FIGURE 22–18 Common Birth Defects** (*a*) Hydrocephalus, or "water on the brain," is caused by an enlargement of the brain's ventricles. (*b*) Scoliosis, a lateral curvature of the spine.

miscarriage—that is, a spontaneous abortion. Two of every three of these miscarriages occur before a woman is even aware that she is pregnant. Why such a high rate? Biologists hypothesize that early miscarriage is nature's way of "discarding" defective embryos. In other words, it is an evolutionary mechanism that helps ensure a healthier population.

Because nature generally eliminates defects, you might expect that most children born into the world would be free of defects. Unfortunately, that is not the case. Humans experience a surprisingly high rate of **birth defects,** physical or physiological abnormalities present in newborns. By various estimates, 10% to 12% of all newborns have some kind of birth defect, ranging from minor biochemical or physiological problems, which are not even noticed at birth, to gross physical defects (▻ Figure 22–18).[5]

The study of birth defects is called **teratology.** The word teratology comes from the Greek *teratos,* meaning "monster" and reflects some of the more gruesome or disfiguring birth defects. Some experts, however, point out that the term is an unfair characterization of all birth defects. Many defects are minor, and people with them are clearly not monsters.

Although there is much to be learned about the causes of birth defects, scientists believe that most arise from chemical, biological, and physical agents, collectively known as **teratogens.** Table 22–3 lists known and suspected teratogens in humans. For example, the virus that produces rubella (German measles) is a known teratogen. So are alcohol, radiation, and megadoses of vitamins A and D.

The effect of a teratogenic agent on the developing embryo is related to three factors: (1) the time of exposure, (2) the nature of the agent, and (3) the dose. Consider time first. Because the organ systems develop at different times, the timing of exposure determines which systems are affected by a given teratogen. Organ systems are usually most sensitive to potentially harmful agents early in their development, as indicated by the yellow bars in Figure 22–13. The central nervous system, for example, begins to develop during the third week of **gestation,** or pregnancy. Because most women do not know they are pregnant for three to four weeks after fertilization, the central nervous system is at risk. In contrast, the teeth, palate, and genitalia

[5]About 2% of all births show gross malformations.

TABLE 22–3 Known and Suspected Human Teratogens

KNOWN AGENTS	POSSIBLE OR SUSPECTED AGENTS
Progesterone	Aspirin
Thalidomide	Certain antibiotics
Rubella (German measles)	Insulin
Alcohol	Antitubercular drugs
Irradiation	Antihistamines
	Barbiturates
	Iron
	Tobacco
	Antacids
	Excess vitamins A and D
	Certain antitumor drugs
	Certain insecticides
	Certain fungicides
	Certain herbicides
	Dioxin
	Cortisone
	Lead

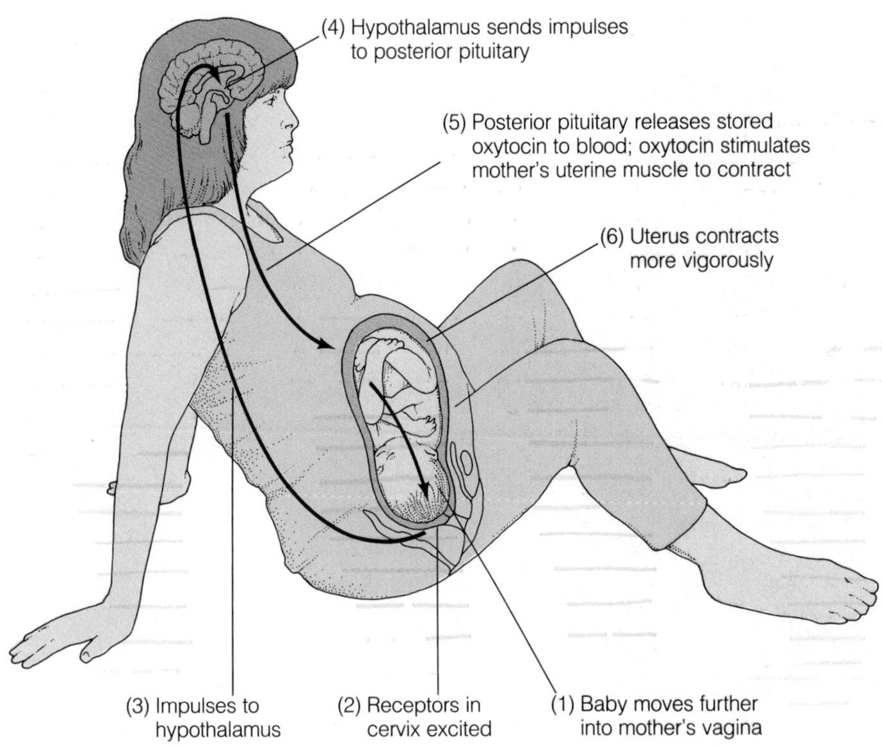
(4) Hypothalamus sends impulses to posterior pituitary
(5) Posterior pituitary releases stored oxytocin to blood; oxytocin stimulates mother's uterine muscle to contract
(6) Uterus contracts more vigorously
(3) Impulses to hypothalamus
(2) Receptors in cervix excited
(1) Baby moves further into mother's vagina

▻ **FIGURE 22–20 Oxytocin Positive Feedback Mechanism in Birth** The mechanism continues to cycle until interrupted by the birth of the baby.

Emotional and physical stress in the mother may also play a role in triggering childbirth. Uterine contractions and discomfort create stress, which is believed to trigger the release of maternal oxytocin. Maternal oxytocin, in turn, augments muscle contractions that are already underway (▻ Figure 22–20). Thus, a positive feedback mechanism is triggered. As uterine muscle contraction increases, maternal oxytocin release increases. This stimulates even stronger contractions, resulting in additional oxytocin release, a cycle that continues until the baby is delivered.

Childbirth also requires the hormone **relaxin,** produced by the ovary and the placenta. Released near the end of gestation, relaxin has two effects. First, it softens the fibrocartilage uniting the pubic bones. This allows the pelvic cavity to widen and thus greatly facilitates childbirth. Relaxin also softens the cervix, allowing it to expand to allow the baby to pass. Without this remarkable adaptation, delivery would be a problematic event indeed!

Childbirth Occurs in Three Stages

▻ Figure 22-21 illustrates the three stages of childbirth.

During the Dilation Stage, the Baby Is Pushed Against the Cervix, which Causes Cervical Dilation. Stage 1, the **dilation stage,** gets its name from the dilation of the cervix. This phase begins when uterine contractions start and generally lasts 6 to 12 hours, but sometimes much longer. At the beginning of stage 1, uterine contractions typically last only 30 seconds and may come every half an hour or so. As time passes, however, uterine contractions become more frequent and powerful. Uterine contractions generally rupture the amnion early in the dilation phase, causing the release of the amniotic fluid, an event commonly referred to as "breaking the water."

Uterine contractions push the infant's head against the relaxin-softened cervix, causing it to stretch and become thin (Figure 22–21b). By the end of stage 1, the cervix has dilated to about 10 centimeters (4 inches), approximately the diameter of a baby's head.

The baby is pushed downward by the uterine contractions and descends into the pelvic cavity. When the head is "locked" in the pelvis, the baby is said to be **engaged.**

During the Expulsion Stage, the Baby Is Pushed Out of the Uterus, Through the Cervix and Vagina. Stage 2, the **expulsion stage,** begins after the cervix is dilated to 10 centimeters and the baby is engaged (Figure 22–21c). At this time, uterine contractions usually occur every 2 or 3 minutes and last 1 to 1.5 minutes each. For most women having their first child, 50 to 60 minutes are required for delivery. If a woman is delivering her second child, only 20 to 30 minutes of uterine contraction are required to push the infant's head through the cervix and the vagina. The expulsion stage ends when the child is pushed through the vagina into the waiting hands of a doctor or midwife.

To facilitate the delivery, physicians or midwives often make an incision to widen the vaginal opening. This procedure, called an **episiotomy,** is performed when the baby's head enters the vagina in stage 2 of labor. The incision enlarges the vaginal orifice, prevents tearing, and allows the infant to pass quickly. The incision is stitched up immediately after the baby is born.

Once the baby's head emerges from the vagina, the rest of the body slips out almost instantly. However, the baby

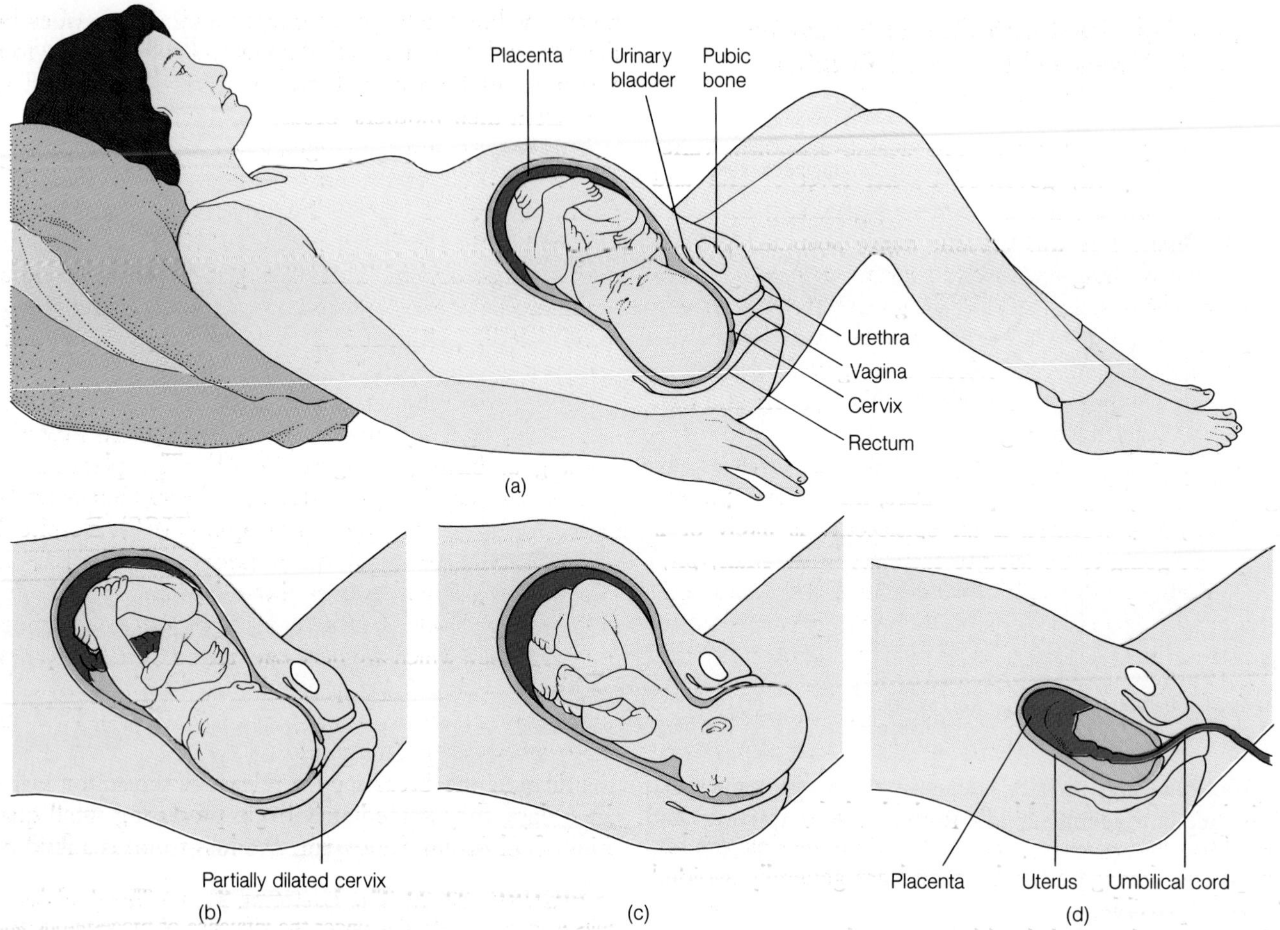

▻ **FIGURE 22–21 Stages of Labor** (*a*) Position of the fetus near birth. (*b*) Dilation stage. Uterine contractions push the fetal head lower in the uterus and cause the relaxin-softened cervix to dilate. (*c*) Expulsion stage. Fetus is expelled through the cervix and vagina. (*d*) Placental stage. Placenta is delivered.

remains attached to the placenta via the umbilical cord, which continues to deliver blood to the baby for a minute or so. Consequently, many health-care workers wait a minute or two to allow the blood remaining in the placenta to be pumped into the newborn before they tie off and cut the umbilical cord.

Most babies (95%) are delivered head first with their noses pointed toward the mother's tailbone (Figure 22–21). Occasionally, however, babies may be oriented in other positions. This makes delivery more difficult, time-consuming, and hazardous for the mother and the baby. The most common alternative delivery is the **breech birth,** in which the baby is expelled rear-end first. Breech births require more time and may cause extreme fatigue in the mother and brain damage in the baby. The umbilical cord sometimes wraps around the baby's neck, cutting off its supply of blood. To avoid complications, breech babies are often physically turned by physicians before birth by applying pressure to the woman's abdomen. If a fetus cannot be turned, the baby is usually delivered by a **cesarean section,** a horizontal incision through the abdomen just at the pubic hair line. Cesarean sections may be performed for other reasons as well—for example, if labor is prolonged or if the mother has an active infection caused by herpes or some other sexually transmitted disease that might be transferred to her child as it passed through the vagina.

The Placental Stage Is Marked by the Delivery of the Placenta. The final stage of delivery is the **placental stage** (Figure 22–21d). The placenta, sometimes called the afterbirth, remains attached to the uterine wall for a short while, then is expelled by uterine contractions, usually within 15 minutes of childbirth. After the placenta is expelled, the uterine blood vessels clamp shut, preventing hemorrhage, although the mother continues to lose some blood for 3 to 6 weeks after delivery.

The uterus gradually returns to its normal size after delivery. Uterine involution (shrinkage) results from the rapid decline in estrogen and progesterone and is accelerated by oxytocin released by the posterior pituitary when a woman breastfeeds her baby. Complete involution in women who breast-feed usually occurs within four weeks of pregnancy. In women who do not, the process usually takes six weeks.

The Pain Associated with Childbirth Can Be Relieved by Drugs and by Special Birthing Methods that Seek to Reduce Tension

The level of pain a woman feels during childbirth varies greatly and is partly governed by her level of fear and tension. The more tense a woman is, generally, the more pain she feels. For this reason, many hospitals provide comfortable birthing rooms and relaxation training to expectant mothers. Drugs can also be given to reduce tension and pain, but they must be administered well before the birth of the baby to be effective. If given just before delivery, the drug offers little relief to the mother and may retard the baby's breathing.

Painkilling drugs can be injected into the wall of the vagina with a syringe. This procedure, known as a **pudendal block,** is performed if an episiotomy is likely or if forceps are going to be used to facilitate birth. Many physicians routinely offer **epidural anesthesia.** An anesthetic agent that temporarily deadens the sensory nerves in the vagina and elsewhere is injected into the bony canal that houses the spinal cord, just outside the dura mater. The drug blocks pain in the nerves that supply the lower body, stopping all sensations. This anesthetic may also block motor nerve impulses to the muscles of the abdomen, rendering a woman unable to push the baby out. In such cases, forceps may be required to facilitate the delivery, but forceps can damage a newborn and are generally avoided whenever possible.

Many couples and health-care workers believe that drugs may be harmful to mothers and their babies. This belief and a desire to do things "naturally" have spawned the **natural childbirth** movement. Natural childbirth means different things to different people. In general, it refers to drug-free deliveries. Couples receive training in relaxation and special breathing techniques.

The most popular natural childbirth method today is known as the Lamaze technique. Parents who are going to use this method attend childbirth classes in which they learn special breathing techniques. Shallow breathing, for example, is used when uterine contractions begin; it keeps the diaphragm from pressing down on abdominal organs, reducing tension and pain. It also ensures an adequate supply of oxygen to the fetus. Other breathing techniques are also learned.

The second most popular method is the Bradley method. The Bradley method teaches women how to relax during uterine contractions, thus reducing pain and the need for drugs. No special breathing is required. The woman helps push, as in the Lamaze method, but only near the end of childbirth.

Milk Production, or Lactation, Is Controlled By the Hormone Prolactin

A baby emerges into a novel environment at birth. No longer connected to the placental lifeline, the newborn must now breathe to get its oxygen and dispose of carbon dioxide, and it must also find a new way of acquiring food. For most of human evolution, newborns have been fed milk from their mothers' breasts.

The breasts of a nonpregnant woman consist primarily of fat and connective tissue interspersed with milk-producing glandular tissue. The glands are drained by ducts that lead to the nipple (▻ Figure 22–22). As noted earlier, during pregnancy the ducts and glands proliferate under the influence of placental and ovarian estrogen and progesterone. Milk production is induced by prolactin and a placental hormone, HCS (described below).

Prolactin secretion in the mother begins during the fifth week of pregnancy and increases throughout gestation, peaking at birth (▻ Figure 22–23). The placenta also secretes a mildly lactogenic (milk-producing) hormone, **human chorionic somatomammotropin (HCS).** Despite the presence of these lactogenic hormones, milk production does not begin until two or three days after birth. Why? Because high levels of estrogen and progesterone throughout pregnancy, which are necessary for breast development, inhibit the action of the prolactin and HCS. When estrogen and progesterone levels decline after birth, HCS and prolactin can exert their effect.

Although the breasts do not produce milk for two or three days, they immediately begin producing small quantities of colostrum. **Colostrum** (co-loss-trum) is a fluid rich

▻ FIGURE 22–22 The Lactating Breast The glandular units enlarge considerably under the influence of progesterone and prolactin. Milk is expelled by contraction of musclelike cells surrounding the glandular units. Ducts drain the milk to the nipple.

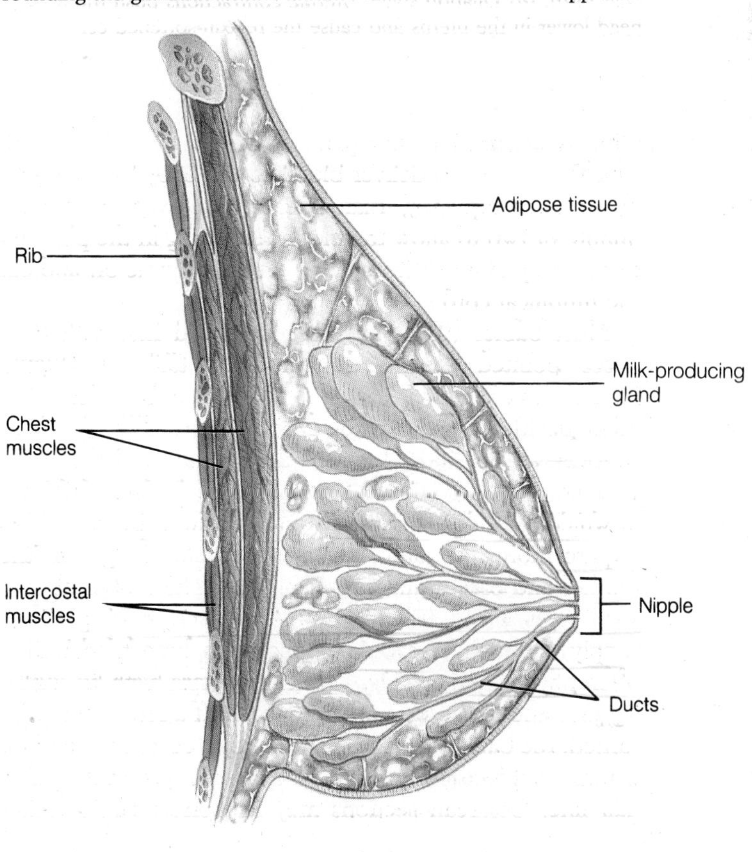

▻ **FIGURE 22–23 Hormone Levels Before and After Birth** Prolactin, estrogen, and progesterone secretion prior to birth and 36 weeks after birth in a lactating woman.

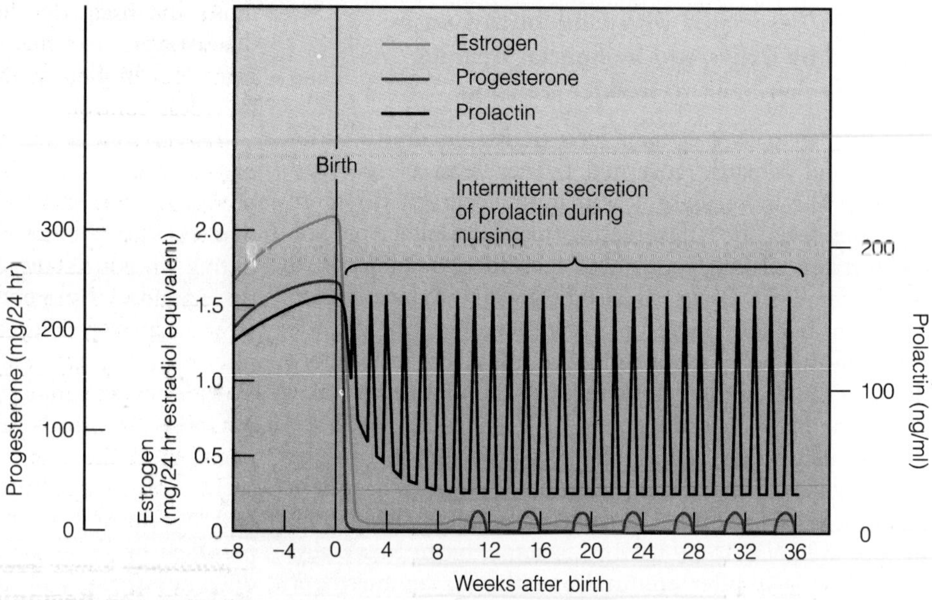

in protein and lactose, but lacking fat. It contains antibodies that protect the infant from bacteria. A newborn subsists on colostrum for the first few days. (Health Note 15–1 describes the benefits of colostrum and breast feeding in general.)

Milk production begins in earnest when estrogen and progesterone levels plummet (Figure 22–23). Thus, within two to three days, the breasts start to produce large quantities of milk. Milk production is facilitated by growth hormone, cortisol, and parathyroid hormone.

Prolactin Secretion Is Maintained After Childbirth by Suckling. Besides stimulating oxytocin release, suckling causes a surge in prolactin secretion. Prolactin levels remain elevated for approximately one hour after each feeding. As a result, each surge stimulates milk production needed for the next feeding. Milk production continues as long as the baby suckles. If nursing is interrupted for three or four days, however, the breasts stop producing milk. In most women, milk production begins to decline by the seventh to ninth month of lactation, a time at which her baby is beginning to feed on semisolid food and breastfeeds less frequently.

Prolactin Secretion Is Controlled By a Neuroendocrine Reflex. Suckling generates nerve impulses that travel to the hypothalamus, causing the release of **prolactin releasing hormone (PRH).** PRH, in turn, stimulates prolactin secretion. When suckling ends, PRH secretion stops.

In order for milk to reach the nipple, it must be actively propelled from the glandular units and through the ducts. This is the function of oxytocin, the release of which is also controlled by a neuroendocrine reflex, described in Chapter 20. Neuroendocrine reflexes are important adaptations that provide services through hormones on demand. Such systems help conserve energy and nutrients and contribute to the evolutionary success of multicellular animals.

INFANCY, CHILDHOOD, ADOLESCENCE, AND ADULTHOOD

Describing the changes that occur from birth to adulthood in one section of one chapter is a little like trying to define the universe in a paragraph. Nevertheless, there are some important milestones along this exciting journey worth noting. Much of this information will be relevant if you have, or are planning to have, children.

Infancy Lasts From Birth to the End of the Second Year of Life

At birth, the infant emerges from its warm, relatively secure uterine home—where its needs were catered to automatically—to the home environment, where it must play a much more active role to satisfy its needs. The human newborn survives in its new environment largely because of parental care and because it comes equipped with several built-in reflexes. One of the most important is the **suckling reflex.** Within minutes of birth, the baby begins to suckle the breast (or anything else put in its mouth).

Babies also come equipped with a **rooting reflex,** which helps them find the nipple. Pressure from a nipple (or any stimulus) touching the cheek causes infants to turn their head toward the stimulus. Interestingly, crying is also a reflex in the newborn. Babies cry when they are hungry, hurt, or uncomfortable. The crying reflex alerts parents that something is wrong. Usually, a simple remedy will cure the crying and restore quiet—at least for a while.

During infancy, the child undergoes a dramatic physical and mental transformation. Although children develop in predictable stages (▻ Figure 22–24), the timetable varies from one child to the next. One infant, for example, may learn to crawl at 5 months but another may not crawl until its 7th or 10th month. This variability is natural and to be expected.

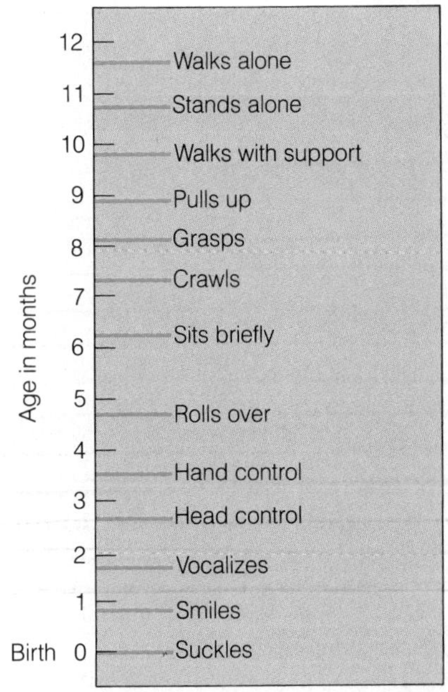

▻ **FIGURE 22–24 Motor Development in the Year after Birth**

Within the first year, the child's body weight triples. Even though growth continues at a fairly fast pace, however, the rate of growth begins to decline. Rapid physical growth is replaced by a growth in sensory, motor, and intellectual capacity.

At birth, the child is fairly immobile but capable of moving its arms and feet in joy, protest, or play. By the 5th month, the baby learns to roll over, a maneuver that provides surprising mobility. Next, the infant learns to crawl, often slithering along the floor like a snake. During the 7th month, the baby begins to crawl on its hands and knees. By the 11th or 12th month, the child stands upright. From an upright position, the child quickly masters the art of walking. A parent's life is forever changed. Throughout this period, the brain is changing as myelin is laid down on axons.

During infancy, the child also becomes more aware of its environment and more and more interactive with parents and peers. Child development involves important emotional and intellectual achievements. One of the first emotional milestones of normal human development is **bonding,** the establishment of an intimate relationship between the infant and one (or preferably both) of its parents or caregivers. Some psychologists think that bonding begins within an hour or so of birth. The initial contact, they say, is crucial to developing the bond. Others disagree. They think that bonding may require a period of several months.

What is important is for children to develop a close relationship with caregivers and to learn that they can be comforted by another human and that they can trust another person explicitly. Successful bonding or attachment is important for subsequent development and may lead to emotional security later in life.

As the brain develops, infants become more curious. They begin to remember things and show rudimentary signs of reasoning. With these changes come improvements in motor control.

Research suggests that intelligence, like motor skills, improves with use. Children are eager explorers, ready to tear apart drawers or closets to explore the objects of the adult world. Parents are encouraged to foster their children's curious nature, to let them explore, and to provide opportunities for a rich and varied experience. A child's innate curiosity, then, is a seed of intellectual development that a parent can nurture. Parents are also advised to provide ample stimuli for their children—toys, books, and personal interaction. Parents should read to their baby even before the infant can talk, should point out and name objects, and should talk to the child whenever possible.

Childhood Lasts From Infancy to Puberty, the Beginning of Adolescence

Growth continues throughout childhood, but as in infancy, the rate of growth continues to decline. During childhood, children display increased motor abilities. Language skills improve enormously. Curiosity abounds, and mental capacity increases remarkably. In early childhood (even during infancy), children start to assert their own will. They may resist or even disobey parents. Asserting one's will and developing an identity are important and natural phases of growth, and parents should be careful not to quash a child's developing will. That task requires the judicious use of discipline, and, of course, lots of love.

Adolescence Lasts From Puberty, or Sexual Maturation, to Adulthood

Puberty, or sexual maturation, marks the onset of **adolescence,** often a rather tumultuous period of growth and development that lasts until adulthood (age 18 to 20). For girls, puberty begins about age 10 or 11. For boys, it begins at age 12 or 13. As in all developmental processes, individual variation is to be expected.

Sexual maturation, discussed briefly in Chapter 21, results from the production of sex steroid hormones in both boys and girls. Research suggests that the pituitary, hypothalamus, and gonads begin functioning long before puberty, but the hypothalamus is extremely sensitive to circulating levels of sex steroid hormones. These hormones inhibit gonadotropin release through a negative feedback mechanism. Thus, the hypothalamus's high degree of sensitivity to sex steroid hormones inhibits gonadotropin release. As an individual ages, however, the hypothalamus becomes less sensitive to the negative feedback influence of these hormones. GnRH release increases, and the pituitary output of gonadotropins climbs. This rise, in turn, increases gonadal hormone production, inducing puberty.

Tables 22–5 and 22–6 summarize some of the impor-

TABLE 22-5 Physical Development in Adolescent Girls

PHYSICAL CHANGE	AVERAGE AGE WHEN CHANGE BEGINS	AVERAGE AGE WHEN NOTICEABLE CHANGE USUALLY STOPS	REMARKS
Increase in rate of growth	10 to 11	15 to 16	If conspicuous growth fails to begin by 15, consult your physician.
Breast development	10 to 11	13 to 14	Noticeable development of breasts (one of which may begin to "bud" before the other) is usually the first sign of puberty. If changes do not occur by 16, there may be cause for concern.
Emergence of body hair	Pubic hair: 10 to 11 Underarm hair: 12 to 13	13 to 14 15 to 16	Age at first appearance of body hair is extremely variable. Pubic hair usually darkens and thickens as puberty progresses.
Development of sweat glands	12 to 13	15 to 16	The apocrine sweat glands are responsible for increased underarm sweating, which causes a type of body odor not present in children.
Menstruation	11 to 14	45 to 55	Menstruation often begins with extremely irregular periods, but by 17 a regular cycle (3 to 7 days every 28 days) usually becomes evident. If menstruation begins before 10 or has not begun by 17, consult your physician.

SOURCE: From *The American Medical Association Family Medical Guide.* Copyright © 1982 by The American Medical Association. Reprinted by permission of Random House, Inc.

tant physical changes taking place during puberty. The physical changes induced by ovarian and testicular steroids during puberty were discussed in Chapter 21. Effects on bone were discussed in Chapter 19.

Adolescents also undergo a tremendous change in mental capacity. They become capable of dealing with abstract concepts, and their thinking skills often improve considerably. Rudiments of critical thinking are apparent. A teenager may, in fact, use the scientific method to test hypotheses.

Adolescence is a time of emerging identity. That identity may contrast sharply with a parent's. It may also run counter to a parent's desire for the child, leading to conflict. Accepting the emerging personality and maintaining a healthy relationship require that parents learn to tolerate, better yet appreciate, differences. Love, support, and understanding are as important now as at any time in human development.

Adulthood Begins When Adolescence Ends

In adults, physical growth stops, but for many people social and psychological development continues throughout adult life. Not everyone continues to grow during adulthood, however. Many people seem to congeal, remaining more or less the same throughout their adult life. They develop fixed patterns of behavior and live accordingly—often to the frustration of those around them.

In their 20s, many adults begin to show signs of aging. With age, wrinkles may begin to appear, weight may increase, and gray hairs may develop.

AGING AND DEATH

Aging is part of the life process, but it remains one of the great mysteries of biology. **Aging** is a progressive deterioration of the body's structure and function. One function of extreme importance is homeostasis. As homeostatic mechanisms falter, the body becomes less resilient. Healing may take longer. Individuals may become more susceptible to disease.

The most notable changes occurring with age are physical changes—wrinkling, loss of hair, and stooping. Less obvious is the gradual deterioration of the function of body organs. As one ages, vision and hearing both decline. Muscular strength also ebbs as a result of a decrease in the number of myofibrils (bundles of contractile filaments) in the skeletal muscles. Bones tend to thin, and joints often show signs of wear. Aging is also accompanied by a gradual reduction in cardiac output and pulmonary function. The decrease in pulmonary function results in a reduction in oxygen absorption and the amount of air that can be inhaled and exhaled. The number of functional nephrons in the kidney declines, as does renal function. The immune system also becomes less able to respond to antigens. The

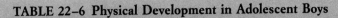

TABLE 22–6 Physical Development in Adolescent Boys

PHYSICAL CHANGE	AVERAGE AGE WHEN CHANGE BEGINS	AVERAGE AGE WHEN NOTICEABLE CHANGE USUALLY STOPS	REMARKS
Increase in rate of growth	12 to 13	17 to 18	If conspicuous growth fails to begin by 15, you should consult your physician.
Enlargement of genitals	Testicles and scrotum: 11 to 12 Penis: 12 to 13	16 to 17 15 to 16	As the testicles grow, the skin of the scrotum darkens. The penis usually lengthens before it broadens. Ability to ejaculate seminal fluid usually begins about a year after the penis starts to lengthen.
Emergence of body hair	Pubic hair: 11 to 12 Underarm hair: 13 to 15	15 to 16 16 to 18	Development of hair is extremely variable and largely dependent on genetic inheritance. The spread of hair up the abdomen and onto the chest usually continues into adulthood.
Development of sweat glands	13 to 15	17 to 18	See remarks in the accompanying table for girls.
Voice change	Enlargement of the larynx, or voice box, begins at 13 to 14, and the voice deepens at 14 to 15	16 to 17	Growth of the larynx may make the adam's apple more prominent. The voice may change rapidly or gradually. If childhood voice persists after 16, consult your physician.

SOURCE: From *The American Medical Association Family Medical Guide*. Copyright © 1982 by The American Medical Association. Reprinted by permission of Random House, Inc.

nervous system does not escape the aging process. Memory deteriorates, and reaction time decreases. Although these changes are part of the aging process, not all of them begin at the same time.

Aging May Be Brought About By a Decline in Cell Numbers and a Decline in the Function of Cells

The decline in homeostatic systems and general body function results from at least two factors: a decrease in the number of cells in the organs and a decline in the function of existing cells. What causes the decline in cell number?

Cell number is determined by the balance between cell division and cell death. Laboratory experiments show that body cells grown in culture divide a certain number of times, then stop. The number of divisions a cell undergoes is directly related to the age of the organism from which the cell is taken: the older the organism, the fewer divisions are possible.

Laboratory studies such as these have also uncovered a correlation between the number of divisions in culture and the life span of the species from which the cells were taken. In other words, cells from species with long life spans undergo more divisions than cells from species with short life spans. These data and others suggest that the end of cell division and, therefore, aging, is genetically programmed.

If cells are programmed to divide a given number of times, why don't all humans live to be the same age? One answer is that people differ genetically. Genetic differences may account for the difference in life span. In addition, people live markedly different lifestyles and eat very different diets. Research shows that the better one lives, the longer the life span (see Health Note 22–1). Diet, exercise, and stress management are keys to a healthy life.

Scientists are beginning to discover why proper diet and exercise help increase the life span. Laboratory studies, for instance, show that by manipulating the chemical environment of cells, one can produce more cell divisions than normal. Vitamin E in large quantities, for example, increases the number of cell divisions in tissue cultures; that is, it increases the cell's life span. Whether vitamin E will extend a person's life is not known.

To understand aging, scientists are looking at the cell itself to see what causes its function to deteriorate. Studies show that the decline in cell function results from problems arising in the DNA, RNA, and cell proteins. Cells are exposed to many potentially harmful factors in the course of a lifetime. Natural and anthropogenic (produced by human activities) radiation and other potentially harmful agents, such as chemicals in the home and work environments, may damage cells and impair their function. As the damage accumulates, body function gradually deteriorates.

Although the decline in cell number and the loss of cell function are thought to be the two leading causes of aging, some researchers think that aging is largely the result of a gradual decline in immune system function caused by a reduction in cell number and a loss of cell function. Although the immune system does falter, this decline surely cannot explain the other signs of aging.

HEALTH NOTE 22–1

CAN WE REVERSE THE PROCESS OF AGING?

Suppose a drug were invented that would help you live to be 100. Would you take it? If you are like most people, you probably would, as long as your additional years would be healthy and fairly trouble free.

Because most of us would like to beat this aging thing that nature imposes on us, we secretly hope for a medical breakthrough that would prolong our lives. Unfortunately, the search for a "cure" for old age has been fraught with frustration. Over the years, numerous treatments have been tried and have failed. As pointed out in this chapter, although the average U.S. life span has increased, we really are not living much longer than our grandparents.

There is some encouraging news, however. Scientists recently discovered a protein called **stomatin,** which they isolated from fibroblasts that had stopped dividing in tissue culture. Scientists hope that stomatin will stop other cells from dividing as well. If this turns out to be true, researchers may have discovered an important clue in the aging puzzle, the signal that ends cell division. Moreover, they may have found a way to prolong human life.

Their reasoning goes as follows: If this protein signals the end of cellular division, there must be a gene that controls its production. If the stomatin gene can be located, researchers may be able to find a way to inactivate the gene. A drug or genetically engineered chemical, for example, could be injected in people to block the gene. The tissues of these people, researchers think, might continue to regenerate beyond the genetically programmed life span.

Efforts to reverse the aging process have met with some success in another arena as well—slowing down the aging of skin. Because age-related changes in the skin mirror those occurring in other aging tissues, scientists have long studied the skin to expand their understanding of the aging process in general. Their studies have shown that skin aging results from two processes: intrinsic, chronological aging, which may be genetically programmed, and extrinsic aging, resulting from accumulated environmental damage from sunlight and other factors.

Many skin scientists doubt whether anything can be done about intrinsic aging. Their studies suggest that after a skin cell has lived out its lifetime, its plasma membrane receptors become insensitive to growth factors that stimulate DNA replication and cell division.

New research suggests, however, that extrinsic aging, especially that induced by sunlight, is preventable and reversible. Unknown to many, sunlight damages epithelial cells of the epidermis and the fibroblasts in the underlying dermis. When exposed to sunlight, molecules inside various skin cells become electrically excited, and the energy they absorb from the sun may drive reactions with other molecules in the cell. These reactions, in turn, may lead to the dissociation of chemical bonds, which may injure important molecules essential for cellular function. Plasma membranes, in fact, can be damaged within a few minutes of exposure to sunlight (▻ Figure 1). Sunlight energy can also be absorbed by oxygen in the tissue, resulting in the formation of oxygen free radicals, highly destructive chemicals that are primarily responsible for extrinsic aging.

▻ **FIGURE 1 Beware: Skin Damage**

In 1988, medical researchers found that Retin-A, a derivative of vitamin A used to treat severe cases of acne, reduces wrinkling of the skin induced by sunlight. A study of Retin-A that followed patients for 22 months or more after treatment showed that the drug reduces wrinkles, age spots, and roughness for at least 22 months as long as it is applied regularly.

Researchers do not know how the drug works, but studies suggest that it does indeed have several beneficial effects. For example, Retin-A stimulates the growth of new blood vessels in the dermis, which may nurture the regeneration of damaged skin cells. It also detoxifies oxygen free radicals in tissues and can inhibit the destruction of collagen fibers in the dermis. Some researchers think that it may regulate genes that play a role in cell growth and differentiation.

Despite the discovery of stomatin and Retin-A, there is as yet no evidence that medical scientists can prevent or even retard the overall rate of aging. The best way to live a long and healthy life is to live well: learn ways to reduce stress, eat well, exercise regularly, take alcohol in moderation or not at all, and avoid harmful practices such as smoking.

Aging Is Often Associated With Certain Diseases, But the Likelihood of Contracting a Disease of Old Age Depends in Large Part on Lifestyle

Many diseases are associated with old age. Osteoporosis, arthritis, atherosclerosis, and cancer are examples. Although these diseases are more prevalent in older people, they are emphatically not the inevitable consequence of aging. That is, they are not unavoidable signs of aging like gray hair. As previous chapters have shown, the likelihood of developing these diseases can be greatly reduced by exercise, diet, and other lifestyle adjustments. In other words, people can age in a healthy manner by living healthy lives.

A Careful Analysis of the Data on Longevity Shows that Modern Medicine Is Not Letting Us Live Much Longer

"Thanks to improvements in medicine, people are living longer." You have heard this statement dozens of times in one form or another. The assertion is that advances in medicine are actually increasing how long you and I will live. This statement is so often repeated that most of us believe it implicitly. But is it true?

Medical scientists measure longevity by a statistic called **life expectancy at birth.** Life expectancy at birth is the number of years, on average, a person lives after he or she is born. As we all know, life expectancy at birth has increased dramatically in the last century. In 1900, for example, on average, white American females lived only 50 years. Today, life expectancy is 81 years. For males, a similar trend is observed. In 1900, for example, the life expectancy of a white American male was 47 years; today, it is 74 years.

Most people take these statistics to mean that men and women are actually living to a much older age. In reality, something very different is happening: the increase in the life expectancy at birth is largely the result of declining infant mortality. That is, thanks to improvements in medicine, more children are living through the first year of life. In the early 1900s, 100 of every 1000 babies died during the first year of life. Today, that number has fallen to 12. This reduction has greatly affected average life expectancy.

This is not to say that all of the gain in life expectancy results from a decline in infant mortality. Medical advances have increased life expectancy after infancy, but these changes are small in comparison to those brought about by lowering infant mortality. In fact, about 85% of the increase in life span in the last century is the result of decreased infant mortality.

Death Is the Final Act of Living

Aging results in a deterioration of function that eventually leads to death. But death can also result from traumatic injury to the body—for example, severe damage to the brain or a sudden loss of blood—and many other causes.

When I began to research this topic for the book, I thought the task would be easy. After all, what can you say about death? It is when a person is no longer alive, isn't it? What I found was that death is difficult to define precisely. Although we can all agree that death is the cessation of life, trouble begins when people try to define it in cases requiring life-support systems. In 25 states, a person is considered dead when his or her heart stops beating and breathing ceases. In the remaining 25 states, however, death is defined more liberally as an irreversible loss of brain function. A woman in a coma, for example, often shows little or no brain activity other than that needed to keep the heart beating and the lungs working. Is this woman alive? In some states, she is; in others, she is not. What is your opinion on this issue? If you were lying in a coma with little or no brain function, what would you want your family to do?

As you can well imagine, maintaining a person on life support is extremely costly and can be financially and emotionally crippling to a family and to society, if, for example the patient is receiving federal benefits through Medicare or Medicaid. The money spent on maintaining a life could arguably be used to save dozens of lives through preventive measures. These issues lead to another biomedical dilemma, the controversy over euthanasia, a word derived from the Greek meaning "easy death." **Euthanasia,** an act or method of causing death painlessly, may be either passive or active. Passive euthanasia consists of deliberate actions and decisions to withhold treatment that might prolong life. Active euthanasia consists of actions and decisions that actively shorten a person's life—for example, injecting a lethal substance to terminate a patient's life.

Opinions on euthanasia vary widely, and the controversy will, no doubt, be debated for many years. (For a debate on physician-assisted suicide, see the Point/Counterpoint in this chapter.) As a final note, individuals who want to save their relatives emotional and legal turmoil can sign a living will. This is a legal document stipulating conditions under which doctors should allow a person to die.

ENVIRONMENT AND HEALTH: VIDEO DISPLAY TERMINALS

Computer monitors, or video display terminals (VDTs), produce numerous types of radiation with frequencies ranging from X-rays to radio waves. Protective shields prevent most of this radiation from escaping. It is what escapes the shield, however, that has some health officials concerned.

Most scientists agree that small amounts of middle- and high-frequency radiation escaping from the protective shielding are not a threat to health. Lower-frequency radiation (radio frequencies) is another story. Laboratory experiments show that electromagnetic fields generated by extremely low-frequency radiation can alter fetal development in chickens, rabbits, and swine.

In addition, researchers in Oakland, California, recently published results of a medical study on the incidence of miscarriage in nearly 1600 women. The study showed that women who sit in front of a computer VDT for more than 20 hours a week during the first three months of pregnancy are nearly twice as likely to miscarry as women in similar jobs not using computers.

What causes these miscarriages? Researchers suspect that they result from radiation emitted from the VDTs. However, other factors may also be involved. For example,

THE RIGHT TO A PHYSICIAN WHO WILL NOT KILL

Rita Marker

Rita L. Marker is Director of the International Anti-Euthanasia Task Force and the author of Deadly Compassion.

Should physician administered euthanasia—the direct and intentional killing of a patient by a physician—be legalized? Common sense has always said NO. And laws have wisely banned euthanasia.

Now, however, an effort is underway to change these laws. Those who seek to legalize euthanasia have framed their arguments in terms of personal rights and freedom. They've clearly recognized the importance of words in molding public opinion: carefully crafted verbal engineering is being employed to transform the appalling crime of mercy killing into an appealing matter of patient choice.

Proposals appearing in state after state have such benign titles as "Death With Dignity Act," and the deceptively soothing term "aid-in-dying" is often substituted for "euthanasia."

As efforts of euthanasia activists increase, careful examination of what is at stake has become increasingly important. This examination should include answers to key questions.

Who Will Be the Recipients of New "Rights" If Euthanasia Is Legalized?

The law would benefit doctors, not patients. Competent adults already have the legal right to refuse medical treatment, as well as the right to make their wishes known about such interventions for the future. Additionally, neither attempting nor carrying out the very personal and tragic decision to commit suicide is a criminal offense. Bluntly put, people can refuse medical care, or can kill themselves without physician intervention.

Doctors, however, are prevented by law from killing their patients. Passage of the Death With Dignity Act would give doctors both the right to directly and intentionally kill patients and the promise of immunity. Legalization of euthanasia would transform an action which is previously considered homicide into a "medical service."

What Types of Safeguards Are Contained in the Death With Dignity Act?

Initial efforts to pass the Death With Dignity Act have failed. In the Fall of 1991, it was turned down by Washington voters. A subsequent unsuccessful effort placed it on California's ballot, where its advocates claimed that the measure contained adequate safeguards. However, the so-called safeguards had as little substance as the Emperor's new clothes. For example:

- A signed request for euthanasia, made years before the diagnosis of a terminal condition, could have been put into effect without any additional witnessed documentation.
- No requirement existed that the family be notified before a loved one was killed.
- No waiting period or counseling about care and symptom alleviation was required.
- The mental health or competency of a patient who was to be killed could not have been assessed unless the patient approved a psychological evaluation.
- Physicians could have offered euthanasia as an "option" and encouraged their patients to "choose" it.
- Only the most general information (such as the number of patients killed) would need to have been reported, thus preventing any serious investigation of abuses.

What Has Happened Where Euthanasia Has Been Practiced?

Euthanasia is technically illegal in Holland, but is nonetheless widely practiced as a result of court decisions which set forth guidelines under which it may be administered. Even with these guidelines, which are far tighter than the "safeguards" proposed in the Death With Dignity Act, a 1991 Dutch government study released horrifying information. The study found that in tiny Holland—a country with only one-half the population of the State of California—Dutch physicians deliberately and intentionally end the lives of more than 11,000 people each year by lethal overdoses or injections. And, the study found that more than half of those killed had *not* requested euthanasia.

What Will Happen If Euthanasia Is Legalized?

While Death With Dignity proposals pose a threat to everyone, they are particularly dangerous for the many people who lack health insurance. Euthanasia as a medical option may present a choice for the rich. For the poor, however, it may become the only affordable "medical treatment."

Expansion of the Death With Dignity Act would also be a problem. Indeed, expansion is already being discussed among euthanasia leaders. For example, in *Final Exit,* Derek Humphry wrote that when the Death With Dignity Act becomes law everywhere, it will provide a means of handing the dilemma of "terminal old age." He has also called for expansion of euthanasia to include those who are physically and mentally disabled.

Those who think that euthanasia, once unleashed, can be controlled are making a deadly mistake. Euthanasia is not a means of dying with dignity, nor is it a matter of liberty. It does not legislate compassion. It only legalizes killing.

THE RIGHT TO CHOOSE TO DIE

Derek Humphry

Derek Humphry was the principal founder of the Hemlock Society and its executive director from 1980 to 1992. He is the author of four books on euthanasia, including the best-seller Final Exit (1991).

If we are truly free people and if our bodies belong to ourselves and not to others, then we have the right to choose when and how to die. Death comes to us all in the end, although modern medicine can often help us improve and extend our life spans. Thus, given today's high technology and ruinous cost of medicine, it is wise if we all give advance thought to the manner of our dying. Such decisions should be transferred to paper (a living will and durable power of attorney for health care), because 46 states now legally recognize one's wishes in respect to the withdrawal of life-support equipment.

Advance-declaration documents, of course, deal only with the legal and ethical problems of medical equipment and treatments. Less than half of dying people are connected to such equipment. For many more patients, technology does nothing for their terminal illness, so—in effect—there is no "plug to pull." Therefore, the cutting edge of the right-to-die debate in the 1990s centers on assisted death and voluntary euthanasia.

Hemlock Society supporters feel not only that hopelessly sick people should have an unfettered right to accelerate their end to save them pain, distress, and indignity but also that willing doctors should be able to help them die. (To clarify some terms: *self-deliverance* (*suicide*) is ending your own life to be free of suffering; *assisted suicide* is helping another die; *euthanasia* is the direct ending of another's life by request. Currently, suicide in any form, for any reason, is not illegal, but assisted suicide and euthanasia are technically crimes.)

Most people at the close of life do not wish to suffer, desiring a quick and painless demise. The sophisticated pain-management drugs now available, when properly administered, control some 90% of terminal pain. But they do nothing to alleviate indignities, psychic pain, and loss of quality of life associated with some debilitating terminal illnesses.

A complaint that any hastening of the end interferes with God's authority can be answered for many through a belief that *their* God is tolerant and charitable and would not wish to see them suffer. Other people have no faith in God. Pious people differ with this view, of course, and I respect that.

Numerous opinion polls in the United States and other Western countries indicate that two-thirds of people want the right to have a doctor lawfully assist them to die. People in the states of Washington, Oregon, and California are engaged in political action to achieve this law reform.

The proposed law in these West Coast states is called the Death With Dignity Act, and the broad outline of its purpose is as follows:

1. The patient wanting physician-assisted dying would have to be a mature adult suffering from a terminal illness likely to cause death within about six months.
2. People with emotional or mental illness (especially depression) could not get help to die under this law.
3. The request would have to be in writing, and the signature would have to be witnessed by two independent persons.
4. The physician could decline to help the patient die on grounds of conscience but would then cease to be the treating physician. The patient could then seek a physician who was willing.
5. The family would have to be informed and its views, if any, taken into account. But the family could neither promote nor veto the patient's request to die.
6. The physician would have to be satisfied that the patient was fully aware of his or her condition, had been informed of all possible alternatives, and was competent to make this request.
7. If the physician was unsure of the patient's mental state, a mental-health professional could be called to make an evaluation.
8. The time and manner of the assisted dying would have to be negotiated between patient and physician, with the patient's wishes paramount.
9. At any time the patient could orally or in writing revoke the request for assisted dying.
10. Any person who pressured another person to get assistance in dying, who forged such a request, or who ignored a revocation would be subject to prosecution.
11. After helping the patient die, the physician would have to report the action in confidence to a state health agency.

The right-to-die movement feels that these conditions, plus others in the Death With Dignity Act too numerous to describe here, are an intelligent and humane approach to euthanasia with the necessary safeguards against abuse.

Dying on one's own terms is not only an idea whose time has come but also the ultimate civil and personal liberty.

SHARPENING YOUR CRITICAL THINKING SKILLS

1. Summarize the key points of each author and their supporting arguments.
2. What is the basis for their fundamental disagreement?
3. Which viewpoint corresponds to yours? Why?
4. How did your values and experiences affect your opinion in this matter?

the stress of using a computer may be the cause. Only more research will tell. Preferring to err on the conservative side, some scientists advise pregnant women to minimize their exposure to radiation from computer monitors.

Should the link between radiation from VDTs and miscarriage be substantiated by further research, it would once again illustrate the importance of a healthy environment to overall human health and reproduction.

SUMMARY

AN OVERVIEW OF ANIMAL DEVELOPMENT

1. The sperm and egg of sexually reproducing animals unite to form a zygote, a process called fertilization.
2. In vertebrates and invertebrates, sperm contact the plasma membrane of eggs and are phagocytized by them. The egg and sperm cell nuclei then fuse to form a diploid zygote.
3. Fertilization is followed by mitotic division, or cleavage. This process produces a multicelled structure, the morula.
4. In many species, such as frogs and sea urchins, a hollow cavity forms in the morula. The resulting structure is a blastula.
5. When cells of the blastula invaginate, gastrulation begins. Three distinct layers of cells form in the gastrula: the endoderm, mesoderm, and ectoderm. Each layer gives rise to specific organs during organogenesis.
6. All animal embryos go through much the same process of development, although there are some notable differences. In chickens and other birds, embryonic cells form a small, disklike structure atop a mass of yolk. It differentiates into three layers that give rise to the organs.

HUMAN REPRODUCTION AND DEVELOPMENT

7. In humans, fertilization usually occurs in the upper third of the uterine tube.
8. Sperm deposited in the vagina reach the site of fertilization with the aid of muscular contractions in the walls of the uterus and uterine tube.
9. Sperm bore through the zona pellucida and contact the plasma membrane. The first one to contact the membrane fertilizes the oocyte. Further sperm penetration is blocked.
10. Sperm are engulfed by the oocyte. The chromosomes of the sperm and oocyte duplicate and merge in the center of the cell, where mitosis begins.
11. Human development during gestation is divided into three stages: pre-embryonic, embryonic, and fetal.
12. Pre-embryonic development begins at fertilization and ends at implantation. The zygote undergoes rapid cellular division and forms a morula.
13. Next, a cavity forms in the morula, converting it to a blastocyst. The blastocyst consists of a clump of cells, the inner cell mass, which becomes the embryo, and the trophoblast, which gives rise to the embryonic portion of the placenta.
14. The morula arrives in the uterus three to four days after fertilization and is converted into a blastocyst, which implants in the uterine wall two to three days later.
15. Soon after implantation begins, the trophoblast differentiates into two layers. The outer layer invades the endometrium. Cavities form in this layer and fill with blood.
16. Fingerlike projections of the inner layer carrying fetal blood vessels then invade the trophoblast, forming the placental villi. The villi are bathed in maternal blood and provide a means of acquiring oxygen and nutrients from the mother and disposing of embryonic wastes.
17. While the placenta is forming, a layer of cells from the inner cell mass of the blastocyst separates from it and forms the amnion. The amnion fills with fluid and enlarges during embryonic and fetal development, eventually surrounding the entire embryo and fetus. The amniotic fluid protects the embryo and fetus during development.
18. After the amnion forms, the cells of the inner cell mass differentiate into the three germ cell layers: ectoderm, mesoderm, and endoderm. The formation of the three primary germ layers marks the beginning of embryonic development.
19. The organs develop from the three basic tissues during organogenesis. Table 22–2 lists the organs and tissues formed from each of the layers.
20. Fetal development begins eight weeks after fertilization. Because most of the organ systems have developed or are under development, fetal development is primarily a period of growth.
21. The placenta produces several hormones that play an important part in reproduction. Human chorionic gonadotropin maintains the corpus luteum during pregnancy. Progesterone and estrogen stimulate uterine growth and the development of the glands and ducts of the breast.
22. About 10% to 12% of all newborns enter the world with some form of birth defect. Birth defects arise from chemical, biological, and physical agents known as teratogens.
23. The effect of teratogenic agents is related to the time of exposure, the nature of the agent, and the dose. A defect is most likely to arise if a woman is exposed to a teratogen during the embryonic period when the organs are forming.
24. Many physical and chemical agents are toxic to the human fetus and when present in sufficient quantities can kill a fetus or retard its growth.
25. During pregnancy a woman's body undergoes incredible change. The uterus and breasts enlarge considerably, blood volume increases, respiration rate climbs, and urination increases.

CHILDBIRTH AND LACTATION

26. Labor consists of intense and frequent uterine contractions that are believed to be caused by the release of small amounts of fetal oxytocin prior to birth. Fetal oxytocin stimulates the release of prostaglandins by the placenta. Oxytocin and prostaglandins stimulate contractions in the sensitized uterine musculature.
27. Emotional and physical stress in the mother may trigger maternal oxytocin release, augmenting muscle contractions.
28. As uterine contractions increase, they cause more maternal oxytocin to be released, which stimulates stronger contractions and more oxytocin release, a positive feedback that continues until the baby is born.
29. Labor consists of three stages, the dilation, the expulsion, and the placental.
30. The breasts of a nonpregnant woman consist primarily of fat and connective tissue interspersed with milk-producing glandular tissue and ducts.

31. During pregnancy, the glands and ducts proliferate under the influence of placental and ovarian estrogen and progesterone.
32. Milk production is induced by maternal prolactin and human chorionic somatomammotropin but does not begin until two to three days after birth.
33. Before milk production begins, the breasts produce small quantities of a protein-rich fluid, colostrum. A newborn can subsist on colostrum for the first few days and derives antibodies from it that help protect it from bacteria.
34. Suckling causes a surge in prolactin secretion. Each surge stimulates milk production needed for the next feeding.

INFANCY, CHILDHOOD, ADOLESCENCE, AND ADULTHOOD

35. The newborn survives because of parental care and because it comes equipped with several built-in reflexes, including the suckling reflex, rooting reflex, and crying reflex.
36. During infancy, the child transforms physically and mentally. Children develop in predictable stages, although the timetable varies considerably.
37. During infancy, sensory, motor, and intellectual capacity increase dramatically. Children develop emotionally as well. One of the first emotional milestones of normal human development is bonding, the establishment of an intimate relationship between the infant and its caregivers. Successful bonding is important for later development and may facilitate emotional security later in life.
38. Research suggests that intelligence, like motor skills, improves with use. Parents are encouraged to foster their children's curious nature, to let them explore, and to provide opportunities for a rich and varied experience.
39. Childhood lasts from infancy to puberty. During childhood, children undergo dramatic improvements in motor skills and in the use and comprehension of language.
40. Puberty, or sexual maturation, marks the onset of adolescence, which lasts until adulthood (age 18 to 20). Sexual maturation results from the production of sex steroids.
41. Adolescence is also a time of emerging identity. Conflict between teenagers and their parents is common during this period, but it can be reduced if parents learn to accept the adolescent's personality and maintain a supportive relationship.
42. Adulthood begins when adolescence ends. Physical growth stops, but, for many people, social and psychological development continues.

AGING AND DEATH

43. Aging is the progressive deterioration of the body's homeostatic abilities and the gradual deterioration of the function of body organs.
44. These changes result from at least two factors: a decrease in the number of cells in the organs and a decline in the function of existing cells.
45. Death results from aging, traumatic injury, or infectious disease.

EXERCISING YOUR CRITICAL THINKING SKILLS

The frequency of Down syndrome babies increases with maternal age. Because 95% of Down syndrome babies can be attributed to maternal chromosomal nondisjunction, it has long been thought to be the result of the age of the oocytes in a woman's ovaries. New research suggests, however, that the reason older women have a higher percentage of Down syndrome babies may be that they are more likely to carry one to term than younger women.

In an international study conducted by 19 scientists, researchers found that the extra chromosome 21 that is responsible for this genetic disorder is most often incorporated during the first meiotic division. Much to their surprise, however, the frequency of nondisjunction in older and younger women was nearly equal.

If the frequency of nondisjunction is the same in mothers of all ages, one possible reason why older mothers have a higher incidence of Down syndrome babies is that their bodies fail to recognize an abnormal embryo as well as those of younger mothers. As a result, they are more likely to carry a Down syndrome baby to term.

Further research is needed to determine whether this hypothesis is valid. Can you think of any way to test it?

After thinking about it for a while, consider this suggestion: one direct way of testing the hypothesis, of course, would be to compare the ages of women who spontaneously abort Down syndrome fetuses.

TEST OF TERMS

1. In humans, the ovum and sperm unite to form a(n) ______________ during fertilization, which takes place in the upper third of the ____________ ____________.
2. Residence in the female reproductive tract for a period of six to seven hours renders sperm membranes fragile, allowing the sperm to release enzymes from the ____________.
3. When a sperm cell contacts the plasma membrane of the oocyte, it triggers an immediate change in the membrane that prevents other sperm from entering. This is called the ____________ block to____________.
4. During pre-embryonic development, the cells first multiply to form a solid ball, the ____________. In humans, a cavity soon forms in this structure, converting it to a blastocyst. The blastocyst consists of a clump of cells, the ____________ ____________

____________, which gives rise to the embryo, and a ring of flattened cells, the ____________, which will give rise to the embryonic portion of the placenta.

5. The blastocyst attaches to the uterine lining, and digests its way into it in a process called ____________, which begins ______ to ______ days after fertilization.
6. The placenta forms from embryonic and maternal tissue. Blood vessels invade the forming placenta with the inner layer of the trophoblast. Fingerlike projections form and are bathed by maternal blood. These structures are called __________ __________.
7. Early in embryonic development, the ____________ forms from the cells of the blastocyst that will eventually give rise to the embryo proper. It eventually forms a complete fluid-filled sac that surrounds the fetus.
8. The organs form during embryonic development; this process is called ____________.
9. Ectoderm forms the neural tube, which later develops into the ____________ and ____________ ____________.
10. Blocks of mesoderm along the neural tube, called ____________, form the muscles of the neck and trunk as well as the ____________.
11. The lining of the intestines comes from embryonic ____________.
12. The umbilical ____________ carries blood rich in oxygen and nutrients from the placenta to the fetus. The ____________ ____________ shunts blood from this vessel to the ______ ____________ ____________, thus bypassing the liver.
13. The hole in the interatrial septum shunts blood from the ____________ to the ____________ and is called the ____________ ____________.
14. The placental hormone ____________ ____________ ____________ maintains the corpus luteum and stimulates production of ____________ and ____________ by the CL.
15. The study of birth defects is called ____________.
16. ____________, a hormone produced by the placenta and ovaries of pregnant women, loosens the connective tissue in the pubic symphysis and softens the cervix in preparation for childbirth.
17. Stage 1 of childbirth is also known as the ____________ phase. During this phase, the baby descends into the pelvic cavity, and the ____________ dilates.
18. Childbirth in which the baby is born buttocks first is called a(n) ____________ birth.
19. Muscular contractions in the wall of the uterus are stimulated by the hormone ____________, secreted by the ____________ ____________ gland.
20. The final stage of childbirth is the ____________ stage.
21. An anesthetic injected into the vagina during childbirth is called a(n) ____________ block.
22. Lactation, the production of milk from the breasts, is stimulated by the maternal hormone ____________, secreted from the anterior pituitary gland.
23. Expulsion of the milk by the glands in the breast is induced by the hormone ____________, whose release is stimulated by suckling.
24. The protein-rich secretion produced by the breasts during the first few days after birth is called ____________.
25. A baby turns its head toward a nipple that brushes against its cheek. This is called the ____________ reflex.
26. One of the first emotional milestones of normal human development is ____________, the establishment of an intimate relationship between the infant and one (or preferably both) of its parents or caregivers.
27. Sexual maturation occurs during ____________.
28. ____________ ____________ at birth is the average life span of an individual.

Answers to the Test of Terms are found in Appendix B.

TEST OF CONCEPTS

1. Give an overview of the process of development in animals. Describe each stage and what is happening at it.
2. The formation of the gastrula results in the formation of endoderm, mesoderm, and ectoderm and the rearrangement of cells. Why is this process so important to the future development of the embryo?
3. Describe the process of fertilization in humans in detail. Use drawings to elaborate your points, and label all drawings.
4. Define the terms *inner cell mass* and *trophoblast.*
5. Where does the human embryo acquire nutrients before the placenta forms?
6. Describe the formation of the placenta in humans.
7. What are the major functions of the human placenta?
8. Describe the flow of blood from the placenta to the fetus and back. What is the role of the various shunts?
9. List and describe the function of the five placental hormones.
10. Describe the changes that occur in pregnant women. Describe the factors that contribute to maternal weight gain during pregnancy.
11. Discuss how the nature of a teratogenic agent and the time of exposure affect teratogenesis, the production of birth defects.
12. What factors trigger labor?
13. What factors are responsible for cervical dilation during labor?
14. Design a humane experiment to test the two major hypotheses about bonding presented in this chapter. Describe how you would go about the experiment and the special considerations that would be necessary to eliminate bias. Could such an experiment be objective? Could such an experiment be ethical?
15. Describe some of the reflexes of the newborn and their importance to the infant's survival.
16. Define the term *aging,* and describe some of the hypotheses that attempt to explain it.
17. Thanks to modern medicine, men and women are living longer, says a friend. Using your understanding of average life expectancy data and the reason for an increase in life expectancy in the last century, explain why this statement is not quite true. 22

APPENDIX A

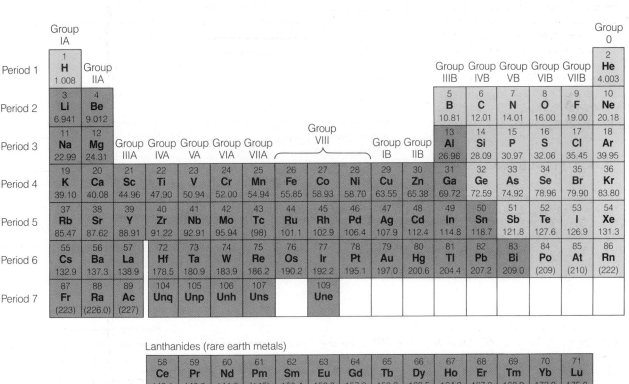

Periodic Table of Elements

	Group IA	Group IIA	Group IIIA	Group IVA	Group VA	Group VIA	Group VIIA	Group VIII			Group IB	Group IIB	Group IIIB	Group IVB	Group VB	Group VIB	Group VIIB	Group 0
Period 1	1 **H** 1.008																	2 **He** 4.003
Period 2	3 **Li** 6.941	4 **Be** 9.012											5 **B** 10.81	6 **C** 12.01	7 **N** 14.01	8 **O** 16.00	9 **F** 19.00	10 **Ne** 20.18
Period 3	11 **Na** 22.99	12 **Mg** 24.31											13 **Al** 26.98	14 **Si** 28.09	15 **P** 30.97	16 **S** 32.06	17 **Cl** 35.45	18 **Ar** 39.95
Period 4	19 **K** 39.10	20 **Ca** 40.08	21 **Sc** 44.96	22 **Ti** 47.90	23 **V** 50.94	24 **Cr** 52.00	25 **Mn** 54.94	26 **Fe** 55.85	27 **Co** 58.93	28 **Ni** 58.70	29 **Cu** 63.55	30 **Zn** 65.38	31 **Ga** 69.72	32 **Ge** 72.59	33 **As** 74.92	34 **Se** 78.96	35 **Br** 79.90	36 **Kr** 83.80
Period 5	37 **Rb** 85.47	38 **Sr** 87.62	39 **Y** 88.91	40 **Zr** 91.22	41 **Nb** 92.91	42 **Mo** 95.94	43 **Tc** (98)	44 **Ru** 101.1	45 **Rh** 102.9	46 **Pd** 106.4	47 **Ag** 107.9	48 **Cd** 112.4	49 **In** 114.8	50 **Sn** 118.7	51 **Sb** 121.8	52 **Te** 127.6	53 **I** 126.9	54 **Xe** 131.3
Period 6	55 **Cs** 132.9	56 **Ba** 137.3	57 **La** 138.9	72 **Hf** 178.5	73 **Ta** 180.9	74 **W** 183.9	75 **Re** 186.2	76 **Os** 190.2	77 **Ir** 192.2	78 **Pt** 195.1	79 **Au** 197.0	80 **Hg** 200.6	81 **Tl** 204.4	82 **Pb** 207.2	83 **Bi** 209.0	84 **Po** (209)	85 **At** (210)	86 **Rn** (222)
Period 7	87 **Fr** (223)	88 **Ra** (226.0)	89 **Ac** (227)	104 **Unq**	105 **Unp**	106 **Unh**	107 **Uns**		109 **Une**									

Lanthanides (rare earth metals)

58 **Ce** 140.1	59 **Pr** 140.9	60 **Nd** 144.2	61 **Pm** (145)	62 **Sm** 150.4	63 **Eu** 152.0	64 **Gd** 157.3	65 **Tb** 158.9	66 **Dy** 162.5	67 **Ho** 164.9	68 **Er** 167.3	69 **Tm** 168.9	70 **Yb** 173.0	71 **Lu** 175.0

Actinides

90 **Th** 232.0	91 **Pa** (231)	92 **U** 238.0	93 **Np** (244)	94 **Pu** (242)	95 **Am** (243)	96 **Cm** (247)	97 **Bk** (247)	98 **Cf** (251)	99 **Es** (252)	100 **Fm** (257)	101 **Md** (258)	102 **No** (259)	103 **Lr** (260)

KEY:

16 — Atomic number
S — Symbol of element
32.06 — Atomic mass

Metals
Nonmetals
Metalloids
Noble gases

APPENDIX B

Answers to the Test of Terms

CHAPTER 10

1. ectoderm, mesoderm, endoderm; 2. extracellular material; 3. muscle, nervous, epithelial, connective; 4. exocrine; 5. epidermis, epithelial; 6. connective; 7. fibroblast; 8. hyaline cartilage; 9. fibrocartilage; 10. central, canaliculi, osteocytes; 11. compact; 12. plasma; 13. red blood cells, platelets; 14. smooth, actin, myosin; 15. neuron; 16. organ system; 17. set point; 18. reflexes; 19. integration; 20. effectors; 21. hormone, paracrine; 22. circadian; 23. suprachiasmatic; 24. jet lag.

CHAPTER 11

1. macronutrients; 2. carbohydrates (glucose) and fat (triglycerides); 3. fat; 4. essential, complete; 5. fiber, colon; 6. Vitamins, water-soluble, water-insoluble; 7. trace minerals; 8. amylase; 9. taste buds; 10. pharynx; 11. peristalsis; 12. gastroesophageal; 13. chyme; 14. pepsin, pepsinogen; 15. pyloric; 16. liver, bile salts; 17. sodium bicarbonate, digestive enzymes; 18. circular folds, villi, microvilli; 19. lymphatic (lacteals); 20. cholecystokinin (CCK).

CHAPTER 12

1. pulmonary, systemic; 2. right atrium; 3. aorta, elastic; 4. semilunar; 5. atrioventricular; 6. sinoatrial, right atrium; 7. atrioventricular bundle; 8. tunica intima, tunica media, tunica adventitia; 9. Elastic fibers; 10. systolic; 11. capillaries, venules; 12. Varicose veins; 13. edema; 14. valves; 15. lymphatic capillaries, interstitial (tissue); 16. Lymph nodes.

CHAPTER 13

1. plasma; 2. immunoglobulins or antibodies; 3. RBC, hemoglobin; 4. red bone; 5. erythropoietin, RBCs; 6. porphyrin; 7. anemia; 8. neutrophils, monocytes; 9. lymphocyte; 10. Leukemia; 11. platelet; 12. Fibrinogen; 13. Plasmin; 14. Hemophilia; 15. Carbon monoxide.

CHAPTER 14

1. Emphysema; 2. pharynx; 3. trachea, bronchi; 4. bronchioles; 5. mucous; 6. surfactant; 7. dust or alveolar macrophage; 8. vocal cords; 9. olfactory epithelium or membrane; 10. hemoglobin, bicarbonate ions; 11. brain, breathing center; 12. inhalation, diaphragm, intercostal muscles; 13. exhalation; 14. chronic bronchitis; 15. tidal, expiratory reserve.

CHAPTER 15

1. capsid; 2. inflammatory response; 3. Pyrogens; 4. Interferons; 5. antigen; 6. immunocompetence; 7. active; 8. memory; 9. immunoglobulins; 10. neutralization; 11. agglutination; 12. cytotoxic T; 13. vaccine.

CHAPTER 16

1. urinary, renal; 2. ureters; 3. urethra; 4. cortex; 5. renal pelvis; 6. glomerulus, renal tubule; 7. afferent, Bowman's space; 8. podocytes; 9. peritubular capillaries, tubular secretion; 10. internal sphincter; 11. dialysis; 12. ADH, increases; 13. Caffeine, diuretic; 14. aldosterone; 15. nephron.

CHAPTER 17

1. central, peripheral; 2. autonomic; 3. dendrites, axon; 4. myelin sheath; 5. terminal boutons, neurotransmitters; 6. resting (membrane) potential, -60; 7. sodium ions, action potential; 8. synaptic cleft, postsynaptic membrane; 9. Acetylcholine, acetylcholinesterase; 10. spinal nerves, ventral; 11. interneuron; 12. primary motor; 13. gyri, sulci; 14. primary sensory, central sulcus; 15. meninges, dura mater, pia mater; 16. association; 17. limbic; 18. cerebellum; 19. hypothalamus, nuclei; 20. reticular activating; 21. cerebrospinal, ventricles; 22. electroencephalogram, EEG; 23. sympathetic; 24. parasympathetic; 25. short-term, seconds, hours.

CHAPTER 18

1. pain; 2. encapsulated receptors; 3. Merkel's disks; 4. Pacinian; 5. Meissner's corpuscle; 6. muscle spindle; 7. Golgi tendon organs; 8. Adaptation; 9. special senses; 10. taste buds, papillae; 11. olfactory, bipolar; 12. sclera, cornea; 13. pigmented, choroid, ciliary body; 14. retina, cones, night; 15. ganglion; 16. fovea centralis, optic disk; 17. lens, ciliary body, cataracts; 18. vitreous humor; 19. Glaucoma, aqueous humor; 20. refraction, change velocity; 21. extrinsic eye; 22. nearsightedness, elongated eyeball, strong lens; 23. radial keratotomy; 24. astigmatism; 25. rhodopsin; 26. sex-linked; 27. external auditory, eardrum; 28. ossicles, oval window; 29. eustachian tube or auditory tube; 30. ampullae, semicircular, endolymph; 31. maculae; 32. organ of Corti, vestibular; 33. conduction; 34. temporary threshold.

CHAPTER 19

1. vertebrae, axial; 2. appendicular; 3. long, diaphysis, compact, marrow; 4. spongy; 5. synovial, synovial, joint capsule, tendons; 6. flexion, extension; 7. arthroscope; 8. osteoarthritis; 9. Rheumatoid, synovial; 10. prosthesis; 11. hyaline cartilage, primary center; 12. osteoclasts, osteoblasts; 13. epiphyseal plate; 14. calcitonin, thyroid, parathormone or PTH; 15. osteoporosis; 16. nuclei, striated; 17. endomysium; 18. epimysium; 19. myofibril; 20. sarcomere, A, I; 21. calcium, sarcoplasmic, troponin, actin; 22. myosin; 23. acetylcholine, T tubules; 24. Muscle fatigue, lactic acid; 25. twitch, all-or-none; 26. wave, motor; 27. slow-twitch, myoglobin; 28. fast-twitch, myosin ATPase; 29. anabolic steroids.

CHAPTER 20

1. hormone, target; 2. receptors; 3. trophic; 4. steroids, amines; 5. second messenger, ATP, adenylate cyclase, phosphates; 6. two-step, genes or DNA; 7. pituitary, hypothalamus; 8. anterior pituitary, releasing, neurosecretory, hypothalamus; 9. Growth, hypertrophy, amino acids; 10. Hyposecretion; 11. ACTH; 12. gonadotropins; 13. prolactin, suckling; 14. neuroendocrine; 15. neuroendocrine, ADH, oxytocin; 16. thyroxine, calcitonin; 17. thyroglobulin; 18. goiter; 19. parathormone or PTH, parathyroid, calcitonin, thyroid; 20. Insulin, glucose, glycogen; 21. gluconeogenesis; 22. autoimmune; 23. insulin-dependent, obesity, diet; 24. adrenalin or epinephrine, noradrenalin or norepinephrine; 25. adrenal cortex, gluconeogenesis, immune; 26. Aldosterone, adrenal cortex.

CHAPTER 21

1. scrotum, seminiferous; 2. epididymes, vasa deferentia; 3. sex accessory; 4. germinal, spermatogonia, primary spermatocytes, secondary spermatocytes; 5. spermatids, 23; 6. acrosome; 7. interstitial, testosterone, luteinizing hormone or LH; 8. foreskin, circumcision; 9. erectile; 10. ejaculation, semen; 11. uterus, uterine tubes; 12. vagina, cervical; 13. labia majora; 14. clitoris; 15. ovulation, menstrual; 16. tertiary or antral, zona pellucida; 17. first, polar, 23, double; 18. corpus luteum or CL, progesterone, estrogen; 19. follicular, estrogen, FSH; 20. endometrium, menstruation; 21. premenstrual syndrome; 22. menopause, estrogen; 23. vasectomy, uterine tubes, tubal ligation; 24. estrogen, progesterone; 25. Pap smear; 26. intrauterine device or IUD; 27. diaphragm, spermicidal; 28. condom; 29. rhythm or natural; 30. temperature; 31. sexually transmitted diseases; 32. Nonspecific urethritis; 33. Herpes.

CHAPTER 22

1. zygote, uterine tube or Fallopian tube; 2. acrosome; 3. fast, polyspermy; 4. morula, inner cell mass, trophoblast; 5. implantation, 5, 7; 6. placental villi; 7. amnion; 8. organogenesis; 9. brain, spinal cord; 10. somites, spine or vertebrae; 11. endoderm; 12. vein, ductus arteriosus, inferior vena cava; 13. right, left, foramen ovale; 14. human chorionic gonadotropin, estrogen, progesterone; 15. teratology; 16. Relaxin; 17. dilation, cervix; 18. breach; 19. oxytocin, posterior pituitary; 20. placental; 21. pudendal; 22. prolactin; 23. oxytocin; 24. colostrum; 25. rooting; 26. bonding; 27. puberty; 28. Life expectancy.

APPENDIX C

The Metric System

In the United States, the metric system is a system of measurement used principally by scientists. In our day-to-day lives, though, most Americans use the English system of measurement—miles, inches, feet, pounds, tons, and so on.

Many countries like New Zealand use the metric system for weights and other measures. If you travel abroad, road signs will list distance in kilometers rather than miles. Other linear measurements will be given in meters and centimeters instead of yards, feet, and inches. The weight of objects will be expressed in kilograms or grams instead of pounds and ounces. Liquids will be measured in liters rather than quarts or gallons.

It can be confusing if you don't know how to convert from one system of weights and measures to another. The lists below show some of the most common units you will encounter in biology and other sciences and compare them to their English equivalents.

Most Common English Units and the Corresponding Metric Units

	ENGLISH UNIT	METRIC UNIT
Weight	tons	metric tons
	pounds	kilograms
	ounces	grams
Length	miles	kilometers
	yards	meters
	inches	centimeters
Square Measure	acres	hectares
	square miles	square kilometers
Volume	quarts and gallons	liters
	fluid ounces	milliliters

—See next page for more Metric System tables

Converting English Units to Metric Units

	ENGLISH UNIT		METRIC UNIT
Weight	1 ton (2000 pounds)	=	0.9 metric tons
	1 pound	=	0.454 kilograms
	1 ounce	=	28.35 grams
Length	*1 mile	=	1.6 kilometers
	1 yard	=	0.9 meters
	*1 inch	=	2.54 centimeters
Square Measure	1 acre	=	0.4 hectares
	1 square mile	=	2.59 square kilometers
Volume	1 quart	=	0.95 liters
	*1 gallon	=	3.78 liters
	1 fluid ounce	=	29.58 milliliters

*Most useful conversions to know.

Converting Metric Units to English Units

	METRIC UNIT		ENGLISH UNIT
Weight	*1 metric ton	=	2204 pounds
	1 metric ton	=	1.1 tons
	*1 kilogram	=	2.2 pounds
	1 gram	=	28.35 ounces
Length	*1 kilometer	=	.6 miles
	1 meter	=	1.1 yards
	1 centimeter	=	.39 inches
Square Measure	*1 hectare	=	2.47 acres
	1 square kilometer	=	0.386 square miles
Volume	1 liter	=	1.057 quarts
	1 liter	=	.26 gallons
	1 milliliter	=	0.0338 fluid ounces

*Most useful conversions to know

Glossary

≈

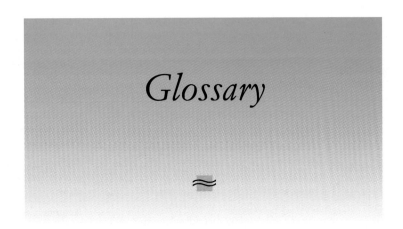

Abiotic factors Physical and chemical components of an organism's environment.

Abyssal zone Deepest waters of the ocean. Characterized by complete darkness.

Accommodation Change in the shape of the lens caused by contraction or relaxation of the smooth muscle of the ciliary body. Through accommodation, the lens adjusts the degree to which incoming light rays are bent, permitting objects to be focused on the retina.

Acetylcholine Neurotransmitter substance in the central and peripheral nervous systems of humans.

Acetylcholinesterase Enzyme that destroys the neurotransmitter acetylcholine in the synaptic cleft.

Achondroplasia Genetic disease that results from an autosomal dominant gene. Individuals with the disease have short legs and arms, but a relatively normal body size.

Acid deposition Deposition of sulfuric and nitric acids in the atmosphere onto the Earth's surface. Damages buildings, lakes, streams, crops, and forests. Acids come from sulfur dioxide and nitrogen dioxide produced during the combustion of fossil fuels.

Acid precursors Sulfur dioxide and nitrogen dioxide gases that combine with water and oxygen to form sulfuric and nitric acids in the atmosphere.

Acrosome Enzyme-filled cap over the head of a sperm. Helps the sperm dissolve its way through the corona radiata and zona pellucida.

Actin microfilaments Contractile filaments made of protein and found in cells as part of the cytoskeleton. Especially abundant in the microfilamentous network beneath the plasma membrane and in muscle cells.

Action potential Recording of electrical change in membrane potential when a neuron is stimulated.

Activation energy The energy needed to force the electron clouds of reactants together before the formation of products; also, the energy needed to break internal chemical bonds.

Active immunity Immune resistance gained when an antigen is introduced into the body either naturally or through vaccination.

Active site The region of an enzyme molecule that binds substrates and performs the catalytic function of the enzyme.

Active transport Movement of molecules across membranes using protein molecules and energy supplied by ATP. Moves molecules and ions from regions of low to high concentration.

Adaptation Genetically based characteristic that increases an organism's chances of passing on its genes.

Adaptive radiation Process in which one species gives rise to many others that occupy different environments. Also known as divergent evolution.

Adenosine diphosphate (ADP) Precursor to *ATP*; consists of adenine, ribose, and two phosphates; *see also* Adenosine triphosphate.

Adenosine triphosphate (ATP) A molecule composed of ribose sugar, adenine, and three phosphate groups. The last two phosphate groups are attached by "high-energy bonds" that require considerable energy to form but release that energy again when broken. ATP serves as the major energy carrier in cells.

Adenylate cyclase Enzyme bound to the inner surface of the plasma membrane. Linked to plasma membrane hormone receptors. Responsible for the conversion of ATP to cyclic AMP (a second messenger).

Adipose tissue Type of loose connective tissue containing numerous fat cells. Important storage area for lipids.

Adolescence Period of human life from puberty until adulthood. Characterized by sexual maturity.

Adrenal cortex Outer portion of the adrenal gland. Produces a variety of steroid hormones, including cortisol and aldosterone.

Adrenal gland Endocrine gland located on top of the kidney. Consists of two parts: adrenal cortex and medulla, each with separate functions.

Adrenal medulla Inner portion of the adrenal glands. Produces epinephrine (adrenalin) and norepinephrine (noradrenalin).

Adrenalin (epinephrine) Hormone secreted under stress. Contributes to the fight-or-flight response by increasing heart rate, shunting blood to muscles, increasing blood glucose levels, and other functions.

Adrenocorticotropic hormone (ACTH) Polypeptide hormone produced by the anterior pituitary. Stimulates the cells of the adrenal cortex, causing them to synthesize and release their hormones, especially glucocorticoids.

Aerobic exercise Exercise, such as swimming, that does not deplete muscle oxygen. Excellent for strengthening the heart and for losing weight.

Age-structure diagram (population histogram) Graphical representation of the number or percentage of males and females in various age groups in a population.

Agglutination Clumping of antigens that occurs when antibodies bind to several antigens.

Aging Inevitable and progressive deterioration of the body's function, especially its homeostatic mechanisms.

AIDS Acquired immune deficiency syndrome. Fatal disease caused by the HIV virus, which attacks T helper cells, greatly reducing the body's ability to fight infection.

Albinism Genetic disease resulting in a lack of pigment in the eyes or the eyes, skin, and hair. An autosomal recessive trait.

Aldosterone Steroid released by the adrenal cortex in response to a decrease in blood pressure, blood volume, and osmotic concentration. Acts principally on the kidney.

Alga (algae, plural) Heterogeneous group of aquatic plants consisting of three major groups: green, red, and brown. Important producer essential to aquatic food chains.

Alleles Alternative form of a gene.

Allergen Antigen that stimulates an allergic response.

Allergy Extreme overreaction to some antigens, such as pollen or foods. Characterized by sneezing, mucus production, and itchy eyes.

Allosteric inhibition Enzyme regulation in which an inhibitor molecule binds to an enzyme at a site away from the active site, changing the shape or charge of the active site so that it can no longer bind substrate molecules.

Allosteric site Region of an enzyme where products of metabolic pathways bind, changing the shape of the active site. In some enzymes this prevents substrates from binding to the active site; in others, it allows them to bind. Thus, allosteric sites can either turn on or turn off enzymes.

Alpine tundra A relatively cold, treeless region similar to the Arctic tundra but found on the tops of mountains.

Alternation of generations Phenomenon exhibited by all land plants and many algae where the plant alternately exists as a haploid generation (produces gametophytes) and as a diploid generation that produces spores (sporophytes).

Altitudinal biome Distinct region on a mountain, characterized by an assemblage of organisms resulting from the climatic conditions at an altitutde.

Alveoli Tiny, thin-walled sacs in the lung where oxygen and carbon dioxide are exchanged between the blood and the air.

Ameboid motion Cellular locomotion common in single-celled organisms and some cells in the human body. The cells send out slender cytoplasmic projections that attach to the substrate "ahead" of the cell. The cytoplasm flows into the projections, or pseudopodia, advancing the organism.

Amenalism Relationship between two organisms in which one organism is negatively impacted while the other is unaffected.

Amniocentesis Procedure whereby physicians extract cells and fluid from the amnion surrounding the fetus. The cells are examined for genetic defects, and the fluid is studied biochemically.

Amnion Layer of cells that separates from the inner cell mass of the embryo and eventually forms a complete sac around the fetus.

Amniotic fluid Liquid in the amniotic cavity surrounding the embryo and fetus during development. Helps protect the fetus.

Amoeboids Group of protozoans that move by way of amoeboid motion.

Amphibians Transitional class of vertebrates that spend part of their lives on land and part in the water. Sexual reproduction generally occurs through fertilization of fish-like eggs in the external environment.

Ampulla Enlarged area of each semicircular canal that houses receptor cells for movement.

Amylase Enzyme in saliva that helps break down starch molecules.

Anabolic reaction A reaction in which substances are formed; for example, the synthesis of glucose during photosynthesis is an anabolic reaction.

Anabolic steroids Synthetic androgen hormones that promote muscle development.

Anaerobe An organism that does not require oxygen to survive. Some anaerobes are actually killed by oxygen.

Analogous structures Anatomical structures that function similarly but differ in structure; for example, the wing of a bird and the wing of an insect.

Anaphase Phase of mitosis during which the chromatids of each chromosome begin to uncouple and are pulled in opposite directions with the aid of the mitotic apparatus.

Androgen-binding protein Cytoplasmic receptor protein that binds to and concentrates testosterone within the Sertoli cell. Production of ABP is stimulated by FSH.

Androgens Male sex steroids such as testosterone produced principally by the testes.

Anemia Condition characterized by an insufficient number of red blood cells in the blood or insufficient hemoglobin. Often caused by insufficient iron intake.

Aneuploidy Describes a genetic condition in which there is an abnormal number of chromosomes.

Aneurysm Ballooning of the arterial wall caused by a degeneration of the tunica media.

Angiosperms Recently evolved and widely distributed land plants that produce seeds contained within specialized structures known as fruits. Commonly called flowering plants.

Anoxia Lack of oxygen.

Antagonistic Refers to hormones or muscles that exert opposite effects.

Anterior pituitary Major portion of the pituitary gland, which is controlled by hypothalamic hormones. Produces seven protein and polypeptide hormones.

Anthropoids Monkeys, the great apes, and humans.

Antibodies Proteins produced by immune system cells that destroy or inactivate antigens, including pollen, bacteria, yeast, and viruses.

Anticodon loop Part of the transfer RNA molecule that bears three bases that bind to the three bases of the codon on messenger RNA.

Anticodon Sequence of three bases found on the transfer RNA molecule. Aligns with the codon on messenger RNA and helps control the sequence of amino acids inserted into the growing protein.

Antidiuretic hormone (ADH) Hormone released by the posterior pituitary. Increases the permeability of the distal convoluted tubule and collecting tubules, increasing water reabsorption.

Antigens Any substance that is detected

as foreign by an organism and elicits an immune response. Most antigens are proteins and large molecular weight carbohydrates.

Anvil (incus) One of three bones of the middle ear that helps transmit sound waves to the receptor for sound in the inner ear.

Aorta Largest artery in the body; carries the oxygenated blood away from the heart and delivers it to the rest of the body through many branches.

Appendicular skeleton Bones of the arms, legs, shoulders, and pelvis. Contrast with axial skeleton.

Aquatic life zones Ecologically distinct regions in fresh water and salt water.

Aqueous humor Liquid in the anterior and posterior chambers of the eye.

Aquifer Porous underground zone containing water.

Arachnids Class of arthropods that includes mites, ticks, scorpions, and spiders.

Arctic tundra Massive biome north of taiga characterized by low precipitation and cold temperatures.

Arteries Vessels that transport blood away from the heart.

Arteriole Smallest of all arteries; usually drains into capillaries.

Arthropods Phylum of invertebrate animals with jointed appendages and chitinous exoskeletons such as insects, arachnids, and crustaceans. Many arthropods inhabit freshwater and marine ecosystems, but the vast majority live on land.

Arthroscope Device used to examine internal joint injuries.

Asexual reproduction Reproductive strategy common in single-celled organisms, such as the amoeba. Reproduction occurs by cell division.

Association cortex Area of the brain where integration occurs.

Association neurons Nerve cells that receive input from many sensory neurons and help process them, ultimately carrying impulses to nearby multipolar neurons.

Aster Array of microtubules found in the cell in association with the spindle fibers during cell division.

Asthma Respiratory disease resulting from an allergic response. Allergens cause histamine to be released in the lungs. Histamine causes the air-carrying ducts (bronchioles) to constrict, cutting down airflow and making breathing difficult.

Astigmatism Unequal curvature of the cornea (sometimes the lens) that distorts vision.

Atom Smallest particles of matter that can be achieved by ordinary chemical means, consisting of protons, neutrons, and electrons.

Atomic mass units Unit used to measure atomic weight. One atomic mass unit is 1/12 the weight of a carbon atom.

Atomic weight Average mass of the atoms of a given element, measured in atomic mass units.

Atrioventricular bundle Tract of modified cardiac muscle fibers that conduct the pacemaker's impulse into the ventricular muscle tissue.

Atrioventricular node (AV node) Knot of tissue located in the right ventricle. Picks up the electrical signal arriving from the atria and transmits it down the atrioventricular bundle.

Atrioventricular valves Valves between the atria and ventricles.

Auditory (eustachian) tube Collapsible tube that joins the nasopharynx and middle ear cavities and helps equalize pressure in the middle ear.

Auricle (or pinna) Skin-covered cartilage portion of the outer ear.

Autocrines Chemical substances produced by cells, which affect the function of the cells producing them.

Autoimmune reaction Immune response directed at one's own cells.

Autonomic nervous system That part of the nervous system not under voluntary control.

Autosomal dominant trait Trait that is carried on the autosomes and is expressed in heterozygotes and homozygote dominants.

Autosomal recessive trait Trait that is carried on the autosomes and is expressed only when both recessive genes are present.

Autosomes All human chromosomes except the sex chromosomes.

Autotrophs Organisms such as plants that, unlike animals, are able to synthesize their own food.

Auxins Group of plant hormones responsible for cell enlargement. Plays a key role in phototropism. Together with cytokinin, auxin regulates the transformation of undifferentiated tissues into stems and roots.

Axial skeleton The skull, vertebral column, and rib cage. Contrast with appendicular skeleton.

Axon Long, unbranched process attached to the nerve cell body of a neuron. Transports bioelectric impulses away from the cell body.

Basal body Organelle located at the base of the cilium and flagellum. Consists of nine sets of microtubules arranged in a circle. Each "set" contains three microtubules.

Basilar membrane Membrane that supports the organ of Corti in the cochlea.

Batesian mimicry Strategy in which one species evolves a warning coloration similar to a species with active defense mechanisms such that the imitating species benefits from the outward similarity.

Bathyal zone Region of semidarkness in salt waters where photosynthesis does not occur; located below the euphotic zone but above the abyssal zone.

Behavior An individual's response to environmental stimuli such as other individuals of the same species, members of other species, or some aspect of the physical environment.

Benign tumor Abnormal cellular proliferation. Unlike a malignant tumor, the cells in a benign tumor stop growing after a while, and the tumor remains localized.

Benthic zone Bottom layer of a lake; supports life forms that can survive in water with very low oxygen levels.

Beta cells Insulin-producing cells of the islets of Langerhans in the pancreas.

Bile Fluid produced by the liver and stored and concentrated in the gallbladder.

Bile salts Steroids produced by the liver, stored in the gallbladder, and released into the small intestine where they emulsify fats, a step necessary for enzyme digestion.

Binary fission Bacterial cellular division.

Bioelectric impulse Nerve impulse resulting from the influx of sodium ions along the plasma membrane of a neuron.

Biomass The dry weight of living material in an ecosystem.

Biomass pyramid Diagram of the amount

of biomass at each trophic level in an ecosystem or, more commonly, a food chain.

Biome Terrestrial region characterized by a distinct climate and a characteristic plant and animal life.

Biorhythms (biological cycles) Naturally fluctuating physiological process.

Biosphere Region on Earth that supports life. Exists at the junction of the atmosphere, lithosphere, and hydrosphere.

Biotic factor Biological components of ecosystems.

Bipedal Refers to the ability to walk on two legs.

Birds Class of vertebrates characterized by feathers and endothermy (warmbloodedness). Most, but not all, of these species retain the ability to fly.

Birth control Any method or device that prevents conception and birth.

Birth control pill Pill generally taken to inhibit ovulation. The most commonly used pills contain both synthetic estrogen and progesterone.

Birth defects Physical or physiological defects in newborns. Caused by a variety of biological, chemical, and physical factors.

Blastocyst Hollow sphere of cells formed from the morula. Consists of the inner cell mass and the trophoblast.

Blood Specialized form of connective tissue. Consists of white blood cells, platelets, red blood cells, and plasma.

Blood clot Mass of fibrin containing platelets, red blood cells, and other cells. Forms in walls of damaged blood vessels, halting the efflux of blood.

B-lymphocytes Type of lymphocyte that transforms into a plasma cell when exposed to antigens.

Bonding Process in which an infant establishes an intimate relationship with one or both of its parents or caretakers. Essential for emotional health.

Bone (organ) Structure comprised of bone tissue. Provides internal support, protects organs, and helps maintain blood calcium levels.

Bone (tissue) Tissue consisting of a calcified extracellular material with numerous cells (osteocytes) embedded in it.

Bony fishes Class of vertebrates with more elaborate bone development than jawless fishes. Contains more species than all other vertebrates combined.

Bowman's capsule Cup-shaped end of the nephron that participates in glomerular filtration.

Bowman's space Cavity between the inner and outer layer of Bowman's capsule.

Brain stem Part of the brain that consists of the medulla and pons. Houses structures, such as the breathing control center and reticular activating systems, that control many basic body functions.

Braxton-Hicks contractions Contractions that begin a month or two before childbirth. Also known as false labor.

Breathing center Aggregation of nerve cells in the brain stem that controls breathing.

Breech birth Delivery of a baby feet first.

Bronchi Ducts that convey air from the trachea to the bronchioles and alveoli.

Bronchioles Smallest ducts in the lungs. Their walls are largely made of smooth muscle that contracts and relaxes, regulating the flow of air into the lung.

Bryophytes Small plants such as mosses and figworts that lack vascular tissue, which severely restricts water transport to distant tissues. Most bryophytes require moist habitats to live and reproduce.

Bulimia Eating disorder characterized by recurrent binge eating followed by vomiting.

Bundle sheath cells Cells surrounding veins in leaves. They participate in the C_4 cycle by breaking down malic acid, producing carbon dioxide that can be used in the C_3 cycle.

C_3 cycle The cyclic series of reactions whereby carbon dioxide is fixed into carbohydrates, using energy from the light-dependent reactions. Also called the Calvin-Benson cycle.

C_4 cycle The series of reactions in certain plants that fixes carbon dioxide into organic acids for later use in the C_3 cycle of photosynthesis.

Calcitonin (thyrocalcitonin) Polypeptide hormone produced by the thyroid gland that inhibits osteoclasts and stimulates osteoblasts to produce bone, thus lowering blood calcium levels.

Calvin-Benson cycle See C_3 cycle.

Canal of Schlemm Network of channels located at the junction of the sclera and cornea that drains aqueous humor from the anterior chamber of the eye.

Canaliculi Tiny canals in compact bone that provide a route for nutrients and wastes to flow to and from the osteocytes.

Cancer Disease characterized by the uncontrollable replication of cells.

Capacitation Process in which the outer protective coat of the sperm is dissolved away in the female reproductive tract. Makes the membrane fragile and disruptible, allowing the acrosome to break down and release its enzymes.

Capillaries Tiny vessels in body tissues whose walls are composed of a flattened layer of cells that allow water and other molecules to flow freely into and out of the tissue fluid.

Capillary bed Branching network of capillaries supplied by arterioles and drained by venules.

Capsid Protein coat of a virus.

Capsomere Globular proteins that make up the capsid of viruses.

Carbohydrate Organic compound consisting of carbon, hydrogen, and oxygen. A structural component of plant cells, it is used principally as a source of energy in animal cells.

Carbon fixation The initial steps in the C_3 cycle, in which carbon dioxide reacts with ribulose bisphosphate to form two stable molecules of 3-phosphoglyceric acid, or PGA.

Carbonic anhydrase Enzyme found in red blood cells that catalyzes the conversion of carbon dioxide to carbonic acid.

Carcinogens Cancer-causing agents.

Cardiac muscle Type of muscle found in the walls of the heart; it is striated and involuntary.

Carnivores Meat eaters; organisms that feed on grazers and other animals.

Carotenoid A family of pigments, usually yellow, orange, or red, found in chloroplasts of plants and serving as accessory light-gathering molecules in thylakoid photosystems.

Carrier proteins Class of proteins that transport smaller molecules and ions across the plasma membrane of the cell. Are involved in facilitated diffusion.

Carriers Individuals who carry a gene for a particular trait that can be passed on to their children, but who do not express the trait.

Carrying capacity The number of organisms an ecosystem can support on a sustainable basis.

Cartilage Type of specialized connective tissue. Found in joints on the articular surfaces of bones and other locations.

Catabolic reaction A reaction in which molecules are broken down; for example, the breakdown of glucose is a catabolic reaction.

Catalyst Class of compounds that speed up chemical reactions. Although they take an active role in the process, they are left unchanged by the reaction. Thus, they can be used over and over again. Catalysts lower the activation energy of a reaction. *See also* Enzyme.

Cataracts Disease of the eye resulting in cloudy spots on the lens (and sometimes the cornea) that cause cloudy vision.

Cell body Part of the nerve cell that contains the nucleus and other cellular organelles; the center of chemical synthesis.

Cell culture Glass bottle or shallow dish containing nutrient medium and designed to permit cells to grow in the laboratory under controlled conditions.

Cell cycle Repeating series of events in the lives of many cells. Consists of two principal parts: interphase and cellular division.

Cell division Process by which the nucleus and the cytoplasm of a cell are split, creating two daughter cells. Consists of mitosis and cytokinesis.

Cellular respiration The complete breakdown of glucose in the cell, producing carbon dioxide and water. Comprised of four separate but interconnected parts: glycolysis, the intermediate reaction, the citric acid cycle, and the electron transport system.

Central nervous system The brain and spinal chord.

Centriole Organelle consisting of a ring of microtubules, arranged in nine sets of three. Structurally identical to basal bodies, but associated with the spindle apparatus. Gives rise to the basal body in ciliated cells.

Centromere Region on each chromatid that joins with the centromere of its sister chromatid.

Cerebellum Structure of the brain that lies blow the cerebral cortex. It has many important functions including synergy.

Cerebral cortex Outer layer of each cerebral hemisphere, consisting of many multipolar neurons and nerve cell fibers.

Cerebral hemisphere Convoluted mass of nervous tissue located above the deeper structures, such as the hypothalamus and limbic system. Home of consciousness, memory, and sensory perception; originates much conscious motor activity.

Cervical cap Birth control device consisting of a small cup that fits over the tip of the cervix.

Cervix Lowermost portion of the uterus; it protrudes into the vagina.

Cesarean section Delivery of a baby via an incision through the abdominal and uterine walls.

Chemical equilibrium The point at which the "forward" reaction from reactants to products proceeds at the same rate as the "backward" reaction from products to reactants, so that there is no net change in chemical composition.

Chemical evolution Formation of organic molecules from inorganic molecules early in the history of the Earth.

Chemiosmosis Production of ATP in the chloroplast's and the *mitochondrion's* electron transport systems. H^+ flowing through pores in ATPase molecules drives the endergonic synthesis of ATP.

Chemosynthetic organisms Cells that lack chlorophyll and acquire electrons from inorganic molecules.

Childhood Period of human life that lasts from infancy to puberty.

Chitin Complex, waterproof, chemical-resistant molecule made of both protein and carbohydrate. Forms the rigid cell wall of fungi and the exoskeleton of insects and other arthropods.

Chlamydia Bacterium that causes nonspecific urethritis, a type of sexually transmitted disease.

Chlorofluorocarbons (CFCs) Chemical substances often used as spray can propellants (outside the United States and several other countries) and refrigerants that drift to the upper stratosphere and dissociate. Chlorine released by CFCs reacts with ozone, thus eroding the ozone layer.

Chlorophyll A green pigment that acts as the primary light-trapping molecule for photosynthesis.

Cholecystokinin (CCK) Hormone produced by cells of the duodenum when chyme is present. Causes the gallbladder to contract, releasing bile.

Chordae tendineae Tendinous chords that anchor the atrioventricular valves to the inner walls of the ventricles.

Chordates Phylum that has three features in common: a hollow dorsal nerve cord, a supporting, cartilaginous notochord, and gill slits in the neck region.

Chorionic villus biopsy Medical procedure to detect genetic defects; involves removing a small portion of the villi and then examining it for chromosomal abnormalities.

Choroid Middle layer of the eye that absorbs stray light and supplies nutrients to the eye.

Chromatid Strand of the chromosome consisting of DNA and protein.

Chromatin Long, threadlike fibers containing DNA and protein in the nucleus.

Chromosome-to-pole fibers Microtubules of the spindle that extend from the centriole to the chromosome where they attach. They play a crucial role in separating the double-stranded chromosomes during mitosis.

Chronic bronchitis Persistent irritation of the bronchi, which causes a mucus buildup, coughing, and difficulty breathing.

Chyme Liquified food in the stomach.

Ciliary body Portion of the middle layer of the eye that is located near the lens. Contains smooth muscle that constricts, thus helping control the shape of the lens and permitting the eye to focus.

Ciliates The most structurally complex of all animal-like protists. They get their name from the many cilia projecting from their surface.

Circadian rhythm Biorhythm that occurs on a daily cycle.

Circulatory system Organ system consisting of the heart, blood vessels, and blood.

Circumcision Operation to remove the foreskin of the penis. Generally performed on newborns.

Cisterna Channels of the endoplasmic reticulum and Golgi.

Citric acid Six-carbon compound produced in the first reaction of the citric acid or Krebs cycle. Formed when oxaloacetate reacts with acetyl Coenzyme A.

Citric acid cycle A cyclic series of reactions in which the pyruvates produced by glycolysis are broken down to CO_2, accompanied by the formation of *ATP* and electron carriers. Occurs in the matrix of mitochondria.

Class Subgroup of a phylum.

Clitoris Small knot of tissue located where the labia minora meet. Consists of erectile tissue.

Cloning Technique of genetic engineering whereby many copies of a gene are produced.

Closed system System that receives no materials from the outside.

Cnidarians Phylum characterized by distinct tissues and primitive nervous systems, coordination, and symmetry. Includes jellyfish, corals, and sea anemones. Reproduction is asexual.

Coacervates Microscopic globules that selectively incorporate molecules from their environment. Coacervates existing on Earth over 3 billion years ago may have given rise to the earliest cells.

Cochlea Sensory organ of the inner ear that houses the receptor for hearing.

Codominant Refers to two equally expressed alleles.

Codon Three adjacent bases in the messenger RNA that code for a single amino acid.

Coelom Body cavity (a space between the body wall and internal organs) lined by mesoderm.

Coevolution Process in which two or more species act as selective forces on each other, resulting in anatomical, behavioral, and functional changes in each other.

Collecting tubules Tubules in the kidney into which nephrons drain. They converge and drain into the renal pelvis.

Color blindness Condition that occurs in individuals who have a deficiency of certain cones. The most common form involves difficulty in distinguishing between red and green.

Colostrum Protein-rich product of the breast, produced for two to three days immediately after delivery.

Commensal Relationship in which one species benefits while the other is unaffected.

Community All of the plants, animals, and microorganisms in an ecosystem.

Compact bone Dense bony tissue in the outer portion of all bones.

Competition Struggle by two or more individuals for the same limited resource.

Complement Group of blood proteins that circulate in the blood in an inactive state until the body is invaded by bacteria; then they help destroy the bacteria.

Complementary base pairing Unalterable coupling of purine adenine to pyrimidine thymine, and purine guanine to pyrimidine cytosine. Responsible for the accurate transmission of genetic information from parent to offspring.

Conditioning Type of learning in which an animal learns to make a connection (association) between a new stimulus and a familiar one—that is, a previous stimulus-response relationship is transferred to a novel stimulus.

Condom Birth control device, consisting of a thin latex rubber sheath or other material that is rolled onto the erect penis. Prevents sperm from entering the vagina and helps prevent the spread of sexually transmitted diseases.

Conduction deafness Loss of hearing that occurs when the conduction of sound waves to the inner ear is impaired. May be caused by ruptured eardrum or damage to ossicles.

Cone Type of photoreceptor that operates in bright light; is responsible for color vision.

Conjugation Process in which bacteria exchange genetic material; considered a primitive form of sexual reproduction.

Connective tissue One of the primary tissues. It contains cells and varying amounts of extracellular material and holds cells together, forming tissues and organs.

Connective tissue proper Name referring to loose and dense connective tissue; supports and joins various body structures.

Consumers Organisms that eat plants and algae (producers) and other consumers.

Contact inhibition Cessation of growth that results when two or more cells contact each other. A feature of normal cells but absent in cancer cells.

Continental shelf Gradually sloping ocean bottom next to continents.

Contraceptive Any measure that helps prevent fertilization and pregnancy.

Contractile vacuole Vacuole in amoeboid protists that collects excess water from the cytoplasm and, when full, contracts, voiding the water through a temporary opening in the plasma membrane.

Convergence Inward turning of the eyes to focus on a nearby object.

Coral reef Biologically rich life zones, found in relatively warm and shallow waters in the tropics or nearby regions consisting of calcium carbonate or limestone produced by calcarous red and green algae and by colonies of organisms called stony corals.

Corepressor Molecule that binds to a repressor protein, allowing it to bind to the operator site, which, in turn, shuts down the structural genes by blocking RNA polymerase.

Cornea Clear part of the wall of the eye continuous with the sclera; allows light into the interior of the eye.

Cork cambium Secondary meristem interposed between the cork and the phloem in the bark that gives rise to new cork, part of the bark.

Coronary bypass surgery Surgical technique used to reestablish blood flow to the heart muscle by grafting a vein to shunt blood around a clogged coronary artery.

Corpus luteum (CL) Structure formed from the ovulated follicle in the ovary; produces estrogen and progesterone.

Cortical granules Secretory vesicles lying beneath the plasma membrane of the oocyte that are released when a sperm cell contacts the oocyte membrane. They block additional sperm from fertilizing the ovum.

Cortisol Glucocorticoid hormone that increases blood glucose by stimulating gluconeogenesis. Also stimulates protein breakdown in muscle and bone.

Cotyledon Part of the embryo that is commonly, but not exclusively, used for the storage of food needed in the germination of an embryo.

Coupled reactions A pair of reactions, one exergonic and one endergonic, that are linked so that the energy produced by the exergonic reaction provides the energy to drive the endergonic reaction.

Cowper's gland Smallest of the sex accessory glands; empties into the urethra.

Cranial nerves Nerves arising from the brain and brain stem.

Creatine phosphate High-energy molecule in muscle.

Cristae Folds formed by the inner membrane of the mitochondrion.

Cro-Magnons Earliest known members of *Homo sapiens sapiens.*

Cross bridges Part of the myosin molecule that attaches to and pulls actin molecules inward causing the sarcomere to shorten.

Crossing over Exchange of chromatin by homologous chromosomes during prophase I. Results in considerably more genetic variation in gametes and offspring.

Crustaceans Class of arthropods that comprise most of the zooplankton as well as crabs, shrimp, and lobster.

Crystallin Protein inside the lens that may denature, causing cataracts.

Culture The ideas, customs, skills, and arts of a given people in a given time that can change or evolve over time.

Cushing's disease Disease that results from pharmacologic doses of cortisone usually administered for rheumatoid arthritis or allergies.

Cyanobacteria The most abundant of the photosynthetic bacteria; once called blue-green algae.

Cyclic AMP Nucleotide derived from ATP. Its synthesis is stimulated when protein and polypeptide hormones bind to the plasma membrane of cells. In the cytoplasm, it activates protein kinase, which, in turn, activates other enzymes.

Cyclic photophosphorylation Production of ATP in plants and photosynthetic protists when $NADP^+$ levels are low; instead of being passed to $NADP^+$, electrons are transferred to the electron transport system, producing ATP. Chief means of ATP production in photosynthetic bacteria.

Cyclosporine Drug used to suppress graft rejection.

Cystic fibrosis Autosomal recessive disease that leads to problems in sweat glands, mucus glands, and the pancreas. Pancreas may become blocked, thus reducing the flow of digestive enzymes to the small intestine. Mucus buildup in the lungs makes breathing difficult.

Cytokinesis Cytoplasmic division brought about by the contraction of a microfilamentous network lying beneath the plasma membrane at the midline. Usually begins when the cell is in late anaphase or early telophase.

Cytoplasm Material occupying the cytoplasmic compartment of a cell. Consists of a semifluid substance, the cytosol, containing many dissolved substances, and formed elements, the organelles.

Cytoskeleton A network of protein tubules in the cytoplasmic compartment of a cell. Attaches to many organelles and enzyme molecules and thus helps organize cellular activities, increasing efficiency.

Cytotoxic cells Type of T cell (T-lymphocyte) that attacks and kills virus-infected cells, parasites, fungi, and tumor cells.

Daughter cells Cells produced during cell division.

Decibel Unit used to measure the intensity of sound.

Deciduous Referring to trees that shed their leaves during the fall.

Decomposer food chain Series of organisms that feed on organic wastes and the dead remains of other organisms.

Defibrillation Procedure to stop fibrillation (erratic electrical activity) of the heart.

Deletion Loss of a piece of a chromosome.

Demographer Scientist who studies populations.

Dendrite Short, highly branched fiber that carries impulses to the nerve cell body.

Dense connective tissue Type of connective tissue that consists primarily of densely packed fibers, such as those found in ligaments and tendons.

Deoxyribonuclease Pancreatic enzyme that breaks RNA and DNA into shorter chains.

Depth perception Ability to judge the relative position of objects in our visual field.

Dermis Layer of dense irregular connective tissue that binds the epidermis to underlying structures.

Desert Biome characterized by low rainfall and a hot climate. Contains organisms well adapted to these conditions.

Detrivores Organisms that feed on animal waste or the remains of plants and animals.

Diabetes insipidus Condition caused by lack of ADH. Main symptoms are polydipsia (excessive drinking) and polyuria (excessive urination).

Diabetes mellitus Insulin disorder either resulting from insufficient insulin production or decreased sensitivity of target cells to insulin. Results in elevated blood glucose levels unless treated.

Dialysis Procedure used to treat patients whose kidneys have failed. Blood is removed from the body and pumped through an artificial filter that removes impurities.

Diaphragm (birth control) Birth control devise consisting of a rubber cup that fits over the end of the cervix. Used in conjunction with spermicidal jelly or cream.

Diaphragm (muscle) Dome-shaped muscle that separates the abdominal and thoracic cavities.

Diaphysis Shaft of the long bones. Consists of an outer layer of compact bone and an inner marrow cavity.

Diastolic pressure The pressure at the moment the heart relaxes. The lower of the two blood pressure readings.

Diatom Plantlike protist found in fresh and salt water. It has a unique cell wall composed of silica, a glassy material that is often fashioned into exquisite designs.

Dicots Angiosperms that contain two cotyledons.

Differentiation Structural and functional divergence from the common cell line. Occurs during embryonic development.

Dihybrid cross Procedure where one plant is bred with another to study two traits.

Dinoflagellate Photosynthetic protists with two flagella, the beating of which cause them to spin like tops.

Diplopia Double vision. May occur when the eyes fail to move synchronously.

Distal convoluted tubule Section of the nephron that connects the loop of Henle to the collecting tubule. Site of tubular reabsorption.

Divergent evolution Process in which organisms evolve in different directions due to exposure to different environmental influences.

Diverticulitis Expansion of the large intestine due to obstruction.

DNA polymerase Enzyme that helps align the nucleotides and join the phosphates and sugar molecules in a newly forming DNA strand.

Dominant Adjective used in genetics to refer to an allele that is always expressed in heterozygotes. Designated by a capital letter.

Dorsal root (of a spinal nerve) Inlet for sensory nerve fibers to the spinal cord.

Double helix Describes the helical structure formed by two polynucleotide chains making up the DNA molecule.

Doubling time Time it takes a population to double.

Down syndrome Genetic disorder caused by an additional chromosome 21 that results in distinctive facial characteristics and mental retardation. Also known as Trisomy 21.

Dryopithecus Genus of apelike creatures that is thought to have given rise to the gibbons, gorillas, orangutans, and chimpanzees.

Ductus arteriosus Shunt that lies between the pulmonary artery and the aorta, helping divert blood from the lungs.

Ductus venosus Shunt that connects directly from the umbilical vein to the inferior vena cava.

Duodenum First portion of the small intestine; site where most food digestion and absorption takes places.

Dust cell Cell found in and around the alveoli; phagocytizes particulate matter that has entered the lung.

E. coli Common bacterium that lives in the large intestine of humans and digests leftover glucose and other materials from food. Used in much genetic research.

Echinoderms Phylum of radially symmetric marine animals with a spiny skin, a complete digestive system, and an internal skeleton composed of shell-like plates.

Ecological niche An organism's habitat and all of the relationships that exist between that organism and its environment.

Ecological system (ecosystem) System consisting of organisms and their environment and all of the interactions that exist between these components.

Ecology Study of living organisms and the web of relationships that binds them together in the economy of nature. The study of ecosystems.

Ecosystem balance Dynamic equilibrium in ecosystems. Maintained by the interplay of growth and reduction factors.

Ectoderm One of the three types of cells that emerges in human embryonic development. Gives rise to the skin and associated structures, including the eyes.

Ectoparasites Parasites that live on the surface of the host.

Edema Swelling resulting from the buildup of fluid in the tissues.

Effector General term for any organ or gland that is controlled by the nervous system.

Ejaculation Ejection of semen from the male reproductive tract.

Elastic arteries Arteries that contain numerous elastic fibers interspersed among the smooth muscle cells of the tunica media.

Elastic cartilage Type of cartilage containing many elastic fibers found in regions where support and flexibility are required.

Electron Highly energetic particle carrying a negative charge that orbits the nucleus of an atom.

Electron carrier A molecule that can reversibly gain and lose electrons. Electron carriers generally accept high-energy electrons produced during an exergonic reaction and donate the electrons to acceptor molecules that use the energy to drive endergonic reactions.

Electron cloud A region surrounding the nucleus of an atom where electrons orbit.

Electron transport system Series of protein molecules in the inner membrane of the mitochondrion that pass electrons from the citric acid cycle from one to another, eventually donating them to oxygen. During their journey along this chain of proteins, the electrons lose energy, which is used to make ATP. *See also* Chemiosmosis.

Elements Purest form of matter; substances that cannot be separated into different substances by chemical means.

Emphysema Progressive, debilitating disease that destroys tiny air sacs in the lung (alveoli). The fastest growing cause of death in the United States.

Endergonic reaction A chemical reaction that requires an input of energy to proceed.

Endocrine glands Glands of internal secretion. They produce hormones that are secreted into the bloodstream.

Endocrine system Numerous, small, hormone-producing glands scattered throughout the body.

Endocytosis Process by which cells engulf solid particles, bacteria, viruses, and even other cells.

Endoderm One of the three types of cells that emerges during embryonic development. Gives rise to the intestinal tract and associated glands.

Endolymph Fluid inside the semicircular canals that deflects the cupula, signaling rotational movement of the head and body.

Endometrium Uterine endothelium or lining.

Endoparasites Parasites that live within their host.

Endoplasmic reticulum Branched network of channels found throughout the cytoplasm of many cells. Formed from flattened sheets of membrane derived from the nuclear membrane.

Endosymbiotic evolution Theory that accounts for the development of the first eukaryotes. Says that free-living bacteria-like organisms were engulfed by other cells and became internal symbionts. Internal symbionts later became the organelles of eukaryotes.

Endothelium Single-celled lining of blood vessels.

End-product inhibition The inhibition of an enzyme by a product of the chemical reaction it catalyzes or by the product of a series of chemical reactions of which the enzyme is a part; may result from binding of the end product to the allosteric site or active site of the enzyme.

Energy The capacity to do work.

Energy carrier A molecule that stores energy in "high-energy" chemical bonds and releases the energy again to drive coupled endergonic reactions. ATP is the most common energy carrier in cells; *NAD* and *FAD* are others.

Energy pyramid Diagram of the amount of energy at various trophic levels in a food chain or ecosystem.

Enhancer Segment of DNA that increases the activity of nearby genes several hundred times.

Envelope Protective membrane of some viruses; lies outside the capsid.

Enzyme A protein catalyst that speeds up the rate of specific biological reactions.

Epicotyl Part of the embryo that gives rise to the shoot system.

Epidermis Outermost layer of the skin that protects underlying tissues from drying out and from bacteria and viruses.

Epididymal duct Duct within the epididymis; site where sperm are stored until ejaculation.

Epididymis Storage site of sperm. Located on the testis, it consists of a long, tortuous duct.

Epiglottis Flap of tissue that closes off the trachea during swallowing.

Epilimnion The warm-water layer of a lake located along the surface.

Epiphyseal plate Band of cartilage cells between the shaft of the bone and the epiphysis. Allows for bone growth.

Epiphysis Expanded end of the long bones.

Epiphytes Plants that obtain their moisture and food from the air and rain rather than from the plants on which they grow.

Episiotomy Surgical incision that runs from the vaginal opening toward the rectum. Enlarges the vaginal opening, easing childbirth.

Epithelium One of the primary tissues. Forms linings and external coatings of organs.

Erectile tissue Spongy tissue of the penis that fills with blood during sexual excitement, making the penis turgid.

Erythropoietin Hormone produced by the kidney when oxygen levels decline. Stimulates red blood cell production in the bone marrow.

Esophagus Muscular tube that transports food to the stomach.

Essential amino acid One of nine amino acids that must be provided in the human diet.

Estuarine zone Estuary and coastal wetland. One of the richest coastal life zones.

Ethology Study of animal behavior. Encompasses both the comparative psychology approach of experimental studies of learned behavior and the classical approach of careful observation of instinctive behavior.

Euchromatin Metabolically active chromatin.

Euphotic zone The open waters of the ocean, extending to the depth at which sunlight no longer penetrates.

Euglenoids Protists with characteristics of both plants and animals. They get their name from the best-known of their kind, *Euglena*.

Evolution Process that leads to structural and functional changes in species, making them better able to survive in their environment; also leads to the formation of new species. Results from natural genetic variation and environmental conditions that select for organisms best suited to their environment.

Exhalation Expulsion of air from the lungs.

Exergonic reaction A chemical reaction that releases energy.

Exocrine gland Gland of external secretion; empties its contents into ducts.

Exocytosis Process by which cells release materials stored in secretory vesicles. The reverse of endocytosis.

Exon Expressed segment of DNA.

Experiment Test performed to prove or disprove a hypothesis.

Exponential growth Type of growth that occurs when a value grows by a fixed percentage and the increase is applied to the base amount.

Extension Movement of a body part (limbs, fingers, and toes) that opens a joint.

External auditory canal Channel that directs sound waves to the eardrum.

External genitalia External portion of the female reproductive system consisting of the clitoris, labia minor, and labia majora.

External sphincter of the bladder Voluntary muscular valve that controls urine release under conscious control. Formed by a flat band of muscle that forms the floor of the pelvic cavity.

Extrinsic eye muscles Six muscles located outside the eye that are responsible for eye movement.

Facilitated diffusion Process in which carrier proteins shuttle molecules across plasma membranes. The molecules move in response to concentration gradients.

Feces Semisolid material containing undigested food, bacteria, ions, and water; produced in the large intestine.

Feedback mechanism A mechanism in which the product of one process regulates the rate of the process, either turning it on or shutting it off.

Fenestrae Minute openings in capillary walls that help permit movement of molecules to and from the capillary.

Fermentation Process occurring in eukaryotic cells in the absence of oxygen, during which pyruvic acid is converted to lactic acid. Also occurs in prokaryotes.

Fertilization Union of sperm and ovum.

Fiber Any of the indigestible polysaccharides in fruits, vegetables, and grains.

Fibrillation Cardiac muscle spasms occurring during heart attacks due to a loss of synchronized electrical signals.

Fibrin Fibrous protein produced from fibrinogen, a soluble plasma protein. Helps form blood clots.

Fibrinogen Protein in plasma that forms fibrin.

Fibroblast Connective tissue cell, found in loose and dense connective tissues that produces collagen, elastic fibers, and a gelatinous extracellular material; responsible for repairing damage created by cuts or tears to connective tissue.

Fibrocartilage Type of cartilage whose extracellular matrix consists of numerous bundles of collagen fibers. Principally found in the intervertebral disks.

Fimbriae Fingerlike projections of the end of the oviduct that sweep the oocyte into the oviduct.

First law of thermodynamics Law stating that energy can be converted from one form to another but is never created or destroyed.

First polar body Cast-off nuclear material produced during the first meiotic division during oogenesis.

Fitness Measure of reproductive success of an organism and, therefore, the genetic influence an individual has on future generations.

Fixed action patterns (FAPs) Stereotyped behavior or response to specific stimuli.

Flagellum Long, whiplike extension of the plasma membrane of certain protozoans and sperm cells in humans. Used for motility.

Flatworms Phylum characterized by bilateral symmetry, cephalization, and three distinct tissue layers. Most flatworms are hermaphrodites (contain both male and female sex organs), but reproduction occurs through cross-fertilization.

Flavine adenine dinucleotide (FAD) A molecule that carries high-energy electrons from one chemical reaction to another; found in the citric acid cycle.

Flexion Movement of a limb, finger, or toe that involves closing a joint.

Follicle (ovary) Structure found in the ovary. Each follicle contains an oocyte and one or more layers of follicle cells that are derived from the loose connective tissue of the ovary surrounding the follicle.

Follicle (thyroid) Structure found in the thyroid gland. Consists of an outer layer of cuboidal cells surrounding thyroglobulin.

Follicle-stimulating hormone (FSH) Gonadotropic hormone that promotes gamete formation in both men and women.

Food chain Series of organisms in an ecosystem in which each organism feeds on the organism preceding it.

Food vacuole Membrane-bound vacuole in a cell containing material engulfed by the cell.

Food web All of the connected food chains in an ecosystem.

Foramen ovale Hole in the interatrial septum of the embryonic heart that diverts blood from the right atrium to the left atrium, reducing the flow of blood to the pulmonary arteries and lungs.

Foreskin Sheath of skin that covers the glans penis.

Fossil Remains or imprints of organisms that lived on Earth many years ago, usually embedded in rocks or sediment.

Fovea centralis Tiny spot in the center of the macula of the eye that contains only cones. Objects are focused onto the fovea for sharp vision.

Fruit Structure derived from the ovary of a flower. Some types of fruit contain seeds and some do not. Some fruits, called grains, fuse with their seeds and remain inseparable.

Fusion Joining of two atoms, which releases large amounts of energy.

Gallbladder Sac on the underside of the liver that stores and concentrates bile.

Gametoyphyte Haploid generation of a plant exhibiting alternation of generations. Produces gametes.

Gastrin Stomach hormone that stimulates HCl production and release by the gastric glands.

Gastroesophageal sphincter Ring of muscle located in the lower esophagus that opens when food arrives, allowing food to pass into the stomach, and then closes to keep food and stomach acid from percolating upward.

Gene Segment of the DNA that controls cell structure and function.

Gene flow Introduction of new genes into a population when new individuals join the population.

Gene pool All the genes of all of the members of a population or species.

Genera Plural of genus.

Genome Genes of an organism.

Genotype Genetic makeup of an organism.

Genus Subgroup of a family of organisms.

Geographic isolation Physical separation of a population by some barrier. Sometimes results in reproductive isolation and the formation of new species.

Geotropism Growth response of a plant or seed to gravity—that is, plant stems bend upward and plant roots bend downward.

Germinal epithelium Germ cells in the wall of the seminiferous tubule that give rise to sperm.

Gestation The period of pregnancy.

Glans penis Slightly enlarged tip of the penis.

Glaucoma Disease of the eye caused by pressure resulting from a buildup of aqueous humor in the anterior chamber.

Glomerular filtration Movement of materials out of the glomeruli into Bowman's capsule in the kidney.

Glomerulus Tuft of capillaries that make up part of the nephron; site of glomerular filtration.

Glucagon Hormone released by the pancreas that stimulates the breakdown of glycogen in liver and muscle and the release of glucose molecules, thus increasing blood levels of glucose.

Glucocorticoids Group of steroid hormones produced by the adrenal cortex that stimulate gluconeogenesis.

Gluconeogenesis Synthesis of glucose from fatty acids and amino acids. Takes place in the liver where amino acids and fatty acids are stored.

Glycogenolysis Breakdown of glycogen, releasing glucose.

Glycolysis Metabolic pathway in the cytoplasm of the cell, during which glucose is split in half, forming two molecules of pyruvic acid. The energy released during the reaction is used to generate two molecules of ATP.

Glycoproteins Proteins that have carbohydrate attached to them.

Goiter Condition in which the thyroid gland enlarges due to lack of dietary iodide.

Golgi complex Organelle consisting of a series of flattened membranes that form channels. It sorts and chemically modifies molecules and repackages its proteins into secretory vesicles.

Golgi tendon organs Special receptors found in tendons that respond to stretch. Also known as neurotendinous organs.

Gonadotropin General term for FSH and LH, which are produced by the anterior pituitary and target male and female gonads.

Gonadotropin-releasing hormone (GnRH) Hormone produced by the hypothalamus that controls the release of FSH (ICSH in males) and LH.

Gonorrhea Sexually transmitted disease caused by a bacterium.

Granum A stack of thylakoid disks found in chloroplasts.

Gray matter Gray, outermost region of the cerebral cortex.

Grazer Herbivorous organism.

Grazer food chain Food chain beginning with plants and grazers (herbivores).

Greenhouse gas Gas, such as carbon dioxide and chlorofluorocarbons, that traps heat escaping from the Earth and radiates it back to the surface.

Growth factor Any biotic or abiotic factor that causes a population to grow.

Growth hormone A protein hormone produced by the anterior pituitary that stimulates cellular growth in the body, causing cellular hypertrophy and hyperplasia. Its major targets are bone and muscle.

Growth rate (of a population) Determined by subtracting the death rate from the birth rate.

Gymnosperms Vascular land trees such as conifers that produce seeds in cones or conelike structures. Also includes cycads and other plants.

Habitat Place in which an organism lives.

Habituation Condition where sensory receptors stop generating impulses, even though a stimulus is still present.

Hammer (malleus) One of three bones of the middle ear. Abuts the tympanic membrane and helps transmit sound from the eardrum to the inner ear.

Helper cell Type of T-lymphocyte that stimulates the proliferation of T and B cells when antigen is present.

Heme group Subunit of the hemoglobin molecule. Consists of a porphyrin ring and a central iron ion to which oxygen binds.

Hemoglobin Protein molecules inside red blood cells; binds to oxygen.

Hemophilia Disease caused by a gene defect occurring on the Y chromosome. Results in absence of certain blood-clotting factors.

Herbicide Any chemical applied to crops to control weeds.

Herbivore Any organism that feeds directly on plants. Also known as a grazer.

Herpes One of the most common sexually transmitted diseases; caused by a virus.

Heterochromatin Inactive chromatin that is slightly coiled or compacted in the interphase nucleus.

Heterotrophic fermenter Evolutionarily probably one of the first cells. Absorbed glucose from the environment and broke it down by anaerobic glycolysis.

Heterotrophs Organisms such as animals that, unlike plants, are unable to synthesize their own food. Consume plants and other organisms.

Heterozygous Adjective describing a genetic condition in which an individual contains one dominant and one recessive gene in a gene pair.

High-density lipoproteins (HDLs) Complexes of lipid and protein that transport cholesterol to the liver for destruction.

Histamine Potent vasodilator released by certain cells in the body during allergic reactions.

Histone Globular protein thought to play a role in regulating the genes.

Homeostasis A condition of stability or equilibrium within any biological or social system. Achieved through a variety of automatic mechanisms that compensate for internal and external changes.

Hominid First humanlike creatures.

Hominoids Subgroup of anthropoids.

Homo sapiens neanderthalensis The Neanderthals. Subspecies of *Homo sapiens.*

Homo sapiens sapiens Species of modern humans that emerged about 400,000 years ago.

Homologous structures Structures thought to have arisen from a common origin.

Homozygous Adjective describing a genetic condition marked by the presence of two identical alleles for a given gene.

Hormone Chemical substance produced in one part of the body that travels to another where it elicits a response.

Human chorionic gonadotropin (HCG) Hormone produced by the embryo that stimulates the corpus luteum to produce estrogen.

Humoral immunity Immune reaction that protects the body primarily against viruses and bacteria in the body fluids via antibodies produced by plasma cells.

Hyperglycemia High blood glucose levels.

Hyperopia (farsightedness) Condition that occurs when the eyeball is too short or the lens is too weak, resulting in poor focus on nearby objects.

Hypertension High blood pressure.

Hypertonic Adjective describing a solution with a higher solute concentration than the cell's cytoplasm, causing the cell to shrivel.

Hyphae Microscopic filaments or strands of which most fungi are composed.

Hypocotyl Part of the embryo that gives rise to root system.

Hypolimnion The coldest water of a lake, which lies below the thermocline.

Hypothalamus Structure in the brain located beneath the thalamus. It consists of many aggregations of nerve cells and controls a variety of autonomic functions aimed at maintaining homeostasis.

Hypothesis Tentative and testable explanation for a phenomenon or observation.

Hypotonic Adjective describing a solution with a solute concentration lower than the cell's cytoplasm, resulting in a swelling of the cell.

I gene Gene that controls blood type through the synthesis of glycoproteins on the plasma membrane of the red blood cell.

Immune system Diffuse system consisting of trillions of cells that circulate in the blood and lymph and take up residence in the lymphoid organs, such as the spleen, thymus, lymph nodes, and tonsils, as well as other body tissues. Helps protect the body against foreign cells, such as bacteria and viruses, and protects against cancer cells.

Immunity Term referring to the resistance of the body to infectious disease.

Immunocompetence Process in which lymphocytes mature and become capable of responding to specific antigens.

Immunoglobulins Antibodies.

Implantation Process in which the blastocyst embeds in the uterine lining.

Impotency Inability of a male to achieve an erection.

Imprinting Attachment of young birds to their mothers or artificial substitutes during a critical period.

Innate behaviors Genetically programmed behaviors—literally in-born behavior—ready to be put to use when needed.

Instinctive behaviors Genetically programmed responses to stimuli that are fixed, automatic, and independent of previous experience.

In vitro Term referring to any procedure carried out in a test tube or petri dish, such as *in vitro* fertilization.

Incomplete dominance Partial dominance. Occurs when an allele exerts only partial dominance over another allele, resulting in an intermediate trait.

Incontinence Inability to control urination.

Induced abortion Deliberate expulsion of a fetus or embryo.

Inducer Chemical substance that activates inducible genes.

Inducible operon Set of genes that remains inactive until needed. Activated by inducers.

Infectious mononucleosis White blood cell disorder caused by a virus. Characterized by a rapid increase of monocytes and lymphocytes.

Inferior vena cava Large vein that empties deoxygenated blood from the body below the heart into the right atrium of the heart.

Infertility Inability to conceive; can be due to problems in either the male or the female.

Inflammatory response Response to tissue damage including an increase in blood flow, the release of chemical attractants, which draw monocytes to the scene, and an increase in the flow of plasma into a wound.

Inhalation Process of air being drawn into the lungs.

Inhibin Substance produced by the seminiferous tubules that inhibits the production of FSH by the anterior pituitary.

Inhibiting hormone Hormone from the hypothalamus that inhibits the release of hormones from the anterior pituitary.

Initiator codon Codon found on a messenger RNA strand that marks where protein synthesis begins.

Inner cell mass Cells of the blastocyst that become the embryo and amnion.

Insects Most successful class of arthropods with nearly a million species named to date. A complex life cycle, a symbiotic relationship with angiosperms, an ability to fly, and seasonal migration have contributed to their phenomenal success.

Insulin Hormone that stimulates the uptake of glucose by body cells, especially muscle and liver cells. Stimulates the synthesis of glycogen in liver and muscle cells.

Insulin-dependent diabetes Type of diabetes that can only be treated with injections of insulin. May be caused by an autoimmune reaction. Also known as early-onset diabetes.

Insulin-independent diabetes Type of diabetes that often occurs in obese people. In most patients, it can be controlled by diet. Also known as late-onset diabetes.

Integral protein Large protein molecules in the lipid bilayer of the plasma membrane.

Integration Process of making sense of various nervous inputs so that a meaningful response can be achieved.

Intercostal muscles Short, powerful muscles that lie between the ribs. Involved in inspiration and active exhalation.

Interferon Protein released from cells infected by viruses that stops the replication of viruses in other cells.

Interleukin 2 Chemical released by helper cells that activates T and B cells, stimulating cell division.

Internal sphincter (of the bladder) Involuntary muscular valve that relaxes reflexively, releasing urine. Formed by a smooth muscle in the neck of the bladder at the junction of the bladder and the urethra.

Internodes Segments of the axon between nodes of Ranvier.

Interphase Period of cellular activity occurring between cell divisions. Synthesis and growth occur in preparation for cell division.

Interstitial cells Cells located in the loose connective tissue between the seminiferous tubules of the testes. Produce testosterone.

Interstitial cell stimulating hormone (ICSH) Luteinizing hormone in males. Regulates testosterone secretion.

Interstitial fluid Fluid surrounding cells in body tissues. Provides a path through which nutrients, gases, and wastes can travel between the capillary and the cells.

Intervertebral disks Shock-absorbing material between the bones of the spine.

Intrauterine device (IUD) Birth control device that consists of a small plastic or metal object with a string attached that is inserted into the uterus through the cervix. Prevents implantation.

Intron Segment of DNA that is not expressed. Lies between exons (expressed segments).

Invertebrates Animals without a spinal column.

Ion Atom that has gained or lost one or more electrons. May be either positively or negatively charged.

Ionic bond Weak bond that forms between oppositely charged ions.

Iris Colored segment of the middle layer of the eye visible through the cornea.

Irritability Ability to perceive and respond to stimuli.

Islets of Langerhans Group of endocrine cells found in the pancreas that produce insulin and glucagon.

Isotonic Having the same solute concentration as a cell or body fluid.

Isotope Alternative form of an atom; differs from other atoms in the number of neutrons in the nucleus.

Jawless fishes Earliest known vertebrates. Slow-moving, bottom-feeders. Not dominant life forms in the ecosystems they occupy.

Joint capsule Connective tissue that connects to the opposing bones of a joint and forms the synovial cavity. The inner layer of the joint capsule produces synovial fluid.

Kidney Organ that rids the body of wastes and plays a key role in regulating the chemical constancy of blood.

Kin selection Altruistic behavior that increases the likelihood that an individual's genes will be transmitted to future generations.

Klinefelter syndrome Genetic disorder that results from an XXY genotype.

Labia majora Outer folds of skin of the external genitalia in women.

Labia minora Inner folds of the external genitalia in women.

Labor The process or period of childbirth.

Lactation Milk production in the breasts.

Laparoscope Instrument used to examine internal organs through small openings made in the skin and underlying muscle.

Larynx Rigid but hollow cartilaginous structure that houses the vocal cords and participates in swallowing.

Learning Process in which stored information can lead to changes in innate behavior.

Lens Transparent structure that lies behind the iris and in front of the vitreous humor. Focuses light on the retina.

Leukemia Cancer of white blood cells.

Leukocytosis An increase in the concentration of white blood cells, which often occurs during a bacterial or viral infection.

Lichen Association resulting from the symbiotic relationship between a fungus and a unicellular alga or cyanobacterium.

Life expectancy Average length of time a person will live.

Ligament Connective tissue structure that runs from bone to bone, located alongside and sometimes inside the joint. Offers support for joints.

Light-independent reactions The second state of photosynthesis, in which the energy obtained by the light-dependent reactions is used to fix carbon dioxide into carbohydrates; occurs in the stroma of chloroplasts.

Limbic system Array of structures in the brain that work in concert with centers of the hypothalamus. Site of instincts and emotions.

Limiting factor One factor that is most important in regulating growth in an ecosystem.

Limnetic zone The region of a lake commonly called the "open water"; extends downward to the point at which light no longer penetrates and is the main photosynthetic body of a lake.

Lipase Enzyme that removes some of the fatty acids from the glycerol molecule, forming a monoglyceride. Produced by the salivary glands and the pancreas.

Lipid Commonly known as fats. Water-insoluble organic molecules that provide energy to body cells, help insulate the body from heat loss, and serve as precursors in the synthesis of certain hormones. A principal component of the plasma membrane.

Liposuction Technique used to remove subcutaneous fat.

Littoral zone The shallow waters at the margin of a lake where rooted vegetation can grow.

Liver Organ located in the abdominal cavity that performs many functions essential to homeostasis. It stores glucose and fats, synthesizes some key blood proteins, stores iron and certain vitamins, detoxifies certain chemicals, and plays an important role in digestion by producing bile.

Long bones Bones of the skeleton that form parts of the extremities.

Loose connective tissue Type of connective tissue that serves primarily as a packing material. Contains many cells among a loose network of collagen and elastic fibers, especially cells that help protect the body from foreign organisms.

Low-density lipoproteins (LDLs) Complexes of protein and lipid that transport cholesterol, depositing it in body tissues.

Lungs Two large saclike organs in the thoracic cavity where the blood and air exchange carbon dioxide and oxygen.

Luteinizing hormone (LH) Hormone that stimulates gonadal hormone production. In men, LH stimulates the production of testosterone, the male sex steroid. In women, LH stimulates estrogen secretion.

Lymph Fluid contained in the lymphatic vessels. Similar to tissue fluid, but also contains white blood cells and may contain large amounts of fat.

Lymph node Small nodular organ interspersed along the course of the lymphatic vessels. Serves as a filter for lymph.

Lymphatic system Network of vessels that drains extracellular fluid from body tissues and returns it to the circulatory system.

Lymphocyte Type of white blood cell. *See also* B-lymphocyte and T-lymphocyte.

Lymphoid organs Organs, such as the spleen and thymus, that belong to the lymphatic system.

Lymphokine Chemical released by suppressor T cells that inhibits the division of B and T cells.

Lysosome Membrane-bound organelle that contains enzymes. Responsible for the breakdown of material that enters the cell by endocytosis. Also destroys aged or malfunctioning cellular organelles.

Lysozyme Enzyme produced in saliva that dissolves the cell wall of bacteria, killing them.

Macronutrients Nutrients required in relatively large amounts by organisms. Includes water, proteins, carbohydrates, and lipids.

Macrophage Phagocytic cell derived from monocytes that resides in loose connective tissues and helps guard tissues against bacterial and viral invasion.

Macula lutea Region of the retina located lateral to the optic disc where cones are most abundant.

Maculae Receptor organs in the saccule and utricle that play a role in position sense.

Malignant tumor Structure resulting from uncontrollable cellular growth. Cells often spread to other parts of the body.

Mammals Class of vertebrates with a single jaw bone, specialized teeth, hair, mammary glands, a four-chambered heart, and an advanced brain. Reproduction is the most complex of animals.

Marfan's syndrome Autosomal dominant genetic disorder that affects the skeletal system, the eye, and the cardiovascular system.

Marrow cavity Cavity inside a bone containing either red or yellow marrow.

Mast cell Cell found in many tissues, especially in the connective tissue surrounding blood vessels. Contains large granules containing histamine.

Matrix Extracellular material found in cartilage. Also the material in the inner compartment of the mitochondrion.

Matter Anything that has mass and occupies space.

Medulla Term referring to the central portion of some organs; for example, the adrenal medulla.

Megakaryocyte Large cell found in bone marrow that produces platelets.

Meiosis Type of cell division that occurs in the gonads during the formation of gametes. Requires two cellular divisions (meiosis I and meiosis II). In humans, it reduces the chromosome number from 46 to 23.

Meiosis I First meiotic division.

Meiosis II Second meiotic division.

Meissner's corpuscle Encapsulated sensory receptor thought to respond to light touch.

Membranous epithelium Refers to any sheet of epithelium that forms a continuous lining on organs.

Memory cells T or B cells produced after antigen exposure. They form a reserve force that responds rapidly to antigen during subsequent exposure.

Menopause End of the reproductive function (ovulation) in women. Usually occurs between the ages of 45 and 55.

Menstrual cycle Recurring series of events in the reproductive functions of women. Characterized by dramatic changes in ovarian and pituitary hormone levels and changes in the uterine lining that prepare the uterus for implantation.

Menstruation Process in which the endometrium is sloughed off, resulting in bleeding. Occurs approximately once every month.

Meristematic tissues Tissues responsible for the production of all other tissues. Located either at the tips of the stem (primary meristems) or along the entire length of the stem and root (secondary meristems). Meristems remain embryonic and give rise to new cells throughout the life of the plant.

Merkel disk Light touch receptor. Consists of dendrites that end on cells in the epidermis.

Mesoderm One of the three types of cells that emerge in human embryonic development. Lies in the middle of the forming embryo and forms muscle, bone, and cartilage.

Messenger RNA (mRNA) Type of RNA that carries genetic information needed to synthesize proteins to the cytoplasm of a cell.

Metabolic pathway Series of linked chemical reactions in which the product of one reaction becomes the reactant in another reaction.

Metabolic water Water produced during cellular respiration by the addition of protons (hydrogen ions) and electrons to oxygen.

Metabolism The chemical reactions of the body, including all catabolic and anabolic reactions.

Metaphase Stage of cellular division in which chromosomes line up in the center of the cell.

Metarterioles Arterioles that serve as circulatory short cuts, connecting arterioles with venules in a capillary bed. Also known as thoroughfare channels.

Metastasis Spread of cancerous cells throughout the body, through the lymph vessels and circulatory system or directly through tissue fluid.

Microfilament Solid fiber consisting of contractile proteins that is found in cells in a dense network under the plasma membrane. Forms part of the cytoskeleton.

Micronutrients Nutrients required in small quantities. They include two broad groups, vitamins and minerals.

Microspheres Small globules consisting of protein that may have been precursors of the first cells. Also known as proteinoids.

Microsurgery Type of surgery performed under dissecting microscopes. Used to reconnect axons, blood vessels, and other small structures.

Microtubules Hollow protein tubules in the cytoplasm of cells that form part of the cytoskeleton. Also form spindles.

Microvilli Tiny projections of the plasma membranes of certain epithelial cells that increase the surface area for absorption.

Middle ear Portion of the ear located within a bony cavity in the temporal bone of the skull. Houses the ossicles.

Migration Movement of an animal from one region to another often over long distances.

Mineralocorticoids Group of steroid hormones produced by the adrenal cortex. Involved in electrolyte or mineral salt balance.

Mitochondrion Membrane-bound organelle where the bulk of cellular energy production occurs in eukaryotic cells. Houses the citric acid cycle and electron transport system.

Mitosis Term referring specifically to the division of a cell's nucleus. Consists of four stages: prophase, metaphase, anaphase, and telophase.

Mitotic spindle Array of microtubules constructed in the cytoplasm during prophase. Microtubules of the mitotic spindle connect to the chromosomes and help draw them apart during mitosis.

Mollusks Phylum that exhibits four common characteristics: a mantle, a radula, a foot, and a shell.

Monerans Kingdom containing prokaryotic organisms, bacteria.

Monocots Angiosperms that contain one cotyledon.

Monocyte White blood cell that phagocytizes bacteria and viruses in body tissues.

Monohybrid cross Procedure in which one plant is bred with another to study the inheritance of a single trait.

Monosomy Genetic condition caused by a missing chromosome.

Morning sickness Nausea that often occurs in the first two to three months of pregnancy.

Morula Solid ball of cells produced from the zygote by numerous cellular divisions.

Motor unit Muscle fibers supplied by a single axon and its branches.

Mucus Thick, slimy material produced by the lining of the respiratory tract and parts of the digestive tract. Moistens and protects them.

Mullerian mimicry Phenomenon of several species, each with toxic characteristics, evolving similar color patterns.

Multipolar neuron Motor neuron found in the central nervous system. Contains a prominent, multiangular cell body and several dendrites.

Muscle fiber Long, unbranched, multinucleated cell found in skeletal muscle.

Muscle spindles Stretch receptors found in skeletal muscle. Also known as neuromuscular spindles.

Muscle tone Inherent firmness of muscle, resulting from contraction of muscle fibers during periods of inactivity.

Muscular artery Any one of the main branches of the aorta. Tunica media consists primarily of smooth muscle cells.

Mutation Technically, a change in the DNA caused by chemical and physical agents. Also refers to a wide range of chromosomal defects.

Mutualism Symbiotic relationship in which both species benefit.

Mycelia Aggregations of hyphae.

Mycorrhiza Fungus root; a symbiotic relationship between certain soil-dwelling fungi and root cells of many vascular plants; acts as root hairs.

Myelin sheath Layer of fatty material coating the axons of many neurons in the central and peripheral nervous systems. Formed by Schwann cells.

Myofibril Bundle of contractile myofilaments in skeletal muscle cells.

Myoglobin Cytoplasmic protein in muscle cells that binds to oxygen.

Myometrium Uterine smooth muscle.

Myopia (nearsightedness) Visual condition that results when the eyeball is slightly elongated or the lens is too strong. In the uncorrected eye, light rays from distant images come into focus in front of the retina.

Myosin Protein filament found in many cells in the microfilamentous network. Also found in muscle cells.

Myosin ATPase Enzyme found in the myosin cross bridges that splits ATP during muscle contraction.

Naked nerve ending Unmodified dendritic ending of the sensory neurons. Responsible for at least three sensations: pain, temperature, and light touch.

Natural childbirth Childbirth without the use of drugs.

Natural selection Evolutionary process in which environmental abiotic and biotic factors "weed" out the less fit—those organisms not as well adapted to the environment as their counterparts.

Nephron Filtering unit in the kidney. Consists of a glomerulus and renal tubule.

Neritic zone Shallow, biologically rich coastal life zone overlying the continental shelf.

Nerve Bundle of nerve fibers. May consist of axons, dendrites, or both. Carries information to and from the central nervous system.

Nerve deafness Loss of hearing resulting from nerve or brain damage.

Nervous tissue One of the primary tissues. Found in the nervous system and consists of two types of cells: conducting cells (neurons) and supportive cells.

Neural groove Ectodermal groove that forms early in embryonic development and runs the length of the embryo, later forming the neural tube.

Neural tube Tube of ectoderm that arises from the neural groove and will become the spinal cord.

Neuroendocrine reflex A reflex involving the endocrine and nervous systems.

Neuron Highly specialized cell that generates and transmits bioelectric impulses from one part of the body to another.

Neurosecretory neurons Specialized nerve cells of the hypothalamus and posterior pituitary that produce and secrete hormones.

Neurotransmitter Chemical substance released from the terminal ends (terminal boutons) of axons when a bioelectric impulse arrives. May stimulate or inhibit the next neuron.

Neutron Uncharged particle in the nucleus of the atom.

Neutrophil Type of white blood cell that phagocytizes bacteria and cellular debris.

Nicotine adenine dinucleotide (NAD) A molecule that carries high-energy electrons from one chemical reaction to another.

Nicotinamide adenine dinucleotide (NAD) Electron acceptor molecule that shuttles energetic electrons from glycolysis, the transition reaction, and the citric acid cycle to the electron transport system.

Nitrogen fixation Process in which bacteria and a few other organisms convert atmospheric nitrogen to nitrate or ammonia, forms usable by plants.

Node of Ranvier Small gap in the myelin sheath of an axon; located between segments formed by Schwann cells. Responsible for saltatory conduction.

Noncyclic photophosphorylation Production of ATP using photosystem I and II and the electron transport system.

Nondisjunction Failure of a chromosome pair or chromatids of a double-stranded chromosome to separate during mitosis or meiosis.

Nonpoint source (of pollution) A source that does not release pollutants via an easily identifiable route. *See also* Point source.

Nonspecific urethritis (NSU) One of the most common sexually transmitted diseases. Caused by several different bacteria.

Nonvascular plants Land plants that not only lack vascular tissue, but also lack roots, stems, and leaves. All nonvascular plants are bryophytes.

Noradrenaline (norepinephrine) Hormone produced by adrenal medulla and secreted under stress. Contributes to the fight-or-flight response.

Nuclear envelope Double membrane delimiting the nucleus.

Nuclear pores Minute openings in the nuclear envelope that allow materials to pass to and from the nucleus.

Nucleoli Temporary structures in the nuclei of cells during interphase. Regions of the DNA that are active in the production of RNA.

Nucleus (atom) Dense, center region of the atom that contains neutrons and protons.

Nucleus (cell) Cellular organelle that contains the genetic information that controls the structure and function of the cell.

Nutrient cycle Circular flow of nutrients from the environment through the various food chains back into the environment.

Olfactory membrane Receptor for smell; found in the roof of the nasal cavity.

Olfactory nerve Nerve that transmits impulses from the olfactory membrane to the brain.

Omnivores Organisms that feed on both plants and animals.

Ontogeny Development of an organism starting with fertilization.

Oogenesis Production of ova.

Oogonium Germ cell in ovary that contains 46 double-stranded chromosomes. Forms primary oocytes.

Operant conditioning Type of associative learning in which the repeated use of a reinforcing stimulus (punishment or reward) elicits a desired behavior.

Operator site Region of the DNA molecule adjacent to the structural genes that acts as a switch to turn the operon on or off.

Operon Functional unit of the DNA of bacteria. Consists of structural and regulatory genes.

Optic disk Site in the retina where the optic nerve exits. Also known as the blind spot.

Optic nerve Nerve that carries impulses from the retina to the brain.

Order Taxonomic term that refers to a subgroup of a class.

Organ Discrete structure that carries out specialized functions.

Organ of Corti Receptor for sound; located in the inner ear within the cochlea.

Organ system Group of organs that participate in a common function.

Organogenesis Organ formation during embryonic development.

Osmosis Diffusion of water across a selectively permeable membrane.

Osmotic pressure Force that drives water across a selectively permeable membrane. Created by differences in solute concentrations.

Ossicles Three small bones inside the middle ear that transmit vibrations created by sound waves to the organ of Corti.

Osteoarthritis Degenerative joint disease caused by wear and tear that impairs movement of joints.

Osteoblast Bone-forming cell; secretes collagen.

Osteoclast Cell that digests the extracellular material of bone. Stimulated by the parathyroid hormone.

Osteocyte Bone cell derived from osteoblasts that has been surrounded by calcified extracellular material.

Osteoporosis Degenerative disease resulting in the deterioration of bone. Due to inactivity in men and women and loss of ovarian hormones in postmenopausal women.

Outer ear External portion of the ear.

Oval window Membrane-covered opening in the cochlea where vibrations are transmitted from the stirrup to the fluid within the cochlea.

Ovary Enlarged, rounded basal portion of the flower pistil containing one or more ovules, each with a sporangium. Produces the female gametophyte containing an egg and several other cells.

Overpopulation Condition in which a species has exceeded the carrying capacity of the environment.

Ovulation Release of the oocyte from ovary. Stimulated by hormones from the anterior pituitary.

Ovum Germ cell containing 23 single-stranded chromosomes. Produced during the second meiotic division.

Oxaloacetate Four-carbon compound of the citric acid cycle. It is involved in the very first reaction of the cycle and is regenerated during the cycle.

Oxidation The loss of hydrogens or electrons from a substance.

Oxytocin Hormone from the posterior pituitary hormone. Stimulates contraction of the smooth muscle of the uterus and smooth-muscle-like cells surrounding the glandular units of the breast.

Ozone O_3 Molecule produced in the stratosphere (upper layer of the atmosphere) from molecular oxygen. Helps screen out incoming ultraviolet light. *See also* Ozone layer.

Ozone layer Region of the atmosphere located approximately 12 to 30 miles above the Earth's surface where ozone molecules are produced. Helps protect the Earth from ultraviolet light.

Pacinian corpuscle Large encapsulated nerve ending that is located in the deeper layers of the skin and near body organs. Responds to pressure.

Pancreas Organ found in the abdominal cavity under the stomach, nestled in a loop formed by the first portion of the small intestine. Produces enzymes needed to digest foodstuffs in the small intestine and hormones that regulate blood glucose levels.

Pap smear Procedure in which cells are retrieved from the cervical canal to be examined for the presence of cancer.

Papillae Small protrusions on the upper surface of the tongue. Some papillae contain taste buds.

Paracrines Chemicals released by cells that elicit a response in nearby regions.

Parasitism Process in which an individual feeds (usually without killing) on another larger individual—the host.

Parasympathetic division (of the autonomic nervous system) Portion of the autonomic nervous system responsible for a variety of involuntary functions.

Parathyroid glands Endocrine glands located on the posterior surface of the thyroid gland in the neck. Produce parathyroid hormone.

Parathyroid hormone (PTH) Hormone that helps regulate blood calcium levels. Stimulates osteoclasts to digest bone, thus raising blood calcium levels. Also known as parathormone.

Parturition Childbirth.

Passive immunity Temporary protection from antigen (bacteria and others) produced by the injection of immunoglobulins.

Penis Male organ of copulation.

Pepsin Enzyme released by the gastric glands of the stomach. Breaks down proteins into large peptide fragments.

Pepsinogen Inactive form of pepsin.

Perforin Chemical released by cytotoxic cells that destroys bacteria. Binds to plasma membrane of target cells, forming pores that make the target cells leak and die.

Perichondrium Connective tissue layer surrounding most types of cartilage. Contains blood vessels that supply nutrients to cartilage cells.

Periodic table of elements Table that lists elements by ascending atomic number. Also lists other vital statistics of each element.

Peripheral nervous system Portion of the nervous system consisting of the cranial and spinal nerves and receptors.

Peristalsis Involuntary contractions of the smooth muscles in the wall of the esophagus, stomach, and intestines, which propel food along the digestive tract.

Peritubular capillaries Capillaries that surround nephrons. They pick up water, nutrients, and ions from the renal tubule, thus helping maintain the osmotic concentration of the blood.

Permafrost Permanently frozen subsoil in the Arctic tundra.

Permanent threshold shift Permanent hearing loss caused by repeated exposure to noise. Results from damage to hair cells of the organ of Corti.

Pharynx Chamber that connects the oral cavity with the esophagus.

Phenotype Outward appearance of an organism.

Phloem Type of vascular tissue in gymnosperms and angiosperms that transports food through stem, petioles, and leaves.

Phonation Production of sound.

Photoelectron transport system Protein carriers that transport electrons from photosystem II to photosystem I and produce ATP from the energy released by the electrons.

Photophosphorylation Production of ATP by the electron transport system.

Phosphoglyceraldehyde A three-carbon monosaccharide produced during glycolysis by the splitting of the glucose molecules; also produced in the light-independent reactions of photosynthesis.

Phosphofructokinase Enzyme in the glycolytic pathway that catalyzes the conversion of glucose-6-phosphate to fructose-6-phosphate; regulated by ATP and citric acid.

Photoreceptors Modified nerve cells that respond to light. Located in the retina of humans and other animals.

Photosynthesis The series of chemical reactions in which the energy of light is used to synthesize high-energy organic molecules, usually carbohydrates, from low-energy inorganic molecules, usually carbon dioxide and water.

Photosystem In thylakoid membranes, a light-harvesting complex and its associated electron transport system.

Phototropism Growth response of a plant to light coming from one direction.

Phylogeny Evolutionary development of a species.

Phylum (phyla, plural) Major grouping of organisms. Animal kingdom contains about two dozen primary phyla, nearly all of which are represented today.

Phytoplankton Microscopic, free-floating organisms, principally algae, which capture sunlight energy, using it to produce carbohydrates from carbon dioxide dissolved in the water.

Pineal gland Small gland located in the brain that secretes a hormone thought to help control the biological clock.

Pistil Portion of flower that consists of terminal stigma, basal ovary, and style of a flower that connects the stigma and ovary.

Pituitary gland Small pea-sized gland located beneath the brain in the sella turcica. It produces numerous hormones and consists of two main subdivisions: anterior and posterior pituitary.

Placenta Organ produced from maternal and embryonic tissue. Supplies nutrients to the growing embryo and fetus and removes fetal wastes.

Plasma Extracellular fluid of blood. Comprises about 55% of the blood.

Plasma cell Cell produced from B-lymphocytes (B cells); synthesizes and releases antibodies.

Plasma membrane Outer layer of the cell. Consists of lipid and protein and controls the movement of materials into and out of the cell.

Plasmids Small circular strands of DNA found in bacterial cytoplasm separate from the main DNA.

Plasmin Enzyme in the blood that helps dissolve blood clots.

Plasminogen Inactive form of plasmin.

Plasmodium Single-celled parasite responsible for malaria.

Platelet Cell fragment produced from megakaryocytes in the red bone marrow. Plays a key role in blood clotting.

Podocyte Type of cell forming the inner lining of Bowman's capsule of the nephron. Part of the filtration mechanism in the glomerulus.

Point source (of pollution) A discrete, easily identifiable source, usually releasing pollutants directly into waterways via pipes.

Polar body Discarded nuclear material produced during meiosis I and meiosis II of oogenesis.

Pole-to-pole fibers Type of microtubule found in the spindle. Extend from one centriole to the other.

Pollination Transfer of pollen grains (male gametophytes) to the ovulate cone in gymnosperms and to the ovaries of flowers in angiosperms.

Polygenic inheritance Transmission of traits that are controlled by more than one gene.

Polyploidy Term referring to a genetic disorder caused by an abnormal number of chromosomes. Includes tetraploidy and triploidy.

Polyribosome Also known as polysome. Organelle formed by several ribosomes attached to a single messenger RNA. Synthesizes proteins used inside the cell.

Population Group of like organisms occupying a specific region.

Porphyrin ring Part of the chlorophyll molecule that absorbs sunlight.

Portal system Arrangement of blood vessels in which a capillary bed drains to a vein, which drains to another capillary bed.

Posterior chamber Posterior portion of the anterior cavity of the eye.

Posterior pituitary Neuroendocrine gland that consists of neural tissue and releases two hormones, oxytocin and antidiuretic hormone.

Precapillary sphincters Tiny rings of smooth muscle that surround the capillaries arising from the metarterioles.

Predation Process in which an individual kills and feeds on another smaller individual.

Premature birth Birth of a baby before 37 weeks of gestation.

Premenstrual syndrome (PMS) Condition that occurs in some women in the days before menstruation normally begins. Characterized by a variety of symptoms such as irritability, depression, fatigue, headaches, bloating, swelling, and tenderness of breasts, joint pain, and tension.

Premotor area Region of the brain in front of the primary motor area. Controls muscle contraction and other less voluntary actions (playing a musical instrument).

Presbyopia Visual impairment caused by aging. Lens becomes stiffer, making it more difficult to focus on nearby objects.

Primary center of ossification Region in the interior of a cartilage mass that first becomes bone.

Primary electron acceptor Protein molecule in the assembly that accepts the energized electron from chlorphyll a in the reaction center.

Primary follicle Structure in the ovary consisting of a primary oocyte and a complete single layer of cuboidal follicle cells.

Primary motor area Ridge of tissue in front of a central groove (the central sulcus). Controls voluntary motor activity.

Primary oocyte Germ cell produced from oogonium in the ovary. Undergoes the first meiotic division.

Primary response Immune response elicited when an antigen first enters the body.

Primary sensory area Region of the brain located just behind the central sulcus. The point of destination for many sensory impulses traveling from the body into the spinal cord and up to the brain.

Primary spermatocyte Cell produced from spermatogonium in the seminiferous tubule. Will undergo first meiotic division.

Primary succession Process of sequential change in which one community is replaced by another. Occurs where no biotic community has existed before.

Primary tissue One of major tissue types, including epithelial, connective, muscle, and nervous tissue.

Primary tumor Cancerous growth that gives rise to cells that spread to other regions of the body.

Primates An order of the kingdom Animalia. Includes prosimians (premonkeys), monkeys, apes, and humans.

Primordial follicle Structure in the ovary that consists of a primary oocyte surrounded by a layer of flattened follicle cells. Gives rise to the primary follicle.

Primordial germ cells Cells that originate in the wall of the yolk sac and eventually become either spermatogonia or oogonia.

Principle of independent assortment Mendel's second law. Hereditary factors are segregated independently during gamete formation. Occurs only when genes are on different chromosomes.

Principle of segregation Mendel's first law, which states that hereditary factors separate during gamete formation.

Producers Generally refers to organisms that can synthesize their own foodstuffs. Major producers are the algae and plants that absorb sunlight and use its energy to synthesize organic foodstuffs from water and carbon dioxide.

Profundal zone The deepest water of a lake, into which very little light penetrates.

Prolactin Protein hormone under control of the hypothalamus. In humans, it is responsible for milk production by the glandular units of the breast.

Promoter Region of the operon between the regulator gene and operator site. Binds to RNA polymerase.

Pronuclei Name of the ovum and sperm cell nuclei shortly after fertilization occurs. Each contains 23 chromosomes.

Prophase First phase of mitosis during which chromosomes condense, the nuclear membrane disappears, and the spindle forms.

Proprioception Sense of body and limb position.

Prosimians Premonkeys; tarsiers and lemurs.

Prostaglandins Group of chemical substances that have a variety of functions. Act on nearby cells.

Prostate gland Sex accessory gland that is located near the neck of the bladder and empties into the urethra. Produces fluid that is added to the sperm during ejaculation.

Proteinoids Spherical structures composed of small amino acid chains formed when amino acids are heated in air. May have been an early precursor of the first cells.

Protists Unicellular eukaryotic organisms such as protozoans and amoebae. Some are autotrophic and some are heterotrophic.

Proton Subatomic particle found in the nucleus of the atom. Each proton carries a positive charge.

Proto-oncogenes Genes in cells that, when mutated, lead to cancerous growth.

Protozoans Animal-like protists, including zooflagellates, ameoboids, ciliates, and sporozoans.

Puberty Period of sexual maturation in humans.

Pulmonary circuit (or circulation) Short circulatory loop that supplies blood to the lungs and transports it back to the heart.

Pulmonary veins Veins that carry oxygenated blood from the lungs to the left atrium.

Punctuated equilibrium Hypothesis explaining how evolutionary change occurs. States that long periods of relatively little change are broken up by briefer periods of relatively rapid evolution.

Pupil Opening in the iris that allows light to penetrate deeper into the eye.

Purine Type of nitrogenous base found in DNA nucleotides. Consists of two fused rings.

Purkinje fiber Modified cardiac muscle fiber that conducts bioelectric impulses to individual heart muscle cells.

Pus Liquid emanating from a wound. Contains plasma, many dead neutrophils, dead cells, and bacteria.

Pyloric sphincter Ring of smooth muscle cells in the lower portion of the stomach where it joins the duodenum. Serves as a gate valve. Opens periodically after a meal, releasing spurts of chyme (liquified, partially digested food) into the small intestine.

Pyramid of numbers Diagram of the number of organisms at various trophic levels in a food chain or ecosystem.

Pyrimidine One of two types of nitrogen base found in DNA nucleotides. Consists of one ring.

Pyrogen Chemical released primarily from macrophages that have been exposed to bacteria and other foreign substances. Responsible for fever.

Radial keratotomy Procedure to correct nearsightedness. Numerous, small superficial incisions are made in the cornea, flattening it and reducing its refractive power.

Radioactivity Tiny bursts of energy or particles emitted from the nucleus of some unstable atoms. Results from excess neutrons in the nuclei of some atoms.

Radionuclide Radioactive isotope of an atom.

Range of tolerance Range of conditions in which an organism is adapted.

Reaction center In the light-harvesting complex of a photosystem, containing chlorophyll a, a molecule to which light energy is transferred by the other pigment molecules.

Receptor Any structure that responds to internal or external changes. Three types of receptors are found in the body: encapsulated, nonencapsulated (naked nerve endings), and specialized (e.g., the retina and semicircular canals).

Recessive Term describing an allele of a gene that is expressed when the dominant factor is missing.

Recombinant DNA technology Procedure in which scientists take segments of DNA from an organism and combine them with DNA from other organisms.

Recombination Process of crossing over during meiosis, resulting in new genetic combinations.

Red blood cells (RBCs) Enucleated cells in blood that transport oxygen in the bloodstream.

Red bone marrow Tissue found in the marrow cavity of bones. Site of blood cell and platelet production.

Reduction The addition of hydrogens or electrons to a molecule.

Reduction factor Any of the factors that cause populations to decline.

Reflex Automatic response to a stimulus. Mediated by the nervous system.

Refraction Bending of light.

Regulator gene Gene that codes for the synthesis of repressor protein in an operon.

Relaxin Hormone produced by the corpus luteum and the placenta. It is released near the end of pregnancy and softens the cervix and the fibrocartilage uniting the pubic bones, thus facilitating birth.

Releasers *See* sign stimuli.

Releasing hormone Any of a group of hormones that stimulates the release of other hormones by the anterior pituitary.

Renal pelvis Hollow chamber inside the kidney. Receives urine from the collecting tubules and empties into the ureter.

Renal tubule That portion of the nephron where urine is produced.

Renewable resources Resources that replenish themselves via natural biological and geological processes, such as wind, hydropower, trees, fish, and wildlife.

Replacement-level fertility Number of children that will replace a couple when they die.

Repressible operon Operon whose genes remain active unless turned off. Found in bacteria and may be present in eukaryotes as well.

Repressor protein Protein produced by a regulator gene. Binds to a region of the DNA molecule (the operator site) adjacent to the structural genes. Blocks RNA polymerase from transcribing structural genes.

Reproductive isolation Condition in which two groups of similar organisms derived from the same parent stock lose the ability to interbreed. Often due to geographic isolation.

Reptiles Class of vertebrates in which fertilization occurs internally. Includes snakes, lizards, and turtles.

Respiratory distress syndrome Disease of premature babies that results from an insufficient amount of surfactant in the infant's lungs, causing alveoli to collapse. Also known as hyaline membrane disease.

Resting potential Minute voltage differential across the membrane of neurons. Also known as the membrane potential.

Restriction endonuclease Enzyme used in recombinant DNA technology. Cuts off segments of the DNA molecule for cloning and splicing.

Reticular activating system (RAS) Region of the medulla that receives nerve impulses from neurons transmitting information to and from the brain. Impulses are transmitted to the cortex, alerting it.

Retina Innermost, light-sensitive layer of the eye. Consists of an outer pigmented layer and an inner layer of nerve cells and photoreceptors (rods and cones).

Retrovirus Special type of RNA virus that carries an enzyme enabling it to produce complementary strands of DNA on the RNA template.

Reverse transcriptase Enzyme that allows the production of DNA from strands of viral RNA.

Rheumatoid arthritis Type of arthritis in which the synovial membrane of the joint becomes inflamed and thickens. Results in pain and stiffness in joints. Thought to be an autoimmune disease.

Rhodopsin (visual purple) Pigment contained in the rods.

Rhythm method Birth control method in which a couple abstains from sexual intercourse around the time of ovulation. Also known as the natural method.

Ribosomal RNA (rRNA) RNA produced at the nucleolus. Combines with protein to form the ribosome.

Ribosome Cellular organelle consisting of two subunits, each made of protein and ribosomal RNA. Plays an important part in protein synthesis.

RNA polymerase Enzyme that helps align and join the nucleotides in a replicating RNA molecule.

RNA replicase Enzyme produced by RNA viruses in host cells that allows them to produce complementary strands of RNA from an RNA template.

Rod Type of photoreceptor in the eye. Provides for vision in dim light.

Root nodule Swelling in the roots of certain plants (legumes) containing nitrogen-fixing bacteria.

Rough endoplasmic reticulum (RER) Ribosome-coated endoplasmic reticulum. Produces lysosomal enzymes and proteins for use outside the cell.

Saccule Membranous sac located inside the vestibule. Contains a receptor for movement and body position.

Salivary gland Any of several exocrine glands situated around the oral cavity. Produces saliva.

Saltatory conduction Conduction of a bioelectric impulse down a myelinated neuron from node to node.

Saprophyte Organism that releases enzymes that digest food materials externally. Smaller food molecules generated in the process are absorbed by the organism. Includes most fungi and nonphotosynthetic bacteria.

Sarcomere Functional unit of the muscle cell. Consists of the myofilaments, actin, and myosin.

Sarcoplasmic reticulum Term given to the smooth endoplasmic reticulum of a skeletal muscle fiber. Stores and releases calcium ions essential for muscle contractions.

Schwann cell Type of neuroglial cell or supportive cell in the nervous system. Responsible for the formation of the myelin sheath.

Science Body of knowledge on the workings of the world and a method of accumulating knowledge. *See also* Scientific method.

Scientific method Deliberate, systematic process of discovery. Begins with observation and measurement. From observations, hypotheses are generated and tested. This leads to more observation and measurement that supports or refutes the original hypothesis.

Sclera Outermost layer of the eye.

Scrotum Skin-covered sac containing the testes.

Sebum Oil excreted by sebaceous glands onto the surface of the skin.

Second law of thermodynamics Law stating that no energy conversion is ever 100% efficient.

Second messenger mechanism Describes how protein hormones and others effect intracellular change by binding to a receptor, activating adenylate cyclase, which leads to the production of cyclic AMP. Cyclic AMP, the second messenger, then activates a cytoplasmic enzyme, protein kinase, which activates or inactivates other enzymes.

Secondary center of ossification Region of bone formation that occurs in the ends (epiphyses) of bones.

Secondary response Generally, a powerful, swift immune system response occurring the second time an antigen enters the body. Much faster than the primary response.

Secondary sex characteristics Distinguishing features of men and women resulting from the sex steroids. In men, includes facial hair growth and deeper voices. In women, includes breast development and fatty deposits in the hips and other regions.

Secondary succession Process of sequential change in which one community is replaced by another. It occurs where a biotic community previously existed, but was destroyed by natural forces or human actions.

Secondary tumor Cancerous growth formed by cells arising from a primary tumor.

Secretin Hormone produced by the cells of the duodenum. Stimulates the pancreas to release sodium bicarbonate.

Secretory vesicles Membrane-bound vesicles containing protein (hormones or enzymes) produced by the endoplasmic reticulum and packaged by the Golgi complex of some cells. They fuse with the membrane, releasing their contents by exocytosis.

Selective permeability Control of what moves across the plasma membrane of a cell.

Sensitization Type of associative learning in which an individual learns to pay heightened attention to stimuli.

Semen Fluid containing sperm and secretions of the secondary sex glands.

Semicircular canal Sensory organ of the inner ear. Houses the receptors that detect body position and movement.

Semilunar valve Type of valve lying between the ventricles and the arteries that conduct blood away from the heart.

Seminal vesicles Sex accessory glands that empty into the vas deferens. Produce the largest portion of ejaculate.

Seminiferous tubule Sperm-producing tubule in the testis.

Sertoli cell Cell in the germinal epithelium of the seminiferous tubule. Houses spermatogenic cells as they develop.

Sex accessory gland One of several glands that produce secretions that are added to sperm during ejaculation.

Sex chromosomes X and Y chromosomes that help determine the sex of an individual.

Sex-linked trait Trait produced by a gene carried on a sex chromosome.

Sex steroid Steroid hormones produced principally by the ovaries (in women) and testes (in men). Help regulate secretion of gonadotropins and determine secondary sex characteristics.

Sexually transmitted diseases (venereal diseases) Infections that are transmitted by sexual contact.

Sickle-cell anemia Genetic disease common in African Americans that results in abnormal hemoglobin in red blood cells, causing cells to become sickle shaped when exposed to low oxygen levels. Sickling causes cells to block capillaries, restricting blood flow to tissues.

Sign stimuli Specific environmental antecedents (stimuli) to particular fixed action patterns.

Sinoatrial node The heart's pacemaker. Located in the wall of the right atrium, it sends timed impulses to the heart muscle, thus synchronizing muscle contractions.

Skeletal muscle Muscle that is generally attached to the skeleton and causes body parts to move.

Skeleton Internal support of humans and other animals. Consists of bones joined together at joints.

Sliding filament mechanism Sliding of actin filaments toward the center of a sarcomere, causing muscle contraction.

Slime molds Fungi that move about by ameobid motion, engulfing food from the soil.

Smooth endoplasmic reticulum (SER) Endoplasmic reticulum without ribosomes. Produces phosphoglycerides used to make the plasma membrane. Performs a variety of different functions in different cells.

Smooth muscle Involuntary muscle that lacks striations. Found around circulatory system vessels and in the walls of such organs as the stomach, uterus, and intestines.

Somite Block of mesoderm that gives rise to the vertebrae, muscles of the neck, and trunk.

Special sense Vision, hearing, taste, smell, and balance.

Speciation Formation of new species resulting from geographic isolation.

Species Group of organisms that is structurally and functionally similar. When members of the group breed, they produce viable, reproductively competent offspring. Also a subgroup of a genus.

Specificity The property of an enzyme allowing it to catalyze only one or a few chemical reactions.

Spermatogenesis Formation of sperm in the seminiferous tubules.

Spermatogonia Sperm-producing cells in the periphery of the germinal epithelium of the seminiferous tubules.

Spermatozoan Sperm.

Spinal nerve Nerve that arises from the spinal cord.

Sponges Phylum of simple, immobile animals that live in colonies, mostly in saltwater. Demonstrate somewhat more complexity than one-celled organisms but less than organisms with distinct tissues.

Spongy bone Type of bony tissue inside most bones. Consists of an irregular network of bone spicules.

Spores (bacterial) Resistant structures produced when environmental conditions become unfavorable. They house the circular chromosome and a tiny amount of cytoplasm and are encased in a thick cell wall.

Sporophyte Diploid generation of a plant exhibiting alternation of generations that produces asexual spores.

Sporozoans Nonmotile, animal-like protists so named because of their ability to produce infectious spores.

Stamen Filamentous structure of a flower that ends with a small bulbous structure known as the anther that produces the male gamete.

Sterilization Procedure to render a man or woman sterile or infertile. In men, the method is generally a vasectomy; in women, it is usually tubal ligation.

Stoma (plural, *stomata*) Adjustable openings in plant leaves. Most gas exchange between leaves and the air occurs through the stoma.

Stirrup (stapes) One of three bones of the middle ear that conducts vibrations from the eardrum to the inner ear.

Stroma The semifluid medium of chloroplasts, in which the membranous grana are embedded; the site of the light-independent reactions.

Structural gene Any gene of an operon that codes for the production of enzymes and other proteins.

Subatomic particles Electrons, protons, and neutrons. Particles that can be separated from an atom by physical means.

Substrate Molecule that fits into the active site of an enzyme.

Succession Process of sequential change in which one community is replaced by another until a mature or climax ecosystem is formed. *See also* Primary succession and Secondary succession.

Sulcus Indented region or groove in the cerebral cortex between ridges.

Suppressor cell Cell of the immune system that shuts down the immune reaction as the antigen begins to disappear.

Suprachiasmatic nucleus Clump of nerve cells in the hypothalamus. Thought to play a major role in coordinating several key functions and several other control centers. Sometimes called the "master clock."

Surfactant Detergent-like substance produced by the lungs. Dissolves in the thin watery lining of the alveoli; helps reduce surface tension, keeping the alveoli from collapsing.

Suspensory ligament Zonular fibers that connect the lens to the ciliary body.

Sustainable-Earth ethic Ethic based on

three tenets: (1) the world has a limited supply of resources that must be shared with all living things, (2) humans are a part of nature and subject to its rules, and (3) nature is not something to conquer, but rather a force we must learn to cooperate with.

Sustainable society Society that lives within the carrying capacity of the environment.

Symbiosis Refers to a number of different types of biotic interaction within a community, including mutualism and commensalism.

Sympathetic division Division of the autonomic nervous system that is responsible for many functions, especially those involved in the fight-or-flight response.

Synapse Juncture of two neurons.

Synaptic cleft Gap between an axon and the dendrite or effector (e.g., gland or muscle) it supplies.

Synergy Coordination of the workings of antagonistic muscle groups.

Synovial fluid Lubricating liquid inside joint cavities. Produced by the synovial membrane.

Synovial membrane Inner layer of the joint capsule.

Syphilis Potentially serious, sexually transmitted disease caused by a bacterium.

Systemic circulation System of blood vessels that transports blood to and from the body and heart, excluding the lungs.

Systolic pressure Peak pressure at the moment the ventricles contract. The higher of the two numbers in a blood pressure reading.

Taiga The northern coniferous forests biome.

Taste bud Receptor for taste principally found in the surface epithelium and certain papillae of the tongue.

Taxis (taxes, plural) Orientation of an animal in relation to some stimulus. An innate mechanism found in less complex organisms.

T cell *See* T-lymphocyte.

Telophase Final stage of mitosis in which the nuclear envelope reforms from vesicles and the chromosomes uncoil.

Temperate deciduous forest Biome that in the United States lies east of the Mississippi River and is characterized by broad-leafed trees.

Temporary threshold shift Temporary loss of hearing after being exposed to a noisy environment.

Tendons Connective tissue structures that generally attach muscles to bones.

Teratogen Chemical, biological, or physical agent that causes birth defects.

Teratology Study of birth defects.

Terminal boutons Small swellings on the terminal fibers of axons. They lie close to the membranes of the dendrites of other axons or the membranes of the effectors, and transfer bioelectric impulses from one cell to another.

Terminator codon Codon found on each strand of messenger RNA that marks where protein synthesis should end.

Testes Male gonads. They produce sex steroids and sperm.

Testosterone Male sex hormone that stimulates sperm formation and is responsible for secondary sex characteristics, such as facial hair growth and muscle growth.

Tetraploidy Condition in which an individual is endowed with two complete sets of chromosomes. Instead of having 46 chromosomes, he or she has 92.

Theories Principles of science—the broader generalizations about the world and its components. Theories are supported by considerable scientific research.

Thermocline A layer of water between the epilimnion and hypolimnion; characterized by rapid temperature change.

Thoracic duct Duct carrying lymph to the circulatory system. Empties into the large veins at the base of the neck.

Thoroughfare channel Vessel that connects the arterioles with the venules, thus allowing blood to bypass a capillary bed. Also known as a metarteriole.

Thylakoid A disk-shaped, membranous sac found in chloroplasts, the membranes of which contain the photosystems and ATP-synthesizing enzymes used in the light-dependent reactions of photosynthesis.

Thyroid gland U- or H-shaped gland located in the neck on either side of the trachea just below the larynx. Produces three hormones: thyroxin, triiodothyronine, and calcitonin.

Thyroid-stimulating hormone (TSH) Hormone produced by the pituitary gland. Stimulates production and release of thyroxine and triiodothyronine by the thyroid gland.

Thyroxin Hormone produced by the thyroid gland that accelerates the rate of mitochondrial glucose catabolism in most body cells and also stimulates cellular growth and development.

Tissue Component of the body from which organs are made. Consists of cells and extracellular material (fluid, fibers, and so on).

T-lymphocyte Type of lymphocyte responsible for cell-mediated immunity. Attacks foreign cells, virus-infected cells, and cancer cells directly. Also known as T cell.

Total fertility rate Number of children a woman is expected to have during her lifetime.

Trachea Duct that leads from the pharynx to the lungs.

Transcription RNA production on a DNA template.

Transfer RNA (tRNA) Small RNA molecules that bind to amino acids in the cytoplasm and deliver them to specific sites on the messenger RNA.

Transformation Conversion of a normal cell to a cancerous one.

Transition reaction Part of cellular respiration in which one carbon is cleaved from pyruvic acid, forming a two-carbon compound, which reacts with Coenzyme A. The resulting chemical enters the citric acid cycle.

Translation Synthesis of protein on a messenger RNA template.

Translocation Process in which a segment of a chromosome breaks off but reattaches to another site on the same chromosome or another one.

Transpiration Process by which water evaporates from leaf surfaces. Creates a tension (reduced pressure) in the xylem that pulls columns of water from the roots to the stems.

Trichocysts Harpoonlike projections found in ciliates and used to capture prey.

Triiodothyronine Hormone produced by the thyroid gland. Nearly identical in function to thyroxin.

Triploidy Genetic disorder in which cells have 69 chromosomes instead of 46.

Trisomy Genetic condition characterized by the presence of one extra chromosome.

Trophic hormones Hormones that stimulate the production and secretion of other hormones. Also known as tropic hormones.

Trophic level Feeding level in a food chain.

Tropism Plant response to stimuli, such as light, gravity, and touch, in which it bends toward (positive response) or away from (negative response) the stimulus.

Trophoblast Outer ring of cells of the blastocyst that form the embryonic portion of the placenta.

True fungi Organisms with distinct cell walls, including single-celled yeasts and the multicellular fungi but not slime molds.

TSH-releasing hormone (TSH-RH) Hormone secreted by the posterior lobe of the pituitary gland. Stimulates thyroxin secretion by the thyroid gland.

T tubules (transverse tubules) Invaginations of the plasma membrane of skeletal muscle fibers that conduct an impulse to the interior of the cell.

Tubal ligation Sterilization procedure in women. Uterine tubes are cut, preventing sperm and ova from uniting.

Tubular reabsorption Process in which nutrients are transported out of the nephron into the peritubular capillaries.

Tubular secretion Process in which wastes are transported from the peritubular capillaries into the nephron.

Tumor Mass of cells derived from a single cell that has begun to divide. In malignant tumors, the cells divide uncontrollably and often release clusters of cells or single cells that spread in the blood and lymphatic systems to other parts of the body. Benign tumors grow to a certain size, then stop.

Tundra Northernmost biome with long, cold winters and a short growing season.

Turner syndrome Genetic disorder in which an offspring contains 22 pairs of autosomes and a single, unmatched X chromosome. Phenotypically female.

Twitch Single muscle fiber contraction.

Tympanic membrane (eardrum) Membrane between the external auditory canal and middle ear that oscillates when struck by sound waves.

Type I diabetes Form of diabetes that occurs mainly in young people and results from an insufficient amount of insulin production and release. Brought on by damage to insulin-producing cells of the pancreas. Also known as early-onset diabetes.

Type II diabetes Form of diabetes that occurs chiefly in older individuals (around age of 40) and results from a loss of tissue responsiveness to insulin. Also known as late-onset diabetes.

Type II alveolar cell Cell found in the lining of the alveoli. Produces surfactant.

Umbilical artery One of two arteries in the umbilical cord that carries blood from the embryo to the placenta.

Umbilical vein Vein in the umbilical cord that carries blood from the placenta to the fetus.

Ureter Hollow, muscular tube that transports urine by peristaltic contractions from the kidney to the urinary bladder.

Urethra Narrow tube that transports urine from the urinary bladder to the outside of the body. In males, it also conducts sperm and semen to the outside.

Urinary bladder Hollow, distensible organ with muscular walls that stores urine. Drained by the urethra.

Urine Fluid containing various wastes that is produced in the kidney and excreted out of the urinary bladder.

Uterus Organ that houses and nourishes the developing embryo and fetus.

Utricle Membranous sac containing a receptor for body position and movement. Located inside the vestibule of the inner ear.

Vaccine Preparation containing dead or weakened bacteria and viruses that, when injected in the body, elicits an immune response. *See also* Active immunity.

Vagina Tubular organ that serves as a receptacle for sperm and provides a route for delivery of the baby at birth.

Vagus nerve Nerve that terminates in the stomach wall and stimulates HCl production by cells in the gastric glands.

Variation Genetically based differences in physical or functional characteristics within a population.

Varicose vein Vein whose wall balloons out because the flow of blood downstream is obstructed.

Vascular plants Land plants having a specialized conducting tissue (xylem and phloem) for the transport of food and water.

Vascular cambium Tissue in woody plants that produces new xylem and phloem.

Vascular tissues Responsible for the transport of materials within a plant, including water, minerals, and food. Includes xylem and phleom.

Vas deferens Duct that carries sperm from the testis to the urethra. Contracts during ejaculation.

Vasectomy Contraceptive procedure in men in which the vas deferens is cut and the free ends sealed to prevent sperm from entering the urethra during ejaculation.

Vasopressin Also known as anti-diuretic hormone, which in high concentrations increases blood pressure.

Vena cava One of two large veins that empty into the right atrium of the heart.

Vein Type of blood vessel that carries blood to the heart.

Venule Smallest of all veins. Empties into capillary networks.

Vertebrates Animals with a spinal column (backbone).

Villi Fingerlike projections of the lining of the small intestine that increase the surface area for absorption.

Virus Nonliving entity consisting of a nucleic acid—either DNA or RNA—core surrounded by a protein coat, the capsid. Viruses are cellular parasites, invading cells and taking over their metabolic machinery to reproduce.

Visible light Electromagnetic radiation visible to humans and other animals.

Vitamin Any of a diverse group of organic compounds. Essential to many metabolic reactions.

Vitreous humor Gelatinous material found in the posterior cavity of the eye.

Vocal cords Elastic ligaments inside the larynx that vibrate as air is expelled from the lungs, generating sound.

Warning coloration Passive means of avoiding predation by warning potential predators, usually backed up with some active defense such as a toxin.

Watershed A region drained by a river.

White blood cells (WBCs) Cells of the blood formed in the bone marrow. Principally involved in fighting infection.

White matter The portion of the brain

and spinal cord that appears white to the naked eye. Consists primarily of white, myelinated nerve fibers.

Xylem Vascular tissue that transports water and minerals in vascular plants from roots to leaves or needles. Comprises the bulk of wood in a tree.

Yellow marrow Inactive marrow of bones in adults containing fat. Formed from red marrow.

Yolk sac Embryonic pouch formed from endoderm. Site of early formation of red blood cells and germ cells.

Zero population growth Condition in which a population stops growing.

Zona pellucida Band of material surrounding the oocyte.

Zonular fibers Thin fibers that attach the lens to the ciliary body.

Zooflagellates Animal-like protists that live in water and in moist soils and contain one or sometimes several flagella.

Zooplankton Crustaceans and protozoans in aquatic ecosystems that feed on phytoplankton.

Zygote Cell produced by a sperm and ovum during fertilization. Contains 46 chromosomes.

Index

A

E

F

M

N

O

P

T

Fig 15-6, page 352: see text credit electronic conversion by Precision Graphics
Fig 15-7, page 358: Darwen and Vally Hennings
Fig 15-8, page 359: John and Judy Waller
Fig 15-9, page 360: layout by Rolin Graphics and computer conversion by Publications Services
Fig 15-10, page 361: Georg Klatt
Fig 15-11, page 363: John and Judy Waller
Fig 15-12, page 364: John and Judy Waller
Fig 15-13, page 365: John and Judy Waller
Fig 15-16, page 370: Cyndie Wooley
Fig 15-17, page 372: Elizabeth Morales-Denney
Fig 15-18, page 373: Elizabeth Morales-Denney
Fig 15-20, page 375: John and Judy Waller

Chapter 16
Fig 16-2, page 383: Pat Rossi
Fig 16-3, page 383: Pat Rossi
Fig 16-4, page 384: Publications Services
Fig 16-5, page 385: Cyndie Wooley
Fig 16-6, page 386: Wayne Clark
Fig 16-7b, page 387: John and Judy Waller
Fig 16-8b, page 388: John and Judy Waller
Fig 16-9, page 389: layout by John and Judy Waller and reflective piece by Elizabeth Morales-Denney
Fig 16-11, page 391: Elizabeth Morales-Denney
Fig 16-12, page 393: John and Judy Waller
Fig 16-13, page 394: John and Judy Waller
Fig 16-14, page 395: John and Judy Waller
Fig 16-15, page 396: layout by John and Judy Waller and computer conversion by J/B Woolsey Associates
Fig 16-16, page 399: layout by John and Judy Waller and computer conversion by J/B Woolsey Associates

Chapter 17
Fig 17-2, page 406: see text credit
Fig 17-3, page 407: Cyndie Wooley
Fig 17-4, page 408: layout by John and Judy Waller and computer conversion by Publications Services
Fig 17-5b, page 409: John and Judy Waller
Fig 17-6, page 410: Cyndie Wooley
Fig 17-7a, page 411: Wayne Clark
Fig 17-7b, page 411: Wayne Clark
Fig 17-7d, page 411: Wayne Clark
Fig 17-7e, page 411: Cyndie Wooley
Fig 17-8, page 412: John and Judy Waller
Fig 17-11, page 413: John and Judy Waller
Fig 17-12, page 414: John and Judy Waller
Fig 17-13, page 415: John and Judy Waller
Fig 17-14b, page 417: John and Judy Waller
Fig 17-15, page 419: Carlyn Iverson
Fig 17-16, page 420: Cyndie Wooley
Fig 17-17, page 421: Cyndie Wooley
Fig 17-18a, page 422: Darwen and Vally Hennings
Fig 17-18b, page 422: Darwen and Vally Hennings
Fig 17-19, page 423: Darwen and Vally Hennings
Fig 17-20a, page 426: layout by Darwen and Vally Hennings and new reflective piece by Carlyn Iverson
Fig 17-20b, page 426: layout by Darwen and Vally Hennings and new reflective piece by Carlyn Iverson
T17-1, page 424: Darwen and Vally Hennings
Fig 17-22a, page 428: Darwen and Vally Hennings
Fig 17-22b, page 428: Darwen and Vally Hennings
Fig 17-23, page 429: Darwen and Vally Hennings
Fig 17-24, page 430: Darwen and Vally Hennings
Fig 17-25, page 430: Carlyn Iverson
Fig 17-26, page 431: Carlyn Iverson
Fig 17-27, page 432: Carlyn Iverson
Fig 17-29, page 433: Publications Services
Fig 17-30, page 434: Publications Services

Chapter 18
Fig 18-1, page 444: Cyndie Wooley
Fig 18-2, page 444: Darwen and Vally Hennings
Fig 18-3, page 445: Cyndie Wooley
Fig 18-6, page 448: Darwen and Vally Hennings
Fig 18-7, page 449: Darwen and Vally Hennings
Fig 18-8, page 450: Darwen and Vally Hennings
Fig 18-9, page 451: Cyndie Wooley
Fig 18-10, page 452: Carlyn Iverson
Fig 18-11, page 453: Carlyn Iverson
Fig 18-12, page 453: Sandra McMahon
Fig 18-13a, b, page 454: Darwen and Vally Hennings
Fig 18-14, page 455: Darwen and Vally Hennings
Fig 18-15, page 456: layout by Darwen and Vally Hennings and new reflective piece by Pat Rossi
Fig 18-17, page 457: layout by John and Judy Waller and new reflective piece by Pat Rossi
Fig 18-18, page 458: layout by John and Judy Waller and new reflective piece by Pat Rossi
Fig 18-19, page 458: John and Judy Waller
Fig 18-20, page 459: layout by John and Judy Waller and new reflective piece conversion by Pat Rossi
Fig 18-21, page 460: Publications Services
Fig 18-22, page 461: layout by Rolin Graphics and computer conversion by Publications Services
Fig 18-26, page 463: Darwen and Vally Hennings
Fig 18-28a, page 464: Carlyn Iverson
Fig 18-29, page 465: Carlyn Iverson
Fig 18-30, page 466: John and Judy Waller
Fig 18-31a, page 467: John and Judy Waller
Fig 18-31b, page 467: John and Judy Waller
Fig 18-32, page 468: John and Judy Waller
Fig 18-35, page 470: Darwen and Vally Hennings
Fig 18-36, page 471: Cyndie Wooley
Fig 18-37, page 472: Pat Rossi
Fig 18-38, page 473: Pat Rossi

Chapter 19
Fig 19-1, page 479: Cyndie Wooley
Fig 19-2, page 480: Carlyn Iverson
Fig 19-4, page 481: Carlyn Iverson
Fig 19-6, page 483: Darwen and Vally Hennings
Fig 19-7b, c, page 484: Darwen and Vally Hennings
Fig 19-9, page 485: Darwen and Vally Hennings
Fig 19-10, page 486: Darwen and Vally Hennings
Fig 19-11, page 486: Darwen and Vally Hennings
Fig 19-15b, page 489: Darwen and Vally Hennings
Fig 19-16, page 490: Darwen and Vally Hennings
Fig 19-17, page 490: John and Judy Waller
Fig 19-19, page 491: John and Judy Waller
Fig 19-21, page 493: Darwen and Vally Hennings
Fig 19-23, page 495: Darwen and Vally Hennings
Fig 19-24, page 497: Sandra McMahon
Fig 19-25, page 498: John and Judy Waller
Fig 19-26, page 499: John and Judy Waller
Fig 19-27, page 501: layout by Rolin Graphics and computer conversion by Publications Services
Fig 19-28, page 501: John and Judy Waller
Fig 19-29, page 502: layout by Rolin Graphics and computer conversion by Publications Services

Chapter 20

Fig 20-1a, page 510: Cyndie Wooley
Fig 20-1b, page 510: Cyndie Wooley
Fig 20-2, page 511: John and Judy Waller
Fig 20-3, page 511: layout by Georg Klatt and computer conversion by Publications Services
Fig 20-4, page 513: layout by Rolin Graphics and computer conversion by Publications Services
Fig 20-5, page 514: John and Judy Waller
Fig 20-6, page 514: Georg Klatt
Fig 20-7, page 515: John and Judy Waller
Fig 20-8a, b, page 515: layout by Darwen and Vally Hennings and new reflective piece by Carlyn Iverson
Fig 20-8c, page 515: layout by John and Judy Waller and new reflective piece by Carlyn Iverson
Fig 20-9, page 517: Publications Services
Fig 20-12, page 518: layout by John and Judy Waller and computer conversion by J/B Woolsey Associates
Fig 20-13, page 519: layout by John and Judy Waller and computer conversion by J/B Woolsey Services
Fig 20-14, page 520: layout by John and Judy Waller and computer conversion by J/B Woolsey Associates
Fig 20-16, page 521: layout by John and Judy Waller and new reflective piece by Carlyn Iverson
Fig 20-17, page 521: layout by John and Judy Waller and computer conversion by J/B Woolsey Associates
Fig 20-18, page 522: layout by John and Judy Waller and computer conversion by J/B Woolsey Associates
Fig 20-19, page 522: Cyndie Wooley
Fig 20-22, page 524: layout by Georg Klatt and computer conversion by Publications Services
Fig 20-23, page 525: layout by John and Judy Waller and modification by Cyndie Wooley
Fig 20-25, page 526: layout by John and Judy Waller and computer conversion by J/B Woolsey Associates
Fig 20-26, page 526: layout by John and Judy Waller and computer conversion by Publications Services
Fig 20-27, page 529: Darwen and Vally Hennings
Fig 20-30, page 532: Pat Rossi
Fig 20-31, page 532: Pat Rossi

Chapter 21

Fig 21-6, page 542: Wayne Clark
Fig 21-7, page 542: Cyndie Wooley
Fig 21-9, page 543: Cyndie Wooley
Fig 21-11b, page 545: John and Judy Waller
Fig 21-12, page 546: John and Judy Waller
Fig 21-15, page 547: Elizabeth Morales-Denney
Fig 21-16, page 548: John and Judy Waller
Fig 21-17, page 549: Wayne Clark
Fig 21-18, page 550: layout by John and Judy Waller and modification by Cyndie Wooley
Fig 21-19 top, page 551: John and Judy Waller
Fig 21-19 bottom, page 551: John and Judy Waller
Fig 21-21, page 553: Darwen and Vally Hennings
Fig 21-21, page 553: Rolin Graphics
Fig 21-22, page 554: layout by Wayne Clark and modification by Cyndie Wooley
Fig 21-23, page 555: layout by John and Judy Waller and modification by Cyndie Wooley
Fig 21-24, page 558: Elizabeth Morales-Denney
Fig 21-25, page 559: layout by John and Judy Waller and modification by Elizabeth Morales-Denney
Fig 21-26, page 560: Elizabeth Morales-Denney
Fig 21-28, page 562: layout by John and Judy Waller and computer conversion by Publications Services
Fig 21-29, page 562: layout by John and Judy Waller and computer conversion by Publications Services
Fig 21-30, page 563: Wayne Clark
Fig 21-34, page 566: Rolin Graphics
Fig 21-38, page 567: John and Judy Waller
Fig 21-39, page 568: layout by John and Judy Waller and computer conversion by Publications Services

Chapter 22

Fig 22-2, page 578: Elizabeth Morales-Denney
Fig 22-3, page 578: Elizabeth Morales-Denney
Fig 22-4, page 579: Elizabeth Morales-Denney
Fig 22-5, page 579: John and Judy Waller
Fig 22-6, page 580: John and Judy Waller
Fig 22-7, page 581: John and Judy Waller
Fig 22-8, page 582: John and Judy Waller
Fig 22-9, page 582: John and Judy Waller
Fig 22-10, page 583: John and Judy Waller
Fig 22-11, page 584: layout by John and Judy Waller and modification by Cyndie Wooley
Fig 22-12, page 585: layout by Rolin Graphics and computer conversion by Publications Services
Fig 22-13, page 586: layout by John and Judy Waller and computer conversion by Publications Services
Fig 22-14, page 587: John and Judy Waller
Fig 22-15, page 588: John and Judy Waller
Fig 22-16, page 588: John and Judy Waller
Fig 22-17, page 589: Cyndie Wooley
Fig 22-19, page 594: Cyndie Wooley
Fig 22-20, page 595: Cyndie Wooley
Fig 22-21, page 596: Cyndie Wooley
Fig 22-22, page 597: Sandra McMahon
Fig 22-23, page 598: layout by John and Judy Waller and computer conversion by Publications Services
Fig 22-24, page 599: John and Judy Waller